Chemical Risk Analysis

Chemical Risk Analysis
A Practical Handbook

BERNARD MARTEL

English language edition consultant editor:
Keith Cassidy

Published by Penton Press,
an imprint of Kogan Page London

This work has been published with the help of
the French Ministère de la Culture - Centre national du livre

Publisher's note
Every possible effort has been made to ensure that the information contained in this handbook is accurate at the time of going to press, and the publishers cannot accept responsibility for any errors or omissions, however, caused. All liability for loss, disappointment, negligence or other damage caused by the reliance of the information contained in this handbook, of in the event of bankruptcy or liquidation or cessation of trade of any company, individual; or firm mentioned, is hereby excluded.

Apart from any fair dealing for the purposes of research or private study, or criticism or review, as permitted under the Copyright, Designs and Patents Act, 1988, this publication may only be reproduced, stored or transmitted, in any form, or by any means, with the prior permission in writing of the publisher, or in the case of reprographic reproduction in accordance with the terms of licences issued by the Copyright Licensing Agency. Enquiries concerning reproduction outside those terms should be sent to the publishers at the undermentioned address:

Penton Press
Kogan Page Ltd
120 Pentonville Road
London N1 9JN
www.kogan-page.co.uk

Original French language edition, Guide d'analyse du risque chimique, © Dunod 1997

© English language edition Penton Press 2000, an imprint of Kogan Page

British Library Cataloguing in Publication Data

A CIP record for this book is available from the British Library

ISBN 1 8571 8028 3

Designed and typeset by Susanne Harris.
Printed and bound in Great Britain by Bell & Bain Ltd, Glasgow

Contents

Publishers note	15
Preface	17
Preface to English language edition	21
Glossary of symbols and abbreviations	25
Introduction – Analysis of risk, prevention, protection	29

PART I: METHODOLOGY

I • Flammability — 35

- **1.1. Vapour pressure** — 36
  - 1.1.1. Problems posed by experimental uncertainty — 36
  - 1.1.2. Estimation of vapour pressures — 37
  - 1.1.3. Calculation of liquid/vapour equilibrium concentration — 48
- **1.2. Limits of flammability** — 50
  - 1.2.1. Definitions, applicability — 50
  - 1.2.2. Accuracy of experimental data — 50
  - 1.2.3. Estimating limits of flammability — 51
  - 1.2.4. Examples — 54
  - 1.2.5. Limits of flammability of inflammable mixtures — 55
  - 1.2.6. Influence of temperature on lower explosive limit — 55
- **1.3. Flashpoints** — 56
  - 1.3.1. Definitions, applicability — 56
  - 1.3.2. Measurement of flashpoint — 56
  - 1.3.3. Precision of measurement, sources of inaccuracy — 57
  - 1.3.4. Methods of estimation — 61
  - 1.3.5. Examples of calculation — 64
  - 1.3.6. Estimation of lower explosive limit — 67
  - 1.3.7. Mixture flashpoints — 68
- **1.4. Autoignition temperatures (AIT)** — 71
  - 1.4.1. Definition — 71
  - 1.4.2. Mode of operation, apparatus — 71
  - 1.4.3. Effect of different physical variables — 72
  - 1.4.4. Effect of chemical structure — 73
  - 1.4.5. Rules of group additivity; estimation of AIT — 74

1.5. Fire hazard risk; methods of classification	**80**
1.5.1. NFPA flammability code	81
1.5.2. Labour Code	83
1.5.3. SAX Code	84
1.5.4. Physical and other factors	84
1.5.5. A new classification: degree of fire hazard risk	87
1.5.6. Case study: 1,2,4-trimethyl benzene	90

2 • Stability   93

2.1. Experimental methods of determination	**94**
2.1.1. Mechanical sensitivity tests	94
2.1.2. Thermal sensitivity tests	94
2.1.3. Shockwave sensitivity tests	95
2.1.4. Ballistic mortar tests	95
2.1.5. Determination of critical diameter	95
2.1.6. Other tests	96
2.2. Qualitative approach to instability	**96**
2.2.1. 'Internal' structural factors	96
2.2.2. External aggravating/attenuating factors	100
2.3. Quantitative approach: The CHETAH programme	**101**
2.3.1. Estimation of enthalpy of formation	101
2.3.2. Criterion C1: enthalpy of decomposition	110
2.3.3. Criterion C2: propensity for combustion	113
2.3.4. Criterion C3; internal redox measure; oxygen balance	115
2.3.5. Criterion C4: effect of mass	116
2.3.6. Search for a unique criterion, a variant of CHETAH	116
2.4. Other quantitative approaches	**119**
2.5. Classification of instability risk	**120**
2.5.1. NFPA grading; reactivity	120
2.5.2. Grading according to Labour Code	122
2.5.3. Comparative instability indicators	123

3 • Toxicity   125

3.1. Substance toxicity: evaluation parameters	**126**
3.1.1. Toxic or lethal concentrations or doses	126
3.1.2. Limiting values and means of exposure	127
3.2. Establishing level of toxic risk of substances	**129**
3.2.1. NFPA Code	130
3.2.2. SAX Code	130
3.2.3. Hodge-Sterner Code	130

3.2.4. Labour Code		130
3.2.5. The transport of dangerous materials: Code of toxicity		133
3.3. Problems posed by the determination of toxicity risk levels		**133**
3.3.1. Level of error of measurement		133
3.3.2. Inadequacy of available parameters		134
3.4. Quantitative estimation methods		**134**
3.4.1. Safety and risk factors		135
3.4.2. Estimation of LC and LD50		136
3.5. Case study		**138**
3.6. Proposal for quantitative index for toxic risk		**140**

4 • Dangerous reactions: methodology — 143

4.1. Dangerous reactions: information parameters — 144
    4.1.1. NFPA Code — 145
    4.1.2. Labour Code — 145
    4.1.3. Code for transport of dangerous materials — 146

4.2. Qualitative approach — 146
    4.2.1. Identification of structural risk factors — 146
    4.2.2. External risk factors: an approach to their prevention — 148

4.3. Quantitative estimation of dangerous character of a reaction — 155
    4.3.1. Physical and other factors — 155
    4.3.2. The Stull method — 157
    4.3.3. The CHETAH programme and proposed variant — 157

PART II: LIST OF DANGEROUS REACTIONS

5 • Dangerous reactions of inorganic chemicals — 163

Introductory note – a management of elements		163
lithium		164
beryllium		164
boron	oxidation-reduction reaction; instability; other dangerous reactions	164
carbon		165
nitrogen	instability of nitrogenous compounds; oxidation-reduction reactions other dangerous reactions	165
oxygen	the element; ozone	169
fluorine	the element; fluorine derivatives	170

Contents — List of dangerous reactions

neon		172
sodium	the metal; hydroxides and carbonates; peroxide	172
magnesium	the element; perchlorate	174
aluminium	the element; aluminium derivatives	176
silicon	the element and its alloys; silane; silica	177
phosphorus	the element; phosphine; phosphorus halides pentasulphide; pentoxide; phosphoric acid	178
sulphur	the element; halogen derivatives; sulphides; oxides; sulphuric and sulphamic acid ; oxygen-containing salts;	181
chlorine	the element ; hydrochloric acid; chlorine dioxide; perchloric acid; oxygen-containing salts; sodium chloride	186
argon		192
potassium	the metal; carbonate; hydroxide	192
calcium	the metal; calcium compounds; carbonate; hydroxide	195
scandium		198
titanium	the metal ; derived substances	198
vanadium	the metal ; compounds	199
chromium	chromium and chromium(II) salts: reduction reactions chromium metal and chromium (II) salts oxidation reactions of chromium (III) derivatives and of chromium (VI); halides	199
manganese	the metal;oxidation reactions of permanganates and oxide of manganese (IV); manganese(III) derivatives	201
iron	the metal oxides; sulphides; hexacyanoferrates; chlorides	203
cobalt		205
nickel		206
copper	the metal; derived substances	206
zinc	the metal; derived substances	208
gallium		210
germanium		210
arsenic	the element; derived substances	210
selenium		211
bromine	the element; derived substances	213
rubidium		215
strontium		215
zirconium	the element; tetrachloride	217
niobium		218
molybdenum		218

ruthenium		219
rhodium		219
palladium		219
silver	the metal; halides and cyanide; nitrate; oxide; dangerous reactions	220
cadmium		212
indium		222
tin	the metal; derived substances	222
antimony	the element; halides; trisulphide; stibine	223
tellurium		225
iodine	the element; hydrogen iodide and metallic iodides; iodates	225
xenon		228
barium	the metal; oxides; nitrates and chlorates; other salts	228
mercury	the metal; derived substances	230
lead	the metal; nitride; oxidising salts; other compounds	231
bismuth	the metal; derived substances	233

6 • Dangerous reactions of organic chemicals — 235

6.1. Hydrocarbons — 235
- 6.1.1. Instability of hydrocarbons — 235
- 6.1.2. Polymerisation of unsaturated hydrocarbons — 237
- 6.1.3. Halogenation of hydrocarbons — 239
- 6.1.4. Oxidation of hydrocarbons — 240
- 6.1.5. Other dangerous reactions — 245

6.2. Alcohols and glycols — 248
- 6.2.1. Substitution reactions of the active hydrogen atom — 248
- 6.2.2. Hydroxyl substitution reactions — 250
- 6.2.3. Reactions of nitric acid and nitrates with alcohols — 251
- 6.2.4. Oxidising reactions — 252
- 6.2.5. Dangerous properties of carbon chains — 255

6.3. Phenols — 258
- 6.3.1. Effect of ring on OH group — 258
- 6.3.2. Effect of hydroxyl group on aromatic ring — 258
- 6.3.3. Hydroxyl-related reactions — 259
- 6.3.4. Ring-related reactions — 259
- 6.3.5. Oxidation reactions — 260

6.4. Ethers	**260**
6.4.1. Reactions related to the mobility of the α hydrogen atom	261
6.4.2. Basicity of active oxygen	264
6.4.3. Rupture of C-O bond; ether instability	265
6.4.4. Oxidation reactions	269
6.4.5. Reactions involving hydrocarbon chains	270
6.4.6. Other dangerous reactions	272
6.5. Halogen derivatives	**272**
6.5.1. Nucleophilic substitution and elimination reactions	272
6.5.2. Effect of metals	276
6.5.3. Oxidation reactions	277
6.5.4. Carbon chain reactions	280
6.5.5. Miscellaneous dangerous reactions	283
6.6. Amines	**285**
6.6.1. Reactions related to basicity of amines	285
6.6.2. Diazotization reactions, instability of diazonium salts and amines	287
6.6.3. Oxidation reactions	289
6.6.4. Halogenation of amines	290
6.6.5. Miscellaneous dangerous reactions	291
6.6.6. Hydrazine reactions	291
6.7. Nitrated derivatives	**292**
6.7.1. Instability	292
6.7.2. Base action	296
6.7.3. Acid action	299
6.7.4. Reactions involving side chain and complex group	301
6.7.5. Oxidation and/or nitration reactions	303
6.7.6. Reactions caused by oxidising nature of nitrated derivatives	304
6.7.7. Miscellaneous dangerous reactions	305
6.8. Aldehydes, ketones and acetals	**307**
6.8.1. Oxidation reactions	307
6.8.2. Dangerous polymerisations	310
6.8.3. Dangerous reactions involving carbonyl group	311
6.8.4. Reactions related to the mobility of the α hydrogen atom	312
6.8.5. Miscellaneous dangerous reactions	313
6.8.6. Acetal reactions	313

6.9. Carboxylic acids — 315
- 6.9.1. Oxidation reactions — 315
- 6.9.2. Instability — 317
- 6.9.3. Reactions related to acidic characteristics — 318
- 6.9.4. Polymerisation of unsaturated α, β acids — 319
- 6.9.5. Miscellaneous dangerous reactions — 319

6.10. Esters — 321
- 6.10.1. Instability — 321
- 6.10.2. Oxidation reactions — 322
- 6.10.3. Reduction reactions — 322
- 6.10.4. Unsaturated ester reactions — 322
- 6.10.5. Miscellaneous dangerous reactions — 325
- 6.10.6. Reactions involving complex esters — 327

6.11. Acid anhydrides and chlorides — 327
- 6.11.1. Reactions involving breaking of CO-Σ bond — 327
- 6.11.2. Oxidation reactions — 331
- 6.11.3. Instability — 332
- 6.11.4. Miscellaneous dangerous reactions — 332

6.12. Nitriles — 334
- 6.12.1. Instability, 'functional polymerisation', additions to triple bond — 334
- 6.12.2. Oxidation reactions — 336
- 6.12.3. Reactions involving the carbon chain and complex nitriles — 336
- 6.12.4. Miscellaneous dangerous reactions — 338

6.13. Amides and analogues — 338
- 6.13.1. Violent interaction of amides and analogues with halogens, halogen derivatives and acid chlorides — 339
- 6.13.2. Oxidation reactions — 341
- 6.13.3. Reduction reactions — 342
- 6.13.4. Instability of amides — 343
- 6.13.5. Miscellaneous dangerous reactions — 343
- 6.13.6. Dangerous isocyanate reactions; Bhopal accident — 343

6.14. Miscellaneous substances — 345
- 6.14.1. Sulphur derivatives — 345
- 6.14.2. Phosphorus compounds — 349
- 6.14.3. Silicon compounds — 350

PART III: TABLES

Introduction 355
Published sources; abbreviations employed; presentation of data in tables

7 • Inorganic elements and their compounds 359

hydrogen; helium; lithium; beryllium; boron; carbon; nitrogen; oxygen; fluorine; neon; sodium; magnesium; aluminium; silicon; phosphorus; sulphur; chlorine; potassium; calcium; scandium; titanium; vanadium; chromium; manganese; iron; cobalt; nickel; copper; zinc; gallium; germanium; arsenic; selenium; bromine; krypton; rubidium; strontium; zirconium; niobium; molybdenum; technetium; ruthenium; rhodium; palladium; silver; cadmium; indium; tin; antinomy; tellurium; iodine; xenon; barium; mercury; lead; bismuth.

8 • Organic compounds 385

hydrocarbons: alkanes; cycloalkanes; alkenes; cycloalkenes; alkynes; aromatic compounds 385

alcohols-polyols: saturated alcohols; unsaturated alcohols; aromatic alcohols 392

phenols 397

ethers: aliphatic ethers; glycollic ethers and aliphatic ether-alcohols; aromatic ethers and ether-phenols; heterocyclic ethers and heterocyclic ether-alcohols 400

halogenated derivatives: alkyl halides; halogenated alkenes and alkynes; aromatic halogenated derivatives; halogenated alcohols; halogenated phenols; halogenated ethers 403

amines: primary amines; secondary amines; polyamines; tertiary amines; heterocyclic amines; hydrazines; amino alcohols; amino phenols; amino ethers; halogenated amines. 412

nitrated, nitrite, nitrate derivatives: nitrated aliphatic derivatives; aliphatic nitrites: aliphatic nitrates; nitrated aromatic derivatives; nitro alcohols; nitro phenols; nitroethers; nitrated halogenated derivatives; nitro amines and other complex nitrogenous compounds 422

aldehydes, ketones: aliphatic aldehydes; aromatic aldehydes and complex groups; aliphatic ketones; aromatic ketones and complex groups; acetals. 426

carboxylic acids: aliphatic saturated acids; aliphatic unsaturated acids and aromatic acids; diacids; acids with complex groups. 434

esters, lactones: aliphatic saturated monoesters; aliphatic unsaturated monoesters; aliphatic diesters; aliphatic and aromatic carbonates; aromatic esters, diesters; lactones; esters with complex groups. 438

acid anhydrides, acid chlorides: acid anhydrides; acid chlorides	446
nitriles: aliphatic saturated nitriles and dinitriles; unsaturated aliphatic and aromatic nitriles; nitriles with complex groups	449
amides, isocyanates: amides, urea	452
miscellaneous compounds: sulphur derivatives; phosphorus derivatives; silicon derivatives; boron derivatives.	453

9 • Tables of chemical names and corresponding formulae   459

9.1 Inorganic substances   459
    9.1.1. equivalent empirical formulae/names   459
    9.1.2. names and CAS numbers   465

9.2 Organic substances   470
    9.2.1. empirical formulae/names   470
    9.2.2. names and CAS numbers   481

Publications and Internet Site Bibliography   493
Chemicals Index   495
Subject Index   527

Publishers Note

As will be evident to the reader of *Chemical Risk Analysis*, this volume is packed with valuable information and analysis, representing the work of an expert specialist, who has done a remarkable and unique job of compilation and explanation.

In making this material available to a wider, English-speaking readership it is the publisher's hope that the work of this author, which includes substantial original work, will gain the readership it merits.

In the process of translation and publication the question of international context and applicability has been of a major concern, and in the extended foreword by the English language edition consultant, himself a leading and well-published specialist expert, this concern is addressed, as it is in minor additions to the text. This includes references to internationally used safety codes. The major fabric of the book remains unchanged, however.

The unique and important ingredient of reports of original research and testing carried out by the author himself, often in his own laboratory, has been retained in full, but, for reasons of clarity, has been set in italics or boxes.

The onerous work of translation has been carried out by Olivia Uttenweiler, and that of production design by Susanne Harris.

Preface

Safety is a young 'science' and, as such, has passed through an essentially descriptive phase. In the development of risk chemistry most of the published work to date is descriptive and directed mainly to the production of safety data material.

Safety is often associated with the sometimes dramatic context of accidents. It is then dominated by the mandatory regulations emanating from the European Union and other equivalent bodies, which impose more and more legal constraints.

Producing a publication that can bring together these two interests, descriptive and juridical, does not appear to me a natural way to progress thinking and practice in risk chemistry.

In the descriptive domain, the material offered by the numerous data banks, by researching the internet, and that available arising from the obligations on suppliers to provide freely complete safety publications for the user (but which are sometimes open to criticism, as in this book) makes it, it seems to me, superfluous to offer yet another similar contribution of that kind.

So the object of this book is to attempt a scientific and technical approach to risk chemistry which, without ignoring regulatory aspects, nevertheless does not subjugate itself to their logic and constraints. To realise a scientific approach assumes, first, a critical attitude to the data-providing techniques themselves, and secondly, the research of quantification of the notion of level of risk in order to dispense with the traditional trilogy of 'very dangerous, moderately dangerous, a little dangerous'.

This book is, then, dominated first of all by a new concept of data concerning the relative safeties of chemical products (hence the quest for exhaustivity which in this book, leads to the inclusion of safety data for more than 1900 substances, plainly different from the updating concept employed by most authors).

It emphasises the need to research quantitative evaluation risk models and thus to provide for the reader means to define a level of danger, in a rigorous fashion, taking into account in his own way work already carried out in this domain. The model proposed ought to be considered as a route and a clue to a way of thinking and not necessarily as definitive.

My wish is that, on the basis of the thinking that might come from reading this book, the reader will be able to forge appropriate tools for the appreciation of risk situations with which he is confronted. There are, in this book, several places where the reader is invited to work out for himself his quantification model and it happens each time that the model is narrowly based on a particular work situation. This is the case with the invitation to the reader to adapt the Dow Chemical method of analysis, developed for industrial chemical installations, to ordinary work situations.

I have also wished to contribute to the bringing to bear of 'order' into the domain of chemical reactions termed 'dangerous'. In this field, the contributions of various authors (eg IRNS, Bretherick) have been in the nature of lexicons or dictionaries. In these, each product is treated as unique without taking account of their belonging to, in the case of organic chemicals, series, or classes of cations and anions for inorganic compounds. But must one install order into dangerous chemical reactions which are often manifestations of disorder: disorders of operating mode, environmental conditions, man-made error, molecular disorder?

I have attempted to organise the presentation of this subject by treating series of chemical products; the chapters dealing with dangerous chemical reactions each virtually constitute a book within a book. However, I have only been able to make some sort of approach to my objective in the domain of organic chemistry and this part is the only one in which I have approached the problem of prevention.

My design here is to bypass reactions declared dangerous merely because they have provoked an accident, in order to provide the means of predicting the potential danger of a reaction, by virtue of global treatment of products that have structures in common.

This area, is in fact, the weak point in most studies of risk chemistry, and also notably, in the regulatory safety data publications mentioned above.

Two objectives have dominated the thinking behind the dangerous reaction chapters: the means to supplement the often inadequate safety documents available in this area and to provide for university teachers (and other lecturers and trainers) an aid which will enable them to integrate the study of safety into their chemical reactivity courses. Safety is often missing from this type of exercise, although there are exceptions.

For whom is this book intended? For those involved in industry, particularly nonchemical operations), who have, for a long while (since the application of the Labour Code) had much experience in safety matters, but for whom my experience as a trainer in the department of Hygiene, Safety and Environment of the IUT allows me to say how inadequately prepared they are when confronted with certain aspects of risk chemistry. Public organisations, curiously, had, until recently, no obligation to administer the Labour Code and that which concerned hygiene and workers' safety. This book is concerned with all these activities but is not addressed to all safety officers because it presupposes a basic knowledge of chemistry. For all that, the chemist may not be at ease with this book, which he may find difficult. For this reason, numerous examples are provided to illustrate the methods studied, and assist in their application, and to permit him to identify the limits.

Meanwhile, this book will not suffice to provide total comprehension of chemical risk and needs to be used in conjunction with other material.

For situations of high chemical risk, this book is at the level of 'preliminary analysis of risk' (hence its title), that is to say, at the first stage of risk inventory, these linked to chemical products and their implicit chemical reactions. It brings together substances and procedure. On the other hand, a deep understanding of work routines, implicit in risks other than chemical, is necessary. Certain areas of activity are excluded from this book, notably related to phytosanitary products and equally, the interaction of chemical products with the environment have been ignored.

Finally, it is hoped that this book will facilitate the broadening of perspective in the domain of safety research, for example, to enable research of the correlations existing between quantitative expression of risk and recorded accidents.

Note that many original contributions are presented in this book (on mixture flash points and autoignition temperatures).

It is my hope that this book will be the starting point for a challenging debate with my readers, (bmartel@imaginet.fr).

B.Martel

Preface to
the English Language Edition

by Keith Cassidy, Consultant Editor, English Language edition.

Life inevitably brings with it the prospect, or the risk, of harm; and it is one of life's ironies that those branches of science and technology which have been prime movers in the improvement of many aspects of our lives – in particular the chemical and engineering industries – are held in greater disregard because of the hazards and risks of their products and artefacts, rather than for the benefits which they have brought.

The tensions which arise between the thrust for technological progress and the concerns of those who might be affected by any adverse impact are not easily resolved. In many cases the science or technology involved is far outside general public experience or understanding; and indeed public perception of the hazards or risks may well be driven by issues which are more in the domains of the social rather than the physical sciences. It is in this context that the approach of 'risk management' has been developed and applied, to try to reach a consensus of 'tolerability' around which the various stakeholders – regulators, manufacturers, consumers,etc, – can operate. The principles of risk assessment and management are central to international initiatives such as the European Union 'Framework Directive' on health and safety, and to more focussed directives such as the so-called 'Seveso' directives on major industrial risks. Similarly, they underpin the Risk Management Programs and Process Safety Management programs of the Occupational Safety and Health Administration and the Environmental Protection Agency in the USA. And they are at the core of regulatory criteria such as 'reasonable practicability', best environmental option', best practicable means' etc.

Not surprisingly, there is no agreed definition of the 'risk management' process; the issues and interactions can be very complex, and much effort is going into defining how they can be characterised but it is possible to sketch the overall process into a coherent architecture, based on the principles of:

IDENTIFICATION	– the recognition and location of any problem;
ASSESSMENT	– the bounding and dimensioning of any problem;
CONTROL	– the limiting of the scale of any potential problem, by prevention or avoidance;
MITIGATION	– the amelioration of the residual elements of the potential problem.

Measures used to parametrise, or to limit, the component elements may vary between different types of hazard and risk, between different components of the overall environment, or between different economic and cultural systems; but the underpinning logic of the approach remains a transparent and potent taxonomy.

At the heart of a risk management approach lies an essentially simple set of critical questions. The core questions are:

WHAT IF?

WHAT THEN?

THEN WHAT?

SO WHAT?

'What if?' requires a combination of technical expertise, experience, and a degree of imaginative insight. 'What then?' and 'Then what?' are essentially the techniques and practices of risk assessment, usually quantified, at least in part. 'So what?' is the area of judgement, informed but not constrained by the earlier inputs. It is a decision process, often rigorous, which involves:

a) dimensioning of likely risk with an understanding of the uncertainties inherent in the assessment process (and answering the question: 'How much of what kind of what risk to whom or what?')

b) reference to the likely benefits generated and the associated political, social and economic considerations;

c) judgements as to tolerability or acceptability for groups directly or indirectly involved, and for any other stakeholders; and

d) sometimes, decisions as to further reduction in risk, taking cost (including effort, and available technology) into account.

It is, in short, a process which is essentially economic and political, and technically informed. And because the process relates to the human condition, or to the well-being of other parts of the environment, it is ethics-rich.

Currently, there is little consensus on the definitions of terms central to the above issues. In the field of the so-called 'major chemical hazards', the nomenclature suggested by the UK Institution of Chemical Engineers has gained a wide recognition, but in some respects are already being overtaken by further developments. All such definitions are currently expressed in essentially social terms; but they are in principle equally applicable in economic or financial contexts, with however increases in the uncertainties inherent in the process. At the international level, such difficulties are further compounded by greater differences of approach, culture and sometimes language; and reaching consensus overall may be a lengthy process. But there is an increasing movement internationally towards a harmonisation of the technical methodologies of assessment, hopefully shadowed by similar movement in the definition, meaning and understanding of terms and approaches; and the recent completion of the OECD initiatives to create and put in place:

a) a glossary of terms used in the risk assessment of chemicals; and

b) a dictionary/thesaurus of risk assessment terms used in the regulation of, and in emergency planning and preparedness for, major chemical accidents. These are important early initiatives in this field. Indeed, these very examples point up the need for consensus, or at least common understanding; for in the former, the focus is on the uncertainty of dose response, whereas in the latter, the critical uncertainty is in the probability of exposure per se.

This brings us to the issue of risk assessment methodologies. Risk assessment is an interactive process which addresses (using risk assessment jargon) the source, the source-term, the dispersion, the dose, and the impact; a process in which there are many common components, including:

– in IDENTIFICATION, the use of substance/threshold approaches;

– in ASSESSMENT, the classical approaches to consequence and probabilistic assessment, including:

a) comparative methods (checklists, audit/review, relative ranking, indices, PHAs, etc;

b) fundamental methods (HAZOP, 'What if analysis', FMEA, FMECA, etc);

c) logic diagram methods (FTA, ETA, CCA, HRA, success trees, etc;)

- in CONTROL, technical, operational, and legal standards, the use of information and justification packages (safety reports and safety cases), and licensing or other permissioning activity

- in MITIGATION, emergency planning, public information and involvement.

Risk assessment, and particularly risk assessment, has the capacity for dimension, to rank, to focus and to test the interdependence and interactive responses of the component elements.

There have been significant developments over the last 20 years in modelling techniques for hazard and risk assessment, (although the assessment of direct environmental and indirect human hazard and risk is currently less well developed and therefore more uncertain than that of direct risk to man). Most of the scientific and technical bases of the approaches are described in the literature, (and this book by Professor Martel is an important addition to that literature); and many computerised packages are available for general use (often as proprietary software). Indeed a significant part of this work focuses on the use of one such package – CHETAH – to apply the methodology described. Generally, such methodologies use mathematical models to estimate the source term, the dispersion, the dose and the impact to estimate hazard ranges or concentrations, to which is added a frequency/probability figure to enable a risk figure to be derived. A criterion for level of harm, or for probability or frequency of exposure, can then be added for risk management purposes. There remain, of course, many uncertainties in both the technical and the social components of this process but their criticality can often be tested and demonstrated by the transparency of the process, and by elements such as sensitivity testing, which can be used to identify critical areas of control, to add insight to debate on tolerability, or to validate the relevance of any criteria which may have been applied.

As already stated, this is a substantial addition to what is already a considerable technical literature. Its particular values lie in the novelty of some of its approaches, in the relevance of the worked examples, in its very useful collation and compilation of substantial amounts of information normally only available in fragmented or dispersed form, and especially at the practical level at which it approaches a very complex technical subject. It is not the purpose of this Foreword to the English Language Edition to expand significantly on the contents of the book. Professor Martel himself provides many additional references for the reader, both in the body of the text, and especially in his introduction to Part Three, where he cites his main sources and gives further worldwide web addresses. However, it may be of assistance to the reader of the English edition if a few further references and sources of information are provided, and these are listed at the end of this Foreword. They are all references to standard works of recognised standing, and will guide the reader to several thousand further references should more detailed information be required; and a few important website addresses of particular relevance to UK and US readers are also given.

This then is a practical handbook of chemical risk assessment, and will be of particular value for such assessors at the practitioner level. It is aimed principally at:

– safety specialists in industry concerned with the safe manufacture, manipulation and transport of chemicals;

– process and safety specialists evaluating chemical hazards and risks;

– those involved in the development and use of analytical procedures in industry and the universities;

– lecturers and trainers in chemical hazard and risk assessment as an essential reference work, and as a practical working manual.

I commend it to the reader.

Glossary

Symbols and abbreviations glossary

AEV	Average Explosive Value
AIT	Autoignition Temperature
ASTM	American Society for Testing and Materials
C	Cleveland (apparatus measuring flash point)
CAS No	Chemical Abstracts Service No
cc	Closed cup
C_{eq}	Equilibrium concentration
Ccc	Cleveland closed cup
Coc	Cleveland open cup
C_s	Concentration in air (% vol)
cu	Cutaneous
cu-m	Cutaneous: mouse
cu-r	Cutaneous:rat
cu-ra	Cutaneous:rabbit
CV	Coefficient of Variation
dec	Decomposition
DTA	Differential Thermal Analysis
Eb	Boiling Point
FER	Fire Explosion Risk
g	Gaseous State
H_{vap}	Enthalpy of Vapourisation
IDLH	Immediately dangerous to life or health
II	Inflammation Index
INRS	Institut National de Recherche et de Sécurité
ip	Intraperitoneal
ip-m	Intraperitoneal:mouse
ip-r	Intraperitoneal:rat
ip-ra	Intraperitoneal:rabbit
IR	Index of Reactivity
IT	Index of Toxicity
iv	Intravenous
iv-m	Intravenous:mouse
iv-r	Intravenous:rat

iv-ra	Intravenous:rabbit
l	Liquid state
LC50	Lethal Concentration 50
LD50	Lethal Dose 50
LEL	Lower Explosive Limit
LEV	Limiting Explosive Value
LVE	Limiting Value of Exposure
m	Mouse
M	Molar Mass
MAK	*Maximale Arbeitsplatzkonzentration*
MLC	Minimum Lethal Concentration
MLD	Minimum Lethal Dose
MVE	Minimum Value of Exposure
NFPA	National Fire Protection Association
o	Orally
oc	Open cup
o-m	Orally:mouse
o-r	Orally:rat
o-ra	Orally:rabbit
Pa	Atmospheric Pressure
PCA	Principal Component Analysis
Pfla	Flash point
PM	Pensky-Martens (apparatus measuring flash point)
PMcc	Pensky-Marten closed cup
PMoc	Pensky-Marten open cup
Pvap	Vapour pressure
r	Rat
ra	Rabbit
RHI	Reaction Hazard Index
s	Standard deviation
S	Setaflash (flash pt measuring apparatus)
scu	Subcutaneous
scu-m	Subcutaneous:mouse
scu-r	Subcutaneous:rat
scu-ra	Subcutaneous:rabbit
Sd	Standard Deviation
SF	Safety Factor
STEL	Short Time Exposure Limit
subl	Sublimation
Svap	Entropy of Vapourisation

Tcc	Tag closed cup
TLV	Threshold Limit Value
Toc	Tag open cup
Tr	Labour Code Risk Degree (for inorganic data)
TR	Transport Code risk degree for dangerous materials
TWA	Abbreviation for TLV/TWA:(Threshold limit value/Time weighted) average
UEL	Upper Explosive Limit

Introduction

Analysis of risk, prevention, protection

Assuring preventive measures must be the main aim of all those involved in potentially hazardous activity, from those responsible for fixing the objectives and details of a task, to the operatives that carry out the work. This includes the setting of rules to enable the operatives to avoid accidents on the job.

Nevertheless, however carefully these rules are developed and applied, an accident cannot always be avoided, the 'zero accident' possibility is fiction beyond reach. Apart from *prevention* rules, it is also necessary to conceive protection rules, intended to minimise the consequences of an accident that could not be prevented.

Prevention consists of developing a very safe mode of operation for a reaction that is capable of being dangerous and explosive. Protection consists of interposing a barrier between reactor and operator. Only in the case of an incident will such a barrier play a part in protecting the operator. To quote further examples, to plan ahead and install very elaborate ventilation systems in an automatic decanting system for highly inflammable liquids is a prevention measure. Installing an automatic incendiary extinguishing system is a protective measure.

However, security management cannot be confined to prevention and protection alone. A preliminary analysis, with a view to determining the nature and level of risk is a prime requirement.

Prevention and protection have costs that are not restricted to the directly financial – the measures involved can lead to more complicated modes of operation and to increased operating discomfort, resulting in longer working routines and increased fatigue. The operator will constantly look for means to rationalise his job, to make it more efficient, less time consuming, less tiring. Preventive measures that get in the way of this are not always carried out.

It is therefore imperative to find a balance, with great care, between the extra restraints placed on the operator, the cost to the firm and the actual level of risk involved.

A low risk situation, with relatively harmless consequences, would carry minimal restriction at little cost; plainly the converse will apply if the risk is high. But there is a psychological dimension, too. If the preventive measures are disproportionately heavy, this might provoke mistrust in them on the part of the operators and render them inoperable. The need to avoid alarming the operators needlessly can play a big role.

The preliminary analysis appropriate to the particular product will determine the level and choice of measures chosen.

For example, simple natural ventilation would be appropriate for weak toxicity through inhalation. If the substance is harmful, it would be necessary to place the operation in a ventilation-assisted area. If it is toxic, a well-ventilated fume hood would be required; if very toxic, it might be necessary to place the work in an airtight environment under negative pressure, with the atmosphere filtered at the exit. For cases of serious chronic toxicity, telemanipulation devices might be required.

For each degree of complication, the costs obviously will increase.

The object of this book is to assist workers involved in chemical processes to determine level of risk wherever the handling of a substance exposes the participants. To do this, 'tools' of analysis are presented in Part One (*Methodology*), and available data in Parts Two and Three. The material provided is intended to provided the means for this analysis, whatever the difficulties encountered – incoherence in known experimental data, insufficient or absent information etc.

The developed methodology is underlined, but this is not to underestimate the importance and interest of the prescribed (ie mandatory) approach. This methodology conforms to technical and scientific logic, and takes into account the *quality of criteria*, in the statistical sense of the term. The approach presented tends to evaluate the levels of risk, as defined, quantitatively rather than qualitatively, in progressive steps.

But our approach must not conflict with the prescribed approach – where this occurs in the text, the reader's attention is drawn to it. The technical approach must complement the prescribed approach and not replace it.

Confronted with a chemical substance or group of compounds the reader has to consider the four potential categories of risk – inflammability, instability, toxicity and dangerous reactions resulting from contact with other substances and materials. But procedure does not only relate to chemical substances. It concerns other materials (eg asbestos), and modes of operation and introduces most often a 'manipulator'. These processes occur in an environment which also can have significant role.

Therefore this book does not claim to provide answers to all the problems that a process throws up. It is only concerned with the substance, and is only one of a number of tools required.

Once the preliminary analysis is completed, it is necessary to integrate it into a more complete analysis in which it will play an essential part. The results of this analysis are formalised by creating a *risk profile* or *risk map*.

In the risk profile is constructed a task procedure description in which are identified different timetabled stages, the successive substances involved, the environmental restrictions deriving from the operative or the climate, as well as the effect of other tasks being carried out in neighbouring operations, with their possible interactions.

The object of this exercise is to identify the dangerous steps in the sequence of events.

In considering a chemical reaction, for example, the sequence will be divided into the various phases: assembly of apparatus, the weighing and manipulation of original reactants, solvents and catalysts; the effect of mixing the various reagents – heat generated (is it above ignition point?), the cooling down process, stopping the reaction, its extraction, distillation, manipulation, decantation and storage. Or the operation might, for example, concern the treatment of surfaces when a series of baths will be used to remove grease, to clean the surfaces, or to treat them with phosphates or by oxidation etc.

Substances can change in the course of an operation, but chronology is difficult to establish when there are operations occurring simultaneously ,eg when different car chassis parts are being treated separately in different baths in the same workshop, one of many examples.

After the four ranks for levels of risk for fire, instability, toxicity and reactivity are established, other ranks will appear – dedicated to the operators involved (are they able to cope with potential dangers?), the peculiarities involved in setting up the apparatus (detecting weak points eg), effect of environmental conditions (lighting, supervision etc), and any rank likely to be involved in the global security of the process.

A risk map has similarities to a noise map. On a plan of the workspace, the different levels of risk pertaining to different activities are represented by symbols and colour codes. This gives an indication as to whether coincidences of activity create extra risk or difficulties with evacuation in the case of an accident. Employing the different levels of risk – fire, instability, reactivity and toxicity, it is possible, using appropriate software, to create a linear multiple regression. This will implement geographical siting using rectangular coordinates x and y with the third 'z' dimension employed if a volumetric consideration is involved. The use of such methods emphasises the need for the expression of risk to be a quantitative one.

Having thus obtained the surface 'isorisk' maps, reconfiguration of the work areas can be carried out, possibly with partial compartmentalisation and the introduction of warning and detection systems.

In this way, the map provides a complete view of the risk situation on a site – the example cited above of surface treatment of surfaces would certainly lend itself to such treatment, as would storage of chemicals.

Using the map, the risks involved in the placing of chemical substances on particular shelves, with respect to proximity to exits and automatic extinguishing equipment can be assessed, and can help form decisions as to whether to reposition the substance, or replace it with a lower risk but equivalent compound.

By producing transparent versions of the maps and superimposing copies related to different classes of risk it is possible to synthesise a map that helps identify 'danger zones' – areas of high risk density.

Having completed this preliminary risk analysis stage, the protection/prevention process can commence. This must take the form of a collective dialogue involving all the potential players, including management, who must not give way to their tendency to delegate this kind of work.

Compromises have to be struck between the different requirements and the restrictions implied. A deep knowledge of the actual task is essential, and for this reason it not possible to provide prescriptive prevention measures; such measures have to be worked out by those involved in the actual processes of a task. The 'security' person appointed can only act in an advisory role, although it is likely to be his responsibility to initiate the preliminary risk evaluation.

The prevention rules must be arrived at by consensus, then be drafted, passed by senior management and signed by all the personnel involved; this should ensure that the measures agreed will be taken seriously. The consideration of the rules need not be confined to this group however; the works inspector and security inspector should be brought in if necessary.

The work itself can now begin, but that is not the end of the security procedure, because it must be maintained and reviewed constantly in the light of unforeseen dangerous situations. Any incident must be reported and the procedure employed should be analysed in order to draw lessons that can be incorporated in a revised pre-vention/protection plan.

Frequent visits from inspectors and senior managers will drive home to members of the team that the safety of all personnel is their business.

The task, being completed, is finished through a listing grouping together safety features and the measures taken, and to propose preventive actions to come. A reconvening of all the participants would enable an exchange of general information and to incorporate them into new sets of rules.

This step is an idealised one; too heavy if a benign operation is envisaged, it might be too succinct if the ensuing risks are serious. Only the personnel involved are in a position to judge the level of constraint after having established the level of risk.

PART ONE
Methodology

1 • FLAMMABILITY

One of the most frequent causes of accidents involving chemicals is linked to their high inflammability. Most organic products are inflammable. This chapter is concerned with these products but it is also relevant to some inflammable inorganic substances as well.

In order to identify and then establish the risk level of a chemical, one uses a certain number of risk parameters. The purpose of this chapter is to consider these parameters in detail. It is also to enable the reader to submit these parameters to critical analysis if values are available, or to estimate them if they are unknown. The user of this book should then be able to offer an evaluation of the risk level of inflammability of a particular chemical. This is necessary even if the chemical is not in the tables included later in this book.

The reader will then be able to take this evaluation into account when considering the prevention and protection measures required.

What are the risk parameters?

The combustion of a chemical substance takes place in the gaseous phase except with metals and metalloids where combustion takes place in the solid phase. This implies that a solid or a liquid inflammable chemical has the ability to vapourise in order to build an inflammable vapour–air mixture. The two indicative parameters are the **boiling point** and, most important, the **vapour pressure** of the liquid.

But that is not sufficient. The ability to vapourise has to be such that the vapour concentration in the air reaches a value that enables propagation of the flame into the gaseous mass from the point where the ignition occurred. If the concentration is insufficient, it is said that the mixture is 'too poor', or, at the other extreme, 'too rich'. The limits of the range within which combustion is possible are called *lower explosive limit* (LEL) and *upper explosive limit* (UEL).

Flashpoints (the reason for the plurality is explained in paragraph 1.3) have the advantage of being linked to the boiling point, the pressure and the lower explosive limit. This is the reason why flashpoints are such important parameters in the evaluation of the inflammability risk of a liquid or a solid. The measurement of flashpoints implies the existence of an ignition flame for the gaseous mixture. Nevertheless, contact of a suitable substance–air mixture with a hot surface can be sufficient to start the combustion of the mixture. The *autoignition temperature* is the parameter that determines the possibility that an inflammable material will combust in contact with a hot substance without the presence of a flame.

After studying these parameters, we will describe and discuss the most important classification methods before suggesting a new quantitative classification method for the fire hazard of a chemical substance.

1 • Flammability   1.1 Vapour pressures

1.1 Vapour Pressure

The vapour pressure of a liquid provides an essential safety parameter and it is mandatory that *safety sheets* contain these values (when they are known). This parameter is taken into account in some classification methods of inflammability risk. It enables one to determine the equilibrium vapour concentration of a liquid in air. This concentration can then be used to ascertain whether a working environment presents an *inflammability risk* (by reference to the inflammability limits) or a *toxicity hazard* (by comparison with the exposure values).

But it will also be seen that vapour pressure estimation methods provide critical analysis of all parameters involved in a fire hazard and thus allow refinement of the methods leading to a quantification of this risk.

1.1.1 Published data; problems posed by experimental uncertainty

There are two disadvantages to the existing vapour pressure tables. First of all, like any experimental data, there is no agreement between sources. This is worsened by the decision to take only one value into account for each chemical substance. This fact may encourage the user to take on trust the figure proposed, which is sometimes unjustified. Secondly, these values are given for a temperature that does not always correspond to the thermal conditions in which the chemical substance will be handled. Some references, to overcome this difficulty, offer several values. For instance, Weka[1] most often gives three values, 20, 30 and 50°C, and the coefficients A, B, C in Antoine's equation can thus be calculated:

$$\log P = A - \frac{B}{T + C}$$

where P is the vapour pressure at temperature T

There are to be found lists of chemical substances in handbooks for each of which log P = f (T), and whose coefficients are to be inserted, are given. These lists are limited but nevertheless provide solutions for the most common chemical substances. When there are several experimental estimates of vapour pressures it is possible to estimate the importance of the *experimental uncertainty* from the standard deviation of the measurements. The relevance of the values can be verified from a series of different sources (to be rigorous, checking that it is a Gaussian sequence would be required).

This method will be fully exploited later but cannot be used here. It is to be noted that when it is attempted to estimate the standard deviation of the different available values at the same temperature the resulting standard deviations are not 'constant' and are strongly correlated to the average of the values. In other words, the higher the average, the higher the standard deviations. This is the phenomenon of *heteroscedasticity*. It does not provide any possibility of estimating uncertainty of measurement.

The table below illustrates this property with a few hydrocarbons described in Part Three. There is one exception that runs contrary to this observation: the 6.99/3.18 value for P_{vap} refers to the average vapour pressure and s the standard deviation.

---

1. Dangerous Products Safety Sheets, Weka ed., updated in 1995.

1 • Flammability

1.1 Vapour pressures

Heteroscedasticity of vapour pressures of hydrocarbons

$\overline{P_{vap}}$	8 421	698	558	10.57	9.28	7.2	6.99	6.4	4.65	2.43
s	685	28.92	22.37	2.06	1.58	0.7	3.18	0.36	0.37	0.4

The diagram below shows in graphical form the same property for alcohols. The two lines joining the four extreme values show the funnel-shaped configuration of values. It is the standard way of spotting the values of heteroscedasticity.

Heteroscedasticity of vapour pressures

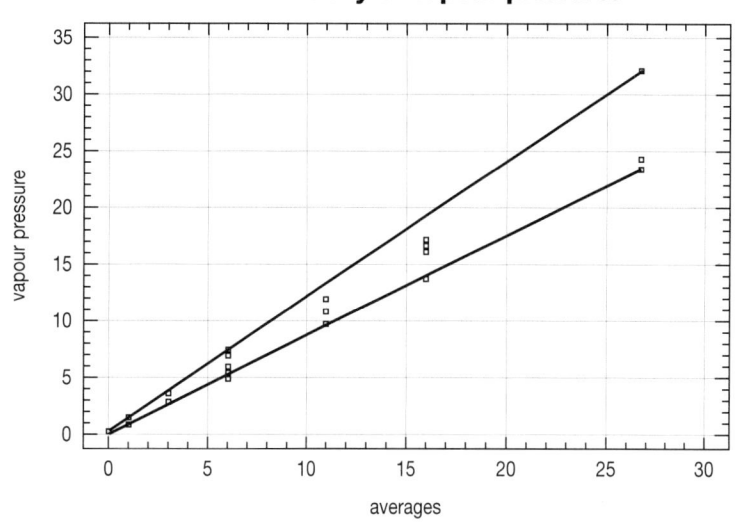

(graph: x-axis "averages" from 0 to 30; y-axis "vapour pressure" from 0 to 35)

In these conditions it is impossible to estimate the uncertainty of measurement that would enable unrealistic experimental points to be noted. It would be interesting to analyse the origin of this heteroscedasticity.

1.1.2 Estimation of vapour pressures

The previous comments help to show why many authors have shown an interest in developing mathematical models likely to result in satisfactory estimations of vapour pressures at any temperature. There are many reasons to make such an estimate:
- They allow the estimation of a value when there is no experimental data or when the available values do not relate to a particular temperature.
- They allow the choice of the most relevant value when the available data presents large contradictions between sources.
- They simply provide a check on a single experimental value to prevent the use of an unrealistic figure. Experience shows that the sole values are the ones to be most wary of.
- As will be seen later, other technical and indirect applications can be derived from these models.

1 • Flammability

1.1 Vapour pressures

Antoine's and Clapeyron's equations offer two possible solutions, and there is also the model developed by Hass and Newton[1].

In order to obtain boiling point corrections for distillation under reduced pressure the following expressions are used:

$$\Delta t = \frac{(273.1 + t)(2.8808 - \log P)}{S + 0.15(2.8808 - \log P)}$$

$$S = \frac{\Delta H_{vap}}{2.303 \times R \times Eb} = \frac{\Delta S_{vap}}{2.303 \times R}$$

S being the entropic factor;

ΔH_{vap} and ΔS_{vap} are respectively the enthalpy and entropy of vapourisation of the liquid;

R the gas constant;

$\log P$ is the decimal logarithm of the pressure in mmHg;

t the boiling point at pressure P in °C;

Δt is the boiling point variation between the standard pressure and pressure P. Knowing the boiling point Eb at the normal pressure and a given temperature t, P is the vapour pressure of the liquid at temperature t.

The equation can then be written:

$$\log P_{vap} = \frac{786.7465 + 3.3129 \times t - 0.4321 \times Eb - S \times Eb + S \times t}{273.1 - 0.15 \times Eb + 1.15 \times t} \quad (1)$$

where

t is the temperature at which P_{vap} is estimated;
Eb is the boiling point of the chemical substance under normal pressure in °C;
S is the *entropic factor*, which depends on Eb and the chemical substance structure;
P_{vap} is obtained in mmHg.

■ Estimation of S using the Hass method

To calculate S, Hass suggests a chart displaying eight straight lines $S = f(Eb)$ corresponding to the groups in which the chemical substances can be classified. It is preferable to present this data in the form of eight equations $S = A + B \times Eb$ which are more suited to calculator or computer programming.

The coefficients of these equations for the eight groups are:

Coefficients of equation S = A + B X Eb

Group	A	B	Group	A	B
1	4.30	0.00205	5	5.22	0.00233
2	4.54	0.00259	6	5.44	0.00230
3	4.76	0.00216	7	5.67	0.002307
4	5	0.00222	8	5.9	0.002307

---

1. H.B. Hass, R.F. Newton, *Handbook of Chemistry and Physics*, 74th edition, 1993-1994, p.15-16 to 16-17, CRC Ed. The author has used this model with students for twenty years and it has proved to be very reliable, apart from a few adjustments.

1 • Flammability

1.1 Vapour pressures

Then the correct group only has to be attributed to the substance. In order to do so, Hass proposes a table bringing together substances and structural groups.

Vapour pressure 'groups' (Hass)

Compound or Functional Group	Group	Compound or Functional Group	Group
Acetaldehyde	3	Glycol diacetate	4
Acetone	3	Dibenzylketone	2
Acetophenone	4	Dimethylamine	4
Acetic acid	4	Dimethyloxalate	4
Benzoic acid	5	Dimethylsilicate	2
Butyric acid	7	Water	6
Formic acid	3	**Esters**	3
Heptanoic acid	7	Ethanol	8
Isobutyric acid	6	**Ethers**	2
Isocaproic acid	7	Ethylamine	4
Propionic acid	5	Ethylene glycol	7
Valeric acid	7	Fluoromethane	3
Benzyl alcohol	5	Methyl formate	4
Isoamyl alcohol	7	Methane	1
Isobutyl alcohol	8	Methanol	7
n-Amyl alcohol	8	Methy ethyl ketone	2
n-Propyl alcohol	8	Methyl ethyl ether	3
Amines	3	α and β Naphthols	3
Acetic anhydride	6	Nitrobenzene	3
Phthalic anhydride	2	Nitromethane	3
Anthracene	1	Nitrotoluenes (o,m,p)	2
Anthraquinone	1	Nitrotoluidines (o,m,p)	2
Benzaldehyde	2	Carbon monoxide	1
Methyl benzoate	2	Ethylene oxide	3
Benzonitrile	2	Methyl oxide	3
Benzophenone	2	Phenanthrene	1
Camphor	2	Phenol	5
Chloroanilines (m,p)	3	Phosgene	2
Cyanogen chloride	3	Quinoline	2
Cresols (o,m,p)	4	Methyl salicylate	2
Cyanogen	4	Methyl silicate	1
Hydrogen cyanide	3	Carbon suboxide	2
1-Hexene	1	**Sulphurs**	2
Hydrocarbons	2	Tetranitromethane	3
Halogen derivative	*	Trichloroethylene	1

*If atoms of halogen (X) are present the same group is attributed as results from the replacement of the halogen by H in the product

From this chart, there are three possible situations that have to be confronted. These lead to three rules:

Rule 1. *The chemical substance appears in the table. The group attribution is straightforward and unambiguous.*

1 • Flammability

1.1 Vapour pressures

Rule 2. *The substance does not appear in the table but contains one of the functional groups in the chart (in bold). The substance is attributed to the group for this structural class.*

Rule 3. *The substance does not fit in either of the previous situations. The compound in the chart that most resembles the substance is attributed.*

The efficacy of the estimation decreases going down this sequence and this is where (but not only where) the weakness of this approach is apparent.

■ Case study

This method will now be applied to a few examples to demonstrate use of this model in the three situations above, choosing chemical substances for which experimental vapour pressures are given in order to be able to make comparisons. When several experimental values are mentioned the distribution range is quoted. When a range of boiling points is given, the values for the minimum and maximum of this range are calculated (but later the highest value will be systematically used). All comparisons are made at 20°C. The data used is as in Part Three. Note that the calculations are tedious but easy to program on the calculator.

The procedure is first to group estimations that proved to be satisfactory, then group examples of failure, which could be due to an imperfect estimate or incorrect experimental values, sometimes without any means of control.

The substance appears in the table

Compound (Hass grouping)	P_{vap} experimental at 20°C (mbar)	P_{vap} estimated (mbar)
Acetic acid (group 4)	14.7 – 16	14.23 – 15.65
Acetone (group 3)	233 – 256	238.38 – 248.22
Ethanol (group 8)	53 – 59.5	56.81 – 57.34
Nitrobenzene (group 3)	0.2	0.17 – 0.18

The substance does not appear in the list, but contains a functional group that is mentioned in the Hass table

Compound (class, n° of group)	P_{vap} experimental at 20°C (mbar)	P_{vap} estimated (mbar)
Benzene (hydrocarbons, group 2)	99 – 101	96.65 – 100.7
Cyclohexane (hydrocarbons, group 2)	90 – 94	85.42 – 92.76
Nonane (hydrocarbons, group 2)	4.26 – 5	4.44
Tetrahydrofuran (ethers, group 2)	180 – 200	170.27 – 191.82
Isopropyl oxide (ether, group 2)	160 – 180	157.19 – 184.36
Thiophene (sulphurs, group 2)	80	84.02 – 92.76
Methyl sulphide (sulphurs, group 2)	532 – 537	522.5 – 563.1

The substance and its compounds do not appear in the list

Substance (compound it most resembles)	P_{vap} experimental at 20°C (mbar)	P_{vap} estimated (mbar)
Benzyl cyanide (benzonitrile, group 2)	0.07	0.06 – 0.07
Acrylic acid (propionic acid, group 5)	4 – 5.33	3.48 – 4.06

Other (doubtful) cases

The examples below illustrate cases in which the chemical substance or the group it belongs to appear in the table.

1 • Flammability
1.1 Vapour pressures

Substance (group or structural type in the Hass table)	P_{vap} experimental [temperature °C] (mbar)	P_{vap} estimated (mbar)
Anthracene (group 1)	1.33 [145]	1.88 – 3.16 [145]
Isoamyl oxide (ethers, group 2)	4.8 [40]	5.78 [40]
1,2-Dimethoxyethane (ethers, group 2)	64 – 81.6 [20]	87 – 92 [20]
Acetic anhydride (group 6)	4.7 – 5.33 [20]	3.11 – 3.46 [20]

In the two last cases the experimental values found in the literature promote some confidence for the range within which correct estimation is expected, so these two last examples question the grouping suggested by Hass.

The possible causes of imperfections in the model are:
- the group may have been attributed by Hass on the basis of erroneous experimental values;
- the attribution of a sole group to a functional group is a compromise that can prove to be inapplicable to substances showing structural particularities that separate them from most substances in this class.

These unsatisfactory (easily replicated) results have led to a search for other approaches to the calculation of the *entropic* factor.

■ Estimation of *S* from boiling points under reduced pressure

The first approach is based on the boiling point under reduced pressure data, which can easily be found in the technical literature[1,2]. The manufacturers' catalogues for substances also give these values. The tables that appear in Part Three systematically mention this data whenever available. They do not appear in any of the safety databases in the *safety sheets* produced by the chemical industry for users.

From the equation used to estimate P_{vap} (called P in the equation below) at temperature t, S can be found:

$$S = \frac{786.7465 + t(3.3129 - 1.15 \times \log P) + Eb \, (0.15 \times \log P - 0.4321) - 273.1 \times \log P}{Eb - t} \quad (2)$$

  Eb is the boiling point at normal pressure in °C;

  t, the boiling point in °C at pressure P in mmHg.

This equation can be used in two different ways:
- to calculate S and incorporate it in equation (1) for vapour pressure;
- to also determine $S_i = A_i + B_i \times Eb$ corresponding to the eight groups defined by Hass and to select the group by looking for the value of S_i closest to the estimated S, i denoting the group.

The advantage of this approach is that it uses data that is both available and accurate. Such boiling points are obtained during processes of purification, so the given data often refers to a pure substance and this constitutes sound data.

---
1. Merck Index, 17th edition, 1989, Merck & Co, Inc.
2. *Guide de la Chimie international*, 1994-1995, Chimedit.

1 • Flammability

1.1 Vapour pressures

In addition, it is quite usual to have a series of about ten boiling points for several pressures for a given substance. The curve of log $P = f(t)$ for the entire pressure and temperature range required can then be drawn.

This approach nevertheless has an important disadvantage, which is of a 'mathematical' nature. The value of S is extremely sensitive to any minute variation of pressure or temperature in equation (2).

The slightest inaccuracy, especially in pressure (manometers are not especially accurate) can lead to high estimation fluctuations.

■ Estimation of S from vapour pressure

A variant of this calculation method for S consists in using equation (2) by inserting the experimental vapour pressure values, provided there is serious reason to believe that these experimental values are reliable. In the opposite case, one would inevitably suffer the consequences of the disadvantage mentioned above.

The examples below will be used to verify the reliability of the experimental data. It can appear not to make much sense to estimate S from the values sought. This way of proceeding will respond to two specific situations; one corresponding to the estimation from this model of parameters other than vapour pressure as will be explained later, the other to a diagnosis of the coherence of a group of vapour pressures.

■ Examples of calculations of the entropic factor

- EXAMPLE 1: CASE OF ACETIC ANHYDRIDE

The relevant data in Part Three is:

$C_4H_6O_3$ • M(g/mol) = 102.09 • CAS no.:108-24-7

Physical data: ΔH^0_f (KJ/mol) = –575.72(g); –624.00(l)

Eb (°C/mmHg) = 138-140; 65/50; 36/10

P_{vap} (bar) (°C) = 4.7; 5; 5.33(20); 13.3(36); 300(100)

By incorporating the two boiling point values for normal and reduced pressure in equation (2), the following four estimates are obtained:

Estimates of entropic factor for acetic anhydride

Eb (°C)	t (°C)	p (mmHg)	S calculated	P_{vap} estimated from S	Group	P_{vap} estimated from group
138	65	50	5.296	5.45	4	5.395
140	65	50	5.150	5.74	3	6.271
138	36	10	5.417	4.84	4	5.395
140	36	10	5.308	4.90	4	4.887

This table shows a preference for the group number 4 and a better coherence in the estimates when using directly calculated S rather than the group. The following calculation rules can be derived (rules 1 to 3 are due to Hass).

Rule 4. Calculation of S. *When several estimates are possible, ones that most often lead to the same group are retained.*

1 • Flammability

1.1 Vapour pressures

Rule 5. Calculation of S. *For the calculation of P_{vap} preference is given to the different values of S corresponding to the group in the majority rather than the group itself.*

> By anticipating rule 6, which gives preference to the boiling point values under reduced pressure that come closer to the particular temperature (in this case, 20°C), the last two values are chosen. Hence, the suggested range of values is:
> $$4.84 \leq P_{vap} \text{ est} \leq 4.90$$
> in practice situated in the middle of the experimental distribution, far from the Hass approximation.
>
> • EXAMPLE 2: CASE OF DIMETHOXYETHANE
> It was seen that the attribution of group 2 of ethers did not provide a satisfactory match with the experimental value (87.2-92.76 mbar at 20°C).
> The data obtained from Part Three are:
> 1,2-Dimethoxyethane: $C_4H_{10}O_2$ • M(g/mol) = 90.12 • CAS no.: 110-71-4
> Physical Data: ΔH^0_f (kJ/mol) = –376.64(l) • Eb (°C/mmHg) = 82-83.5; 20/61.5; 16/50; –14/10 • P_{vap} (bar) (°C) = 64; 80; 81.6(20); 125(30); 270(50)
>
> The procedure is exactly the same as with the previous example and leads to the table below. It should be noted nevertheless that the value 20/61.5 (81.98 mbar) corresponding to the value of the boiling point is an experimental value that refers to this case. However, this value will be treated just like any other experimental value, which needs to be incorporated in the calculation.
>
> **Estimation of entropic factor of dimethoxyethane**
>
> | Eb (°C) | t (°C) | p | S calc | P_{vap} estim from S | Group | P_{vap} estim from group |
> |---|---|---|---|---|---|---|
> | 82 | 20 | 61.5 | 4.998 | 81.981 | 3 | 84.531 |
> | 83.5 | 20 | 61.5 | 4.876 | 81.984 | 3 | 79.308 |
> | 82 | 16 | 50 | 5.000 | 81.90 | 3 | id |
> | 83.5 | 16 | 50 | 4.884 | 81.65 | 3 | id |
> | 82 | –14 | 10 | 4.794 | 90.84 | 2 | 92.763 |
> | 83.5 | –14 | 10 | 4.716 | 89.03 | 2 | 87.205 |
>
> Applying rule 4, this substance must be considered as belonging to group 3 and not 2 as suggested by Hass. Applying rule 5, this leads to the values shown.

Rule 6 can now be stated:

Rule 6. Calculation of S. *When there are several boiling point values for reduced pressure, preference is for estimates from data obtained at temperatures that come closer to the temperature at which the estimate is made.*

Rule 6 reinforces rule 5 and leads to rejection of the two last estimates in the table. It can be resolved by using the following estimate, which also leads to rejection of the 64 mbar figure. It might well be that this figure is the result of mistaken units during transcription to mbar from a value expressed in mmHg.

$$81.65 \leq P_{vap} \text{ estim} \leq 81.984$$

1 • Flammability   1.1 Vapour pressures

> • EXAMPLE 3: CASE OF DIMETHYLDICHLOROSILANE
>
> The available values are :
> Dimethyldichlorosilane : $C_2H_6Cl_2Si$ • M((g/mol) = 129.06 • CAS No 75-78-5
> Physical data: ΔH_f^0 (KJ/mol) = –461.3(g) • Eb (°C/mmHg) = 69-70 •
> P_{vap} (bar) (°C) = 145; 173(20); 230(30); 490(50).
>
> The substance does not appear in any of the structural groups in Hass's table. The substance that most 'resembles' it, ethyl silicate, is found in Group 1, which does not give any indication that is at all in line with the experimental pressure values at 30 and 50°C. Finally, there is a great discrepancy between the values at 20°C, indicating a need for a more thorough study. However, none of the boiling points are given under reduced pressure. The only available values are thus the pressures whose single value is unacceptable. It is an ideal case on which to test the method. So s is calculated from the available values.
>
> The table below contains the figures obtained and reached using estimates based on application of rule 5:
>
> **Estimation of entropic factor of dimethyldichlorosilane**
>
> | Eb(°C) | t (°C) | p (mmHg) | S calc | P_{vap} est from S | Group |
> |---|---|---|---|---|---|
> | 69 | 20 | 108.75 | 4.924 | 145.956 | 3 |
> | 70 | 20 | 108.75 | 4.823 | 144.955 | 3 |
> | 69 | 20 | 129.75 | 4.477 | 172.936 | 1 |
> | 70 | 20 | 129.75 | 4.385 | 172.947 | 1 |
> | 69 | 30 | 172.5 | 4.909 | 229.891 | 3 |
> | 70 | 30 | 172.5 | 4.783 | 229.952 | 2 |
> | 69 | 50 | 367.5 | 5.319 | 489.809 | 5 |
> | 70 | 50 | 367.5 | 5.050 | 489.857 | 3 |
>
> By attribution of the group in the majority, ie group 3 and above, all the calculations based on rule 5 lead to a result that is perfectly in line with the values at 30 and 50°C and to the rejection of the 173 mbar value at 20°C, the 145 mbar value being perfectly consistent with the other values at different temperatures. This technique also enables determination of so-called 'reliable' pressure values.

The following rule can now be stated :

Rule 7. *An experimental distribution of pressure values will be regarded as reliable when attribution of a same group is made when calculating S using these values, or when different calculations of S lead to values that are very close to this.*

There are still two cases that need to be considered:

• the case of substances whose boiling point is too high to allow distillation under atmospheric pressure. This factor prevents estimation of the vapour pressure. It might be thought that these substances are not really worthy of study, being of such low volatility. This is true if based on inflammability alone but would not be the case if they were highly toxic since *limiting or average exposure values are demanded by regulation*. Even compounds of low volatility (for instance, mercury) can easily exceed these values. There is also the case of substances with very high boiling points for which there are contradictory boiling point values;

1 • Flammability   1.1 Vapour pressures

- when there is no other data other than the boiling point under atmospheric pressure or only under reduced pressure and in which the substance and the structural group it belongs to do not appear in the Hass table. The first case will be considered, followed by an actual example.

■ Estimation of boiling point under normal pressure

The equations (1) and/or (2) allow determination of this value if the group, and the boiling point under reduced pressure (or reliable pressures) are known. From (1) or (2):

$$\alpha . Eb^2 + \beta . Eb + \gamma = 0 \quad \text{in which:}$$

$\alpha = B$ (this is the constant B that enables one to calculate S according to the Hass method);

$\beta = A - B \times t + 0.4321 - 0.15 \times \log P$ (A is the other Hass constant);

$\gamma = 273.1 \times \log P + 1.15 \times t \times \log P - 3.3129 \times t - 786.7465 - A \times t$

The solution

$$Eb = \frac{-\beta + \sqrt{\beta^2 - 4\alpha\gamma}}{2\alpha}$$ is the only one that is physically feasible.

In order to estimate Eb only the boiling point 't' (in °C) under pressure 'P' (in mmHg) and the group need be known.

There is no easy example to illustrate this method. The substances found in Part Three, triphenylphosphate or pentaerythritol, require preliminary research.

- EXAMPLE: TRIETHYLENETETRAMINE

This case illustrates the most frequently found situation. In this specific case there is the boiling point under normal pressure, but possibly because of the distillation difficulties caused by this heavy substance, there are huge contradictions amongst the published boiling point data. So this boiling point has the value range 267-287°C.
The data from the Part Three is :

Triethylenetetramine: $C_6H_{18}N_4$ • M(g/mol) = 146.24 • CAS no.: 112-24-3
Eb (°C/mmHg) = 267; 277; 284-287; 183/50; 144/10 •
P_{vap} (bar) (°C) = < 0.013; 0.013(20); 15(50).

Without looking for a more accurate method, group 3 of amines is attributed to this substance following Hass rule 2.

Then:
- with $t = 183°C$ and $P = 50$ mmHg, $Eb = 280.253°C$;
- with $t = 144°C$ and $P = 10$ mmHg, $Eb = 282.775°C$.

The result allows clear rejection of the 267°C value and possibly allows retention of either the estimated 280-282 value or the 284-287°C experimental interval. In any case, to go further, it will be seen the $P_{vap} < 0.013$ value is the only one in line with the method shown. The 15(50) pressure value is obviously unrealistic as simple comparison with the boiling point under 10 mmHg (ie under 13.3 mbar) pressure demonstrates.

1 • Flammability　　　　1.1 Vapour pressures

■ Methodology in difficult cases; case study

In the situation when there is no information apart from the boiling point under atmospheric pressure or even under vacuum, there is the paradoxical situation of having to reconstitute a set of estimated data without having any means whatsoever to check the suitability of the approach.

Therefore a case for which data is available will be considered, but it is as if nothing is known about the substance. In this way the method may be tested.

- EXAMPLE: 2-PENTANOL

Ignore everything about this substance except its boiling point under atmospheric pressure:

$$Eb = 116\text{-}119.3°C.$$

Also, assume ignorance of the Hass table.

Assuming that Hass's grouping applies to all substances which look like the substance under consideration from a structural point of view, are they all the same as assumed by Hass? All structures that come close to 2-pentanol will be studied in turn; **for these, there is the information required to find the entropic classification**. This can then be applied to this substance.

Nevertheless the case chosen complicates the task. Indeed, the interesting series here is: 2-propanol, 2-butanol, 2-pentanol, 2-hexanol and 2-heptanol.

Unfortunately, the last ones are not helpful at all since there is no data for 2-hexanol and the available value of vapour pressure for 2-heptanol leads to an unrealistic group for this series (group 3, none of the aliphatic alcohols have a group inferior to 7). Group 8 is obtained for the first two substances. Indeed, by applying rule 6:

- With 2-propanol for $t = 23.8°C$, $P = 40$ mmHg, $Eb = 83°C$, $S = 6.221$ → group 8 ($S_8 = 6.091$);
- With 2-butanol for $t = 20°C$, $P = 12.975$ mmHg, $Eb = 100°C$, $S = 6.211$ → group 8 ($S_8 = 6.131$).

It is necessary to use values of vapour pressure in order to be able to deal with the last stage since boiling point values under reduced pressure are not available. The selected vapour pressure value is the only one that is coherent since the others would provide a value of S very far outside the entropic domain. The calculated value of S is already greater than the maximal value S_8 (as in the case of 2-propanol), a point that will be returned to later.

The work would stop here given the previous comments regarding compounds C_6 and C_7; for the reason that the comparative study will be extended to the case of primary alcohols:

1-propanol, 1-butanol, 1-pentanol, 1-hexanol and 1-heptanol.

The comparisons of 1 and 2-propanol, on the one hand, and 1 and 2-butanol, on the other, will enable quantification of the effect of positional change of the alcohol group. If there is no effect, the comparison of 1 and 2-pentanol becomes legitimate and allows a conclusion to be drawn on the substance.

- with 1-propanol, for $t = 20°C$, $P = 15$ mmHg, $Eb = 98°C$, $S = 6.150$ → group 8 ($S_8 = 6.126$) (approach through vapour pressure);
- with 1-butanol, for $t = 20°C$, $P = 5.475$ mmHg, $Eb = 118°C$, $S = 6.086$ → group 8 ($S_8 = 6.172$) (approach through vapour pressure).

It can thus be concluded that the change from secondary to primary structure does not modify the entropic group. Classified 1-propanol can be noted under group 8, and the superior primary alcohols can be studied.

- with 1-pentanol, for $t = 48°C$, $P = 10$ mm Hg, $Eb = 138°C$, $S = 6.428$ → group 8 ($S_8 = 6.218$);
- with 1-hexanol, for $t = 24,4°C$, $P = 1$ mm Hg, $Eb = 158°C$, $S = 5.983$ → group 7 ($S_7 = 6.035$);
- with 1-heptanol, for $t = 42,4°C$, $P = 1$ mm Hg, $Eb = 176°C$, $S = 6.371$ → group 8 ($S_8 = 6.306$).

1 • Flammability
1.1 Vapour pressures

The fact that the group remains the same during the positional change of the alcohol group is confirmed in two cases out of three and particularly for the C_5 alcohol (in line with the indications in the Hass table in this last case). It can thus be safely concluded that 2-pentanol belongs to group 8.

The fact that this conclusion is verified is extremely interesting since the results, acceptable but only moderately satisfactory, will give way to an unexpected perspective.

Indeed, by attributing group 8 to 2-pentanol the following comparative results are obtained:

Estimation of P_{vap} assuming group 8 for 2-pentanol

Temperature (°C)	P_{vap} experimental (mbar)	P_{vap} est (mbar)
20	5.3	6.33
30	11.2	12.65
50	40	43.36

It will be noticed that the estimate is systematically higher than the corresponding experimental value whereas the latter is obtained with group 8, which gives the highest possible estimate. The boiling point under reduced pressure allows calculation of S for 2-pentanol, obtaining:
- for $t = 62°C$, $P = 60$ mmHg and $Eb = 119.3°C$, $S = 6.283$ → group 8 ($S_8 = 6.175$).

That the value is systematically too high is due to the group attribution and leads to an entropic factor which is too low (and thus increases the estimated P_{vap}). Using the value obtained from equation (2) results in a better match with experimental data being obtained, as shown in the table below:

Estimation of P_{vap} with $S = 6.283$ for 2-pentanol

Temperature (°C)	P_{vap} experimental (mbar)	P_{vap} est (mbar)
20	5.3	5.79
30	11.2	11.72
50	40	41.04

This reasoning is based on faith in experimental data. This is justified by the values resulting from rule 7, which give a remarkable consistency in the entropic factor calculated by using equation 2 (respectively 6.392; 6.347 and 6.334 at 20. 30 and 50°C).

These results show that $S > S_8$ is most often verified in the series considered, as shown in the following table:

substance	S_8	S calc
1-Propanol	6.126	6.150
2-Propanol	6.091	6.221
1-Butanol	6.172	6.086
2-Butanol	6.131	6.211
1-Pentanol	6.218	6.428
2-Pentanol	6.175	6.283
1-Hexanol	$S_7 = 6.035$	5.983
1-Heptanol	6.306	6.371

It might be possible to suggest a new group 9 for alcohols.

It may be concluded that vigilance is needed in order to get the most out of this method.

The method presented explains why the substances in Part Three are categorised by structural similarity rather than in alphabetical order.

There is also the situation in which there is no available data, the substance not being in the database. In a case that involves all inflammability parameters in the study, (see para 1.5.6). This situation can be resolved using information in this book.

1.1.3 Calculation of liquid/vapour equilibrium concentration

Having obtained vapour pressure values, which are of minimum reliability, the vapour equilibrium concentration of the substance can be obtained and compared with various risk parameters or regulatory criteria in order to estimate a potentially dangerous situation. Depending on the work to be done, one has to proceed to a selection of appropriate units.

If Pa is the atmospheric pressure, the equilibrium concentration C_{eq} can be expressed as volume percentage in the form:

$$C_{eq} = \frac{100 \times P_{vap}}{Pa} \tag{3}$$

The value obtained will allow direct comparison with the limits of inflammability that are always given in percentage form and enable us to deduce whether the substance presents the risk of building an explosive mixture with air. Note that if the chosen unit is the millibar, the value P_{vap} is to be divided by ten in order to obtain the concentration. This expression will be used in the determination of 'II' code (infammability index; (see para 1.5.5).

The concentration can be expressed in ppm (parts per million) and thus make a direct comparison with the limiting or average explosive values (LEV or AEV) that are often given in ppm. This calculation will enable us to know if the regulations and recommendations are conformed to for the substances.
Then:

$$C_{eq} = \frac{10^6 \times P_{vap}}{Pa} \tag{4}$$

Here again use of the mbar allows direct calculation; multiply the value of P_{vap} by one thousand to obtain C_{eq} in ppm.

Finally, unit mg/m³ is also used to express LEV and AEV. Furthermore, g/m³ or mg/l units are used in regulations to specify the criteria of toxicity by inhalation of substance. The equation below expresses C_{eq} in g/m³ or mg/l:

$$C_{eq} = \frac{1000 \times M \times P_{vap}}{Pa \times Vm} \tag{5}$$

M is the molar mass and Vm the molar volume expressed in litres, which, if compared with the vapour produced by a perfect gas, gives, depending on the temperature at which C_{eq} is measured:

$$Vm = 22.4 \times \frac{t + 273.1}{273.1} \tag{6}$$

t being the temperature in °C.

1 • Flammability

1.1 Vapour pressures

> **EXAMPLE:**
> The average exposure value of nitrobenzene (see Part Three) (maximum authorised concentration for 8h). Does the presence of an open receptacle in a poorly ventilated room comply with regulations?
> Nitrobenzene belongs to group 3 and from previous study of this substance it was noticed that the value 0.2 mbar was relevant. Then:
>
> $$C_{eq} = 1000 \times 0.2 = \sim 200 \text{ ppm} \gg 1 \text{ ppm}$$
>
> Nitrobenzene may not be handled in these conditions according to the regulations; an extractor or fume hood is necessary as well as closed recipients, if possible. But, even worse: C_{eq} reaches the value of concentration that is immediately dangerous to health (IDLH).
> These conclusions are drawn from the following data quoted in Part Three:
> **Nitrobenzene**: $C_6H_5NO_2$ • M(g/mol) = 123.12 • CAS No: 98-95-3
> ΔH_f^0 (KJ/mol) = +15.90 (l) • Eb (°C/mmHg) = 210-211; 83-84/10 • P_{vap} (bar) (°C)
> = <u>0.2(20)</u>; 0.4(25); 1.33(44.4); 1.9(50); 66.7(120); 500(185) • Tr: 26/27/28-33 •
> NFPA: 3 TR: 60 • LVE 'MVE (F): – (1) • TWA (USA): 1; MAK(D): 1 • STEL: 2;
> IDLH: <u>200</u>

In conclusion, the rules developed are summarised:

The Hass table of group attribution of the entropic factor is first considered:

1. *The substance appears in the table. The group attribution is straightforward and unambiguous.*

2. *The substance does not appear in the table but belongs to one of the structural groups in the chart (in bold). The group mentioned for this functional group is attributed to the substance.*

3. *The substance does not fit into any of the previous situations. It is attributed to the group in the chart that most resembles the substance.*

Data either for boiling points under reduced pressure or reliable vapour pressures at different temperatures are sought and S then calculated:

4. *When several estimations of S are possible, those that lead most often to the same group are retained.*

5. *For the calculation of P_{vap} preference is given to the different values of S corresponding to the group in the majority rather than the group itself.*

6. When there are several boiling point values under reduced pressure or experimental pressure values, preference is given to the estimates from data measured at temperatures that come closer to the temperature at which the estimation is carried out.

7. An experimental distribution of vapour pressure values is regarded as reliable when there is attribution of a same group when calculating S with these values or when different calculations of S lead to values that are very close to this factor.

1.2 Limits of flammability

1.2.1 Definitions, applicability

Mixtures become inflammable when the vapour concentration lies between two limits called the *lower explosive limit* (LEL) and the *upper explosive limit* (UEL). These limits depend on numerous factors, especially temperature and pressure.

LEL is the most important of the two limits. It is mostly useful when inflammable substances are handled in confined spaces (reservoirs, painting cabins, ovens etc). Details of limits of inflammability are kept by chemical substance manufacturers who are required to mention them on safety sheets that have to be put at clients' disposal. When compared with the equilibrium concentration determined as indicated before, LEL allows determination of whether a working environment presents a risk of explosion in the presence of a source of ignition.

Despite all of this, LEL is not regarded as an important inflammability parameter. In fact, as will be seen in paragraph 1.3, flashpoints are considered to be more convenient in the evaluation of fire hazard of chemical substances.

But as will be seen in paragraph1.3.3 LEL presents an indirect factor of interest in the estimation of flashpoints and in this respect will be considered as an essential parameter in the study of evaluation criteria of fire hazard. This point of view will be reinforced when using LEL as a parameter in the calculation of the 'Il' code of fire hazard presented in paragraph 1.5.5.

The object of these comments is, first of all, to draw attention to the fact that experimental limits of inflammability present a big experimental error and have to be handled with caution. The methods enable an estimation of the relevance of these experimental values to be made, and when these are not sufficiently reliable or are unknown – an estimation based on calculation models is made.

1.2.2 Accuracy of experimental data

A brief study of the available data related to limits of inflammability in Part Two shows that these parameters are subject to high experimental uncertainty. For a large number of substances, the experimental values are widely dispersed. When they are submitted to quality estimation using statistical tools, in many cases they reveal that it is impossible to use them with confidence. The examples of difficulties raised by the statistical analysis of the LEL data can be multiplied.

> • EXAMPLE 1: BUTYL ACETATE
> The LEL data given for this substance in the 'Esters' entry of Part Three is the following:
>
> 1.2   1.38   1.4   1.7   3
>
> The test of normality (Shapiro and Wirk) indicates that this distribution is not Gaussian; the test by Dixon indicates that the maximum value is not realistic.
>
> • EXAMPLE 2: ETHANOL
> The LEL are the following:
>
> 3.2   3.3   3.3   3.3   3.5   4.3

As before, the distribution is not Gaussian and Dixon's test concludes that the maximum value is not realistic.

- EXAMPLE 3: ACETYLENE

$$1.5 \quad 2.2 \quad 2.3 \quad 2.5 \quad 2.5 \quad 3$$

This distribution belongs to a standard model, Dixon's test does not give rise to an unrealistic value (hence the normality). The calculation of the confidence range of the individual values in this sequence then leads to:

$$1.07 < \text{LEL} \leq 3.59$$

that shows that the fluctuations in the experimental values are so high that the statistical tool does not allow any unrealistic hypothesis on the real level of LEL.

- EXAMPLE 4: 1-CHLORO-1,1-DIFLUOROETHANE

There are four values:

$$4.4 \quad 6.2 \quad 6.2 \quad 9$$

the distribution is normal; the confidence interval of the individual range is:

$$0.408 < \text{LEL} \leq 12.49$$

The last two calculations of the confidence interval indicate in which range there are 95 out of 100 chances of finding an experimental LEL for this substance. The only legitimate experimental approach to the measurement of LEL is the repetition of measurements and the calculation of the average. The first two sequences show that the fluctuations in LEL in both of these cases cannot be considered to be linked to the *uncertainty of measurement*. There is a predictable cause, which cancels out all interest in this data.

Finally, a sample of 21 substances and the analysis of the standard deviations of measurements of LEL show that these standard deviations are not equal and therefore reflect different causes of variation. Without making the statistical analysis worse, the experimental values are very unstable and therefore hardly useful.

These considerations, along with the fact that this parameter is hardly mentioned in any available table, have led several authors to research models that incorporate estimations of limits of inflammability.

1.2.3 Estimating limits of flammability

Two methods for estimating lower limits of inflammability will be considered. The first involves data on flashpoints which will be analysed in the next paragraph; the other, which is also used for the estimation of UEL, is due to Hilado[1], in which the stoichiometric concentration of a substance has only to be multiplied by 0.5 to 0.55 to estimate its LEL approximately (and by about 3.2 for the UEL). Hilado has perfected this approach. For the present writer LEL and UEL are estimated with the help of the following equations:

$$\text{LEL} = a \times C_s$$
$$\text{UEL} = b \times C_s$$

---

1. Hilado, G, Li, J., *Fire and Flammability*, 8, 34-40, 1977??

1 • Flammability 1.2 Limits of flammability

in which *a* and *b* are constants characteristic of the structure of the substance. The stoichiometric concentration, Cs, is the concentration of a substance in the air, expressed in percentage volume, which enables it to burn completely using all available oxygen. There are various tables due to the author, some more precise than others. Here is the most precise data due to Hilado:

Limits of inflammability (Hilado)

Functional Group	a	b
1 - Saturated linear hydrocarbons	0.555	3.10
2 - Cycloalkanes	0.567	3.34
3 - Alkenes	0.475	3.41
4 - Aromatic hydrocarbons	0.531	3.16
5 - Alcohols – Glycols	0.476	3.12
6 - Ethers	0.537	7.03
7 - Epoxides	0.537	10.19
8 - Esters	0.552	2.88
9 - Other C, H, O compounds	0.537	3.09
10 - Monochlorinated derivatives	0.609	2.61
11 - Dichlorinated derivatives	0.716	2.61
12 - Brominated derivatives	1.147	1.5
13 - Amines (N $\rightarrow$ NO$_2$)	0.692	3.58
14 - Compounds containing sulphur	0.577	3.95

a and *b* being chosen with the help of the table; the stoichiometric concentration needs to be calculated.

This is given by the expression:

$$Cs = \frac{100}{1 + 4.773 \times z} \qquad (7)$$

in which *z* represents the number of necessary moles of oxygen to obtain complete combustion of the substance.

The figure 4.773 is the ratio $\frac{100}{[O_2]}$ in which $[O_2]$ is the percentage concentration of oxygen in air.

Although the most energetic equation of the combustion of a nitrogenous substance implies that nitrogen has to combine to form a nitrogen molecule, the estimation of LEL is more accurate, according to Hilado, if nitrogen is combined in the form of nitrogen dioxide NO_2.

z can easily be calculated in the case of a compound with a general formula $C_cH_hO_oN_nX_xS_s$ in which X represents any halogen.

The equation for the complete combustion of this substance is:

$C_cH_hO_oN_nX_xS_s + zO_2 \rightarrow cCO_2 + yH_2O + nNO_2 + sSO_2 + xHX$

y can be calculated by writing down the equivalent numbers of atoms of hydrogen in both parts of the equation:

$$h = 2y + x \rightarrow y = \frac{h - x}{2}$$

1 • Flammability

1.2 Limits of flammability

The equivalent numbers of the oxygen quantities between both parts of the equation can now be written by using the value calculated above for y:

$$o + 2z = 2c + 2n + 2s + \frac{h-x}{2}$$

Hence:

$$z = c + \frac{h-x}{4} + n + s - \frac{o}{2}$$

This calculation enables one to program easily the stoichiometric concentration, using a small calculator. If the molecule contains other atoms, silicon, tin, manganese, lead, etc, the most stable oxides thermodynamically are sought perhaps by using enthalpies of formation data listed for inorganic substances in Part Two.

The calculation of a and b is carried out by simple linear regression. If it is accurately carried out, it can give access to the coefficient of linear correlation, the 'Fischer', the residual standard deviations and those forecasts which provide the advantage of knowing the quality of the correlation and forecasting and with which accuracy of the estimation is carried out. The Hilado approach does not make this possible and gives reassuring estimated figures since there is no mistake in estimation. The user can make his calculations without reservation. Taking the author's calculations for LEL only, with all the data available, led to differences compared with Hilado, but there is no objective basis for preference; the alternative approach accepts that regression only rarely leads to very accurate forecasts. The table below constitutes a counter proposal, which has the sole advantage of providing an appreciation of the forecast.

The uncertainties of forecast have been calculated from the standard deviations of forecast 'awarding' five points. The uncertainty subtracted and added to the estimated value gives a confidence interval of 95% for the LEL.

During this analysis it was noticed, contrary to Hilado, that it did not appear judicious to convert nitrogen into nitrogen dioxide but rather into nitrogen. n has to be taken out of the above expression for

$$z = c + \frac{h-x}{4} + s - \frac{o}{2}$$

Here is the table obtained:

Revised estimations of lower limits of inflammability

Functional Group	a	Confidence Interval
Saturated hydrocarbons	0.5448	± 0.048
Unsaturated hydrocarbons	0.4265	± 0.177
Aromatic hydrocarbons	0.5133	± 0.043
Saturated alcohols – glycols	0.5015	± 0.104
Non cyclic ethers	0.4743	± 0.171
Oxygen-containing heterocyclics	0.4294	± 0.128

Functional Group	a	Confidence Interval
Glycol ethers	0.4683	± 0.093
Monochlorinated derivatives	0.6114	± 0.233
Dichlorinated derivatives	0.7051	± 0.331
Amines (N → N_2)	0.5726	± 0.155
Aldehydes	0.4424	± 0.233
Ketones	0.4979	± 0.087
Carboxylic acids	0.4852	± 0.242
Esters	0.5529	± 0.110

These different approaches will make possible:
- either an estimate of an unknown LEL;
- or a decision between various contradictory values in favour of those which are the most convincing.

1.2.4 Examples

- EXAMPLE 1: 3-OCTANOL

There is no data on this compound but 1-octanol, which has the same C_s, has an experimental LEL of 0.8%. A value for C_s of 1.72% is obtained.

$$LEL_{est} = 1.72 \times 0.476 = 0.82 \text{ (H: Hilado)}$$

$$LEL_{est} = 1.72 \times 0.5015 = 0.86 \pm 0.10 \text{ (M: author)}$$

- EXAMPLE 2: BUTYL ACETATE

This compound has been considered but its experimental value was found to be unrealistic.

$$C_s = 2.55\%;$$

$$LEL = 1.41\% \text{ (H); } LEL = 1.41 \pm 0.11 \text{ (M)}.$$

This estimate indicates that values 1.2; 1.7 and 3% are unrealistic.

- EXAMPLE 3: ETHANOL (already considered)

$$C_s = 6.53\%; LEL = 3.11\% \text{ (H); } LEL = 3.27 \pm 0.10 \text{ (M)}.$$

Here again the maximum value is to be rejected.

- EXAMPLE 4: PHENYLAMINE

This is a contentious compound since Hilado's approach and the author's use a different calculation of stoichiometric concentration:

$$\text{Hilado: } C_s = 2.34\% \text{ (N →} NO_2\text{); } LEL = 2.34 \times 0.692 = 1.62\%;$$

$$\text{Author: } C_s = 2.63\% \text{ (N →} N_2\text{); } LEL = 1.51 \pm 0.17\%.$$

Our estimate is a compromise between the experimental values and Hilado's apparently slightly high value. This comparative analysis of the two approaches will be continued within the paragraph that deals with flashpoints since there will then be available better evaluation tools for both methods. The comparison between both tables shows that the range of values is higher than the author's. In particular, sulphur-containing compounds were not considered. The regression conducted for this substance was of mediocre quality because of the small amount of data, so an equation was not proposed.

1.2.5 Limits of inflammability of inflammable mixtures

It is only in a few cases that the limits of inflammability of a vapour mixture can be calculated, thanks to the limits of inflammability of the components in the pure state of a mixture[1]. In fact, Le Châtelier's law only applies to mixtures of saturated aliphatic hydrocarbons and to two or three other mixtures of inorganic substances (CO, H_2) alone or with methane.

This law can be written in the following general form:

$$LEL_m = \frac{100}{\sum_{i=1}^{n} \frac{p_i}{LEL_i}}$$

in which LEL is the lower explosive limit of 'i' in its pure state, p_i its percentage in the mixture of n substances and LEL_m the LEL of the mixture. In this formula Σp_i = 100. Apart from the cases mentioned before, there would be differences between the LEL calculated in this way and the experimental LEL. It appears that this equation is hardly useful.

1.2.6 Influence of temperature on lower explosive limit (LEL)

An empirical formula has been suggested that allows estimation of an LEL at a temperature t $(LEL)_t$ knowing LEL (LEL_0) at temperature t_0 to be:

$$LEL_t = LEL_0 \left(1 - \frac{t - t_0}{600 - t_0}\right) \qquad (8)$$

This equation is to be tested by comparing a few experimental values of LEL at high temperature with the values estimated from this equation for the same compounds. The values of LEL, which can be found in Part Three, are supposed to be obtained at 20°C unless otherwise stated. In actual fact, none of the sources mention this temperature of 20°C. On the other hand, there is sometimes more detail on measurements carried out at higher temperatures. There is some confusion in these values since it is not rare for different authors to give the same values; one of them giving a temperature that is different from 20°C whereas the other does not mention anything. There is limited confidence in this data. Values were chosen only when these were without specified temperatures (eg 20°C) higher than the value at high temperature, because when the temperature rises, LEL decreases and UEL increases, which this equation does not take into account though it can be applied to UEL according to the author's sources.

Here is the result of this comparative work:

Compound	LEL (20°C)	LEL (t°C) experimental	LEL_{calc} (equation 8)
t-Butylbenzene	0.8	0.70 (100)	0.69
Tetrahydronaphthalene	0.8	0.50 (150)	0.62
Biphenyl	0.7	0.60 (100)	0.60
2-Ethylhexanol	1.1	0.88 (104)	0.94
Cyclohexanol	1.3	1.10 (100)	1.12
Isoamyl acetate	1.1	1.00 (100)	0.95

---

1. *Les mélanges explosifs*, Brochure n° 335, INRS, 1980.

Bearing in mind the high levels of error of measurement of this parameter there is a temptation to conclude that the effect of temperature remains small compared with that due to the level of experimental error.

This paragraph is closely linked to the study of flashpoints. In order to be able to get an overall view on uses of lower explosive limits, one will need to refer to this paragraph and the next. To the user, LEL is characterised by the high level of experimental error that goes with its measurement. But, for a purely mathematical reason, this level of error will not affect the use of this parameter greatly. It is thus undeniably useful and will be helpful in the difficult task of measuring flashpoints.

1.3 Flashpoints

1.3.1 Definitions, applicability

Flashpoint is the temperature at which an inflammable liquid builds enough vapour so that this, together with air, forms an inflammable mixture in the presence of an igniting flame. The inflammation has to be very brief when this parameter is measured. If the combustion lasts for longer than five seconds, this temperature is defined as *fire point*. Fire point is never used because it is really difficult to obtain an accurate value. *Flashpoint* is the most important parameter in fire hazard. It plays an essential role in the determination of risk criteria related to the inflammability of a substance.

1.3.2 Measurement of flashpoint

Measurement of flashpoint in practice is based on a simple principle. A liquid substance is placed in a cup. This is slowly heated (approximately one degree per minute) and regularly (every minute) a 'pilot light' equipped with a little flame (about 5 mm long) is placed on the surface of the liquid. Flashpoint is found when the gaseous volume situated above the liquid combusts very quickly giving the impression of a flash. The cup can be either open or closed. In the first case, flashpoint is called 'open cup' (oc) in our tables and in the second 'closed cup' (cc). In this latter case, the lid covering the cup has windows that open during the test in order to enable the flame to be in contact with the gaseous volume situated in the free space of the cup. All windows need to open, which indicates that the complete contents of the cup has been burned. In each of the 'open' or 'closed' types of procedure there are often two types of apparatus depending on whether the substance examined has a flashpoint greater or lower than 70-80°C.

These four main types of apparatus being defined, (scientists and manufacturers have let their imagination go in order to create apparatus). There are now about ten models, which differ by the volume of liquid used (from 2 cm^3 to about 70 cm^3), the metal used for the cup (brass, aluminium), the heating mode (water bath, Bunsen burner, electrical), the type of gas used by the pilot light (natural gas, butane), the level of complexity of automatic controls; some apparatus equipped with several cups can actually be programmed in order to make measurements automatically without the help of the operator. The liquid can be shaken manually or, thanks to an electrical motor, the ignition can be manual or automatic.

The apparatus have various names, Abel-Pensky, Pensky-Martens (PM in our tables), Tag (T), Cleveland (C), Setaflash (S), Luchaire. These abbreviations can be associated with the 'oc' and 'cc' abbreviations. Each country has its own preferences, and, plainly, this situation will not enable flashpoints to be obtained with minimum required coherence.

For this reason a thorough analysis of the level of error of flashpoint measurements is required. It is interesting to note that this situation has been criticised by a lot of authors quoted in the following pages and that all of them have recommended a world-wide experimental approach. Moreover, they all suggested the need for an estimation method while waiting for a standardisation in tests. This standardisation has not happened yet and will be less likely to since there is no apparatus that is better than the rest and it would imply ignoring the manufacturers' commercial interests.

1.3.3 Precision of measurement, sources of inaccuracy

The situation just described creates real confusion in the experimental data of flashpoints and seriously complicates their use in the evaluation of the risk level of a given substance. Nevertheless, how measured flashpoints depend on the type of apparatus is known.

Thus Coffee[1] logically indicates that flashpoint 'oc' is higher than flashpoint 'cc'. For flashpoints 'oc' this same author thinks that the Cleveland apparatus gives flashpoints higher than the Pensky-Martens apparatus, the latter giving higher flashpoints than Tag's. This information can facilitate the choice of values that are likely to be made for the purpose of a safety analysis. Texts on regulations about chemical hazard prevention demand that the method of measurement and the open or closed cup aspects as well as the apparatus must appear on safety sheets, which have to be put at users' disposal by manufacturers. But this partly contradicts the French legal requirement (decree of April 20th 1994[2]) on labelling, which states that it is compulsory to provide the flashpoint of a substance, but does not seem to have this similarly compulsory character with regard to the make of apparatus.

It is certainly this lack of clarity which leads to the current situation. Flashpoints are usually given without any mention of either the open or closed cup aspects or the make of apparatus. Amongst the thousand or so organic substances listed in Part Two, more than one hundred of them mention 'oc' and 'cc', which enables comparisons to be made. Nevertheless, a study of the data indicates that the difference between experimental values can reach 56°C for the same substance (for instance, butadiene). It happens quite often that for flashpoints lower values 'oc' than 'cc' values for the same substance are found. The nature of the sources of the level of measurement error in flashpoints can easily be guessed at.

The causes of fluctuations are:
- nature of 'oc' or 'cc' of flashpoint ('cup' effect);
- type of apparatus ('apparatus' effect);

---

1. R.D. Coffee. Safety in the chemical laboratory, vol. 2, *J. Chem. Educ*. 1971, 51.
2. Decree of April 20th 1994 on declaration, classification, packaging and labelling of substances, *J.O.* of 8th May 1994, p. 6753.

- variant 'high temperature' or 'low temperature' usage for different apparatus and this constitutes another aspect of the apparatus effect;
- experimental level of error itself (called 'operator' effect and 'laboratory' effect);
- purity of substance of which flashpoint is measured ('substance' effect);
- physical conditions (pressure, temperature, humidity... 'environmental' effect).

In an attempt to make a partial analysis of these different causes of fluctuations; the 'substance' effect will be discussed in paragraph 1.3.7.

■ 'Cup' effect

The systematic rereading of the database that appears in Part Three, devoted to organic substances, enables us to identify one hundred and ten substances for which 'oc' and 'cc' are mentioned. It is thus possible to estimate the average of differences between oc and cc. y_c is this value and s_d the standard deviation of these differences, then

$$\overline{y_c} = 5.16°C; s_d = 7.2°C$$

The high value of the standard deviation shows how the situation changes from one substance to the next. The experiment with thousands of measurements accumulated by the author's students shows that when the Cleveland's apparatus is used for flashpoint 'oc' and Setaflash for 'cc', the difference between flashpoints is between 7 and 15°C, depending on the substance. When there is such a difference between two experimental values of a flashpoint, it can then, without too much risk of error, be put down to the 'cup' effect.

The study of this data enables selection of twenty seven substances for which there is extra information since flashpoints Toc and Tcc are given for these substances (as indicated before, 'T' is the abbreviation for Tag in the tables).

Then :

$$\overline{y_c} = 6.22°C; s_d = 7.29°C$$

The small differences between these parameters allow us to infer that the given flashpoints mainly come from the Tag apparatus, which is popular with American experimenters.

■ 'Apparatus' effect

There is hardly any experimental data needed to assess the effect of differences in the apparatus. There are only isolated values for flashpoint PMoc (Pensky-Martens open cup), only two pairs of values for Tcc and PMcc and only three pairs of values for Toc and Coc (Coc: Cleveland open cup).
With these three pairs:

$$\overline{y_c} = 8.467°C; s_d = 9.93°C \quad \text{is obtained.}$$

These figures cannot be used for statistical purposes. However, they do not contradict the indications from Coffee.

■ 'Operator' effect

This effect has not been found in the source measurements carried out by the same or different operators in the same or different laboratories, with the same make of apparatus. None of the levels of error of measurement ever relates to an experimental value of flashpoint. This is incredible since over the past few years techniques of statistical control of testing and experimental planning have been developed which are now compulsory for some activities under the ISO 9000 quality standard.

In order to estimate part of this level of error, data from the author's laboratory was used where the students calculated flashpoints of varied substances using the Cleveland apparatus 'oc' and Setaflash 'cc' that work below 70-80°C. Every year about thirty groups of two students analyse flashpoints of about ten substances. Measurements are thus repeated many times by each group, which enables estimations of standard deviations within a group (intragroup standard deviation) to be made and also between groups (intergroup standard deviation).

Using the results of 1994-1995 the table below contains the list of compounds for that year, the results obtained with the global standard deviation (intra- and intergroup) and the range of values of Part Three. The substances were chosen for their low toxicity, flashpoints that allowed coverage of the required domain of measurement of the apparatus and their low cost (it was a donation). The choice of apparatus for flashpoints was also prosaic. The 'Cleveland' is rather cheap, the 'Setaflash' has the advantage of using only 2 cm^3 per test.

Substance flashpoints

Compound	Indicative values °C (published)	Experimental results ± s (number of values)
1-Butanol	29 - 38	37.00 ± 1.222 (25)
1-Pentanol	33 - 57	50.62 ± 2.725 (25)
Cyclohexanone	44 - 63	45.20 ± 1.323 (30)
p-Isopropyltoluene	47 - 53	49.31 ± 1.440 (30)
1,2,4-Trimethylbenzene	Absent	50.40 ± 1.775 (15)
Cyclohexanol	68	62.97 ± 1.734 (15)
Ethyl malonate	73 - 99	62.46 ± 4.070 (35)

Data for 1,2,4-trimethylbenzene does not appear in the tables as it will be analysed in the case study at the end of this chapter, to show what to do when faced with this type of situation (substance that does not appear in the tables or does appear but has hardly any data).

The Bartlett test gives an 'abnormal' standard deviation for the last compound. This figure can be easily explained by postulating that it is a partially hydrolysed substance, containing ethanol with a low flashpoint, which must have been altered by numerous decantations. Despite this difficulty an analysis of the multiple variance to determine the importance of the level of experimental error, split into intra- and intergroup effects, and the effect relating to the nature of the substance was carried out, which is precisely the only effect sought since it allows measurement of the applicability of the flashpoint to differentiate substances from the point of view of fire hazard. None of the details regarding the statistical aspect will be given here.

The following conclusions can be drawn from the table on variance analysis:
- *2.18% of total variance can be explained by the intragroup effect (merged standard deviation of the five measurements carried out by each group for each substance);*
- *39.97% of total variance can be explained by the intergroup effect (merged standard deviation obtained by all groups working on each substance);*
- *57.84% of this total variance is due to differences in inflammability of the seven substances (standard deviation of averages – indicated in the above table – obtained from all measurements carried out on each of the seven substances).*

It can be seen that everything has been done to optimise the effect of the nature of the substance on the global variance (by choosing substances whose flashpoints cover the majority of the experimental domain allowed by the substance); the effect relating to the experimenters plays an important part in this variance (total of 42.15%).The second observation is the small effect of the intragroup compared with the intergroup. The effect of environmental conditions on the experiments may be one of the reasons. Indeed, intragroup measurements were carried out during the same session, whereas there is a gap of several days between the tests performed by different groups and a gap of six months between the first and last sessions. Therefore, quantification of the 'operator' effect must be modified by adding that it also contains the 'environmental' effect and, as seen with ethyl malonate, another 'substance' effect due to a possible chemical change of the substance stock used during this period of six months. This does nevertheless seem to be applied to malonate only since the other substances seem to be stable enough.

■ 'Environmental' effect

Nothing was done to highlight the environmental effect when the students' work was planned, since it was not thought it would be a relevant factor. Only the flashpoint was corrected to take into account the local pressure because it is imposed by the standard of the Setaflash apparatus. This correction is very minute (0.1 to 0.4°C). The only source of an 'environmental effect' could be climatic – the whole year's work was partially done over three seasons centred on winter. The measurements, classified in chronological order, were submitted to an autocorrection test by Von Newman for each substance. It obviously (very low level of significance in favour of H_o) led to the conclusion that there is a chronological tendency for six substances out of seven. However, a graphical study of this tendency did not lead to a conclusion of a coherent link with a climatic change.

These were three bell curves with a maximum in winter for hydrocarbons and cyclohexanol. There are two monotone decays for two alcohols (pentanol and cyclohexanol). Malonate has a more complex time evolution (two maxima, two minima).1-butanol is the only one that has this unpredictable nature to be expected from a variable whose fluctuations are due to the error of measurement.

This long assessment of the analysis of the level of error of measurement that goes with flashpoint will be completed later (see para 1.3.7) by considering the effect of impurities that can be found in substances at their flashpoint. Nevertheless, it is sufficient to prove that it is not possible to have any confidence in the data of flash-points that can be found in the technical literature, especially when the safety expert has unique data only. To the author's knowledge, there were not until now any documents or databases putting the majority of flashpoint data published on each substance at the users' disposal; in particular works like 'Weka' apparently regularly update their sheets by exchanging rather than adding data. This implies that they are not aware of the status of the experimental measurement. This is the reason why it was decided to mention all the available data.

This choice of exhaustiveness is meant to, at best, analytically 're-use' the data in order to statistically determine the level of confidence one can have in it, or at worst to make a diagnosis of incoherence and opt for a theoretical estimation. The latter approach will be used.

1.3.4 Methods of estimation

In the minds of all authors who favour the estimation of flashpoints based on a theoretical model rather than experimental results this approach was temporary and only supposed to be used during the period used by commissions of experts to lay down a standard technique for the determination of flashpoints. As has already been seen, it is less likely that this method will be used in the near future. This is the reason why we think estimation techniques have to be part of the priority tools of risk analysis in work on chemical risk prevention. Why is such work on estimation important? We will see later that flashpoint is the crucial parameter in order to establish the level of fire hazard of a substance.

The prevention plan, and in particular symbols and warnings on labels and packaging of a substance, will depend on this risk level. In order to make sense, the risk classification has to take into account the inflammability level of a substance, to be on a coherent risk scale and not the mere result of more or less unpredictable fluctuations, particularly those due to the choice of apparatus or working method. The aim of estimation is to be able to identify substances on a scale where their position directly indicates their level of inflammability risk.

■ From boiling points

The relationship between flashpoints and boiling points has been long noted. The first approach to the estimation of flashpoints was the following.

The analysis of experimental results by simple linear regression provides an equation from which the estimation is straightforward. Nevertheless, to obtain an accurate model, an equation for each structural type is needed. Thus, for hydrocarbons, which are one of the best examples for this approach, an equation for linear saturated hydrocarbons is required, one for the branched ones, and one for the cyclic compounds. The same is needed for unsaturated, then aromatic compounds etc. The more the study is based on a precise structural type, the better the linear adjustment and the better the forecast standard deviation but at the same time there will be fewer points with which to calculate the model and the forecast standard deviation will be higher. It is not simple to find a compromise and it was decided to give up on this approach as soon as the relevance of the Hass model was noted.

In order to allow a better appreciation of this approach it will be illustrated in the case of a saturated hydrocarbon whether it is linear, branched or cyclic. In order to get the best possible adjustment only the experimental values listed in Part Three, which were obviously in line with the expected result and rejected flashpoints 'oc', were used.

1 • Flammability

1.3 Flashpoints

The following model for flashpoints 'cc':

$$P_{fla} = 0.6712 \times Eb - 71.1 \quad \text{was obtained}$$

where Eb is the boiling point of the substance at standard pressure expressed in °C and P_{fla} the estimated 'cc' flashpoint.

The residual standard deviation is $s_r = 5.558$, the forecast standard deviation is $s_0 = 5.63$ in the middle of the experimental domain. The coefficient of the linear correlation is $r = 0.9966$. The table below gives some examples of estimations compared with the extreme experimental values found in literature. The confidence range at 95% of the forecast is added in order to show the quality of the estimate.

Examples of experiments and estimated flashpoints

Compound	P_{fla} (°C) exp	P_{fla} est (°C)	Confidence interval at 95%
Ethane	−183; −93	−130.16	−141.97 < P_{fla} ≤ −118.36
Cyclopropane	?	−93.25	−104.8 < P_{fla} ≤ −81.67
Butane	−138; −60	−71.44	−82.92 < P_{fla} ≤ −59.85
Pentane	−49; −40	−46.87	−58.29 < P_{fla} ≤ −35.46
2-Methylbutane	−56; −51	−50.97	−62.39 < P_{fla} ≤ −39.54
Cyclopentane	−42; −7	−37.88	−49.28 < P_{fla} ≤ −26.48
Hexane	−26; −11	−24.12	−35.50 < P_{fla} ≤ −12.74
Cyclohexane	−26; −18	−16.74	−28.12 < P_{fla} ≤ −5.36
Octane	8; 15	13.46	2.06 < P_{fla} ≤ 24.86

The accuracy of the estimate is only of the order of about 10 °C. Despite this, it enables one to have an idea of an unknown flashpoint (in the case of cyclopropane) or to reject absurd flashpoints or identify 'oc' flashpoints. This is the case for ethane for which points -183 and -93 do not appear to be relevant or cyclopentane for which points -20 and -7 are, one or the other, probably 'oc' flashpoints.

This approach is not accurate enough, especially to provide a risk level, but from a practical point of view and without looking for any model, a simple tabular study can be of some help in the choice of experimental values. Indeed, substances are classified in the tables in ascending order of number of carbon atoms (thus by ascending order of boiling points). For the same number of carbon atoms the classification gives in ascending order of carbon atom arrangement the linear, branched (boiling point drop) then cyclic substances (increase of boiling point). In certain cases, when there is a sequence of experimental flashpoint values, it will be possible to estimate an approximate flashpoint.

EXAMPLE: CASE OF CYCLOHEXANOL

There was a discrepancy between the published flashpoint value of this substance (68°C) and that obtained using a Setaflash apparatus (63°C), Could this discrepancy be explained by a systematic difference due to the use of two different apparatus (the published flashpoint indicates the use of a Tag apparatus) or is it due to authors' inaccuracy, including this author's? It would be quite surprising if it were this author's, since the figure is an average of fifteen tests. The study of the available values in Part Three for alcohols whose boiling points – those of cyclohexanol and (in particular) 1-pentanol, cyclopentanol, 1-hexanol, 1-heptanol and above all 2-heptanol, which has the same boiling point as cyclohexanol, leads us to think that the value 63°C is more relevant than the published experimental value (we obviously have to ignore flashpoints 'oc' in this analysis).

1 • Flammability 1.3 Flashpoints

So this approach cannot be ignored, but it can only be complementary to the analysis and cannot replace the estimation technique, which will be preferred and presented and discussed below.

■ From the Hass equation

A lot of authors such as Coffee (already quoted) Gooding[1], Gmehling[2] and Thorne[3] have put forward the following hypothesis, which is another way of defining flashpoints:

- *flashpoint 'cc' is the temperature at which pressure reaches a value so that the resulting equilibrium concentration equals the lower explosive limit.*
- *in the same way flashpoint 'oc' can be defined as the temperature to which one has to heat a liquid so that its vapour gives an equilibrium concentration equal to its stoichiometric concentration.*

These hypotheses are now generally recognised, especially the first one. The resulting reasoning immediately gives way to the estimation model.

Let us call LEL the lower explosive limit of a given substance in percentage, 'Pi' the vapour pressure that allows an equilibrium concentration equal to LEL and 'Pa' the atmospheric pressure

Then:
$$C_{eq} = LEL \, \frac{100 \times Pi}{Pa}$$

$$Pi = 7.6 \times LEL$$

in which Pi and Pa are expressed in mmHg.

By inserting Pi, Eb, and S or the entropic group in the Hass equation (1) given in paragraph 1.1.2, after rearrangement, 't' can be calculated using equation (9), which will be the estimated flashpoint 'cc'.

$$t_{fla} = \frac{273.1 \times \log Pi + 0.4321 \times Eb - 0.15 \times Eb \times \log Pi + S \times Eb - 786.7465}{S - 1.15 \times \log Pi + 3.3129} \quad (9)$$

In the same way $P_{cs} = 7.6 \times C_s$ provides estimation of flashpoint 'oc'. The accuracy of the equation will here depend on two unequal factors:

- quality of the calculation of S and/or group. There are numerous previous examples;
- quality of the experimental value of the lower explosive limit (LEL) and of C_s.

Two comments can be made on the second point. For a simple mathematical reason mistakes made with the LEL value are of little consequence to the calculated value of flashpoint 'cc'. Indeed, this mistake is not that significant since there is a logarithm involved. Secondly, in theory no mistake is made with the stoichiometric concentration (except for nitrogenous compounds where there is an ambiguous aspect with regard to the nitrogen reaction). This second approach (with C_s) can thus provide preliminary control of the model parameters (S or the group) and there

---
1. C.H. Gooding, *Chemical Engineering*, 1983, p.88.
2. J. Gmehling, P. Rasmussen, *Ind. Eng. Chem. Fundam.*, 1982, 21, p.186-88.
3. P.F. Thorne, *Fire and materials*, 1976, p. 134-39.

is good reason to think that experimental flashpoint 'oc' is reliable (several independent, hardly dispersed, values).

Finally: do calculated flashpoints 'oc' and 'cc' preferentially estimate values given by particular makes of apparatus?

This can be answered by carrying out an extended case study in which substances are chosen with often varied complex structures from a functional point of view and for which there are a large number of experimental values for which a named apparatus are usually mentioned. As far as possible a triple approach is used, employing experimental LEL, calculated LEL using Hilado's method and/or the suggested variant.

> The author also included, for a reason that will be given later, the three alcohols used in the author's laboratory by the students. We excluded ethyl malonate, which caused problems. We will not go into detail concerning calculations of S and group determination since these aspects were largely covered previously. We will not pass any judgement on the different approaches, letting the reader be his own judge.

1.3.5 Examples of calculation

EXAMPLE 1: 2-BUTOXYETHANOL

Here is the available data from Part Three:
2-Butoxyethanol: $C_6H_{14}O_2$ • M(g/mol) = 118.18 • CAS No.: 111-76-2
Physical data: Eb (°C/mmHg) = 168.4-172; 94/50; 62/10 • P_{vap} (bar) (°C) = 0.8; 0.9; < 1.33(20); 1.1(25); 8(50); 400(140) • LEL/UEL = 1.1/10.6; 1.1/10.7; 1.1/12.7; 1.9/10.3; 1/10.6 • P_{fla} = 60cc; 65.6Tcc; 71.1PMcc; 71oc; 74oc; 85Coc • AIT = 230; 240; 244

The situation given by the distribution of flashpoints is not particularly clear. 11 and 14°C respectively separate the extreme values of flashpoints 'cc' and 'oc'. The maximum flashpoint 'cc' is equal to the minimum flashpoint 'oc'. The evidence is inconclusive.

Choice of group or calculation of S. The Hass table does not provide the answer. The author's method gives S = 5.448 (group 4).

Choice of LEL. The median experimental value (1.1%) seems to be the most relevant. Hilado's method is problematic; the choice is between:
a = 0.476 (alcohols) and a = 0.537 (ethers)

This type of function is dealt with easily using the author's method.
C_s = 2.406; LEL = 1.127 (0.90 ≤ LEL ≤ 1.35) taking the confidence interval into account.
Eb = 172°C is chosen.

Then, for flashpoint 'cc': with LEL = 1.1, P_{fla} = 58.789°C; is obtained with the interval resulting from the author's approach: P_{fla} = 59.22 (55.27 ≤ P_{fla} ≤ 62.475);
For flashpoint 'oc':
C_s = 2.406, P_{fla} = 73.79°C.

EXAMPLE 2: ETHANOLAMINE

Ethanolamine: C_2H_7No • M(g/mol) = 61.10 • CAS no.: 141-43-5
Physical data: Eb (°C/mmHg) = 170-172; 100/50; 68/10; 69-70/10 • P_{vap} (bar) (°C) = 0.27; 0.5; 0.64; 0.7; 5.2(20); 8(60); 20(80); 55(100) • LEL/UEL = 2.5/17; 5.5/17 • P_{fla} = 85cc; 93oc; 96PMcc; 104Coc • AIT = 385; 779

1 • Flammability 1.3 Flashpoints

The flashpoint experimental data 'overlap' each other. Notice in anticipation that there is a limit of 93°C, which leads to a change in the 'NFPA' classification code (inflammability from 1 to 2) showing how sometimes the level of error can have important consequences on risk evaluation. This example shows up another concern, which is in considering a comparison of the approach by Hilado with the author's variant, since there is a 'conflict' between the methods on the calculation of stoichiometric concentration.
The two experimental data contradict each other badly and will not be used.

Hilado's method: C_s = 4.698; LEL = 3.25.
Author's variant: C_s = 6.056; LEL = 3.407 (2.407 ≤ LEL ≤ 4.406).

It is difficult to choose between groups when using the Hass table; the substance can belong to group 4 (ethylamine) or 7 (ethylene glycol) or even 8 (ethanol).
Using the author's approach that gives S = 6.044 (group 7)
Then:
flashpoint 'cc' according to Hilado:
LEL = 3.25; P_{fla} = 86.593° C
flashpoint using the author's variant;
LEL = 3.407; P_{fla} = 87.513 deg C (80.87 ≤ P_{fla} ≤ 92.623).

The author's method which gives an interval of confidence shows that we have ≤ 95% chance to be right when asserting the NFPA code. 1 can not be attributed to this substance (this will be explained; p.81):

    flashpoint 'oc' according to Hilado:
    C_s = 4.698; P_{fla} = 93.925°C;
    flashpoint 'oc' following the author's variant:
    C_s = 6.056; P_{fla} = 99.189°C.

EXAMPLE 3: BUTANONE
Butanone: C_4H_8O • M(g/mol) = 72.11 • CAS no.: 78-93-3
Physical data: ΔH_f^0 (KJ/mol) = –235.39(g); –273.17(l) • Eb (°C/mmHg) = 79-80; 25/100; 6/40 • P_{vap} (bar) (°C) = 32.4(0); 56.6(10); 94.64; 95; 99; 103; 104; 105(20) • LEL/UEL = 1.8/9.5; 1.8/10.1; 1.8/11.5; 1.8/12.5; 2/10 • P_{fla} = –9Tcc; –6oc; –7Tcc; –6Tcc; –4Toc; –3; –1Toc • AIT = 403; 505; 514; 516; 550-615

The Hass table as well as the author's approach leads to group 2 for this substance.
By using the two experimental values LEL = 1.8% and 2%, Hilado shows that 2% is a legitimate value:
C_s = 3.67; LEL = 1.97.

The author's approach is more centred on 1.8%: LEL = 1.827 (1.51 ≤ LEL ≤ 2.146).

So, for flashpoint 'cc':
using LEL experimental values: P_{fla} = -11.219°C and P_{fla} = -9.524
    for Hilado P_{fla} = -9.769°C;
    for us P_{fla} = -10.981°C (-13.987 ≤ LEL ≤ -8.376);
For flashpoint 'oc' P_{fla} = 0.874°C.

EXAMPLE 4: ACRYLIC ACID
Acrylic acid: $C_3H_4O_2$ • M(g/mol) = 72.06 • CAS No.: 79-10-7
Physical data: ΔH_f^0 (KJ/mol) = –336.23(g); –384.09(l) • Eb (°C/mmHg) = 139-142; 122/400; 103.3/200; 86.1/100; 69/50; 66.2/40; 39/10; 27.3/5 • P_{vap} (bar) (°C) = 4; 4.13; 5.33(20); 53.3(39.9); 53.3(60) • LEL/UEL = 2/13.7; 2.0/8.0; 2.4/8; 5.3/26; P_{fla} = 46; 50Tcc; 52oc; 54Toc; 62; 68Coc • AIT = 374; 395; 412; 429; 541

It is difficult to use experimental data with this substance; discrepancy exists between the pressure values at 20°C, a figure apparently unreasonable in the lower and upper explosive limits, and 16°C between the extreme values of flashpoint 'oc', not to mention autoinflammation temperatures that will be discussed in paragraph 1.4 (the reader will notice that the relevant value is 395°C when using our autoinflammation estimation method). The choice of group according to Hass would lead us to group 5 if we consider that propionic acid is the most similar (same number of carbon atoms). Our approach gives $S = 5.387$ (group 4).

The LEL according to Hilado's method gives $C_s = 6.528$; LEL = 3.51%.
A verification of the coherence of both approaches consists in estimating vapours at 20°C; group 5 gives $P_{vap} = 3.48$ mbar whereas $S = 5.387$ gives $P_{vap} = 4.11$ mbar, thus is better.

According to the author's variant we obtain LEL = 3.17% but with high inaccuracy (1.6 ≤ LEL ≤ 4.73). In both cases, value 5.3 is rejected. $Eb = 142°C$ is obtained:
For flashpoint 'cc':
according to Hilado: $P_{fla} = 58.41°C$;
according to the author: $P_{fla} = 54.438°C$ (41.97 ≤ P_{fla} ≤ 62.40);
For flashpoint 'oc':
according to Hilado: $P_{fla} = 70.841°C$; according to author: $P_{fla} = 69.114°C$.

EXAMPLE 5: ACETIC ANHYDRIDE
This example has already been studied regarding vapours and an important discrepancy observed between Hass (group 6) and the author's method (group 4; $S = 5.308$). Do we find the same discrepancy with flashpoints?

Acetic anhydride: $C_4H_6O_3$ • M(g/mol) = 102.09 • CAS No.: 108-24-7
Physical data: ΔH^0_f (KJ/mol) = –575.72(g); –624.00(l) • Eb (°C/mmHg) = 138-140; 65/50; 36/10 • P_{vap} (bar) (°C) = 4.7; 5; 5.33(20); 13.3(36); 300(100) • LEL/UEL = 2/10; 2/10.2; 2.7/ 10.3; 2.8/12.4; 2.9/10.3; 3/10 • P_{fla} = 49Tcc; 52; 54cc; 64Toc • AIT = 315; 331; 380; 385; 389; 392

Both methods will be used with the experimental LEL (2 ≤ LEL ≤ 3), according to Hilado ($C_s = 4.977$; LEL = 2.673; the available LEL data did not allow a regression to be conducted in order to calculate 'a'; so Hilado's figures are used). $Eb = 140°C$ is obtained:
For flashpoint 'cc':

with experimental LEL and group 6:	50.653 ≤ P_{fla} ≤ 57.915
with experimental LEL and $S = 5.308$,	44.88 ≤ P_{fla} ≤ 52.453;
with group 6:	$P_{fla} = 54.172°C$;
with $S = 5.308$:	$P_{fla} = 48.697$;

For flashpoint 'oc':

with group 6:	$P_{fla} = 65.86°C$;
with $S = 5.308$:	$P_{fla} = 60.917$.

EXAMPLE 6: ESTIMATION OF FLASHPOINTS OF SUBSTANCES STUDIED IN AUTHOR'S LABORATORY
Ethyl malonate was excluded because of the comments already made.

These estimates are reformulated in the form of the confidence interval at 95%.
Estimations of flashpoints are carried out using LEL and S or groups calculated according to the author's method.

Comparison of experimental and published flashpoints

Compound	Flashpoint (°C)	Experiment results Confidence interval	Estimated confidence interval P_{fla}
1-Butanol	29 - 38	$36.50 < P_{fla} \leq 37.50$	$29.12 \leq P_{fla} \leq 35.69$
1-Pentanol	33 - 57	$49.50 < P_{fla} \leq 51.74$	$44.96 \leq P_{fla} \leq 51.52$
Cyclohexanone	44 - (63 co?)	$44.71 < P_{fla} \leq 45.70$	$35.28 \leq P_{fla} \leq 41.77$
p-Isopropyltoluene	47 - 53	$48.77 < P_{fla} \leq 49.85$	$47.74 \leq P_{fla} \leq 50.71$
1,2,4-Trimethylbenzene	Absent	$49.42 < P_{fla} \leq 51.38$	see para 1.5.6
Cyclohexanol	68	$62.01 < P_{fla} \leq 63.93$	$51.02 \leq P_{fla} \leq 58.32$

There is an apparent discrepancy between the two compounds, cycloaliphatic cyclohexanone and cyclohexanol. For the latter there is a discrepancy anyway between the three columns, which makes testing difficult (the industrial substance, olone, is a mixture of these two substances; is our substance pure? This would explain the discrepancy between the first two columns but not the one of the theoretical estimation). We do not have enough confidence in the experimental results found in the literature to reject our estimates, especially when we have only one value.

1.3.6 Estimation of lower explosive limit

The possibilities offered by the lower explosive limit of estimating a flashpoint 'cc' have already been discussed in detail. It would be possible, vice versa, to estimate a lower explosive limit given a flashpoint 'cc' in which there is some confidence. To do so one simply needs to calculate, using equation (1) in paragraph 1.1.2, the vapour pressure of a substance at the temperature of the flashpoint, then for its equilibrium concentration C_{eq}, in these conditions C_{eq} = LEL. But an error in P_{fla} has big consequences on C_{eq} and thus on LEL. This is the reason why this approach is less reliable than the other and can only be used when there is certainty about the flashpoint.

EXAMPLE 1:
One can take as an example the flashpoint method used in the author's laboratory. The measured flashpoints were the subject of repetitions, which gave an accurate confidence interval which was relatively narrow. Therefore these figures can be used with confidence. The following is obtained.

Comparison between published experimental and estimated values of LEL (%)

Compound	LEL (published)	Experimental results Confidence interval	LEL est.
1-Butanol	1.4	$36.50 < P_{fla} \leq 37.50$	$2.03 \leq LEL \leq 2.17$
1-Pentanol	1.2	$49.50 < P_{fla} \leq 51.74$	$1.66 \leq LEL \leq 1.91$
Cyclohexanone	$1.1 \leq LEL \leq 1.3$	$44.71 < P_{fla} \leq 45.70$	$1.81 \leq LEL \leq 1.90$
p-Isopropyltoluene	0.7	$48.77 < P_{fla} \leq 49.85$	$0.76 \leq LEL \leq 0.81$
1,2,4-Trimethylbenzene	Absent	$49.42 < P_{fla} \leq 51.38$	see para 1.5.6
Cyclohexanol	$1.25 \leq LEL \leq 2.4$	$62.01 < P_{fla} \leq 63.93$	$1.84 \leq LEL \leq 2.04$

There is a big gap compared with the majority of experimental data, which are themselves rarely consistent with the experimental flashpoints compared with the accepted hypothesis based on the relationship between C_{eq} and LEL at the temperature of flashpoint 'cc'.

EXAMPLE 2: MESITYLENE
Apparently only the lower explosive limit of 100°C is available for this substance (see Part Three). The empirical formula linking the LEL to both temperatures could be used; this would give an LEL of 0.93% at 20°C. What result does this approach provide, if in accordance with Hass, this substance is considered as belonging to entropic group 2 (hydrocarbons) and determine its pressure at the flashpoint temperature?
There are three flashpoint values: 44, 46, 54°C that are known, but flashpoints 'cc' and 'oc' lead to 54°C as an 'oc' value, keeping the two others
  with P_{fla} = 44: LEL = 0.976%;
  with P_{fla} = 46: LEL = 1.091% are obtained.

Are the hypotheses regarding the choice of group and flashpoint 'oc' coherent? Calculating flashpoint 'oc' with C_s = 1.716%; the 'oc' estimated, P_{fla} = 54.47°C is obtained, which is perfectly consistent. This approach shows that confidence based on a satisfactory coherence of two values of P_{fla} 'oc' separated by only 2°C that confirm each other, and on a coherence between the estimated and experimental values of P_{fla} 'oc'. The result is not inconsistent with the experimental LEL at 100°C given the inaccuracy of LEL and the empirical model linking the LELs at different temperatures.

Plainly, instinct plays a significant part in conducting this method, personal experience playing an important role and each substance requiring a specific approach.

EXAMPLE 3: LEL of 2,4-DICHLOROTOLUENE
There is no value available for LEL and two flashpoint values, which lead to flashpoints of 79 ('cc') and 87 a flashpoint 'oc'. S = 5.071 (group 2) is obtained for the group, which is consistent with the Hass table that indicates that the group of this substance is the same as the one resulting from an exchange of halogens for hydrogen atoms.
This leads to toluene, which is a hydrocarbon (group 2).
Calculate P_{vap} and C_{eq} with t = 79°C; S = 5.071 and Eb = 200°C; P_{vap} = 14.73 mbar; C_{eq} = 1.454 = LEL. Is this result consistent with the group hypothesis and flashpoint 'oc' given to value 87? Calculating this flashpoint with C_s = 2.55%; P_{fla} = 90.80°C is obtained. This proves to be coherent.

1.3.7 Mixture flashpoints

It is not crucial in regard to safety to devise an estimation model for mixture flashpoints given the complexity of the issue. Generally speaking, in most cases it will be sufficient to know the inflammability parameters of the pure compounds, at least for mixtures of inflammable substances.

■ The Thorne and Gmehling models

There are several authors who have shown interest in estimating mixture flash-points from the point of view of studying mixtures of inflammable substances, and mixtures of an inflammable substance with one or several incombustible substances. This latter aspect was of particular interest because this type of mixture is sometimes used to improve solvent qualities and reduce the inflammability.

These incombustible substances are halogenous derivatives whose extinction properties are well known and some molecules were used as extinction agents before it was realised they were toxic. Paint stripper containing methanol and dichloromethane is an example of this type of mixture that is currently marketed. The interest in mixtures of inflammable liquids in the context of this book is that it would enable us to identify the effect of impurities on the flashpoint of a substance. Indeed, the question was asked during the discussion over the origin of the level of measurement error if the important discrepancies between the different bibliographical sources could not be explained by variable quantities of various impurities.

Thorne, already quoted (note p.63), tried to estimate the possibilities of theoretical forecasts of flashpoints by using hydrocarbon mixtures in a study that was carried out in order to model fuel flashpoints and to study the effect of halogenous compounds on these hydrocarbons.

The study is based on four linear hydrocarbons (in C_1, C_6 to C_8) and the model uses Antoine and Clapeyron's equations. The flashpoints used by the author do not take into account all experimental values that are currently available; the correlation coefficients obtained during multiple linear regression adjustments between experimental and estimated values are very bad (0.90 to 0.98; see the huge errors obtained from a correlation study concerning flashpoints for which the present writer still has a coefficient of 0.9966). The model can be used if differences between pure cmpounds are still low regarding boiling and flashpoints.

Another estimation method of mixture flashpoints was suggested by Gmehling (note p.63). The method uses the forecast technique of activity coefficients of liquid mixtures called 'UNIFAC' that would therefore enable calculation of the vapour pressure of the mixtures and, thanks to Le Châtelier equation, calculate the temperature to which the mixture has to be heated so that its equilibrium concentration reaches the lower explosive limit.

The author gives an example of a study concerning a mixture of ethanol, toluene and ethyl acetate. The case is presented in the form of a Scheffe plan for which choice of compound quantities are not optimised to obtain a good matrix as shown in the matrix of effects correlation; there is no point repetition in the middle of the matrix, which thus excludes the quantification of the level of error of measurement that can only be estimated by the residual standard deviation of the regression. Finally, the author uses flashpoints of pure substances from partial experimental data. The available data give 9 to 13°C for ethanol (the author 12.8), 2 to 9°C for toluene (5.56) and -4 to -2°C for ethyl acetate.

The inaccuracies of the approach lead the present writer to disagree with the author's conclusions on the UNIFAC approach, which according to him is better than the one of comparing the mixture with an ideal mixture.

This criticism is made in order to make the reader understand the highly unpredictable aspect of experimental approaches concerning flashpoints. If the approach is experimental, all statistical tools need to be employed to provide conclusions that include a calculated error level.

■ An experimental approach

In 1996 the present writer offered the students a study of the 1-butanol/1-pentanol/cyclohexanol mixture using Scheffe's plan (mixture plan) which used a standard experimental matrix offered by 'LUMIERE' software. The substance choice takes into account the limited possibilities of the Setaflash apparatus, the low toxicity of the chosen substances and their reciprocal chemical inertia. These had been

1 • Flammability
1.3 Flashpoints

studied for years and the level of measurement error with this type of apparatus and the experimenters were known. The mixture plans give very useful graphical triangular representations but have the inconvenience of having very high forecast standard deviations especially in the middle of the experimental domain.

The figures obtained are shown in the table below, which as well as giving the values obtained, includes the estimations from the regression model with their confidence interval at 95%:

Flashpoint of mixture (deg C)

Molar fractions of alcohols C4(1), C5(2), C6(3)			P_{fla}	P_{fla} est. using model	Confidence interval of estimation
x_1	x_2	x_3			
1	0	0	37.5	36.20	31.40 - 41.00
0	1	0	48.8	48.18	43.81 - 52.55
0	0	1	61.1	60.68	56.64 - 64.72
0.5	0.5	0	32.5	32.68	25.79 - 39.52
0.5	0	0.5	31.5	30.70	23.66 - 38.05
0	0.5	0.5	53.1	54.43	47.77 - 60.44
0.66	0.17	0.17	29.0	30.16	22.74 - 35.32
0.17	0.66	0.17	42.4	41.95	36.11 - 48.61
0.17	0.17	0.66	44.0	45.33	39.66 - 52.22
0.33	0.33	0.33	31.5		
0.33	0.33	0.33	37.0		
0.33	0.33	0.33	38.0	36.00	30.45 - 42.75
0.33	0.33	0.33	38.0		
0.33	0.33	0.33	38.5		
0.33	0.33	0.33	37.0		

The calculation of the model using multiple linear regression gives:

$P_{fla} = 36.20[C_4H_{10}O] + 48.18[C_5H_{12}O] + 60.68[C_6H_{12}O] - 38.04[C_4H_{10}O] \times [C_5H_{12}O] - 70.96[C_4H_{10}O] \times [C_6H_{12}O]$

Tests on coefficients were carried out with the help of the residual standard deviation.

Thanks to this model it is possible to calculate the quantity of 1-butanol by percentage and weight from which the flashpoint of cyclohexanol decreases significantly. It is found to be 0.14% of butanol (molar fraction: 0.002). This calculation was made supposing there is no pentanol in the mixture and a flashpoint target at 56.64°C, the lower limit of the confidence interval was at 95% of pure cyclohexanol.

The aim was to demonstrate the considerable effect of an impurity of 'low' flashpoint on the flashpoint of a substance. This effect is obvious even though a flashpoint had to be chosen that did not facilitate the demonstration (the butanol flashpoint is not as low as expected; for instance, what would be the result for ethyl oxides after extraction of a compound with ether?)

This study also provided the opportunity to compare the flashpoints of these three alcohols with those obtained two years previously (see para 1.3.3) by other students on other stocks of substances. One can question whether the differences are not precisely related to the differences in purities of some of them. These differences show once more the high levels of measurement error of flashpoints whatever their origin.

Finally, mention should be made of the two effects of interaction of the mathematical model whose negative coefficients show minima of flashpoints for the binary butanol/cyclohexanol and butanol/pentanol combinations. Can they be explained by the presence of azeotropes in these substances? The tables examined did not list these mixtures and there was no time to do an experimental check with the students.

1.4 Autoignition temperatures (AIT)

This last inflammability parameter presents problems. After stating its definition it will be seen that measuring autoignition temperature proves to be a difficult exercise because its measurement is sensitive to the experimental conditions, even more sensitive than for flashpoints. Worse, this parameter seems to be controlled by kinetic factors far more complex to master than the thermodynamic factors that probably control flashpoints (in fact it is a liquid/vapour equilibrium). So whilst the influence of the nature of the cup metal on a flashpoint has never been demonstrated, this demonstration was easily made with autoignition temperatures.

1.4.1 Definition

'Autoignition temperature is the temperature at which a substance burns up in the absence of any inflammation source (flame, spark).'

Hilado[1] gives another version of this definition.
'Autoignition temperature is the lowest temperature at which a substance starts to heat itself at a speed that is sufficient to trigger its ignition.'

This second definition explains why autoignition appears after a certain time and that according to the set period of time there can be a different AIT for the same substance.

1.4.2 Mode of operation, apparatus

The most commonly used apparatus is ASTM 2155-69[2]. Its European version is described in the AFNOR NF T 20-037 standard. Its simplified operating method is the following: a small sample of substance placed in a hypodermic syringe of volume less than 1 cm^3 is injected into a container made of borosilicate glass containing 200 cm^3 air and is heated. AIT is the lowest container temperature at which there is ignition, which is observed when a flame appears inside the container placed in semi-darkness. This ignition has to occur within less than five minutes after injection. For tests conducted a complete range of their apparatus was offered, which varied by the volume and nature of the container materials, the air being static or on the contrary circulating in the apparatus. The effect of these variations on conditions is disastrous, at least if one wants to get figures that are characteristic

---

1. C.J. Hilado, S.W. Clark, *Chem. Eng.*, 1972, pp. 75-80.
2. American Soc. for Testing and Materials, ASTM 2155-69, ASTM Std., 17, Nov. 1970, pp. 724-727.

of substances and can be repeated and reproduced. It does not seem that these effects can be quantified since they depend on the AIT level. Some figures providing approximate values are to be found in Hilado's publication.

1.4.3 Effect of different physical variables

■ Nature of the container material

The highest autoinflammation temperatures are obtained when the container is made of glass. With a metal container not only are the values low but they also depend on the state of the metal surface and especially on the potential presence of oxides that play a catalytic role. A drop in temperature can then reach at least 50°C with substances that have AIT greater than 290°C. The effect of the material of the container varies hugely depending on the substance. Thus the AIT results are the same for toluene with glass and metal whereas with methyl acetate they are respectively 502 and 454°C. But is it due to the presence of oxides?

■ Container volume

Tests were carried out with apparatus volumes that varied between 8 and 12000 cm^3. AIT decreased and the self-ignition waiting time increased when the volume increased. The following AIT extreme values and waiting time (τ) for the three compounds below were:

toluene: AIT = 649°C (0 cm^3, extrapolated); AIT = 482°C (12000 cm^3);

τ = 48 s (125 cm^3); τ = 220 s (12000 cm^3);

ethanol: AIT = 495°C, τ = 5 s (8 cm^3); AIT = 363°C, τ = 160 s (12.000 cm^3);

acetone: AIT=676°C(8cm^3);AIT=467°C(12000cm^3);

τ = 26 s (200 cm^3); τ = 85 s (12000 cm^3).

Making a detailed comparison of the figures given by the bibliography and all the results compiled by Hilado on the volume effect on the AIT of the three above substances, the 1 to 2 °C differences that appear between some values in both sions from °F to °C.

Effect of container volume on AIT

Compound	AIT (°C) [volume in cm^3]	AIT (°C) [Part 3]
Toluene	649 [8]	
	584 [35]	
	568 [125]	
	553 [150]	552
	536 [200]	536
	519 [1000]	
	482 [12000]	480

1 • Flammability
1.4 Autoignition temperatures (AIT)

Compound	AIT (°C) [volume in cm³]	AIT (°C) [Part 3]
Ethanol	495 [8]	
	445 [35]	
	426 [125]	427
	421 [150]	425
	402 [200]	
	391 [1000]	
	363 [12000]	361; 363 [to 425]; 371 [to 427]
Acetone	676 [8]	
	570 [35]	
	561 [88]	560
	636 [125]	
	633 [150]	
	519 [200]	538; 540
	491 [1000]	
	467 [12000]	465 to 560

■ Conditions of air/vapour contact

AIT values would be lower in stagnant rather than in circulating air.

1.4.4 Effect of chemical structure

It now has to be ascertained whether, despite the numerous causes mentioned previously, AIT remains a sufficiently 'discriminating' parameter to render it suitable as an analysis criterion of the level of inflammability risk of a substance.

By way of introduction to this subject the effect of some structural characteristics on this parameter will be set out (but without going into any detail of how to handle the statistical tool) by limiting the study to hydrocarbons in order to simplify it. The above table shows that the database always contains the lowest AIT value obtained by using a 12 l container. In order to simplify the study even more the lowest AIT values for each substance are used.

So for hydrocarbons:
- the effect of the hydrocarbon chain length and of the chain ramification was studied by limiting the study to only one lateral chain, the effect of the saturated, ethylenic or aromatic nature of hydrocarbon was considered.
- The entire study was carried out by analysing the multiple variance and multiple linear regression of 55 values. These two approaches show that the effects of these three factors are very important.

The main results are as follows.

■ Effect of chain length

Autoignition temperature of hydrocarbons decreases with the length of the hydrocarbon chain.

By limiting the study to structures for which there are several compounds the average AIT value changes from 651 ± 36°C for substances with two carbon atoms to 284 ± 16°C for hydrocarbons with ten carbon atoms.

■ Effect of the 'shape' of the carbon skeleton

On the basis of the analyses carried out it can be maintained that the presence of branching or an aliphatic cycle increases AIT. The analysis does not allow differentiation between the effects of branching or cyclization. The AIT averages of the linear, branched and cyclic compounds are respectively: 364 ± 10°C, 437 ± 17°C and 424 ± 11°C.

■ Effect of unsaturation

Saturated and ethylenic compounds have the same AIT averages; on the other hand the presence of an aromatic cycle significantly increases this parameter. AIT averages for these groups are respectively: 339 ± 9°C, 347 ± 14°c and 538 ± 16°C.

1.4.5 Rules of group additivity; estimation of AIT

These results thus show that whereas the flashpoint was only moderately influenced by the compound structure (their chemical functionality but especially their atomic composition and vapour), autoignition temperatures seem to be closely linked to the structural factors that affect the chain. So additivity rules for estimation of AIT should be sought. Every time a chemical or physical property is highly influenced by the structure, chemists tried to establish rules that enable one to reduce a molecule to characteristic groups for which the contribution to the value of this property is known. This was done for instance by Kinney[1] for boiling points and Benson[2] for thermochemical properties.

Kinney's method was used in the author's lectures for several years but did not prove to be exploitable since it does not allow estimation of boiling points of many substances such as halogenous derived substances. On the other hand, Benson's method proved to be very useful and will be discussed in Chapter 2 ('Stability').

The result of studies performed in the laboratory in the search for additivity rules from the substance formula are shown below. This study was carried out using multiple linear regression, the software being limited to the analysis of the main effects of fourteen factors, the study had to be subdivided into groups of compounds starting with hydrocarbons.

This constraint leads to the conclusion that the contribution of a hydrocarbon group is not influenced by the proximity of a particular grouping (under certain conditions).

The identified groups are the primary, secondary, tertiary and quaternary saturated hydrocarbon groups then the ethylenic, aromatic, etc. groups. The Benson's group definitions provided a lead but complete changes had to be made to this approach every time a group did not have hydrogen atoms. Thus the carbonyl group contribution cannot be determined since it leads to incoherent values from one substance to the other. Nevertheless, the effect can be easily quantified when

---

1. Contribution in Dean, *Lange's Handbook of Chemistry*, 13th ed., 1985, McGraw Hill.
2. S.W. Benson and al., *J. Chem. Phys.*, **29**, 1958, 546 and previous publications by same author.

1 • Flammability   1.4 Autoignition temperatures (AIT)

studying the contribution of hydrocarbon groups linked to a carbonyl group eg the changes caused by the presence of a carbonyl group on the hydrocarbon group.

In the same way it was found through analysis incorporating second-order interaction coefficients that the aldehyde AIT is not conditioned by the CHO effect on the close hydrocarbon groups as was the case with ketonic carbonyl, but by the effect of close groups on aldehydic C-H.

The methodology proceeded as follows:

For hydrocarbons. *The different group constituent of each substance, for which at least one AIT is known, is identified (if several AIT are known, the lowest is retained). Thus 4-ethyltoluene has two methyl groups, one methylene, two aromatic carbon atoms surrounded by three carbon atoms and four aromatic carbon atoms that have a hydrogen atom. Moreover if there is a 'para' arrangement and two AITs for this compound; the lowest ie 475°C is retained. For all usable substances, using the data in Part Three, a multiple linear regression on the table is used, choosing a statistical model that will allow us to know what the best regression models are. After choosing the most suitable one from a chemical point of view a model is obtained and calculated AIT values to experimental values compared. If the calculated value is more than three times the standard deviation of the experimental value, another, more appropriate experimental value is considered. If it does not exist, a check is made for a low represented structure in the list or finally if by some analogies found in the list we cannot come to a conclusion, then the value can be regarded as unacceptable. Experience with this data helps one to act with as much discernment as possible. After 'correction' the final model is completed and leads to 'contributions of self-ignition of groups'.*

For functional compounds, eg alcohols. *The assumption is made that the contributions of hydrocarbon groups are the same as in the previous compounds except for groups directly linked to hydroxyl.*

The procedure is as follows:

1. *Three types of arrangements are identified, CH_2OH (primary alcohols), CHOH (secondary) and COH (tertiary) groups.*
2. *The hydrocarbon groups contribution values from chosen experimental AIT are subtracted. Thus propane-1,2-diol has one CH_2OH, one CHOH and one CH_3; the chosen AIT value is 371°C (see Part Three).*
3. *The methyl group contribution (199.5°C) is subtracted; 171.5 is the contribution of the two previously mentioned alcohol functions.*

This approach is obviously simplified but can only be improved when there is much more data available than in this specific case. This enriched approach was not possible given the poor bibliographical source. It was made possible because of availability of current databases. The rules presented here are nevertheless sufficient since the published AIT are not very accurate. The table below gives the contributions of groups that could be determined. It is impossible to have access to all types of structures since some of them are not well documented enough for the time being.

1 • Flammability
1.4 Autoignition temperatures (AIT)

The table below gives the 'increments of self-ignition of the groups' that can be calculated along with the estimated standard deviation (the forecast standard deviation is far too important).

Increments in autoignition of groups and estimated standard deviation

N°	Group	Increment in AIT	Divergence in estimate
1	CH_3-	199.5	9.6
2	$-CH_2-$	−27	3.9
3	$>CH-$	−229	23
4	$-C-$	−373	40.3
5	$=CH_2$	204	16.5
6	$=CH-$	−19.2	13.55
7	$=C-$	−282.9	25.7
8	$HC_{Ar}-$	68.5	5.7
9	$C_{Ar}-$	−58.5	21.4
10	Aliphatic cycle	460	31
11	Ortho	−95.5	43.9
12	Meta	−50	35
13	Para	−55	37.7
Alcohols			
14	$-CH_2OH$	202	3.8
15	$-CHOH-$	−3.41	7.6
16	$>COH$	−158	21.2
Ethers			
17	CH_3-O	133.7	19
18	$-CH_2O$	−61.5	6.8
19	$>CH-O$	−207	26.4
20	$=CH-O$	−79	39
21	$C_{Ar}-O$	−57	29
22	Epoxy cycle	492	37
23	Other heterocycles	438.3	41

1 • Flammability
1.4 Autoignition temperatures (AIT)

N°	Group	Increment in AIT	Divergence in estimate
	Chlorinated derivatives		
24	$-CH_2Cl$	219.5	18.3
25	$-CHCl-$	139	51
26	$\smash{\diagdown\!\!\!\diagup}\!C-Cl$	0	
27	$-CHCl_2$	216.5	62.6
28	$=\overset{(H)}{C}-Cl$	204	34
29	$=CCl_2$	222	64
30	$C_{AR}-Cl$	181.3	31
31	Ortho Correction	−89.3	67.3
	Amines		
32	CH_3N	66.4	11.3
33	CH_3NH	177.3	22.14
34	$-CH_2NH_2$	183.5	10.4
35	$-CH_2NH$	−21.6	8.4
36	$-CH_2N$	−108.6	6.4
37	$\diagdown\!\!\!\diagup CHNH_2$	−26.8	28.5
38	$\diagdown\!\!\!\diagup CHNH$	−241.5	16.5
39	$\diagdown\!\!\!\diagup CNH_2$	−222	49
40	$C_{Ar}-NH_2$	147.2	18.5
41	$C_{Ar}-NH$	−44.2	20.2
42	$C_{Ar}-N$	−175.4	37
43	Heterocycle	410.5	24
44	$C_{Ar}-N(-C_{Ar})$ heterocyclic	−78.8	17
	Ketones		
45	$CH_3-(CO)$	208	12.6
46	$-CH_2-(CO)$	42	10
47	$\diagdown\!\!\!\diagup CH-(CO)$	−163.5	22
48	$(OC)-CH_2-(CO)$	−75.7	57

N°	Group	Increment in AIT	Divergence in estimate
49	=CH–(CO)	45	34
50	C_{Ar}–(CO)	0	
51	Cyclic correction	405.6	37
Esters			
52	H–CO(OR)	0	
53	CH_3–CO(OR)	0	
54	–CH_2CO(OR)	–206.7	19.5
55	\CH–CO(OR)	–341.7	36.6
56	=CHCO(OR)	–266.3	21
57	=C–CO(OR)	–440	26.5
58	C_{Ar}–CO(OR)	–306	19.6
59	(RCO)–OCH_3	426.5	17.7
60	(RCO)–PCH_2–	236.8	101
61	(RCO)–OCH<	0	
62	(RCO)–OCH=	180	49
63	Cyclic correction (lactones)	432.3	39
Acids			
64	–CH_2CO_2H	271	16
65	\CH$CO_2H$	60	32
66	=CHCO_2H	192.5	39
67	=C–CCO_2H	–38	56
68	C_{Ar}–CO_2H	228.5	56

The missing structures and compounds could not be studied through lack of sufficient data. The atoms mentioned define the group except the three elements in brackets. In this last case, the content of the bracket helps identify the group. Besides, the free linkages indicate a linkage with a hydrocarbon group.

1 • Flammability
1.4 Autoignition temperatures (AIT)

Examples of AIT estimations using additivity method

In order to choose the 'worst' conditions, possible substances that are not listed in Part Three were taken, ie substances that were not used for self-ignition increment calculations by regression. Indeed, choosing as examples substances from the database, the results are almost too 'good' to be used in a demonstration. These new substances come from the publication by Hilado quoted before. The table below gives the list of compounds, the estimated AIT_{est}, the AIT found in the publication AIT_{exp}, the list of group numbers taken into account for the calculation as well as cyclic corrections and corrections of potential positions. There is no ketone in the list since they were mentioned in the text.

Formula and name	AIT_{est} (°C)	AIT_{exp} (°C)	Structural group N°(number)
CH₃CH₃ | | CH₃–CH–CH–CH₂–CH₃ 2,3-Dimethylpentane	313	335; 338	1(4); 2(1); 3(2)
CH₃ CH₃ CH₃ | | | CH₂ = C – CH–CH–CH₃ 2,3,4-Trimethyl-1-pentene	261.1	257	1(4); 3(2); 5(1); 7(1)
1,2,4-Trimethyl benzene	483	521	1(3); 8(3); 9(3); 11(1); 12(1) or 13(1)
C₂H₅ | HOCH₂–CH–CHOH–(CH₂)₂–CH₃ 2-Ethylhexane-1,3-diol	287.6	360	1(2); 2(3); 3(1); 14(1); 15(1)
Styrene oxide	507.5	498; 538	8(5); 9(1); 18(1); 19(1); 22(1)
CH₂=CH–O–CH(CH₃)₂ Isopropylvinylether	317	272	1(2); 5(1); 19(1); 20(1)
CH₃–CH–(CH₂)₂–CH–CO₂H | CH₃ 2-Methylpentanoic acid	405	433	1(2); 2(2); 64(1)
CH₂=CH–CO₂–CH₂–⟨O⟩ Glycidyl acrylate	415	398	2(1); 5(1); 18(1); 19(1); 55(1); 59(1)
Glycerol Triacetate	473.6	433	52(3); 59(2); 60(1)
CH₃–CH₂–CHCl–CH₂Cl 1,2-dichlorobutane	531	275	1(1); 2(1); 24(1); 25(1)

First it is noticed that there is no AIT for 2-methylpentanoic acid in Part Three, which is the result of a study of the databases listed on p.355; this was quoted in Hilado's publication.

Finally, note that 1,2,4-trimethyl benzene will be considered in paragraph 1.5.6. The previous table thus already solves some of the forthcoming problems.

Compared with the experimental values for which was noted a high level of measurement error, a level of agreement was found that is not worse than the disparities found for a lot of compounds, which were the subject of independent measurement. Note in particular the good estimates obtained with two compounds that have relatively complex structures, such as styrene oxide and glycidyl acrylate. Nevertheless, there are two estimates that seem sufficiently different from the experimental values to require explanation.

- For 2-ethylhexane-1,3-diol it is possible that the lack of data regarding diols biased the model. Simply add that the difference observed (~82°C) is unacceptable compared with the 49°C difference between the two extreme experimental values of a substance as well known as propane-1,2-diol. Nevertheless there is a huge disadvantage to this model, which can be observed when a hydrocarbon chain becomes long and that may happen here. Indeed, taking for example a chain in which CH_2 groups are progressively added, the increment of this group (n° 2) being negative will lower the AIT value excessively. Thus with dodecane an unacceptable value of 129°C is obtained. Consideration has therefore to be restricted to compounds with less than five to six groups of this nature or not include increments for group n°2 after more than five or six repetitions. Finally, apart from aldehydes and ethers, a few substances have an AIT lower than 200°C.

- For 1,2-dichlorobutane one only has to compare the compounds that come close to this in the table (1,2-dichloropropane for instance, 555 to 557°C) to realise that there is, surprisingly, such a fall in the experimental value for the compound, which differs from 1,2-dichloropropane by one CH_2 group only. A wall effect for this substance is thus postulated since the AIT due to Hilado was measured in a metallic container apparatus whose surface condition could be seriously modified by the presence of hydrogen chloride produced by decomposition and combustion of chlorinated substances during previous tests.

This illustrative exercise completes the presentation of the different parameters that can be taken into account in the search for the fire hazard risk level of a substance. The methods available will now be discussed followed by a case study.

1.5 Fire hazard risk; methods of classification[1]

1. English language editor's note
Throughout this book the regulatory regime is referred to as 'Labour Code', a translation of the French 'Code du Travail'. All countries have national regulatory codes, some of which reflect wider imperatives (eg EU directories, ILO conventions, UN standards). In adddition, there are many technical and other codes of practice or procedure produced by national or international professional, technical and industrial organisations. Examples of the former are the regulatory codes of the HSE and The EA in the UK, and of OSHA and EPA in the USA; and of the latter, the NFPA, which is referred to extensively in this book.

The purpose of these classifications is to enable the user to know the risk level of a substance, and for those that are related to regulations, to codify danger signals that are compulsory in order to inform users as well as carriers. There are four classifications:

- *National Fire Protection Association* (NFPA, United States) classification.
- Labelling code used by the European Union and labour regulations in France.
- Inflammability code used by SAX in his work (see beginning of Part Three).
- Physical and other factors used by American industry meant for risk calculation of industrial chemical installations.

1.5.1 NFPA flammability code

This classification system allocates to substances a number between 0 (minimum danger) and 4 (maximum danger) defined by NFPA[1] as follows:

4; substances that quickly or completely evaporate under normal pressure and at ambient temperature or that can be easily dispersed in the air and easily burnt. This level includes gas, cryogenic liquids, all liquids with a flashpoint less than 22.8°C (73°F) and boiling point less than 37.8°C (100°F). In addition, there are substances like inflammable dust that can form suspensions in air leading to explosive mixtures, as well as aerosols. These mixtures can be formed in certain environmental and activating conditions.

3; Liquid or solid substances that can combust in conditions close to the ambient temperature. This level includes liquids with flashpoints of less than 22.8°C and boiling points greater than or equal to 37.8°C and those with flashpoints greater or equal to 22.8°C and any boiling point. It also includes solids, which can form suspensions that can combust in the air without exploding (this is what differentiates them in this particular case from solids classified 3 and 4). Fibrous solids such as cotton substances that can combust quickly can also be included. Finally, there are substances that combust spontaneously and substances for which their *oxygen balance* (para 2.3.4) turns them into substances that can combust violently (nitrocellulose or organic peroxides in their dry state) without any oxygen supply.

2; Substances that need to be preheated to give off inflammable gaseous mixtures. This is so for liquids with flashpoints greater or equal to 37.8°C or less than 93.3°C and solids that easily provide inflammable vapour.

1; substances that need to be heated to a high temperature to enable them to combust. These are liquids with a flashpoint greater than or equal to 93.3°C and solids that spontaneously combust after being heated to 815.6°C (1500°F) for a maximum of five minutes. This class also includes all inflammable ordinary solids (wood, paper, cloth).

0; substances that do not burn when heated at 815.6°C for more than five minutes.

---

1. http://www.orcbs.msu.edu/chemical/nfpa/flammability.

1 • Flammability

1.5 Fire hazard; methods of classification

So far as liquids are concerned, the diagram below sums up this classification:

Attribution criteria for NFPA flammabilty Code

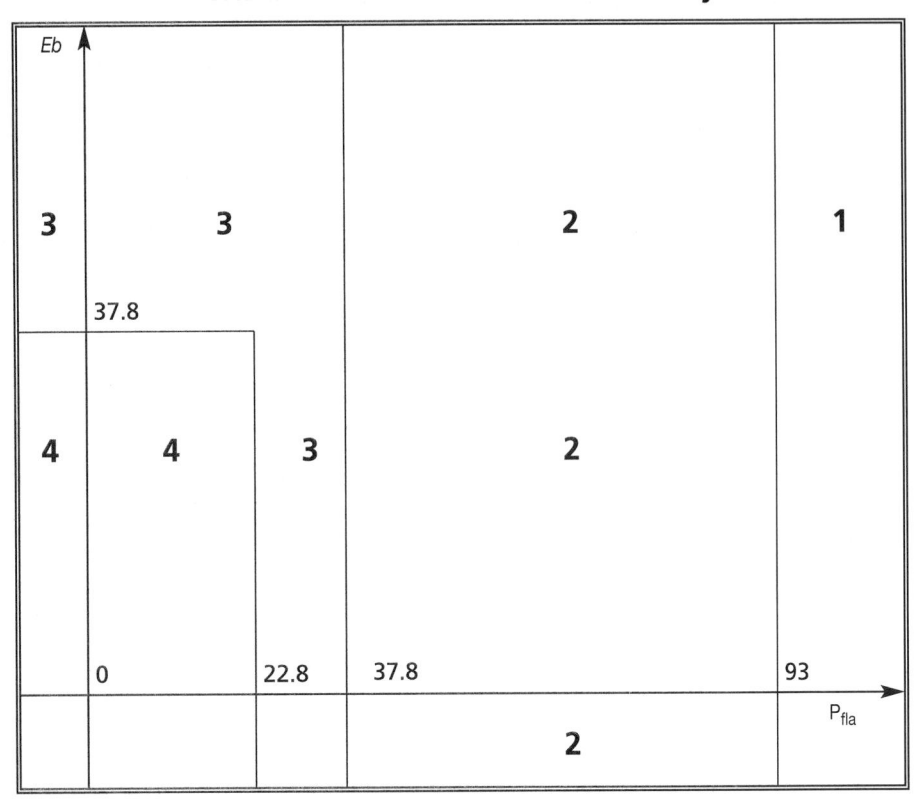

E_b				
3	3		2	1
37.8				
4	4	3	2	
0	22.8	37.8		93 P_{fla}
			2	

The NFPA code is represented in a diamond containing 4 sectors, respectively: toxicity, inflammability, reactivity and 'special risks'. A coloured code that will appear on glass labels, at the back of transport vehicles, room doors etc enables the danger to be better noted. It is used by American companies although some French companies have also adopted it and it appears to be an efficient device.

Below is a picture of this diamond that comes from an American source[1]. This source allows immediate and up-to-date access to the current NFPA codes.

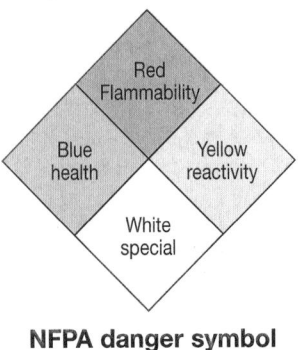

NFPA danger symbol

(Red Flammability, Blue health, Yellow reactivity, White special)

---

1. http://www.orcbs.msu.edu/chemical/nfpa/nfpa.html.

1.5.2 Labour code

This classification system is compulsory throughout Europe, although other classification methods are also used. Like the NFPA code it is based on boiling and flashpoint parameters. It classifies substances into five categories that are defined as follows:

Attribution criteria for flammability Code

Code	Risk description	Classification criterion
12	Extremely inflammable	Boiling point equal to or less than 35 °C, flashpoint less than 0°C. substances and gaseous preparations which, at ambient temperature and pressure, are inflammable in air.
13	Liquefied gas extremely inflammable	Same criterion as above for flashpoint. The substance is stored liquefied because of its low boiling point.
11	Highly inflammable	Flashpoint equal to or greater than 0 °C and less than 21 °C. Boiling point greater than 35°C.
10	Inflammable	Flashpoint greater than or equal to 21 °C and less than or equal to 55 °C.
–	Combustible liquid	Flashpoint greater than 55 °C.

Next to the risk number and sentence on the label there is a symbol representing a black flame on an orange background and the letter 'F'. This labelling is described in a decree published in the official French journal already mentioned (see note, p.58).

The symbol used with this text on regulations is shown below:

F: Highly inflammable

So far as risk 10 is concerned, fire regulations indicate that compounds that have the above mentioned inflammable properties could be classified as not inflammable if they could not 'in any way encourage combustion...'.

These classification criteria are summarised in the diagram below:

Attribution criteria for flammability danger according to safety code

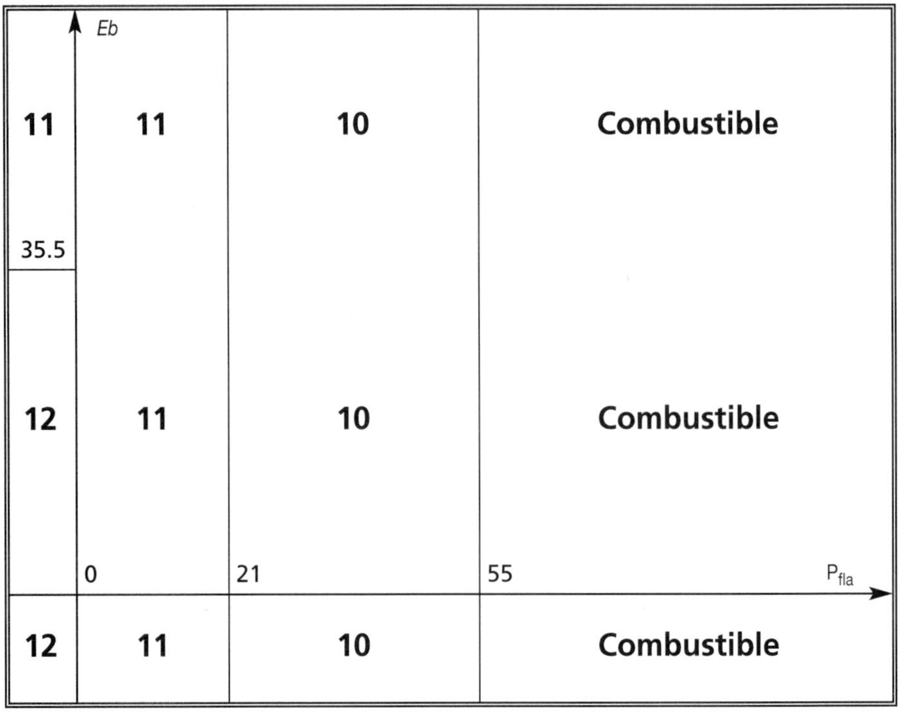

Eb			
11	11	10	Combustible
35.5			
12	11	10	Combustible
0 21		55	P_{fla}
12	11	10	Combustible

There are numerous types of classification, based on flash and boiling point criteria in the fire regulations.

In this case of classified installation classification the risk clause is replaced by a risk category number or transport regulations for dangerous substances, which is displayed on the back of tankers or on substances packaging as codes 30 (corresponding to 10), 33 (corresponding to 11 and 12) or 233 (corresponding to 13).

1.5.3 SAX code

In this book each substance is given a hazard code between 1 ('slightly dangerous') and 3 ('highly dangerous'). There are two disadvantages to this approach:
- 'inflammability' criteria 'are qualitative';
- this single number indicates inflammability hazard as well as toxicity and stability. Except for substances for which the two risk levels are really distinctive, there can be a great confusion in the identification of the figure's origin.

1.5.4 Physical and other factors

A complex evaluation system is required to take into account not only risk of substance but also of physical conditions in which the substances are used as well

1 • Flammability

1.5 Fire hazard; methods of classification

as all used parameters (for instance, quantity). One was suggested by the Dow Chemical Co[1] in order to enable the integration of the preliminary study into the other approaches when it comes to designing an industrial workshop. This subject goes far beyond the purpose of this book. Its disadvantage is that it is aimed at something really specific. Its interest lies in the fact that it can be used by a safety expert as guidelines to work out his own approach. For inflammability risk this method is based on the NFPA code classification criteria. These two methods differ by the number chosen, reflecting the risk level.

This technique will now be illustrated in the case study that follows (see para 1.5.6).

This type of classification has the advantage of offering a more suitable scale for risk quantification (this is precisely why it was chosen). It can be compared with a quantitative variable. Besides, it takes into account four main factors of risk aggravation:

1. insolubility in water, which does not enable the automatic 'flooding' of substances in an incident;
2. working methods that require a temperature higher than the substances' flashpoints;
3. the process of adding a factor that takes into account an 'alarm zone' of 80% of flashpoint, though it is not really relevant (there is available a better approach), takes a typical observation into account ie when thermal conditions come closer to the flashpoint temperature there is a higher risk. Indeed, numerous studies show that in very different situations than those used to determine a flashpoint substances combusted although they were heated below their flashpoint. This can be explained by the differences between handling conditions of flashpoints during flashpoint measurement and real conditions in which liquid vapourisation can be increased (for instance, spreading on clothes);
4. finally this approach takes into account autoinflammation temperature.

It also has another advantage as will be seen later. The classifications given here can be modified according to:

- other dangerous properties of substance (oxidising properties, dangerous reactions to water etc.);
- conditions of use (continuous processes, multiple reactions in the same equipment, etc.);
- quantities used.

---

1. Dow Chemicals Co, *Chem. Eng.* Progress, 62 (8 and 9), 93-127

1 • Flammability
1.5 Fire hazard; methods of classification

Criteria and classification of flammability depending upon physical factors*

Criterion	Physical factor
Gases	
Gas of low enthalpy of combustion and of low limit of elevated inflammability ($\leq$ 17 kJ/g; LEL > 12 %)	6
Gas of elevated enthalpy of combustion and LEL < 10 %	18
Unstable gases capable of spontaneous explosion	20
Liquids	
I - Liquids with flashpoint greater than or equal to 260 °C	
Applications at a temperature below flashpoint:	3
Appications at a temperature above or at 80 % of flashpoint:	10
II - Liquids with flashpoint greater than or equal to 60 °C and less than 260 °C	
Applications at temperature lower than flashpoint:	5
Applications at temperature greater than or at 80 % of flashpoint :	10
III - Liquids with flashpoint greater than or equal to 22.8 °C but below 60 °C	
Applications at temperature below flashpoint soluble in water:	7
Applications at temperature greater than 80% of flashpoint and soluble in water:	8.5
Applications at temperature above flashpoint and soluble in water:	10.5
IV - Liquids with flashpoint greater than or equal to 22.8 °C and below 60 °C	
Applications at a temperature less than flashpoint and non soluble:	10
Applications at temperature greater than 80 % of flashpoint and non soluble:	12
Applications at temperature greater than flashpoint and non soluble:	15
V - Liquids with flashpoint below 22.8 °C and boiling point greater than 37.8 °C	
Soluble in water:	12
Insoluble in water:	15
VI - Liquids with flashpoint and boiling point less than 22.8 and 37.8 °C respectively:	18
VII - Liquids with auto-ignition temperature less than or equal to 190 °C and spontaneously inflammable liquids	20
Solids	
Non combustible solids:	1
'Ordinary' combustible solids (wood, paper):	3
Solids with flashpoint corresponding to liquid criterion :I	
temperature < $0.8 \times P_{fla}$:	5
$0.8 \times P_{fla}$ < temperature < P_{fla}:	7.5
temperature > P_{fla}:	10
Easily burnable solid, but easily extinguished by water:	10
Finely divided solids capable of forming explosive suspensions in air:	10
Spontaneously inflammable solids not extinguishable by water:	16

* The disadvantage for gases is the ambiguity that is related to the lack of accurate threshold values. The values that are mentioned here come from examples found in the Dow publication.

1.5.5 A new classification: degree of fire hazard II

■ Characteristics

The disadvantages of all classification methods so far considered are the same for all qualitative classifications.

- They cannot be part of a mathematical model whose purpose would be to turn the classification into a *continuous quantitative* variable. In particular, the example of physical factors illustrates this. Whereas for the highest degree criteria are the same as those of the NFPA code, the simple fact of wanting to add in physical factors to these calculation models forced the originators of this technique to forget about the NFPA code.
- Their 'value' does not refer to a real degree of the risk level variation. Therefore a substance in NFPA code 4 cannot be considered as presenting a risk that is twice as high as of a code 2 substance. This indicates the limits of physical factors, which are only a qualitative variable masked by an arbitrary increase of modalities that nevertheless improve the accuracy of classification.
- They do not have any physical meaning. It is impossible to establish an accurate and quantified relationship between their level and physical quantities defining risk. Perfect knowledge of each definition is the only way to enable a relationship between code and property.
- It is difficult for them to cover a risk situation that depends on different parameters since a same risk level might cover situations that have nothing in common. So a comparison using physical factors of amyl oxide and vinyl oxide gives code 20 for the first because its autoinflammation temperature is less than 190°C though it has a flashpoint greater than 57°C, and 18 for the second for which the flashpoint is less than 0°C.
- In addition, these codes are ambiguous at the limit and suffer artificial threshold effects. So ethyl oxide and pentane are respectively classified 12 and 11 in the regulations although they have the same flashpoints but 2°C difference in their boiling points on each side of the crucial value of 35°C. In the same way code 11 is placed by labour regulations on substances as different as pentane ($P_{fla} \sim -40°C$) and octane ($P_{fla} \sim 15°C$). Finally, even a minor mistake can cause a swing from one level to another, if when close to the limits, whereas a quantitative classification would only suffer a slight variation. An example is given below.

In seeking a classification that does not present the disadvantages mentioned above, this classification cannot be replaced by those fixed by regulation. Texts on regulations have the disadvantage of presenting an important 'viscosity'. The purpose of this book is not to suggest counter-proposals but to look for approaches that answer technical and scientific concerns. This will be returned to several times.

■ Definition, physical meaning

It is suggested then that 'fire hazard index', II (inflammation index), is defined by the following equation:

$$II_t = \frac{100 \times C_t}{C_{fp}}$$

in which C_{fp} is the substance equilibrium concentration at flashpoint temperature 'cc' and C_t the substance equilibrium concentration at temperature 't' defined as $t = 21°C$. If the thermal conditions are not determined, it is given as 'II'. In the opposite case, when t, temperature is known, the index is symbolised as 'II$_t$', the temperature being in subscript. There is a theoretical figure expressed in %. Note that C_{fp} = LEL.

Note that:
- II = 100% corresponds to code 11 in the labour regulations;
- II = 80% constitutes the 80% limit of LEL, which is taken into account by explosimeters and some automatic atmospheric monitors in order to define the alarm zone of an environment likely to become explosive.

In the same way:
- II$_t$ = 100% indicates that the working temperature is equal to the substance flash-point;
- II$_t$ = 80% also indicates the alarm index.

If the substance studied is a gas that combusts at 21°C or at temperature 't', write down C_t = 100. II code can lose all its meaning in confined environments (since UEL is not taken into account).

II code (or II$_t$) gives the vapour equilibrium concentration as a percent of LEL. It has a clear physical meaning and its value can be directly linked to the risk. We know that above 100% the equilibrium concentration 'enters' the inflammability zone.

Nevertheless, it should not be concluded that any substance with a degree greater than 100% creates an inflammable environment. There is an environmental factor that was not taken into account here, which is the quantity of substance handled, the ventilation rate of the premises and the vapourisation speed of the liquid. This last factor is recommended by regulations but there are few figures available. These values are determined in conditions that cannot be compared with each real condition and are related to substances of different natures that cannot allow any direct comparison.

Nevertheless this risk degree is defined in such a way only when the difference between AIT and the temperature is greater than 300°C.

In the opposite case following "penalties" are listed:

If AIT-t ≤ 200°C   II$_t$* = II$_t$ x 1.2;

If 200° < AIT-t ≤ 250°   II$_t$* = II$_t$ x 1.1;

If 250° < AIT-t ≤ 300°   II$_t$* = II$_t$ x 1.05.

If temperature 't' is unknown, calculate II* by replacing AIT-t by AIT.

■ Calculations

II and II$_t$ are estimated degrees only; there is no experimental value involved except for AIT and calculation of the entropic factor from equation (1) by Hass. The process is indicated in the chart below:

1 • Flammability

1.5 Fire hazard; methods of classification

```
           Determination of Cs
                   ↓
           Determination of LEL
                   ↓
        Determination of entropic group
                   ↓
       Determination of C_eq at 't' or 21°C
                   ↓
              Calculation of II
                   ↓
             Take into account AIT
                   ↓
    Calculation of II_t* if AIT−t ≤ 300°C
         or of II_t* if AIT ≤ 300°C
```

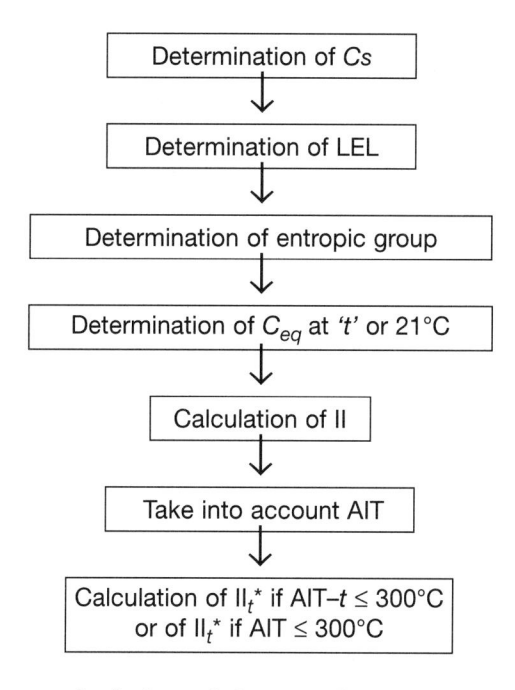

Calculation of degree of fire risk II

In order to calculate LEL use the lowest value obtained from one of the two estimations by Hilado or the variant. For AIT use the lowest value obtained either by using the author's source or by calculating it with the help of additivity rules.

■ Examples

The table below lists some of the substances dealt with in previous examples and compares different numbers.

Example of different classes of risk

Compound	C_s %	LEL est	Group or S	II (%)	AIT[1]	Labour Code	NFPA Code
Propane	4.022	2.19	group 2	4566.21	372	13	4
Acetaldehyde	7.73	3.42	group 3	3508.77*	140	12	4
Ethyl oxide	6.528	3.10	group 2	2356.27*	160	12	4
Pentane	2.552	1.39	4.576	4379.55*	260	11	4
Cyclohexane	2.275	1.24	4.607	952.27*	245	11	3
Toluene	2.275	1.17	4.685	258.44	480	11	3
Ethylbenzene	1.956	1.00	4.786	98.14	430	11	3
Styrene	2.052	1.05	4.804	62.58	*468.8*	10	3
Propylbenzene	1.716	0.88	4.833	39.50	450	10	3
Tetrahydronaphthalene	1.586	0.87	4.619	4.62	384	–	2
Diphenylmethane	1.293	0.66	4.861	0.36	436	–	1

1. The estimated AIT is shown in italics.

One can make two comments:
- whereas ethyl oxide is classified as more dangerous than pentane (the latter boils above 35°C, ether below) II* classification ranks it second because its LEL is significantly higher than pentane's.
- with II < 100 ethylbenzene should be ranked '10' by labour regulations. In fact, the flashpoint taken into account by these regulations is 15°C whereas our table gives the 15-21 interval and our calculation gives 21.3°C. With the last two figures favoured by our calculation the code would indeed be 10. A possible mistake here causes a big change using the labour regulations but a slight modification with the II degree.

The latter comment raises a problem that should be tackled as follows. In a technical analysis, conclusions should never be in contradiction with the regulations in force. If there is such a contradiction, one should opt for the solution offered by regulation if it leads to a classification harsher than the technical approach. If on the contrary the technical approach leads to the opposite, approval is needed before an aggravated classification can be used. In our case our approach has nothing to do with this notion of threshold. In practical terms, the analyst deals with a scale that makes it easy to compare these substances with each other. It will be his knowledge about the situation in which the work with these substances would be like which will enable him to choose the measures according to the different II and II* levels more than the code dictated by regulations.

Indeed our scale makes it possible to make a more subtle comparison of risk levels, which allows us to classify substances in ascending risk order and can be directly interpreted (an II value of 4.62% for instance means that with a vapour equilibrium concentration at 21°C in air this concentration cannot exceed 4.62% of LEL value, which is equivalent to the reading of the dial of an explosimeter). Note in particular that it is clearly seen by examining the tables for these substances that the same code '11' set by the regulations or '3' by NFPA hide very different risk situations, which is well shown by II code. Vice versa, the threshold effects of these codes tend to overestimate insignificant risk differences (for instance between ethylbenzene and styrene).

1.5.6 Case study: 1,2,4-trimethyl benzene

To conclude, here is an example that will enable the reader to familiarise himself with another approach and that corresponds to a situation in which the substance is not mentioned in this book (or any other). A case with no available data will be selected which is hardly realistic since it is always possible to find the boiling point of a substance in a supplier's catalogue, and then consider the state of knowledge about this substance through the available sources. In this case a substance analysed by the students in the author's laboratory, and that was not included in the tables in Part Three, is considered.

Estimation of boiling point

'Hydrocarbons' listed in Part Three are analysed and the effect of substitution of a methyl group for a hydrogen atom of an aromatic cycle determined. The change from benzene to toluene that increases the boiling point by 31°C is noted and the change from toluene to xylenes gives an increase of 32.5°C for the *ortho* position and of 27°C for the other positions (in fact 27 and 26.5). If this increase seems too modest, by adding a third group methyl in meta of m-xylene 139 + 27 =

1 • Flammability
1.5 Fire hazard; methods of classification

166°C is obtained for mesitylene. Yet the tables give 165°C. So the method has satisfactory reliability. It can be used with the 'unknown' substance by adding increment 27 or 26.5 to the boiling point of o-xylene ie $Eb = \sim 171°C$.

Estimation of entropic factor

This analysis uses factors S calculated with toluene, xylene and mesitylene by using boiling point under reduced pressure data. This analysis leads to the conclusion that these compounds belonging to group 2 enable S to be modelled in the equation $S = 4.363 + 0.003085 \times Eb$. This gives $S = 4.892$, which enables P_{vap} to be estimated at different temperatures:

at 20°C: 1.88 mbar;
at 30°C: 3.62 mbar;
at 50°C: 11.6 mbar.

Estimation of LEL and flashpoints

$Cs = 1.716\%$; LEL = 0.911 (according to Hilado); LEL = 0.88% (according to the author's variant); Hence: P_{fla} 'cc' = 45.8°C and P_{fla} 'oc' = 57.736°C.

Hazard codes

NFPA: **2**;
Labour regulations: **10**;
Physical factors code: **10**;
II code: **22.59%**;

Estimation of autoignition temperature

As in the previous paragraph:
AIT = 483°C
To summarise by filling the cells below:
1,2,4-Trimethylbenzene: C_9H_{12} • M(g/mol) = 120.21 • CAS No.: 95-63-6
Physical data: Eb (°C/mmHg) = 171 • P_{vap} (bar) (°C) = 1.88(20); 3.62(30); 11.6(50) • LEL/UEL = 0.88; 0.911 • P_{fla} = 46cc; 57oc • AIT = 483

Now follows an analysis of published material on this subject (see list of sources at the beginning of Part Three):

From Weka: Eb = 169-171°C; P_{vap} = 15 mbar (50°C); 30 mbar (65°C).

P_{fla} = 48°C (nothing was mentioned about the apparatus used). LEL = 0.8%. Labour regulations: 10; AIT = 485°C;

Merck index: Eb = 169-171°C;

SAX (CD ROM version): Eb = 169°C; P_{fla} = 54.4°C (no further information); AIT = 515°C;

MSDS Sigma-Aldrich (CD ROM): Eb = 169°C; P_{vap} = 48°C (no further information); LEL = 0.9%; P_{vap} = 6 mbar (37.8°C); 9.33 mbar (44.4°C); AIT = 514°C;

Guide de la chimie: Eb = 169.3°C; 145.4/400; 102.8/100; 79.8/40; 13.6/1.

P_{fla} = 51-54°C (no information). LEL/UEL = 1/6. AIT = 520°C;

Canadian databases: substance is not mentioned in the tables.

A few values of P_{vap} are given at other temperatures. They were therefore calculated under the same conditions as previously, ie with the estimations done at the beginning without any data. All results are listed below. The author's estimated data is underlined:

1 • Flammability

1.5 Fire hazard; methods of classification

1,2,4-Trimethylbenzene: C_9H_{12} • M(g/mol) = 120.21 • CAS No.: 95-63-6

Physical data: ΔH_f^0 (KJ/mol) = –61.88(l); –15.96(g); –57.54(l)
- Eb (°C/mmHg) = 169-171; 171; 145.4/400; 102.8/100; 79.8/40; 13.6/1
- P_{vap} (bar) (°C) = 1.88(20); 3.62(30); 5.8; 6(37.8); 8.5; 9(44.4); 11.6; 15(50); 25; 30(65);
- LEL/UEL=0.8;0.88;0.9;0.911;1
- P_{fla}=46cc;48;51-54;54;57oc• AII=483;485;515;521

This summary table demonstrates the coherence in the estimates carried out without having any data to start with. Added in anticipation is the estimate of enthalpies of formation from the additivity tables by Benson as in chapter 2 on stability.

Notice that boiling point data under reduced pressure enable us now to calculate factor S more accurately than before. We estimated S at 4.892. We can now find the 'exact' figure, obtaining $S = 4.815$. When using this figure we obtain better P_{vap} estimates but a worse estimation of flashpoint for closed cup than that compared with experimental values. The reader will be able to observe by carefully comparing the examples dealt with previously, that it is a general observation. It is certainly due to the poor original hypothesis according to which the equilibrium concentration at flashpoint temperature is equal to LEL.

2 • STABILITY

Any chemical product, which, when subjected to an external *physical* effect, undergoes a more or less violent decomposition, is considered to be *unstable*.

Temperature, shock, shockwaves, friction and light may be the physical agency of instability. Unsaturated organic substances can sometimes undergo violent chemical transformations under the influence of some of these but do not come within the above definition. In these specific cases, dangerous chemical reactions, which often involve catalytic impurities, are the cause and are treated in chapter 4 as 'dangerous reactions'.

An unstable substance is not necessarily dangerous. The dangerous characteristic involves intense mechanical and thermal effects linked to, in the first category, the formation of large quantities of gaseous substances, in the second one, the exothermicity of the degradation process. Nevertheless, these characteristics have to be accompanied by an appropriate degradation speed which is the main factor of the destructive characteristic of an unstable substance. The rate at which the temperature increases is more important than the level it reaches; eg petrol releases more thermal energy than nitroglycerine. It is the same principle with the formation of decomposition gas; the speed at which gas is formed (ie speed of growth of local pressure and thus the resulting shockwave) is more important than the total volume of gas. These considerations enable us to understand why quantitative evaluation of instability risk is difficult. Indeed, it proves to be more complicated to work out kinetic forecast models than thermodynamic ones.

The aim of this chapter is to enable the reader to evaluate danger related to the instability of any chemical substance, despite these difficulties, in order to work out the prevention measures needed to minimise the risk of accidental instability.

It is useful to first relate the methods of determining a substance sensitivity to the different physical causes likely to create decomposition. Then a qualitative approach of how to forecast instability risk by listing all different factors that have an impact on the stability of a molecule will be presented. The notion of 'plosophoric' groups, which proves to be suitable despite its simplistic form, will be used. Then, the most important part of this chapter, which deals with the analysis of the CHETAH programme, with examples, will be discussed. Although it is a thermodynamic approach, it proves to be perfect for forecasting substance instability especially when using our suggested variant.

Finally, the methods used in implementing labour regulations (NFPA) will be described as well as public labour regulations, which will be compared with each other and with the quantitative evaluation methods.

2.1 Experimental methods of determination

Most chemical substance manufacturers systematically submit their new substances (or preparations) to tests[1] that enable them to evaluate the decomposition risks. There are many types of apparatus that are used to test the effect of the different physical causes of instability. The most important are the mechanical and thermal sensitivity tests. The methods listed below are simply intended to give an idea of the available experimental possibilities.

2.1.1 Mechanical sensitivity tests

How easily a substance can spontaneously decompose or explode under the influence of shock or friction is analysed. These are some of the most frequent causes of explosive decomposition of unstable substances related to common handling: stirring, breaking, sieving, etc, or the accidental dropping of a substance or the opening of a flask etc.

■ **Impact sensitivity test**

A mass of 5 to 20 kg is allowed to fall from a given height, according to the type of apparatus, on a quantity of substance of about 20 mg placed in a steel recipient equipped with a 'piston'. The French call this test *mouton de choc*. This process is repeated by placing the test mass at different heights and by doing a series of fifty tests at each height. Thus the energy that leads to a probability of decomposition of 50% is determined; m-dinitrobenzene used as a reference gives a probability of 0% with a mass of 5 kg falling from a height of 1 metre (~5Nm). (m-Dinitrobenzene is the first substance that warrants being called 'explosive' according to the dangerous substance transport regulations.)

■ **Friction sensitivity test**

A small quantity of substance is placed on a porcelain plate. Then an hemispheric mass of porcelain is placed on the substance with a weight that can be adjusted and then proceed to a displacement of the porcelain plate. The weight that is necessary to obtain a probability of decomposition of 50% by doing a series of thirty tests is determined; m-nitrobenzene gives 0% at 36 kgf (327N).

2.1.2 Thermal sensitivity tests

■ **Violent heating test**

200mg of substance is put in a test tube, which is kept open, and which is then placed vigorously in an oil or melted metal bath. The temperature at which the substance is burnt up or explodes is thus determined. With this test m-dinitrobenzene combusts at 400°C.

---

1. L. Médard, *Les explosifs occasionnels*, volume 1, p. 377, 1987.

2 • Stability   2.1 Experimental methods of determination

■ Heating under containment test

20 g of the substance is placed in a cylindrical steel socket of diameter of 24 mm and height 75 mm. This socket is closed with a lid equipped with a 'light' that has a calibrated diameter. It is placed in an enclosure, which contains four burners; three laterally and one on the lower part. The socket is then heated vigorously and the limiting diameter of the light with which an explosion of the socket is obtained after three tests is determined. The limiting diameter with m-dinitrobenzene is found to be less than 1 mm.

2.1.3 Shockwave sensitivity tests

■ Detonatability test

400 g of substance is placed in a steel cylinder of 40 mm diameter on a lead plate, and a detonator placed on the substance. After triggering the detonator, if on close examination of the lead plate, it is jagged, it is concluded that the substance had detonated.

■ Card gap test

The difference between this test and the previous one is that layers of screens are added between the substance and the explosive in order to measure the shockwave effect of the detonator on the substance.

A detonator, in the form of explosive plates made with tetryl then a screen made of cellulose acetate plates is placed in a wooden container. A 26 mm diameter cylinder full of the substance to be analysed is placed on the screen, and finally,,a steel plate on top of the cylinder is added. If the substance transmits the detonation, the steel plate will be pierced and not projected. Piercing serves as an indicator of detonation transmission. The number of cellulose acetate disks needed between the sample and the detonator to prevent the detonation from being transmitted is found. Only one is needed for most chemical substances, but with m-dinitrobenzene, 240 are required.

2.1.4 Ballistic mortar tests

This is a performance test for explosives. In this respect it only meets specific requirements related to the use of so-called 'intentional' explosives.

2.1.5 Determination of critical diameter

The purpose of this test is the same as the previous one. The critical volume is the sphere diameter of a substance, below which it is impossible to obtain detonation conditions under the influence of a firing blasting charge. These tests can only be conducted on shooting ranges and sometimes with large substance quantities.

2.1.6 Other tests

In addition to these standardised test methods set by regulation (in particular the transport regulations of dangerous substances), there are laboratory methods that can provide more details regarding substance behaviour. In particular, there is differential thermal analysis (DTA), thermal gravimetric analysis, calorimetry and thermomanometry, which will not be described here.

2.2 Qualitative approach to instability

The purpose of this section is to enable the reader to identify the potential stability properties of a chemical substance by simply analysing its structural formula. This will be made possible by listing the structural properties of unstable molecules. The reader will see the need to identify two types of structural properties; those that bear the 'hallmarks' of an unstable property and those whose presence can increase or alternatively reduce the risk of violent decomposition. Since the latter are the only ones in the molecule, they do not represent any danger for it in terms of stability. But when performing qualitative analysis, one also has to take into account the conditions under which the substance is handled. In addition to the structural properties, the analyst will have to carry out a study on the 'external' risk factors.

2.2.1 'Internal' structural factors

The following are relevant to fragile and 'incompatible' parts of a molecule.

■ Low energy bonds

The presence of weak bonds in a molecule will create instability. The decomposition seat will involve this bond, which will tend to break easily under the influence of a physical cause, sometimes of low intensity. The table below lists the weakest bonds.

Energies of weakest bonds in polyatomic molecules

Bond	Energy of bond (kJ/mol)	Bond	Energy of bond (kJ/mol)
O-O	146.44	N-O	221.75
N-N	163.18	S-S	226.44
O-F	180.58	C-S	271.96
N-Cl	192.46	N-F	271.96
CO-Br	200.83	C-Br	284.51
C-I	213.38	C-N	304.60
O-Cl	217.57	P-H	318.28

Common observation and accident reports often refer to molecules with these bonds, but especially the first eight in the list. Carbon-metal bonds are to be added to this list, in particular the ones involving transition metals.

Endothermic property

Besides the weak bonds listed in the previous table, there are other multiple bonds that endow the molecules in which they are situated with a positive enthalpy of formation. Such compounds are termed 'endothermic' compounds. The danger they represent does not necessarily come from the fact that they are unstable, but is related to the exothermicity of their decomposition reaction. The most convincing examples are the acetylenic compounds, and in particular, acetylene. It is also the case for ethylene, aromatic compounds, imines and nitriles.

The following table illustrates some cases.

Enthalpies of formation

Product	ΔH_f (kJ/mol)	Product	ΔH_f (kJ/mol)
Diborane	+ 31.3 (g)	Silane (SiH_4)	+ 34.3 (g)
Hydrogen azide	+ 294.14 (g)	Phosphine (PH_3)	+ 22.8 (g)
Sodium azide	+ 21.7 (s)	Carbon disulphide	+ 116.6 (g)
Hydrazine	+ 95.40 (g)	Arsine (AsH_3)	+ 66.4 (g)
Nitrous oxide (N_2O)	+ 82.05 (g)	Stibine (SbH_3)	+ 145.1 (g)
Nitric oxide (NO)	+ 90.25 (g)	Hydrogen iodide	+ 24.9 (g)
Nitrogen dioxide (NO_2)	+ 33.18 (g)	Lead nitride	+ 476.14 (s)
Ozone	+ 142.20 (g)	Lead thiocyanate	+ 200.8 (s)
Ethylene	+ 52.3 (g)	Methylhydrazine	+ 94.35 (g)
Propene	+ 20.42 (g)	1,1-Dimethylhydrazine	+ 49.37 (l)
Allene	+ 192.13 (g)	1,2-Dimethylhydrazine	+ 55.65 (l)
Isoprene	+ 75.73 (g)	Phenylhydrazine	+ 142.39 (l)
Cyclopentene	+ 32.93 (g)	Hydrogen cyanide	+ 135.14 (g)
Cyclopentadiene	+ 133.89 (g)	Acetonitrile	+ 87.86 (g)
Acetylene	+ 226.73 (g)	Propionitrile	+ 50.63 (g)
Propyne	+ 185.43 (g)	Butyronitrile	+ 34.06 (g)
Benzene	+ 82.93 (g)	Isobutyronitrile	+ 25.4 (g)
Toluene	+ 50.02 (g)	Cyanogen	+ 307.9 (g)
Styrene	+ 147.36 (g)	Malononitrile	+ 186.61 (l)
Naphthalene	+ 75.31 (s)	Acrylonitrile	+ 184.93 (g)

In parentheses, physical state: gas (g), liquid (l) or solid (s).

It is more difficult to interpret these tables in terms of risk related to instability. If some of these compounds are indeed unstable (phosphine, arsine, azides, hydrazines [inorganic compounds]; acetylene, hydrazines [organic substances]), other compounds, although endothermic, are stable (benzene, naphthalene, ethylene and in general non-acetylenic hydrocarbons). It will be seen later that the thermodynamic approach to instability faces the same difficult; the limit of the pertinence of the quantitative estimation of instability according to the CHETAH programme will force us to suggest a variant of this method.

2 • Stability — 2.2 Qualitative approach to instability

■ **'Internal redox' property, oxygen balance**

This is the most remarkable situation created by the simultaneous presence in a molecule of two incompatible structural sites, as safety experts usually put it. These are structural entities which have a strong tendency to interact violently with each other. This happens especially when two sites of the molecule have atoms of extreme states of oxidation; one with the maximum reduction state, the other with the maximum oxidation state. The best examples can be found in inorganic chemistry and are in general currency. Ammonium nitrate and dichromate are the best known. In the first, there is a 'confrontation' between the nitrogen (V) of the nitrate ion and the nitrogen (-III) of the ammonium ion. In the second, chromium (VI) is next to the ammonium ion. This is the reason why there are such big stability differences between these two compounds and corresponding alkali salts, sodium nitrate and potassium dichromate, for instance. The quantity of oxygen in a molecule can be compared with the quantity of oxygen that is necessary to oxidise it completely. This is called *oxygen balance*, which is a way of judging the oxidising property of a molecule.

■ **Aggravating/weakening structural effects**

When comparing various molecules for which instability is known for some of the above mentioned characteristics, big differences in respect of risk level are to be noticed according to the characteristics of the structural sites that are close to these instability factors.

They include:

- *Effects of chain length and substitution*. Comparing the stability of alkyl nitrates or aliphatic nitrated derivatives, in spite of instability to the extent of making it hazardous to handle them, stability increases rapidly with the number of alkyl groups. So nitroethanes, especially nitropropanes and nitrobutanes, are commonly used industrial raw materials.

  The same is true for peroxides and peroxy acids; so performic and peracetic acids or acetyl peroxide are particularly dangerous, whereas perstearic acid (eighteen carbon atoms) is perfectly stable.

  There is a similar situation related to substitution in a nearby unstable site. Thus tert-butyl peroxide can be handled whereas butyl peroxide cannot. These are steric effects, not to be confused with mass effects (see later).

- *Effects of functional group repetition and close groups*. The presence of a sole site with an unstable characteristic often leads to more or less stable molecules. Nevertheless, its repetition puts an end to this stability. Thus, if nitrobenzene is a satisfactorily stable compound, this is not the case for m-dinitrobenzene, trinitrotoluene (TNT) or trinitrophenol, which are all strong explosives, especially TNT.

  It can be the same with compounds with complex groups. The presence of another compound that is not unstable can increase or in some cases decrease the instability of a molecule. This can be observed if propene or 1-propanol are compared with allylic alcohol, for which behaviour is aggravated, compared with the first two compounds that each have one function only.

- *Effects of cyclic tension*. Cyclic compounds including six bonds present maximum stability. This rapidly decreases when the number of bonds increases

2 • Stability
2.2 Qualitative approach to instability

or decreases. So cycles with five, four and three bonds, especially the last two, are more or less unstable due to strain caused by valency angles and steric effects between groups and hydrogen atoms. Thus epoxides and aziridines are very unstable.

- *Mass effects due to some ions in salts*. It is generally observed that there is a greater instability amongst compounds containing heavy atoms compared with elements in the first periods of the periodic table. This can be observed by analysing enthalpies of formation of ammonia, phosphine, arsine and stibine (see previous table for the last three). In the same way, it is easier to handle sodium azide than lead azide, which is a primary explosive for detonators. It is exactly the same with the relatively highly stable zinc and cadmium thiocyanates and the much less stable mercury thiocyanate.

- These last examples illustrate the effects of heavy cations on anionic stability. The opposite case of an anionic effect is also possible. Thus diazonium salts are hardly stable but not dangerous when the anion is a chloride ion, whereas they become dangerous when the anion is a sulphide or carboxylate.

■ Plosophoric groups

All characteristics described in the previous four paragraphs can be attributed to structural entities which make molecules unstable. These groups are called 'plosophoric groups', probably by analogy with the chromophoric groups used by dyers. This expression does not add much, but is frequently used. Bretherick (see reference at the beginning of Part Three) gave them the English name: 'explosophors' and gives the complete list in the table below.

Plosophoric groups

Name of group	Formula	Name of group	Formula
Acetylenic	$-C \equiv C-$	N-Nitroso	$>N-N = O$
Acetylide	$-C \equiv C-$ Metal	N-Nitro	$>N-NO_2$
Acetylene halide	$-C \equiv C-X$	Azoic	$>C-N = N-C<$
Diazo	$>CN_2$	Peroxide	$>C-O-O-C<$
		Hydroperoxide	$>C-O-O-H<$
Nitroso	$>C-N = O$	Azide	$-N_3 ; N_3^-$
Nitro	$>C-NO_2$	Sulphur, carboxylates of diazonium	$>C-N_2^+ S^-$ $>C-N_2^+ RCO_2^-$
Nitrite	$>C-O-N = O$ NO_2^-	Halogenamine	$>N-X$

Name of Group	Formula	Name of Group	Formula
Nitrate	$\diagdown\!C\!-\!O\!-\!NO_2$ ⁄	N, N-difluoramin	$-NF_2$
		NO_3^-	
Epoxide	$\overset{O}{\triangle}$	Hypochlorite	$-O\!-\!Cl$
			ClO^-
Fulminate	$CN\!-\!O\!-\!Metal$	Chlorite, chlorate, perchlorate	$-ClO_x$ $x = 2, 3, 4$
Amide	$\diagdown\!N\!-\!Metal$ ⁄ $C\!=\!O$	Chlorate	$-ClO_2$ $-ClO$

2.2.2 'External' aggravating/attenuating factors

■ **Quantity**

The same unstable substance will behave according to the quantity that is handled. For sufficiently small quantities, the substance cannot decompose further by detonating. This was mentioned with regard to the tests carried out to determine the critical diameter. This phenomenon can be easily understood by postulating that the decomposition process is governed by the thermal exchanges between the decomposing system and the exterior. So long as the heat produced by decomposition is compensated by the energy loss by conduction and convection, the substance's temperature does not increase, and decomposition remains slow. Nevertheless, heat loss depends on substance-air contact. When substance volume increases, surface/volume ratio decreases, which is a disadvantage for the thermal exchange with the exterior and thus unbalances it in favour of heat production. The decomposition thus accelerates, which aggravates the phenomenon. In addition, the 'core' of the system, which is protected by the superficial layers of the substance, cannot dissipate the heat that is produced. There are some examples cited in Part Two, especially in the formation of butadiene peroxidized polymers.

■ **Dilution, desensitisation**

Dilution or simple mixing with a stable compound is sufficient to stabilise an unstable substance. In the case of a simple mixture with a neutral substance, this stabilisation process is called 'desensitisation'. Thus hardeners such as benzoyl peroxide are normally in the form of suspensions in heavy esters or oils. This peroxide is mixed with 30% of water by weight. Dynamite is nitroglycerine stabilised with the help of a neutral material. In all these cases, heat that is produced by the potential beginning of decomposition is absorbed by the inert substance.

■ **Confinement**

A lot of substances, even those that are relatively unstable, decompose often by heating up or spontaneously combusting without any destructive effects, or decompose in closed containers by detonation. Gunpowder is a typical example.

Confinement volume plays a huge role but cannot be modelled. The solidity of the closure does not perform any role. So a sheet of paper placed on the test tube is enough to transform the deflagration decomposition process of benzoyl peroxide into a detonation, which pulverises the test tube (this accident happened to the author but proved to be unrepeatable during a later test). So it is not a transition -al effect of pressure increase that plays an aggravating role.

2.3 Quantitative approach: the CHETAH programme[1]

The CHETAH programme is the most 'popular' process of quantitatively estimating the risks related to the instability of a compound. It stands for *Chemical Thermodynamic and Energy Release Program*. This program was developed by the National Institute of Standards (NIST)[2]. The 7-0 version is not distributed in France.

The program sets four criteria, leading to a three-level qualitative classification: *low risk, medium, high* for each of them. Each criterion quantifies an aspect of the decomposition risk. So these four classifications need to be taken into account to arrive at a final estimation. Some workers[3] have tried to use a sole criterion, which mathematically combines the four criteria, but failed. Three out of these four criteria involve calculating the enthalpies of decomposition and combustion of the particular compound. In order to do so it is necessary to know the enthalpies of formation of the compound and of the decomposition and combustion products. A lot of these values are inevitably absent in Part Three, so it was thought necessary to include estimation methods for enthalpies of formation as well as for enthalpies of vapourisation/condensation, since in many cases there is only available the value for the physical state of the compound that is not always appropriate.

2.3.1 Estimation of enthalpy of formation

■ **Benson's tables**

Benson carried out a very important study in order to establish a set of additivity rules for groups, relating to thermochemical data (see reference p.102). These tables enable one to determine the enthalpy of formation of a group in its gaseous phase by identifying in its structural formula characteristic groups for which the contribution to the enthalpy value is given. Benson did this work in different stages and the nature of the groups as well as their representation changed as he progressed. The latest available results are shown in Table I but for some structures, previous approaches seemed to give more reliable results. So it was decided to add Table II regarding functional aromatic derivatives for which estimates given in the first table are not always satisfactory.

---

1. The CHETAH programme is one of many and varied semiquantitative approaches to the estimation of risk. These approaches have proliferated in recent years in academia, commerce, industry and regulatory circles. One of the advantages of CHETAH is that it avoids some of the complexities and uncertainties of more complex models.
2. http://www.nist.gov/srd/o_spec16.htm ; e-mail : srdp@enh.nist.gov.
3. The INRS suggested a sole criterion, which the author used to use with his students until it was suggested not to use it anymore.

2 • Stability
2.3 Quantitative approach: the CHETAH programme

The enthalpy of formation of the substance that is obtained is for the gaseous state. The enthalpy that corresponds to the gas → liquid change of state therefore needs to be added. The enthalpy corresponding to the liquid → solid change is to be ignored if the enthalpy of formation of a substance in the solid state is required, since there is not much difference between the two. For the enthalpy of formation in the liquid state the enthalpy of condensation is calculated from the equation that bonds it to the entropic factor discussed in Chapter 1, ie:

$$S = \frac{\Delta H_{vap}}{2.303 \times R \times Eb}$$

The ΔH_{vap} sign has to be changed to obtain the enthalpy of condensation. R equals 8.31 Jmol^{-1}K^{-1}. Eb is given in °Kelvin.

Advice regarding the reading of tables. The group representations used here are different from Benson's, which are difficult to read. The group is usually represented in bold. In brackets are close structural elements that give a particular value to this group but are not taken into account for the contribution value. The contribution of these groups in brackets can be found in the appropriate section of the table. Finally, when closeness of a group is present to one or several saturated aliphatic groups, this closeness is represented by simple bonds.

The CHETAH programme used kcal/mole units whereas kJ/mole is used here. Nevertheless, numerous published documents give results using the old units. This is the reason it was decided to provide these group values in both units to be able to make calculations using both systems for comparative purposes. kJ is therefore retained.

Benson table (I)

No.	Group (in bold)	ΔH_f^0 kcal/mol	ΔH_f^0 kJ/mol	No.	Group (in bold)	ΔH_f^0 kcal/mol	ΔH_f^0 kJ/mol
	Hydrocarbons				**Hydrocarbons**		
1	H$_3$C–	–10.08	–42.19	14	(=C)–CH$_2$–(C=)	–4.29	–17.96
2	–CH$_2$–	–4.95	–20.72	15	(=C)–CH$_2$–(C$_{Ar}$)	–4.29	–17.96
3	>CH–	–1.90	–7.95	16	–CH$_2$–(C≡)	–4.73	–19.80
4	>C<	0.5	2.09	17	–CH$_2$–(C$_{Ar}$)	–4.86	–20.34
5	=CH$_2$	6.26	26.20	18	>CH–(C=)	–1.48	–6.20
6	=CH–	8.59	35.96	19	CH–(C≡)	–1.72	–7.20
7	=C<	10.34	43.28	20	>CH–(C$_{Ar}$)	–0.98	–4.10
8	=CH–(CH=)	6.78	28.38	21	C–(C=)	1.68	7.03
9	=C–(CH=)	8.88	37.17	22	C–(C$_{Ar}$)	2.81	11.76
10	=CH–(C$_{Ar}$)	6.78	28.38	23	≡CH	26.93	112.73
11	=C–(C$_{Ar}$)	8.64	36.17	24	≡C–	27.55	115.32
12	=CH–(C≡)	6.78	28.38	25	≡C–(C=)	29.20	122.23
13	–CH$_2$–(C=)	–4.76	–19.93	26	≡C–(C$_{Ar}$)	29.20	122.23

2.3 Quantitative approach: the CHETAH programme

No.	Group (in bold)	ΔH_f^0 kcal/mol	ΔH_f^0 kJ/mo	No.	Group (in bold)	ΔH_f^0 kcal/mol	ΔH_f^0 kJ/mo
	Hydrocarbons				**Hydrocarbons**		
27	$C_{Ar}H$	3.30	13.81	56	Z-Cyclononene	9.9	41.44
28	$C_{Ar}-$	5.51	23.06	57	E-Cyclononene	12.8	53.58
29	$C_{Ar}-(C=)$	5.68	23.78	58	Spiropentane	63.5	265.81
30	$C_{Ar}-(C\equiv)$	5.70	23.86	59	Bicyclo[1,1,0-]butane	67	280.46
31	$C_{Ar}-(C_{Ar})$	4.96	20.76				
32	Correction configuration **Z**	1.00	4.19	60	Bicyclo[2,1,0-]pentane	55.3	231.49
33	Correction **ortho**	0.57	2.39	61	Bicyclo[3,1,0-]hexane	32.7	136.88
	Correction of cyclic tension			62	Bicyclo[4,1,0-]heptane	28.9	120.98
34	Cyclopropane	27.6	115.53				
35	Methylene-cyclopropane	40.9	171.21	63	Bicyclo[5,1,0-]octane	29.6	123.91
36	Cyclopropene	53.7	224.79		**Oxygen compounds**		
37	Cyclobutane	26.2	109.67	64	$-\mathbf{CO}-(CO)$	−29.2	−122.23
38	Cyclobutene	29.8	124.74	65	$(=C)-\mathbf{CO}-(O)$	−33.5	−140.23
39	Cyclopentane	6.3	26.37	66	$(C_{Ar})-\mathbf{CO}-(O)$	−46.0	−192.56
40	Cyclopentene	5.9	24.70	67	$-\mathbf{CO}-(O)$	−33.4	−139.81
41	Cyclopentadiene	6.0	25.12	68	$\mathbf{HCO}-(O)$	−29.5	−123.49
42	Cyclohexane	0	0	69	$\mathbf{HCO}-(C=)$	−31.7	−132.70
43	Cyclohexene	1.4	5.86	70	$(CAr)-\mathbf{CO}-(C_{Ar})$	−39.1	−163.67
44	1,3-Cyclohexadiene	4.8	20.09	71	$-\mathbf{CO}-(C_{Ar})$	−37.6	−157.39
45	1,4-Cyclohexadiene	0.5	2.09	72	$\mathbf{HCO}-(C_{Ar})$	−31.7	−132.70
46	Cycloheptane	6.4	26.79	73	$-\mathbf{CO}-$	−31.5	−131.86
47	Cycloheptene	5.4	22.60	74	$\mathbf{HCO}-$	−29.6	−123.91
48	1,3-Cycloheptadiene	6.6	27.63	75	$\mathbf{H_2CO}$ (formaldehyde)	−27.7	−115.95
49	1,3,5-Cyclohepta-triene	4.7	19.67	76	$(CO)-\mathbf{O}-(CO)$	−50.9	−213.07
50	Cyclo-octane	9.9	41.44	77	$(CO)-\mathbf{O}-(O)$	−19.0	−79.53
51	Z-Cyclo-octene	6	25.12	78	$(CO)-\mathbf{O}-(C=)$ or $(CO)-\mathbf{O}-$	−41.3	−172.88
52	E-Cyclo-octene	15.3	64.05	79	$(CO)-\mathbf{OH}$	−60.3	−252.42
53	1,3,5-Cyclo-octa-triene	8.9	37.26	80	$-\mathbf{O}-(O)$	−4.5	−18.84
54	1,3,5,7-Cycloocta-tetraene	17.1	71.58	81	$-\mathbf{O}-(OH)$	−16.27	−68.11
				82	$(=C)-\mathbf{O}-(C=)$	−32.8	−137.30
55	Cyclononane	12.8	53.58	83	$(=C)-\mathbf{O}-$	−31.3	−131.02

2.3 Quantitative approach: the CHETAH programme

No.	Group (in bold)	ΔH_f^0 kcal/mol	ΔH_f^0 kJ/mo	No.	Group (in bold)	ΔH_f^0 kcal/mol	ΔH_f^0 kJ/mo
	Oxygen compounds				**Oxygen-containing compounds**		
84	(CAr)–**O**–(CAr)	−22.6	−94.60		*Correction of cyclic tensions*		
85	(CAr)–**OH**	−37.9	−158.65	111	Epoxy (ring)	27.6	115.53
86	–**O**–	−23.7	−99.21	112	Oxacyclobutane	26.4	110.51
87	–**OH**	−37.88	−158.57	113	Tetrahydrofuran	6.7	28.05
88	=**C**(CO)(O)	6.3	26.37	114	Pyran	2.2	9.21
89	=**C**–(CO)	9.4	39.35	115	1,3-Dioxan	3.5	14.65
90	=**CH**–(CO)	7.7	32.23	116	1,4-Dioxan	5.4	22.60
91	=**C**(C=)(O)	8.9	37.26	117	1,3,5-Trioxan	3.4	14.23
92	=**C**–(O)	10.3	43.12	118	Furan	−6.2	−25.95
93	=**CH**–(O)	8.6	36.00	119	1,2-Dihydropyran	2.5	10.47
94	**C**$_{Ar}$–(CO)	9.7	40.60	120	Cyclopentanone	6.0	25.12
95	**C**$_{Ar}$–(O)	−1.8	−7.53	121	Cyclohexanone	3.4	14.23
96	(CO)–**CH$_2$**–(CO)	−7.2	−30.14	122	Cyclic anhydride 5 bonds	1.1	4.60
97	>**C**–(CO)	1.58	6.61	123	Cyclic anhydride 6 bonds	1.4	5.86
98	>**CH**–(CO)	−1.83	−7.66	124	Structure 122 unsaturated	4.6	19.26
99	–**CH$_2$**–(CO)	−5	−20.93		**Nitrogen compounds**		
	Oxygen-containing compounds			125	**CH$_3$**–(N)	−10.08	−42.19
100	**CH$_3$**–(CO)	−10.08	−42.19	126	–**CH$_2$**–(N)	−6.6	−27.63
101	**C**(O)(O)	−16.8	−70.32	127	>**CH**–(N)	−5.2	−21.77
102	H**C**(O)(O)	−17.2	−72.00	128	>**C**–(N)	−3.2	−13.40
				129	–**CH$_2$**–(N=N)	−5.5	−23.02
103	=H$_2$**C**(O)(O)	−17.7	−74.09	130	>**CH**–(N=N)	−3.3	−13.81
				131	>**C**–(N=N)	−1.9	−7.95
104	(C$_{Ar}$)–**CH$_2$**–(O)	−6.6	−27.63	132	–**NH$_2$**	4.8	20.09
105	(=C)–**CH$_2$**–(O)	−6.9	−28.88	133	>**NH**	15.4	64.46
106	>**C**–(O)	−6.60	−27.63	134	–**N**<	24.4	102.14
				135	(N)–**NH$_2$**	11.4	47.72
107	>**CH**–(O)	−7.00	−29.30	136	(N)–**NH**–	20.9	87.49
108	–**CH$_2$**–(O)	−8.5	−35.58	137	(N)–**N**<	29.2	122.23
109	**CH$_3$**–(O)	−10.08	−42.19	138	(C$_{Ar}$)–**NH**–(N)	22.1	92.51
110	Correction >C–O–C<	8.4	35.16	139	(C)=**NH**	0	0
				140	(C)=**N**–	21.3	89.16
				141	(C)=**N**–(C$_{Ar}$)	16.7	69.91

2 • Stability

2.3 Quantitative approach: the CHETAH programme

No.	Group (in bold)	ΔH_f^0 kcal/mol	ΔH_f^0 kJ/mol	No.	Group (in bold)	ΔH_f^0 kcal/mol	ΔH_f^0 kJ/mol
	Nitrogen compounds				**Nitrogen compounds**		
142	(N=)**NH**	25.1	105.07	176	$(C_{Ar})_2$**N**–(N)	36.3	151.95
143	(N=)**N**–	32.5	136.05	177	(C=)**N–N**(=C)	21.6	90.42
144	(C_{Ar})–**NH$_2$**	4.8	20.09	178	(C=)**N**–(CAr)	14.1	59.02
145	(C_{Ar})–**NH**–	14.9	62.37	179	**C$_{Ar}$**–(N=CH–)	–0.5	–2.09
146	(C_{Ar})–**N**<	26.2	109.67	180	–**CH$_2$–NO**	19.1	79.95
147	(C_{Ar})–**NH**–(C_{Ar})	16.3	68.23	181	>**CH–NO**	22.2	92.93
148	(N aliphatic)–**C$_{Ar}$**	–0.5	–2.09	182	∖**C–NO**∕	23.2	97.12
149	**N$_{Ar}$** (pyridine)	16.7	69.87	183	**H$_2$N**–(O)	4.8	20.09
150	(N=)**N**–(N)	23	96.28	184	–**NH**–(O)	12.2	51.07
151	(N)–**CH=O**	–29.6	–123.91	185	**HO**–(N)	–14.1	–59.02
152	(N)–**CO**–	–32.8	–137.30	186	–**O**–(N)	–0.9	–3.77
153	(CO)–**NH$_2$**	–14.9	–62.37	187	(–CH=)**N**–(OH)	–5	–20.93
154	(CO)–**NH**–	–4.4	–18.42		*Correction of tensions of nitrogen cycles :*		
155	(CO)–**N**<	0	0	188	Aziridine	27.7	115.95
156	(CO)–**NH**–(Ar)	0.4	1.67	189	Pyrrolidine	6.8	28.46
157	(CO)–**NH**–(CO)	–18.5	–77.44	190	Piperidine	1.0	4.184
158	(–CO–)$_2$**N**–	–5.9	–24.70		**Halogen compounds** (except fluorides)		
159	(–CO–)$_2$**N**–(C_{Ar})	–0.5	–2.09	191	= **CCl$_2$**	–1.8	–7.53
160	–**CH$_2$–CN**	22.5	94.19	192	= **CBr$_2$**	0	0
161	>**CH–CN**	25.8	108.00	193	= **CClBr**	0	0
162	∖**C–CN**∕	0	0	194	= **CHCl**	2.1	8.79
163	>**C(CN)$_2$**	0	0	195	= **CHBr**	12.7	53.14
164	= **CH–CN**	37.4	156.56	196	= **CHI**	24.5	102.51
165	= **C(CN)$_2$**	84.1	352.04	197	**C$_{Ar}$–Cl**	–3.8	–15.90
166	= **CH–NO$_2$**	0	0	198	**C$_{Ar}$–Br**	10.7	44.77
167	**C$_{Ar}$–CN**	35.8	149.86	199	**C$_{Ar}$–I**	24.0	100.42
168	≡ **C–CN**	63.8	267.07	200	**C$_{Ar}$–CH$_2$–Br**	–6.9	–28.87
169	–**CH$_2$–NO$_2$**	–15.1	–63.21	201	**C$_{Ar}$–CH$_2$–I**	8.4	35.15
170	>**CH–NO$_2$**	–15.8	–66.14	202	ortho correction		
171	∖**C–NO$_2$**∕	0	0	203	**C$_{Ar}$Cl–C$_{Ar}$Cl**	2.2	9.20
172	–**CH(NO$_2$)$_2$**	–14.9	–62.37	204	**C$_{Ar}$**(C) –**C$_{Ar}$Cl**	0.6	2.51
173	–**O–NO**	–5.9	–24.70	205	–**CCl$_3$**	–20.7	–86.61
174	>**N–NO$_2$**	17.8	74.51	206	–**CHCl$_2$**	–18.9	–79.08
175	–**CH$_2$–N$_3$**	70	293.02	207	–**CH$_2$Cl**	–15.6	–65.27

2.3 Quantitative approach: the CHETAH programme

No.	Group (in bold)	ΔH_f^0 kcal/mol	ΔH_f^0 kJ/mol	No.	Group (in bold)	ΔH_f^0 kcal/mol	ΔH_f^0 kJ/mol
	Halogen compounds (except fluorides)				**Sulphur compounds**		
208	>**CCl$_2$**	−19.5	−81.59	241	(=C)−**CH$_2$**−(SO)	−7.35	−30.75
209	>**CH−Cl**	−12.8	−53.56	242	**C$_{Ar}$**−(SO)	2.3	9.62
210	>**C−Cl**	−12.8	−53.56	243	−**SO**−	−14.41	−60.29
211	−**CBr$_3$**			244	(C$_{Ar}$)−**SO**−(C$_{Ar}$)	−12.0	−50.21
212	−**CH$_2$−Br**	−5.4	−22.59	245	**CH$_3$**−(SO$_2$)	−10.08	−42.19
213	>**CH−Br**	−3.4	−14.23	246	−**CH$_2$**−(SO$_2$)	−7.68	−32.13
214	>**C−Br**	−0.4	−1.67	247	>**CH**−(SO$_2$)	−2.62	−10.96
215	−**CH$_2$−I**	7.95	33.26	248	>**C**−(SO$_2$)	−0.61	−2.55
216	>**CH−I**	10.7	44.77	249	(=C)−**CH$_2$**−(SO$_2$)	−7.14	−29.87
217	>**C−I**	13.0	54.39	250	(C$_{Ar}$)−**CH$_2$**−(SO$_2$)	−5.54	−23.18
218	−**CHBrCl**			251	**C$_{Ar}$**−(SO$_2$)	2.3	9.62
	Sulphur compounds			252	−**SO$_2$**−	−69.74	−291.79
219	**CH$_3$**−(S)	−10.08	−42.19	253	−**SO$_2$**−(C$_{Ar}$)	−72.29	−302.46
220	−**CH$_2$**−(S)	−5.65	−23.64	254	(C$_{Ar}$)−**SO$_2$**−(C$_{Ar}$)	−68.58	−286.94
221	>**CH**−(S)	−2.64	−11.05	255	(=C)−**SO$_2$**−(C$_{Ar}$)	−68.58	−286.94
222	>**C**−(S)	−0.55	−2.30	256	(=C)−**SO$_2$**−(C=)	−73.58	−307.86
223	(C$_{Ar}$)−**CH$_2$**−(S)	−4.75	−19.87	257	(C$_{Ar}$)−**SO$_2$**−(SO$_2$)	−76.25	−319.09
224	(=C)−**CH$_2$**−(S)	−6.45	−26.99	258	−**CO**−(S)	−31.56	−132.05
225	**C$_{Ar}$**−(S)	−1.8	−7.53	259	(CO)−**SH**	−1.41	−5.90
226	=**CH**−(S)	8.56	35.82	260	−**S−CN**	37.18	155.56
227	=**C**−(S)	10.93	45.73	261	(N)−**CS**−(N)	12.78	53.47
228	−**SH**	4.62	19.33	262	(S)−**S**−(N)	−4.90	−20.5
229	(C$_{Ar}$)−**SH**	11.96	50.04	263	>**N**−(S)	29.9	125.10
230	−**S**−	11.51	48.16	264	(N)−**SO**−(N)	−31.56	−132.05
231	(=C)−**S**−	9.97	41.71	265	>**N**−(SO$_2$)	−20.4	−85.35
232	(=C)−**S**−(C=)	−4.54	−19.00		*Cyclic correction*		
233	(C$_{Ar}$)−**S**−	19.16	80.17	266	S (3-ring)	17.7	74.06
234	(C$_{Ar}$)−**S**−(C$_{Ar}$)	25.90	108.37	267	S (4-ring)	19.37	81.04
235	−**S**−(S)	7.05	29.5				
236	(C$_{Ar}$)−**S**−(S)	14.5	60.67	268	S (5-ring)	1.73	7.24
237	(S)−**S**−(S)	3.04	12.72	269	S (6-ring)	0	0
238	**CH$_3$**−(SO)	−10.08	−42.19				
239	−**CH$_2$**−(SO)	−7.72	−32.30	270	S (7-ring)	3.89	16.28
240	>**C**−(SO)	−3.05	−12.76				

2 • Stability

2.3 Quantitative approach: the CHETAH programme

No.	Group (in bold)	ΔH_f^0 kcal/mol	ΔH_f^0 kJ/mol	No.	Group (in bold)	ΔH_f^0 kcal/mol	ΔH_f^0 kJ/mol
	Sulphur compounds				**Phosphorus compounds**		
				300	>**N**–(P)	32.2	134.72
271	S-S rings	5.07	21.22	301	>**N**–(PO)	17.8	74.48
				302	–**P**=**N**–	0.5	2.09
272	ring-**SO₂**	5.74	24.02	303	$(C_{Ar})_2$ **P**=**N**–	−25.7	−107.53
273	S ring	1.73	7.24		**Boron compounds**		
				304	**CH₃**–(B)	−10.08	−42.19
	Phosphorus compounds			305	–**CH₂**–(B)	−2.22	−9.29
274	**CH₃**–(P)	−10.08	−42.19	306	>**CH**–(B)	1.1	4.60
275	–**CH₂**–(P)	−2.47	−10.33	307	=**CH**–(B)	15.6	65.27
276	**CH₃**–(PO)	−10.08	−42.19	308	>**B**–	0.9	3.77
277	–**CH₂**–(PO)	−3.4	−14.23	309	>**BCl**	−42.7	−176.02
278	**CH₃**–(P=N)	−10.08	−42.19	310	>**B**–(O)	29.3	122.59
279	–**CH₂**–(P=N)	19.4	81.17	311	(–O) **B**–(O–) (–O)	24.4	102.09
280	C_{Ar}–(P)	−1.8	−7.53				
281	C_{Ar}–(PO)	2.3	9.62	312	(–O)–**BCl**–(O–)	−19.7	−82.42
282	C_{Ar}–(P=N)	2.3	9.62	313	(–O)–**BCl₂**	−61.2	−256.06
283	>**P**–	7.04	29.46	314	(–O)–**BH**–(O–)	19.9	83.26
284	–**PCl₂**	−50.1	−209.62	315	(N–)₃ **B**	24.4	102.09
285	$(C_{Ar})_3$ **P**	28.3	118.41	316	(N–)₂ **BCl**	−23.8	−99.58
286	(–O–)₃ **P**	−66.8	−279.49	317	(N)–**BCl₂**	−67.9	−284.09
287	(N–)₃ **P**	−66.8	−279.49	318	(B)–**OH**	−115.5	−483.25
288	–**PO**	−72.3	−302.5	319	(B)–**O**–	−69.43	−290.5
289	–**POCl₂**	−123.0	−514.63	320	–**B**<	−9.93	−41.55
290	(–O)–**PO** (–O) (phosphonates)	−99.5	−416.31	321	(–S–)₃ **B**	24.4	102.09
				322	–**S**–(B)	−14.5	−60.67
291	**OP**(–O–)₃	−104.6	−437.65	323	(C_{Ar})–**S**–(B)	−7.8	−32.64
292	$(C_{Ar})_3$ **PO**	−52.9	−221.33		**Tin compounds**		
293	(N–)₃ **PO**	−104.6	−437.65	324	**CH₃**–(Sn)	−10.08	−42.19
294	–**O**–(P)	−23.5	−98.32	325	–**CH₂**–(Sn)	−2.18	−9.12
295	**HO**–(P)	−58.7	−245.6	326	>**CH**–(Sn)	3.38	14.14
296	–**O**–(PO)	−40.7	−170.29	327	(Sn)–**C**–	8.16	34.14
297	**HO**–(PO)	−65	−271.96	328	(C_{Ar})–**CH₂**–(Sn)	−7.77	−32.51
298	(PO)–**O**–(PO)	−54.5	−228.03	329	C_{Ar}–(Sn)	5.51	23.05
299	–**O**–(P=N)	−40.7	−170.29	330	=**CH**–(Sn)	8.77	36.69
				331	>**Sn**<	36.2	151.46

2.3 Quantitative approach: the CHETAH programme

No.	Group (in bold)	ΔH_f^0 kcal/mol	ΔH_f^0 kJ/mol	No.	Group (in bold)	ΔH_f^0 kcal/mol	ΔH_f^0 kJ/mol
	Tin compounds				**Zinc compounds**		
332	\-**SnCl**/	-9.8	-41.00	347	**CH$_3$**-(Zn)	-10.08	-42.19
333	>**SnCl$_2$**	-49.2	-205.85	348	-**CH$_2$**-(Zn)	-1.78	-7.45
334	-**SnCl$_3$**	-89.5	-374.47	349	-**Zn**-	33.3	139.33
335	\-**SnH**/	34.8	145.6		**Aluminium compounds**		
336	(=C-)$_4$ **Sn**	36.2	151.46	350	**CH$_3$**-(Al)	-10.08	-42.19
337	(=C-)$_3$ **SnCl**	-8.2	-34.31	351	-**CH$_2$**-(Al)	0.7	2.93
338	(=C-)$_2$ **SnCl$_2$**	-50.7	-212.13	352	>**Al**-	9.2	38.49
339	(=C-)-**SnCl$_3$**	-82.2	-343.92		**Mercury compounds**		
340	\-**Sn**-(C=)/	37.6	157.32	353	**CH$_3$**-(Hg)	-10.08	-42.19
341	(C$_{Ar}$-)$_4$ **Sn**	26.26	109.87	354	-**CH$_2$**-(Hg)	-2.68	-11.2
342	\-**Sn**-(C$_{Ar}$)/	4.93	146.15	355	>**CH**-(Hg)	3.62	15.15
343	\-**Sn**-(Sn)/	26.4	110.46	356	**C$_{Ar}$**-(Hg)	-1.8	-7.53
	Lead compounds			357	-**Hg**-	42.5	177.82
344	**CH$_3$**-(Pb)	-10.08	-42.19	358	-**HgCl**	-2.82	-11.8
345	-**CH$_2$**-(Pb)	-1.7	-7.11	359	(C$_{Ar}$)-**Hg**-(C$_{Ar}$)	64.4	269.45
346	>**Pb**<	72.9	305.01	360	(C$_{Ar}$)-**HgCl**	9.9	41.42

Below is the additional table for aromatic compounds, which is sometimes more useful than the previous one, especially in the case of aromatic nitrated derivatives.

Benson table(II)

No.	Group (in bold)	ΔH_f^0 kcal/mol	ΔH_f^0 kJ/mol	No.	Group (in bold)	ΔH_f^0 kcal/mol	ΔH_f^0 kJ/mol
A	-**C$_6$H$_5$**	21.97	91.92	G	-**Br**	5.50	23.01
B	-**C$_6$H$_4$**-	24.70	103.34	H	-**OH**	-45.50	-190.37
C	>**C$_6$H$_3$**-	27.30	114.22	I	-**O-CH$_3$**	-38.20	-176.56
D	>**C$_6$H$_2$**<	29.90	125.10	J	-**O-CH$_2$**-	-36.20	-151.46
	Substituent in nucleus			K	-**NH$_2$**	-1.20	-5.02
				L	-**NH-CH$_3$**	-0.40	-1.67
E	-**F**	-48.00	-200.83	M	-**N(CH$_3$)$_2$**	0	0
F	-**Cl**	-9.50	-39.75	N	-**NO$_2$**	-6.00	-25.10

■ Examples of estimation of enthalpy of formation

A few compounds will be chosen to test the method with various types of structures. Compounds will be chosen for which enthalpy can, for the purpose of comparison, be found in the tables and the enthalpy calculated in the liquid state only when this value is needed to be known.

2 • Stability
2.3 Quantitative approach: the CHETAH programme

Examples of enthalpy calculations (kJ/mol)

Compound, structure, no. of group	ΔH_f (g) est. kJ/mol	ΔH_f (l) est.	ΔH publ (state)
Ethylene: $H_2C=CH_2$, 5 (2)	52.40		52.30 (g)
Allyl alcohol $CH_2=CH-CH_2-OH$ ↑ ↑ ↑ ↑ 5 6 108 87	−131.99	Group 7 −174.29	−132.04 (g)
1,4-Dioxan 108 (4) 86 (2) Correction cycle no 116	−318.14		−315.06
2-Nitroaniline: $H_2N-C_6H_4-NO_2$ (2nd table) Groups: B (1); K (1); N(1) entropic group; 7	73.22	5.72	−14.43 (s)
N-methylaniline 148, 145, 125 27(5) → —NH–CH$_3$	87.14	S = 5.186 40.49	32.22 (l)
2-Methylpyridine 149, 28, 1 27(4) → —CH$_3$	105.98	S = 5.04 67.20	100.67 (g) 57.86 (l)
1.2-Dichlorobenzene 27(4) → Cl, Cl 197(2) 203 (ortho correction)	32.64 (table I) 23.84 (table II)	S = 5.774 −20.45	29.96 (g)
Isobutyronitrile 161 1 → CH$_3$, CH$_3$ CH–CN	23.62	Group 2 −9.76	25.40 (g)
Thiophene 6(2) 226(2) 232 273 (correction)	112.8		114.00 (g)

109

The Benson tables lack some information about the nitro group bonded with an aromatic cycle, which is one of the most frequent groups amongst unstable compounds. An attempt was made to estimate the C_{Ar}-NO_2 value using the experimental data from Part Three. The results are incoherent and this explains why Benson did not mention anything. As a result, only the approach of the second table could be used, which gives results with a high level of error.

2.3.2 CHETAH Criterion C1: enthalpy of decomposition

Criterion C_1 estimates the enthalpy of decomposition ΔH_d of the particular molecule. If the equation for this decomposition is known, the enthalpy of formation of the product is subtracted from the sum of the enthalpies of formation of the degradation products.

If the equation of the decomposition reaction is not known, this calculation is carried out using the equation of the hypothetical reaction that releases the strongest energy theoretically possible.

The comparison of the experimental and estimated enthalpies of decomposition according to CHETAH shows that the latter are almost always overestimated. This is why ΔH_d is calculated using ΔH_f of the substance in the physical state at 25°C.

C_1 is given here in kJ/g.

Criteria regarding risk levels are as follows:

- $|C_1| < 1.254$ Low risk;
- $1.254 \leq |C_1| < 2.926$ Medium risk;
- $|C_1| \geq 2.926$ High risk.

The complicated values chosen as a threshold can be surprising. This can be explained by the choice of units. Originally CHETAH chose the threshold values 0.3 and 0.7 kcal/g.

■ Method of writing for most energetic reaction

The main problem is knowing how to write the equation of the most energetic hypothetical reaction. Here are some rules that can be applied in the simplest case in which the molecule contains C, H, O, N, X (X = halogen) atoms:

1. Having X and H, as many HX as possible are formed.
2. If after 1. H and O are left, as many H_2O as possible are formed.
3. If after 2. O and C are left, as many CO_2 as possible are formed.
4. If after 2. C, H and N are left, as many CH_4 as possible are formed.
5. If after 4. H and N are left, as many NH_3 as possible are formed.
6. If after 4. there are any C and Cl left, as many CCl_4 as possible are eliminated.
7. If after 3. or 4. N and C are left, both atoms will create the two elements in the fundamental state N_2 and C.

The reasoning that leads to 2., 3. and 4. is the following.

If H, O, C and N are present, the oxygen onset in the form of H_2O is more energetic than in the form of CO_2. Indeed, in the first case the enthalpy due to the oxygen atom is -241.60 kJ/mol (Part Three) and in the second case:

$$\frac{-393.34}{2} = -196.67 \text{ kJ/mol (Part Three)}.$$

In the same way, the formation of water produces an energy (per hydrogen atom) of:

$$\frac{-241.60}{2} = -120.8 \text{ kJ/mol}$$

whereas in CH_4 form, we obtain:

$$\frac{-74.82}{4} = -18.71 \text{ kJ/mol}.$$

Following the same reasoning, the energies produced by carbon in the form of CO_2 and CH_4 are compared, and in the same way the energies by hydrogen in the form of CH_4 and NH_3 are compared.

This reasoning is used systematically in the most complex cases by looking for the most energetic compounds that are likely to be formed. It leads to the descending order of priority of the degradation substances; this is taken into account in the writing of the equation.

■ Examples of equations and calculations

In the previous examples, the calculated enthalpies of decomposition are taken. The enthalpies of formation of the decomposition substances come from the corresponding chapters in Part Two. The published values of enthalpies of formation are favoured and use of the values estimated is only made when there is no experimental data. A few inorganic compounds have been added which are noted for their instability: eg ammonium dichromate and ammonium nitrate.

Ethylene

$$CH_2=CH_2 \longrightarrow CH_4 + C$$

$\Delta H_d = \Delta H_f (CH_4) - \Delta H_f (C_2H_4) = -74.82 - 52.3 = -127.12$ kJ/mol
$C_1 = -4.53$ kJ/g → high risk

Allyl alcohol

$$C_3H_6O \longrightarrow H_2O + CH_4 + 2\,C$$

$\Delta H_d = \Delta H_f (H_2O) + \Delta H_f (CH_4) - \Delta H_f (C_3H_6O) = -241.8 - 74.82 + 174.29 = -142.33$ kJ/mol
$C_1 = -2.45$ kJ/g → moderate risk

1,4-Dioxan

$$C_4H_8O_2 \longrightarrow 2\,H_2O + CH_4 + 3\,C$$

$\Delta H_d = 2 \times \Delta H_f (H_2O) + \Delta H_f (CH_4) - \Delta H_f (C_4H_8O_2) = -241.8 \times 2 - 74.82 + 353.42 = -205$ kJ/mol
$C_1 = -2.327$ kJ/g → moderate risk

2-Nitroaniline

$$C_6H_6N_2O_2 \longrightarrow 2\,H_2O + 0.5\,CH_4 + N_2 + 5.5\,C$$

$\Delta H_d = 2 \times \Delta H_f (H_2O) + 0.5 \times \Delta H_f (CH_4) - \Delta H_f (C_6H_6N_2O_2) = -241.8 \times 2 - 0.5 \times 74.82 + 14.43 = 506.58$ kJ/mol
$C_1 = -3.67$ kJ/g → high risk

N-Methylaniline

$$C_7H_9N \longrightarrow 2.25\ CH_4 + 0.5\ N_2 + 4.75\ C$$

$\Delta H_d = 2.25\ \Delta H_f\ (CH_4) - \Delta H_f\ (C_7H_9N) = 2.25 \times (-74.82) - 32.22 = -200.565$ kJ/mol
$C_1 = -1.87$ kJ/g $\rightarrow$ moderate risk

2-Methylpyridine

$$C_6H_7N \longrightarrow 1.75CH_4 + 0.5N_2 + 4.25C$$

$\Delta H_d = 1.75 \times \Delta H_f\ (CH_4) - \Delta H_f\ (C_6H_7N) = 1.75 \times (-74.82) - 57.86 = -188.795$ kJ/mol
$C_1 = -2.03$ kJ/g $\rightarrow$ moderate risk

1,2-Dichlorobenzene

$$C_6H_4Cl_2 \longrightarrow 2\ HCl + 0.5CH_4 + 5.5C$$

$\Delta H_d = 2 \times \Delta H_f\ (HCl) + 0.5 \times \Delta H_f\ (CH_4) - \Delta H_f\ (C_6H_4Cl_2) = -92.3 \times 2 - 0.5 \times 74.82 + 20.45$
$= -202.06$ kJ/mol
$C_1 = -1.37$ kJ/g $\rightarrow$ moderate risk

Isobutyronitrile

$$C_4H_7N \longrightarrow 1.75CH_4 + 0.5N_2 + 2.25\ C$$

$\Delta H_d = 1.75 \times \Delta H_f\ (CH_4) - \Delta H_f\ (C_4H_7N) = 1.75 \times (74.82) + 9.76 = -121.175$ kJ/mol
$C_1 = -1.75$ kJ/g $\rightarrow$ moderate risk

Thiophene

This compound contains a sulphide atom. The above rules for the writing of the equation of the most energetic reaction are not applied. The reasoning mentioned above will have to be applied to know if this substance favours H_2S or CH_4.
The formation of hydrogen sulphide produces -20.6/2 kJ ie -10.3 kJ per hydrogen atom whereas the formation of methane produces -74.82/4 kJ ie -18.705 kJ per hydrogen atom. Maximum methane has then to be formed in this hypothetical reaction. To be thorough, the potential formation of carbon disulphide has to be analysed, but this substance is 'endothermic'.

Then:

$$C_4H_4S \longrightarrow CH_4 + 3\ C + S$$

$\Delta H_d = \Delta H_f\ (CH_4) - \Delta H_f\ (C_4H_4S) = -74.82 - 80.2 = -155.02$ kJ/mol
$C_1 = -1.84$ kJ/g $\rightarrow$ moderate risk

Ammonium dichromate

$$(NH_4)_2Cr_2O_7 \longrightarrow Cr_2O_3 + 4\ H_2O + N_2$$

$\Delta H_d = 4 \times \Delta H_f\ (H_2O) + \Delta H_f\ (Cr_2O_3) - \Delta H_f\ ((NH_4)_2Cr_2O_7) = -241.8 \times 4 - 1139.7 + 1806.45 =$
-300.45 kJ/mol
$C_1 = -1.19$ kJ/g $\rightarrow$ low risk

Ammonium nitrate

$$NH_4NO_3 \longrightarrow 2\ H_2O + N_2 + 0.5O_2$$

$\Delta H_d = 2 \times \Delta H_f\ (H_2O) - \Delta H_f\ (NH_4NO_3) = -241.8 \times 2 + 365.56 = -118.04$ kJ/mol
$C_1 = -1.47$ kJ/mol $\rightarrow$ moderate risk

Most substances which appear in the examples of this chapter are analysed In Part Two and their enthalpy of decomposition determined experimentally. This is because most of them are considered hardly stable. This is one of the reasons for assigning no 'low risk' in the suggested classifications. But it is also indisputable that criterion C_1 overestimates the instability risk. It is the case for all aromatic compounds that are generally very stable. In the examples above, N-methylaniline, dichlorobenzene and 2-methylpyridine are of medium risk although 'low risk' would be more appropriate. The only way to identify risk accurately is by 'associating' the qualitative approach with the CHETAH one. Thus none of the three substances mentioned have an explosophoric group. One way to get a better estimate is by analysing the aliphatic compound whose structure 'comes closer' to the particular aromatic compound. Another method will be used later in order to obtain a better correspondence between forecast and reality.

2.3.3 CHETAH Criterion C_2: propensity for combustion

■ **Definition and criteria**

Criterion C_2 takes into account criterion C_1 as well as the enthalpy of combustion of the compound. This criterion is the result of the observation that compounds of similar enthalpies of decomposition and combustion are the only ones that present a risk bonded to instability.

Classification is carried out by writing down the co-ordinates of the compound in kJ/g on the graph below, which defines the zones corresponding to each risk degree:

- C_1 is on the x-axis;
- The $\Delta H_c - \Delta H_d$ is on the y-axis in which ΔH_c is the enthalpy of the combustion reaction that leads to substances with maximum degree of oxidation of the mol-ecule.

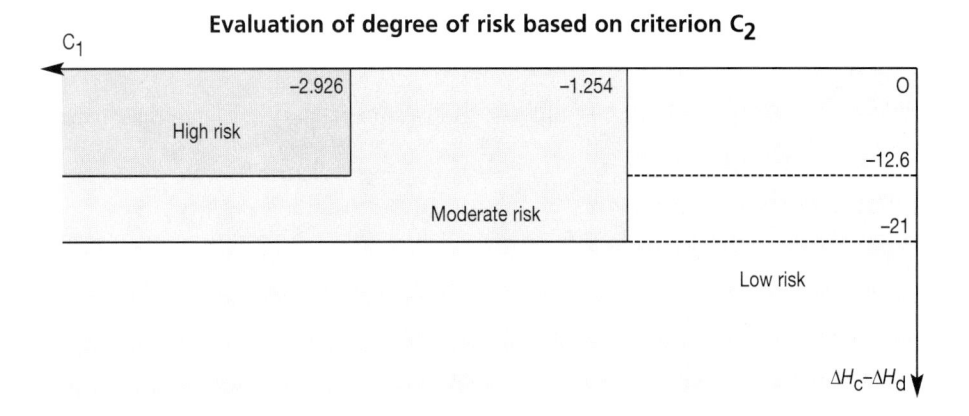

Evaluation of degree of risk based on criterion C_2

C_1

| | −2.926 | | −1.254 | | 0 |

High risk

−12.6

Moderate risk

−21

Low risk

$\Delta H_c - \Delta H_d$

■ **Examples**

The same examples as before are considered and the equations of combustion written, without giving all the calculation details.

Ethylene

$$C_2H_4 + 3\,O_2 \longrightarrow 2\,CO_2 + 2\,H_2O$$

$\Delta H_c = -1322.90$ kJ/mol $= -47.72$ kJ/g; $\Delta H_c - \Delta H_d = -42.72$ kJ/g → low risk

Allyl alcohol

$$C_3H_6O + 4 O_2 \longrightarrow 3 H_2O + 3 CO_2$$

$\Delta H_c = -1731.61$ kJ/mol $= -29.81$ kJ/g; $\Delta H_c - \Delta H_d = -27.36$ kJ/mol → moderate risk

1,4-Dioxan

$$C_4H_8O_2 + 5 O_2 \longrightarrow 4 CO_2 + 4 H_2O$$

$\Delta H_c = -2187.78$ kJ/mol $= -24.83$ kJ/g; $\Delta H_c - \Delta H_d = -22.50$ kJ/g → low risk

2-Nitroaniline

$$C_6H_6N_2O_2 + 6.5 O_2 \longrightarrow 6 CO_2 + 3 H_2O + N_2$$

$\Delta H_c = -3071.97$ kJ/mol $= -22.24$ kJ/g; $\Delta H_c - \Delta H_d = -18.57$ kJ/g → moderate risk

N-Methylaniline

$$C_6H_9N + 8.25 O_2 \longrightarrow 6 CO_2 + 4.5 H_2O + 0.5 N_2$$

$\Delta H_c = -3481.32$ kJ/mol $= -32.48$ kJ/g; $\Delta H_c - \Delta H_d = -30.61$ kJ/g → low risk

2-Methylpyridine

$$C_6H_7N + 7.75 O_2 \longrightarrow 6 CO_2 + 3.5 H_2O + 0.5 N_2$$

$\Delta H_c = -3265.16$ kJ/mol $= -35.06$ kJ/g; $\Delta H_c - \Delta H_d = -33.02$ kJ/g → low risk

1,2-Dichlorobenzene

$$C_6H_4Cl_2 + 6.5 O_2 \longrightarrow 6 CO_2 + H_2O + 2 HCl$$

$\Delta H_c = -2766.95$ kJ/mol $= -18.82$ kJ/g; $\Delta H_c - \Delta H_d = -17.45$ kJ/g → moderate risk

Isobutyronitrile

$$C_4H_7N + 5.75 O_2 \longrightarrow 4 CO_2 + 3.5 H_2O + 0.5 N_2$$

$\Delta H_c = -2478.16$ kJ/mol $= -35.86$ kJ/g; $\Delta H_c - \Delta H_d = -34.11$ → low risk

Thiophene

$$C_4H_4S + 6 O_2 \longrightarrow 4 CO_2 + 2 H_2O + SO_2$$

$\Delta H_c = -2434.63$ kJ/mol $= -28.95$ kJ/g; $\Delta H_c - \Delta H_d = -27.11$ kJ/g → low risk

Ammonium dichromate

The compound is incombustible; the combustion reaction, ie the oxidation reaction, is the same as the decomposition one.

So $\Delta H_c = \Delta H_d$ and $\Delta H_c - \Delta H_d = 0$ → *low risk*.

Ammonium nitrate

The same is true as for dichromate: $\Delta H_c - \Delta H_d = 0$ → *medium risk*.

The results obtained using this criterion are very close to reality. Two of the compounds that are known to be unstable and appear in this series, ie nitroaniline and ammonium nitrate, which have an explosophoric group without necessarily being noted for being explosive, are classified 'medium risk'. There are still two anomalies: the far too severe classification for 1,2-dichlorobenzene, which is obviously due to the endothermic nature of the aromatic cycle (it would be better to analyse 1,2-dichlorocyclohexane using the technique mentioned before); and on the other hand, the underestimated risk of ammonium dichromate, which is, incidentally, overestimated in the regulations as will be seen later.

2.3.4 Criterion C_3: 'internal redox' measure; oxygen balance

■ Definition and criteria

Oxygen balance is described by the equation:

$$C_3 = \frac{3200}{M} \times z$$

where z is the number of oxygen molecules that are necessary to balance the equation of the reaction of the most energetic complete combustion.

If oxygen is in the left part of the equation, z bears the - sign, if it is on the right, z is positive.

M is the molar mass of the compound.

C_3 is in grams per cent (g%).

The oxygen balance is, when negative, the quantity of oxygen 'internal' to the molecule that is missing to ensure complete oxidation of this molecule. z is given directly by the coefficient of O_2 in the equation of combustion.

Then:

- $C3 < -240$ Low risk.
- $-240 \leq C3 < -120$
  or $80 < C3 \leq 100$ Medium risk.
- $-120 \leq C3 \leq 80$ High risk.

The following table lists the above examples and their classification following this criterion.

Internal redox character of various compounds

Compound	M	z	C_3 (g%)	Degree of risk
Ethylene	28.06	−3.00	−342.12	Low
Allyl alcohol	58.09	−4.00	−220.34	Moderate
1,4 Dioxan	88.11	−5.00	−181.59	Moderate
2-Nitroaniline	138.14	−6.50	−150.57	Moderate
N-Methylaniline	107.17	−8.25	−246.34	Low
2-Methylpyridine	93.14	−7.75	−266.27	Low
1.2-Dichlorobenzene	147.00	−6.50	−141.50	Moderate
Isobutyronitrile	69.11	−5.75	−266.24	Low
Thiophene	84.14	−6.00	−228.19	Low
Ammonium dichromate	252.10	0.00	0	High
Ammonium nitrate	80.04	0.50	+ 19.99	High

Oxygen balance plays a role on its own within the CHETAH programme and its interpretation proves to be difficult. Indeed, is it possible that the very similar oxygen balance values of nitroaniline and dichlorobenzene refer to instability behaviours of the same nature? There is a real internal redox danger for the first one, but certainly not for the second. It is obvious that dichlorobenzene is overestimated from a risk point of view criteria C_3. For the rest, C_3 is essentially a way of measuring the oxidising power and in organic chemistry a way of measuring the compound propensity for behaving like a propellant.

2.3.5 CHETAH Criterion C_4: effect of mass

$$C_4 = 10 \times C_1^2 \times \frac{M}{n}$$

M is the molar mass and n the total number of atoms that are in the particular molecule.

- $C_4 < 525$ Low risk.
- $525 \leq C_4 < 1926$ Medium risk.
- $C_4 \geq 1926$ High risk.

Using again the example above:

Classification of compounds according to criterion C_4

Compound	M	C_1 (kJ/g)	C_4	Degree of risk
Ethylene	28.06	−4.53	959.69	Moderate
Allyl alcohol	58.09	−2.45	348.69	Low
1,4-Dioxan	88.11	−2.33	341.67	Low
2-Nitroaniline	138.14	−3.67	1162.87	Moderate
N-Methylaniline	107.17	−1.87	220.45	Low
2-Methylpyridine	93.14	−2.03	274.16	Low
1.2-Dichlorobenzene	147.00	−1.37	229.92	Low
Isobutyronitrile	69.11	−1.75	176.37	Low
Thiophene	84.14	−1.84	316.52	Low
Ammonium dichromate	252.10	−1.19	187.89	Low
Ammonium nitrate	80.04	−1.47	192.18	Low

When analysing the results it is noticed this time that criterion C_4 is the least severe of the four CHETAH criteria. It emphasises the unstable property of nitroaniline but underestimates the instability of ammonium nitrate and ammonium dichromate for which there is no indication of danger whatsoever.

2.3.6 Search for a unique criterion, a variant of CHETAH

■ **Why a unique criterion?**

The table below lists all CHETAH results for the substances used as examples (medium and high classifications are in bold).

2 • Stability

2.3 Quantitative approach: the CHETAH programme

CHETAH criteria for various compounds

Compound	C_1 (kJ/g)	C_2	C_3 (g%)	C_4
Ethylene	**−4.53**	Low	−342.12	**959.69**
Allyl alcohol	−2.45	Low	−220.34	348.69
1,4-Dioxan	−2.33	Low	−181.59	341.67
2-Nitroaniline	**−3.67**	**Medium**	−150.57	**1162.87**
N-Methylaniline	−1.87	Low	−246.34	220.45
2-Methylpyridine	−2.03	Low	−266.27	274.16
1.2-Dichlorobenzene	−1.37	**Medium**	−141.50	229.92
Isobutyronitrile	−1.75	Low	−266.24	176.37
Thiophene	−1.84	Low	−228.19	316.52
Ammonium dichromate	−1.19	Low	0	187.89
Ammonium nitrate	−1.47	**Medium**	+19.99	192.18

*The medium and high risks are in bold.

It is noted that 2-nitroaniline and ammonium nitrate are the two compounds that give the most consistent results. These are the two most unstable compounds of this series. All authors who showed some interest in this approach ruled out the possibility of looking for a unique criterion which would synthesise these four approaches, a point of view not shared by the present author.

First, it is thought that the CHETAH approach generally overestimates instability risk. Indeed, if these compounds are analysed, they are all, apart from nitroaniline and ammonium nitrate, known as not being dangerous; but they are all mentioned as 'medium risk' at least once.

Examining criterion C_1; it obviously overestimates risk since all compounds except one have one medium risk level at least.

It is to be noticed that three criteria depend on this criterion C_1, which seems somehow redundant. It is particularly true in the case of C_4 and C_1, which from a thermodynamic point of view are exactly the same. Knowing the dangerous nature of these compounds, it would seem that C_4 is more representative of instability risk than C_1.

Besides, the fact that C_1 depends on the $\Delta H_c - \Delta H_d$ difference makes the case for a general property for explosives not possible (in the case of a weak C_1) or limited (in the case of a medium C_1). This effect emphasises the importance on stability of a small difference between enthalpies of combustion and degradation.

Finally, internal redox is closely bonded to its potential thermal effects by adding an element of fragility to the molecule.

Based on these comments, that are reinforced by the difficulties the users of this method have encountered in doing their own summary of their observations, a unique criterion was sought. It had to have two characteristics:

1. Be quantitative and continuous to avoid qualitative criteria, which can hardly be bonded to accurate physico-chemical properties and distort estimates close to thresholds.

2 • Stability   2.3 Quantitative approach: the CHETAH programme

2. Give results as far as possible in line with the 'on the job' observations. There is no point in claiming that a compound, which is well known as being stable, being listed in the regulations as an unstable compound, especially the ones 'subjected to regulations'. These will be analysed later and before returning in paragraph 2.5.3 to the bond between the instability hazard codes and the author's and other's indicators. But, in advance, after defining our unique criterion, an attempt will be made to show this bond with the compounds used as examples.

■ Suggested criterion: instability index II_s

Criterion C_4 is offered as a unique criterion, which will be modified if necessary by corrective coefficients that take into account the redefined criteria C_2 and C_3. Criterion C_1 is only used as a calculation element of C_4 and not as a decision element.

■ Redefinition of criterion C_2

Whatever C_1, it is accepted that:

- $|\Delta H_c - \Delta H_d| > 21.6$ kJ/g is a low risk.
- $12.6 < |\Delta H_c - \Delta H_d| \leq 21$ is a medium risk.
- $|\Delta H_c - \Delta H_d| \leq 12.6$ is a high risk.

The C_3 risk levels remain the same.

■ Corrective coefficients

It is accepted in what follows that $C_i = C_j$, meaning that the two criteria mentioned have the same risk level according to CHETAH. The 'greater than' (>) and 'less than' (<) signs will not need to be explained.

1. If $C_4 = C_2 = C_3$, instability degree IIs (ie C_4) corresponds to the unchanged CHETAH criterion.

2. If $C_2 > C_4$, criterion C_4 will be increased by 50%, if there is a risk degree between both criteria and by 100% if there are two levels between the two.

3. If $C_2 < C_4$, there will be a fall of respectively -25 and -50%. Criterion C_4 will thus be altered and is written as IIs*.

4. If $C_3 > C_4$ or $C_3 >$ IIs*, the same increasing or decreasing rate is used as before.

5. If $C_3 < C_4$ or $C_3 <$ IIs*, do the same as with C_2.

The IIs* symbol is kept whatever the nature and number of amendments to IIs criterion.

It is important to notice that, if after first increasing or decreasing IIs, there is a change of risk level, which brings it to the same level as the C_3 risk, there is no need for a second amendment. This happens when $C_4 < C_2 = C_3$ and $C_4 > C_2 = C_3$.

Applying this approach to these particular compounds; the following table was obtained:

Index of stability of various compounds

Compound	C_4 initial	$\Delta H_c - \Delta H_d$	C_3 (g%)	II_s	Indicative degree	Safety code
Ethylene	**959.69**	–40.56	–342.12	539.82*	**Medium**	
Allyl alcohol	348.69	–27.36	**–220.34**	523.04*	Low	
1,4-Dioxan	341.67	–22.50	**–181.59**	512.51*	Low	
2-Nitroaniline	**1162.87**	–13.93	**–150.57**	1162.87	**Medium**	
N-Methylaniline	220.45	–30.61	–246.34	220.45	Low	-
2-Methylpyridine	274.16	–33.02	–266.27	274.16	Low	
1.2-Dichlorobenzene	229.92	**–17.45**	**–141.50**	517.32*	Low	
Isobutyronitrile	176.37	–34.11	–266.24	176.37	Low	
Thiophene	316.52	–27.11	**–228.19**	474.78*	Low	
Ammonium dichromate	187.89	0	0	751.56*	**Medium**	**
Ammonium nitrate	192.18	0	+19.99	768.72*	**Medium**	

*Denotes modification of C_4 parameters, taking into account C_2 and C_3 parameters of CHETAH programme.
**Unstable compound according to labour regulations.
If nothing is mentioned, this means that codes other than the instability code was attributed to the compouund.
Medium and high risks according to CHETAH are in bold.

The choice of a unique criterion according to the author's approach is more in line with reality, especially if the 'indicative degree' is regarded as a simple application of the conclusions suggested by CHETAH. This qualitative classification, which is not recommended, has all the disadvantages of the effect of threshold that exists with ethylene, allyl alcohol, 1,4-dioxan and 1,2-dichlorobenzene, which are all the same so far as the value is concerned, and not the threshold of 526 that artificially separates ethylene from the other compounds. The only contradiction would be the low value given to C_4^* of ammonium dichromate, which is regarded as an explosive in labour regulations, but nevertheless considered valid.

The reader will be able to read in Chapter 5 about this compound, and convince himself that the author's estimate makes more sense than that used in the regulations, which is contrary to general observation. On the other hand, the CHETAH programme completely underestimates this compound.

So far as nitroaniline is concerned, the absence of an estimate in the regulations does not seem wise as the entries on nitrated derivatives in Chapter 6 will demonstrate.

These comments seem to justify our approach, which to us remains technical.

2.4 Other quantitative approaches

Despite the adjustments made to the CHETAH method, there is still the disadvantage of only taking into account thermodynamic factors of instability. Stull[1] suggested a method, which attempts to combine the thermodynamic and kinetic factors.

---

1. D.R. Stull, *J. Chem. Educ.*, 51(1), pp. A21-A25, 1974.

For the thermodynamic factors Stull takes into account the 'decomposition temperature'. This is defined as the temperature reached by the decomposition compounds of the particular substance when the latter decomposes into these constituent elements. It is therefore calculated using the standard enthalpy of formation of the compound.

For the kinetic factors he chooses the activation energy of the decomposition reaction. The maximum theoretical hazard is defined as that of a reaction of activation energy of 0 kJ/mol and decomposition temperature of 3000 K. Vice-versa the minimum risk is respectively defined by 418.4 kJ/mol and 0 K. These extreme values are placed on two vertical axes respectively graduated between 0 and 3000K and 418.4 and 0 kJ/mol. A diagonal bonds the origins '0' of the two axes and is graduated in ten degrees. The intersection of a line bonding the decomposition parameters of a given substance with the diagonal thus defines the reaction hazard index (RHI). Calculated for eighty chemical substances this RHI varies from 0.88 for methane to 7.05 for acetylene.This method is the most elaborate stage of the technique invented by Stull. He abandons in this method another approach he used on the thermodynamic aspect in 1971[1] in which he used the concept of flame temperature calculated from the enthalpy of combustion reaction.

The limits of this method are the difficulty in having access to the activation energy of this decomposition reaction. Besides, if one analyses Stull's results, one realises that he chooses enthalpies of formation either in the liquid or gaseous state according to the data available to him. This makes it difficult to compare substances. The second difficulty is the choice of regarding the decomposition reaction as a reaction that leads to the simple elements within the molecule and in this respect the CHETAH approach seems to be more realistic (although postulating for the formation of carbon monoxide rather than carbon dioxide in the decomposition reaction would be more like reality). We will come back to this RHI when comparing it with the modified CHETAH and hazard codes.

2.5 Classification of instability risk

Following is the technique used by the NFPA and European Union and labour regulations in order to classify instability hazard before coming back to the technical approaches.

2.5.1 NFPA grading; 'reactivity'

■ **Definitions**

With regard to inflammability, NFPA coding classifies 'reactivity' hazard into five degrees from 0 (no danger) to 4 (maximum danger), defined[2] as follows:

4: Substances, which, by themselves, can very easily detonate or cause an explosive decomposition reaction at normal pressure and temperature. This

---

1. D.R. Stull, *J. Chem. Educ.*, 48(3), pp. A173-A183, 1971.
2. Hazard identification signals XLVII, *J. Chem. Educ.*, 45(5), pp. A413-A422, 1968 and http://www.orcbs.msu.edu/chemical/nfpa/reactivity.

classification also includes substances sensitive to mechanical and thermal shocks under these same conditions.

3: Materials or substances, which can, by themselves, detonate or cause an explosive reaction, but need a source that initiates these decompositions or reactions or need to be heated in containment. This includes substances that are sensitive to mechanical or thermal shocks at high temperature and/or pressure or that react explosively to water.

2: Materials or substances, which, by themselves, are unstable and easily cause a violent chemical reaction without causing an explosion. This includes all substances that can easily engage in very exothermic transformations at high temperature and pressure. It also includes all substances that react violently with water or that can produce mixtures potentially explosive to water.

1: Substances, which, by themselves, are normally stable, but that can become unstable at high temperature and pressure. This includes all substances that have an exothermic reaction with water without causing a violent reaction.

0: Substances, which, by themselves, are stable even under fire conditions and do not react to water.

■ **Examples**

In Part Two will be found a presentation of NFPA 'reactivity' codes (when these are different from 0). The table below gives twenty substances which have the different NFPA degrees. These will later be used as examples when comparing the different types of classification of instability hazard.

Examples of substances bearing different NFPA codes

NFPA degrees	Examples
4	acetylene; nitromethane; peracetic acid; ethyl nitrate
3	ethylene oxide; nitroethane; 1-nitropropane; tert-butyl peroxide
2	ethylene; butadiene; styrene; acetaldehyde
1	propylene; isopropyl oxide; acetic anhydride ; propionitrile
0	cyclohexane; t-butanol; butanone; ethyl acetate

■ **Comments**

When analysing the previous table, it shows the ambiguity of NFPA 'reactivity' codes vis-à-vis instability. It is not so much an instability code but rather, like its name indicates, a code related to dangerous chemical reactions. Degrees 3 and 4 are the only ones that are more or less usable for defining an instability level, with the exception of ethylene oxide. So, even ethylene oxide's main hazard is not its explosive decomposition but its very violent polymerization caused by catalytic impurities (see chapter 6).

In degree 2 only reactivity degrees are treated vis-à-vis exothermic polymerization in particular and addition reactions on the double bond (ethylene, butadiene, styrene, propylene), easy peroxidation (isopropyl oxide, acetaldehyde), hydrolysis (acetic anhydride). Possibly only propionitrile and substances with code 0 have an actual NFPA 'stability' code. Every time one has to deal with the NFPA code one has to interpret it after carefully reading the paragraphs in Part Two.

2.5.2 Grading according to Labour code

■ **Definitions**

Labour regulations, based on the European regulations, set a real instability code, which has to appear on the labelling of chemical substance containers. This code defines the following risk indications (decree from 10.10.1983):

R 1: Explosive when dry.

R 2: Risk of explosion by shock, friction, fire or other sources of ignition.

R 3: Extreme risk of explosion by shock, friction, fire or other sources of ignition.

...

R 5: Heating may cause explosion.

R 6: Explosive with or without contact with air.

...

R44: Risk of explosion if heated under confinement.

The labelling symbol below appears next to these codes.

E: Explosive

■ **Examples**

These codes are mentioned for all the substances that are subjected to regulations in Part Three. Amongst the twenty substances chosen previously only four of them have such a code. They are nitromethane and peracetic acid (code 5), acetylene (codes 5 and 6) and ethyl nitrate (code 2).

■ **Comments**

When analysing the labelled contents, apart from code 6, which is ambiguous since the risk is bonded to the presence of air, the other codes are obviously instability codes. Nevertheless, their definitions seem to lack accuracy and coherence. Indeed, what does code 5 add compared to codes 2 or 3? Why do most of the previously mentioned substances not have code 44, which in all likelihood seems to apply to them? Can we really distinguish codes 1 and 2 on one hand, and 2 and 3 on the other? Finally, it will be seen that this code tends to be applied in the obvious cases of instability, which could be almost identified by the qualitative approach alone, and does not mention anything for the intermediate unstable substances. To put it in a nutshell it seems that it is not used enough and thus hardly usable except in noted cases of instability.

2.5.3 Comparing the different instability indicators

To summarise the information already presented and to enable an opinion on the possibilities of forecasting instability risk to be formed, it is interesting to make a comparison of the different reactivity/instability risk classifications. The table below sums up these different approaches. The substances are classified in descending order according to the NFPA coding and are taken from Stull's publication already mentioned (note, p.120):

Comparison of different indicators of instability for various substances

Compound	C_1	C_2	C_3	C_4	CHETAH modified IIs or IIs*	RHI	NFPA	Safety Code
Acetylene	High	High	High	High	6693.6	7.05	4	5 – 6
Nitromethane	High	High	High	High	2833.6	5.97	4	5
Ethyl nitrate	High	High	High	High	3010.7	6.36	4	2
Peracetic acid	High	High	High	Medium	2421.0	5.04	4	5
Ethylene oxide	High	Medium	Medium	Medium	1656.1	3.81	3	–
Nitroethane	High	Medium	High	Medium	2660.0*	4.62	3	–
1-Nitropropane	High	Medium	Medium	Medium	1197.0	4.22	3	–
t-Butyl peroxide	Medium	Weak	Weak	Weak	400.4	4.31	3	–
Ethylene	High	Weak	Weak	Medium	539.82*	4.19	2	–
Butadiene	High	Weak	Weak	Medium	515.6*	2.94 or 5.72	2	–
Styrene	Medium	Weak	Weak	Weak	384.4	6.33	2	–
Acetaldehyde	Medium	Medium	Medium	Weak	549.5*	3.76	2	–
Propylene	High	Weak	Weak	Weak	464.0	2.70	1	–
Isopropyl oxide	Weak	Weak	Weak	Weak	62.1	2.72	1	–
Acetic anhydride	Weak	Medium	Medium	Weak	173.2*	4.34	1	–
Propionitrile	Medium	Weak	Weak	Weak	236.1	2.93	1	–
Cyclohexane	Weak	Weak	Weak	Weak	30.7	2.60	0	–
t-Butanol	Weak	Weak	Weak	Weak	9.4	2.54	0	–
Butanone	Weak	Weak	Weak	Weak	69.6	2.72	0	–
Ethyl acetate	Weak	Weak	Medium	Weak	76.5*	3.38	0	–

* Is not mentioned in the labour regulations.

There is a reasonably satisfactory correlation between the different indicators. Our criterion combines the simplicity of calculations (compared in particular with RHI) with the advantage of having a continuous and quantitative scale, which thus makes it easy to establish a risk classification (such as RHI). In NFPA, labour regulations and to a lesser extent in the standard CHETAH the disadvantages of a qualitative classification are to be found.

In closing this Chapter it can be seen that instability risk remains relatively easy to forecast, but more difficult to estimate on a risk scale. The approach that seems most suitable could consist in:

- Looking for and identifying explosophoric groups and potential aggravating-/attenuating structural factors.
- If the physical conditions of use are normal, carry out the CHETAH then 'changed CHETAH' analyses according to our variant for the substances that have a potential instability risk only. If conditions are harsher, extend the analysis to all substances.

Substances are considered as dangerous if they are:

1. Dangerous under normal conditions: if IIs or IIs* $\geq$ 1926.
2. Dangerous at high temperature and/or pressure (up to 250 °C; $P > 1.5$ bar)
3. If $525 \leq$ IIs or IIs* < 1926;

or

Not dangerous under the conditions below (and $T < 500°C; P < 10$ bar).

if IIs or IIs* < 525.

Under harsher physical conditions there are systems that are not considered in this analysis.

Over-precise 'recipes' have to be handled with caution especially when dealing with thresholds.

3 • TOXICITY

The most difficult problem of risk evaluation linked to chemicals will be discussed in this Part. This is primarily a medical problem, which therefore comes within the competence of the company medical officer and epidemiologists, but nevertheless need not only be dealt with by them. The person in charge of safety control in a place where chemicals are handled also has to tackle this problem. This person will have to take into account the level of toxicity risk of a substance. This will determine the constraint level of the measures to be taken, its favoured means of penetration, which depends on the activity, and its penetration properties specific to the organism. The physical properties of the substance (which will determine the nature of the precautions to be taken) and also the values of toxicity parameters have to be taken into account. He has to check the container labelling and know how to interpret and explain the toxicity instructions on this labelling.

For numerous reasons it is very difficult to use the information provided. These are:

- The effect of a harmful substance obviously depends on the quantity the person is exposed to. This quantity is expressed in air concentration and exposure time for a gas and the subject weight for liquids and solids.
- The effect depends on species. Animals used for experimental purposes do not all react in the same way and with the same sensitivity to toxic substances. The reaction depends on the species. Within the same species, members have different degrees of resistance; this has been particularly observed in situations of human poisoning.
- The nature and intensity of the effect highly depend on the way toxic substances penetrate into the organism. Three main means of penetration are identified: *inhalation, skin contact, ingestion*. Some penetration means are favoured under identical exposure conditions depending on the substance's properties.
- The effect differs according to time and place. So far as the time effect is concerned, there is a need to distinguish acute effect, which appears a short while after the substance penetration, from the long term or chronic effect, for which effects can be identified after several years of exposure. The action can be local, ie contact point with the substance, or 'systemic', reaching organs that are distant from the penetration point. The local effect affects skin and eyes and/or mucous membranes, especially the inhalation ones. The local effects are irritant and sensitive.
- In addition, 'cumulative' effects can occur when the elimination and/or detoxification speed of the substance is low compared with its absorption quantity and rate. The concept of dosage is not useful in this context.
- It can be thus appreciated how uncertain the information from these experimental observations and estimates can be.

3 • Toxicity

3.1 Evaluation parameters for product toxicity

In describing the main parameters used to estimate toxicity, the previous factors will be taken into account for some types of estimates. It will then be seen how difficult it is to interpret these estimates illustrating the difficulties raised by their use. Finally, a possible way of estimating toxic risk parameters will be suggested and applied to a few practical cases, emphasising how cautious one has to be when using it.

3.1 Substance toxicity; evaluation parameters

The parameters below come either from animal experiments or epidemiological observation.

3.1.1 Toxic or lethal concentrations or doses

Lethal concentration 50 (LC50) is the vapour concentration of a substance in air, which kills 50% of the animals exposed. This estimate comes from a protocol that was statistically controlled. This value depends on the animal chosen for the experiments and exposure time. The three animals that are most commonly used are in descending order *rat, mouse and rabbit*. It is a parameter that estimates risk level by inhalation, which is the most important means of penetration involving toxic substances in the work place.

The lethal dose 50 (LD50) is defined as the dose which kills 50% of a batch of animals. This dose depends on the animal weight, space and means of penetration. The animals that are most commonly used are those mentioned before. Cat, dog, monkey, birds, frog, guinea-pig, etc. are not represented well enough to give interesting LD50 data. The means of penetration that are most commonly used are in an approximate descending order: *oral, intraperitoneal, skin, intravenous and subcutaneous*. Data obtained with the other means of penetration: intramuscular, rectal etc cannot be used because the data are sparse. LD50 are given per subject weight unit to be able to compare, as much as possible, values from one species to another.

The units are preferably ppm (parts per million in volume) for LC50; mg/l for gas and vapour; mg/m^3 or g/m^3 for aerosols or non-volatile solids that can form suspensions in air. For gas and vapour mg/l is used for the reason that will be explained later. For LD50 use of mg/kg is customary. It goes without saying that the animal genus must be named in all cases.

There are variants that need to be added to these parameters which can hardly be used at a non medical level. These are minimum toxic doses or concentrations and *minimum lethal concentrations* or *doses (MLC, MLD)*. The last two are mentioned without using them later on, but given the important level of error regarding LC and LD50 measurements, MLC and MLD can nevertheless be used as a reference.

These last values are the only ones available for humans. In this case these are the lowest values published following accidents. After collecting these values it was decided not to keep them in the tables after realising that they were useless in the context of this book.

Lethal concentrations and doses are parameters of acute, systemic poisoning. There are also specific protocols to evaluate the irritant and corrosive effects on skin, eyes and mucous membranes.

3.1.2 Limiting and mean values of exposure

Exposure values are vapour, aerosol or particle concentrations not to be exceeded in working spaces. For *limiting values (LVE)* the exposure time is fifteen minutes. The exposure time for *mean values (MVE)* are the daily working hours (eight). These parameters indirectly give the toxic risk by inhalation or in some cases the transmission risk through skin. Thus they have the same units as LC50; gas being usually given in ppm; dusts and aerosols given in mg/m^3.

They are the result of a choice made by epidemiologists following toxicological tests on animals and epidemiological campaigns. These values are regularly updated. Some values are sometimes given on a temporary basis and 'put to the test' for one to two years. The fact that they keep changing and take into account toxic effects on humans gives them an advantage over LC and LD. Their disadvantage lies in the long and tedious analyses they require and thus only offer figures for the most common substances. These values are not necessarily the same from one country to the another. INRS[1] publishes daily figures for the United States, Germany and France.

One finds it rather disappointing when looking for a correlation between LC50 and LVE (or MVE) values using the statistical tools. INRS nevertheless did it and found out that there is a normal log law which governs this data. For an oral LD50 or a LC50 for rat an estimate of LVE can be obtained by using linear regression. Correlation coefficients are around 0.7, which seems a little bit low. This link leads to the two equations:

$$LVE_{est} = 0.004 \times [LC50] \text{ (in ppm)}$$

and $\quad LVE_{est} = 0.025 \times [LD50]$

(LVE given in mg/m^3; LD50 given in mg/kg).

INRS finds it very useful when giving advice to epidemiologists who look for an approximate value for a substance, which has not yet been analysed (called *calculated limiting value of exposure*).

IDLH (*immediately dangerous to life and health*) is an index of exposure used in the United States. If such a concentration is reached, there is a need for emergency evacuation of buildings or use of an insulating mask.

Finally, STEL (*short term exposure limit*) corresponding to the definition of LVE and TLV/TWA (*threshold limit value/time weight average*) are the equivalent of the French MVE. There are sometimes differences in values between STEL and LVE on the one hand, and TWA and MVE on the other hand.

---

[1]. Institut National de Recherche et de Sécurité.

3 • Toxicity

3.1 Evaluation parameters for product toxicity

Comparison between different values of toxicity

Compound	LC50 r mg/l (ppm)	LVE/STEL (ppm)	MVE/TWA (ppm)	IDLH (ppm)
Acetaldehyde	15.6 (8714)	-	100	10000
Ethyl acetate	5.76 (1600)	-	400	10000
Isopropyl acetate	133.6 (31840)	300	250	16000
Methyl acetate	97 (32000)	250	200	10000
Pentyl acetate	27.69 (5200)	150	100	4000
Acetonitrile	12.66 (7550)	60	40	4000
Acetic acid	39.296 (16000)	10	-	1000
Formic acid	7.4 (3931)	5	-	1000
Acrolein	0.151 (66)	0.1	-	5
Ethyl acrylate	9.75 (2381)	15	5	2000
Methyl acrylate	4.832 (1372)	15	10	1000
Furfurylic alcohol	0.934 (232)	15	10	250
Ammonia	2.94 (4228)	50	25	500
Acetic anhydride	4.17 (998)	5	-	1000
Benzene	31.9 (9981)	5	-	2000
Butanol	24.21 (7984)	50	-	8000
Chloroform	47.7 (9769)	50	5	1000
Chloromethane	17.14 (2566)	100	-	1000
Chlorotoluenes	90.54 (17487)	2	-	10
Hydrogen cyanide	0.534 (484)	10	2	60
Cyclohexane	70 (20331)	375	300	10000
Dichloromethane	52 (14969)	500	100	1000
Diethylamine	11.97 (4000)	10	-	2000
Dimethylamine	8.374 (4539)	10	-	2000
Dioxan	46 (12764)	40	10	2000
Nitrogen dioxide	0.165 (88)	3	-	50
Sulphur dioxide	2.66 (1015)	5	2	100
Carbon bisulphide	0.079 (25)	25	10	500
Furfural	0.687 (174)	2	-	250
Isophorone	10.40 (1840)	5	-	800
Isopropanol	39.26 (15969)	400	-	20000
Methyl methacrylate	15.61 (3812)	200	100	4000
Methanol	83 (63378)	1000	200	25000
Methylamine	0.57 (450)	10	-	4000
Ethylene oxide	1.44 (800)	10	5	800
Pyridine	12.94 (4000)	10	5	3600
Hydrogen sulphide	0.627 (451)	10	5	300
Tetrachloromethane	50.39 (8000)	10	-	300
Trichloroethylene	155 (28845)	200	75	1000
Triethylamine	4.138 (1000)	10	-	1000
Xylenes	19.72 (4540)	150	100	10000

The table above shows the links between LC50 for rat, LVE, MVE and IDLH. There does not seem to be any logical link between the different choices.

3.2 Establishing level of toxic risk of substances

3.2.1 NFPA code

Here again there is a classification system with five degrees, already mentioned for fire hazard and reactivity risks. The definition of these five degrees is qualitative. They run from 4 to 1 ie:

- **4**: Substances, which after very limited exposure cause death or major irreversible effects, even if prompt medical treatment is undertaken. Also included in this category are substances that easily go through protective rubber clothing and those that, under normal conditions or in a fire, release extremely dangerous vapour, whether they are toxic or corrosive by inhalation or skin contact.

- **3**: Substances, which after short exposure can cause serious effects that are likely to be irreversible, even if prompt medical treatment is undertaken. People who have been exposed need to be kept away from these. Also included in this category are materials which give highly toxic combustion substances and corrosive substances.

- **2**: Substances which can cause temporary disabilities or possible permanent injuries following intense or continuous exposure if no prompt medical treatment is undertaken. Included are all substances that require a breathing insulating mask, substances that give off toxic combustion gas. In addition to these are highly irritant combustion substances and those that can release toxic substances, which cannot be identified.

- **1**: Substances, which after exposure can cause irritation leading to minor injuries only, even if no treatment is undertaken. They only require use of a breathing mask with cartridge. Also included are materials that give off irritant fumes in fire conditions and those that are an irritant to skin without causing any destruction of the tissues.

- **0**: Substances and materials which do not give off any substances that are dangerous to health in fire conditions.

This code is displayed in the diamond represented in chapter 1 called 'Inflammability' (p.82). It is in a blue section of this diamond. The table below gives the list of the most toxic substances according to NFPA.

Most toxic substances according to NFPA

Compound	NFPA toxicity	Compound	NFPA toxicity
Acrylonitrile	4	Methyl sulphate	4
Bromine	4	Methyl sulphide	4
Beryllium and salts	4	Methylparathion	4
Chloroacetic acid	4	Nickel tetracarbonyl	4
Chloropicrin	4	Nicotine	4
Cyanogen	4	Parathion	4
Cyanohydrin acetone	4	Proiolactone(β-)	4
Dinitrobenzene (m-)	4	Propionitrile	4
Ethyl fluoroacetate	4	Propargyl bromate	4
Fluorine	4	Sulpho-nitric mixture	4
Hydrogen cyanide	4	Trifluoromethylbenzene	4
Hydrogen fluoride	4		

3.2.2 SAX Code

This code is of limited usefulness (see para 1.5.3) (it is supposed to contain in one figure only the level of the three factors: inflammability, toxicity and stability). So far as toxicity risk is concerned, the definition of the three degrees is clear and makes it easy to choose between the different risk levels. It is defined by LD50 values (given in mg/kg) and LC50 (given in ppm).

Toxicity according to SAX code

Code	Risk description	Classification criterion
3	High risk	LD 50 < 400 and/or LC 50 < 100
2	Medium risk	400 ≤ LD 50 < 4000 and/or 100 ≤ LC 50 < 500
1	Low risk	4000 ≤ LD 50 < 40000 and/or 500 ≤ LC 50 < 5000

These degrees are then used by SAX to define toxic risk according to the nature of penetration. So a substance might be 'hardly toxic by intraperitoneal penetration and moderately toxic when it penetrates subcutaneously'.

3.2.3 Hodge-Sterner Code

The code, cited by Sax, proposes six degrees:

Toxicity according to Hodge-Sterner Code

Dangerously toxic	LD 50 < 1 mg/kg
Seriously toxic	1 ≤ LD 50 < 50 mg/kg
Very toxicity	50 ≤ LD 50 < 500 mg/kg
Moderately toxic	500 ≤ LD 50 < 5000 mg/kg
Weakly toxic	5000 ≤ LD 50 ≤ 15000 mg/kg
Extremely low toxicity	LD 50 > 15000 mg/kg

3.2.4 Labour Code

In France these correspond to what is decided by the European Union[1]. The codes used for inflammable, unstable substances and for some risks linked to reactivity have already been described. There are numerous codes for toxicity and corrosiveness and these have the skull as a symbol for toxic substances, an 'X' for harmful substances (Xn) or irritant ones (Xi), and a burnt hand for corrosives. In addition to these symbols there is a sentence about risk and some cautionary advice that bear numbers preceded with the letter R, and sentences about risk and S for cautionary advice. Everything has to appear on the container labels. Cautionary advice has never seemed coherent or sufficiently exhaustive and only risk codes are mentioned in Part Three. Notes on risk appear in the following table.

---

1. Journal officiel des Communautés européennes

3 • Toxicity

3.2 Establishment of toxicity risk levels

Toxicity according to Labour Regulations (European Union)

Acute systemic toxicity

20	Harmful by inhalation	28	Very toxic if swallowed
21	Harmful in contact with skin	29	Contact with water liberates toxic gas
22	Harmful if swallowed		
23	Toxic by inhalation	30	Can become highly flammable in use
24	Toxic in contact with skin		
25	Toxic if swallowed	31	Contact with acids liberates toxic gas
26	Very toxic by inhalation		
27	Very toxic in contact with skin	32	Contact with acids liberates toxic gas

Chronic toxicity

33	Danger of cumulative effects	60	May impair fertility
39	Danger of very serious irreversible effects	61	May cause harm to the unborn child
40	Possible risks of irreversible effects	62	Possible risk of impaired fertility
45	May cause cancer	63	Possible risk of harm to unborn child
46	May cause heritable genetic damage		
48	Danger of serious danger to health by prolonged exposure	64	May cause harm to breastfed babies
49	Can cause cancer by inhalation		

Irritant and corrosive nature

34	Causes burns	41	Risk of serious damage to eyes
35	Causes severe burns	42	May cause sensitization by inhalation
36	Irritating to eyes		
37	Irritating to respiratory system	43	May cause sensitization by skin contact
38	Irritating to skin		

Codes 39, 40 and 48 linked to acute toxicity codes indicate chronic effect of these risks. So 48/20 indicates that the substance is harmful by inhalation in the case of prolonged exposure.

Codes 50 and 59 deal with environmental risks; these are beyond the scope of this book.

These codes can be combined when a substance has multiple risks. Codes are then separated by a slash (for different means of penetration) or a dash (for risks of different nature). Thus 20/21/22 means that a substance is harmful by inhalation, skin contact and ingestion. If the substance is inflammable as well, it is 10-20/21/22. A compound can have different acute and chronic toxicity levels. For example, for phenylamine: 20/21/22-40-48/23/24/25, ie: harmful by inhalation, skin contact, ingestion; possibility of irreversible effects; risk of toxic effects by inhalation, skin contact and prolonged ingestion.

3 • Toxicity
3.2 Establishment of toxicity risk levels

■ Classification criteria

These take into account the systemic effects of lethal concentrations and doses (LC and LD50). However, when epidemiological observation has shown that effects for humans are different from those observed with experiments on animals, the observations on humans are the ones that will be taken into account.

Here are the criteria, at this stage confined to considerations that seem to concern lay (ie non-medical) workers.

Criteria for Health and Safety Code classification

Code	Parameter taken into account	Criteria for classification
R 20	LC 50 inhalation by rat	1 < LC 50 ≤ 5 mg/l/4 h (aerosols and particles) 2 < LC 50 ≤ 20 mg/l/4 h (gas and vapours)
R 21	LD 50 cutaneously (skin), rat or rabbit	400 < LD 50 ≤ 2000 mg/kg
R 22	LD 50 orally by rat	200 < LD 50 ≤ 2000 mg/kg
R 23	LC 50 through inhalation by rat	0.25 < LC 50 ≤ 1mg/l/4 h (aerosols and particles) 0.5 < LC 50 ≤ 2 mg/l/4 h (gas and vapours)
R 24	LD 50 cutaneously(skin), rat or rabbit	50 < LD 50 ≤ 400 mg/kg
R 25	LD 50 orally by rat	25 < LD 50 ≤ 200 mg/kg
R 26	LC 50 through inhalation by rat	LC 50 ≤ 0.25 mg/l/4 h (aerosols and particles) LC 50 ≤ 0.50 mg/l/4 h (gas and vapours)
R 27	LD 50 cutaneously (skin), rat or rabbit	LD 50 ≤ 50 mg/kg
R 28	LD 50 orally by rat	LD 50 ≤ 25 mg/kg

These criteria can be used to put forward a hypothesis on the potential toxicity risk level of a substance, which is not listed in labour regulations. The other criteria are either qualitative or take into account biological tests that go beyond the sphere of non-medical staff. It can be interesting to compare LC and LD50 data with clauses concerned with risk offered by the regulations and that appear in our tables. Here are a few examples of common substances, for which there is better information than for the others.

When analysing the table on LC and LD the following comments are relevant:
- based on the previous criteria, phenol in its vapour state should bear code R26, in the state of aerosols code R23 for inhalation and code R22 for ingestion (ie: 26-22-34);
- dioxan is in accordance with the criteria (no acute toxic effects);
- the corrosive character of acid seems to be favoured over codes R20/22-24-34, which are the logical result of the values given by literature;
- propionitrile is in accordance with the criteria.

Examples of labelling codes and of corresponding LC and LD 50

Compound	LC 50 (rat) mg/l; LD 50 o-r (oral-rat); cu-ra (skin-rabbit) mg/kg	Labelling Code
Phenol	LC 50: 0.32; LD 50 o-r: 423; LD 50 cu-ra: 740	24/25--4
Dioxan	LC 50: 46; LD 50 o-r: 4855; LD 50 cu-ra: 7700	36/37-40
Acrylic acid	LC 50: 10.1; LD 50 o-r: 321; LD 50 cu-ra: 289	34
Propionitrile	LC 50: 1.12; LD 50 o-r: 39; LD 50 cu-ra: 210	23/24/25

3.2.5 The transport of dangerous materials; code of toxicity

The toxic and harmful degree bears code 6, the highly toxic degree, code 66. For corrosive materials, codes will be 8 and 88. If there is no secondary code, the minimum degree of danger will bear the figures 60 and 80 respectively. These codes appear at the rear of vehicles and on packaging.

3.3 Problems posed by the determination of toxic risk levels

3.3.1 The level of error of measurement

The quality of the determination of risk levels set by regulation depends on the quality of measurements of lethal concentrations and doses (LC and LD50). From safety data in Part Three it is seen that this quality is far from being reached.

With data from different sources for the same animal and means of penetration the dispersion of results is to be taken into account.

For Part Three multiple data was obtained with rat orally by choosing examples that provided at least three values. Forty seven substances were considered. Applying the determination criteria for the labelling code for ingestion, two different codes can be allocated, depending on the LD50, to twenty of these substances. It is even possible to allocate three different codes to one of them.

So acrylic acid would bear R25 (LD50 o-r: 34 mg/kg) R22 (LD50 o-r: 235; 340; 353; 355) or no code (LD50 o-r: 2590). Benzene, toluene, 1-propanol, dichloromethane, etc., can be either R22 or have no code by ingestion depending on the values (labour regulations actually chose not to allocate any acute toxicity code to them).

When analysing the standard deviation value, which measures the dispersion of measurements, the effect of heteroscedasticity, already discussed in connection with the measurement of vapour pressure, is noted ie the dependency between standard deviation and average (the higher the average, the greater the dispersion of measurements). One way to make this unfortunate property obvious when it comes to analysing data is to calculate the coefficient of variation for each distribution (CV). If it is more or less constant, there is heteroscedasticity.

The coefficient of variation (CV) is defined by the equation:

$$CV\% = 100 \frac{s}{\bar{y}}$$

where s and $\bar{y}$ are the standard deviation and the average respectively. For distributions of LD50 o-r the average coefficient of variation is about 50% (standard deviation is generally equal to half the average value). This leads to the conclusion that where there is a unique LD50 value it is wise to assume that the real value of this LD is in the range:

$$0.20 \times \bar{y} \leq LD\ 50 \leq 1.80 \times \bar{y}$$

This range is a confidence interval of the average of two values with a Student's t value of 2, which corresponds to a free degree of forty six with which the average coefficient of variation is known. So for benzene the only value of LD50 provided by Sax (3306 mg/kg) would lead to the following range:

$$661.2 \leq LD\ 50\ \text{o-r} \leq 5950.8$$

Note that this large range contains the value 930 mg/kg, which comes from another source and would have justified application of code R22. This example thus illustrates the fact that one has to be careful in this particularly serious context of toxicity risk. This approach to estimation by range is another reason why the approach to estimation by regression as seen in paragraph 3.4 should be used.

3.3.2 Inadequacy of available parameters

Determinations of LC and LD50 are very often made under experimental conditions that are very different from the ones concerned with the classification of dangers for chemical substances. The drug industry in particular is one that requires most analyses in this field and in this case the choice of animals and means of penetration are not necessarily the same as those demanded by regulation. This is the reason why when having a closer look at LC and LD50, many LD50 values do not correspond to the ones sought.

There are only two animals that are used in the search for criteria: rat (intoxication by inhalation, skin and orally) and rabbit (by skin). The only means of penetration that can be used are inhalation, skin and orally. When a substance is not subjected to regulations, the absence of one or several animal/means of penetration combinations prevents the proposal of any level of danger for the substance labelling and also any suitable prevention measures.

3.4 Quantitative estimation methods

With the other risks the emphasis is placed on the importance of having available a quantitative evaluation method. Although it seems difficult, it would also be useful for toxicity. The following paragraph illustrates an approach, which used to be used in France by the old 'PUCK'[1]. This is followed by an introduction to the approach suggested in paragraph 3.6.

---

1. PUCK: *Produits chimiques Ugine-Kuhlmann.*

3 • Toxicity
3.4 Quantitative estimation methods

3.4.1 Safety and risk factors

Safety and risk factors evaluate approximately the speed at which a toxic substance reaches a toxic vapour concentration in air. An accurate way to do this would be to know the vapourisation speed for this substance and the air renewal rate of the room in which it is handled. This is why regulations recommend measurement of the vapourisation speed for a particular substance and include it in safety sheets. One can hardly use this figure since it is rarely mentioned. The only substances which were subjected to such measurements are the most commonly used although these figures only are remotely linked to the real conditions. So it was decided to suggest[1] a method derived from the vapour pressure of the substance, which is a factor the vapourisation speed depends on precisely.

The *safety factor* corresponds to:

$$SF = \frac{C_{eq}}{LVE}$$

in which C_{eq} is the vapour equilibrium concentration of a liquid at 20°C and LVE the limiting value of exposure; both having the same unit. The safety factor measures the propensity of this substance to reach the limiting value of exposure. *Risk factor*, which will not be used, is the ratio of the safety factor of a particular substance to the safety factor of a reference substance, which is generally acetone. It is used to compare solvents with each other; acetone is the most frequently used.

Following are some examples of safety factors selected by choosing from the comparative table of LC50, LVE, MVE and IDLH substances, which are hardly, moderately and highly toxic. The vapour pressures of the substances come from the tables in Part Three, the estimation techniques in paragraph 1.1.2 should be applied, if need be.

The safety factor scale takes into account the volatility as well as the toxicity of the substance. It is an acute intoxication factor. As a result, the long term toxicity of benzene, which is carcinogenic, is hardly taken into account. This is a variant of the author's approach.

Examples of safety factors for various compounds

Compound	P_{vap} (mbar)	C_{eq} (ppm)	LVE (ppm)	SF
Acetonitrile	97	95755	60	1596
Acrolein	286	282330	0.1	2820000
Benzene	100	98720	5	987
Chloroform	210	207305	50	4146
Hydrogen cyanide	832	821323	10	82132
Cyclohexane	112	110563	375	295
Dichloromethane	465	459032	500	918
Dioxan	39	38500	40	963
Methanol	128	126357	1000	126
Trichloroethylene	80	78973	200	395
Xylenes	8	7897	150	53

1. Note Tt 401, *Centre d'applications de Levallois, Produits chimiques Ugine-Kuhlman*, non dated.

3.4.2 Estimation of LC and LD50

When analysing the level of measurement error of LC and LD50 it was realised that the set of data was difficult to use since it is hardly reliable, and therefore of questionable coherence amongst all the figures. In order to find an answer to this a sample of the LC and LD50 values were submitted to an analysis based on principal components (PCA). It would take far too much time to describe this method, besides this goes beyond our subject. Its purpose is to look for and classify the different types of information contained in a complex table of quantitative data.

The method replaces a 'space' of particular variables with a simplified space defined by axes. Variables, which, if they are correlated to each other, have a multiple piece of information that is difficult to identify, and are replaced with independent axes, which are supposed to have a piece of information 'decomposed' into elementary elements. The arrangement of these variables in this new space enables establishment of their reciprocal relationships.

From this it can be decided whether it makes sense to identify toxicity risk using LD values determined for a means of penetration different from inhalation, skin or orally. If it does make sense, it will be possible to estimate risk although regulations did not provide any criteria. Note that on top of the space of variables mentioned above it is possible to have in this space 'bodies', which can be of help in interpreting axes.

The only difficulty in this method (in addition to the calculations, which are easily carried out using computers) is the fact that it is impossible to analyse tables with values that are missing, so there is a need to choose substances for which there are a whole range of LC and LD values. Since this is impossible, three tables were used, which all have in common the LD50 variables for rat and mouse, orally and by intraperitoneal means of penetration, so that the coherence of the three tables and a strong enough relationship between them could be established. The purpose was to determine, if, in the absence of one of the classification criteria set by regulation, it was possible to choose another available criterion to determine the risk level of toxicity.

The three tables contained respectively 39, 24 and 31 specimens. In addition to the four common variables mentioned already, the first table enabled analysis of the LD50 for rat and mouse by the intravenous means of penetration. The second table enabled analysis of the LC50 for rat and mouse and LD50 by skin for rabbit. The third table deals with subcutaneous means of penetration and the LD50 for rabbit by skin. Another statistical analysis was carried out to be able to compare the LD50 for rat and mouse by the subcutaneous means of penetration. The correlation matrix for these three tables leads to the only satisfactory correlations:

3 • Toxicity
3.4 Quantitative estimation methods

Correlation matrix for toxicity data for rat and mouse

	LC r	LC m	o-r	o-m	ip-r	ip-m	iv-r	iv-m	scu-r	scu-m
LC m	0.91									
o-r										
o-m	0.73	0.78	0.90							
ip-r			0.91	0.87						
ip-m			0.77	0.78	0.83					
iv-r						0.73				
iv-m						0.76	0.86			
scu-r										
scu-m			0.82	0.74	0.85					
cu-ra			0.78		0.78				0.78	

The abbreviations can be checked in the list of abbreviations at the beginning of Part III.

Although there are other correlations, only the ones with a correlation coefficient greater than 0.7 were kept.

This confirms the tacit hypothesis which states that the effect of a toxic substance on an organism is of the same nature whatever the animal (provided they are all mammals) and the means of penetration. Yet the effective intensity, ie toxic dose, is obviously influenced by the means of penetration since the different mechanisms of penetration make it more or less difficult for the toxic substance to enter the organism.

The intensity of this effect was estimated by simple linear regression.

The previous correlation matrix was used so that the equations, which enable estimation of LC50 r; LD50 o-r and LD50 cu-r from the other available LC and LD could be chosen. Priority here is given to these three criteria. There are a few other equations, in addition to these, which allow estimation of the criteria set by regulations in different stages. The tables of variance analysis that come with regressions allow favouring of an estimate by range for the coefficient of the general equation below:

$$LC_{est} \text{ or } LD_{est} = a \times LC \text{ or } LD \text{ (used)}$$

Estimation of toxicity

Estimated LC or LD	Interval of coefficient 'a'	LC or LD employed
LC 50 r	$0.90 \leq a \leq 1.30$	LC m
LC 50 r	$0.003 \leq a \leq 0.007$	LD 50 o-m
LD 50 o-r	$0.7 \leq a \leq 1.10$	LD 50 o-m
LD 50 o-r	$1.05 \leq a \leq 1.40$	LD 50 ip-r
LD 50 o-r	$1.10 \leq a \leq 1.80$	LD 50 ip-m
LD 50 cu-ra	$1.10 \leq a \leq 1.96$	LD 50 o-r
LD 50 cu-ra	$1.98 \leq a \leq 3.15$	LD 50 ip-r
LD 50 cu-ra	$1.43 \leq a \leq 2.25$	LD 50 scu-m
LD 50 o-m	$0.96 \leq a \leq 1.96$	LD 50 ip-r
LD 50 o-m	$1.15 \leq a \leq 1.82$	LD 50 ip-m
LD 50 o-m	$0.82 \leq a \leq 1.37$	LD 50 scu-m

Estimated LC or LD	Interval of coefficient 'a'	LC or LD employed
LD 50 ip-r	$0.53 \leq a \leq 1.00$	LD 50 ip-m
LD 50 iv-r	$0.40 \leq a \leq 0.66$	LD 50 ip-m
LD 50 iv-r	$0.85 \leq a \leq 1.19$	LD 50 iv-m
LD 50 iv-m	$0.38 \leq a \leq 0.60$	LD 50 ip-m
LD 50 scu-r	$0.40 \leq a \leq 1.50$	LD 50 scu-m

This approach leads to a simple if not accurate way of tackling a particular situation (but is the experimental determination better?). Assume a substance, which is not subjected to regulations and for which there are none, two or three of the LC and LD50 that are essential for determining the classification criteria by inhalation, skin or ingestion, but for which we have available some of the LC and LD50 as mentioned above. We then make the estimates using the available figures and keep the harshest. We apply the criteria set by labour regulations. We then need to obtain agreement of the company doctors and government inspectors on safety labelling for the substance.

So far as toxicity risk through inhalation is concerned, which is the most common under normal working conditions, a quantitative 'index of toxicity' is suggested on p.140.

3.5 Case study

To illustrate this with a few cases to demonstrate the possibilities and limits of this method, chemicals will be chosen for which there are numerous values of LC and/or LD50 available to be able to check the quality of the estimates. For each of these the values that are listed in the safety data in Part Three will be given.

Potassium chloride
LD 50 o-r: 2600; 3020; LD 50 o-m: 383; LD 50 ip-r: 660; LD 50 ip-m: 1181; LD 50 iv-r: 39; 142; LD 50 iv-m: 117:
- estimation of LD 50 o-r using LD 50 o-m, $LD\ 50_{est} = 268 - 421$
- estimation of LD 50 o-r using LD 50 ip-r, $LD\ 50_{est} = 693 - 924$
- estimation of LD 50 o-r using LD 50 ip-m, $LD\ 50_{est} = 1300 - 2126$

The estimate does not seem realistic and gives inconsistent experimental values (compare values for rat and mouse orally).

Propionaldehyde
Labour regulations: 36/37/38; NFPA: 2 TR: 33; LC 50 r: 19.319/4h; LC 50 m: 21.8/2h; LD 50 o-r: 1410; MLD o-m: 800; LD 50 scu-r: 820; LD 50 scu-m: 680; LD 50 cu-ra: 5040:
- estimation of LC 50 o-r using LC 50 m, $LC\ 50_{est} = 19.62 - 27.47$
- estimation of LC 50 o-r using MLD o-m, $LC\ 50_{est} = 2.4 - 5.6$
- estimation of LD 50 o-r using MLD o-m, $LD\ 50\ o\text{-}r_{est} = 560 - 880$
- estimation of LD 50 cu-ra using LD 50 o-r, $LD\ 50v = 1551 - 2763.6$

First note that the determination criteria for risk level and nature set by labour regulations are not respected. It should be R20/22. Our approach gives R20/21/22.

Dimethylacetamide
Labour Regulations: 20/21-36; NFPA: 2 TR: 30; LVE (MVE) (F): - (10); TWA (USA): 10; MAK(D): 10;

STEL: 20; LC 50 r: 8.81/1h; LC 50 m: 7.2/?; LD 50 o-r: 4300; 4930; 5000; 5400; LD 50 o-m: 4620; LD 50 ip-r: 2750; LD 50 iv-r: 2640; LD 50 ip-m: 2800; LD 50 iv-m: 3020; LD 50 w-s: 9600; LD 50 cu-ra: 2240:
- estimation of LC 50 r using LC 50 m, $\quad$ LC 50_{est} = 6.48 - 9.072
  (Note that the time of exposure for the experimental value is not that advocated in the Labour Regulations).
- estimation of LC 50 r using LD 50 o-m, $\quad$ LC 50 r: 13.86 - 32.34
- estimation of LD 50 o-r using LD 50 o-m, $\quad$ LD 50_{est} = 3234 - 5082
- estimation of LD 50 o-r using LD 50 ip-r, $\quad$ LD 50_{est} = 2887 - 3850
- estimation of LD 50 cu-ra using LD 50 o-r, $\quad$ LD 50_{est} = 4730 - 10584
- estimation of LD 50 cu-ra using LD 50 ip-r, $\quad$ LD 50_{est} = 5445 - 8662

Note that by simply analysing the available data of LD50 by skin for rabbit code R21 could not apply. Thus this code is due to other considerations (on the other hand, code R20 and the absence of R22 are justified). Our observations do not contradict this comment and provide R20.

Acetonitrile
Labour regulations: 23/24/25; NFPA: 3; 2 TR:336; VLE(VME) (F): - (40); TWA (USA): 40; MAK(D): 40; STEL: 60 IDLH: 4000; LC 50 r: 12.66/4h; LC 50 m: 4.52/4h; LC 50 l: 4.74; LD 50 o-r: 2460; 2730; 3800; LD 50 o-m: 2690; LD 50 o-l: 50; LD 50 ip-r: 850; LD 50 ip-m: 175; LD 50 iv-r: 1680; LD 50 iv-m: 612; LD 50 scu-r: 3500; LD 50 scu-m: 4480; LD 50 cu-ra: 1250:
- estimation of LC 50 r using LC 50 m, $\quad$ LC 50_{est} = 4.27 - 5.97
- estimation of LC 50 r using LD 50 o-m, $\quad$ LC 50 r: 8.07 - 18.83
- estimation of LD 50 o-r using LD 50 o-m, $\quad$ LD 50_{est} = 1883 - 2959
- estimation of LD 50 o-r using LD 50 ip-r, $\quad$ LD 50_{est} = 893 - 1190
- estimation of LD 50 o-r using LD 50 ip-m, $\quad$ LD 50_{est} = 192.5 - 315
- estimation of LD 50 cu-ra using LD 50 o-r, $\quad$ LD 50_{est} = 2706 - 4822
- estimation of LD 50 cu-ra using LD 50 ip-r, $\quad$ LD 50_{est} = 1683 - 2678
- estimation of LD 50 cu-ra using LD 50 scu-m, $\quad$ LD 50_{est} = 6407 - 10080

If we confine ourselves to the criteria set by labour regulations, the analysis of the experimental figures would provide codes R20/21. Our approach gives R23/25-21, which are better in line with the official figures.

Propylene oxide
Labour regulations: 45-20/21/22-36/37/38; NFPA: 2; TR: 33; VLE (VME) (F): – (20); TWA (USA): 20; MAK(D): 20; LC 50 r: 9.16/4h; LC 50 m: 4/4h; LD 50 o-r: 380; 930; 1140; LD 50 o-m: 440; LD 50 ip-r: 150; 364; LD 50 ip-m: 175; LD 50 cu-ra: 1500; 1245:
- estimation of LC 50 r using LC 50 m, $\quad$ LC 50_{est} = 3.6 - 5.04
- estimation of LC 50 r using LD 50 o-m, $\quad$ LC 50 r: 1.32 - 3.08
- estimation of LD 50 o-r using LD 50 o-m, $\quad$ LD 50_{est} = 308 - 484
- estimation of LD 50 o-r using LD 50 ip-r, $\quad$ LD 50_{est} = 158 - 210
- estimation of LD 50 o-r using LD 50 ip-m, $\quad$ LD 50_{est} = 192.5 -510
- estimation of LD 50 cu-ra using LD 50 o-r, $\quad$ LD 50_{est} = 418 - 1596
- estimation of LD 50 cu-ra using LD 50 ip-r, $\quad$ LD 50_{est} = 297 - 1147

Note that the 'estimation' approach would lead to a more severe classification: R23/24/25, which is in line for inhalation with that for ethylene oxide.

These results are disappointing. However, the comments made after analysing some examples show that this approach usefully complements the sole observation of experimental figures and codes offered by the regulations. But what seems most important here is the fact that the work carried out using regression can serve as a diagnosis. Its poor quality only reflects the incoherence of experimental data. Both experimental and estimation approaches complement each other and should be considered as two essential elements in risk analysis.

3.6 Proposal for quantitative index for toxicity risk

It is evident that a 'safety factor' for toxicity risk by inhalation is needed. Like the index for fire hazard, it differs from a risk indicator, in this case LVE being a physical quantity linked to the volatility of the substance. The safety factor thus has a significance and can be interpreted so as to give it a physical and biological meaning. So the safety factor for acrolein is about two million, ie at liquid/vapour equilibrium the concentration of this vapour can reach a value which is two million times greater than its limiting value of exposure. This means that it is absolutely essential to efficiently ventilate the work place. However, it is more difficult to interpret code 26, which refers to the 'very toxicity by inhalation', provided by labour regulations for this substance. This code is also ascribed to potassium cyanide, which does not release any vapour (but whose dusts can undoubtedly represent the same risk level). The whole point of the quantitative approach lies in this difference.

The disadvantage of a safety factor is linked to the one of LVE:
- few chemical substances have a LVE;
- there is no accurate quantifiable relationship between LVE and all other similar indicators such as MVE, LC50 r and IDLH;
- unlike flashpoint, which once reached, will be able to accidentally create fire risk, it is common knowledge that LVE is a non dangerous dose. Its comparison with the other values shows that it is lower by an average factor greater than ten compared with the doses from which toxicity effects develop among humans. As a consequence, it leads to 'SF' figures that can be unnecessarily alarming. Experts are aware of the excessively alarming risk parameters.

LC50 r does not have the disadvantages previously noted. It is indeed a commonly used figure that can be more or less directly calculated from the equations of regression previously offered. Besides, the comparative table between LC50 r and IDLH shows similar values in many cases. LC50 r is indisputably a toxicity concentration for rats but that is difficult to accept in human toxicityology. There is also a close relationship with the toxicity degrees from labour regulations hence guaranteeing a correlation between both factors. Finally, the use of this LC will lead to an 'index for toxicity risk' lower in value than SF.

These considerations are illustrated in the following table with the calculation of this 'index of toxicity' (IT) using the previous examples.

Calculation of toxicity index

Compound	P_{vap} (mbar)	C_{eq} (ppm)	LC50 r (ppm)	SF	IT
Acetonitrile	97	95755	7550	1596	12.68
Acrolein	286	282330	66	2820000	4277.73
Benzene	100	98720	9981	987	9.89
Chloroform	210	207305	9769	4146	21.22
Hydrogen cyanide	832	821323	484	82132	1696.95
Cyclohexane	112	110563	20331	295	5.44
Dichloromethane	465	459032	14969	918	30.66

Compound	P_{vap} (mbar)	C_{eq} (ppm)	LC50 r (ppm)	SF	IT
Dioxan	39	38500	12764	963	3.02
Methanol	128	126357	63378	126	1.99
Trichloroethylene	80	78973	28848	395	2.74
Xylenes	8	7897	4540	53	1.74

The objective is attained from:
- numerical values of the same importance as the other indices thus providing possibilities to look for a unique quantitative criterion that will include the different risks enabling work on a new safety measure to be initiated:
- total respect for the toxicityities related to inhalation that were given by 'SF';
- more marked difference between the values for the two most toxic substances; indeed the SF for these two substances seemed to be too different.

There is an anomaly in the higher index value for dichloromethane compared with chloroform although dichloromethane is noted for being the less toxicity halogenous solvent. But this figure only reflects the extreme volatility of this substance since its LC50 is lower than the one for chloroform. There is probably a similar problem for acetonitrile, which is underestimated by this approach.

As is needed for all potential risks for chemical substances, an index of toxicity enables quantification of risk. Nevertheless, it only applies to risk by inhalation, which is yet the most common as well as insidious risk under normal working conditions with chemical substances. However, this approach should be treated with caution because of the difficulties inherent in toxicological risks. At this stage of the analysis it is essential to work in collaboration with the company doctor, whose total agreement is necessary.

4 • DANGEROUS REACTIONS: methodology

Many accidents resulting from dangerous reactions have a history of repeating themselves over the years, without the lessons of history being properly drawn. For example, the accident at Seveso, arising from difficulties in controlling the reaction temperature of sodium hydroxide with 1,2,4,5-tetrachlorobenzene, had already happened three times a few years before. The symptoms due to acute intoxication caused by dioxin were already known.

When analysing such accidents the same patterns keep repeating themselves, and this could be due to a systematic lack of bibliographic work.

In this section an attempt will be made to identify some of the reasons that can explain this situation.

A first reason is linked to the way one identifies a dangerous reaction. Most often a reaction is only regarded as dangerous when it has caused an accident, details of which are then published. But the experimental descriptions in publications rarely mention any accident and in some countries this kind of information remains 'top secret'. Dealing with industrial injury from a legal rather than a technical or scientific point of view does not help deliver the information. As a result, the absence of information on potential risk caused by a chemical reaction has often been interpreted as no danger at all.

A second reason is due to the absence of preliminary risk analyses in everyday activities involving chemical substances. Only 'sensitive' activities are subjected to analyses (mainly because of what is at stake and for financial reasons). This is due to, on the one hand, the absence of literature on how to diagnose *a priori* risks and, on the other hand, the difficulty of such an analysis,(the chemistry being difficult). Finally, preliminary risk analyses are not part of managerial experience and work practice.

There are only a few publications that deal with risks of dangerous reactions. Whereas the texts quoted refer to inflammability (with all the reservations made in the first chapter) and regulations, there is hardly any mentioning of risk of dangerous mixtures. However, whenever this is the case, the authors only list compounds known as being 'incompatible' without any explanatory comments, Bretherick's publication apart, although its layout makes it difficult to draw any general rules regarding the substances' chemical behaviours.

A third reason is due to the influence of external factors, which often play a more important role than the factors that characterise the chemical nature of the reaction. All reactions can be regarded as dangerous (or not dangerous) at this stage. This means that most of the time it is the ambient conditions that make chemical reactions uncontrollable.

However, if a reaction is dangerous in itself, it is much more difficult to determine the working mode than for a reaction in which there is no chemical factor, which would make us think that there is a danger. This is the reason why it is important to address this question.

The most important part of the approach adopted in this book is concerned with the analysis of 'dangerous reactions'. In this chapter a forecast analysis of risks that can be caused by an interaction between two or several chemicals, or between chemicals and materials, is proposed. It will be a general approach, and its purpose is to develop a range of tools in order to forecast and quantify risks due to substance reactivity in deliberate or accidental contact.

The aim of Part Two, which lists the dangerous reactions of inorganic and organic chemicals, is to enable the reader to carry out an overall analysis on how the principal classes of compounds behave in order to recognise the reaction properties that are likely to lead to a danger diagnosis for a planned chemical reaction. Both chapters in this Part are intended to produce a compilation of dangerous chemical reactions, which will allow us to gather together pieces of information in a coherent way with explanatory attempts or mnemotechnic links, when the explanation seems out of reach. They give priority to the structural factors, which are likely to give an explanation for the dangers. However, the explanations remain mainly qualitative. The level of risk is not presented in a systematic way and is often subjective. Also, factors linked to failures in determining operating modes are not classified by types of factor.

First, the risk codes offered by regulations to inform the user of risks linked to potential dangerous reactions caused by chemical substances will be listed.

Then it will be shown how the texts of the two chapters following can be used to carry out a first qualitative study on risk.

Finally, an attempt will be made to show how the identified qualitative risk can be quantified to adapt the level of constraint of the operating modes.

4.1 Dangerous reactions: information parameters

Commercial and public regulations made a deliberate effort to place 'signals' on containers carrying labelling and packaging to enable the user to identify the potential risks of dangerous reactions of a substance.These types of labels focus on four types of explicit dangerous reactions only:
- risk of reaction with water in two different kinds of situations: potential outdoor storage and problems posed by water used as an extinguisher;
- risk of reaction with oxygen, in particular air, a risk to which all substances are inevitably exposed ;
- risk of polymerisation, which can be spontaneous due to ambient effects that are difficult to control (especially heat) or impurities;
- risk of violent oxidation.

4 • Dangerous reactions: methodology

4.1 Dangerous reactions: information parameters

4.1.1 NFPA Code

There is nothing to add to the description of 'reactivity' risk levels in the NFPA code described in paragraph 2.5.1. It was seen in particular that degrees 1, 2 and 3 could be compared respectively with ascending levels of danger of chemicals with water. To reinforce this aspect, the code adds in these cases the symbol W̶ indicating to the fire service that it is forbidden to use water as an extinguisher for this type of substance. This symbol is positioned on the white diamond of the NFPA label illustrated in paragraph 1.5.1. The reactivity code is systematically given for inorganic substances and only if it is greater than '0' for organic substances. This includes '**OXY**', which indicates strong oxidants.

4.1.2 Labour Code

These are quoted in Part Three and in chapter 6 with reference to dangerous reactions of organic chemicals. Here the codes concerning dangerous reactions are compiled. They start with the letter R and are followed by a number; together they correspond to 'risk clauses':

R4: Forms very sensitive explosive metallic compounds.

R6: Explosive with or without contact with air.

R7: May cause fire.

R8: Contact with combustible material may cause fire.

R9: Explosive when mixed with combustible materials.

R15: Contact with water liberates extremely flammable gases.

R16: Explosive when mixed with oxidising substances.

R19: May form explosive peroxides.

R30: Can become highly flammable in use.

R4 refers to the 'real' acetylenic compounds, which form salts, in particular copper salt, and are highly sensitive explosives. R6 remains as ambiguous as when it was first noted in paragraph 2.5.2. R30 remains a mystery to the author.

How can such codes be attributed?

For code 'R8' it commences with a solid with the help of AFNOR standards 'NF T 20-035'. Handling consists in preparing mixtures of variable compositions of an oxidant to be classified as cellulose. Both substances have to have a definite particle size distribution. The composition which gives the fastest combustion on a 'moulding' of the mixture at a distance of 20 cm is established. This speed is compared with the one of the mixture used as a reference, which has an imposed composition of barium nitrate and cellulose. If the combustion speed of the particular substance is higher than that of the reference, it will bear R8.

Two years of experiments using this technique with the author's students have shown us that these measurements have a very high level of experimental error.

4 • Dangerous reactions: methodology

4.1.3 Code for transport of dangerous materials

When transporting dangerous chemical substances the carrier has to affix at the rear of the vehicle and/or substance packaging plates and labels, a danger label and/or a substance number. The risk number contains at least two figures; the first being the main risk, the second the secondary risk. 0 indicates the absence of a secondary risk. If the main risk number is divided into two, this indicates an aggravated risk. The numbers that are of interest here are:

1: Detonatable material.

4: Flammable solid or self-heating solid materials.

5: Oxidant or peroxide (encourages fire).

9: Danger of violent reaction resulting from spontaneous decomposition or polymerisation.

X: The substance must not, in any case, be in contact with water.

The other numbers are 0 (absence of secondary danger), 2 (gas), 3 (flammable liquid already seen), 6 (toxic material), 7 (radioactive material), 8 (corrosive materials). When different figures are put together it leads to more or less complicated risk clauses. Thus 265 refers to oxidising toxic gases. The code is a stability as well as a reactivity one, as can be seen.

4.2 Qualitative approach

The aim is to collect all information likely to identify the different types of dangerous chemical reactions a substance can be involved in. This is relatively easy in organic chemistry due to the role played by the functional chemical group in reactivity, whether dangerous or not. This point is considered in the chapter on dangerous reactions of organic substances. In inorganic chemistry, the classification mode chosen for Part 2 requires additional work. In both cases, there is a precise place to describe any type of substance. Thus, when there is a dangerous reaction between two chemicals, the chapter in which the method is described is always the most distant one in the presentation order of the substances. Two general examples corresponding to the potentially dangerous reactions of an organic and inorganic substance will be considered.

4.2.1 Identification of 'structural' risk factors

As an example a violent reaction that brings together, in an industrial process, two substances responsible for explosive reactions will be discussed because it is no longer in use ie.:

■ **2-hydroxy-2-methylpropanoic acid**

has the following chemical formula:

$$\begin{array}{c} CH_3 \diagdown \quad \diagup OH \\ C \\ CH_3 \diagup \quad \diagdown CO_2H \end{array}$$

It can therefore have the dangerous characteristics of both the alcohol and acid functions. Besides, methyl groups can be involved in reactions that affect the C-H linkage. Starting with the acid function discussed in chapter 6 there are the following factors to take into account:

1. The presence of a complex function is an aggravating factor.

2. Hydrogen peroxide and sodium can lead to the formation of a peracid which is likely to be unstable. This is the case with peroxides in general.

3. Chromium trioxide represents a potential risk vis-à-vis acetic acid so possibly regarding this acid also. The contact with dichromate or permanganate when hot should also be avoided. This indicates that other oxidising salts such as nitrites, nitrates, persulphates need to be handled with care regarding this acid.

4. Nitric acid is indicated as being particularly dangerous with carboxylic acids. It is interesting to note that an accident is reported with the nitric acid/lactic acid mixture combined with hydrogen fluoride; in actual fact, lactic acid has a structure which is very close to that of this particular compound. The latter will certainly be less reactive due to the steric nature of the second methyl group.

5. Sodium hypochlorite could also be dangerous when hot, and alkaline chlorites also; in this case however it is not guaranteed that the acid character of this example is sufficient to form chlorine dioxide. Caution is still required when handling these mixtures to which chlorates and perchlorates can be added.

Other reactions will not be dealt with here.

At this stage, remember that carboxylic acids represent risks vis-à-vis esters, acid chlorides and anhydrides, amides or various sulphur-containing or phosphorated groups.

This will be dealt with in later chapters.

- With anhydrides and acid chlorides dangerous reactions with alcohols are to be noted.

Beginning with alcohols and then with acids, note:

- the possibility of interaction of the alcohol group with highly reducing metals (so care must be taken not to have any type of this metal present in the apparatus).
- the dangers due to nitric acid, as well as all the oxidants mentioned previously;
- when analysing the chemical groupings it is found that the majority of the dangers linked to nitric acid and generally of all nitrogen oxides are explicitly quoted for nitriles, amines and nitrated derivatives.

This last comment is a warning on the potential risk that dinitrogen tetroxide presents as a reagent vis-à-vis the particular alcohol-acid. The paragraph on 'hydrocarbons' confirms this. This oxide has even caused accidents with saturated hydrocarbons (these can be due to impurities). In any case, there is every chance that oxide will activate the methyl groups of this compound.

■ Dinitrogen tetroxide

The tetroxide, which forms an equilibrium with nitrogen dioxide, is a very strong oxidant and an endothermic compound. To analyse all the dangerous reactions of

4 • Dangerous reactions: methodology

4.2 Qualitative approach

this substance, refer not only to the paragraph about nitrogen but also all the following paragraphs, using all potential summary tables.

Paragraphs to be consulted regarding the dangerous reactions of dinitrogen tetroxide

Paragraph	Dangerous reactions N_2O_4/NO_2 with:	Paragraph	Dangerous reactions of N_2O_4/NO_2 with:
Fluorine	Fluorine	Calcium	Calcium
Magnesium	Magnesium	Chromium	Chromium
Aluminium	Aluminium	Manganese	Manganese
Phosphorus	Phosphorus;phosphine	Iron	Iron
Potassium	Potassium	Iodine	Hydrogen iodide

It can be assumed that the dangerous reactions mentioned here should also be applied to all substances which are similar to those listed from a reactivity point of view. So it is likely that the dangerous reactions between these nitrogen oxides and magnesium also exist with calcium, strontium and barium, which belong to the same group of the periodic table. This is not necessarily the case with beryllium, which is located somewhere completely different in this classification. When following a similar reasoning it can be postulated that interactions are also dangerous with alkalis, which are more reactive than alkaline earths and for which potassium was the only one quoted. It can be the same in column V: phosphorus, arsenic, antimony (apart from nitrogen and bismuth, which are far too different from the others in the same group).

An hypothesis of close behaviours between all nitrogen oxides can be made: monoxide, dinitrogen oxide and even nitric acid, which is a very strong oxidant, are similar to the nitrogen oxides. It can be then postulated that the dangerous reactions of strong oxidants, chlorates, perchlorates will possibly apply to dinitrogen tetroxide. Other additional lists can be drawn up to create a record of all situations of potential risk. This is thus the first stage when analysing the dangers linked to the potentially hazardous chemical interactions. A first selection that consists in getting rid of the most hazardous reactions can lead to the development of an operating process. In order to do so, the following factors, which list some examples of accidents linked to mistakes made at different stages of the operating modes should be considered.

4.2.2 'External' risk factors: an approach to their prevention

The contents of Part Two have been drawn on to classify all accident factors related to operating modes using the examples listed below. From these, prevention rules have been drawn. This corresponds to the standard approach used by workers involved in safety.

4 • Dangerous reactions: methodology
4.2 Qualitative approach

■ **Thermal effects**

Examples
- Accident during hydrogenation of acetylene at a temperature accidentally set at 400°C.
- Explosion of propadiene accidentally heated to 95°C.
- Excessive heating of ethylenic compounds, which caused their polymerisation.
- Violent polymerisation of styrene accidentally heated to 95°C.
- Action of bromine on acetylene, action of nitric acid on aromatic hydrocarbons. The accidents involving both reactions are very specific. They are the result of application of an unsuitable temperature. If it is too low, the reaction is too slow and causes an accumulation of reagent that is not converted and whose concentration increases, causing an acceleration of the reaction, i.e. a rise in temperature. The temperature then becomes too high and causes a more or less violent speeding-up of the reaction.
- Accidents due to thermal shock: mixtures of cryogenic liquids with water at ambient temperature. There will be an explosion if the difference in temperature of the mixture of molten metals with water is greater than 90°C.
- Heating to an excessive temperature (50°C) of a nitrogen tetroxide/petroleum ether mixture due to exceptional climatic conditions.
- Overheating of a propene/nitrogen tetroxide mixture at 200°C due to pump failure.
- Change in the way a reaction occurs because of inappropriate temperatures. At 20°C nitric acid converts mesitylene into 3,5-dimethylbenzoic acid without danger, but at 115°C the explosive 1,3,5-tris(nitromethyl)benzene is formed. Excessive heating of thiophosphoryl trichloride with pentaerythritol at 160°C leads to the formation of phosphine, which combusts spontaneously.
- Ethanol/potassium permanganate/sulphuric acid mixture detonates when heated.
- A reaction of oxygen with hydroquinone, usually carried out at 70°C and 90 bar, caused a fire and detonated after reaching 90°C and 110 bar.
- Heat, sometimes ambient, accelerates the speed of formation of peroxide by oxidation in air.
- The action of benzyl chloride on potassium cyanide without any temperature control led to an explosion.
- The reaction of vinyl chloride becomes violent at 90°C and above when oxygen is present.
- Explosion of three tons of an ethylene oxide/glycerol mixture when heated to 200°C instead of 115-125°C.

Prevention checklist
1. Carefully check the storage and handling conditions of sensitive chemical substances.
2. Carry out a bibliographical analysis and an estimate of the risk level according to the methods discussed earlier if the thermal sensitivity or thermal conditions of a reaction are unknown.

3. Clearly indicate on the labels of sensitive substances the temperature not to be exceeded.
4. If there is uncertainty, carry out tests using small quantities in a thermal analysis apparatus.
5. Have efficient refrigeration systems ready to try to stop reactions likely to speed up or for which the preliminary analysis showed them to be exothermic.
'Flooding' systems might be needed in extreme situations and if the compounds are incompatible with water.

■ Effects of pressure

Examples

A pressure effect was mentioned in the previous paragraph on thermal effects (reaction of oxygen with hydroquinone) eg:
- explosion of ethylene when submitted to a dramatic rise in pressure from 1 to 88 bar.
- same accident when propene is submitted to a dramatic change at 4860 bar.
- allene detonates at 29 bar. Acetylene detonates below pressures that are even lower.
- ethane/ethylene/chlorine mixtures detonate at 10 bar.
- the sudden change from 1 to 11 bar of ethylene oxide/oxygen mixture causes its decomposition and ignition.
- oxygen/trichloroethylene mixtures detonate at 27 bar.

Prevention checklist
6. Ensure means of controlling pressure and alarm systems, which go off when the thresholds are exceeded, allowing an automatic decrease in pressure.
7. Isolate the apparatus that work under pressure.
8. Place guard screens in front of all the fragile apparatus that are likely to be submitted to pressures they are not going to withstand.
9. Carry out tests in well-protected microbombs.
10. Proceed by progressive rises in pressure, ie in stages.

■ Effects of concentration

Examples
- Two examples have been given for the temperature/concentration action in the earlier reference to thermal effects (actions of bromine on acetylene and nitric acid on aromatic hydrocarbons).
- A methane/chlorine mixture detonates if the chlorine concentration is greater than 20%.
- A 95/5 nitric acid/ethanol mixture made in error (instead of 5/95) caused it to explode.
- Explosion, due to unclear instructions; use of concentrated nitric acid instead of acid at 2.6% in water with isopropanol.

4 • Dangerous reactions: methodology
4.2 Qualitative approach

- Accident caused when using a solution at 15% of nitric acid in ethanol. These concentrations should never exceed 10%. These concentration effects are due to an inversion in the order in which reagents should be incorporated.
- When incorporating sulphuric acid into a hydrogen peroxide/alcohol mixture there will be an explosion whereas a sulphuric acid/peroxide mixture poured onto alcohol would not cause any fire. Everything here is linked to the concentration of peroxomonosulphuric acid in alcohol.
- It is exactly the same with the 'haloform' reaction. If chloroform is poured onto the acetone/sodium hydroxide mixture, there can be an explosion whereas incorporating hydroxide in the other two substances would not be dangerous. In this case the risk factor is the high hydroxide concentration.
- Ethylene oxide becomes dangerous in oxygen at a concentration greater than 10%.
- When preparing 4-methoxybenzaldehyde by the action of oxygen on 4-methylanisol, a dramatic output of oxygen caused the reaction to accelerate.
- All vapourisation processes of solutions made of unstable substances are dangerous because the concentration of the unstable substance increases. In this category the heterogeneous reactions can be grouped together; they lead to accidents because of compounds with too thin a particle size distribution. So it is possible to control the reaction of phenyllithium by using thick pieces of lithium.
- The fact that Würtz reactions take a long time to start with halogenous derivatives and sodium is due to a concentration effect which can lead to an acceleration of the reaction once the concentration of reagents is critical. It is aggravated by thermal phenomena.
- The dangers linked to unstable substances that are insoluble compared with those that are soluble are also concentration effects. This is the case for many polyperoxides. Even those that are soluble can detonate if their concentration exceeds a critical value eg 15% for the peroxide of vinylidene chloride.
- An inappropriate mixture of compounds can cause excessive concentrations of reagents, which are hardly miscible because of huge differences in densities. This is why a stirrer breakdown caused an accident when stirring started again in a reaction between 3-phenyl-1-propanol and the very heavy phosphorus tribomide, which accumulated in the lower part of the reactor.

Prevention checklist
11. Use solvents as far as possible when conducting a chemical reaction.
12. If it is impossible to use solvents, incorporate one of the reagents, carefully chosen, slowly into the other.
13. Analyse the order in which substances should be incorporated, if necessary using small quantities.
14. Make sure suitable stirring and means of controlling the whole process is used.
15. Anticipate any precipitation of a reaction substance and give instructions to the technician.

4 • Dangerous reactions: methodology

4.2 Qualitative approach

16. The beginning of the reaction needs to be carefully monitored. For reactions that prove difficult to start technicians should have at their disposal a list of indications that show that the reactions start properly. Generally, the absence of moderate heating indicates that the reaction is not starting.
17. For heterogeneous reactions in the solid state, tests on small quantities with different particle size distributions can help the technician make the right decisions.
18. For reactions in the gaseous phase, dilution by inert gases can enable one to control the reactions that are considered violent. Nitrogen and carbon dioxide are often used in this context. The latter is used to dilute ethylene oxide.

■ Effect of light

Examples

- Light catalysed the violent polymerisation of propene.
- The action of chlorine on ethylene can be dangerous in ambient light.
- The formation of peroxides is catalysed by light.

Prevention checklist

19. Avoid working surfaces and extraction and fume hoods that are exposed to bright lighting.
20. Use opaque containers and/or reactors. Yellow glass or black or grey plastic containers need to be systematically used.
21. Protect laboratories from direct lighting from solar rays.

■ Mass effects

Examples

- Butadiene polyperoxide has a critical mass, which corresponds to a sphere of diameter of 9 cm.
- A few drops of tetrachloromethane on lithium do not represent any danger. Large quantities of chlorinated derivative cause an explosion.
- 140 g of perfluoroiodohexane in contact with seven grams of sodium detonated after being heated for thirty minutes. Under the same conditions, small quantities did not cause fire.

Prevention checklist

22. Work as much as possible with small quantities.
23. Do not change the handling scale without previously analysing risks.
24. For reactions for which there is little information, start with the smallest possible quantities.
25. Increase the quantities by proceeding through progressive tests.

Contact time effects

Examples

'Old' mixtures are dangerous!
- Xylene stored for sixteen years on sodium wire detonated when it was handled.
- A nitric acid solution in ethanol detonated after being stored for a long time.
- A nitric acid/hydrogen fluoride/glycerol mixture used as a chemical metal polish detonated after three days in storage.
- After one week to six months most ethers contain dangerous quantities of peroxides.
- 2-Butanol detonated after being stored for twelve years.
- A mixture of polypropyleneglycol and water stored at a temperature greater than 100°C detonated after fifteen minutes.
- A mixture of lithium tetrahydrogenaluminate and tetrahydrofuran stored for two years combusted when it was handled.

Prevention checklist

26. Affix the date of receipt, of manufacture and of the last time of handling of the stored chemical.
27. For substances that can turn into peroxides, clearly indicate on the bottles the dates of peroxide analyses and the dates on which the substances will have to be destroyed.
28. Monitoring will be more or less rigorous depending on the substances' propensity to peroxidation.

Effect of reagent viscosity

Examples

- Viscous liquids have difficulty in dissipating the heat produced by the beginning of a reaction. The heat can thus cause the substance to reach a temperature that is sufficient to provoke its self-ignition. It occurs in a lot of reactions involving glycerol with potassium permanganate or sodium hydride.

Prevention checklist

29. Viscous liquids need to be diluted.

Effect of apparatus materials

Examples

- There are a lot of accidents involving the handling of hydrogen peroxide in partly rusted iron containers, which catalyse the explosive decomposition of peroxide.
- The use of glass apparatus to carry out reactions sensitive to light causes accidents.

4 • Dangerous reactions: methodology

4.2 Qualitative approach

- The use of steel reactors or containers to carry out reactions involving halogenous compounds can, by causing the formation of hydrogen chloride, and then iron chloride, provoke violent reactions of decomposition or polymerisation.
- Trichloroethylene containing traces of water was stored in a metallic drum. After a while the compound suddenly decomposed (or polymerised).

Prevention checklist

30. The choice of materials, tools and apparatus has to be made only after analysing the potential reactions of the compounds with these materials. It is particularly the case when handling acetylenic compounds, unsaturates, halogenated derivatives.

■ Substance confusion

Examples

Carbon was used by mistake instead of manganese dioxide when preparing oxygen. Both substances look the same.

Prevention checklist

31. One should be able to read the labels from a distance and these should be 'standardised' sufficiently to allow one to quickly identify without any doubt important information regarding potential dangers.
32. Always think about possible mix-ups between compounds that look similar. The workplace needs to be properly laid out, especially from an ergonomic point of view, taking into account substance incompatibility. Analyse how far or close bottles are from each other to avoid contact between them, which would cause reactions between compounds. Draw up a map of risks with the different workplaces using chemical substances.
33. Check the workplaces to make sure chemicals no longer needed are put away and on a time basis, which varies according to the potential risk.

By analysing the risk factors linked to the 'environment' of a chemical reaction the aim is to analyse the factors which cause an accident before providing elements of prevention. The normal procedure consists in drawing up a 'tree of causes' whose aim is to describe the sequence of events, which takes place before the accident. Then one needs to analyse the different ways of breaking this sequence. Most of the time the accident would not have happened if the sequence was broken. These breaks planned after the event should prevent the same accidents from happening again. A simple example is given here since the problems caused by dangerous reactions are hardly mentioned in the existing literature on safety. The approach is incomplete, for instance it does not deal with any aspect of staff training. This post-event approach has to be preceded by an *a priori* approach. This chapter (as well as the others) will concentrate on the latter approach.

4.3 Quantitative estimation of dangerous character of a reaction

4.3.1 Physical and other factors

Physical factors favouring inflammability were analysed in paragraph 1.5.4, and the physical factors that apply to unstable compounds were also mentioned. Also underlined was that this classification method was aimed at carrying out quantitative risk analyses. It is precisely for the analysis of dangerous reactions that this method was suggested. It works as follows:

1. A physical factor is attributed to the mixture of substances by using the table in paragraph 1.5.4 and the tables, which can be found in the publication already quoted (note p.84).
2. 'Penalties' are attributed to this factor, which is above all a function of inflammability. These penalties consist of adding a given percentage if the substances have any specific unstable or reactive properties. The table below indicates the aggravating percentages to be applied.

Aggravating factors linked to substance characteristics	Aggravating percentages (%)
A Materials that react with water to produce combustible gas (eg sodium)	0 to 30
B Oxidising materials (eg oxygen, chlorates)	0 to 20
C Materials subject to explosive decomposition (eg peroxide)	125
D Materials subject to detonation (acetylene...)	150
E Materials subject to spontaneous polymerisation (eg ethylene oxide)	50 to 75
F Materials subject to spontaneous heating up	30
G Materials of resistivity favourable to accumulation of static electricity (hydrocarbons)	30 30

3. Increased coefficients relating to the conditions of preparation and the handling procedure are then applied. The aim is to quantify the risk in industrial plants. This might be where this approach may appear to be too specialised to be of use in small scale reactions. The table below gives these coefficients.

Aggravating factors linked to operating conditions	Coefficient of aggravation 'm'
H Pouring of flammable products into open receptacles	0 to 0.3
I Continuously carried out reactions	0.25 to 0.5
J Reactions discontinued or in successive baths	0.4 to 0.8

4 • Dangerous reactions: methodology
4.3 Quantitative estimation of dangerous character of a reaction

Aggravating factors linked to operating conditions		Coefficient of aggravation 'm'
K	Multi stage reactions in the same equipment	0 to 0.5
L	Processes or reactions difficult to control (nitrations, certain polymerisations)	0.5 to 1
M	High pressures $0 \leq P < 17$ bar	0
	$17 \leq P < 210$ bar	0.3
	$P \geq 210$ bar	0.6
N	Reactions under reduced pressure	0 to 1
O	High temperatures (°C) $T \geq 0.8 \times AIT$	0.5
	$260 \leq T < 537 \rightarrow$ gas	0.1
	$\rightarrow$ liquid	0.2
	$T \geq 537 \rightarrow$ gas	0.2
	$\rightarrow$ liquid	0.3
P	Low temperatures (°C) $-29 \leq T \leq 10$	0.15
	$T < -29$	0.25
Q	Functioning within inflammability limits Filling, emptying of flammable liquids	0.25
	Storage of flammable liquids in old containers with risk of air ingress	0.5
	Chemical reaction conducted near to flammable concentration	1
	Distillation within flammable limits	1.5
R	Functioning outside flammable domain but under instrumental control	0.2 to 2
S	Possibility of formation of aerosols or powders in suspension	0.3 to 0.6
T	Danger of explosion greater than average	0.6 to 2
U	Volumes utilised (in m³) $V < 7$	0
	$7 \leq V < 20$	0.4 to 0.5
	$20 \leq V < 100$	0.5 to 0.75
	$100 \leq V < 200$	0.75 to 1
	$200 \leq V < 1000$	1 to 2
	$V \geq 1000$	2 to 4

4. MF denotes the particular physical factor for the mixture and Σm corresponds to the sum of the increase coefficients applied to each property of the substances and operating conditions.

Fire-explosion risk FER is determined using the equation :

$$FER = (1 + \Sigma m) \times FM$$

5. The activity is classified as follows:

if FER < 20   low risk;

if 20 ≤ FER < 40   moderate risk;

if $40 \leq FER < 90$ important risk;

if $FER \geq 90$ very high risk.

As it stands this method can hardly be applied to activities of the scale of interest. Indeed, regarding temperature, pressure and volume the thresholds used are far too critical. The distinction between continuous and discontinuous techniques does not seem of any use in this context. There are a lot of situations that often arise under smaller scale conditions which are missing; for instance, glass materials, activities outside working hours or unsupervised handling, etc.

It is conceivable that in each large group of activity; eg research, development, analysis laboratory, in workshops (of surface treatment, cleaning, degreasing of metallic parts) people in charge who have gained a good knowledge of safety use this framework to develop their own quantitative classification system.

This seems to be the case for the categories of operating conditions (H to U in the previous table).So far as categories A to G are concerned, they do not seem to be adequate. Except for a few categories, the coefficients that are used are mostly quantified using qualitative methods. It is common practice to make a qualitative variable into a quantitative variable, but this does not give any more physical meaning to the quantitative one than it had before. So it cannot be said that a FER of 92 constitutes a risk that is twice as great as a FER of 46. The physical factor used with the increase in coefficients defined from H to U should be replaced by the risk level suggested later as a variant of CHETAH.

4.3.2 The Stull Method

The technique suggested by Stull and described in paragraph 2.4 obviously can be applied to a chemical reaction with the difficulties in finding the activation energy.

4.3.3 The CHETAH programme and proposed variant

The CHETAH programme would apply to chemical reactions as well as isolated substances. This programme has not been demonstrated with chemical reactions, which leads to the difficulties raised in the previous chapter regarding the difficulty in synthesising the information given by the four criteria and the overestimation of risk by criterion C_1. Therefore we prefer to suggest a direct adaptation of the variant, which was described before and leads to a unique risk criterion, first giving the variant using the same writing conventions as in the previous chapter. It is essential for the reader to study the paragraph about CHETAH (see 2.3) before going any further.

■ **Calculation of enthalpy of reaction ΔH_r**

The enthalpy of decomposition is now replaced by the enthalpy of reaction to analyse the potential danger. Since the danger of a chemical reaction is usually related to a modification in its procedure, which makes it uncontrollable and causes destruction of the molecular groups, it seems to make more sense to write down the most energetic reaction possible. The risk will indicate the maximum potential danger considering the stoichiometry chosen. This approach may be

4 • Dangerous reactions: methodology

4.3 Quantitative estimation of dangerous character of a reaction

considered as artificial. Nevertheless, it corresponds to the standard approach. Let us imagine that we want to make 4-methoxybenzaldehyde from 4-methylanisole by catalytic oxidation by oxygen. The most energetic reaction consists in the following:

$$CH_3O-C_6H_4-CH_3 + O_2 \longrightarrow CH_3O-C_6H_4-CH=O + H_2O$$

and not in the one that would lead to a complete oxidation of the molecule and requires ten moles of oxygen. But, if efficiency optimisation forces one to use for instance a molar quantity that is twice as much as the one given by the theoretical reaction, the calculation of the enthalpy will include the excess oxygen. This calculation needs to be carried out using details of the stoichiometry.

■ Calculation of degree of risk IR (index of reactivity)

Applying the calculation formula given by the CHETAH programme for C_4 by replacing ΔH_d by ΔH_r, which is the enthalpy of reaction defined previously as 'IR', and apply the same classification criteria as with CHETAH:

- if IR < 525   low risk;
- if 525 ≤ IR < 1926   average risk;
- if IR ≥ 1926   high risk is obtained.

■ Potential increases or reductions

1. Calculation of $C_2 = \Delta H_c - \Delta H_r$, ΔH_c actually refers to the complete oxidation reaction of the molecules; indeed, this applies to the reactions between incombustible substances.

- if $|C_2| > 21$ kJ/g   low risk;
- if $12.6 < |C2| \leq 21$ average risk;
- if $|C_2| \leq 12.6$ high risk.

which leads to:

- if $C_2 = IR$ no modification of IR.
- if $C_2 < IR$ IR* = 0.75 x IR, if one risk range IR* = 0.5 x IR (if two risk ranges);
- if $C_2 > IR$ IR* = 1.5 x IR, if one risk range IR* = 2 x IR (if two risk ranges).

2. Calculation of C_3. This is defined as in the CHETAH programme and criteria are the same.

- C3 < -240   low risk;
- -240 ≤ C_3 < -120 or 80 < C_3 ≤ 100   average risk;
- -120 ≤ C_3 ≤ 80   high risk.

4 • Dangerous reactions: methodology

4.3 Quantitative estimation of dangerous character of a reaction

The increased coefficients according to C_3 obtained are then applied. Increases or reductions are applied on IR and IR* in the same way as with C_2.

The final IR or IR* degree of risk implies that the reaction is:

- without danger if IR or IR* < 525 →
  no need for operating precautions;

- of average danger if 526 ≤ IR or IR* < 1926 →
  One needs to check the temperature and the flow rates of the reagents. Tests need to be done on small quantities to determine the effect of pressure, if required;

- of very high danger if IR or IR* ≥ 1926 →
  The operating mode needs to be tested or set on microquantities. It is recommended to carry out a preliminary differential thermal analysis.

One should use the modified CHETAH method every time a task requires the use of substance mixtures for which the qualitative analysis shows signs of a potentially dangerous reaction. It can also be used when carrying out an accident investigation to test the different factors involved which have not been elucidated, or to decide between several scenarios of dangerous reactions that are likely to give an explanation for an incident. Finally, the analysis should also be carried out before attempting any new chemical reaction experiments which are likely to be dangerous.

Ten examples will be considered: six of them are organic reactions and four inorganic. Seven of these reactions caused accidents, mentioned in the corresponding chapters. Two of them were chosen because they are known for not representing any danger and are commonly quoted in lectures.

The following table lists these reactions. Only the stoichiometry is indicated, but more information will be found concerning the circumstances of the accident and the type of reaction in the references concerning the function of the main compound.

N°	Reaction	ΔH_f est.	CHETAH standard criteria				IR		Level of risk
			C_1	C_2	C_3	C_4			
1	$C_6H_5-C≡CH + HClO_4$	209.7(l)	–5.84	–14.6	–118.5	3456.8	2592.6*		High
2	$(CH_3)_2CH-OH + HNO_3$		–6.68	–7.76	–84.5	3231.7	3231.7		High
3	$CH_3O-C_6H_4-CH_3 + O_2$	–156 (l)	–4.18	–23.07	–186.8	1282.7	962.0*		Medium
4	$Ca(OCl)_2 + 2\ CH_3OH$		–2.56	–6.36	–30.9	798.4	1796.4*		Medium
5	$(CH_3)_3C-OH + HCl$		–0.43	–32.26	–259.0	9.1	9.1		Low
6	C₆H₄(OH)(CO₂H) + (CH₃CO)₂O (acid)	–785.84	–0.75	–13.60	–110.61	61.7	185.2*		Low
6	$3\ Na_2SO_4 + 8\ Al \rightarrow 3\ Na_2S$		–5.66	–1.73	–22.43	7091.6	7091.6		High
8	$3\ NaCN + 5\ NaNO_2$		–3.54	0	0	2124.6	2124.6		High
9	$CuO + H_2S \rightarrow H_2O + CuS$		–1.06	–5.25	–42.27	309.9	929.8*		Medium
10	$H_2S + 2\ O_2 \rightarrow H_2O + SO_2$		–6.32	0	0	5458.8	5428.8		High

When analysing this table the following should be noted:
- Reaction (3) has already been mentioned. The estimated risk level makes it probable that it is the use of an unsuitable operating mode that caused the accident. It seems that oxygen was cautiously incorporated.
- Reactions (5) and (6) are the reactions known for not representing any danger.
- The $CuO + H_2S$ reaction caused an accident, which was blamed on either reaction (9) or (10) catalysed by copper oxide. The comparison of the IR risk degrees makes it probable that the second interpretation is the most plausible.

This analysis facilitates an *a priori* study, which enables us to be informed about the potential risks of dangerous reactions. This methodological study is now followed by two chapters on dangerous reactions focusing on first, inorganic, and then organic substances.

PART TWO
List of Dangerous Reactions

5 • DANGEROUS REACTIONS OF INORGANIC CHEMICALS

Arrangement of elements in this Part

For reasons apparent in other parts of this book, the elements in this Part are arranged in the order in which they appear in the periodic table of the elements.

For the additional convenience of the reader an alphabetical index of the elements follows:

• aluminium	176		• molybdenum	218
• antinomy	223		• neon	172
• argon	192		• nickel	206
• arsenic	210		• niobium	218
• barium	228		• nitrogen	165
• beryllium	164		• oxygen	164
• bismuth	233		• palladium	219
• boron	164		• phosphorus	178
• bromine	213		• potassium	192
• cadmium	222		• rhodium	219
• calcium	195		• rubidium	215
• carbon	165		• ruthenium	219
• chlorine	186		• scandium	198
• chromium	199		• selenium	211
• cobalt	205		• silicon	177
• copper	206		• silver	220
• fluorine	170		• sodium	172
• gallium	210		• strontium	215
• germanium	210		• sulphur	181
• iodine	225		• tellurium	225
• indium	222		• tin	222
• iron	203		• titanium	198
• lead	231		• vanadium	199
• lithium	164		• xenon	228
• manganese	201		• zinc	208
• magnesium	174		• zirconium	217
• mercury	230			

Lithium

Lithium is extremely reactive with water and hydrogen peroxide. These are the only reactions which cause accidents. But one can find other dangerous reactions of this class of substances elsewhere in this Part (see p.146).

- Lithium has a moderate reaction with water when cold, but extremely violent when hot. The violent release of hydrogen along with a high temperature rise of the lithium particles sometimes provokes the gas to combust. With powdered lithium, an explosion will occur.
- Above 300°C lithium combusts in contact with hydrogen.
- A mixture of lithium with hydrogen peroxide combusts immediately. It has been used as a propellant for rockets.

Beryllium

See later references.

Boron

Boron has a large range of oxidation states. The element and its hydrides are reducing agents, perborates are strong oxidants. The oxidation-reduction reactions are the only known dangerous reactions of boron compounds. Two compounds amongst the ones analysed here are hardly stable. Two dangerous reactions, which are difficult to classify, will be dealt with separately.

Oxidation-reduction reactions

- When there are oxidants present, boron always reacts violently. It reduces water; and if the temperature is greater than 600°C, the reaction is violent or even explosive.
- Boron carbide can combust spontaneously in air.
- When there are reducing agents present, sodium perborate gives an explosive reaction. It is less dangerous if the perborate is hydrated.

Instability

- Boron hydrides are unstable. Their enthalpy of formation is positive.
- When dry, sodium perborate detonates on friction.

Other dangerous reactions

- Diborane forms complex hydrides with lithium, which combust spontaneously in air.
- When boron trifluoride is heated in the presence of beryllium or lithium, it provokes incandescence.

Carbon

Carbon is a reducing agent; the oxidised forms of carbon are carbon oxides. The dangerous reactions of these will therefore mainly be oxidation-reduction reactions.

- Carbon provokes violent or even explosive decomposition of hydrogen peroxide. However, the real cause of this may be linked to the presence of metallic impurities in carbon.
- Carbon/lithium contact in the presence of lithium battery electrolytes can provoke spontaneous ignition.
- Carbon dioxide reacts slowly with fragmented lithium at 20°C. With the powdered metal, its ignition is instantaneous. It also happens when it is hot. This is the reason why carbon dioxide extinguishers are forbidden for putting out lithium fires (and metals in general; see later).
- Powdered beryllium can also combust in contact with carbon dioxide.
- In one instance, carbon dioxide was incorporated into lithium in solution in liquid ammonia. When oxygen and/or water was later added it led to detonations.

Published sources do not mention any dangerous reactions with carbon suboxide, which is hardly ever used.

Nitrogen

The fact that there are a large number of different types of bonds with nitrogen leads to a great variety of combinations with very different behaviours regarding reactions. These reactions are usually dangerous.

There are two features so far as dangerous reactions involving nitrogenous compounds are concerned; firstly, those involving the instability of certain types of molecular arrangements containing nitrogen bonds with very electronegative elements including nitrogen, oxygen, chlorine etc. Secondly, there are the variety of oxidation states of nitrogen, which can vary between -III and +V.

Depending on the cases, nitrogen compounds can be very strong reducing agents (ammonia; amide ions, cyanide, ammonium; hydrazine, hydroxylamine) or oxidants (nitrite, nitrate ions; nitrogen oxides). Accidents are caused by the molecules containing oxidising as well as reducing nitrogen atoms, which give very unfavourable oxygen balance systems with regard to risk.

The instability of nitrogenous molecules with weak bonds will first be considered, then the dangerous oxidation-reduction reactions. Then, all the reactions which cannot be classified under the two main categories of dangerous reactions will be gathered together.

Instability of nitrogenous compounds

Compounds with nitrogen-metal (nitride) bonds

- In the gaseous phase and in the presence of lithium, nitrogen gives lithium nitride Li_3N $\Delta_f^0 = -197.48$ KJ/mole, which is the only alkali nitride known to explode when heated between 115°C and 293°C.

5 • Reactions of inorganic chemicals

Nitrogen

- Hydrogen azide in the pure state, which is highly endothermic, detonates violently.
- In hot water, sodium azide decomposes violently.
- In the presence of acid, sodium azide leads to hydrogen azide, the dangerous behaviour of which has already been mentioned.

Compounds with nitrogen-nitrogen bonds

- Hydrazine is an unstable substance because of its positive enthalpy of formation. It decomposes when heated. The decomposition can cause an inflammation even in the absence of air. It can also combust spontaneously in the presence of various materials from clothes to soil (see tables in Part Three: the self-ignition temperatures vary according to the materials in contact with hydrazine). Also, violent decomposition of hydrazine in a steel reactor occurs when in a carbon dioxide atmosphere.

Compounds with nitrogen-oxygen bonds

- Nitrogen dioxide forms an equilibrium with dinitrogen tetroxide. When it is cold, the equilibrium favours the second compound. Depending on the thermal conditions, the dangerous reactions will involve one or the other of these two compounds. Their endothermic character makes them hardly stable.
- Under a pressure of 100 kbar dinitrogen tetroxide even detonates at -152°C.
- Nitrous oxide is unstable because it is an endothermic substance. It gives rise to an explosive decomposition if it is heated.
- Hydroxylamine detonates when it is heated.

Compounds with nitrogen-chlorine bonds

- Nitrogen trichloride, which is highly endothermic, is also extremely unstable. All the accidents that are described below and involve this compound are linked to the catalysis of its decomposition. It is the same when it is in the pure state and exposed to a physical agent such as shock, light or ultrasound.
- Nitrogen trichloride detonates in the presence of nitrogen monoxide, dioxide or trioxide and also in contact with ammonia and potassium cyanide.

Oxidation-reduction reactions

Most of the dangerous reactions cited in this paragraph involve nitrogenous molecules, which are both reducing and oxidising, priority being given to the oxidising ones.

Oxidation by nitric acid

- In the presence of nitric acid, finely divided boron combusts spontaneously.
- Nitric acid forms explosive mixtures with boranes, especially diborane.
- Ammonia forms highly unstable mixtures with nitric acid. Their decomposition is extremely violent. In the vapour state, nitric acid causes the spontaneous ignition of ammonia.

- The same applies to hydrazine. In the presence of catalysts, the ignition is instantaneous and violent. This mixture was used as the first propellant for rockets, in particular for V2 and the French rocket called 'Véronique'. In the absence of catalysts, there is a period of induction, but when decomposition occurs it can be explosive.
- Metallic cyanides, sodium azide and hydrogen azide form explosive mixtures with nitric acid.

Oxidation by nitrogen oxides
- In contact with dinitrogen tetroxide, sodamide produces showers of sparks.
- Nitrogen dioxide, tetroxide and monoxide detonate when they are heated with ammonia. Ammonia combusts in the presence of nitrous oxide.
- When a mixture of nitrogen dioxide and hydrogen was heated, it spontaneously detonated; the presence of oxygen traces may have caused the accident.
- Dinitrogen tetroxide has a highly exothermic reaction with boron trichloride.
- Hydrazine combusts spontaneously in contact with dinitrogen tetroxide. As with nitric acid, this property made it suitable as a propellant for rockets from the N_2O_4/N_2NNH_2 system.
- At 20°C boron produces a highly violent lightning flash in contact with nitrogen tetroxide.
- In nitrogen oxide, carbon that is heated at 400°C burns better than in air.
- If hydrogen containing oxygen traces is heated with this oxide it spontaneously combusts or detonates.
- The same oxide detonates in the presence of boron and forms an explosive mixture with carbon monoxide or pure hydrogen.
- Nitrous oxide/hydrazine mixture combusts spontaneously.

Oxidations by nitrite and nitrate ions
- Nitrites are strong oxidants.
- They decompose violently in the presence of lithium; the same is true for boron.
- They cause the ignition of all ammonium salts, in particular ammonium nitrate as will be seen below.

The principal dangerous reaction of cyanides can be observed when these are heated in contact with strong oxidants.

- This is the case with nitrites and alkaline nitrates. Sodium nitrite/potassium cyanide mixture has been suggested as a primary explosive for detonators.
- In the presence of molten sodium nitrite, sodamide forms an explosive substance. It is also the case for potassium nitrite; it detonates violently when heated with sodamide.

Ammonium and alkaline nitrates have the same oxidising character as nitrites.

- A powdered carbon/nitrate mixture heated at about 160°C causes incandescence of carbon followed by explosion of the mixture.
- Sodium and potassium nitrates give an explosive reaction with lithium.

5 • Reactions of inorganic chemicals

Nitrogen

Dangerous reactions of nitrogen and its derivatives

	Nitric acid	Alkaline amides	Ammonia	Nitrogen	Hydrogen azide	Sodium azide	Alkaline cyanides	Hydrazine	Hydroxylamine	Ammonium nitrate	Potassium, sodium nitrate	Potassium, sodium nitrite	Oxides of nitrogen	Nitrogen trichloride
Hydrogen				■									■	
Group I (Li; Na; K)						■		■						
Group II (Be; Mg; Ca; Sr; Ba)														
Group III (B)	■							■						
Group IV (C; Si; Ge)					■									
Group V (N, P, As, Sb)														
Group VI (O$_2$; S; Se; Te)														
Group VII (F$_2$; Cl$_2$; Br$_2$; I$_2$)														
Transition metals														
Other metals (Al...Bi, Sn, Pb)														
Hydrides group III (B$_x$H$_y$)	■													
Hydrides group IV (SiH$_4$...)														
Hydrides group V (NH$_3$) H$_2$NNH$_2$; NH$_4^+$; NH$_2$OH	■										■			
Water		■												
Hydrides group VI (H$_2$S...)														
Hydrogen halides			■											
Metal halides (AlCl$_3$; FeCl$_3$...)														
Non-metal halides (NCl$_3$; P$_x$X$_y$; SOCl$_2$; S$_x$Cl$_y$)		■												
Metal oxides														
Non-metal oxides (O$_3$; CO$_x$; N$_x$O$_y$; P$_x$O$_y$; SO$_x$...)												■		
Peroxides (H$_2$O$_2$; Na$_2$O$_2$)	■													
Nitric acid														
Sulphuric acid and/or sulphates			■											
ClO$_x^-$; HClO$_4$														
MnO$_4^-$, Cr$_2$O$_7^{2-}$														
NO$_3^-$; NO$_2^-$														
Instability			■		■	■							■	
Polymerisation														
Bases														
BCl$_3$		■											■	

- An alkaline nitrate/boron mixture, both in the powdered state, detonate on impact.

Ammonium nitrate has both the dangerous properties of a nitrate ion, which is the same as for metal nitrates, and the dangerous reactions of ammonium ion. It also has the quite rare feature of containing two nitrogen atoms with extreme oxidation states; -III for the ammonium ion and +V for the nitrate ion (positive oxygen balance and equal to 20 g%), which makes it a very dangerous redox system by criterion C_3 in the CHETAH programme. Its instability is due to an intramolecular oxidation-reduction reaction. This is why it was decided not to discuss it in the paragraph about the instability of nitrogenous compounds.

- Ammonium nitrate detonates at temperatures greater than 250°C when it is pure, but at 170° and above in the presence of chloride ions, in particular with ammonium chloride (in the presence of this, a mixture of zinc powder and ammonium nitrate, in contact with a few drops of water, created a spectacular flare-up). Also, if it is heated with 50% of its weight in water, it detonates.
- When it is mixed with potassium nitrite in the melted state (440°C), ammonium nitrate detonates. The presence of rust decreases the temperature of the explosion to 80-120°C. This behaviour is linked to the chemical incompatibility of the nitrite ion and ammonium ion.

Other dangerous reactions

- Liquid nitrogen condenses oxygen and has similar dangerous reactions to those of liquid oxygen. This is how an explosion, which occurred during contact between hydrogen and liquid nitrogen, was interpreted.
- Hydrogen peroxide forms highly sensitive explosive mixtures with nitric acid.
- Sodamide reacts violently with water: $NaNH_2 + H_2O \longrightarrow NaOH + NH_3$

The dangerous behaviour of this substance vis-à-vis various reagents can be explained by the basic properties of ammonia.

- In the presence of boron trichloride or trifluoride (Lewis acids), ammonia is likely to detonate. In contact with boron, it causes incandescence.

Depending on the quantity of ammonia, it stabilises or destabilises ammonium nitrate.

The numerous reactions between nitrogenous derivatives with different states of oxidation are represented in the table (p168), which is more difficult to read than the usual summary tables in which there are less troublesome crossings between lines and groups.

Oxygen

Oxygen and especially ozone are strong oxidants. They form very dangerous redox reactions with all the reducing agents mentioned in the previous chapters.

5 • Reactions of inorganic chemicals — Fluorine

The element

- In the presence of metal oxides, oxygen heated with hydrogen gives rise to explosive mixtures.
- The same goes for carbon monoxide or carbon monoxide/hydrogen (synthesised gas) mixtures.
- Liquid oxygen/carbon or liquid oxygen/liquid carbon monoxide mixtures have been used as explosives. The activated carbon atoms constitute an aggravating factor.
- Powdered or molten lithium combusts as soon as it is in contact with oxygen.
- Diborane heated at 105-165°C with oxygen detonates after a latent period.

All the reduced forms of nitrogen as well as the element itself can give rise to dangerous reactions with oxygen.

- When it is incorporated (or condensed from ambient air) into liquid nitrogen, oxygen causes explosion of the mixture if it is exposed to gamma radiation or neutrons.
- There have been a lot of accidents caused by the ignition of ammonia in oxygen. If ammonia is in the liquid state, the mixture can instantaneously detonate.
- An attempt to grind sodamide was made in the presence of water traces. Sodamide combusted violently in the presence of ambient air. Note that the sodamide/water interaction is dangerous even in the absence of oxygen.
- An oxygen/hydrazine mixture which was heated caused a violent explosion.

Ozone

The same chemicals are involved in accidents with ozone, the difference being that the consequences are worse than with oxygen and occur at lower temperatures.

- Ozone is a highly endothermic substance. Its enthalpy of formation reaches 2.96 KJ/g, which makes it very unstable thermodynamically by CHETAH criterion C_1. In the liquid or solid state, it detonates spontaneously.
- Ozone in the solid state forms explosive mixtures with liquid hydrogen. It also detonates in contact with liquid nitrogen if there are metal impurities present.
- With carbon monoxide, if the temperature is greater than -78°C, or with nitrogen oxide, even at -189°C, ozone forms explosive oxidation reactions.
- The same goes for liquid ammonia above -78°C; the explosion occurs as soon as the compounds are in contact with each other.
- Ozone catalyses the explosive decomposition of nitrogen trichloride.

Fluorine

Fluorine is the most reactive element. It reacts with all other elements and their compounds, even with its own derivatives and usually in an extremely violent way, even at low temperatures. It is the strongest oxidant.

The element

Action of hydrogen and its derivatives

- Fluorine detonates in contact with hydrogen at temperatures starting at -252°C if the molar quantities of both reagents are present.

5 • Reactions of inorganic chemicals — Fluorine

- Water in the solid state forms mixtures that are very sensitive to impact with liquid fluorine, which, if they detonate, have the same disruptive property as trinitrotoluene. If water is in the liquid state, the explosion is instantaneous and equally violent.

Action of lithium and its derivatives
- In contact with fluorine, lithium combusts instantaneously.
- Fluorine causes the ignition of dilithium carbonate. Lithium oxide glows in contact with it.

Action of beryllium and its derivatives
- If heated in the presence of fluorine, beryllium becomes incandescent. The same is true for beryllium oxide.

Action of boron and its derivatives
- With boron there is immediate ignition with fluorine sometimes followed by an explosion. Boron oxide becomes incandescent and so does boron nitride.
- Metal borides become incandescent in the presence of fluorine.
- Boron trichloride combusts in contact with fluorine.

Action of carbon and its derivatives
- When carbon is in the impure form of soot or coal, it combusts in contact with fluorine and then can detonate spontaneously. If the carbon is very pure, the ignition is delayed.
- Graphite detonates violently in the presence of fluorine under a pressure of 13.6 bar.
- The carbon monoxide/fluorine mixture can detonate spontaneously. If oxygen is present and the temperature is greater than 30°C, the explosion is instantaneous. It is attributed to the formation of the following unstable substance:

$$F_2 + 2CO + O_2 \xrightarrow{>30°C} F-\overset{O}{\underset{\|}{C}}-O-O-\overset{O}{\underset{\|}{C}}-F$$

Action of the nitrogenous compounds
- Ammonia combusts spontaneously when it is mixed with fluorine. This system has been considered as a propellant for rockets. If ammonia is in the form of an aqueous solution, the mixture detonates as soon as it is formed.
- The same goes for hydrazine and cyanides.

The most oxidised forms of nitrogen are also incompatible with fluorine. So:
- the action of fluorine on nitric acid causes a violent explosion, which is thought to be due to the following reaction that leads to unstable fluorine nitrate:

$$HNO_3 \xrightarrow{F_2} FNO_3$$

- fluorine detonates when it is mixed with all nitrogen oxides and is incompatible with alkaline nitrates.

Action of oxygen and hydrogen fluoride
- Fluorine reacts violently with oxygen and liquid air.
- Fluorine reacts with hydrogen fluoride.

Fluorine derivatives
- Potassium fluoride forms explosive 'peroxidates' with hydrogen peroxide. They are analogues of hydrates.
- In the liquid state, hydrogen fluoride gives rise to a violent reaction with potassium hexafluorosilicate.
- Hydrogen fluoride reacts violently with ammonia.

Neon

Neon is a perfectly inert element.

Sodium

Sodium is, like all other alkali metals, a very strong reducing agent (more reactive than lithium), which has extremely violent reactions with numerous compounds. It causes a large number of accidents. Sodium peroxide is a very reactive oxidant, which has violent interactions with reducing agents. Carbonates, and especially sodium hydroxide, are bases which react with acids (the reaction is aggravated by the formation of carbon dioxide).

The metal

Actions of boron compounds
- If heated with boron trifluoride, sodium becomes incandescent.

With carbon and its derivatives
- Sodium in the gaseous state gives a highly violent explosion with powdered carbon, in the presence of air.
- It reacts violently with carbon dioxide even if both compounds are in the solid state; carbon dioxide cannot be used as an extinguishing agent for sodium fires.
- In the presence of sodium in liquid ammonia, carbon dioxide gives rise to a violent explosion when the unstable substance formed during the reaction is heated at a temperature of 90°C. The reaction is presumed to be the following:

$$Na + CO \xrightarrow{NH_{3liq}} NaCO$$

5 • Reactions of inorganic chemicals — Sodium

Action of oxygen and its derivatives
- Sodium can combust spontaneously in contact with air.
- Sodium reacts violently with water, forming hydrogen, which usually combusts. Under certain circumstances of containment the interaction is explosive. It is also explosive when the water is in the solid (ice) state. The interaction is equally violent with hydrogen peroxide (in an aqueous solution).
- At a temperature of 500°C sodium has an extremely violent reaction with sodium peroxide:

$$2Na + Na_2O_2 \longrightarrow 2\ Na_2O$$

Action of nitrogenous compounds

Whether nitrogen is in the reduced form (ammonia, ammonium salts, hydrazine) or the oxidised form (nitrites, nitrates), sodium gives rise to dangerous interactions. With the first set of compounds metalation occurs with formation of unstable or highly reducing compounds, whilst, with the second, redox reactions occur.

- A mixture of sodium and ammonium chloride creates a weak explosion on impact.
- In the presence of anhydrous hydrazine, sodium in suspension in ethyl oxide is thought to give the following reaction:

$$H_2NNH_2 + CO \xrightarrow[(C_2H_5)_2O]{Na} NaCO \quad H_2NNHNa \dashrightarrow[air] \text{explosion}$$

The substance obtained detonates in contact with ambient air. If the hydrazine is hydrated, the reaction is highly exothermic and forms ammonia as well as hydrogen.

- The reaction of metal with hydroxylamine varies according to the conditions. In the absence of a solvent, there is immediate ignition. In the presence of a solvent, the substance below for which there is no indication whatsoever regarding risks is obtained.

$$NH_2OH \xrightarrow{Na} NH_2ONa$$

- With nitric acid sodium combusts spontaneously.
- The action of sodium on ammonium nitrate results in a violent explosion. It is assumed to be caused by the decomposition of a hyponitrite formed as follows:

$$NH_4NO_3 \xrightarrow{Na} Na_2N_2O_2$$

- With sodium nitrate, pure sodium (or in liquid ammonia) gives rise to an explosive yellow substance, which may be the same hyponitrite as the one above.

Action of fluorine and its derivatives
- Gaseous fluorine provokes the immediate ignition of sodium.
- Sodium slowly reacts with hydrogen fluoride by forming sodium fluoride. The reaction is not considered to be dangerous. It detonates if hydrogen fluoride is in aqueous solution.

Sodium hydroxides and carbonates

- When they are mixed vigorously with sodium carbonate, acids cause a sudden eruption of the medium, which is due to the massive formation of carbon dioxide. This is how the foam extinguishers work. Pressures of 14 bar are obtained by this reaction in such an extinguisher.
- Lithium burns violently in contact with sodium carbonate. This behaviour explains why use of extinguishing powders containing carbonates (and especially hydrogencarbonates) are forbidden for putting out lithium fires (and all fires of highly reducing metals).
- Fluorine provokes a violent 'combustion' of sodium carbonate.
- Sodium hydrogencarbonate was heated at a temperature of 100°C in the presence of carbon and water. During the handling, stirring was interrupted. When it started again a violent release of carbon dioxide caused the content of the reactor to overflow.
- Since the enthalpy of 'dissolution' of sodium hydroxide is very high, when hydroxide in the solid state is introduced into water this can cause the water to boil violently.

Sodium peroxide

This oxidant reacts with many compounds. In organic chemistry it is an agent which entirely oxidises substances enabling one to measure them. In 'microbombs', the contact causes a small explosion. Thus silicon compounds are converted into silica, which can then be measured by gravimetry. When opening the 'bomb', overpressures can cause corrosive substances (especially peroxide in excess) to be thrown up and cause harm.

- It reacts violently with water in small quantities and large quantities of water can cause the compound to detonate.
- With hydrogen peroxide it is thought to form very unstable peroxydates.
- Mixtures of boron or carbon with this peroxide are explosive. Boron nitride becomes incandescent.

Numerous examples of dangerous reactions involving this substance will be noted later.

Magnesium

Magnesium is a strong reducing agent like all alkali and alkaline earth metals. It has a number of dangerous reactions very similar to the one for sodium. Amongst the magnesium compounds, perchlorate, which is highly oxidising and unstable, is the only one that has ever been involved in an accident.

The element

Action of water and peroxides

- Contact of water with flaming magnesium gives rise to a very violent explosion. The extinguishing agents made with water therefore cannot be used on magnesium fires, which are common in the aeronautical industry.

5 • Reactions of inorganic chemicals

Magnesium

- Magnesium burns in contact with hydrogen peroxide containing traces of manganese.
- With sodium peroxide containing traces of water or combined with carbon dioxide, the mixtures obtained detonate spontaneously.

Action of beryllium derivatives
- The interaction of magnesium with beryllium difluoride is highly exothermic.
- Magnesium violently reduces beryllium oxide.

Action of carbon dioxide and carbonates
- Magnesium combusts above 900°C with carbon dioxide. The ignition takes place at a temperature of 780°C and above if water traces are present. This behaviour thus does not allow use of extinguishers containing carbon dioxide for magnesium fires.
- A magnesium/calcium carbonate mixture gives rise to a violent explosion in the presence of hydrogen. In fact, attempts to extinguish magnesium fires using extinguishing powders containing alkaline hydrogencarbonates caused numerous detonations. This shows that this reaction is general and does not require hydrogen.

Action of nitrogen and its derivatives
- Magnesium reacts very violently with liquid nitrogen.
- If it is heated in the presence of metal cyanides, magnesium starts glowing.
- The metal burns violently in contact with nitric acid in the gaseous state.
- Magnesium sets off a violent reaction or even an explosion with ammonium nitrate around 200°C or in contact with molten sodium nitrate.
- Magnesium wire burns violently in an atmosphere of nitrogen dioxide or di-nitrogen tetroxide.

Action of oxygen
- Magnesium combusts when it is heated in air at a temperature of 520°C and above. The magnesium/liquid oxygen mixture is explosive.

Action of fluorine
- The violent combustion of magnesium in gaseous fluorine in the presence of water traces caused an accident.

Magnesium perchlorate

This compound easily produces perchloric acid in the presence of acid traces and dangerous salts are often created by its combination with traces of perchloric acid. Since this salt is a very efficient dehydrating agent and used very often, there are a very large number of accidents which involve it.

- During an operation which consisted in drying ammonia with magnesium perchlorate, a highly violent explosion took place.

Aluminium

The dangers are the same as with magnesium and sodium, ie those caused by highly reducing metals. Aluminium is less dangerous with sodium.

The element

- Powdered aluminium can create explosive suspensions in air in the presence of an ignition source. It can combust spontaneously if the powder is moist. Aluminium powder detonates spontaneously in liquid oxygen.
- Aluminium powder can (depending on the surface texture of the metal) react with water, forming hydrogen, which can provoke explosions due to the overpressures created if the interaction occurs in a closed container. The same thing happens, whatever the surface texture, if the reaction occurs with an aqueous sodium hydroxide solution.
- Molten aluminium detonates in contact with water.
- The aluminium/Raney nickel mixture can explode in contact with water (see nickel).
- Metal catalyses the explosive decomposition of hydrogen peroxide.
- Powdered aluminium combusts – or even detonates, if it is heated in the presence of carbon dioxide. Ignition takes place at ambient temperature when anaqueous aluminium chloride is present. The same is true for the aluminium chloride-/carbon monoxide mixture.

It is forbidden to use water and carbon dioxide when putting out aluminium fires because of dangerous reactions with these two elements.

- Aluminium is thought to form dangerous mixtures with lithium.
- Aluminium amalgam is thought to combust spontaneously in air.
- Aluminium can detonate when mixed with ammonium nitrate. It has a highly exothermic reaction with metal nitrates if it is heated at 70-135°C when water traces are present.
- The metal forms highly sensitive explosive mixtures with dinitrogen tetroxide.
- Aluminium forms a complex that combusts spontaneously in air with diborane.

The aluminium derivatives

Aluminium borohydride in the gaseous state has a positive enthalpy of formation, so it can be assumed to be rather unstable. It combusts spontaneously in contact with air, but detonates if water traces are present. The same thing happens in contact with water at 20°C.

- Anaqueous aluminium chloride is dangerous. In contact with water, it hydrolyses, forming hydrogen chloride. If it is kept in a closed bottle in which there are water traces, the overpressure that is created can cause the bottle to detonate. There have been numerous accidents illustrating this.

- There was also the case of an accident caused by the explosion of a container due to overpressure. An aluminium chloride/sodium peroxide/powdered aluminium mixture in the container was kept like this for 41 days. Note that these three elements represent great danger when water is present.
- When mineral acids are present, aluminium phosphide rapidly converts into phosphine, which can combust or even detonate. When bases are present the phosphine release is slow.
- There was a rather surprising instantaneous detonation when making an aluminium oxide/sodium nitrate mixture.

Silicon

There are a comparatively limited number of accidents which involve silicon and its compounds.

The element and its alloys

- Powdered silicon can combust when water is present. It creates suspensions in air, which detonate violently when there is an ignition source present. The silicon/lithium alloy can combust spontaneously in air.
- If it is heated in presence of alkaline carbonates, it becomes incandescent and forms carbon monoxide.

The behaviour of silicon with water and carbonates shows that there are similar problems as with the other metals previously cited when it comes to putting out silicon fires.

- Silicon combusts spontaneously in fluorine. The temperature reaches 1400°C.
- Ferrosilicon (Si-Fe alloy) gives a very violent reaction with aluminium.

Silane

- This gas combusts spontaneously in contact with air. Its enthalpy of formation is positive and makes it unstable.
- When mixed with 30% of oxygen it creates a metastable system, which is particularly dangerous if the temperature is greater than 80°C.

Silica

- Silica is reduced violently when heated with the metals from earlier groups of the periodic table. For magnesium this reaction is aggravated when water traces are present.

As a result, sand was recommended for extinguishing fires involving metals. Nevertheless, sand should not be used for metal fires, especially alkali metal fires where the reaction is particularly violent.

- Silica easily occludes ozone, which increases the danger that a concentration of this compound represents.

Phosphorus

The element

The dangerous reactions of this element are mostly oxidation reactions caused by strong oxidants. There are three exceptions with the compounds mentioned in the previous paragraphs.

- Phosphorus combusts spontaneously in contact with air.
- In contact with phosphorus, beryllium incandesces (formation of a phosphide?).
- It catalyses the explosive decomposition of nitrogen trichloride.
- It oxidised violently when it was heated with potassium nitrate. When it was submitted to a pressure of 70 bar and sodium nitrate was present, it detonated violently. With ammonium nitrate in the molten state it combusts, whereas in the solid state it detonates when exposed to impact.
- An air/nitric acid/phosphorus mixture in the gaseous state combusts spontaneously. The same is true for hot phosphorus or in the molten state when nitrogen oxides are present.
- Phosphorus combusts in contact with fluorine.
- It detonates in contact with magnesium perchlorate.

Phosphine

Phosphine has a positive enthalpy of formation, which makes it an unstable compound from a thermodynamic point of view. The author's sources provide figures, which contradict each other, of a factor of 4.18. This shows that there is a mistake in the units employed. It is a strong reducing agent.

- In the liquid state, phosphine can detonate spontaneously.
- If it is pure, phosphine does not combust in contact with air below 150°C. If it is impure, it combusts when it reaches a temperature of -15°C.
- Pure oxygen, even traces, creates an explosive mixture with phosphine. The detonation is immediate when nitrogen oxide is present.
- Phosphine combusts spontaneously in contact with nitrogen oxides.
- It undergoes a very exothermic reaction with boron trichloride.
- It catalyses the explosive decomposition of nitrogen trichloride.

Phosphorus halides

Actions of metals and their oxides

- The mixtures of sodium with phosphorus, phosphoryl trichloride or with phosphorus pentachloride combust spontaneously and/or detonate. There is an instantaneous ignition when powdered aluminium is mixed with trichloride or phosphorus pentachloride.
- When it is heated with magnesium oxide, phosphorus pentachloride incandesces.

Phosphorus

Action of water

- Phosphorus halides violently react with water by producing the corresponding hydrogen halide. In the case of phosphorus trichloride, the following reaction equation is postulated:

$$PCl_3 + H_2O \longrightarrow P_2H_4 + O_2$$

The phosphorus hydride which is formed combusts in contact with the oxygen that is formed. (This equation is hardly plausible.)

Action of oxidants

- Oxygen in contact with phosphorus trichloride between 100 and 700°C forms phosphoryl trichloride. The reaction proved to be dangerous (the nature of the accidents was not mentioned).
- Phosphorus trichloride detonates in contact with nitric acid. The same may occur during the highly exothermic interaction between this chloride and sodium peroxide.
- Trichloride and phosphorus pentachloride combust in contact with fluorine.

Action of hydroxylamine

- Trichloride and pentachloride phosphorus combust when they are mixed with hydroxylamine.

Phosphorus pentasulphide

- Can hydrolyse violently when water is present by releasing hydrogen sulphide, which sometimes combusts.
- Combusts in contact with air if it is subjected to friction or a flame or sparks. Ignition can also take place if it is really dry and submitted to a temperature of 275°C. Finally, in the powdered state, it can detonate when air is present.

Phosphorus pentoxide

- If heated with sodium, phosphorus pentoxide causes the metal to glow.
- If heated with sodium oxide or at ambient temperature when water traces are present, phosphorus oxide reacts very violently.
- It gives rise to an extremely violent reaction with hydrogen peroxide. This is certainly due to the catalysis of the decomposition of peroxide caused by the phosphoric acid produced by water.
- Phosphorus pentoxide reacts violently with hydrogen fluoride at a temperature lower than 20°C.

Phosphoric acid

- There was an explosion in a closed steel container in which phosphoric acid was present with chloride ions. In these conditions, the hydrogen that is produced created an overpressure.

5 • Reactions of inorganic chemicals

Phosphorus

Synopsis of dangerous reactions of phosphorus and its derivatives

Column	Substance
1	Phosphorus
2	Metaphosphoric acid
3	Phosphoric acid
4	Phosphorus pentachloride
5	Phosphorus, phosphoryl trichloride
6	Diphosphorus pentasulphide
7	Diphosphorus pentoxide
8	Ammonium phosphates
9	Phosphates of potassium, sodium
10	Phosphine
11	Diammonium phosphite
12	Sodium polymetaphosphate
13	Potassium pyrophosphate

Rows:
- Hydrogen
- Group I (Li; Na; K)
- Group II (Be; Mg; Ca; Sr; Ba)
- Group III (B)
- Group IV (C; Si; Ge)
- Group V (N; P; As; Sb)
- Group VI (O_2; S; Se; Te)
- Group VII (F_2; Cl_2; Br_2; I_2)
- Transition metals
- Other metals (Al...Bi; Sn; Pb)
- Hydrides group III (B_xH_y)
- Hydrides group IV (SiH_4...)
- Hydrides group V (NH_3; H_2NNH_2; NH_4^+; NH_2OH)
- Water
- Hydrides group VI (H_2S...)
- Hydrogen halides
- BeX_2
- BX_3
- Metal halides ($AlCl_3$; $FeCl_3$...)
- Non-metal halides ($SOCl_2$; S_xCl_y; P_xX_y; NCl_3)
- Metal oxides
- Non-metal oxides (O_3; N_xO_y; P_xO_y; SO_x...)
- Peroxides (H_2O_2; Na_2O_2)
- Nitric acid
- NO_3^-; NO_2^-
- Sulphuric acid and its salts
- CO_3^-
- ClO_x^-; $HClO_4$
- MnO_4^-, $Cr_2O_7^{2-}$
- Instability
- Polymerisation
- Bases
- Cl^-

180

Sulphur

Sulphur can exist in the extreme oxidation states -II and +VI. Besides, some forms have peroxidic catenations, which have a very strong oxidising character and a high instability (peroxodisulphates, peroxomonosulphuric acid) especially with ammonium salts because of their highly unfavourable oxygen balance. Depending on the state of oxidation there are dangerous redox reactions either with strong reducing agents (metals) or oxidants (nitrates, peroxides). The dangerous reactions will be analysed by substance and not by type of reaction.

The element

Sulphur is a strong oxidant although not so strong as oxygen, which is listed before it in group VI of the periodic table. It can thus be oxidised by the elements situated above it, in group VII or by strong oxidants.

Oxidation of sulphur

- Sulphur suspensions in air can detonate when they are submitted to an igniting flame if their concentration lies between 35 and 1400 g/m^3. The air/sulphur/carbon mixture spontaneously combusts if carbon is subjected to great heat. The combustion of sulphur under controlled conditions produces sulphur dioxide, as used in industry.
- Sulphur combusts spontaneously in fluorine at ambient temperature. Different sulphur chlorides (S_2Cl_2 then SCl_2) are formed with chlorine. The reaction is not considered dangerous.
- The element forms explosive mixtures with sodium peroxide.

Oxidation by sulphur

- Sulphur/metal mixtures form metal sulphides. The reaction can be very violent. This is the case for lithium or magnesium when both the metal and sulphur are in the molten state. This is also true for sodium by heating. The reaction is much less violent when sodium chloride is present. It is also less dangerous if the interaction is made in toluene.
- With aluminium powder, sulphur has an extremely violent reaction, which can be explosive if an igniting flame is applied to the mixture. The flame is blinding.
- Boron reacts with more difficulty. When it is heated to 600°C with molten sulphur it becomes incandescent.
- The white phosphorus/sulphur mixture either combusts or detonates when it is heated. It forms diphosphorus pentasulphide. Red phosphorus reacts with more difficulty; there is a need for an ignition flame and as a result of which the mixture combusts violently. This reaction is thought to lead to tetraphosphorus trisulphide.

Sulphur halogen derivatives

The halogen atom of these compounds is highly reactive. Water, bases and metals can react with these compounds violently or even explosively. So disulphur dichloride reacts violently with water the following way:

$$S_2Cl_2 \xrightarrow{H_2O} HCl + SO_2$$

If the interaction takes place in a closed container, it can detonate.

- The same is true for thionyl chloride, which when incorporated into an aqueous solution of ammonia causes violent hydrolysis.
- Thionyl chloride was incorporated in a pipe, which contained water by mistake. This caused a violent reaction along with a violent release of sulphur dioxide and hydrogen chloride in the following way:

$$SOCl_2 + H_2O \longrightarrow SO_2 + 2\ HCl$$

The volume increase caused by the change of liquid into gas corresponds to the ratio 1: 993 at 20°C.

With sulphur dichloride the hydrolysis is particularly exothermic.

- Chlorosulphonic acid gave rise to a bad accident, which was interpreted as the result of a huge and violent hydrolysis. 500 kg of 98% sulphuric acid were incorporated into 520 kg of chlorosulphonic acid without stirring the medium. Two liquid layers were formed, which were mixed vigorously when the stirrer started to work. The highly violent release of hydrogen chloride was explained by the action of the hydrolysis of chlorosulphonic acid by the 2% of water contained in the sulphuric acid.
- Disulphur dichloride reacts violently with sodium peroxide. Other oxidants have given rise to accidents with halogen derivatives. Nitric acid reacts violently with sulphur dichloride, releasing large quantities of gas. With dinitrogen pentoxide, sulphur dichloride gave rise to a violent detonation.
- In contact with aluminium, disulphur dichloride provokes the instantaneous ignition of the metal. Lithium batteries contain thionyl chloride. A large number of explosions of batteries have been explained by the violent interaction of lithium with the chloride, which was assumed to be released through the anode. Sodium combusts in contact with thionyl chloride vapour heated to a temperature of 300°C. Finally, sulphur dichloride gives rise to explosive mixtures on impact with sodium.
- Sulphuryl chloride will detonate in contact with a base.
- When it is heated with red phosphorus, sulphuryl chloride gives rise to a very violent reaction. Similar behaviour between phosphorus and chlorosulphonic acid was observed. Above 25-30°C there is an energetic reaction, which speeds up before detonating violently.
- Finally, a particularly dangerous reaction between chlorosulphonic acid and silver nitrate was also observed. This is probably due to the following reaction, which leads to a particularly unstable acid:

$$AgNO_3 + ClSO_3H \longrightarrow NO_3 - SO_3H$$

5 • Reactions of inorganic chemicals — Sulphur

Sulphides

In these, sulphur is at a high state of reduction. These compounds are therefore particularly oxidisable.

Hydrogen sulphide
- Hydrogen sulphide combusts spontaneously in the presence of oxygen when the mixture is at a temperature of 280-360°. Under controlled conditions this reaction is used to make sulphur from hydrogen sulphide. When soda lime is present, the interaction with pure oxygen is explosive whereas air simply causes soda lime to glow.
- The reaction of hydrogen sulphide with sodium provokes fusion.
- Hydrogen sulphide/fluorine mixture combusts spontaneously.
- Strong oxidants, fuming nitric acid or sodium peroxide, also cause the immediate ignition of hydrogen sulphide.
- Finally, like many other compounds, it catalyses the explosive decomposition of nitrogen trichloride.

Sodium sulphide
- In the presence of moist oxygen, sodium sulphide heats up spontaneously and can reach a temperature of 120°C. In controlled situations, the sodium sulphide/oxygen interaction constitutes a preparation for sodium thiosulphate (that can also be prepared by the action of sulphur on sodium sulphite). When sodium sulphide is exposed to oxygen in the presence of carbon, it reacts violently. Supposedly, the oxidation of carbon is catalysed by sulphide.
- In one reported incident sodium sulphide was incorporated into residuary waters from some pulp wood processing, which contained sodium carbonate. This produced an unexplained explosion.

Carbon disulphide
- The interaction of disulphide with oxygen can be very dangerous. This compound has a very low self-ignition temperature (see tables in Part Three). When rust is present it can cause the mixture to detonate by catalysing the oxidation reaction.
- In the presence of metals, carbon disulphide can combust as with aluminium or create mixtures sensitive to impact eg with sodium.
- Carbon disulphide burns in nitrogen monoxide and this combustion can be explosive. With dinitrogen tetroxide the mixture is supposed to be stable up to 200°C but a shockwave or a spark can cause the mixture to detonate.
- Fluorine at 20°C causes the ignition of carbon disulphide.
- With alkaline azides an explosive dithioformate is formed as follows:

$$N_3^- + S = C = S \longrightarrow N_3 - C \underset{S^-}{\overset{S}{\lessgtr}}$$

5 • Reactions of inorganic chemicals

Sulphur

Oxides of sulphur

Sulphur dioxide

- In the presence of fluorine, sulphur dioxide combusts spontaneously or detonates. The seriousness of the result is supposed to depend on the order in which the compounds are added. Note that the action of chlorine leads to sulphuryl chloride and was not mentioned as being dangerous.
- Sulphur dioxide violently reacts with sodium in the melted state.

Sulphur trioxide

- Made industrially by catalytic oxidation of sulphur dioxide, its interaction with water gives sulphuric acid.
- The interaction of sulphur trioxide with water is highly exothermic. The heat that is generated is substantial and sometimes light is produced as well. This is a cause of accidents involving this oxide.
- With white phosphorus there is ignition after a period of induction if phosphorus is in the gaseous state, or immediately if it is in the liquid state.

Sulphuric and sulphamic acids

- The solution of pure acid in water is extremely exothermic. If water is mixed into acid, this can give rise to release of corrosive material causing many serious accidents.
- Because of its strong acid character sulphuric acid can decompose alkaline carbonates, violently forming a huge amount of carbon dioxide. Emissions are possible if the mixture is made without using any protection. The container can detonate if closed.
- Sodium reacts slowly with pure acid. If it is in an aqueous solution, the reaction is extremely violent.
- White phosphorus reacts violently with sulphuric acid when it is hot.
- A solution of hydrogen peroxide at 50%, in excess, converts sulphuric acid into peroxomonosulphuric acid, which is very unstable and a particularly strong oxidant.
- When hot sodium nitrate or nitrite is treated with sulphamic acid, violent detonations occur.

Sulphur oxygen-containing salts

According to the state of oxidation of sulphur or oxygen and the types of bonds, salts will be reducing, more or less oxidising or unstable. Sulphates are the least reactive and only extreme conditions can lead to dangerous reactions.

- Sodium hydrosulphite is unstable. It decomposes violently at 190°C. The interaction with water is particularly dangerous and can cause its ignition.
- Ammonium sulphamate is also unstable; it hydrolyses violently in an acid medium when it is hot.

5 • Reactions of inorganic chemicals

Sulphur

Synopsis of dangerous reactions of sulphur and derivatives

Sulphur	Sulphamic acid	Sulphuric acid	Chlorides of sulphur (S_xCl_x; $SOCl_2$; SO_2Cl_2; HSO_3Cl)	Sulphur dioxide	Carbon disulphide	Sodium hydrosulphite	Peroxodisulphates (K^+; NH_4^+)	Ammonium sulphamate	Sodium sulphate	Hydrogen sulphide	Alkali sulphur	Thiosulphates (Na^+; NH_4^+)	Sulphur trioxide	
														Hydrogen
■	■	■					■							Group I (Li; Na; K)
■		■												Group II (Be; Mg; Ca; Sr; Ba)
■														Group III (B)
												■		Group IV (C; Si; Ge)
		■										■		Group V (N; P; As; Sb)
														Group VI (O_2; S; Se; Te)
■														Group VII (F_2; Cl_2; Br_2; I_2)
■		■	■											Transition metals
■		■	■											Other metals (Al...Bi; Sn; Pb)
														Hydrides group III (B_xH_y)
														Hydrides group IV (SiH_4...)
		■		■										Hydrides group V (NH_3; H_2NNH_2; NH_4^+; NH_2OH)
■		■	■											Water
														Hydrides group VI (H_2S...)
														Hydrogen halides
														BeX_2
														BX_3
														Metal halides ($AlCl_3$; $FeCl_3$...)
												■		Non-metal halides ($SOCl_2$; S_xCl_y; P_xX_y...)
														Metal oxides
■				■								■		Non-metal oxides (O_3; N_xO_y; PxO_y; SO_x; CO_3^-...)
■		■		■										Peroxides H_2O_2; Na_2O_2
														Nitric acid
	■											■		NO_3^-; NO_2^-
														Sulphuric acid and its salts
			■	■										ClO_x^-; $HClO_4$
														MnO_4^-; $Cr_2O_7^{2-}$
			■									■		Instability
	■													Bases
			■											Alkaline azides (N_3^-)
												■		Acids

185

5 • Reactions of inorganic chemicals — Chlorine

- Sodium thiosulphate is a reducing agent. In the presence of sodium nitrate it leads to explosive mixtures, which are sensitive if they are heated. Under the same conditions the detonation is immediate with sodium nitrite.
- Ammonium thiosulphate reacts in a very dangerous way with hydrogen peroxide.
- Sodium sulphate forms, in the presence of aluminium at 800°C, sodium sulphide. The reaction usually proves to be explosive and the usual method of preparing sodium sulphide from the sulphate uses carbon as a reducing agent.

Peroxodisulphates are particularly dangerous oxidants because of their reactivity. This is particularly true of the ammonium salt, which contains a highly reducing cation together with an oxidant anion. This property gives it a theoretical oxygen balance of 0 g%, which makes it very unstable thermodynamically (see the CHETAH programme, para 2.3). So:

- Potassium peroxodisulphate decomposes violently when it is heated above 100°C by releasing oxygen. The same violent decomposition takes place at temperatures lower than 100°C, if water is present. The salt combusts spontaneously if soda is present.
- Potassium peroxodisulphate reacts violently with hydrazine in a basic medium. The danger is aggravated by the formation of nitrogen, which can represent a risk of overpressure if the apparatus used creates the substances' containment.
- Ammonium peroxodisulphate detonates when it is heated above 75°C, but there is the same result at ambient temperature in the presence of water, carbon dioxide or sodium peroxide.
- Detonation occurs during the reaction of molten aluminium with ammonium peroxodisulphate in the presence of water. However, since the temperature is above 75°C, the presence of water is sufficient to decompose it and the water/molten aluminium interaction has already been mentioned as being explosive.
- Finally, there is an extremely violent reaction during the action of ammonia on ammonium peroxodisulphate in the presence of Ag^+ ions.

Chlorine

Chlorine is one of the strongest oxidants whether it is in the elementary form or as oxidised anions, with oxidation states of +I (hypochlorites) to +VII (perchlorates). The chloride ion with an oxidation state of -I is very stable (octet electronic structure); only hydrochloric acid is dangerously reactive, linked to its strongly acidic character. This explains the nature of the dangerous reactions which have already been described and have caused a large number of accidents. The accidental aspect is aggravated by the fact that the derivatives mentioned in this paragraph are much used.

The element

Chlorine reacts with all previously mentioned elements and their reduced forms as previously described.

- Chlorine detonates in contact with hydrogen if the mixture is submitted to ultraviolet radiation or if a catalyst is present and if the concentration of chlorine is between 5 and 89% in volume. In the dark such an explosion occurs in the presence of 2% of nitrogen trichloride. This is to be expected, given the instability of nitrogen trichloride.
- Chlorine has caused numerous accidents with metals. Beryllium becomes incandescent if it is heated in the presence of chlorine. Sodium, aluminium, aluminium/titanium alloy, magnesium (especially if water traces are present) combust in contact with chlorine, if they are in the form of powder. There was an explosion reported with molten aluminium and liquid chlorine. The same is true for boron (when it is heated to 400°C), active carbon and silicon. With white phosphorus there is a detonation even at -34°C (liquid chlorine).
- The situation does not improve with mixtures with the 'hydrides' of the elements. Thus, a detonation occurred during contact between water and chlorine due to an accidental spark. Phosphine, silane and diborane all combust spontaneously in chlorine (their behaviour is the same in oxygen). With hydrogenated nitrogenous compounds; ammonia, hydrazine, hydroxylamine, ammonium salts (especially ammonium chloride), and also sulphamic acid (these last two in an acid medium) there is ignition or even detonation.
- In almost all cases of nitrogenous derivatives it seems that nitrogen trichloride or more complex compounds with a nitrogen-chlorine bond are formed. Their instability here is the main factor in accidents. The interpretation is the same as with the accidents involving nitrogenous compounds that were present during the electrolysis of sodium chloride. These led to detonations due to the formation of nitrogen trichloride formed from chlorine, which was the substance of the electrolysis.
- A spark provokes the detonation of a fluorine/chlorine mixture if traces of water are present.
- Chlorine was incorporated into carbon disulphide in a steel reactor. This led to a violent explosion which destroyed the installation and which is probably due to the catalytic effect of steel (formation of iron trichloride).
- In the presence of sodium hydroxide, chlorine converts into sodium hypochlorite (bleach). This is a typical way to eliminate chlorine; it has to be in the gaseous state and slowly incorporated into the base. Such an operation with liquid chlorine gives rise to a violent detonation.
- Sulphuric acid is sometimes used to dry out chlorine. There was a detonation during such an operation.

Hydrochloric acid
- Hydrochloric acid generally reacts violently with electropositive metals such as aluminium or magnesium. With sodium the reaction is slow if hydrogen chloride is anaqueous, and explosive with aqueous acid.
- In contact with fluorine there is spontaneous 'ignition' of the mixture.
- Concentrated sulphuric acid is used to obtain anaqueous hydrogen chloride. There was an accident due to the overpressure created by the gas in a confined apparatus. This drying operation was also tried with silica. The exo-

thermicity of the moist HCl/SiO$_2$ interaction makes the operation hazardous and it is recommended to use anaqueous calcium sulphate as a dehydrating agent.
- Hydrochloric acid is thought to give dangerous reactions with metal carbides. There are no further details regarding risks. The dangers may be linked to the acetylene produced.
- Hydrochloric acid catalyses the explosive decomposition of nitrogen trichloride.

Chlorine dioxide

The enthalpy of formation for this compound is positive. So it is hardly stable thermodynamically and it is to be expected that chlorine dioxide will give dangerous reactions, which are often linked to the catalysis of its decomposition.
- A chlorine dioxide/hydrogen mixture detonates if it is made in the presence of platinum or submitted to a spark.
- It also detonates in the presence of phosphorus, sulphur and carbon monoxide. With a chlorine/phosphorus pentachloride mixture the resulting detonation was explained by the hypothetical formation of chlorine monoxide.

Perchloric acid

Perchloric acid is particularly dangerous because of its exceptionally strong oxidising character and instability, especially in the anhydrous state. Therefore all methods of obtaining this acid in the anhydrous state are particularly dangerous. Metal perchlorates and in particular, magnesium perchlorate, are very strong dehydrating agents. However, their transformation into perchloric acid under the influence of acids, which are sometimes present in the form of impurities, make them dangerous. The main dangerous reactions of this acid involve organic compounds. (See chapter 6.) With inorganic substances the following dangers were mentioned:
- A drop of perchloric acid on carbon or charcoal causes a violent explosion.
- The acid reacts with a number of metals when it is hot. The hydrogen produced forms gaseous vapours with the perchoric acid, which spontaneously detonate at these temperatures.
- It reacts with phosphine by forming an explosive salt:

$$PH_3 + HClO_4 \longrightarrow PH_4^+ ClO_4^-$$

It is assumed that fluorine with perchloric acid at 72% forms the following highly explosive substance:

$$F_2 + HClO_4 \longrightarrow F-ClO_4$$

- Note that this halogen reacts violently with water.
- When it is cold, perchloric acid is thought not to react with sodium hypophosphite, when it is hot the mixture immediately detonates.

Chlorine-oxygen containing salts

Hypochlorites, chlorites, chlorates and perchlorates all represent the same dangers, which are linked to the fact that they are strong oxidants. The danger is not directly linked to the importance of the oxidation state of chlorine atom and this is partly for kinetic reasons. The main factors of the accidents described in the technical literature are not the intrinsic properties of each anion, but rather the frequency with which they are used. So chlorates and perchlorates are more often involved in accidents than hypochlorites and especially chlorites, which are hardly used. Thus the classification below does not provide positive indications about the dangerous properties of each substance mentioned.

Hypochlorites

- Anhydrous sodium hypochlorite detonates spontaneously at ambient temperature. It can only be handled in an aqueous solution. It is interesting to point out that calcium hypochlorite can be used in the anhydrous state (see calcium p.196). However, in an aqueous solution the product instability is not surprising. It slowly decomposes by forming oxygen and creating overpressures in storage containers leading to their detonation.

- With ammonium salts, sodium hypochlorite gives nitrogen trichloride, which detonates spontaneously.

Chlorites

- When it is subjected to an impact or a temperature of 200°C, sodium chlorite decomposes violently.

- In the presence of an acid, sodium chlorite gives off chlorine dioxide, which can detonate spontaneously. Almost all oxygen-containing chlorine salts give rise to such a reaction.

- The sodium chlorite/sodium hydrosulphite mixture combusts spontaneously. This is explained by the fact that sulphur-containing salts catalyse the decomposition of chlorite.

Chlorates, perchlorates

Their properties and many of their uses are very similar and their dangerous reactions are the same. Perchlorates seem more stable than chlorates.

- Sodium chlorate stored in closed containers detonated because of its confinement and the heat given off by a nearby fire. For a reason that was not specified, a stock of 80 tonnes of potassium chlorate detonated. Regarding this accident the authors make the assumption that the decomposition reaction of this compound would be the following:

$$KClO_3 \longrightarrow Cl_2 + O_2$$

- Chlorates and perchlorates react violently with metals when they are exposed to heat (the salt is then in molten state), a flame, a spark, friction or impact. The ignition often involves very spectacular blinding flames. The mixtures can detonate; the conditions of the reaction depend on how the interaction is

5 • Reactions of inorganic chemicals

Chlorine

Synopsis of dangerous reactions of chlorine and its derivatives

Chlorine	Chloric acid	Perchloric acid	Alkaline chlorates	Alkaline chlorites	Sodium chloride	Chlorine dioxide	Alkaline hypochlorites	Ammonium perchlorate	
■		■				■			Hydrogen
■		■							Group I (Li; Na; K)
■		■							Group II (Be; Mg; Ca; Sr; Ba)
■		■							Group III (B)
■		■						■	Group IV (C; Si; Ge)
■		■							Group V (N; P; As; Sb)
■		■							Group VI (O_2; S; Se; Te)
		■							Group VII (F_2; Cl_2; Br_2; I_2)
■		■				■			Transition metals
■		■							Other metals (Al...Bi, Sn, Pb)
		■							Hydrides group III (B_xH_y)
■		■							Hydrides group IV (SiH_4...)
■		■							Hydrides group V (NH_3; H_2NNH_2; NH_4^+; NH_2OH)
■									Water
■									Hydrides group VI (H_2S...)
									Hydrogen halides
									Beryllium halides
									Boron halides
									Metal halides ($AlCl_3$; $FeCl_3$...)
■		■							Non-metal halides ($SOCl_2$; S_xCl_y; P_xX_y; NCl_3)
		■							Metal oxides (SiO_2)
		■							Non-metal oxides (O_3; N_xO_y; P_xO_y; SO_x; CO_3^-)
		■							Peroxides (H_2O_2; Na_2O_2)
									Nitric acid
		■							NO_3^-; NO_2^-
		■							N_3^-; CN^-
									Sulphuric acid
		■							Non-oxidant salts of acids (SO_x^{x-}; PO_x^{x-}; CO_3^-)
									ClO_x^-; $HClO_4$
									MnO_4^-, $Cr_2O_7^{2-}$
			■						Instability
									Polymerisation
■		■							Bases
■		■							Acids
■									Carbon sulphide

caused. The cited metals are magnesium as well as aluminium, on its own or combined with aluminium trifluoride, which plays a catalytic role, or combined with titanium dioxide, boron, silicon.

- The same goes for carbon (the accident was caused because carbon was used instead of manganese dioxide, by mistake), sulphur and phosphorus. There was a detonation with carbon. With phosphorus the detonation occurred once the carbon disulphide used to dissolve phosphorus vapourised; red phosphorus behaves the same way. The same happened with the potassium chlorate/sodium nitrate/sulphur/carbon mixture, which led to a violent detonation as well as with the potassium perchlorate/aluminium/potassium nitrate/barium nitrate/water mixture. In the last case the explosion took place after an induction period of 24h.

One can intuitively forecast the dangerous character of these mixtures; in fact, they are pyrotechnic.

All reducing agents can give rise to dangerous reactions with chlorates. The following accidents illustrate this.

- When hot, ammonia and compounds, which contain nitrogen-hydrogen bonds; eg ammonium salts and cyanides react violently with chlorates and alkaline perchlorates. Diammonium sulphate, ammonium chloride, hydroxyl-amine, hydrazine, sodamide, sodium cyanide and ammonium thiocyanate have been cited. So far as hydrazine is concerned, the danger comes from the formation of a complex with sodium or lithium perchlorate, which is explosive when ground. Many of these interactions are explosive but the factors which determine the seriousness of the accident are not known.

Finally, various dangerous reactions with different properties are:

- The action of sulphuric acid on chlorates produces chlorine dioxide and possesses the dangerous characteristics of this substance. Incorporating a drop of sulphuric acid into a mixture for Bengal lights produces spectacular results. The acidic character of sulphur dioxide would be sufficient also to convert chlorates into chlorine dioxide, but a temperature of 60°C is needed to do so.

- Fluorine causes a detonation in contact with potassium chlorate. This is put down to the formation of the following explosive:

$$KClO_3 + F_2 \longrightarrow FClO_4$$

- Ammonium perchlorate has, in addition to the properties mentioned for the previous compounds, a specific instability due to its unfavourable oxygen balance. Friction is enough to make it detonate. Combined with aluminium it has been used as a propellant for rockets. In the presence of carbon and metal salts, it reacts exothermically below 240°C and detonates above this temperature.

Sodium chloride

This compound is perfectly stable. In addition to the accident described on p.187, which happened during electrolysis and was caused by chlorine, there was another accident, which is not really linked to a chemical reaction.

- This is the detonation caused by the incorporation of water into melted sodium chloride heated at 1100°C.

- The only dangerous reaction that was cited deals with the following violent reaction:

$$NaCl + Li \longrightarrow Na + LiCl$$

This reaction shows that it is impossible to put out lithium fires with sodium chloride although it is an extinguishing agent for metal fires.

Argon

- The argon/liquid nitrogen mixture is dangerous because argon solidifies in liquid nitrogen. At minus 196°C solid argon can remain in containers for a very long time, but if they are closed, the vapourisation of solid Ar can cause the apparatus to detonate by overpressure.
- During a drying operation a detonation occurred, which was probably due to an impurity of argon.

Potassium

Potassium has similar properties to sodium, but its reactivity is higher. The same is true for the sodium/potassium alloy. One illustration of this feature is the fact that it is not allowed to destroy potassium with methanol or ethanol as is often the case with sodium; the interaction between primary alcohols and potassium cannot be controlled. Bretherick suggests tert-butanol, which is far less reactive. This suggestion is quite surprising since the potassium tert-butylate formed is so dangerous that this operation only changes the nature of the risk. Thus potassium represents more risk than sodium (but creates fewer accidents since it is hardly used). It is one of the strongest reducing agents.

The metal

Action of water

- Potassium reacts with water like sodium, but more violently. Small pieces of potassium incorporated into a very small amount of water cause a violent release of hydrogen, which combusts. In the same conditions big pieces cause a violent detonation.

Oxidation by oxygen and peroxides

In the absence of humidity potassium can not oxidise even in the molten state.

- However, water traces can rapidly oxidise the metal and the exothermicity can cause its fusion followed by its ignition. There are violent detonations when potassium suspended in benzene is exposed to oxygen.
- In the presence of air, there is very slow oxidation, which forms a layer of potassium superoxide as follows:

$$K + O_2 \longrightarrow KO_2$$

Underneath this layer of superoxide a layer of oxide K_2O is formed. The interaction between potassium and its superoxide is violent and causes the metal to combust; the intermediate layer acts as a protection. Contact between both compounds happens when the metal is cut. This slow oxidation is avoided by keeping potassium under anhydrous xylene. It is possible that many dangerous reactions may be in fact due to the exceptional reactivity of the superoxide.

- Accidents with violent detonations were reported, which involved sodium or potassium peroxides with metal as well as hydrogen peroxide.

Action of carbon and derivatives

- Potassium violently reacts with carbon in most of its allotropic forms. Thus the action of molten metal on graphite results in the formation of some kind of a 'carbide':

$$8C + K \longrightarrow KC_8$$

- In the presence of air, the carbide combusts spontaneously. With water this substance detonates. Another accident involved powdered amorphous carbon mixed with potassium and heated to a high temperature. Ignition followed by detonation was attributed to the presence of the superoxide on the metal surface. If the suggested interpretation is correct, this accident illustrates the comment made previously.

- There are numerous consequences of the reactivity of potassium with carbon. First of all, it is impossible to prepare this metal by electrolysis of melted potassium chloride with the help of graphite electrodes as is the case for sodium; otherwise this causes a detonation. Secondly, it is extremely dangerous to put out potassium fires with extinguishing agents made with graphite, which are commonly used for sodium fires ('Marcalina' powders, 'Graphex CK 23'). Detonations during such attempts have been mentioned. Note that Sax surprisingly recommends the use of graphite to put out potassium fires before mentioning the dangers, which result from potassium/graphite interaction.

- With carbon monoxide at -50°C the danger is thought to come from the formation of dicarbonyl potassium as follows:

$$K + 2CO \longrightarrow K(CO)_2$$

The carbonyl detonates if it is heated to a temperature of 100°C or at ambient temperature, in contact with air or water.

- With carbon dioxide in the solid state the mixture detonates on impact. Therefore, with graphite, carbon dioxide cannot be used as an extinguishing agent for potassium fires. The slow reaction of potassium with gaseous carbon dioxide at ambient temperature gave rise to an accident. Potassium was stored in an aluminium container in a laboratory in contact with carbon dioxide; the formation of potassium carbonate caused the corrosion of the container, which caused potassium to combust on contact with air.

Action of halogens and halides

- In contact with fluorine or dry chlorine, potassium combusts immediately.

5 • Reactions of inorganic chemicals

Potassium

- Anhydrous hydrogen chloride in contact with this metal causes a violent detonation.
- Boron trifluoride, sulphur and disulphur dichlorides, phosphorus trichloride in the liquid state cause potassium to combust. The same is true for phosphorus pentachloride in the solid state. In the latter case the same accident happened with gaseous halide. The same is also true for the bromide analogues of these compounds.
- With dichlorine oxide there is a highly violent detonation.
- Potassium reacts very dangerously with fluoride and aluminium chloride as well as with silicon tetrachloride.

Action of nitrogenous compounds
- With the compounds that contain nitrogen-hydrogen bonds there are highly reactive or unstable metal compounds formed. Thus with hydrazine :

$$K + H_2N - NH_2 \longrightarrow H_2N - NHK$$

is obtained.
- The reaction is explosive. The reaction is also dangerous with ammonia combined with phosphine or in the presence of sodium nitrite (in the latter case, the danger may as well come from the ammonia/nitrite interaction).
- Oxygen-containing derivatives are strong oxidants with which potassium will react violently. The detonation characterises the action of nitric acid, nitrogen tetroxide and nitrogen pentoxide at 20°C. With nitrogen dioxide or monoxide there would be an ignition. Ammonium nitrate creates a detonation.

Action of sulphur, phosphorus and their compounds
- Sulphur as well as phosphorus gives highly violent reactions when they are heated with potassium. With carbon disulphide, potassium creates a mixture, which detonates on impact. The same happens by heating. With sulphur dioxide, potassium in the molten state combusts. Finally, the potassium/sulphuric acid mixture gives rise to detonations.
- With phosphorus pentoxide, potassium becomes incandescent.

Potassium carbonate
- When a 1:1 carbonate/magnesium mixture was heated, this gave rise to a detonation. This accident was explained by the formation of carbonyl potassium.
- With silicon the interaction is highly violent.

Potassium hydroxide
- When hydroxide dissolves in water it gives off a lot of heat. If this operation is carried out without care the effects can be particularly dangerous. Hydroxide pellets were used. Then a stirrer was switched on. A very violent eruption threw out part of the reactor's content all over the surrounding area.
- Potassium hydroxide catalyses the explosive decomposition of nitrogen trichloride or chlorine dioxide.

- Moist potassium peroxodisulphate was accidentally exposed to traces of potassium hydroxide. It combusted spontaneously, causing the installations to catch fire. It is impossible to put out a fire involving this peroxydic compound with carbon dioxide or extinguishing powders although these agents are suitable for fires of chemical substances. Only water, which is usually not recommended in this case, can put out this type of fire.

Calcium

Calcium is a very electropositive metal, which is more reactive than magnesium. The reactivity of the metals listed in this group of the periodic table increases when we go down the group and such is the case for alkali metals, which are more reactive than the alkaline earths. The main danger from calcium is the ignition of this metal. It is, like all electropositive metals, very difficult to quench when it burns. One needs special extinguishing agents to put out any fire. For salts, the dangerous reactions are linked to the behaviour of the anion. In the non-ionic compounds the dangerous reactions are not linked to calcium. The table of dangerous reactions is very similar to what we have already seen.

The metal

Action of water and hydroxides

- Calcium reacts violently with water, releasing hydrogen, which can combust or even detonate.

Mixtures of calcium with lithium or sodium hydroxide sometimes detonate.

Reactions with elements

- Above 300°C calcium can 'combust' violently in an atmosphere of hydrogen.
- It also reacts violently with lithium in the molten state as well as with sodium, causing emissions of flaming metal. The reaction takes place with pure sodium as well as with sodium combined with calcium or sodium oxide.
- When it is hot, in the compact state or at ambient temperature in the powdered state, calcium can combust in air. It also reacts violently in air with sulphur.
- Pieces of calcium combust spontaneously in contact with gaseous fluorine. With chlorine the result is the same with powdered calcium.
- Finally, at high temperatures (~1000°C) calcium reacts violently with silicon after a short induction time. The products incandesce.

Dangerous reductions of metal oxides

- Calcium powder heated in a nitrogen and carbon dioxide atmosphere combusts violently.
- It also reacts with nitrogen oxides. It combusts if it is heated in nitrogen oxide. It detonates in nitrogen tetroxide at ambient temperature. The same happens with phosphorus pentoxide.

- Finally, it reacts dangerously with metal carbonates; attempts to put out calcium fires with powders containing alkaline carbonates give rise to explosions.

Other dangerous reactions
- With ammonia, when it is hot, calcium becomes incandescent. When it is cold, in liquid ammonia, a harmless solution is created. However, the evaporation of ammonia leaves calcium powder, which is extremely reactive. It combusts violently and instantaneously in contact with air.
- There have not been any accidents recorded that have mentioned metal halides (except with 'interhalogens', which are very reactive anyway). Nevertheless, it seems important that calcium should not be in contact with such halides without taking any precautions. It was noticed that calcium starts glowing when it is heated with boron trifluoride.
- Finally, there was a detonation caused by the interaction of molten calcium with asbestos cement (complex silicate).

Calcium compounds

Dangerous reactions involving atoms or atomic groups linked to calcium, whether they are ionic (anions) or not, occur.

Anions and analogues with oxidising properties
- Calcium hypochlorite, which is a strong halogenation agent generating Cl^+ ions creates dangerous reactions (already discussed on p.189) with the compounds that contain hydrogen-nitrogen bonds. Indeed, they lead to nitrogen trichloride. An accident occurred after mixing the hypochlorite with calcium hydroxide. The hypochlorite is relatively stable, much more stable than magnesium hypochlorite. It is assumed that the latter was responsible for detonations of containers filled with calcium hypochlorite and magnesium oxide. It was also suggested that this compound leads to explosive dichlorine oxide by hydrolysis. Indeed, accidents occur after repeated openings of storage containers of calcium hypochlorite, which might have made this hydrolysis possible. Finally, detonation occurs when the hypochlorite is mixed with sulphur (in the presence of humidity) and carbon powder (when hot).
- Calcium permanganate, with hydrogen peroxide, ignites immediately. This phenomenon is used to ignite certain types of rockets.
- Calcium chromate, which contains 20% of boron is a composition used in pyrotechnics.
- Calcium chlorate reacts dangerously as do metal chlorates. This occurs with interactions of this compound with carbon, aluminium, metal sulphides, sulphur and phosphorus.
- Calcium sulphate is reduced violently by aluminium. In the presence of this sulphate, red phosphorus, which was previously treated by potassium nitrate and calcium silicide, combusted violently.
- When heated in the presence of magnesium and hydrogen, calcium carbonate gives rise to a highly violent reaction.

5 • Reactions of inorganic chemicals

Calcium

Reducing anions

- Calcium sulphide creates very violent detonations with chlorate or potassium nitrate.
- When it is heated to 300°C in the presence of oxygen, calcium phosphide becomes incandescent.
- In contact with dichlorine oxide, calcium phosphide heats up, causing explosive decomposition of the chlorinated compound.

Other dangerous reactions

- Calcium halides are relatively stable; there is the possibility of a violent reaction in the presence of a more electropositive alkaline metal; eg detonation on impact of a mixture of calcium bromide and potassium. Calcium chloride has a very high enthalpy of solution in water. When dissolved in large quantities in hot water this causes the solution to boil vigorously, creating emissions.
- When mixed with potassium chlorate calcium dihydrogenphosphate detonates as violently as with decomposition of nitroglycerine. It is probably the result of the explosive decomposition of chlorine dioxide, which is formed because of the presence of acid radicals in the phosphate.
- The dangers presented by calcium carbides are usually linked to the formation of acetylene, due to humidity (see acetylene on p.235). When handling the carbide with a steel tool, which generated a spark, this caused the acetylene in the container to combust. If calcium carbide is heated in a container under air, it strongly occludes the nitrogen present, making the ambient atmosphere richer in oxygen to such an extent that it can cause the acetylene to combust. There have been numerous detonations of acetylene generators as well. In addition, the heating in the presence of chemicals as varied as chlorine, magnesium and hydrogen chloride causes incandescence of calcium carbide. With sodium peroxide the detonation is immediate. Sulphur combusts at 500°C.
- Accidents occur with calcium cyanamide, which are due to the presence of 2% of calcium carbide.
- Calcium oxide hydrates very exothermically. A large number of accidents involving this hydration have been described. If a quantity of water corresponding to 33% of the weight of oxide is added, the temperature can reach 150-300°C, causing the water to suddenly vapourise. In certain situations of confinement the overpressure due to vapour is enough to blow up the installation. Temperatures of 800-900°C were recorded during such accidents. Calcium oxide has to be stored in plastic containers. Incandescence of oxide in contact with liquid hydrogen fluoride occurs. Finally, when calcium oxide was heated with phosphorus pentoxide this led to a very violent reaction. The exothermicity of this reaction causes phosphorus oxide to vapourise. However, these compounds do not seem to react when they are cold.
- Violent combustion of calcium carbonate placed in a fluorine atmosphere has been recorded.

Scandium

This metal is very electropositive. In the powdered state it is extremely inflammable in the presence of a flame or simply when heated. It reacts violently with halogens and oxidants.

Titanium

Titanium is a highly reducing metal. It gives rise to numerous violent reactions, which lead to many accidents. The metals in group I (alkali) and II (alkaline earth) are the only ones that are more reducing than titanium. This property plays a role in all the accidents listed below. Carbide, hydride and titanium trichloride also have reducing properties and cause accidents. Titanium oxide is very stable. The metals in groups I and II are the only ones with dangerous reactions with this oxide.

The metal

All the reactions below involved powdered titanium.

- It is very difficult to handle since it combusts spontaneously in air. The fire is difficult to put out and requires special extinguishing agents. With liquid oxygen the reaction leads to a violent detonation. The surface texture of titanium seems to be an important element in the violence of the effects obtained.
- Titanium is regarded by many authors as the only element that burns in nitrogen where it changes into nitride. This is not exactly correct since the same goes for lithium. Sometimes the combustion is capable of detonation.
- It also reacts with carbon. There are cases of spontaneous ignition with titanium/powdered carbon mixtures prepared for the purpose of producing titanium carbide.
- With chlorine the ignition occurs when it is heated to 350°C. The controlled reaction leads to titanium tetrachloride (the trichloride is prepared from tetrachloride by the action of silver or zinc).
- Titanium burns in carbon dioxide, when pure or diluted by nitrogen. If there is titanium dust in suspension in carbon dioxide, the mixture detonates when it is exposed to a flame or heat.
- Titanium becomes incandescent if it is mixed with alkaline carbonates or alkaline nitrates. But in this latter case if the cation is ammonium, there is a detonation. However, detonation of a device containing potassium nitrate into which powdered titanium had been thrown has been observed.
- When it was heated to 700°C in a mixture of water vapour and air, titanium caused a violent detonation due to the hydrogen formed. Titanium hydride might have formed, which also detonates in air. When it is in contact with hydrogen peroxide, it gives rise to an immediate detonation at ambient temperature.
- With alkaline chlorates, the metal glows when it is heated. Finally, with nitric acid containing less than 1.5% of water and more than 6% of nitrogen tetroxide, the mixture obtained detonates on impact.

Derived substances

- Powdered titanium carbide detonated spontaneously in oxygen. It was thought that the detonation was caused by a spark of static electricity.
- Titanium dioxide is reduced violently by lithium, sodium, potassium, magnesium, calcium and aluminium ('thermite' type of reaction).
- Fine powdered titanium trichloride combusts spontaneously in contact with air or water.
- It gives a very violent reaction with potassium or hydrogen fluoride above 500°C.

Vanadium

Since they are hardly ever handled, vanadium and its derivatives, apart from vanadium pentoxide, give rise to a small number of accidents.

The metal

- In the powdered state, the metal, which is obtained by the action of magnesium on vanadium trichloride, combusts spontaneously in air.
- Liquid chlorine in contact with powdered vanadium causes the mixture to detonate.

Vanadium compounds

- Vanadyl chloride is very hygroscopic. Its high exothermic hydration can cause such temperature rises that water is violently vapourised.
- When sodium or potassium is added too quickly to vanadium chloride, this gives rise to a violent detonation. If it is done without caution, the reaction is violent.
- Divanadium trioxide combusts spontaneously in air. For some authors this behaviour is observed at ambient temperature, for others, heating is necessary.
- Divanadium pentoxide is reduced very violently by lithium at 394-768°C. With calcium containing sulphur in the form of impurities, in the presence of water traces, a violent combustion is observed.

Chromium

Apart from chromium and chromium (II) salts, which are reducing, the table of accidents involving chromium derivatives is mainly one of very reactive oxidants. The effect of this dangerous property is worsened by the fact that the cited substances are commonly used, not only in industry (especially in surface treatment) but also in analysis, research and training laboratories.

Finally, there are halides whose dangerous reactions are partly linked to the reactivity of the Cl⁻ anion.

Chromium and chromium (II) salts: reduction reactions

Compact chromium is not dangerous. The metal powder ('pyrophoric') obtained by the amalgam evaporation is the only one that represents risk. As far as chromium

(II) salts are concerned (not mentioned in Part Three due to absence of data), they are hardly used compared with the compounds with 'normal' valencies (III and VI). They give rise to a limited number of accidents.

Chromium metal and chromium (II) salts

- Chromium causes various accidents with sulphur dioxide or nitrogen oxide (incandescence), carbon dioxide (with which it forms explosive suspensions), hydrogen peroxide (with which it detonates at ambient temperature) and ammonium nitrate (with which it detonates when it is heated to 200°C).
- Chromium (II) salts reduce water into hydrogen. This gradual transformation at ambient temperature has been responsible for accidents of the same nature. A glass bottle, in which chromium dichloride was stored for several years, detonated spontaneously. A sealed tube, in which there was chromium (II) sulphate heptahydrate and water in excess, detonated after being stored in darkness for a year. In both cases the accidents were interpreted as the result of water reduction, forming hydrogen whose pressure eventually caused the explosion of the containers.

Oxidation reactions of chromium (III) derivatives and chromium (VI)

Chromium oxides

- Chromium (III) oxide gave rise to a very violent reaction with lithium heated to 180°C. The temperature exceeds 1000°C. It is a typical 'thermite' reaction during which a more electropositive metal violently reduces an oxide by forming a metal, which is chromium in this particular case. However, the mixture needs to reach a high temperature to be able to react.

    Chromium (VI) oxide proves to be more dangerous.

- Some technicians thought they had to reinforce the oxidising property of this oxide by combining it with nitric acid or metal nitrates. In these conditions the oxidations proved to be uncontrollable, often leading to violent detonations.
- The contact of sodium or potassium causes incandescence of the mixture.
- The interaction between ammonia in an aqueous solution and oxide is very exothermic and has to be managed carefully. If ammonia is in the gaseous state, the exothermicity of the interaction causes incandescence of the oxide, causing serious accidents.
- With even more reducing or unstable nitrogenous compounds such as sodamide (when the mixture is ground) or hydrazine, the substances detonate.
- Sulphur, hydrogen sulphide and metal sulphides also led either to the ignition of the sulphurous substance (S and H_2S) or the incandescence of the mixture. Note that sulphides and cyanides are amongst the most dangerous salts in the presence of oxidants, a fact not to be ignored.
- Molten chromium trioxide (196°C) causes phosphorus to combust.
- Chromyl chloride also has a dangerous oxidising property, which makes it behave in a similar way to chromium trioxide. This goes for ammonia (incandescence), hydrogen sulphide, phosphine, sulphur and disulphur dichloride

(ignition), sodium azide, moist phosphorus, phosphorus trichloride (with which one drop of chromyl chloride is sufficient for detonation) and finally calcium sulphide (highly violent reaction).

Chromates and dichromates

They are strong oxidants especially when they are combined with sulphuric acid. Ammonium dichromate or chromate (less used) have an oxygen balance ($C_3 = 0$ g% for the first one) that makes them likely to be very unstable thermodynamically. This is confirmed by the labelling code '1' (explosive in the dry state) provided by the regulations and attested by a number of accidents.

Ammonium dichromate decomposes at a temperature starting at 170-190°C to give the following reaction:

$$(NH_4)_2Cr_2O_7 \longrightarrow Cr_2O_3 + 4\ H_2O + N_2$$

The enthalpy of decomposition, which is rather low compared with criterion C_1 in the CHETAH programme (see p.117), was determined by DSC and corresponds to 0.76 KJ/g at 230-260°C. In a confined atmosphere the decomposition conditions would be explosive. This reaction has been used many times because of its spectacular aspect. This is why it is called 'Vesuvius fire' by artificers and 'green volcano' by lecturers. These demonstrations should be stopped because of the detonation risks they represent as well as chromium (III) oxide toxicity.

- Boron and silicon, whether they are on their own or combined, are used with sodium or potassium chromate to make pyrotechnic mixtures.
- Hydrazine and hydroxylamine give explosive or inflammable reactions with all chromates and dichromates.

Chromium halides

This type of dangerous reaction are not necessarily (except in one case) linked to the chromium atom but rather to the halogen atoms.

- Therefore chromyl chloride, already mentioned on p.200 for its oxidation reactions, hydrolyses very violently forming hydrogen chloride.
- Alkali metals and especially potassium cause their mixtures to detonate with chromium trifluoride and trichloride. There was also a detonation involving the violent combustion of lithium in contact with chromium trichloride in a nitrogen atmosphere. However, it should be noted that lithium (as well as titanium) is the only alkali metal which can burn in nitrogen. So the chloride implication is not demonstrated in this last case.

Manganese

Like all the transition metals already analysed manganese is a very reducing metal. The same goes, but to a lesser extent, for its oxide with an oxidation state of +II. Its most common combinations are oxidants, in which the oxidation state of manganese is between +III and +VII. Thus they will give rise to dangerous oxidations as in hydrogenated derivatives of nitrogen, metals and some metalloids. Manganese halides will behave in the same way as chromium halides.

The metal

- Powdered manganese combusts easily in air and burns very quickly. It forms dust suspensions in air, giving rise to explosive mixtures.
- The same is true for manganese powder heated in a carbon dioxide, sulphur dioxide or nitrogen oxide atmosphere. The combustion is extremely violent in these three cases.
- It creates deflagrations with nitric acid. When it was mixed with molten ammonium nitrate and heated to 200°C, violent explosions were observed.
- With fluorine, at ambient temperature and with chlorine, when it is hot, it combusts and becomes incandescent. Similar incandescences were also noted with phosphorus.
- It causes the explosive decomposition of hydrogen peroxide.

Oxidation reactions of permanganates and manganese (IV) oxide

Potassium permanganate is dangerous when it is in the form of fine powder. Almost all the reactions below involve this form.

- Potassium permanganate creates violent combustion when it is heated with carbon powder and an explosion in the same conditions with titanium. It causes violent explosions when it is ground with phosphorus or sulphur.
- When it is mixed with small quantities of sulphuric acid and in the presence of water traces, it is thought to form permanganic acid, $HMnO_4$, which is very unstable and a violent oxidant. Thus, this system incandesces in the presence of ammonia. Mixtures of potassium permanganate /peroxomonosulphuric acid have also been used to measure sulphur in the presence of carbon. However, this is not recommended since it is highly likely to detonate.
- The reaction of potassium permanganate with an aqueous solution of hydrogen fluoride at 40% enables one to prepare potassium hexafluoromanganate without danger. However, this reaction becomes dangerous when one uses solutions of hydrofluoric acid at 60-90%. With hydrogen chloride the reaction is explosive and it is thought that its danger is linked to the following reaction:

$$HCl \xrightarrow{KMnO_4} Cl_2 \xrightarrow{KMnO_4} Cl_2O \dashrightarrow \text{Explosion}$$

Another view is that the explosion is due to the following reaction:

$$HCl + KMnO_4 \longrightarrow Cl - MnO_4$$

There was an explosion when sulphuric acid was in contact with a mixture of potassium choride (forming hydrogen chloride) and potassium permanganate.

- Potassium permanganate gives rise to a very violent reaction with hydrogen peroxide. It causes the ignition of anhydrous hydroxylamine and the explosion of hydrazine. A quantity of 0.5% of permanganate was incorporated into ammonium nitrate. It detonated seven hours later. It was assumed that this detonation was due to the decomposition of ammonium permanganate, which is hardly stable (probably because of its oxygen balance).

Manganese dioxide is also a strong oxidant.

- It reacts violently when it is heated with aluminium. It causes hydrogen sulphide to combust. It decomposes hydroxylamine chlorhydrate, hydrogen peroxide (detonation) and potassium azide violently.
- It is used to make oxygen in laboratories. One has to heat a mixture of manganese dioxide and an alkaline chlorate. This reaction, which is known for not being dangerous and often used in lectures, gives rise to anhydrous manganese dichloride which reacts explosively with potassium or zinc. Detonation of the mixture is known to have led to several accidents.
- There was a detonation of a mixture of manganese oxide and peroxomonosulphuric acid at 92%.

Manganese (III) derivatives

- Manganese (III) oxide reacts violently with hydrogen peroxide.
- Anaqueous manganese dichloride reacts explosively with potassium or zinc.

Iron

The metal

- There are 'pyrophoric' forms of iron, which combust spontaneously in air if they are heated above 180°C.
- Iron combusts spontaneously in liquid fluorine. With chlorine, heating is necessary to cause thread iron to combust.
- It combusts in pure hydrogen peroxide or when containing traces of potassium permanganate.
- The iron/potassium perchlorate mixtures react violently with each other. The exothermicity of the reaction is sufficient to melt part of the iron. This reaction is used to weld metal parts.
- The same goes for almost all oxidants and especially molten ammonium nitrate, sodium dichromate (temperatures reach 1100°C), ammonium dichromate in concentrated solution, sodium peroxide (at 240°C, if there is friction) and dinitrogen tetroxide. In all these cases iron combusts in contact with these oxidants.
- Iron reacts with water; the reaction is catalysed by rust and forms hydrogen. The exothermicity of the reaction can cause hydrogen to combust.
- There was a surprising fatal accident during which an iron container with magnesium chloride (probably moist) detonated. It was thought that the magnesium salt had catalysed the interaction between the metal and water.

5 • Reactions of inorganic chemicals

Iron

Iron oxides

Iron (II) oxide is a strong reducing agent. Iron II/III or III oxides are oxidants. The danger of the latter can also result from their catalytic behaviour vis-à-vis substances, which are hardly stable.

Iron (II) oxide
- Iron (II) oxide and especially that made by reducing the other oxides, combusts spontaneously if it is heated to 200°C. It also strongly catalyses the combustion of carbon in air. This behaviour can explain the spontaneous inflammable property of the products of burning iron oxalate, which contain this oxide and carbon. When they are placed on the hand and thrown into the air, they form very spectacular showers of sparks. It combusts in contact with liquid oxygen in the presence of carbon.
- This oxide reduces sulphur dioxide when the mixture is heated. The oxide glows during the reaction. The same happens in contact with nitric acid and hydrogen peroxide.

Iron (III) oxide
- With very electropositive metals this oxide is reduced very violently following thermite type of reactions. Violent reactions of this type happen with lithium, magnesium, aluminium and the Al-Mg-Zn alloy. The iron formed is melted due to the exothermicity of this reaction. This experiment is not recommended for lectures.
- This oxide catalyses the violent or even explosive decomposition of hydrogen peroxide. This reaction explains the numerous accidents mentioned involving the contact of hydrogen peroxide with rusted iron. Two accidents of this nature dealt with mixtures of hydrogen peroxide with ammonia and an alkaline hydroxide The detonations took place after a period of induction of respectively several hours and four minutes. Iron (III) oxide also catalyses the explosive decomposition of calcium hypochlorite.
- When it was heated at 150°C with carbon monoxide, the mixture detonated violently. It was assumed that the accident was due to the formation of unstable iron pentycarboxyl.
- Finally, this oxide gives rise to a very violent reaction with calcium disilicide.

Triiron tetroxide
- Triferric tetroxide gives rise to a highly violent detonation when it was heated with calcium silicide combined with aluminium and sodium nitrate.
- When tetroxide is combined with aluminium and sulphur, it is a compound used in incendiary bombs.

Iron sulphides

These compounds are not listed in the tables in Part III since there is no safety information about them. Like any other sulphide this does not make them less dangerous especially regarding oxidants. The sulphur obtained by the ignition of an iron/sulphur mixture has the FeS formula whereas natural sulphides (in particular, pyrites) have the FeS_2 formula.

- Synthetic sulphide combusts spontaneously in air when it is impure and has caused numerous accidents. The same goes for pyrites in the powdered state and that contain humidity traces. In the absence of humidity, powder pyrites can combust on heating. The temperature of this ignition can be dramatically lowered if it contains carbon.
- Synthetic sulphide reacts violently with lithium at 260°C. The temperature then rises above 900°C.
- Pyrites are likely to catalyse the explosive decomposition of hydrogen peroxide.

Hexacyanoferrates

These possess the dangerous reactions of the CN^- ion, or cyano group. The hexacyanoferrate (III) anion also has oxidising properties. Finally, it is thought to produce an extremely unstable acid in certain conditions. These salts also produce hydrogen cyanide, which is highly toxic, in an acid medium (see p.334).

- Thus in hydrochloric acid medium, potassium hexacyanoferrate (III) is thought to give cyanoferric acid, which has a very endothermic enthalpy of formation (ΔH_f = 534.7 kJ/mol; 2.48 kJ/g), explaining the explosive property.
- When it is heated above 196°C, the mixture of this compound with chromium trioxide combusts. When it is submitted to friction or impact, the same mixture detonates violently. The same happens if it is heated with sodium nitrite. Nitrite gives rise to a detonation with potassium hexacyanoferrate (II) too. The dangerous site of these complex anions is the cyano group.
- The oxidising property of this salt appears in the very violent oxidation of ammonia.

Iron chlorides

- These give rise to the same dangerous reactions as chromium halides. Iron dichloride and trichloride detonate in contact with sodium or potassium. There is also a violent combustion of an iron trichloride/calcium carbide mixture, which leads to the formation of melted iron.

Cobalt

There are only a few accidents reported which involve cobalt and its derivatives. The reason must be because compounds of this metal are hardly ever used. The accidents described are similar to the previously mentioned analogous transition metals.

- Finely powdered cobalt can detonate spontaneously in air. Its behaviour depends on its surface texture. Raney cobalt is much more dangerous than Raney nickel, which is more commonly used (see nickel).
- Cobalt reacts violently or even explosively with ammonium nitrate at a temperature close to its melting point (~200°C).
- Cobalt (II, CoO) and (III, Co_2O_3) oxides catalyse the explosive decomposition of hydrogen peroxide.

- Cobalt dichloride and dibromide react highly violently with sodium and potassium.
- Cobalt nitrate has the typical dangerous reactions of metal nitrates. Two accidents during which violent detonations occurred were reported. One of them happened when a mixture of this nitrate with powder carbon was ground up. The other took place when a tetraammonium hexacyanoferrate/cobalt nitrate mixture was heated to 220°C. It typifies incompatibility of the cyano group with oxidants.
- Tripotassium hexanitrocobaltate is a highly unstable compound, which tends to detonate easily under specific circumstances. This happened during an evaporation attempt of this salt in an aqueous solution. Its detonation was explained by the instability of the dry compound.

Nickel

There are only a small number of dangerous reactions of nickel and its derivatives. The metal proves to be dangerous in the pulverised state and with Raney nickel (prepared by the action of sodium hydroxide on a 50:50 Ni-Al alloy); aluminium is dissolved by hydroxide, which leaves nickel in a very porous state.

- Nickel powder gives rise to dangerous reactions, which has led to accidents with potassium perchlorate (ignition), with chlorine at 600°C (ignition) and with ammonium nitrate at about 200°C (detonation). It catalyses the explosive decomposition of hydrogen peroxide.
- Raney nickel is a hydrogenation catalyst. It easily occludes hydrogen. The subsequent gas desorption can cause its ignition. When this metal is incorporated into a liquid mixture of oxygen and hydrogen, it combusts. When it is mixed with magnesium silicate, it creates a catalyst, which combusts spontaneously in air. When it is mixed with sulphur, it heats up and glows. Finally, an accident was reported which involved a mixture of Raney nickel and water that was heated in an autoclave. A strong release of hydrogen caused the pressure to increase dramatically, reaching 1000 bar.
- Nickel cyanide mixed with magnesium causes the incandescence of magnesium. It is a dangerous reaction, which is typical of the cyanide ion in which the cation does not play a crucial role.
- In contact with fluorine, when it is cold, nickel oxide glows. It reacts violently with hydrogen peroxide (catalysis of its decomposition?). Finally, in contact with a mixture of hydrogen sulphide and air, it glows and causes this gaseous mixture to detonate.
- Nickel fluoride and especially chloride react explosively with potassium.

Copper

The metal

Most dangerous reactions of this metal consist in reactions which lead to its oxidation although it is far less reducing than the metals, previously described.

5 • Reactions of inorganic chemicals — Copper

- Copper reacts exothermically with chlorine dissolved in heptane at 0°C. The compounds combust during the reaction. The same happens when dry gaseous chlorine is in contact with the metal in fine leaf form.
- A violent detonation took place when an attempt was made to synthesise a sulphide mixture of aluminium and copper (I) according to the following reaction:

$$Cu + Al + 2S \longrightarrow CuAlS_2$$

- Copper wire, which when heated to 121°C combusts in fluorine.
- Copper catalyses the explosive decomposition of hydrogen peroxide.
- It violently reacts with strong oxidants such as ammonium perchlorate (ignition in contact with copper pipes), alkaline chlorates (detonation by heating, impact or friction; powdered copper), ammonium nitrate (detonation; molten ammonium nitrate and powdered copper) and potassium superoxide (copper glows).
- Copper glows in an atmosphere of air and hydrogen sulphide.
- Lecturers often demonstrate the reducing property of copper by making it react with concentrated sulphuric acid (this reaction is also used to make sulphur dioxide). Apparently this reaction has given rise to several accidents and this is why it is not recommended for use in lectures.
- Finally, there was a detonation, which involved molten copper and water. It is not a properly dangerous reaction and the consequences of such a thermal 'shock' are rather predictable.

Copper derivatives

With copper salts the dangerous reaction is usually due to the anion atoms and not the copper cation. However, there is an exception, which will be considered before mentioning the other reactions.

- Thus, all copper salts give an explosive reaction with calcium carbide. This is due to the formation of explosive copper acetylide, which is formed from cupric or cuprous cation and the acetylene formed.
- Copper nitrate reacts with sodamide and ammonia by forming explosive copper amides. The oxidising properties of this nitrate have led to violent detonations with ammonium hexacyanoferrates heated to 220°C in the presence of water traces, or dry at the same temperature, but in the presence of an excess of hexacyanoferrate. These accidents illustrate once more the incompatibility between compounds with a cyano group (or cyanide anion) and oxidants. An accident also occurred with a potassium hexacyanoferrate.
- Copper oxides give rise to numerous accidents. When copper (II) oxide was heated with boron, it gave a highly violent reaction, which caused the melting of the Pyrex container. This is true for alkali metals and titanium as well as aluminium. The reactions lead to liquid metal copper. The emissions of glowing compounds make the reaction very dangerous.
- Copper (II) oxide violently oxidises hydrogen sulphide causing its ignition. The same phenomenon was observed with hydrogen. However, in the latter case, it was thought that air was present and that oxide had catalysed the oxygen + hydrogen reaction.

- Particularly violent interactions took place between this oxide and hydrazine or hydroxylamine. In fact, the danger is linked to the extreme in-stability and the highly reducing property of these nitrogenous compounds.
- Cuprous oxide is also reduced violently by electropositive metals as discovered in an accident which occurred with aluminium. With chromium (III) oxide the reaction enables one to make copper chromite, which is a very common catalyst. The activity of copper chromite is such that it frequently combusts at the end of the reaction.
- Copper (II) sulphide is like any other sulphide, ie very reactive vis-à-vis oxidants. Therefore detonations occur during mixtures between this compound and magnesium, zinc and cadmium chlorates.
- Hydroxylamine combusts in contact with anhydrous copper sulphate.
- As with other metals, cupric and cuprous halides can detonate in contact with potassium. In addition, the action of cuprous chloride on lithium azide gives rise to a very violent reaction.
- Finally, when copper dicyanide is heated with magnesium, it causes the mixture to glow, as with nickel cyanide.

Zinc

The metal

All the dangerous reactions involve the metal in the pulverised state and are the result of its reducing property.

- Powdered zinc can combust spontaneously in air. It depends on its surface texture and how fine the powdered is. A detonation occurred during a sieving operation on hot zinc. The presence of zinc chloride traces seems to catalyse the combustion of the metal in air. If a small quantity of water is present, a temperature rise is observed, which is sometimes followed by the ignition of hydrogen. It is assumed that hydrogen is produced when the water is reduced by zinc. Also, zinc alloyed with platinum, iridium or rhodium gives, after acid washing, residues, which detonate when heated in air. This behaviour is probably due to the absorption of oxygen or hydrogen by the noble metals, which became porous after zinc had been 'dissolved' by acid washing.

Many other oxidants give dangerous reactions with zinc. These reactions have already been mentioned by means of previously discussed transition metals.

- At ambient temperature, zinc violently reacts with ammonium nitrate, especially in the presence of small quantities of ammonium chloride. If zinc is hardly mixed with nitrate, one or two water drops are enough to cause the reaction, which in these conditions consists in a strong release of white zinc oxide fumes. If the compounds are carefully mixed, the same operation leads to the ignition and strong combustion of zinc after a couple of seconds, during which fumes are released. Note that zinc oxide which is freshly produced is significantly toxic (see tables in Part Three). Despite its spectacular and enigmatic character (indeed, it is possible to shake one's wet hand above the mixture without the audience noticing the presence of water and giving the impression of a magic trick) we think that this demonstration should be banned in lectures. At 200°C the reaction between ammonium nitrate and zinc can become explosive.

5 • Reactions of inorganic chemicals — Zinc

- A zinc/potassium chlorate mixture is explosive on impact or friction. With potassium superoxide the metal glows. The same thing happens when a mixture of zinc and titanium oxide is heated or when nitric acid vapour is in contact with melted zinc (~400°C).
- With chlorine vapour, zinc in leaf form combusts spontaneously. Zinc glows in contact with fluorine vapour.
- When a sulphur/zinc mixture is heated around 600°C, it combusts violently. The lightning is blinding and its brevity makes it a spectacular experiment, often performed in front of students (using video recordings the author timed lightnings, which lasted for 0.2 seconds with 4 g of mixture). It also has the advantage of demonstrating the fluorescence of zinc sulphide, which is formed during this reaction, especially if zinc is previously treated with copper sulphate traces. Some authors do not recommend carrying out this experiment in lectures. Zinc also gives dangerous reactions with sulphur derivatives, especially carbon sulphide (zinc incandescence) and ammonium sulphide. In the latter case, the danger seems to come from the compoundion of a mixed sulphide made with zinc and ammonium together with the formation of hydrogen sulphide and hydrogen. To illustrate this, there was an accident followed by the detonation of the container, which was due to the overpressure created by the gases formed.
- Zinc gives an explosive reaction with manganese dichloride, whereas with calcium chloride, which was in a galvanised iron container, the detonation is blamed on the overpressure created by the release of hydrogen, which is formed in these conditions.
- A zinc/hydroxylamine mixture can combust or detonate spontaneously.
- Finally, strong bases in concentrated aqueous solutions react very exothermically with zinc as follows:

$$Zn + OH^- \longrightarrow ZnO_2^{2-}$$

With a very small amount of concentrated sodium hydroxide solution mixed with zinc, the temperature rise is such that water vapourises rapidly and the remaining zinc can detonate spontaneously. Attempts to obtain delayed ignitions of the mixture with this reaction have failed.

Zinc derivatives

- When zinc cyanide is heated with zinc, it reacts violently causing incandescence.
- If powdered carbon is thrown on zinc nitrate when it is hot, detonation can take place.
- When zinc oxide and magnesium are heated, they give rise to an extremely violent reaction. It is the typical 'thermite' type of reaction. The same goes for the zinc peroxide/aluminium or zinc mixtures.
- Zinc halides detonate strongly in the presence of potassium.
- Zinc phosphide can combust in contact with water probably because of the formation of phosphine. Finally, it reacts violently with perchloric acid.

Gallium

Two dangerous reactions were reported for this element, which are hardly used.
- A mixture of gallium/hydrogen peroxide in 30% hydrochloric acid detonates if chilled insufficiently.
- Cold chlorine reacts very violently with gallium.

Germanium

- Strong oxidants react violently with germanium. This applies to chlorine (ignition when it is hot; germanium powdered), oxygen (combustion, if germanium is heated), sodium peroxide, nitric acid, sodium chlorate or nitrate (violent or even explosive reaction).
- Germanium glows in molten potassium hydroxide.
- Amongst the germanium compounds chosen, only germanium tetrachloride gave rise to an accident. Indeed, this compound undergoes extremely violent deflagration if it is poured into water.

Arsenic

The element

- There is thought to be an explosive form of arsenic, possibly including a combination of unknown structure. This form is obtained when sulphuric acid, which is strongly charged with arsenic, is in contact with an iron container protected by a lead coating. It detonates on friction or impact. An analysis attempt did not indicate the presence of hydrogen. This case is similar to that of antimony.
- Finely powdered arsenic combusts spontaneously in contact with chlorine.
- Arsenic is violently oxidised by strong oxidants. This applies to potassium superoxide (incandescence of the element), dichlorine oxide (the released heat causes the chlorinated derivative to detonate), chromium (III) oxide (incandescence), potassium or silver nitrates (ignition), potassium permanganate or sodium peroxide (detonations).
- Arsenic catalyses the explosive decomposition of nitrogen trichloride.
- When it is heated with zinc powder, arsenic reacts very exothermically causing a bright light emission. The same goes for aluminium.

Arsenic derivatives

Arsenic (III) oxide
- Zinc wire violently reduces this oxide, sometimes explosively.
- With hydrogen fluoride in the liquid state, it glows.
- A veterinary preparation made with diarsenic trioxide, sodium nitrate and ferrous sulphate combusted spontaneously.

5 • Reactions of inorganic chemicals — Selenium

Arsine

An endothermic compound, which is hardly stable thermodynamically. However, it only decomposes under extreme conditions (very violent shock, high temperature). When it is involved in a reaction, this compound can combust or detonate easily because of the exothermicity of the transformation. Since it is hardly ever handled in the pure state (it forms when compounds containing arsenic are handled) it gives rise to a limited number of accidents.

- Arsine combusts immediately in contact with chlorine, fluorine and potassium chlorate.
- It detonates in the presence of fuming nitric acid. It detonates in contact with potassium in ammonia.

Arsenic sulphides

These have the dangerous reactions of the sulphides, which are very sensitive to oxidants.

- Diarsenic trisulphide produces particularly toxic sulphur dioxide, hydrogen sulphide and arsenic by thermal decomposition or acid treatment.
- Various authors have mentioned the particularly dangerous character of the arsenic trisulphide/hydrogen peroxide interaction without giving any details regarding their appearances.
- When this sulphide is mixed with a concentrated solution of chloric acid, it glows. With potassium chlorate, if both compounds are in the form of fine powdered and the mixture contains at least 30% of weight of chlorate, it detonates on impact. This mixture has been used for pyrotechnic purposes.
- Diarsenic pentasulphide detonates in contact with chlorine, potassium nitrate and alkaline chlorates.

Arsenic trichloride

- Reacts very violently with metals. Accidents have been reported with aluminium (ignition) and potassium (detonation).

Selenium

Only one dangerous reaction was mentioned for selenium derivatives. When it is cold, phosphorus trichloride causes the incandescence of selenium dioxide.

- The mixtures of selenium and zinc, nickel or sodium glow on heating. The danger of the reaction depends on the state of division. With potassium, the ignition is immediate.
- A selenium/phosphorus mixture incandesces, whereas fluorine causes selenium to combust at ambient temperature.
- The element reacts violently with oxidants. There have been incidents reported with alkaline chlorates in the presence of water traces (incandescence), chromium (VI) oxide (violent reaction) and sodium peroxide (very sensitive explosive mixture).

5 • Reactions of inorganic chemicals

Arsenic

Synopsis of dangerous reactions with arsenic and derivatives

Arsenic	Arsenic acid	Calcium arsenate	Potassium arsenate	Sodium arsenate	Sodium arsenite	Gallium arsenide	Arsine	Arsenic oxide (III)	Arsenic oxide (V)	Arsenic pentasulphide	Arsenic trichloride	Arsenic trisulphide	
													Hydrogen
			X				X						Group I (Li; Na; K)
													Group II (Be; Mg; Ca; Sr; Ba)
													Group III (B)
													Group IV (C; Si; Ge)
													Group V (N; P; As; Sb)
													Group VI (O_2; S; Se; Te)
X							X						Group VII (F_2; Cl_2; Br_2; I_2)
X							X						Transition metals
							X						Other metals (Al...Bi; Sn; Pb)
													Hydrides group III (B_xH_y)
													Hydrides group IV (SiH_4...)
				X									Hydrides group V (NH_3; H_2NNH_2; NH_4^+; NH_2OH)
													Water
													Hydrides group VI (H_2S...)
			X										Hydrogen halides
													Beryllium halides
													Boron halides
													Metal halides ($AlCl_3$; $FeCl_3$...)
													Non-metal halides ($SOCl_2$; S_xCl_y; P_xX_y; NCl_3)
													Metal oxides
													Non-metal oxides (O_3; N_xO_y; P_xO_y; SO_x...)
						X							Peroxides (H_2O_2; Na_2O_2)
													Nitric acid
X				X									NO_3^-; NO_2^-
													N_3^-
X													Sulphuric acid
													Non-oxidant salts of acids (SO_x^{x-}; PO_x^{x-}; CO_3^-)
													ClO_x^-; $HClO_4$
X				X									MnO_4^-, $Cr_2O_7^{2-}$
													Instability
													Polymerisation
													Bases
								X					Acids

212

- Finally, there was also spontaneous incandescence of a calcium carbide/selenium mixture and the formation of an explosive compound in the presence of an alkali or alkaline earth amide. In addition, selenium catalyses the explosive decomposition of nitrogen trichloride.

Bromine

Bromine's dangerous reactions are similar to those of chlorine. However, bromine is less reactive. Its physical state (liquid) is another aggravating factor.

The element

- Hydrogen/bromine gaseous mixture represents a risk of explosion, which depends on the pressure and temperature conditions (not quoted by our sources).
- Suspensions of alkali metals in bromine detonate on impact. Accidents have occurred with lithium, sodium and especially potassium. Very violent or even explosive detonations were reported with powdered boron heated to 600°C, gallium (very violent exothermic reaction at -33°C causing a 'flash' at 0°C), germanium (ignition by heating), aluminium and titanium at 360°C (very violent reaction causing incandescence of the metals), scandium and red phosphorus in the presence of water. This last reaction is a synthesis method for hydrogen bromide. It is regarded as too dangerous and the action of bromine on tetrahydrofuran is favoured.
- Bromine creates an explosive reaction with ozone at ambient temperature and pressure.
- There was an incident reported with the calcium carbide/bromine mixture at 350°C. The reaction gave rise to incandescence of the mixture.
- A mixture of bromine in the gaseous state and nitrogen was in contact with sodium azide which detonated, and was explained by formation of $Br-N_3$ bromine azide, which like all compounds that have halogen-nitrogen bonds, is hardly stable. Thus, bromine was mixed with ammonia at ambient temperature. The mixture was then cooled down to -95°C and red oil was formed, which detonated when the mixture was heated and reached -67°C. It formed the following complex:

$$[NBr_3, 6NH_3]$$

This complex is stable in water, but detonates in contact with arsenic or phosphorus. When it is in water, iodine behaves in the same way (p.227).

- Bromine reacts very violently at ambient temperature with silane (SiH_4). This reaction leads to detonations. One needs to use temperatures which do not exceed -30°C to carry out this reaction safely. Note that the same dangers were observed with germanium (GeH_4).
- Bromine reacts with sodium hydroxide in aqueous solution by forming sodium hypobromate. Such a preparation was made without stirring the medium; two layers were formed. A couple of hours later, the stirring was started and caused

5 • Reactions of inorganic chemicals

Selenium

Synopsis of dangerous reactions of selenium and its derivatives

Selenium	Selenium dioxide	Selenium hexafluoride	Sodium selenite	Arsenic selenide	
					Hydrogen
▓					Group I (Li; Na; K)
					Group II (Be; Mg; Ca; Sr; Ba)
					Group III (B)
					Group IV (C; Si; Ge)
▓					Group V (N; P; As; Sb)
					Group VI (O_2; S; Se; Te)
▓					Group VII (F_2; Cl_2; Br_2; I_2)
					Transition metals
					Other metals (Al...Bi, Sn, Pb)
					Hydrides group III (B_xH_y)
					Hydrides group IV (SiH_4...)
▓					Hydrides group V (NH_3; H_2NNH_2; NH_4^+; NH_2OH)
					Water
					Hydrides group VI (H_2S...)
					Hydrogen halides
					Beryllium halides
					Boron halides
					Metal halides ($AlCl_3$; $FeCl_3$...)
▓	▓				Non-metal halides ($SOCl_2$; S_xCl_y; P_xX_y; NCl_3)
					Metal oxides
▓					Non-metal oxides (O_3; N_xO_y; P_xO_y; SO_x; CO_3^-)
▓					Peroxides (H_2O_2; Na_2O_2)
					Nitric acid
					NO_3^-; NO_2^-
					N_3^-
					Sulphuric acid
					Non-oxidant salts of acids (SO_x^{x-}; PO_x^{x-}; CO_3^-)
					ClO_x^-; $HClO_4$
					MnO_4^-, $Cr_2O_7^{2-}$
					Instability
					Polymerisation
					Bases
					Acids
▓					CaC_2

the mixture to erupt. The same handling is known not to be dangerous when it is stirred constantly.

Bromine derivatives

- Whether hydrogen bromide is anhydrous or in an aqueous solution, it 'combusts' spontaneously when in contact with fluorine. It reacts rapidly with ozone even around -100°C. The reaction speeds up before detonating.
- Hydrogen bromide catalyses the explosive decomposition of nitrogen trichloride.
- It causes molten potassium to combust.
- Ammonium bromide gives rise to 'moderate' detonations in the presence of potassium.
- Bromates are strong oxidants, being as strong if not stronger than chlorates (*a comparative test was carried out using the NF T 20-035 standard with our students showing that bromates are more reactive than chlorates vis-à-vis cellulose, see chapter 4*). Since they are not very often handled they cause a limited number of accidents. Two incidents were mentioned: one causing the sodium bromate/sulphur mixture to combust a few hours after being prepared; the other giving rise to a very violent reaction of selenium with potassium bromate in an aqueous solution.

Rubidium

Rubidium is more reactive than potassium. Therefore there is greater risk of dangerous reactions of the same nature. Since it belongs to the category of alkali metals which are less used, like caesium, this explains why there is only a small number of accidents.

- Rubidium reacts violently with water and gives off heat, which ignites the hydrogen formed.
- In air and even more so in oxygen, rubidium combusts spontaneously. There was no mention of any potential formation of superoxide as was the case for potassium.
- Halogens, (fluorine, chlorine and bromine) in the gaseous state, cause rubidium to combust spontaneously. With liquid bromine, it is very difficult to avoid detonation.
- Rubidium reacts with carbon and forms a carbide RbC_8, which combusts spontaneously in air.
- At 60°C, the metal reacts violently with vanadium trichloride.

Strontium

This gives rise to few dangerous reactions since it is hardly used.

- Fine powdered strontium combusts spontaneously in air. It reacts more violently with water than calcium and produces hydrogen and strontium hydroxide.

5 • Reactions of inorganic chemicals

Bromine

Synopsis of dangerous reactions of bromine and its derivatives

Bromine	Hydrogen bromide	Potassium bromate	Sodium bromate	Ammonium bromide	Potassium bromide	Sodium bromide	
■							Hydrogen
■	■						Group I (Li; Na; K)
■							Group II (Be; Mg; Ca; Sr; Ba)
■							Group III (B)
■							Group IV (C; Si; Ge)
■	■						Group V (N; P; As; Sb)
■							Group VI (O_2; S; Se; Te)
■	■	■					Group VII (F_2; Cl_2; Br_2; I_2)
■							Transition metals
■							Other metals (Al...Bi, Sn, Pb)
■							Hydrides group III (B_xH_y)
■							Hydrides group IV (SiH_4...)
							Hydrides group V (NH_3; H_2NNH_2; NH_4^+; NH_2OH)
■							Water
							Hydrides group VI (H_2S...)
							Hydrogen halides
							Beryllium halides
							Boron halides
							Metal halides ($AlCl_3$; $FeCl_3$...)
■							Non-metal halides ($SOCl_2$; S_xCl_y; P_xX_y; NCl_3)
							Metal oxides
■	■						Non-metal halides (O_3; N_xO_y; P_xO_y; SO_x; CO_3^-)
							Peroxides (H_2O_2; Na_2O_2)
							Nitric acid
							NO_3^-; NO_2^-
■							N_3^-
							Sulphuric acid
							Non-oxidant salts of acids (SO_x^{x-}; PO_x^{x-}; CO_3^-)
							ClO_x^-; $HClO_4$
							MnO_4^-, $Cr_2O_7^{2-}$
							Instability
							Polymerisation
■							Bases
							Acids
■							CaC_2

5 • Reactions of inorganic chemicals — Zirconium

- It incandesces with chlorine at 300°C and combusts in contact with bromine vapour at 400°C.
- A strontium/hydrogen mixture can combust above 300°C.
- It incandesces when it is heated in the presence of boron trifluoride.
- Molten lithium reacts violently with strontium and the mixture can combust.

Zirconium

This belongs to the second series of the transition metals. So far as reactivity is concerned, this series is very similar to the first and especially the third. Thus, zirconium and hafnium are so similar that it is very difficult to separate them (except for nuclear purposes where the presence of the second is harmful). The same goes for tantalum and niobium. There are two main differences between the three series within each group.

- In the first series, with high oxidation states (> + III) they are strong oxidants (for instance, $Cr_2O_7^{2-}$). This is not the case for the high oxidation states of the elements in the second and third series.
- In the first series, there are a large number of compounds of elements with a low oxidation state (II or III like MnO_4) whereas there are only a few compounds with oxidation states II or III in the higher series. Compounds with oxidation states IV and VIII are the most stable. This phenomenon can be observed in all the groups of the periodic table.

To illustrate the first remark, no reported accident has ever involved an oxygen-containing compound of zirconium with a high oxidation state.

The element

Zirconium is an extremely active reducing agent. It reduces numerous compounds violently.

- Most accidents involving zirconium are with the powdered metal and its propensity to oxidise in air. The NFPA 'inflammability' code illustrates this property. Over forty accidents describe the ignition, which is usually followed by the detonation of powdered zirconium in air. Humidity (up to 5-10% of water, otherwise there is an inhibition), friction and static electricity are the most common factors. Aqueous hydrogen fluoride, of 1% concentration, is thought to stabilise zirconium vis-à-vis the ignition of electrostatic origin. The Zr-Si and Zr-Ti alloys behave in the same way. Ignition can occur even when air contains carbon dioxide or nitrogen in surplus.
- Zirconium reduces almost all oxygen-containing salts. This is the case for alkali hydroxides (accidents with the lithium, sodium and potassium compounds) and zirconium hydroxide, lithium, sodium and potassium carbonates, alkaline sulphates; sodium tetraborate and copper (II) oxide. This is true especially for oxidising salts such as alkaline chromates and dichromates, chlorates (accident with potassium salt) and nitrates (accident with potassium salt).
- Zirconium adsorbs hydrogen very easily when it is hot. Numerous accidents are linked to the ignition of this adsorbed hydrogen.

- Finally, a mixture of phosphorus and zirconium causes incandescence when heated.

Zirconium tetrachloride

- This reacts very violently with water when it is hot.
- It also reacts dangerously with hydrogen fluoride. There are no details regarding the conditions and consequences of this interaction.
- Finally, there was reported ignition and violent combustion of a mixture of zirconium tetrachloride and lithium that was kept under nitrogen. However, note that lithium gives rise to violent combustions in nitrogen with which it is, together with titanium, the only metal to react.

Niobium

- Niobium combusts spontaneously in fluorine when it is cold and in chlorine at 205°C.
- Niobium pentoxide gives rise to a very violent reaction when it is heated at 200°C with lithium.

Molybdenum

There are only a few accidents reported involving this metal and its derivatives. They are described below.

- Molybdenum combusts spontaneously in fluorine when it is cold.
- When a mixture of 70% of molybdenum and 30% of potassium perchlorate was heated at 330°C, it combusted violently. With sodium peroxide, the mixture with powdered molybdenum caused a violent detonation.
- Molybdenum sulphide, like any sulphide, reacts violently with oxidants. There was an accident reported describing a molybdenum sulphide/potassium nitrate mixture, which detonated.
- Molybdenum trioxide gives rise to a violent reaction when it is heated with lithium during which temperatures sometimes reach 1400°C. With sodium or potassium, under the same conditions, the mixture incandesces whereas with magnesium it detonates.
- There was a fatal accident reported which involved graphite and molybdenum trioxide. When the steel was being prepared, 372 kg of molybdenum trioxide and 340 kg of graphite were incorporated into an electrical oven, which contained steel melted at 1500-1600°C. Some time after adding graphite, a flare emerged from the oven and fatally burnt a technician. This flare was 15 m long and had a diameter of 2 m to 3 m. It was assumed that it was due to a volume of 800 m^3 of carbon monoxide formed by graphite oxidation by molybdenum oxide and combusted in contact with ambient air because of the temperature in front of the oven. It was thought that the combustion was aggravated by the presence of graphite particles.

5 • Reactions of inorganic chemicals

Ruthenium/Rhodium/Palladium

- Molybdates react very violently with zirconium. Detonations were mentioned during these reactions.

Ruthenium

- In the presence of potassium chlorate/nitric acid/hydrochloric acid mixture, ruthenium oxidises explosively. Ruthenium is chemically inert like any 'noble' metal and maybe more than any other metal. Indeed, it is inactive vis-à-vis 'aqua regia', but on this occasion potassium chlorate was combined with aqua regia.
- Despite the previous comments there are dangerous forms of this metal. Thus, the Ru-Zn alloy, when treated by hydrochloric acid leads to zinc 'dissolving' into a very porous ruthenium, which detonates in air spontaneously. The same goes for ruthenium, which is obtained by reduction of its salts by sodium borohydride. It is recommended to reduce ruthenium salts using hydrazine, which is reputed to be 'not dangerous'. However, with ruthenium trichloride this reaction seems to be 'not dangerous' only when hydrazine has a very low molar ratio (0.9 mol per cent). If it is not the case, a huge volume of gas could constitute an important pressure risk.
- Ruthenium tetroxide is unstable. It is thought to decompose explosively above 108°C or 180°C (the figures seem to indicate that the difference between them is the result of an input mistake, but which one is correct?). This oxide is a strong oxidant. This property, as well as the fact that it is less toxic than osmium tetroxide, makes it interesting. Four accidents have been described. This oxide gave rise to ignition followed by a violent combustion after being treated by ammonia at -70°C under 1 mbar. By mixing the reagents slowly, a solid is formed, which detonates at 206°C. It is either nitride or unstable ruthenium oxyamide. Ruthenium tetroxide reacts very exothermically with phosphorus tribromide. Its vapour also detonates when it is in contact with sulphur. Is it due to a destabilisation of oxide by sulphur or a violent oxidation? Perfectly identical results were obtained with carbon powder.

Rhodium

There are little data available. The only piece of information consists in rhodium being obtained by the reduction of its derivatives by hydrogen. The metal obtained occludes large quantities of hydrogen. If it is desorbed from hot rhodium, it combusts in air. Thus, the treatment has to be carried out in inert gas. Rhodium is inert, not even fluorine reacts with it. It occludes oxygen, but only reacts slowly when it is hot. It reacts with halogens at very high temperatures.

Palladium

The only dangerous reactions cited deal with metal palladium, especially when it is in the pulverised state.

- Powdered palladium combusts spontaneously in air. The combustion is even stronger with the Pd/C catalyst. The interaction of palladium with the O_2/H_2

system is particularly dangerous in the 'anthraquinone' method used to prepare hydrogen peroxide and that works with palladium as a catalyst. Indeed, 9,10-dihydroxyanthracene treated by oxygen gives hydrogen peroxide and anthraquinone. The regeneration of diphenol is obtained by hydrogenation of anthraquinone. Palladium is thus alternatively exposed to both gases, which it occludes perfectly. This can create a very dangerous interaction between them. This situation has caused a large number of accidents.

- A dry Pd/C catalyst was incorporated in sodium borohydride. The hydrogen which was released combusted spontaneously in contact with air and the catalyst. The interaction between the catalyst and borohydride would not represent any danger if borohydride was incorporated into the aqueous phase onto the Pd/C and not the other way round.
- Palladium mixed with arsenic when it is hot causes a violent incandescence of the mixture with a very bright light emission. The same goes for sulphur.
- A thermocouple of palladium was put in contact with a molten aluminium bath at around 600°C. This led to a very violent 'flash', which was probably due to the formation of an Al-Pd alloy. Differential thermal analysis showed that an exotherm is formed at 2800°C between these two metals.

Silver

The most characteristic type of dangerous reactions is linked to the propensity of silver derivatives (especially water-soluble salts) to form silver salts or non-ionic silver compounds, which are highly unstable.

The metal

- Silver catalyses the explosive decomposition of hydrogen peroxide. The same goes for peroxomonosulphuric acid at 92%, which has a peroxidic bond.
- Detonations of batteries containing silver, zinc and electrolytes have been reported.

Halides and silver cyanide

- Silver fluoride and boron detonate violently when their mixture is ground up. The same thing happened with the mixture of this fluoride with silicon. When it is mixed with potassium, it gives rise to an extremely violent detonation. At 320°C it causes the incandescence of titanium.
- Silver chloride reacts violently with aluminium. Various accidents, which involved both compounds, have ended up in a violent detonation of the mixture.
- But the main danger regarding this salt lies in the solutions of this compound in ammonia (ie ammoniacal silver chloride). They are not dangerous when they are freshly prepared. But when they are left in air for a prolonged time or heated, they produce a black precipitate, which almost always detonates. It was thought that this black sediment was nitride, amide or imidic silver. These three structures are highly unstable.
- Silver cyanide detonates in contact with fluorine even at low temperatures.

Silver nitrate

- An accident described with ammoniacal silver chloride also occurs with ammoniacal silver nitrate, when kept for too long. Such a solution, which was kept for two weeks, detonated when it was stirred with a glass stirrer. Some authors vapourised such a solution when it was dry and isolated a solid, which proved to be explosive on impact. The structure of this (or these) compound(s), which is (or are) formed is not clearly defined. The authors, who proceeded to add sodium hydroxide containing ammonia to the solid, obtained a solid, which detonated almost immediately after its appearance in the solution. They considered it to be trisilver nitride.

- Most of the other dangerous reactions of silver nitrate are those of nitrates, which are strong oxidants. With powdered magnesium the mixture detonates as soon as it is in contact with a drop of water. A mixture of carbon and silver nitrate also detonates when it is subjected to a very violent impact. With phosphine the type of reaction depends on the conditions. If a strong flow rate of gaseous phosphine is incorporated into an aqueous solution of silver nitrate, the detonation takes place in a violent way and not very long after starting the operations. With a slow flow rate the compounds combust on the surface of the solution. In the latter case, the authors thought silver phosphide was formed. Finally, two dangerous reactions can be interpreted as being linked to the explosive decomposition of the compounds formed. They are represented by the following reactions:

$$Cl\text{-}SO_3H + AgNO_3 \longrightarrow O_3N-SO_3H$$

$$CaC_2 + AgNO_3 \longrightarrow C_2Ag_2$$

The second one is typical and was also cited in the cupric series.

Silver oxide

- Ammoniacal silver oxide is as dangerous as the previous similar compounds. Thus, the clear solution, which is obtained after centrifuging ammoniacal silver oxide leaves a very explosive compound as a residue to which was attributed the Ag_3N_4 formula. It seems that the decomposition of this nitride is inhibited by ammonium salts. A similar situation was created after treating this ammoniacal oxide with silver nitrate until a solid started to precipitate, and this detonated 10 to 14 days after being prepared. It was assumed that this solid was Ag_2NH.

- The oxidising properties of this compound play a role in most other dangerous reactions. For instance, in the presence of aluminium and on heating there is a very violent detonation. This reaction has nothing in common with the usual 'thermite' reactions carried out with aluminium or magnesium. With magnesium the mixture was heated in a sealed tube and the tube detonated. Silver oxide with the Na-K alloy gives rise to a mixture that detonates on impact. When the oxide was ground up with sulphur, selenium and phosphorus it gave rise to a violent ignition. Finally, silver oxide oxidises hydrogen sulphur exothermically by causing its ignition. It behaves in the same way with carbon monoxide. The temperature rise of the medium reaches 300°C:

$$CO + Ag_2O \longrightarrow CO_2 + 2Ag$$

Other dangerous reactions

- A mixture of silver sulphide heated with potassium chlorate causes a very violent reaction, which illustrates the usual incompatibility of sulphides with oxidants.
- A very violent detonation occurred when silver permanganate was dried under vacuum in a dessicator containing sulphuric acid. It was thought that this accident was due to the formation of dimanganese heptoxide, which is extremely explosive through the action of sulphuric acid vapour on silver permanganate.

Cadmium

The specialised literature mentions a few accidents which involve cadmium and its derivatives.

- Cadmium, when finely divided, combusts spontaneously in air.
- It creates violent or even explosive reactions when it is mixed with melted ammonium nitrate.
- Fine powdered cadmium reacts violently with selenium when it is heated. Nevertheless, the reaction is less violent than that of zinc with selenium. Therefore, one has to be careful when preparing cadmium selenide. The particle size of both reagents is a crucial safety factor regarding this reaction.
- Cadmium halides (chloride, bromide, iodide) react explosively with potassium. No accident has been mentioned with fluoride (but this does not prove anything).
- In the presence of magnesium, cadmium cyanide causes the mixture to incandesce when it is heated. The positive enthalpy of formation of this compound leads one to think that it is an unstable compound. No accident involving this instability has ever been reported.
- Cadmium oxide gives rise to a violent 'thermite' reaction with magnesium when it is heated. Detonations occur during this reaction.

Indium

The only dangerous reaction reported is the incandescent interaction between indium and sulphur.

Tin

From tin onwards, the elements in group IV of the periodic table have a stronger metal character. As a result, tin behaves like a relatively reducing metal and has the same dangerous reactions as those with the previous metals. The feature of tin (II) compounds is also their reducing character. For tin (IV) derivatives the oxide has an oxidising character, which is hardly active. However, for all the following compounds the danger is mainly represented by the anion reactivity.

The metal

- An accident resulted from molten tin being incorporated in small quantities (drop by drop) into water, causing the medium to detonate. This cannot be explained by thermal shock, given the low melting temperature of this metal.
- Tin burns in fluorine at 100°C whereas there was a spontaneous ignition of the metal when it was put in contact with liquid chlorine at -34°C.
- It is oxidised very violently by nitrates. The reaction is very violent in molten ammonium nitrate; detonations were also mentioned. Tin in thin leaves and incorporated in an aqueous solution of copper nitrate combusted and threw up showers of sparks all over the surrounding area.
- The same goes for peroxides and superoxides. Thus, tin powdered which was in contact with sodium peroxide in a carbon dioxide atmosphere in the presence of water traces glowed before combusting. With potassium superoxide there is an immediate incandescence of the mixture.
- Tin mixed with sulphur and then heated glows and combusts violently. A very violent reaction occurs between tin and disulphur dichloride.
- A violent reaction takes place with molten lithium. Several accidents have been described and in some instances the reaction caused the compounds to combust.

Tin derivatives

- Tin chlorides have the same dangerous reactions as the chloride ion and, as far as tin (II) derivatives are concerned, as oxidation reactions. Therefore, potassium gives the usual explosive reaction with both chlorides.
- With nitrates, stannous chloride gives rise to frequent detonations. The same happens with stannous fluoride mixed with magnesium nitrate. In the presence of oxygen, stannous fluoride is converted into tin oxydifluoride $SnOF_2$, but this reaction and the resulting compound are not mentioned as dangerous. The reaction proved to be extremely violent with hydrogen peroxide at 3%. This is not a violent decomposition of peroxide because of its dilution, but their actual interaction.
- A calcium carbide/stannous mixture combusted. Could it be the exothermic decomposition of a tin acetylide?
- Tin disulphide has the same dangerous reactions as sulphides, for which sulphide anion is very sensitive to the effect of oxidants. So the effect of chloric acid and chlorates gives rise respectively to violent detonations and the formation of spectacular showers of sparks. Dichlorine oxide detonates, but this is to be expected.
- When it is hot, tin oxide is reduced violently by electropositive metals in 'thermite' reactions. This goes for sodium, potassium, magnesium and aluminium.

Antimony

As with the previous element, the dangerous reactions of antimony derivatives are linked to the behaviour of the anion. The element is reducing and apart from its interaction with two metals it gives rise to dangerous reactions with strong oxidants.

5 • Reactions of inorganic chemicals

Antimony

The element

- Antimony gives rise to a violent reaction when it is heated with aluminium. The same happens with cerium.
- It reacts violently with halogens; it combusts in fluorine, chlorine and gaseous bromine. With liquid bromine the reaction is more violent and can lead to a detonation as with iodine.
- Powdered antimony reacts explosively when it is carefully mixed with alkaline nitrates; alkaline antimonates are formed. The same happens with molten ammonium nitrate, which combusts violently when it is ground up with potassium permanganate (to be compared with arsenic on p.210).
- Mixtures of antimony with chloric acid and dichlorine oxide give rise to detonations.

Antimony halides

- Antimony trichloride vapour in contact with powdered aluminium causes the ignition of aluminium. Note that the aluminium/alkaline chlorates interaction is 'catalysed' by aluminium trifluoride. Is it due to a specific role played by fluoride ion?
- Potassium reacts explosively with all antimony trihalides except trifluoride, as with tin halides (p.223).
- The electrolysis of antimony trihalides forms an antimide, which is considered as 'explosive' (an analysis shows that it contains halogen). But it is not known if it is the same as the one cited in the tables of enthalpy of formation, unless the 'explosive' antimony mentioned in the tables is formed by a method which is similar to 'explosive' arsenic. The documentation is confusing on this point.

Antimony trisulphide

There are the same typically dangerous reactions as with sulphides. Antimony pentasulphide is not mentioned in the specialised literature, but it is unlikely that it behaves in a different way from the trisulphide. Antimony trisulphide can produce dazzling white flames. This is the reason it is used in pyrotechnics.

- Antimony trisulphide combusts spontaneously when it is ground up with silver(I) oxide.
- A pyrotechnic mixture of sulphide/potassium chlorate/aluminium has led to regular detonations. This sulphide incandesces as soon as it is in contact with chloric acid. Mixtures of antimony trisulphide with alkaline nitrates, which are probably used for pyrotechnic purposes, also lead to detonations. 'Bengal lights' has been made with this mixture, which was used in small quantities in mixtures and no accidents were experienced. Finally, dichlorine oxide detonates in contact with this sulphide.

Stibine

The enthalpy of formation is positive. Thus it is a compound which is thermodynamically unstable.

- When it is heated, it gives rise to detonations. The presence of a hot spot on the walls of a container in which stibine is present is enough to cause an explosion.

5 • Reactions of inorganic chemicals — Tellurium/Iodine

An accident which led to a low deflagration also took place when stibine was distilled around -17°C.
- Bases catalyse this decomposition in the presence of ammonia and, when it is gently heated, stibine detonates.
- Stibine is also a strong reducing agent.
- In the presence of chlorine, the reaction is explosive. The same happens with concentrated nitric acid. In the latter case, the reaction is more violent than with arsine (that needs fuming nitric acid). Finally, it detonates when it is put in contact with oxygen containing 2% of ozone even at -90°C.

Other dangerous reactions

- Antimony trioxide has been reported to combust spontaneously when heated in the presence of air. This is surprising since the author has worked on this compound for about ten years without experiencing any problem of this kind.
- Antimony (III) salts are thought to give rise to dangerous reactions with perchloric acid. There is no detail regarding the nature of the dangers.

Tellurium

Only reactions which involve the element itself are reported.
- It incandesces in chlorine on heating moderately, and combusts spontaneously in fluorine. It is interesting to note that it is the only element of this group to react easily with chlorine (in fact, with bromine too but not in a dangerous way).
- It reacts rather violently with numerous metals. With molten sodium, the reaction is strong. With potassium, the mixture incandesces but, if a large quantity of potassium in excess is used, there is deflagration. With zinc, cadmium and tin, there is incandescence. In the latter case, tin telluride, SnTe, forms.

Iodine

The element

There are typical halogenic reactions with this element although this halogen is less reactive than the ones which precede it.
- Iodine reacts dangerously with numerous elements. Thus, with aluminium, magnesium or powdered zinc, the mixture with iodine in contact with a drop of water gives rise to a 'flash', which is extremely violent and blinding. In the very spectacular experiment involving ammonium nitrate and zinc described on p.208 and carried out with students, it is recommended to incorporate a small quantity of iodine before throwing a few drops of water on the mixture.
- With potassium or antimony, the interaction gives rise to a 'small' deflagration. With lithium at 200°C, the reaction is violent. With rubidium, the mixture combusts. The action of iodine on titanium or the Ti-Al alloy enables one to prepare titanium diodide. The reaction, which is carried out above 113°C under reduced pressure or at 360°C under normal pressure is violent and produces showers of sparks.

5 • Reactions of inorganic chemicals

Iodine

Synopsis of dangerous reactions of iodine and its derivatives

Iodine	Potassium iodate	Sodium iodate	Ammonium iodide	Hydrogen iodide	Potassium iodide	Sodium iodide	
							Hydrogen
▓	▓	▓					Group I (Li; Na; K; Rb)
▓							Group II (Be; Mg; Ca; Sr; Ba)
▓							Group III (B)
▓	▓	▓					Group IV (C; Si; Ge)
▓	▓	▓					Group V (N; P; As; Sb)
▓	▓						Group VI (O_2; S; Se; Te)
▓							Group VII (F_2; Cl_2; Br_2; I_2)
▓							Transition metals
▓							Other metals (Al...Bi, Sn, Pb)
							Hydrides group III (B_xH_y)
							Hydrides group IV (SiH_4...)
▓							Hydrides group V (NH_3; H_2NNH_2; NH_4^+; NH_2OH)
							Water
							Hydrides group VI (H_2S...)
							Hydrogen halides
							Beryllium halides
							Boron halides
							Metal halides ($AlCl_3$; $FeCl_3$...)
			▓				Non-metal halides ($SOCl_2$; S_xCl_y; P_xX_y; NCl_3)
							Metal oxides (RuO_4)
			▓				Non-metal oxides (O_3; N_xO_y; P_xO_y; SO_x; CO_x)
			▓				Peroxides (H_2O_2; Na_2O_2)
▓							Nitric acid
							NO_3^-; NO_2^-
							N_3^-
							Sulphuric acid
▓	▓		▓				Non-oxidant salts (SO_x^{x-}; PO_x^{x-}; CO_3^{2-}; S^{2-}; CN^-; X^-)
			▓				ClO_x^-; $HClO_4$
			▓				MnO_4^-, $Cr_2O_7^{2-}$
							Instability
							Polymerisation
							Bases
							Acids
▓							CaC_2

226

- There have been several accidents with metalloids: detonation with fluorine; very violent reaction with boron at 700°C, and ignition with white phosphorus. In the last case, the dangerous character of the reaction of the preparation of hydrogen iodide by distillation of the phosphorus/moist iodide mixture was also mentioned. The formation of phosphonium iodide often causes the conduits of the apparatus to block, which causes the apparatus to detonate due to overpressure. Several accidents involve this factor, which is not due to a reaction that is intrinsically dangerous.
- At 305°C, iodine mixed with calcium carbide produces incandescence in the mixture. Can it be explained by the formation of diodoacetylene followed by its decomposition?
- One of the best known dangerous reactions of iodine is its conversion into a 1:1 ammonia/nitrogen triiodide complex in the presence of ammonia. Iodine incorporated into ammonia and left there for about ten minutes converts into a black solid. It is stable in water provided the water is not hot. When it is left in cold water for a prolonged period, nitrogen triiodide is thought to convert into diiodide, which is as unstable. After filtering and drying, the complex mentioned proves to be one of the most sensitive explosives. It detonates under the effect of the smallest causes; draughts, light, sounds, vibration*.

Hydrogen iodide and metal iodides

Hydrogen iodide is hardly endothermic. This property makes it rather unstable thermodynamically, which explains the exothermicity of the reactions that it gives rise to.

- It gives rise to dangerous reactions with very electropositive metals. With molten potassium, there is immediate ignition. If the metal is solid, it detonates on impact in the presence of hydrogen iodide. Magnesium combusts spontaneously and the ignition is thought to last for a short period of time. Ammonium iodide gives rise to a dangerous reaction with potassium too (this danger was not mentioned with the other iodides). One can assume that the danger comes from hydrogen iodide, which is formed by the decomposition of this hardly stable iodide.
- With phosphorus, there is conversion into phosphine under the effect of hydrogen iodide. Phosphine combusts spontaneously when oxygen is present.
- Hydrogen iodide catalyses the explosive decomposition of nitrogen trichloride.
- Its reducing character makes it dangerous as with strong oxidants. So contact with anhydrous perchloric acid causes the mixture to combust. Alkaline iodides behave the same way as with this acid. Hydrogen iodide in contact with fluorine, fuming nitric acid, nitrogen tetroxide, melted potassium chlorate also combusts. An attempt to prepare ruthenium triiodide by the action of hydrogen iodide on ruthenium tetroxide leads to a violent detonation.
- Finally, metal iodides react violently with diphosphorus pentoxide.

---

* It was detonated by clapping, which delighted the author's students.

Iodates

These have an oxidising character, which is not as strong as with chlorates or bromates, according to comparative analysis carried out using the NF T 20-035 standard, already cited for bromine on p.145.

- They violently oxidise metals such as sodium, potassium and magnesium. In the last case, the mixture corresponds to a typical pyrotechnic preparation. Aluminium and copper also give rise to violent combustions or even detonations.
- With metalloids, the reactions are dangerous too. There are cases of accidents during which violent combustions or detonations occur with carbon, phosphorus, arsenic, sulphur and tellurium (with silver iodate, in this last case). Friction is one of the main factors.
- Finally, anions that are incompatible with oxidants will give rise to violent reactions with iodates. This goes for cyanides, thiocyanates and sulphides. In the last case, arsenic, antimony, copper and tin sulphides were the main ones cited.
- A dangerous reaction took place between potassium iodate and manganese dioxide. This accident occurred with an impure manganese oxide, which may be responsible for this reaction.
- Finally, an accident involved an active carbon filter, which contained potassium iodide and through which gas containing ozone was passing. The detonation of the filter was explained by ozone oxidising iodide into iodate and by the iodate interacting to produce an explosion with the active carbon (see the effect of iodates on metalloids above).

Xenon

Xenon is involved in various fluoride and mixed oxygen-containing combinations. Some of them, such as xenon trioxide, which are extremely unstable, are explosive.

Barium

Barium is the most reactive of all the alkaline earths which are analysed here. As far as its salts are concerned, the danger depends on the nature of anion.

The metal

- Powdered barium combusts spontaneously in air and oxidising gases.
- It reacts with water in a more violent way than the metals, which are situated above it in group II, but less violently than sodium.
- Below 300°C, it combusts in hydrogen.
- It incandesces with boron trifluoride.

Barium oxides

- As with calcium and strontium oxides, barium oxide reacts very exothermically with water, forming hydrogen. The heat which is released is such that it can, like any other oxide of this series, be an ignition source if inflammable materials are present (for instance, wood). If it is perfectly dry, it causes hydroxylamine to combust. In the presence of nitrogen tetroxide, the reaction is very exothermic and causes the oxide to incandesce before melting (F: 1918°C). Finally, it incandesces with sulphur trioxide.

- Barium peroxide is a strong oxidant. Thus it can be expected to give rise to the usual dangerous reactions of this class of compounds. It causes hydrogen sulphide to combust. The same goes for selenium if the mixture of the compounds in the powdered state is at a temperature greater than 265°C. It causes the immediate ignition of hydroxylamine. The same is true for electropositive metals such as aluminium and magnesium as well as the calcium-silicon alloy. The incandescence observed when this peroxide is in contact with sulphur dioxide, and in particular carbon dioxide, is even more unexpected.

Nitrates and chlorates

The oxidising properties of these salts were described previously. It is not surprising to find a table of dangerous reactions for this type of oxidant. The problem may be aggravated by the frequency of these accidents since these salts are used in pyrotechnics. Indeed, they are the raw materials of all fireworks which produce green colours.

- In one instance, a pyrotechnic preparation, which contained aluminium, water traces, potassium chlorate and potassium and barium nitrates detonated violently twenty four hours after being prepared. A mixture of barium nitrate, aluminium and magnesium proves very sensitive to friction or impact (risk of ignition or spontaneous detonation).
- Barium chlorate creates mixtures which are strongly explosive with sulphur or phosphorus. It is incompatible with ammonium salts. Besides, in the presence of sulphuric acid, and like any other chlorate, it forms chlorine dioxide, which detonates at ambient temperature.

Other salts

- Barium sulphate is not a strong oxidant; yet it can give rise to a very violent thermite reaction when it is heated in the presence of a very electropositive metal. This applies to aluminium for which an accident was described. In addition, there was an accident which involved a mixture of this sulphate with red phosphorus although the latter had been previously treated with potassium nitrate and calcium silicide.
- Barium sulphide has the usual dangerous reactions of sulphides (detonations with potassium chlorate and nitrate, combustion with phosphorus pentoxide). It catalyses the explosive combustion of dichlorine oxide.

Mercury

The metal

- Mercury forms amalgams with numerous metals. Usually, this conversion is very exothermic, therefore it can present risks; the reaction can become violent if a metal is added too quickly into mercury. Accidents have been described with calcium (at 390°C), aluminium, alkali metals (lithium, sodium, potassium, rubidium) and cerium. Some of these alloys are very inflammable, in particular the Hg-Zn amalgam.
- Mercury reacts violently with chlorine at 200-300°C and dry bromine. It also combusts spontaneously in chlorine dioxide.
- A strange accident occurred with ammonia. A mercury manometer was in contact with ammonia in the presence of water traces. The grey-brown solid formed detonated when a technician attempted to take it away mechanically. Tests showed that this solid did not form in the absence of water.

Mercury derivatives

Mercury halides

- Apart from the fluoride, mercuric halides react explosively with potassium like all analogues of the other metals already mentioned. With mercurous salts, the reaction seems less violent since with mercurous chloride, molten potassium causes the mixture to incandesce without ever combusting. It is likely that other metals react too; an extreme violent reaction was mentioned between indium and mercuric bromide.

Cyanide and mercuric thiocyanate

- These compounds are endothermic enough not to be sufficiently stable thermodynamically. No accident has been linked to this instability of cyanide. The same is not true for thiocyanate, which decomposes easily at 165°C and not dangerously according to the descriptions about its thermal behaviour and the author's own experience. It gives one of the best shows in chemistry by forming the 'Pharaoh's snake', which is obtained by a fantastic volume increase. This 'show' creates mercury in the gaseous state, which is highly toxic.
- These compounds have the usual incompatibilities of cyanides, especially when oxidants and magnesium are present. Nevertheless, these dangers only give rise to accidents with mercuric cyanide. Thus, accidents have been mentioned with fluorine (ignition, if there is a low temperature rise), sodium nitrite (detonation by heating) and magnesium (detonation). In the last case, the detonation is explained by the formation of cyanogen $(CN)_2$, which then decomposes explosively. The same happens with gold cyanide, but not the other metal cyanides. Pharaoh's snake is explained by the polymerisation of this cyanogen (together with the formation of large quantities of gas). Mercuric cyanide causes the explosive decomposition (or polymerisation) of hydrogen cyanide (see on p.335).

Mercuric oxide

This has a strong tendency to form oxygen and the metal by heating. It can prove to be a dangerous oxidant in certain conditions, but can also catalyse some reactions of decomposition, which is a feature shared by most heavy metal compounds. In the presence of very active oxidants, it can form explosive mercury peroxide.

- This is exactly what happens when hydrogen peroxide, which contains traces of nitric acid, is present. The mercury peroxide obtained detonates on friction or impact. The oxide decomposes hydrogen peroxide explosively without nitric acid.
- Its oxidising character plays a role in all other reactions. Surprisingly, it is thought to form explosive dichlorine oxide with chlorine. It leads to a fast and very exothermic reaction with disulphur dichloride and detonations with metals: potassium, K-Na alloy, magnesium; with phosphorus and anhydrous or hydrated hydrazine.

Other dangerous reactions

- Mercuric and mercurous nitrates have the same oxidising properties as nitrates. The first has been involved in accidents with phosphine (formation of a complex, which detonates on impact) and alkaline cyanides. In the latter case, it is assumed that the danger arises from the formation of mercury nitrite, which is highly unstable; in one particular accident the use of an apparatus with a narrow neck aggravated the effect, causing an effect of confinement. The second nitrate led to more or less strong detonations with carbon (red-hot) and phosphorus (on impact).
- Finally, the sulphide is absolutely incompatible with oxidants, chlorine (incandescence), dichlorine oxide (explosive decomposition), silver (I) oxide (ignition when the mixture is ground up).

Lead

The metal

- Lead, which is formed by reducing lead oxide using furfural and heated at 290°C combusts spontaneously.
- It can combust in melted lithium.
- It catalyses the explosive decomposition of hydrogen peroxide.
- Like many metals, already described, lead reacts violently and sometimes explosively in melted ammonium nitrate at a temperature which is lower than 200°C.

Lead azide

- Its positive enthalpy of formation makes it likely to be highly unstable thermodynamically. Indeed, its explosive decomposition caused numerous accidents. So, 5 g of this compound touched with a metal spatula causes its detonation. Its crystallisation can also cause its detonation and there should be a constant stirring during this operation to avoid the formation of large crystals. In the presence of transition metals, such as copper or brass, it converts into copper or zinc azide, which are much more sensitive than the lead salt. Indeed, lead azide is a detonatable agent, which is far more 'reliable' than most other primary explosives; it has replaced mercury fulminate in detonators.

Oxidising salts of lead

The oxidising salts are chromate, lead oxide and peroxide as well as nitrate. They all give rise to the typical dangerous reactions of this class of compounds.

- Lead chromate gave rise to two accidents, which involved an alkaline (II) ferrocyanide . They are explained by incompatibility between the cyano group (or cyanide anion) and oxidants.
- With sulphur, this chromate forms mixtures, which combust spontaneously. The same is true for titanium.
- Lead oxide reacts violently with numerous metals such as sodium powder (immediate ignition), aluminium (thermite reaction, which is often explosive), zirconium (detonation), titanium, some metalloids, boron (incandescence by heating), boron-silicon or boron-aluminium mixtures (detonation in the last two cases). Finally, silicon gives rise to a violent reaction unless it is combined with aluminium (violent detonation). It also catalyses the explosive decomposition of hydrogen peroxide.
- There is a very spectacular reaction with the charring of lead tartrate. When this residue is freshly prepared, it combusts spontaneously in air by throwing up showers of sparks. When it is rubbed on a sheet of paper, or put in the hand and thrown, the reactions of this residue are more violent. It is thought that this is due to the catalysis by lead oxide of the combustion of carbon formed, which is very porous.
- Lead peroxide is even more active than the oxide. It reacts violently with sulphur and sulphides. When it is ground up with sulphur, the mixture combusts. With hydrogen sulphide, the reaction is very exothermic and causes peroxide to incandesce and hydrogen sulphide to combust. Finally, it reacts violently with calcium, strontium and barium sulphides on heating.
- There are other sulphur derivatives which react violently. Lead peroxide combusts in contact with sulphuric acid. It forms an explosive mixture with sulphuryl chloride and incandesces in sulphur dioxide.
- It gives rise to very violent reactions with metals, and accidents have been mentioned with sodium or potassium (detonation), magnesium or aluminium (ignition), titanium, molybdenum, tungsten (incandescence), zirconium (very sensitive to impact, friction, shockwaves, static electricity). It also reacts with metalloids: boron and white phosphorus (detonations by grinding), red phosphorus (ignition by grinding). Silicon incandesces when it is exposed to a small ignition flame (the temperature of the mixture reaches 1400°C).
- It causes the ignition of hydroxylamine and incandesces in phosphorus trichloride when it is hot.
- Finally, there is a detonation with lead nitrate when it is mixed with carbon and heated until red-hot. With the Ca-Si alloy the mixture combusts.

The other lead compounds
- Three types of accidents have been mentioned involving lead chloride (detonation, when it was heated with calcium), carbonate (violent combustion in fluorine) and sulphate (violent reaction, sometimes detonation with potassium).

Bismuth

The metal
- Bismuth forms an alloy with melted lithium; the conversion is dangerous because of its exothermicity. The aluminium/bismuth mixture combusts spontaneously in air. The same goes for a cerium/bismuth mixture.
- All other dangerous reactions consist of oxidations of bismuth by strong oxidants. Thus, chloric and perchloric acids lead to highly sensitive explosives (probably bismuth chlorate and perchlorate). Fuming nitric acid causes the incandescence of bismuth at ambient temperature whereas a detonation occurs when molten bismuth is mixed with concentrated nitric acid. Finally, a bismuth/molten ammonium nitrate mixture causes a very violent or even an explosive reaction.

Bismuth derivatives
- Bismuth trioxide incandesces with sodium or potassium. Bismuth halogens (chlorinated, brominated, iodated) detonate in contact with potassium.

6 • REACTIONS OF ORGANIC CHEMICALS

A description on how this chapter is organised can be found on p.146.

6.1 Hydrocarbons

The nature of dangerous reactions involving organic chemicals depends on the saturated, unsaturated or aromatic structures of a particular compound. Saturated hydrocarbons are hardly reactive, especially when they are linear. Branched or cyclic hydrocarbons (especially polycyclic condensed ones) are more reactive, in particular as with oxidation reactions. With ethylenic or acetylenic unsaturated compounds, the products are 'endothermic'.

The reactions in which a multiple bond plays a role will be very exothermic and particularly dangerous, especially in the case of the lower ranking and conjugated dienes. In addition, the presence of a multiple bond increases the reactivity of hydrogen atoms in an allylic position. Aromatic compounds possess endothermic character too. Their dangerous reactions will most often involve intermediate or ultimate products, which are highly unstable as in nitration or oxidation. The hydrogen atom in the benzyl position in the aromatic ring is much more reactive. Finally, note that the non-polar simple C-H and C-C bonds can be cleaved homolytically and give rise to free radical reactions that are sometimes difficult to control, in particular, in the case of oxidation and halogenation.

6.1.1 Instability of hydrocarbons

Saturated hydrocarbons are stable. Only cycloalkanes with a tight ring are unstable. Alkenes and alkynes have a strong endothermic character, especially the first homologues and polyunsaturated conjugated hydrocarbons. This is also true for aromatic compounds, but this thermodynamic approach does not show up their real stability very well. Apart from a few special cases, the decomposition of unsaturated hydrocarbons requires extreme conditions, which are only encountered in the chemical industry.

■ **Ethylene**

This decomposes at a high temperature and pressure.
- Hydrogenation of acetylene into ethylene gave rise to an accident (this reaction is no longer carried out on an industrial level). The temperature was fixed at 400°C by mistake. The ethylene reached a temperature of 950°C because of the exothermicity of the reaction. Its decomposition caused the installation to detonate and led to a fire which took four days to put out. It was assumed that the degradation had led to the formation of methane and hydrogen given the pressure that was reached.

- Other sources indicate that decomposition can take place at 350°C under 170 bar and in the absence of air. In this case, the temperature reaches 1350°C and the pressure is multiplied by six. An estimation of the enthalpy of decomposition gives: ΔH_d = -4.33 kJ/g. The violent pressure increase also causes the decomposition of ethylene.
- An accident occurred in a gas pipeline in which the pressure of ethylene radically changed from 1 to 88.5 bar. The explosion which followed caused the gas pipeline to break by increasing the temperature of its walls to 700-800°C. The sudden temperature rise caused by the pressurisation and the fall of temperature at which gas combusts can be another factor (see p. 241).

■ Propylene

- When it was submitted to a pressure of 955 bar and a temperature of 327°C and then violently compressed to 4860 bar, propylene detonated, creating an overpressure, which reached 10000 bar. It was thought that such a pressure rise would lead to the formation of methane and hydrogen.

■ Dienes

Allene decomposes at pressures starting at 29 bar. Butadiene decomposes explosively at a higher pressure.

■ Acetylene

This is hardly stable and it was not until suitable conditions of dilution were found that it was possible to handle it in industry. Even at low temperatures it detonates easily, when it is in the solid or liquid state. Detonations occurred during attempts at liquefaction. Its dilution in nitrogen at -181° stabilises it, but there was an accident under these conditions, which was due to the presence of carborundum that makes it sensitive to impact. In the gaseous state, it detonates at a pressure of 1.4 bar and above. It can only be kept under pressure when it is in a solution of acetone in which it is highly soluble. Alcohols C_1 to C_4, ketones to C_4, diols C_3 and C_4, and carboxylic acids C_1 to C_4 all play the same stabilising role as acetone. The same goes for propane and butane. Nitrogen oxide, hydrogen halides, and vinyl bromide stabilise acetylene, but sulphur dioxide destabilises it.

- Acetylene mixed with acetic acid is stable at 70°C, but an accidental temperature rise in a cylinder containing a solution of acetylene in acetic acid resulted in a temperature rise reaching 185°C on a specific point of the cylinder walls. There was an extreme heating of the cylinder, which could not be controlled by severe external cooling, and caused the cylinder to detonate.

■ Propyne

Commercial propyne contains important quantities of propadiene, this mixture not being sensitive to impact.

- A mixture of propyne containing 30% of propadiene under a pressure of 3.5 bar was accidentally brought to 95°C at a spot in the storage cylinder. It then detonated. Commercial propyne becomes dangerous at 3.4 bar and at 20°C. It is lowered to 2.1 bar if the compound is heated to 120°C. Some authors think that the stability of the commercial mixture is close to that of ethylene.

6 • Reactions of organic chemicals

6.1 Hydrocarbons

■ Vinylacetylene

Probably the least stable industrial hydrocarbon. It has been used less often since it became a compound of minor interest in the chemical industry. The compound in the pure state detonates spontaneously. As a by-compound of an old synthesis of butadiene, which is not used anymore, it used to be mixed with butadiene. As soon as its concentration in such a mixture reaches 50%, it detonates. The speed at which its pressure goes up is extremely high and makes it highly explosive.

6.1.2 Polymerisation of unsaturated hydrocarbons

Polymerisation is the logical conclusion since in many accidents of unsaturated hydrocarbons, which are often followed by 'spontaneous' detonations, the exact reasons for these accidents could not always be identified and the distinction between decomposition and polymerisation could not always be made.

■ Ethylene

Numerous accidents have brought about a violent polymerisation of ethylene. In two examples the conditions were the following:
- Presence of copper, temperature greater than 400°C and pressure greater than 54 bar.
- Presence of zeolites with 5 Å, which were used to dry ethylene under pressure and catalysed the polymerisation. The installation was destroyed. The zeolites with 3 Å do not play any catalytic role.

■ Propylene

When this compound is mixed with lithium nitrate or sulphur dioxide, this leads to its explosive polymerisation. This reaction was carried out in a glass reactor at 20°C. It seems that ambient light played a role.

■ Styrene

- An accidental heating at 95°C was sufficient to cause this compound to polymerise violently and be thrown out of the reactor.
- A large number of accidents due to the polymerisation of styrene in storage containers are caused by a temperature which is too high. It is recommended not to store styrene at a temperature greater than 32°C. However, the presence of polymerisation primers, which are placed deliberately or accidentally, is the most frequent cause of violent polymerisations of styrene.
- In the presence of benzoyl peroxide, the polymerisation got out of control due to an excessive pressure of 3 bar and caused an accident. Similar accidents have occurred with other free radical primers.
- The ionic polymerisation of styrene is as dangerous. Interlaminar compounds of sodium or potassium with graphite catalyse the polymerisation of styrene. This method can usually be controlled. Nevertheless, it gives rise to detonations. It was assumed that in these cases the lamellar structure of graphite is destroyed and the metallic particles dispersed.

- Polymerisations triggered by butyl lithium have often led to detonations, which destroyed the installations. The reaction can be best controlled by first incorporating polystyrene with a low molecular mass.

■ Butadiene

- A temperature of 30-40°C and a moderate pressure are enough to cause a violent polymerisation, which can increase the pressure in the reactor to 1000 - 1200 bar. In storage, a low polymerisation can also be dangerous for a different reason. In this case, polymer precipitates in the form of flakes causing the volume to rise, which can eventually cause the storage tanks to detonate. Butadiene can only be stored if it contains a polymerisation inhibitor, which also plays the role of an oxidation inhibitor. Tert-butylcatechol concentrated at 0.2% is perfect for this use, but rust and water can damage the inhibitor.

■ Cyclopentadiene

- Can polymerise, but its main danger lies in its propensity to dimerise by the Diels-Alder reaction. This reaction takes place at a temperature starting at 0-40°C, under pressure. If the dimerisation is not controlled, the storage equipments' temperature and pressure rise very quickly, which leads to their destruction. Storage temperatures of -80°C have been recommended.
- Cyclopentadiene polymerises very violently by forming large volumes of tar in the presence of potassium hydroxide in an alcoholic solution or sulphuric acid.

■ Acetylene compounds

Rarely involved in dangerous polymerisation reactions. However, phenylacetylene polymerises violently in the presence of palladium (II) acetate. With other palladium (II) salts the polymerisation does not appear to be dangerous.

The NFPA 'reactivity' code states an instability and polymerisation risk. Here are the data (classified in alphabetic order) for hydrocarbons:

Hydrocarbon	NFPA reactivity	Hydrocarbon	NFPA reactivity
Acetylene	3	3- or 4-Methylstyrene	0
1,3-butadiene	2	1,3-Pentadiene	1
Dicyclopentadiene	1	Propylene	1
Ethylene	2	Propyne	2
Isoprene	1	Styrene	2
α-Methylstyrene	1		

These values do not always refer to the same thing. The acetylene and propyne codes indicate the instability. The other compounds have codes which indicate the danger of polymerisation. Comparing the methylstyrene codes with the others raises doubts about the codes chosen.

6 • Reactions of organic chemicals

6.1 Hydrocarbons

6.1.3 Halogenation of hydrocarbons

Halogens can give rise to substitution or addition reactions. Usually, the substitution consists of a free radical reaction catalysed by heat, light or free radical generators. With unsaturated hydrocarbons, there is an ionic addition (electrophilic addition). The reactivity of hydrocarbons is thus linked respectively to the free radicals that are formed and the intermediate carbocations. For halogens, fluorine is so reactive that nearly all its reactions are very dangerous and often lead to complete halogenation of the hydrocarbon molecule where the C-C bonds can be broken. On the other hand, the reactions are very rarely dangerous with iodine. Chlorine and bromine have intermediate reactivities.

■ **Effect of fluorine**

This element can never be controlled. Fluorine reacts violently with solid methane at -190°C. With liquid hydrocarbons at -210°C, the reaction is dangerous. All hydrocarbons react dangerously, from the first homologues to anthracene as well as lubricants. In the gaseous state, there is ignition with small quantities, and detonation with large quantities and when the mixture is made quickly. There is immediate detonation with alkenes and alkynes. With benzene, when fluorine is incorporated bubble by bubble and at a low temperature, this causes ignition on the surface. If the flow rate is substantial, there is immediate detonation.

■ **Effect of chlorine**

The CH_4/Cl_2 mixture is explosive if it contains more than 20% of chlorine. If mercury oxide is present, the reaction is very violent. The accidents mentioned below deal with saturated hydrocarbons.

- When it was heated to 350°C and activated carbon was present, ethane gave rise to a violent detonation. Analysis showed that when carbon dioxide is incorporated as a thinner, this reduces the risks of detonation.
- When petrol was accidentally incorporated into a tube containing liquefied chlorine, this caused a reaction that started slowly and then accelerated.
- A polypropylene filter, which contained zinc oxide, reacted explosively in contact with chlorine and was destroyed. This filter was tested at 300 bar. Zinc chloride, which is formed by the effect of chlorine on zinc oxide, may have catalysed this reaction.
- The same accident with the same consequences happened when chlorine accidentally came into contact with an oil pump diaphragm.
- There was also a detonation when gaseous chlorine was introduced into an apparatus lubricated with Vaseline.
- Chlorine caused a significant temperature rise when it came into contact with polyisobutene.
- When ethylene was radiated by UV or in the presence of mercury, mercury (II) or silver oxide at ambient temperature or lead (II) oxide at 100°C, the reaction was explosive.
- A chloroethane compoundion plant was detonated because of the pressure of ethane/ethylene/chlorine mixture at 80°C and under 10 bar.

- A chlorination of styrene carried out at 50°C and in the presence of iron trichloride led to a detonation after having incorporated 10% of the chlorine. The presence of iron trichloride or even the use of steel reactors is sufficient to make styrene chlorination hazardous. Various authors assume that the danger comes from the catalysis of the styrene decomposition by iron salt rather than the catalysis of the addition of chlorine by this Lewis acid.

■ Effect of halogens and halogenation agents on acetylene and toluene

Acetylene gives rise to dangerous reactions with all halogens including iodine. However, the compounds need to be heated or there has to be light present to create conditions which cause accidents. As with the addition reaction, acetylene and alkynes in general are less reactive than alkenes, in particular chlorine and bromine. Here the danger is linked to the difficulty in finding conditions which allow control of the conversion. If the temperature is too low, the reagents accumulate and the reaction is out of control once the reagents' concentrations are in excess. If the temperature is too high, the reaction is out of control right from the start. The acetylene/chlorine reaction takes place at 100°C. With bromine, the reaction does not represent any danger at the reflux of tetrachloromethane used as a solvent (Eb: 77°C), which leads to tetrabromoethane.

- This reaction was carried out at a lower temperature since it was not properly monitored. As a result, the reaction got out of control and caused the compounds to overflow.

Acetylene as well as its compounds have an acetylenic hydrogen atom with an acid character that can be substituted by a chlorine atom. Thus, with acetylene, calcium hypochlorite and sodium hypochlorite in an acid medium form (this can be formed by calcium carbide hydrolysis) chloroacetylene and dichloro-acetylene, which combust spontaneously in air.

In addition to the hypochlorites, mentioned before, sulphur dichloride has been involved in a dangerous reaction with toluene.

- Toluene was introduced into sulphur dichloride in a steel reactor. A violent reaction occurred, together with a massive release of hydrogen chloride, creating an overpressure, which led to the detonation of the apparatus. It was explained by a catalytic effect due to either iron or iron trichloride. Several possible conversions have been suggested, which all generate hydrogen chloride and produce either chlorotoluenes, polychloromethylbenzenes or toluyl sulphides.

6.1.4 Oxidation of hydrocarbons

■ Combustion

Hydrocarbons burn in air giving off large quantities of heat. The general equation of the complete combustion of a saturated hydrocarbon is the following:

$$C_n H_{2n+2} + \left(\frac{3n+1}{2}\right) O_2 \longrightarrow n\, CO_2 + (n+1)\, H_2 O$$

6 • Reactions of organic chemicals

6.1 Hydrocarbons

If a significant volume of gas (caused by a leak, for example) is exposed to an ignition source and this gas is mixed with air in proportions that are close to stoichiometric, the gas 'cloud' can cause a lot of damage when it gives rise to a detonation. The accident at Flixborough is one example. The lower explosive limit of hydrocarbons is extremely low. If the carbon chain length exceeds 8, the autoinflammation temperature of a linear hydrocarbon is close to 200°C. All these parameters decrease with pressure. The table below shows to which extent pressure influences the AIT of ethylene:

Pressure (bar)	1	68	102
AIT (°C)	492	371	204

But all these parameters are also sensitive to the presence of other compounds. Therefore, the AIT of ethylene is lower when the walls of a reactor are covered with diboron trioxide.

So the handling of hydrocarbons presents serious fire hazards. There are many accidents linked to this in the industrial sector. For instance, a serious accident happened when polyethylene was stored. It appeared to be caused by the diffusion of monomer through the mass of polymer, which created an inflammable atmosphere in the storage container. Incorporating a mixture of oxygen and styrene in a reactor cause spontaneous ignition.

As a consequence, operating conditions that enable one to work in concentration ranges that do not overlap the range set by the limits of inflammability are often sought in industrial synthesis.

For example, the industrial synthesis of vinyl chloride involves a mixture of ethylene, chlorine and oxygen. This is carried out in such a way that hydrogen chloride which forms during the reaction keeps the ethylene/oxygen mixture outside the LEL - UEL range.

The combustion of acetylene is as dangerous. Two accidents, which illustrate the dangers linked to the inflammability of this hydrocarbon are described here:

- An oxygen/acetylene mixture detonated spontaneously when it was introduced into a steel tube (the exact reason for this accident could not be identified).
- The second accident deals with a mixture which contained 54% of acetylene and 46% of oxygen. When it was heated to 270°C under 11 bar, the pressure of the system reached 56 bar in 0.7 seconds ,when the reactor detonated. It had been previously tested at several thousand bars.

■ Effect of gaseous oxygen

It is particularly dangerous to have unsaturated hydrocarbons in contact with oxygen.

- So styrene, which is not stabilised by an inhibitor at ambient temperature, can, especially from 40°C onwards and in the presence of oxygen or air, give rise to polymerisation as well as oxidation, which leads to a polyperoxide. If this polyperoxide is isolated, it detonates spontaneously. However, its solubility in monomer limits the risks. This property probably caused a detonation that occurred in a workshop, which was completely destroyed during oxidation caused by oxygen in contact with 1,1-diphenylethylene at 40-50°C under 100 bar.

- The 'polyperoxidation' of 1,3-dienes is even more dangerous because they are more reactive and some of their polyperoxides are insoluble. With butadiene, the polyperoxidation takes place at temperatures lower than -113°C; the oxygen is 'absorbed' very quickly and forms insoluble polyperoxides that precipitate. It was estimated that at a temperature of 25°C the critical mass of such a compound consists of a sphere of diameter 9 cm. This diameter decreases quickly with the temperature. Isoprene behaves in the same way, but its polyperoxide is soluble. In these conditions, the monomer can absorb any temperature rise which would be caused by the beginning of a decomposition, thus reducing risks. If the monomer evaporates, a gum that detonates at 20°C is formed if the medium is stirred. With cyclopentadiene, polyperoxide is more stable and only detonates at a high temperature.

- With aromatic compounds that have benzylic sites, the peroxidation is easy. An apparatus in which tetrahydronaphthalene was distilled detonated. It was assumed that this accident was linked to the concentration of peroxide formed in contact with oxygen. When using phenols as inhibitors of oxidation (and polymerisation too) for all these compounds, this avoids these risks.

The *Du Pont de Nemours* company attempted to classify the risks of dangerous peroxidation of chemical compounds. This classification includes the hydrocarbons mentioned above. This list is to be found in the entry about ethers (see on p.261).

Effect of liquid oxygen

Liquid oxygen and sometimes liquid air form explosive mixtures with hydrocarbons and organic compounds, which are very sensitive and lead to detonations that are highly catastrophic. George Claude was seriously injured in 1903 after inserting a candle into liquid oxygen. When liquid methane, benzene, traces of oil are in the presence of liquid oxygen they detonate easily. With methane the detonation is even more powerful when aluminium powder is present. Liquid oxygen explosives were used in mines and quarries. Petrol/liquid oxygen and carbon/liquid oxygen mixtures were used as explosives. An accident happened when liquid oxygen was in contact with an aluminium oxide filter which had absorbed oils.

Effect of the ozone/oxygen mixture

Ozonolysis is a reaction used with unsaturated hydrocarbons when preparing aldehydes and ketones, by reducing intermediate ozonide or acids by oxidation.

The reducing agents used include hydrogen in the presence of palladium, and zinc in acid medium.

Oxidation is usually brought about by using the acetic acid/hydrogen peroxide mixture (the active species is peracetic acid). Ozonide is highly dangerous, especially when it is insoluble in the medium.

- Ozonide formed from ethylene is particularly unstable.

6 • Reactions of organic chemicals

6.1 Hydrocarbons

- With isoprene, 2,3-dimethylbuta-1,3-diene and cyclopentadiene, if the ozone concentration in the ozone/oxygen mixture exceeds a certain limit (not stated), the medium immediately combusts when incorporating this mixture at -78°C.
- 1g of isoprene dissolved in heptane was treated by ozone at -78°C. This gave rise to a violent detonation when the cooling was stopped. It was thought that the reason for this accident was the temperature, which was too low, given the dilution of isoprene. Ozonide, which was stable in the dilute state, accumulated, leading to a high concentration, which decomposed violently when it was heated. This could have been avoided by increasing the temperature to let ozonide decompose gradually during its formation.
- With benzene and aromatic hydrocarbons in general (in particular phenylamine), the triozonide formed is insoluble and forms a very unstable gel. There was a violent detonation during the ozonisation of rubber dissolved in benzene. A gelatinous precipitate formed just before the detonation of this reaction.
- Finally, acetylene detonates violently when it comes into contact with an ozone/oxygen mixture in which the quantity of ozone exceeds 50 mg/l.

■ Effect of nitrogen oxides

The NO_2/N_2O_4 system is a particularly dangerous oxidant of unsaturated and aromatic hydrocarbons. A few accidents have been reported with saturated hydrocarbons, but most seemed to be caused by the presence of traces of unsaturated compounds.

- A petroleum ether/N_2O_4 mixture heated at 50°C (due to exceptional climatic conditions) detonated. The same happened with hexane at 28°C.
- A N_2O_4/cyclohexane mixture detonated violently. On the other hand, the NO_2/cyclohexane mixture does not seem to represent any danger.
- The mixtures are particularly dangerous with alkenes, cycloalkenes and dienes. Accidents have been reported with propene, butenes, isobutylenes, 1-hexene, butadiene and cyclopentadiene. However, the reaction below is not thought to be dangerous if one operates at a temperature of 30°C and under two bar.

$$CH_3-CH=CH_2 + N_2O_4 \longrightarrow CH_3-CH(NO_3)CO_2H$$

This reaction gave rise to a detonation following accidental overheating of a pump at 200°C.

- Benzene or toluene detonate when they are in contact with dinitrogen tetroxide, especially when dinitrogen tetroxide is in liquid form.
- The N_2O_4/cyclopentadiene system was used as a propellant for rockets.
- When large quantities of oxygen are introduced into N_2O_4/alkene or cycloalkene mixtures, detonations occur at a temperature starting at 0°C. These detonations have been explained by the formation of peroxynitrates and nitro-peroxynitrates, which are unstable:

$$R-O-O-NO_2 \qquad R-CH(NO_2)OONO_2$$
$$\text{peroxynitrate} \qquad \text{nitroperoxynitrate}$$

6 • Reactions of organic chemicals

6.1 Hydrocarbons

- Other sources say that, in general, nitrogen oxides give rise to relatively slow reactions with alkenes, 1,2-dienes and alkynes, which lead to an ignition of the medium only from 30°C onwards. For 1,3-dienes, the same sources say that the interaction between these reagents should be carried out between -180 and -100°C and becomes dangerous between -35 and -15°C.

The naphthalene/N_2O_5 mixture detonates spontaneously.

■ Effect of sodium nitrite

This salt is used as an inhibitor for the polymerisation of butadiene. It seems to play its role very well when its concentration is lower than 0.5%. If the concentration is greater than 5%, a black precipitate forms which contains 80% of organic polymer containing nitrate and nitrite groups in the proportions of 2:1. This solid combusts spontaneously at 150°C even when there is no air.

■ Effect of nitric acid

Nitric acid and especially 'fuming' acid is a strong oxidant and nitrating agent, especially when it is combined with sulphuric acid (formation of an electrophilic species). The dangers of the reactions which involve this compound are linked to the exothermicity of the reactions and the eventual formation, particularly in an aromatic series, of nitrated species that can be very unstable in some cases.

Most dangerous reactions deal with ethylenic, acetylenic and aromatic hydrocarbons. But there is an example of a dangerous reaction, which brings a saturated hydrocarbon mixed with pentacarbonyl iron into play.

- Fuel and oil compounds detonate on contact with nitric acid, if they contain 15 to 20% of unsaturated compounds.
- Ethylenic and acetylenic hydrocarbons combust spontaneously when they come into contact with nitric acid, within a millisecond.
- Cyclopentadiene reacts explosively in the presence of fuming acid even in inert gas. The danger is aggravated by sulphuric acid.
- Terpenes and in particular, spirit of turpentine, give rise to mixtures which combust spontaneously with nitric acid. Some of the above mentioned mixtures have been used as propellants for rockets.
- A benzene/HNO_3 mixture heated at 170°C reacts by giving off heat, which is sufficient to cause benzene to self-ignite.
- Nitrating toluene with HNO_3/H_2SO_4 mixture led to a violent detonation because it was badly monitored.
- Oxidising p-xylene in terephthalic acid is thought to be potentially dangerous.
- The synthesis of 3,5-dimethylbenzoic acid by oxidation of mesitylene by nitric acid has to be carefully monitored. Nitric acid has to be introduced very slowly so that the temperature of the medium does not exceed 20°C. In one method an acetic acid/acetic anhydride mixture, is used as a solvent (this seems surprising given the dangers of this solvent vis-à-vis nitric acid).
- When this acid was prepared in an autoclave at a temperature of 115°C, a violent detonation occurred, which pulverised the apparatus. This accident was

explained by the formation of 1,3,5-trinitromethylbenzene, which is very unstable.
- There is a synthesis, which is supposed to be safe and consists in using very small quantities of reagents and closely monitoring the temperature. However, the thermal control of the aromatic hydrocarbons/nitric acid reaction usually proves to be very difficult. Indeed, the temperature is either too high and the reaction is out of control and can lead to detonation, or too low and the nitration or oxidation takes place too slowly causing the compounds to accumulate and the reaction to be delayed. The consequences are the same as before.
- This occurred when attempting to prepare pyromellitic acid by the action of nitric acid on hexamethylbenzene. This accident was worsened by the fact that there was no stirring.

■ Effect of barium peroxide

- Propane in the gaseous state and under atmospheric pressure reacts with barium peroxide when it is hot, giving rise to a violent exothermic reaction.

■ Effect of sodium peroxide

- Benzene, which contains traces of water, combusts immediately when it is in contact with sodium peroxide. It is likely that in this case the active component is hydrogen peroxide, which results from the interaction between water and sodium peroxide.

■ Effect of peroxysulphuric acids

- When alkanes come into contact with peroxymonosulphuric or peroxydisulphuric acid, this leads to a slow charring. With benzene the detonation is immediate.

■ Effect of chromyl dichloride (Cl_2CrO_2)

- Spirit of turpentine combusts spontaneously in the presence of this reagent.

6.1.5 Other dangerous reactions

■ Other reactions of alkynes

The dangerous reactions of halogenation, which affect either the acetylenic hydrogen atom or the triple bond have just been described. There are also reactions which lead to accidents and affect one of the acetylenic sites.

So it can be concluded that it is almost always dangerous to substitute a hydrogen atom for a metal since the acetylenic organometallic derivative formed is unstable. The danger is particularly great when this metal is a transition element.

- When acetylene comes into contact with molten potassium it forms potassium acetylide, which detonates spontaneously under thermal conditions.

6 • Reactions of organic chemicals

6.1 Hydrocarbons

Synopsis of dangerous reactions of hydrocarbons

Acetylene, propyne, vinylacetylene...	Allene	Benzene	Butadiene	Cyclopentadiene	Essence of turpentine	Ethylene, propylene...	Methane, ethane... cyclohexane...	Phenylacetylene	Styrene	Toluene, xylenes, mesitylene...	Reagent
		■									Hydrogen
■			■		■						Group I (Li; Na; K...)
											Group II (Be; Mg; Ca...)
											Group III (B)
			■								Group IV (C; Si; Ge)
			–								Group V (N; P; As; Sb)
■		■	■		■						Group VI (O₂)
■		■	■								Group VII (Cl₂; Br₂)
■			■								Transition metals
			■								Other metals (Al... Bi; Sn; Pb)
											Hydrides Group III (BₓHᵧ)
											Hydrides Group IV (SiH₄...)
											Hydrides Group V (NH₃...)
											Water
											Hydrides Group VI (H₂S...)
											Hydrogen halides
					■						Metal halides (AlCl₃; FeCl₃...)
			■								Chromium chloride
	■							■			Non-metal halides (SOCl₂...)
											Metal oxides
■	■	■	■	■	■				■		Non-metal oxides (O₃; N₂O₄; N₂O₅)
■	■	■	■		■				■		Organic peroxides, (H₂O₂; Na₂O₂...)
■		■	■			■					Acid oxidants (HNO₃; H₂SO₅...)
		■			■	■					Acids and salts (H₂SO₄; HClO₄)
■		■									ClOₓ⁻
■	■		■								NO₃⁻; NO₂⁻
			■								MnO₄⁻, Cr₂O₇²⁻
■			■	■				■			Instability
■			■	■					■		Polymerisation
	■										Bases
											Organometallics
			■								Zeolites

246

6 • Reactions of organic chemicals

6.1 Hydrocarbons

- Propyne reacts with ammoniacal silver nitrate by forming a metallic derivative that detonates at 150°C.
- When acetylene and 'real' alkynes are in contact with copper, silver and transition metals, they form acetylides and analogues that are explosive, from ambient temperature upwards for acetylides. For instance, acetylene that was accidentally in the presence of electrical wires whose copper was bare led to a detonation.
- The violent reaction of lithium with ethylene is explained by formation of lithium acetylide; the metal glows.

Other dangerous reactions seem to affect the triple bond.

- It is dangerous to put mercury (II) salts in contact with acetylene in the presence of concentrated sulphuric acid.
- Mixtures of acetylene and dinitrogen oxide used to create flames in atomic absorption apparatus detonate in the presence of perchloric acid.
- Phenylacetylene reacts with anhydrous perchloric acid at -180°C and forms an organic perchlorate, which decomposes violently at a temperature starting at -78°C. The following reaction has been suggested:

$$C_6H_5 - C\equiv CH + HClO_4 \xrightarrow{-180°C} [C_6H_5 - C=CH_2]^{\oplus} ClO_4^{\ominus} \xrightarrow{-78°C} \text{Explosion}$$

■ Dangerous reactions in aromatic series

This paragraph, which is concerned with the oxidation of hydrocarbons, describes a few dangerous reactions involving this type of compound. The catalytic hydrogenation of benzene into cyclohexane, which is the main method of making cyclohexane in industry, does not represent any danger when it is carried out at 210-230°C.

- However, an accident was reported in which a drop in the production of cyclohexane was observed during such an operation. The technicians decided to proceed with a gradual temperature rise (10°C increments) and increase the flow rate of hydrogen. The reaction suddenly got out of control, the temperature reached 280°C and even 600°C, causing the destruction of the installation.
- When benzene is used as a solvent, it causes accidents.
- The effect of potassium methanoate on arsenic pentafluoride does not represent any danger provided the solvent used is trichlorotrifluoroethane. Every time benzene is used instead of the halogenic solvent, a large number of detonations have been reported.
- Benzene and toluene form complexes with some salts; these complexes are often very unstable. With silver perchlorate, benzene gives rise to a complex that leads to very dangerous benzenic solutions. Besides, it detonates when it is ground up. Its enthalpy of formation corresponds to -3.4 kJ/g, which makes it dangerous according to the CHETAH criterion C_1 (see para 2.3.2).

Finally, the following accident was reported.

- After storing sodium in m-xylene for 16 years, a layer of white solid was found at the bottom of the container which detonated when the flask was moved. It seems that the reaction was due to a peroxydic impurity. After this accident, it was recommended not to keep suspensions of alkali metals in hydrocarbons for longer than six months. Their labelling should mention the date they were prepared. In America, once the time limit has passed they are destroyed by combustion.

■ Accidents of cryogenic liquids

This paragraph is not concerned with dangers of a chemical nature. It is concerned with all gaseous compounds which are handled in the liquefied state. Generally speaking, liquefied gases are likely to detonate every time they come into contact with water at ambient temperature, if the difference in temperature between both liquids reaches about a hundred degrees. However, they behave very differently according to their nature.

- Propylene has been involved in such detonations. The detonation is inevitable when the difference between its temperature and that of water lies between 96 and 109°C. The same is not true for ethylene.

6.2 Alcohols and glycols

The alcohols' reactivity is linked to:
- the nucleophilic character of active oxygen atom, which can cleave the O-H or C-O bond; the behaviour depends on the structure of the hydrocarbon chain;
- their reducing property, which can lead to oxidation reactions that are often dangerous;
- the 'activating' influence of the active group on the hydrogen atom in a single or multiple bond, or aromatic rings that are potentially present.

The dangerous reactions of alcohols, apart from the ones that involve the carbon chain, are linked either to the exothermicity of the reactions whose consequences are often aggravated by poor temperature monitoring, or the instability of the intermediate or final compounds formed. The first case often happens with oxidation reactions, the second especially with substitutions of active hydrogen or hydroxyl. Nitric acid will be the subject of special consideration since it can have both characters, without knowing which one played a role during accidents that have involved this compound.

6.2.1 Substitution reactions of the active hydrogen atom

■ Effect of metals, organometallic compounds and strong bases

Reducing metals sometimes react violently with alcohols.
- 1-Butanol gave rise to aluminium tributanoate when it came into contact with equipment containing aluminium. The equipment detonated because of the overpressure created by the hydrogen formed. The exothermicity of this reaction is also a risk factor.
- An accident was caused by an excessive temperature rise due to 2-propanol in contact with a surplus of aluminium.
- There was a similar accident when methanol was added to magnesium turnings; the quantity of methanol was too low and the turnings' heating up was sufficient to cause methanol to combust.

- There are a lot of accidents which occur when ethanol is used without taking enough precautions when destroying sodium waste. With organometallic compounds an alkoxide is formed.
- 2-propanol mixed with triethylaluminium caused a detonation. The same accident happened when the same alcohol was mixed with diethylzinc.
- Organoaluminium derivatives with the formula: R_xAlZ_{3-x} (Z = OR, OH, X) give rise to explosive reactions with methanol and ethanol. With very strong bases, the reaction can be written down as:

$$B^- + R\text{–}OH \longrightarrow BH + R\text{–}O^-$$

In unfavourable operating conditions or with very thick alcohols (polyols), the exothermicity of the reaction always causes alcohol to combust, sometimes after a latency period.

- When potassium tert-butylate in the solid state comes into contact with a few drops of methanol, ethanol, or 1- or 2-propanol it causes the alcohols to combust after a latency period of two, seven, and one minute(s) respectively. If the alcohol is in the vapour state, ignition can also take place. However, if there is a large quantity of alcohol, there is no incident since alcohol in excess absorbs the heat produced.
- Sodium hydride reacts with glycerine forming hydrogen and the corresponding alkoxide.

  If hydride is in the form of granules and glycerine is in the pure state, the alcohol combusts. The fact that its thickness limits the heat dissipation makes it responsible for this behaviour. When glycerine is diluted with tetrahydrofuran, for instance, there is no ignition.
- Sodium borohydride also gives rise to ignition with the same alcohol.
- Ethylene glycol reacts dangerously with sodium hydroxide at a temperature starting at 230°C; a highly exothermic decomposition, which emits a lot of hydrogen occurs. If the reaction takes place in a closed reactor, the apparatus detonates systematically.

■ Effect of halogens and halogenation agents

The dangers linked to reactions of this kind are explained by the formation of hypohalites.

- When chlorine was put in contact with methanol, this led to a deflagration followed by a fire due to the decomposition of methyl hypochlorite.
- A mixture of 15 cm³ of methanol and 9 cm³ of bromine was refluxed. After two minutes there was a very exothermic reaction that could not be kept under control and caused the compounds to overflow. A similar accident was described with ethanol.
- The following reaction was carried out in a school. It was considered as being too dangerous to be used in such an environment.

$$C_2H_5\text{–}OH + Br_2 + P \longrightarrow C_2H_5\text{–}Br$$

- Liquid chlorine and glycerine were heated in a steel bomb at 70-80°C, which detonated.

6 • Reactions of organic chemicals
6.2 Alcohols and glycols

Alcohols react violently with solid calcium hypochlorite, usually causing an ignition. Sometimes very irritant fumes form. Such behaviour was mentioned with methanol, ethanol, glycerol and monoethers of diethyleneglycol. It was assumed that hypochlorites were formed.

- The detonation of a mixture of methanol with an acid solution of sodium hypochlorite was explained in the same way.

■ Effect of phosphorus derivatives

When the phosphorus compound is a phosphorus trihalide or phosphoryl (or thiophosphoryl) trihalide, two types of reactions are possible. For instance, with a phosphorus trihalide:

$$PX_3 + R-OH \begin{cases} (1) \rightarrow (RO)_3P + HX \\ (2) \rightarrow R-X + P(OH)_3 \end{cases}$$

reaction (1), which forms a phosphite, usually occurs with primary alcohols whereas reaction (2), which leads to a halogenated derivative, occurs with tertiary alcohols. These reactions are not dangerous provided a suitable operating mode is used.

- When adding phosphorus tribomide drop by drop on to 3-phenyl-1-propanol, the stirrer broke down causing the formation of a thick layer of phosphorus tribomide. An attempt was made to make the stirrer work manually, which caused the apparatus to detonate because of the massive and sudden formation of hydrogen bromide.
- Glycerol gave rise to a particularly violent reaction with phosphorus triiodide.
- A mole of thiophosphoryl trichloride and a mole of pentaerythritol were heated for four hours at 160°C, when a violent detonation took place. The investigation revealed that the detonation could be linked to the formation of PH_3, phosphine.
- Finally, when tetraphosphorus decasulphide was refluxed with ethylene glycol with hexane used as a solvent (~70°C), this gave rise to an exothermic reaction, which was out of control and raised the temperature to 180°C.

■ Effect of ethanol on butadiene

- The following reaction gave rise to a detonation, which could not be explained:

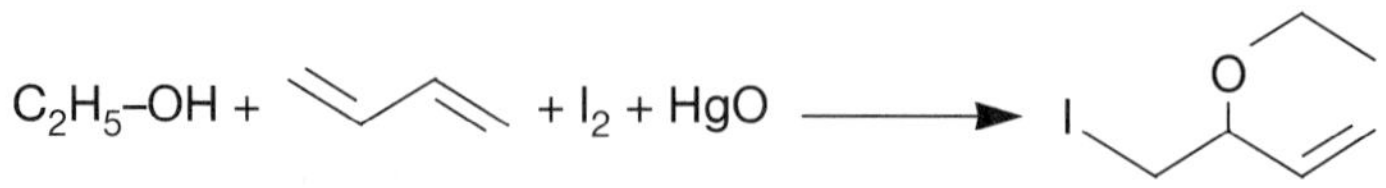

$$C_2H_5-OH + \text{(butadiene)} + I_2 + HgO \longrightarrow \text{(iodo-pyran product)}$$

6.2.2 Hydroxyl substitution reactions

Except with nitric acid, most dangerous reactions are those of alcohols with perchloric acid or its salts (that can form acid). All accidents reported are caused by the formation of highly unstable organic perchlorates.

- Magnesium perchlorate used as a dehydrating agent was exposed to ethanol fumes for several months, while attempting to crumble the salt formed. The detonation which followed was explained by the formation of ethyl perchlorate which is explosive on impact or friciton.

- Similar accidents have happened when barium perchlorate is added to C_1 to C_3 alcohols as well as 1-octanol. The latter alcohol demonstrates the nature of this danger and is a counter-example of the typical observation of stabilisation of unstable species, when the number of carbon atoms increases (whereas it is the case for peroxides and peracids).
- When ethanol was added to potassium perchlorate, formed during a gravimetric measuring of potassium, it caused a very violent detonation.

With perchloric acid, the following dangerous reactions have been reported:

- A lipid was extracted from a methanol/chloroform mixture, the solution obtained was then treated with 1.5 cm^3 of 72% perchloric acid. This led to a violent detonation, which was explained by the decomposition of methyl perchlorate.
- When a few drops of anhydrous perchloric acid are added to ethanol, there is an immediate detonation. With ethylene glycol, glycerine and pentaerythritol, this acid forms liquids that detonate as soon as they are moved from one container to the next. The same is true for glycerine, when handled when in the presence of perchloric acid and lead (II) oxide.
- 2-Butoxyethanol forms highly explosive liquids with perchloric acid at 50-95% at 20°C or 40-90% at 75°C.
- Finally, there was an unexpected accident that happened during hydrogenation of an organic compound by hydrogen in the presence of Raney nickel, under a pressure of 100 bar. Methanol was used as a solvent. The detonation of the installation was explained by the overpressure resulting from the formation of methane. This explanation does not seem very convincing since this reaction does not seem to involve any change in volume (thus in pressure). Calculating the enthalpy of this reaction gives a value of approx -3kJ/g that shows an exothermicity, which is important enough to provide an explanation for the damaging nature of this accident.

6.2.3 Reactions of nitric acid and nitrates with alcohols

There are a plethora of accidents which involve this acid and the reasons put forward are ambiguous. Nitric acid gives rise to three kinds of dangerous reactions, which are represented below:

$$CO_2, H_2O, R-CO_2H, R-CHO \xleftarrow{HNO_3 \; (2)} [R-OH] \begin{array}{c} \xrightarrow{HNO_3 \; (1)} R-ONO_2 \; \text{Nitrates} \\ \xrightarrow{MNO_3} \\ \xrightarrow{HNO_3 \; M \, (3)} M^+[O-N\equiv C]^- \; \text{Fulminates} \end{array}$$

The transformations (1) lead to unstable alkyl nitrates, which can detonate very easily. The reactions (2) lead to more or less complete oxidations of the organic molecule. The formation of aldehydes is purely theoretical since they are more oxidisable than alcohols and therefore are not part of the oxidation process by nitric acid. On the other hand, a ketone can form with a secondary alcohol. With tertiary alcohols, carboxylic acid is the only possible outcome of the partial oxidation, which is caused by the breaking of C-C bonds. When the oxidation is out of control, it is likely to have a complete oxidation. Finally, with heavy metal nitrates (symbolised by a capital M) or with nitric acid contaminated by heavy metals (see equation (3)), it is thought that explosive heavy metal fulminates form. It is not always easy to know which type of transformation is the factor of the accident.

- A methanol/nitric acid mixture was used as a propellant for rockets.
- A technician was asked to make a 5/95 nitric acid/ethanol solution. He made a 95/5 mixture because of a misunderstanding (or a lack of clarity in the instructions given). The mixture detonated immediately.
- Because of unclear instructions, 120 litres of concentrated nitric acid (instead of 2.6%) was fed into a reactor, which contained 5 litres of 2-propanol. The mixture detonated violently and caused the installation to be destroyed. It was assumed that isopropyl nitrate was present.
- A mixture of nitric acid, sulphuric acid and glycerine detonated violently, probably because of the formation of nitroglycerine.
- A 15% nitric acid solution in ethanol was used to clean a bismuth crystal. The solution decomposed violently after this treatment and projected the content of the container into the extraction hood in which it was placed. Mixtures with less than 10% of nitric acid in alcohol are the only ones to remain stable for short periods of time. They should never be stored. In this particular accident, the presence of bismuth suggests a reaction of type (3).
- A mixture of 80 cm^3 of nitric acid, 80 cm^3 of hydrogen fluoride and 240 cm^3 of glycerine was used to polish a metal. It was stored after treatment. It detonated three days later. Did the metal that was found in what was left of this mixture after the accident play a role in the detonation (reaction (3))?
- A metal polishing fluid of a similar composition that contained propane-1,2-diol, nitric acid, hydrogen fluoride and silver nitrate detonated thirty minutes after being used.
- When nitric acid was added to cyclohexanol when preparing cyclohexane-1,2-dione, the reactor detonated, probably because of the exothermicity of the oxidation reaction, which was complete instead of partial (reaction (2)).
- A mixture of solid silver nitrate and ethanol was handled with a metallic spatula. The mixture detonated as soon as it came into contact with the spatula. This detonation could be explained by the accidental formation of silver fulminate. However, witnesses noticed the smell of ethyl nitrate after the accident.

6.2.4 Dangerous oxidising reactions

■ **Effect of oxygen**

Some alcohols peroxidise in air.

- An extremely violent detonation occurred when 2-butanol, which had been stored for ten years, was distilled. Analysis that was carried out after the accident on what was left from butanol showed that it contained 12% of peroxides. An exactly similar accident was mentioned with 2-butanol distilled after being stored for twelve years. The analyses carried out after the accident led to the same level of peroxides. It seems that traces of butanol must have played a role of promoter in the peroxidation.

6 • Reactions of organic chemicals

6.2 Alcohols and glycols

- Five accidents have been described, which all led to a detonation at the end of distillation of 2-propanol stored over a period of about four years.

Experiments were made with pure 2-propanol samples stored for six months in white glass containers, which were half full and exposed to the light. It could be proved that peroxides form in these conditions. The experiment showed that peroxidation is faster when a ketone is present. The fact that this oxidation only takes place with secondary alcohols shows that it affects the hydrogen atom in the α position of the active group, (atom that is more reactive than a primary hydrogen), and not the actual function.

Before showing how direct peroxidation of active group works, it should be mentioned that platinum catalyses oxidation of alcohols at ambient temperature in aldehydes (provided they are primary). The exothermicity of the reaction is sufficient to cause the alcohol to combust by heating a thread of platinum at a high temperature. Note that 2-butanol (but not isopropanol) is classified in list B, producing peroxides that become explosive at a certain concentration in the lists of peroxidable compounds set up by Du Pont de Nemours company (see p.261).

■ Effect of peroxides

The reaction of hydrogen peroxide with an alcohol in the presence of sulphuric acid gives rise to the corresponding hydroperoxide:

$$R-OH + H_2O_2 \longrightarrow R-O-O-H + H_2O$$

The danger of this reaction depends on the order in which the reagents are introduced. If acid is introduced into the alcohol/peroxide mixture, there can be detonation especially when the medium is heterogeneous. It is assumed that the danger of the reaction is linked to the formation of peroxymonosulphuric acid, which has such an oxidising property that it can react with all organic compounds:

$$H_2SO_4 + H_2O_2 \longrightarrow H_2SO_5$$
$$H_2SO_5 + R-OH \longrightarrow [R-O-OH] \longrightarrow \text{or } [R-O-O-R] \longrightarrow \text{Explosion}$$

Similar accidents have been mentioned with $H_2S_2O_8$, peroxydisulphuric acid. If the peroxide/alcohol mixtures are made with concentrated hydrogen peroxide, they either detonate or combust spontaneously. Analysis shows that the self-ignition temperature of 2-propanol is much lower when hydrogen peroxide is present.

- Preparing tert-butyl peroxide by the effect of 50% hydrogen peroxide in the presence of 78% sulphuric acid has led to numerous accidents. They were due to the high exothermicity of the reaction causing a temperature rise and leading to the explosive decomposition of the peroxide formed, if the temperature rise is badly monitored.

$$(CH_3)_3C-OH + H_2SO_4 \text{ (78%)} + H_2O_2 \text{ (50%)} \longrightarrow (CH_3)_3C-O-O-C(CH_3)_3$$

- Methanol gives rise to an explosive mixture with phosphoric acid and hydrogen peroxide if the proportion of peroxide is too high. This mixture is used to polish metals.

- The following reaction can be dangerous depending on the order in which the reagents are introduced:

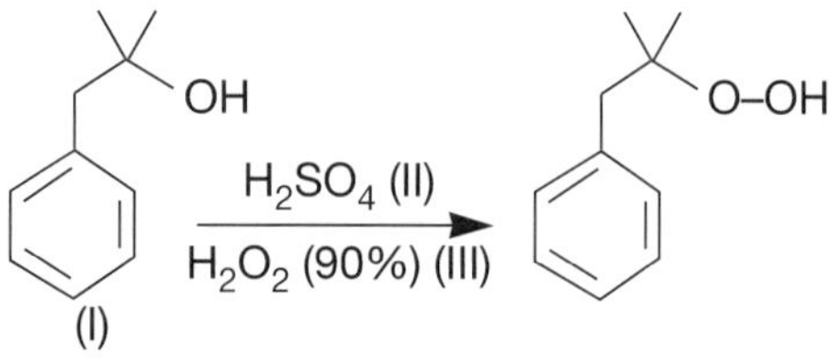

$$\text{PhCH}_2\text{C}(\text{CH}_3)_2\text{OH} \xrightarrow[\text{H}_2\text{O}_2 \text{ (90\%) (III)}]{\text{H}_2\text{SO}_4 \text{ (II)}} \text{PhCH}_2\text{C}(\text{CH}_3)_2\text{O-OH}$$

(I) → product

If (II) is introduced into [(I) + (III)], the medium detonates.
If [(II) + (III)] are introduced into (I), the reaction is regarded as less dangerous.
Alkaline peroxides are equally dangerous.

- Potassium reacts violently with the alcohols used for its destruction. This danger is linked to the fact that the potassium residues to be destroyed are exposed to oxygen and the destruction takes place in air. If the operation is carried out in inert gas and with potassium that is not exposed to air, the reaction is not dangerous. The danger of this reaction comes from the fact that potassium forms a KO_2 superoxide easily, which oxidises the alcohol violently.

It should be noted that the superoxide ion O_2^- and peroxide ion O_2^{2-} are different, but are put together because they behave the same way.

- Sodium peroxide gives rise to accidents of the same nature as potassium superoxide with ethanol, ethyleneglycol, sugars at ambient temperature.

■ Effect of other oxidants

All accidents that have happened with oxygenated salts of transition metals (permanganates, dichromates), metallic oxides (CrO_3), chlorates and chlorites have been listed (perchlorates and hypochorites are described on p.250).

- An explosive detonation occurs when potassium permanganate comes into contact with ethanol in the presence of sulphuric acid, if the medium is heated.

- An accident occurred when 2-propanol reacted with potassium dichromate in a sulphuric medium. The aqueous solution of dichromate and sulphuric acid were added in small quantities. The homogeneous liquid was throwing up glowing particles (CrO_3 particles?). The same reaction proved not to be dangerous when this process was carried out drop by drop.

- Ethylene glycol combusted spontaneously when it was added to potassium permanganate at ambient temperature and at 100°C in the presence of ammonium or potassium dichromate, sodium chlorite or silver chlorate. With potassium dichromate, the medium reached 170°C.

- Ethanol combusts when chromyl dichloride (Cl_2CrO_2) is present.

- Finally, chromium trioxide in the solid state causes alcohols with which it comes into contact to combust immediately. Accidents are reported with methanol,

ethanol, 2-propanol and butanols. With butanols, the order in which they react vis-à-vis this oxide is the following:

$$n\text{-}C_4H_9OH > sCH_4H_9OH > t\text{-}C_4H_9OH$$

Methanol is sometimes used to prepare dichromium trioxide by the reduction of chromium oxide (VI). Since these attempts lead to spontaneous ignition and detonation, it is preferable to use the thermal decomposition of ammonium dichromate, provided it is carried out with caution (see p.200).

6.2.5 Dangerous properies of carbon chains

The OH activates the hydrogen atoms situated with the active carbon. This was established, for example, with the hydroperoxidation by oxygen as described previously. Nor can the halogenation effect of α OH on hydrogen atoms be excluded although the dangerous character of these halogenation reactions is better explained by the formation of hypohalites as seen on p.224. In fact, dangerous interactions usually occur when the carbon chain is unsaturated or has an aromatic ring.

■ Instability of unsaturated alcohols

This property was described for hydrocarbons on p.235. The presence of an alcohol function is an aggravating factor if it is situated close to the multiple bond.

Therefore allylic alcohol is regarded as unstable. At 360-500°C, its enthalpy of decomposition is 0.67 kJ/g. This is rather low according to the CHETAH method (see para 2.3) but high according to T. Grewer, who regards a compound as unstable when $|\Delta H_d| > 0.5$ kJ/g.

The same goes for butyne-1,4-diol, which is unstable, in particular in the presence of strong bases, metal halides or Hg (II) salts in a sulphuric medium.

- Distilling this diol when a base is present leads to violent detonation.

As will be seen on p.256 some accidents of propargyl alcohol can be explained by its easy decomposition although other explanations are given.

■ Polymerisation

With ethylenic hydrocarbons, as seen on p.237, it is rather difficult to know whether a dangerous reaction is due to the instability of the double bond or its polymerisation. Allylic alcohol dehydration is one example. Alcohols treated in a sulphuric acid medium can form, according to the conditions and the alcohol structure, either an ethylenic hydrocarbon or an ether.

- When preparing allyl oxide according to this method, a violent detonation interrupted the operation. It was explained by the alcohol polymerisation catalysed by sulphuric acid. There was a less convincing explanation, which is peroxidation in the allylic position of alcohol or the ether obtained. Indeed, there should have been prolonged storage for this peroxidation.
- There was also a 'polymerisation' reaction during the sulphuric treatment of benzyl alcohol. The reactor in which the reaction of alcohol dehydration was

6 • Reactions of organic chemicals

6.2 Alcohols and glycols

carried out detonated at 180°C, which is the usual temperature for this type of reaction. This accident was explained by an aromatic polysubstitution (improperly called polymerisation) through benzyl carbocation in the following way:

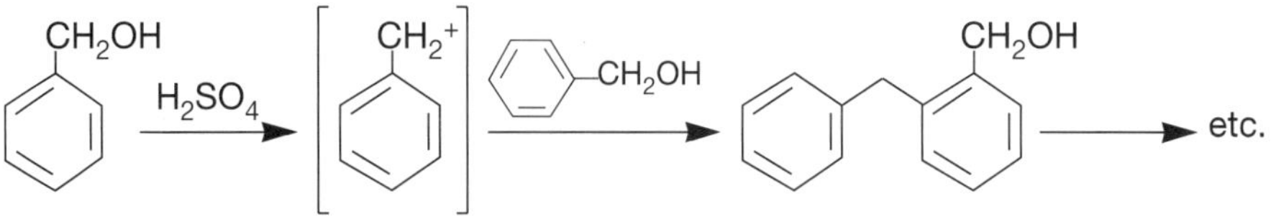

$$C_6H_5CH_2OH \xrightarrow{H_2SO_4} [C_6H_5CH_2^+] \xrightarrow{C_6H_5CH_2OH} \text{dimer} \longrightarrow \text{etc.}$$

The aromatic substitution can go on indefinitely and violently by affecting the ortho and para sites of the different rings that are formed at each stage.

- The same explanation was suggested for a detonation that involved the same alcohol contaminated by 1.4% of hydrogen bromide and 1.1% of ferrous ion. The electrophilic intermediate formed as follows:

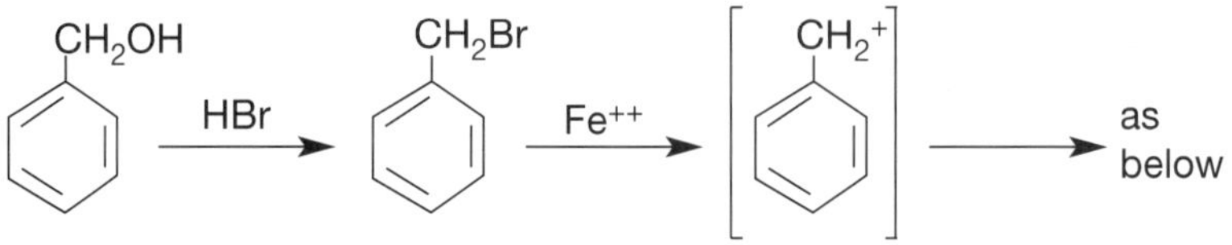

$$C_6H_5CH_2OH \xrightarrow{HBr} C_6H_5CH_2Br \xrightarrow{Fe^{++}} [C_6H_5CH_2^+] \longrightarrow \text{as below}$$

■ Reactions of the carbon chain of propargyl alcohol

This very unstable alcohol, which is usually used in an aqueous solution, gives rise to the dangerous reactions of the acetylene 'functional' group.

- Drying this alcohol by using a base before distillation led to a violent detonation, which was explained by the formation of unstable sodium acetylide.
- An accident of the same nature happened when it was dried with diphosphorus pentoxide. In this case, it is unlikely that the reason for it is the formation of an acetylene salt. The author believes that, since water plays a 'desensitising' role on this compound, it leads to a very unstable pure alcohol when it is removed by strong dehydrating agents. The temperature rise due to the dessicating agent hydration is sufficient to decompose the pure alcohol violently. This interpretation could also be applied to the previous case.
- An aqueous solution of 33% of propargyl alcohol gives rise to an unexpected detonation in a sulphuric medium (acid at 56%). Indeed, this reaction is thought to be safe. It was thought that this accident was either linked to difficulties in cooling or the presence of a heavy metal salt (see next reaction).
- The effect of 6 g of aqueous sulphuric acid in 0.6% mole of aqueous propargyl in alcohol at 30% in the presence of mercury (II) sulphate was used when preparing the corresponding keto-alcohol in the following way:

$$HC\equiv C-CH_2OH + H_2SO_{4\,(aq)} + HgSO_4 \longrightarrow \left[CH_2=\underset{\underset{\displaystyle OH}{|}}{C}-CH_2OH \right] \longrightarrow CH_3-CO-CH_2OH$$

This reaction got out of control at 70°C when the stirrer was switched on, causing the compound to overflow.

6 • Reactions of organic chemicals

6.2 Alcohols and glycols

Synopsis of dangerous reactions of alcohols and glycols

Allyl alcohol	Benzyl alcohol	Propargyl alcohol, butyne-1,4-diol	2-Butanol, 2-propanol...	t-Butanol	Ethylene glycol	Glycerol	Methanol, ethanol, 1-propanol	Pentaerythritol	3-Phenyl-1-propanol	Propane-1,2-diol	Reagent
					■						Hydrogen
											Group I (Li; Na; K...)
											Group II (Be; Mg; Ca...)
											Group III (B)
											Group IV (C; Si; Ge)
					■						Group V (N; P; As; Sb)
		■									Group VI (O_2)
				■							Group VII (Cl_2; Br_2)
											Transition metals
		■									Other metals (Al... Bi; Sn; Pb)
											Hydrides group III (B_xH_y)
											Hydrides group IV (SiH_4...)
											Hydrides group V (NH_3...)
											Water
											Hydrides group VI (H_2S...)
■							■				Hydrogen halides
											Metal halides ($AlCl_3$; $FeCl_3$...)
				■	■						Non-metal halides (PBr_3; PI_3...)
					■						Cl_2CrO_2
■							■				Metal oxides (CrO_3; HgO)
	■										Non-metal oxides (O_3; N_2O_5; P_2O_5...)
						■					P_2S_5
■											Peroxides (H_2O_2; Na_2O_2)
■											Acid oxidants (HNO_3...)
■											Other acids and salts (H_2SO_4; $HClO_4$)
											ClO_x^-
											NO_3^-; NO_2^-
	■										MnO_4^-, $Cr_2O_7^{2-}$
											Instability
■		■									Polymerisation
		■									Bases
							■				$(CH_3)_3C-O^-\ K^+$
							■				Metal hydrides (NaH; $LiAlH_4$...)
											Organometallics
					■						Butadiene

257

6.3 Phenols

Phenols have two structural features, the hydroxyl function and the aromatic ring, for which the dangerous reactions are analysed on p.250. There follows an analysis of the effects of each of these features on the other and their consequences in terms of risk.

6.3.1 Effect of ring on OH group

The ring increases the mobility (acidity) of the hydroxyl hydrogen atom greatly. This is due to the conjugation between the lone electronic pair of oxygen and the ring. Moving this pair of electrons towards the ring makes oxygen become positive, which facilitates the proton's separation. This effect stabilises the anion created by moving its charge.

The first situation is illustrated in the following resonance structures in the case of phenol:

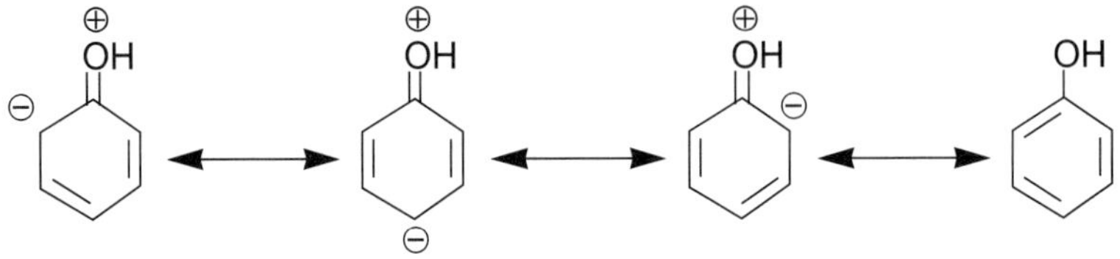

the second is illustrated in the following:

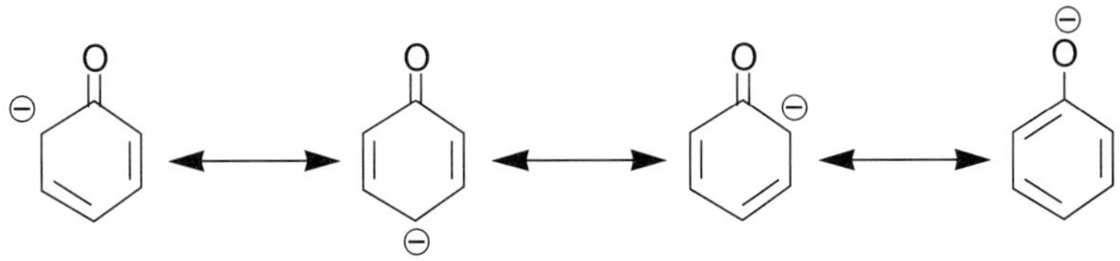

This electronic effect increases the stability of the anion created by the proton loss more than the phenolic one. It may therefore be assumed that the reactions linked to the substitution of the active hydrogen are more dangerous than with alcohols.

6.3.2 Effect of hydroxyl group on aromatic ring

By increasing the negative charge on the nucleus, the hydroxyl radical will activate this ring by means of electrophilic substitution, which is the principal dangerous reaction of the aromatic ring in the case of nitration. Thus, in phenol nitration, the slow reaction involves the formation of a carbocation by adding the electrophilic species NO_2^+. This cation is more stabilised by conjugation than the carbocation of a non substituted ring (benzene):

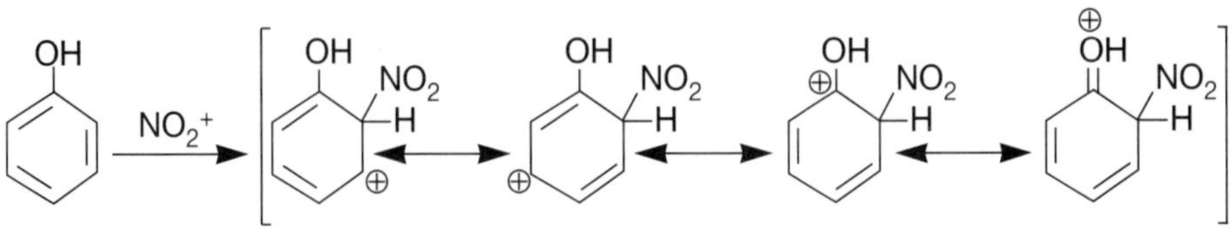

6 • Reactions of organic chemicals

6.3 Phenols

The stabilisation of the intermediate cation explains why the activation energy of the reaction is reduced; this will increase its speed considerably. The transition state resembles the cation. There are numerous consequences so far as risk is concerned: the speed at which heat evolves increases due to the exothermicity of the reaction and there is an easy polysubstitution, which leads to trisubstitutions that can give rise to very unstable compounds.

The molecular properties are the only risk factors that are responsible for accidents. This can explain why there are only a few accidents mentioned in the specialised literature. Here they are listed by subdividing them into three sections:

- dangerous reactions linked to hydroxyl;
- dangerous reactions linked to the ring;
- dangerous reactions of oxidation of phenols.

6.3.3 Hydroxyl-related reactions

- A very violent exothermic reaction occurred when heating a concentrated sodium hydroxide solution with hydroquinone.

6.3.4 Ring-related reactions

- The [CF_3-CO_2H, $NaNO_3$] nitrating mixture gives rise to a dangerous reaction and the formation of tars with phenol.
- Resorcinol was nitrated with nitric acid at 82% in the following way:

[Resorcinol] + HNO_3 (82%) → [2,4-dinitroresorcinol]

In the course of the experiment it was felt that the nitric acid concentration was not sufficient and it was decided to interrupt the reaction. When it was noticed that tars formed, the medium was decanted into another container and more concentrated nitric acid poured on to the tar left at the bottom of the reactor. The medium detonated immediately.

- As soon as phenol is in contact with calcium hypochlorite the mixture flares up, giving off fumes that are highly irritant and identified as chlorophenols and/or dichlorophenols. Comparing this observation with similar accidents obtained between this hypochlorite and alcohols shows an important modification of the reaction process. By analogy with alcohols, one would expect that the danger from this reaction arises from the formation of an aryl hypochlorite following (1) below. In fact, the reaction (2) took place, which shows that Cl^+ ion generated by the hypochlorite gives an electrophilic substitution instead. However, the radical aspect of the mechanism cannot be excluded as it is sometimes put forward for hypochlorites (especially organic).

6.3.5 Oxidation reactions of phenols

- Peroxymono- and disulphuric acids detonate in contact with phenol.
- A mixture of phenol and sodium nitrite detonated when it was heated in a test tube.
- When hydroquinone comes into contact with nitric acid, it combusts spontaneously within a millisecond.
- The effect of oxygen on hydroquinone produces a clathrate. The usual process consists in operating in a solution of hydroquinone in 1-propanol, at 70°C and with oxygen pressure of between 20 and 150 bar. Then, there is a slow cooling without stirring to obtain the formation of large crystals. In order to obtain smaller crystals the process was modified. It was decided to raise the temperature (90°C) and the pressure first to 90 bar then 100 bar. The cooling was carried out by stirring. After 20 minutes, the pressure reached 450 bar and caused the disk of the explosion vent to detonate. Then the detonation damaged the reactor and the equipment in the surrounding area. It was assumed that the accident was due to the violent and very exothermic oxidation of hydroquinone, which brought the medium to a temperature greater than the AIT of 1-propanol (370-415°C).

6.4 Ethers

Ethers can give rise to five different types of dangerous reactions.

1. Those related to the nucleophilic character of the active oxygen atom. The resulting 'acid-base' reactions are only dangerous if the acid-base complex obtained is unstable.
2. The reactions due to C-O bond rupture. They are generally safe due to the low reactivity of this bond in general. Most ethers are stable. This is the reason why they are often used as solvents which can react. The reaction will be dangerous in the following cases: heterocyclic ethers with strained ring systems, in particular epoxides, as well as hetero rings with five chains; if the compound that is in the presence of ether is particularly reactive and forms unstable products; finally, if it is carried out in the presence of an acid, which for reason 1 can weaken the C-O bond by sequestering oxygen.

3. The presence of active oxygen will activate a hydrogen atom in α position. This can give rise to dangerous substitution reactions because of their exothermicity and above all the instability of the products obtained. The danger is even higher if the carbon chain is broken.
4. Like all organic compounds, ethers have a reducing character. Thus they will give rise to dangerous oxidations due to the exothermicity of the reaction.
5. Finally, carbon chains that have an unsaturated bond, an aromatic ring or another group, can cause dangerous reactions which involve these structural elements or groups. In this case, the simultaneous presence of these structural parameters can boost their reactivity due to the electronic effects that they exert on each other.

However, the reaction is often dangerous because the operating conditions are not under control or because reagents, which have a particularly 'violent' chemical behaviour, are used. For these reasons, several sites of the particular molecule can give rise to a reaction; if not, the entire molecular system can be affected by the conversion. This explains why it is difficult and ambiguous to classify the dangerous reactions. The subdivision above will be used by changing the order of the paragraphs following a 'logic of risks', ie according to the importance of their implication in accidents, rather than by chemical logic.

6.4.1 Reactions related to the mobility of the α hydrogen atom

■ Peroxidation

This is caused by the effect of air on an ether. This is classed as a radical reaction as the effects of light, heat and radical sources demonstrate. The general reaction can be written:

$$-\underset{H}{\overset{|}{C}}-O-\underset{H}{\overset{|}{C}}- \xrightarrow[h\nu, \Delta, R^{\bullet}]{O_2} -\underset{OOH}{\overset{|}{C}}-O-\underset{H}{\overset{|}{C}}- \xrightarrow{O_2} -\underset{OOH}{\overset{|}{C}}-O-\underset{OOH}{\overset{|}{C}}- \dashrightarrow \begin{array}{l}\text{Diperoxides}\\ \text{Triperoxides}\end{array}$$

There are several analyses that determine the most dangerous ether structures.

The *Du Pont de Nemours* analysis consists of three lists: A, B and C in the descending order of risk:

- List A lists all compounds that form explosive peroxides, which are extremely sensitive. Amongst them are divinylacetylene for hydrocarbons and isopropyl oxide for ethers.
- List B contains all compounds that form peroxides which become dangerous when they reach a critical concentration. The danger will often become apparent during distillation operations. For hydrocarbons, this is the case for deca- and tetrahydronaphthalene, cyclohexene, dicyclopentadiene, propyne and butadiene. Secondary alcohols such as 2-butanol also form part of this list. Finally, for ethers there are diethyl ethers, ethyl and vinyl ethers, tetrahydrofuran, 1,4-dioxan, ethylene glycol diethers and monoethers.

- List C deals with butadiene and unsaturated compounds, but does not include any ethers. Peroxides that are formed with the compounds in this list catalyse their polymerisation. This is the dangerous reaction.

Ethers bearing the danger code R 19: able to form explosive peroxides

Aliphatic ethers	Glycol ethers	Heterocyclic oxides
Diethyl ether	1,2-Dimethoxyethane	Furan
Ethyl vinyl ether		Dihydrofuran
Dipropyl ether		Tetrahydrofuran
Diisopropyl ether		1,4-Dioxan
Methyl butyl ether		Isopropyl glycidyl
Butyl vinyl ether		ether

Labour regulations have a code 19 with regards to the labelling of chemical compounds: 'Can form explosive peroxides'. Ethers with code 19 are listed in the table below:

NFPA reactivity code for ethers

Ether	NFPA reactivity	Ether	NFPA activity
Ethylene oxide	3	Diisopropyl ether	1
Propylene oxide	2	Tetrahydrofuran	1
Butyl vinyl ether	2	Tetrahydrofurfurylic alcohol	0
Ethyl vinyl ether	2	Anisole	0
Methyl vinyl ether	2	2-Butoxyethanol	0
Divinyl ether	2	Dimethoxyethane	0
Furfuryl alcohol	1	Diethyleneglycol	01
1,4-Dioxan	1	Dihydropyran	0
Furan	1	2-Methoxyethanol	0
2-Methylfuran	1	Dibutyl ether	0
Diethyl ether	1	Dipentyl ether	0
Ethyl methyl ether	1	Triethyleneglycol	0

Finally, the 'reactivity' NFPA code takes this risk into account. The ethers listed in the table below are given in descending order of risk.

When analysing these different lists one realises that the different sources do not agree with each other. So far as the NFPA reactivity code is concerned, codes 2 and 3 have been attributed to epoxides and ethers that are unsaturated. Their purpose is to inform the reader about dangerous polymerisation and not peroxidation risks. The accidents described below also involve compounds such as dibutyl ether that are not considered as dangerous in the regulations or NFPA:

- Extracting a fatty substance and a floor wax with diethyl ether gives rise to a detonation.

- Another extraction using wood pulp led to a detonation too.

In both cases an analysis that was carried out on ether showed that it contained 250 ppm of peroxide. As mentioned in the Du Pont list B, the danger only appears during the vapourisation operation that follows extraction. Peroxide formation in diethyl ether can be avoided by storing it on sodium threads or by incorporating 1 ppm of pyrogallol or 0.05 ppm of ethyldithiocarbamate. If the ether is already oxidised, peroxide can be destroyed by stirring it with a ferrous salt or alkaline sulphite solution. Peroxide is indicated by stirring the ether that is suspected of being peroxidised with an acetic sodium iodide solution when starch is present. The blue colour of the iodine-starch complex indicates the presence of peroxide.

The presence of certain metals in diethyl ether activates peroxide formation. Sulphur catalyses peroxide decomposition.

- Vapourisation of diethyl ether containing sulphur caused a system to detonate.
- Diisopropyl ether that is peroxidised detonated many times: by stirring, unscrewing the cap, impact. All accidents are fatal due to the violence of the detonations. It peroxidises in a few hours and forms complex mixtures, which are highly dangerous. The following peroxides have been identified:

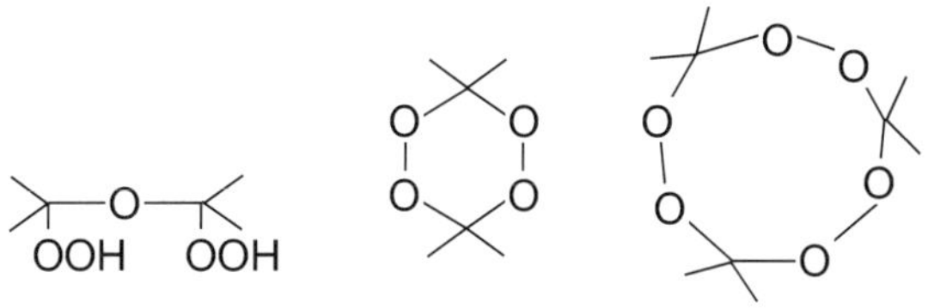

Generally speaking, the danger from ether is not related to its propensity to form a hydroperoxide or to the structure of the compound formed, but to the propensity of this hydroperoxide to form a mono-, di- or triperoxide.

Protecting diisopropyl ether against oxidation requires 16 ppm of N-benzyl-4-aminophenol or 50 ppm of diethylenetriamine.

- Dibutyl ether peroxidises easily, forming very dangerous peroxides. However, these can be destroyed easily by heating them when alumina is present.
- Tetrahydrofuran has a similar behaviour to dibutyl ether.
  This hydroperoxide decomposes slowly, avoiding accumulation. However, if the conditions are ideal for peroxidation (heat, prolonged time exposure to air, solar light), the hydroperoxide converts into extremely dangerous peroxides. Phenolic antioxidants inhibit this peroxidation efficiently. If tetrahydrofuran is peroxidised, it is not possible to destroy peroxides with ferrous salts or sulphites since tetrahydrofuran dissolves in water. Alumina or active carbon (passing over an alumina column or activated carbon at 20-66°C with a contact period of two minutes) are used, or by stirring in the presence of cuprous chloride.
- The explosion of peroxidised tetrahydrofuran in the presence of bases shows how sensitive these compounds are to bases.
- Same dangers of peroxidation and decomposition exist for the compounds formed. However, this peroxidation is inhibited efficiently by N-phenyl-1-naphthylamine.

- Diallyl ether had been exposed to air during two weeks before distillation. The handling was interrupted by a very violent detonation, which was put down to the peroxidation of this ether.
- The same accident and effects occurred during the distillation of 1,2-dimethoxyethane, which had been stored for seven years.
- Ether peroxidation was responsible for the detonation that occurred when dibenzyl ether was treated by the etherate of dichloroaluminium hydride. Note that in this case as well as for diallyl ether there will be easy peroxidation since both sites are of allyl and benzyl nature respectively.

1,4-dioxan peroxidises very easily. A distillation simulation of this compound after being exposed to air for fifteen days was carried out by differential thermal analysis by the author and his students. This indicated the violent decomposition of the peroxides formed at the end of distillation. The peroxides of this ether are destroyed by passage over an alumina column.

Halogenation

- Diethyl ether combusts when it comes into contact with gaseous chlorine. An ethereal solution of chlorine gives rise to deflagration in the presence of light.
- Deflagration took place after bromine had been introduced into diethyl ether.
- Bromine was rapidly added to aqueous tetrahydrofuran when preparing a solution containing 10% of bromine in this ether. This gives rise to a violent reaction as well as a substantial release of hydrogen bromide. This unexpected behaviour was explained by the bright light coming from the extraction hood, which had been freshly painted in white.
- A mixture of sulphuryl chloride and diethyl ether gives rise to a very violent reaction and significant release of hydrogen chloride, probably as follows:

$$CH_3-CH_2-O-C_2H_5 \xrightarrow{SO_2Cl_2} CH_3-CHCl-O-C_2H_5 + HCl + SO_2$$

All dangerous reactions mentioned in this paragraph are radical. Therefore, there will be greater dangers in the presence of light, excessive heat, radical sources (in the last case, organic peroxides are frequently used as catalysts).

6.4.2 Basicity of active oxygen

Ethers can give rise to 'acid-basic' reactions in the presence of acids, in particular Lewis acids. The reaction can prove to be dangerous in two situations:

In the presence of Lewis acids that are particularly reactive:

- The interaction between anhydrous hafnium tetrachloride and tetrahydrofuran is particularly exothermic and violent.
- The same happens with the same ether and titanium or zirconium tetrachloride.

When the acid leads to an acid-base complex, which is particularly unstable:

- Synthesis of the complex below using a boron hydride had produced a molar solution of the complex in tetrahydrofuran that detonated after being stored at 15°C for two weeks.

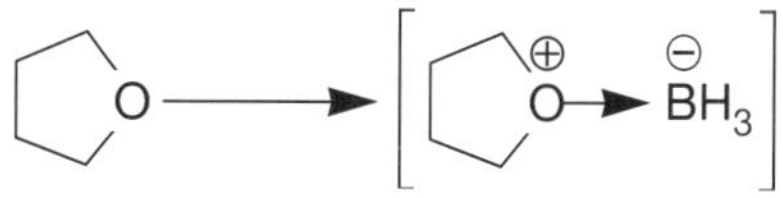

6.4 Ethers

A study showed that 0.05 M solutions stored in a open bottle and in the presence of sodium borohydride (used as a stabiliser) were the only ones that could be stored.

- In the same way, the following scheme represents the preparation of two acid-base complexes of 1,4-dioxan, which detonates when it is stored at 20°C (in the case of the complex with SO_3) or in the dry state (with the aluminium derivative):

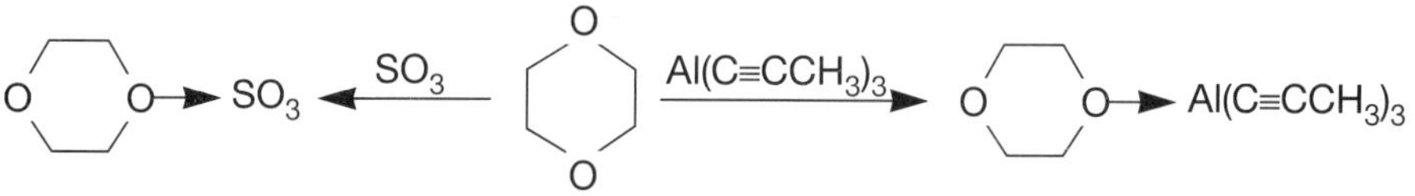

- Finally, the $B_{10}H_{14}$, $(C_2H_5)_2O$ complexes detonate on impact.

6.4.3 Rupture of C-O bond; ether stability

The relative chemical inertia of the C-O bond has already been emphasised. Structural factors can increase this reactivity significantly. This can lead to ruptures that can be dangerous in certain situations. In this case, the higher propensity of this bond to rupture makes the particular ether rather unstable. The following table gives the enthalpies of decomposition for a few ethers.

Enthalpy of decomposition of various ethers

Compound	ΔH_d (kJ/g)	Temperature range (°C)
1,4-Dioxan	0.165	130-200
Epichlorhydrine	0.500	375-500
Phenylglycidyl ether	0.626	360-450
Glycidol	1.365	130-150
Ethylene oxide	1.516	320-490

It is obvious that the unstable character of an ether comes from the epoxy ring. On the other hand, linear ethers are perfectly stable. Obviously, ethers with structures similar to the examples above are those that are mainly responsible for the accidents bonded to the decomposition of ethers.

Dangerous reactions of epoxides

- Pure ethylene oxide in the gaseous state can detonate spontaneously.
- Vigorously compressing a mixture of ethylene oxide and oxygen causes the epoxide to decompose (it was thought that it could be due to the ether polymerisation). If a 5%/95% mixture of ethylene oxide and air is compressed vigourously to 1/11 volume, the epoxide can decompose violently and combust (the 3/100 explosive limits suggest that this ignition can occur in the absence of

6 • Reactions of organic chemicals

6.4 Ethers

air). This sensitivity to decompression varies according to the oxygen pro portion. It has been observed that the dangerous factor in compression is speed.

When this strained ring opens, this can give rise to polymerisation that can be very dangerous due to exothermicity. Note that it will be very difficult to explain an accident by decomposition leading to the destruction of the molecular structure or polymerisation.

- There was a sudden temperature rise after heating ethylene oxide at a moderate temperature. The medium could not be cooled and as a result the equipment detonated. It was assumed that violent polymerisation of epoxide caused the accident.

When used as a sterilizer for medical equipment, ethylene oxide has to be handled using concentrations in carbon dioxide or a 'freon' that are lower than 10% (the mixture would combust in air). Bases catalyse the polymerisation.

- Sodium hydroxide was introduced into a reactor containing 90 kg of ethylene oxide. The reactor detonated eight hours later. A study was carried out to try to understand what caused the accident. It showed that in similar conditions the temperature of the medium reaches 100°C when 13% of oxide is polymerised, then 160°C at 28% of conversion. At this stage, it takes sixteen seconds for the medium to reach a conversion rate of 100% and a temperature that reaches 700°C.

Another study drew a comparison between the polymerisation of ethylene oxides and propylene oxides in similar operating conditions and in the presence of 10% of sodium hydroxide. When the polymerisation reached its maximum speed, the temperature reached 439°C for the former and 451°C for the latter; the pressures obtained are 44.6 and 26.6 bar respectively.

- Propylene oxide was introduced into a container that contained epoxy resins; it detonated. This accident was put down to compound polymerisation catalysed by triamines or superior homologues, which are used to 'harden' resins (for example, triethylenetetramine).

- Propylene oxide detonates very violently when sodium hydroxide is present.

- A road haulier carrying epichlorhydrin and who had covered a distance of 500 km noticed that there was an unusual temperature rise in the tanker. It was probably due to the beginning of polymerisation in the presence of an impurity, which played a catalytic role. The temperature reached 115°C very quickly and later the safety valve opened releasing a large quantity of fumes.

- The catalysts of epoxide polymerisation are strong bases, acids (H_2SO_4), Lewis acids ($AlCl_3$, $FeCl_3$), metals (K, Al, Zn), and metal oxides (Al_2O_3, Fe_2O_3).

The epoxy ring opens easily when nucleophilic reagents are present and releases a large quantity of energy due to the cyclic tension. This leads to reactions that can be hardly controlled.

- Twenty tons of ethylene oxide were contaminated by ammonia accidentally. The tank broke open releasing a fume 'cloud', which gives rise to a devastating explosion. Again, it is rather difficult to interpret this accident. Indeed, it could be a violent polymerisation, which was the result of the catalytic effect of ammonia or a very exothermic reaction:

$$\text{(ethylene oxide)} \xrightarrow{NH_3} N(CH_2CH_2OH)_3$$

It is thought that the quantities of ammonia are low and that the exothermicity of the reaction caused the oxide polymerisation.

- Ethylene oxide was treated with glycerol. The temperature of the reaction should have been about 115-125°C, but an accidental temperature rise caused it to reach 200°C; the reaction got out of control and caused the reactor to break and the equipment to detonate.

The epoxy ring does not open more easily when certain acids are present. When the molecule is unstable, risks are higher as shown in the accident described below:

- Ethylene oxide was added to perchloric acid. The medium detonated probably because the compound formed was unstable.

$$\text{(ethylene oxide)} + HClO_4 \longrightarrow HO-CH_2CH_2-ClO_4$$

■ Rupture of C-O bond of other ethers

Rupture can only happen due to the high reactivity of the compound in the presence of ether and especially the instability of the compound obtained.

Ozone can give rise to such reactions. Oxygen acts upon the α site of oxygen, ozone breaks the C-O bond, by forming an explosive peroxide:

$$C_2H_5-O-C_2H_5 \xrightarrow{O_3} C_2H_5-O-O-C_2H_5$$

- This reaction gives rise to a detonation.
- Diethyl ether detonated when it was added to perchloric acid. This was probably due to the instability of one of the compounds below:

$$C_2H_5-O-C_2H_5 \xrightarrow{HClO_4} C_2H_5-ClO_4 \text{ or } (C_2H_5)_2\overset{\oplus}{O}H \; ClO_4^{\ominus}$$

- A homogeneous mixture of diethyl ether and nitric acid decomposes vigorously after a latency period during which the medium splits into two liquid phases. One of them was attributed to the formation of ethyl nitrate, which is unstable.
- Extracting nitric acid using diethyl ether leads to explosive mixtures.
- Diethyl ether introduced into a nitric acid/2-bromotoluene/water mixture caused a deflagration.
- Laboratory glassware containing diethyl ether residues was cleaned with a sulpho-nitric mixture and caused detonation of the container.

- The epichlorhydrin oxidation reaction by nitric acid using diethyl ether as a solvent had been carried out the following way:

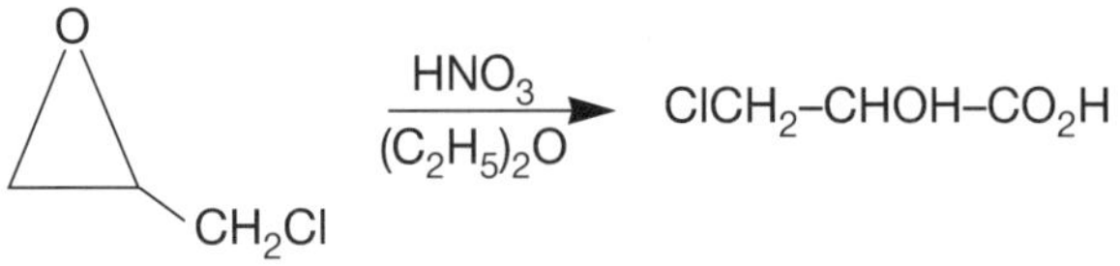

$$\text{epichlorhydrin} \xrightarrow[(C_2H_5)_2O]{HNO_3} ClCH_2-CHOH-CO_2H$$

The medium detonated.

In all these cases the reaction responsible for the accident is very likely to be:

$$C_2H_5-O-C_2H_5 \xrightarrow{HNO_3} C_2H_5-ONO_2$$

This reaction is catalysed by sulphuric acid and its presence constitutes an additional risk.

- Boron trifluoride in aqueous dioxan had been treated by three successive portions of nitric acid. After this treatment the medium was heated to eliminate boron trifluoride by vapourisation. Finally, perchloric acid was added, which caused detonation. It was explained by the decomposition of one of the two compounds below:

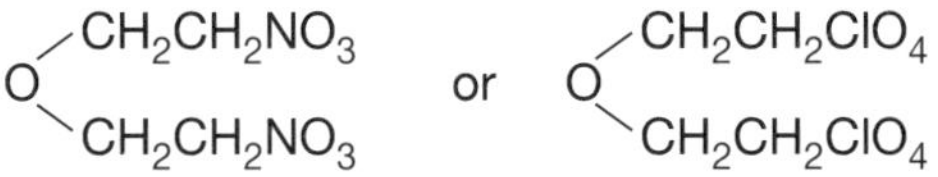

$$O\begin{matrix}CH_2CH_2NO_3\\CH_2CH_2NO_3\end{matrix} \quad \text{or} \quad O\begin{matrix}CH_2CH_2ClO_4\\CH_2CH_2ClO_4\end{matrix}$$

The time sequence that led to the accident makes it likely that the second compound was responsible.

- The effect of dinitrogen pentoxide in a dichloromethane solution on diethyl ether at a temperature lower than 20°C caused a detonation. It was explained by the decomposition of nitroglycol formed as follows:

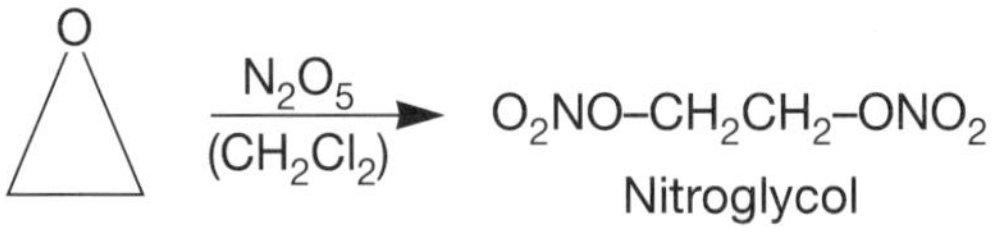

$$\text{epoxide} \xrightarrow[(CH_2Cl_2)]{N_2O_5} O_2NO-CH_2CH_2-ONO_2$$
$$\text{Nitroglycol}$$

- Sodium tetrahydrogenaluminate was prepared by the action of hydrogen on a suspension of aluminium and sodium in tetrahydrofuran. A detonation caused by the following reaction interrupted the handling:

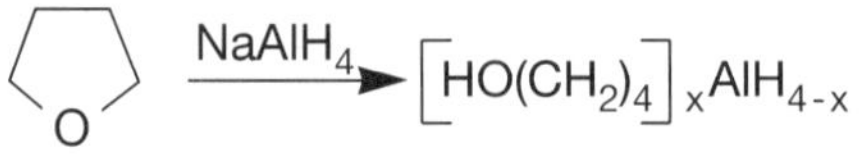

$$\text{THF} \xrightarrow{NaAlH_4} \left[HO(CH_2)_4\right]_x AlH_{4-x}$$

- Tetrahydrofuran treated with thionyl chloride at 60°C gives rise to a highly exothermic or even explosive reaction, depending on the reagents' proportions, assumed to happen in the following way:

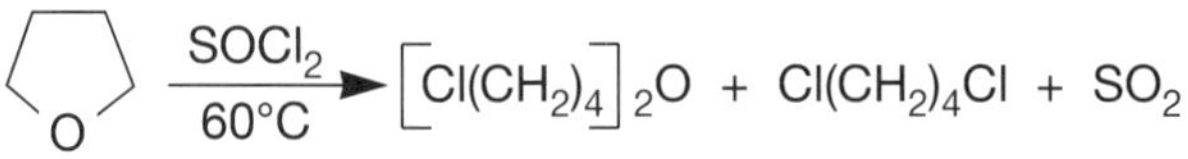

$$\text{THF} \xrightarrow[60°C]{SOCl_2} \left[Cl(CH_2)_4\right]_2O + Cl(CH_2)_4Cl + SO_2$$

6.4.4 Oxidation reactions

A large number of reactions mentioned in the previous paragraphs are oxidation reactions, which are preferably classified according to other criteria. Here are listed the reactions, which do not form any identifiable compounds because of the violence of the combustion reactions.

■ Oxidation by oxygen

- Liquid air is a strong oxidant of diethyl ether with which it detonates on contact.
- The autoinflammation temperature of a 50/50 mixture of ethylene oxide and air that is around 460°C lowers to between 416-251°C when ethylene oxide comes into contact with certain insulation materials (that can be found in sterilisation chambers where this oxide is used).
- An attempt to oxidise 4-methylanisole was carried out in the following way:

$$CH_3O-\langle \rangle- \xrightarrow[60\ bar]{O_2\ 115°C} CH_3O-\langle \rangle-CHO$$

Acetic acid was the solvent; the temperature and pressure used are mentioned in the equation. Heavy metals were present. An extremely violent reaction took place. An investigation was conducted after the accident which revealed that the accident was caused by a flow rate of oxygen that was too fast.

- Polypropylene glycol was stored in the presence of water at a temperature greater than 100°C. Ten to fifteen hours later, there was a violent reaction as well as a substantial gas release. This accident was interpreted by polyether oxidation by air (peroxidation?).

■ Oxidation by hydrogen peroxide and alkaline peroxides

- Evaporating a hydrogen peroxide solution in diethyl ether gives rise to a residue that detonates when it comes into contact with a glass stirrer or a platinum spatula.
- When furfuryl alcohol comes into contact with hydrogen peroxide at 85% concentration this leads to an ignition within a second.
- A 1/1,2/1 mixture of 2-methoxyethanol/polyacrylamide + hydrogen peroxide/toluene respectively volatilised slowly and intermittently by heating over a period of four weeks. The detonation was explained by the ether-alcohol oxidation by the peroxide.
- A mixture of sodium peroxide and diethyl ether combusts when water is present. This accident is caused by ether oxidation by hydrogen peroxide, which is formed by the effect of water on alkaline peroxide.

Peroxides, which are formed by the effect of hydrogen peroxide on hydrogen atom in α of active oxygen or on the alcohol group, can come into play in any of these reactions.

■ Oxidation by various agents

- Divinyl ether combusts in the presence of nitric acid within a millisecond.

6 • Reactions of organic chemicals

6.4 Ethers

- Potassium permanganate or an alkaline peroxydisulphate in an acid medium violently oxidises diethyl ether and causes detonations.
- Once made, the chromium dichloride/diethyl ether mixture combusts immediately.

6.4.5 Reactions involving hydrocarbon chain and miscellaneous

■ **Aromatic chains**

The Ar-O-R aromatic ethers can give rise to reactions of electrophilic substitution that can be dangerous; not only because of the reagents' nature but also the activation effect due to the active oxygen. The activation mechanism can be described in the same way as for phenols.

The same goes for the furan ring, which has a π system with six conjugated electrons, hence is aromatic.

- Chlorosulphonic acid reacts very violently with diphenyl ether if the medium reaches a temperature greater than 40°C. The reaction is the following:

$$\text{Ph-O-Ph} \xrightarrow{ClSO_3H} ClO_2S\text{-C}_6H_4\text{-O-C}_6H_4\text{-}SO_2Cl$$

If the reaction is carried out when fatty acids or nitrogenous compounds are present, it appears to be less dangerous.

- Furfuryl alcohol in an acid medium gives rise to reactions of polycondensation; reactions of successive electrophilic substitutions involving furan molecules. This reaction is identical to the reaction described for benzyl alcohol on p.256 and represents the same dangers. It is carried out under the same conditions, ie in a sulphuric medium. The electrophilic species that comes into play is very similar to the benzyl cation.

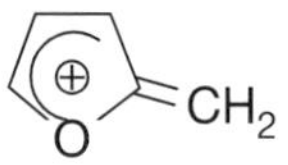

$$\text{[furan cation]}=CH_2$$

The reaction seems less dangerous when sulphuric acid is replaced by 1,3-phenylenediamine or epichlorhydrin.

■ **Acetylene chain**

Acetylene ethers have the same instability and dangerous reactions as acetylene compounds.

- Ethoxyacetylene detonates violently when it is heated to 100°C in a sealed tube.
- The effect of ethylmagnesium iodide on ethoxyacetylene leads to a metal acetylide, which detonate when the medium is stirred.

$$C_2H_5O-C\equiv CH + C_2H_5MgI \longrightarrow C_2H_5O-C\equiv C^- MgI^+ \dashrightarrow \text{Explosion}$$

The interpretation put forward is not entirely satisfactory since the reaction does not seem to be dangerous when it is carried out with ethyl magnesium bromide.

6.4 Ethers

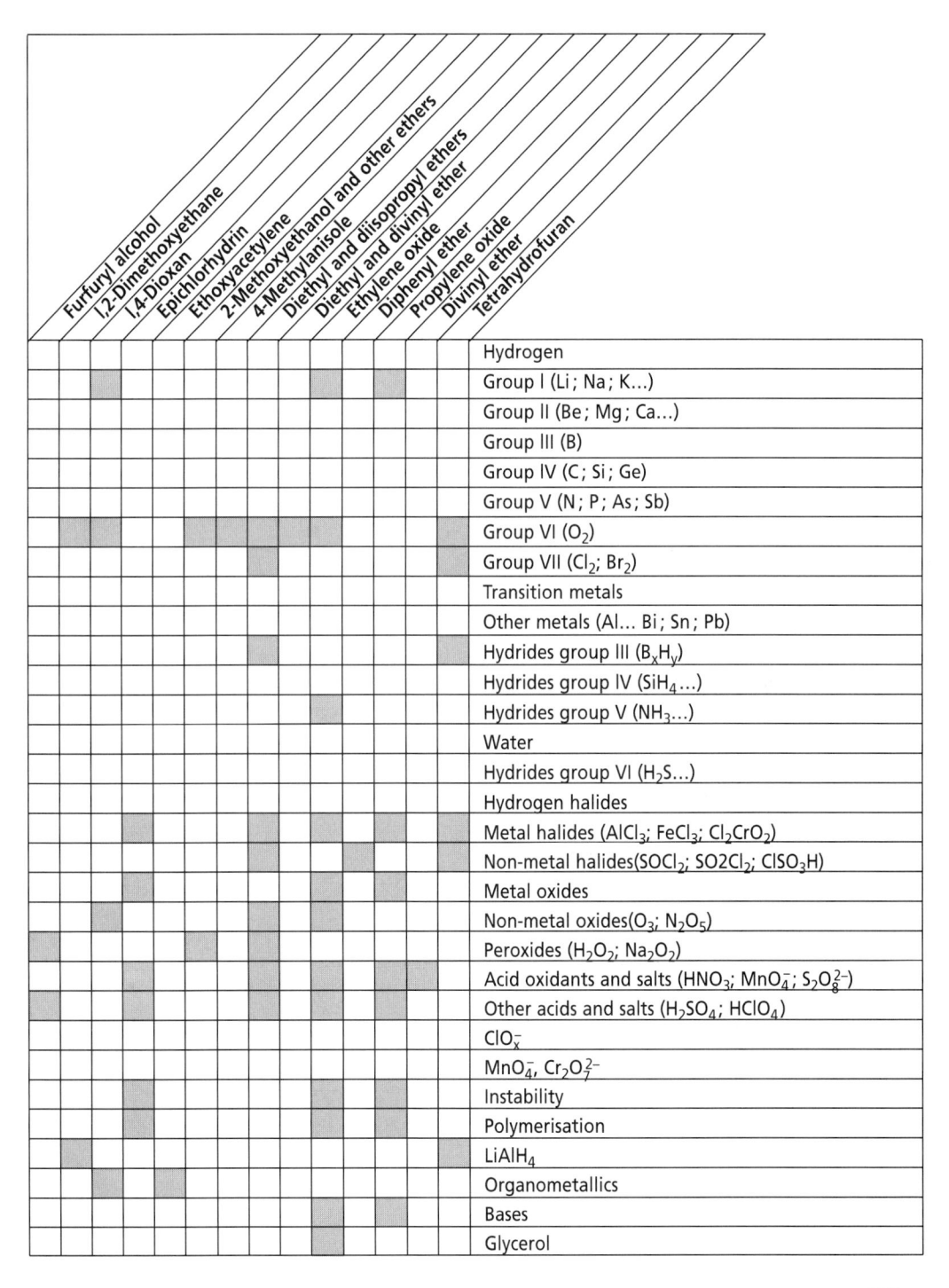

Synopsis of dangerous reactions of ethers

Columns (diagonal headers, left to right):
Furfuryl alcohol; 1,2-Dimethoxyethane; 1,4-Dioxan; Epichlorhydrin; Ethoxyacetylene; 2-Methoxyethanol and other ethers; 4-Methylanisole; Diethyl and diisopropyl ethers; Diethyl and divinyl ether; Ethylene oxide; Diphenyl ether; Propylene oxide; Divinyl ether; Tetrahydrofuran

Hydrogen
Group I (Li; Na; K...)
Group II (Be; Mg; Ca...)
Group III (B)
Group IV (C; Si; Ge)
Group V (N; P; As; Sb)
Group VI (O$_2$)
Group VII (Cl$_2$; Br$_2$)
Transition metals
Other metals (Al... Bi; Sn; Pb)
Hydrides group III (B$_x$H$_y$)
Hydrides group IV (SiH$_4$...)
Hydrides group V (NH$_3$...)
Water
Hydrides group VI (H$_2$S...)
Hydrogen halides
Metal halides (AlCl$_3$; FeCl$_3$; Cl$_2$CrO$_2$)
Non-metal halides (SOCl$_2$; SO2Cl$_2$; ClSO$_3$H)
Metal oxides
Non-metal oxides (O$_3$; N$_2$O$_5$)
Peroxides (H$_2$O$_2$; Na$_2$O$_2$)
Acid oxidants and salts (HNO$_3$; MnO$_4^-$; S$_2$O$_8^{2-}$)
Other acids and salts (H$_2$SO$_4$; HClO$_4$)
ClO$_x^-$
MnO$_4^-$, Cr$_2$O$_7^{2-}$
Instability
Polymerisation
LiAlH$_4$
Organometallics
Bases
Glycerol

6.4.6 Other dangerous reactions

Four accidents will be described in which it is more difficult to implicate the ether grouping.

- 1,2-Dimethoxyethane had been distilled on lithium tetrahydrogenaluminate powder under nitrogen and reduced pressure to obtain anhydrous ether. When distillation stopped, air entered under atmospheric pressure, which caused the apparatus to detonate (violent oxidation of lithium aluminate, which was nearly 'dried up' by oxygen?).
- A suspension of lithium tetrahydrogenaluminate in tetrahydrofuran had been stored for two years. After this period, a new portion of aluminate was introduced into the medium, which heated up and combusted for an unknown reason. This ignition was put down to the hydrogen produced.
- After being in contact with diethyl ether, silver perchlorate detonated when ground up. It does not appear that ether was involved in this explosion.
- Prussian blue (iron hexaferrocyanate) came into contact with ethylene oxide at 20°C. The reaction was very violent and the residue formed combusted spontaneously in air.

6.5 Halogen derivatives

The electronic character of halogens gives electropositive character to the carbon atom, which endows an 'acid' character to the α hydrogen atom. In the presence of a nucleophilic reagent, a halogen derivative can give rise to either nucleophilic substitution reactions or elimination reactions of hydrogen halide. Usually these reactions do not cause any incidents except when the nucleophile is extremely reactive, or the halogen derivative has several halogens on the same carbon, or when the substitution or elimination compound is unstable. Nucleophilic reagents are bases, organometallic, ethylene or aromatic compounds.

Like any organic compound they will have a reducing property that will make them likely to cause dangerous oxidations aggravated by the tendency to peroxidise in some situations (due to the mobility of hydrogen atom of active carbon). Finally, if reactive structural elements (nonsaturation, aromaticity) are present in the hydrocarbon chain, this can give rise to dangerous reactions, which may be aggravated by the presence of halogen atoms.

As usual the dangerous reaction has most effect on the compounds which are most often handled, ie chlorinated derivatives. If the operating methods are not well executed, this will be a determining factor in an accident.

6.5.1 Nucleophilic substitution and elimination reactions

The main danger factor is the polyhalogenation on one carbon site only. The substitution (1) or elimination (2) reaction can then lead to an intermediate 'carbene' (or to a transition state that has this character). Its reactivity and instability will create accidents.

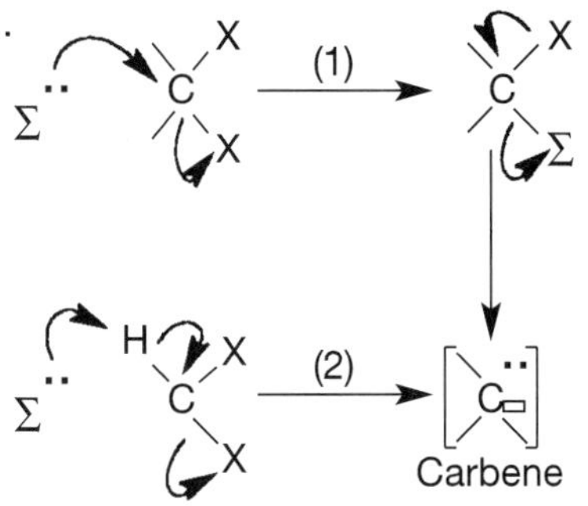

6 • Reactions of organic chemicals
6.5 Halogen derivatives

Effect of bases

- The effect of sodium, potassium or calcium hydroxide on trichloromethane or bromoform in the presence of acetone gives rise to the following reaction, which is very exothermic and caused numerous accidents with detonations:

$$HCX_3 + CH_3-CO-CH_3 \xrightarrow{OH^-} X_3C-\underset{\underset{CH_3}{|}}{\overset{\overset{OH}{|}}{C}}-CH_3$$

If the solvent is a cyclic ether, the reaction does not seem to be dangerous, but detonations have been mentioned when hydroxide is in excess (trihalomethane introduced into a hydroxide/acetone mixture) and when cyclohexane is used as solvent.

If sodium in an alcoholic medium is used, the active species is alkoxide and there can be a substitution.

- In the following reaction:

$$CHCl_3 + Na + CH_3OH \longrightarrow CH(OCH_3)_3$$

the result was not the formation of methyl orthoformate but the detonation of the apparatus, which was probably due to inadequate cooling of the apparatus.

With a dihalomethane, hydrogen atoms are not 'acid' enough to give rise to this reaction with hydroxides.

If a hydrogen atom is present in the β position of the halogen, a β-elimination reaction becomes possible. If the unsaturated compound obtained is particularly unstable, the reaction will be dangerous.

- Thus, 1,1,2,2-tetrachloroethane treated by potassium hydroxide forms dichloroacetylene, which is inflammable spontaneously and is very unstable (see p.240).

$$Cl_2CH-CHCl_2 \xrightarrow{KOH} Cl-C\equiv C-Cl$$

It is very likely that this reaction takes place in two stages, the second being the dehydrohalogenation of trichloroethylene formed during the first stage.

- Indeed, trichloroethylene treated by concentrated sodium hydroxide produces dichloroacetylene, which combusts spontaneously.

- When epichlorohydrin was in the presence of trichloroethylene with chloride ion traces, there was a deflagration. Similar accidents have been described with other epoxidic substances.

- According to the same principle, 1,2-dichloroethylene reacts with sodium hydroxide in concentrated solution or in the solid state by giving rise to a spontaneous ignition, which is probably due to chloroacetylene.

The nucleophilic property of water seems sufficient to explain the following accident:

- distilling trichloroethylene containing water led to a violent detonation caused by a massive hydrogen chloride formation, which created an overpressure which the distillation apparatus could not contain.
- In the same way, the detonation of moist trichloroethylene, which had been stored in a metal container, was explained by the hydrogen chloride formed. In this case it is possible to suggest another cause, which would involve iron trichloride forming by the interaction of hydrogen chloride with rust traces, and the catalysis by this salt of a polymerisation or degradation of the chlorinated derivative (see the similar case of aluminium chloride on p.281).
- Potassium tert-butylate leads to ignition of halogen derivatives, which is more or less immediate, depending on their nature and physical state: two minutes with liquid dichloromethane and gaseous chloroform; one minute with liquid carbon tetrachloride and epichlorhydrin; immediate ignition with gaseous chloroform.
- The interaction of triisopropylphosphine with chloroform is extremely violent, if in the absence of solvent.

■ Effect of metal salts

- Chloroform gives rise to the following reaction with potassium sulphide:

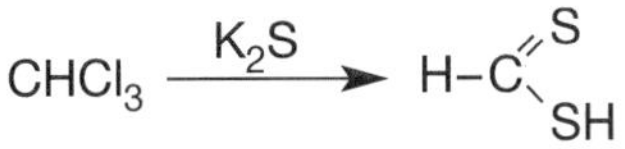

$$CHCl_3 \xrightarrow{K_2S} H-C\begin{smallmatrix}S\\SH\end{smallmatrix}$$

The reaction got out of control in two attempts out of three causing the apparatus to detonate. When it is carried out in the presence of methanol, it appears to be dangerous.

- When an alkali azide is present, dichloromethane and chloroform form explosive organic azides.
- An attempt to prepare phenylacetonitrile by the effect of sodium cyanide on benzyl chloride in a methanol medium led to the explosion of the reactor. This was explained by insufficient cooling of the medium.
- The effect of silver chlorite on iodomethane in the absence of any solvent gives rise to an immediate explosion. During another attempt it was observed that the presence of a solvent delays but does not prevent the explosion from happening. This accident can be interpreted in two different ways: instability of the methyl chlorite formed, or violent oxidation of the iodised compound by silver chlorite, which is a strong oxidant.

■ Effect of organometallic compounds

- Carbon tetrachloride/triethylaluminium/aluminium chloride mixture is explosive at a temperature of 20°C.
- Dimethylzinc leads to the same result with 2,2-dichloropropane if the mixture is not prepared carefully. The reaction is less dangerous if the dichlorinated deriv-

ative is introduced in very small quantities at the beginning of the operation before the reaction starts.
- It is not recommended to use carbon tetrachloride as an elution solvent in borane chromatographic separation since serious accidents have been reported during such operations.
- When attempting to put out a diborane fire using carbon tetrachloride (now prohibited) there was a violent detonation.
- A carbon tetrachloride/decaborane ($C_{10}H_{14}$) mixture detonates on impact.
- Calcium disilicide detonates when it comes into contact with carbon tetrachloride.
- Carbon tetrabromide or hexabromoethane give rise to explosive mixtures with hexahexyldilead in air and in the absence of solvent.

■ Effect of ethylene and aromatic compounds

- The following reaction was considered safe:

$$BrCCl_3 + CH_2=CH_2 \xrightarrow[51 \text{ bar}]{120°C} Br-CH_2-CH_2-CCl_3$$

Three operations were carried out without causing any incident and by sticking to the operating rules and quantities recommended in the literature. A fourth attempt caused the equipment to detonate.

- An identical reaction to the previous one can be carried out with carbon tetrachloride either at 25-105°C and under 30-80 bar or in the presence of a radical initiator (peroxide):

$$CH_2=CH_2 + CCl_4 \longrightarrow Cl-CH_2-CH_2-CCl_3$$

These different methods have given rise to detonations, which have destroyed the equipment due to a six-fold pressure rise.

- The effect of chloromethane on ethylene when aluminium chloride, a catalyst made of nickel, and nitromethane under 30-60 bar are present, gives rise to a highly violent detonation. The fact that there are so many compounds present makes it difficult to come up with a simple explanation. Ethylene (polymerisation) and nitromethane may have caused this accident (as can be seen in the paragraph about nitrated derivatives).
- The reaction:

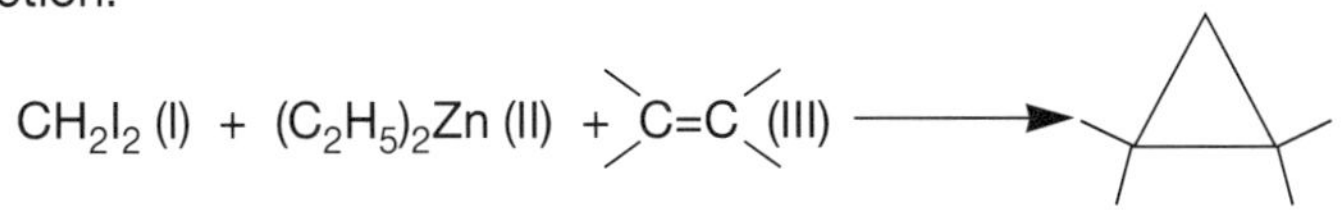

$$CH_2I_2 \text{ (I)} + (C_2H_5)_2Zn \text{ (II)} + \text{C=C (III)} \longrightarrow \triangle$$

gives rise to a detonation, if (II) is poured onto (I + III) but apparently it is not dangerous when (I) is poured onto (II + III).

- Finally, in an aromatic series, the effect of allyl chloride on benzene or toluene in the presence of ethyl aluminium dichloride (Friedel-Crafts catalysts) at -70°C is very violent and has led to a large number of accidents. It is thought that the exothermicity of the reaction below (in the case of benzene) caused these accidents, but one can not exclude a violent polymerisation of allyl chloride.

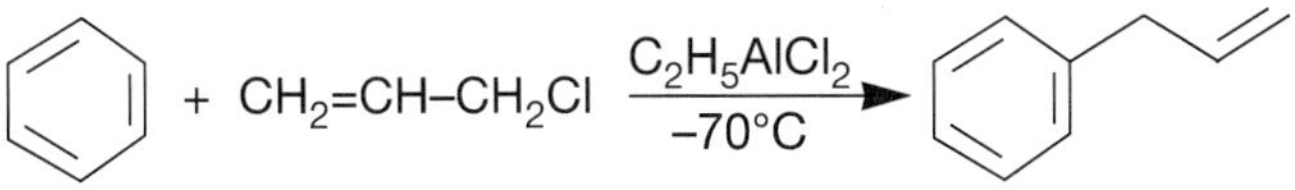

$$\text{C}_6\text{H}_6 + CH_2=CH-CH_2Cl \xrightarrow[-70°C]{C_2H_5AlCl_2} \text{C}_6\text{H}_5\text{-CH}_2\text{-CH=CH}_2$$

6.5.2 Effect of metals

In the simplest case, the reactions can be written down as follows:

$$R\text{–}X + 2\ MR \longrightarrow M + MX$$
$$R\text{–}X + R\text{–}M \longrightarrow R\text{–}R + MX$$

Depending on the operating conditions, one can stop at the organometallic stage or carry out the second stage (Wurtz reaction). Two different halogen derivatives can be used, together with an inorganic halogen derivative, from group IV (Si, Ge, Sn) or V (in particular phosphorus halides) for example. When the starting compound is a polyhalogen the reaction turns out to be complex. These reactions are always dangerous because they are difficult to start, which means that reagents may accumulate. This reaction is aggravated when the organometallic intermediate is unstable or the halogen compound is polyfunctional.

A few examples of dangerous reactions of monohalogen compounds will now be listed ending with the polyhalogens. The table on p.278 provides a general idea.

- Bromobenzene was added to lithium in an ethereal medium to obtain phenyl-lithium.
- The method described recommended use of lithium in rough fragments. Instead, the experimenter used lithium that had been finely divided. The reaction got out of control after thirty minutes and detonated.
- The same happened with the reaction of bromobenzene with finely divided sodium. The less risky method consists in using sodium wire in benzene or toluene.
- The Wurtz reaction between bromobenzene and 1-bromobutane, sodium being the metal, needs to be carried out between 15 and 30°C. It is difficult to start the reaction below 15°C; starting compounds will accumulate and the delayed start in a halogen medium, which is too concentrated, will lead to a detonation. Above 30°C the reaction is too violent and cannot be controlled.
- The same goes for temperature monitoring when preparing butyl-sodium from 1-dichlorobutane and sodium in a diethyl ether medium. Below -23°C the reaction only starts after a dangerous accumulation of chlorinated compound. Above this temperature the reaction is feasible, but difficult to control.
- It is prohibited to dry chloromethane, chloroethane and dichloromethane on sodium mirror since it is too dangerous.
- Iodomethane was introduced into sodium in toluene too quickly. This led to a reaction which went out of control and dispersed the compounds. Sodium particles were in contact with air and combusted causing the ignition of toluene and a fire.
- There is a high explosion risk in the preparation of triphenylphosphine as in the reaction below. This risk can be reduced by adding a mole per cent of a light alcohol.

$$C_6H_5\text{–}Cl + PCl_3 \xrightarrow{Na,\ toluene} (C_6H_5)_3P$$

- The effect of chloromethane on aluminium powder in the presence of aluminium chloride traces forms trimethylaluminium. Its spontaneous ignition has caused numerous accidents. The same is true for zinc.

- Bromoethane forms the same organometallics, which are inflammable with aluminium, zinc and also magnesium. There was a very serious industrial accident that caused a warehouse to be completely destroyed, and this was explained by the effect of bromoethane on an aluminium pipe. Methylaluminium bromide formed and combusted in contact with air, causing the fire. A red cloud formed due to the bromoethane combustion (this was used as an extinguishing agent before being prohibited).

The accidents that are described below deal with chlorinated, brominated, iodised and mixed polyhalogen derivatives; the polyfluoridated derivations are much more stable.

- Mixtures of lithium with polychlorinated, polybrominated, polyiodised and mixed derivatives are extremely sensitive to impact.
- Carbon tetrachloride was used to 'rinse' lithium which had been kept in Vaseline oil. It detonated when the metal was cut with a knife. It is better to use hexane for this type of operation.
- A few drops of carbon tetrachloride in contact with blazing lithium do not represent any danger. A quantity of 25 cm^3 causes a very violent explosion.
- If hexachlorocyclopentadiene is stirred for a couple of minutes with sodium, it detonates.
- 140 g of tridecafluoroiodohexane in the presence of 7 g of sodium detonated after being heated for thirty minutes. If the quantities of reagent are small, the reaction does not appear to be dangerous.
- With polyhaloalkanes, potassium forms mixtures that detonate on impact. The potassium/carbon tetrachloride mixture is hundred and fifty to two hundred times more sensitive than mercury fulminate. A simple door slam can cause its detonation.
- The addition of trichloro- or tetrachloroethylene to aluminium components in dry cleaning equipments is responsible for many accidents. The effect of the carbon tetrachloride/methanol mixture in the 1/9 proportion of aluminium, magnesium or zinc causes the dissolution of these metals, whose exothermicity makes the interaction dangerous. There is a period of induction with zinc, which is cancelled out when copper dichloride, mercury dichloride or chromium tribromide is present.
- The tetrachloroethylene/aluminium/zinc oxide mixture was used as a military 'flare'.
- The reaction of barium with trichloroethylene has an enthalpy of -2.6 kJ/g. According to the CHETAH method (see para 2.3) this reaction is thought to be hazardous.

6.5.3 Oxidation reactions

■ **Inflammability**

Halogen derivatives are hardly inflammable. Their behaviour depends on the halogen 'rate' that is present in the molecule and its nature (brominated derivatives are less inflammable than chlorinated) as well as the temperature to which they are exposed. So some of them that used to be used as extinguishing

6 • Reactions of organic chemicals
6.5 Halogen derivatives

Common accidents reported as a result of interaction between metals and halogen derivatives

Compound	Li	Na	K	K-Na	Be	Mg	Ba	Al	Ti	Zr	Zn	Sm	Pu
CH_3-Cl		■	■			■		■			■		
CH_2Cl_2	■	■				■							
$CHCl_3$	■	■					■	■					
CCl_4	■	■	■		■	■	■	■			■		■
C_2H_5-Cl		■		■									
$Cl-CH_2-CH_2-Cl$	■	■						■					
$C_2H_3Cl_3$		■	■										
$Cl_2CH-CCl_3$		■	■										
Cl_3C-CH_3						■		■					
Cl_3C-CCl_3								■			■		
CH_3-Br						■		■			■		
CH_2Br_2			■	■									
$CHBr_3$	■	■	■										
CBr_4	■												
$Br-CH_2-CH_2-Br$						■							
C_4H_9-Br		■											
CH_2I_2	■		■	■									
CI_4	■												
$CFCl_3$	■						■						
CF_2Cl_2							■	■					
Cl_3C-CF_3							■						
Cl_2FC-CF_2Cl	■						■	■				■	
$C_2H_2FCl_3$								■					
ClF_2C-CF_2Cl								■					
$(ClFC-CF_2)_n$								■					
$BrCF_3$								■					
$Cl_2C=CHCl$	■						■						
$Cl_2C=CCl_2$	■				■	■	■	■					
C_6H_5-Cl		■									■		
$C_6H_4Cl_2$								■			■		
C_6H_5-Br	■	■											
Chlorinated rubber								■			■		

agents (bromomethane) combust when they are heated and/or submitted to an ignition source that emits a lot of energy (sparks). The perfect extinguishing agent, which is either a compound or a mixture will be a compromise between efficiency, costs, physical properties (state, boiling point, vapour pressure at ambient temperature) and innocuousness. Two agents corresponded to these requirements in France: bromotrifluoromethane and chlorobromodifluoromethane. They are about to be prohibited because they are environmentally unfriendly.

- Bromoethane was used in sterilisation equipment to make diethyl ether atmospheres fireproof. This method is inefficient since 31% brominated derivative should be used. Inflammability data regarding bromoethane vary from one author to the next. It is most often agreed that this compound has very narrow inflammability limits (13.5-14.5%). The range is more critical under pressure and is thought to reach 8.6-20%.

- The methanol/dichloromethane mixture has been used as a solvent when pickling paints and varnishes. The fact that it is hardly inflammable is thought to be one of its qualities. In actual fact, this mixture is inflammable in air when the rate of methanol reaches 0.5% and above.

- 1,1,1-trichloroethane is considered to be a non-flammable solvent. In fact, it is inflammable when the ignition source is energetic. Its inflammability limits are much debated (see inflammability data in the tables in Part Three). An accident has been described during a welding operation on a warehouse roof, with a diameter of 34 m in which this compound had been stored: 65% of the roof surface was blown out by the detonation of the air/halogen derivative mixture.

- There have also been fires of dichloroethylene, which is well known for being incombustible. (It may be that the ignition was due to the formation of chloroacetylene).

Peroxidation

A large number of ethylene halogen derivatives peroxidise easily at ambient temperature, by forming peroxides and polyperoxides that are particularly dangerous.

- With chlorotrifluoroethylene the following peroxidation reaction was identified:

$$ClFC=CF_2 \xrightarrow{O_2} \text{(cyclic peroxide with } O\text{-}F_2, O, F_2, F_2, F_2\text{)}$$

- 2-chloro-1,3-butadiene auto-oxidises quickly, even at 0°C. The polyperoxide formed is insoluble, hence particularly dangerous. This peroxidation is quicker than with butadiene. However, the polyperoxide can be easily destroyed by an aqueous solution containing 20-30% of sodium hydroxide.

- The same goes for vinylidene chloride between -40 and 25°C. If the rate of the peroxide formed reaches 15%, the compound can detonate as soon as it is disturbed. Phenols that contain bulky groups are good inhibitors of such compound peroxidation.

The paragraphs about the peroxidation of ethylene hydrocarbons (on p.242), secondary alcohols (on p.253) and ethers (on p.200) can be referred to. In this last paragraph, the list created by *Du Pont de Nemours* lists the most dangerous peroxidisable compounds. Tetrafluoroethylene and vinyl chloride are mentioned in list C; vinylidene chloride in list A, which is the most dangerous one. It is rather surprising that the EU labelling code does not mention the peroxidation risk of ethylene halogen derivatives.

■ Other oxidations involving oxygen

Numerous accidents demonstrate the effect of oxygen on halogen derivatives. It is not demonstrated that peroxides do not intervene.
- There was an explosion during the reaction of oxygen with iodomethane at 300-500°C. It was attributed to the formation of methyl periodate, an explosive.
- A mixture of 1,1,1-trichloroethane/oxygen, that is heated to 100°C under a pressure of 54 bar, detonates after three hours. Similar accidents have been reported with chloro- and bromotrifluoroethylene.
- The trichloroethylene/oxygen mixture detonates at 20°C under 27 bar.
- Vinyl chloride reacts violently with oxygen at 90-100°C.
- Liquid oxygen/dichloromethane, 1,1,1-trichloroethane, trichloroethylene mixtures detonate violently. Under the same conditions, carbon tetrachloride detonates with less intensity and some fluoridated derivatives do not detonate.

■ Dangerous reactions with oxidising agents

- A trichloroethylene/potassium mixture detonates around 100°C apart from when the superoxide KO_2 layer that covers potassium is removed beforehand. The danger is therefore assumed to come from the oxidising property of this oxide. However, it seems surprising that pure potassium is inactive vis-à-vis this halogen derivative at such a temperature.
- When carbon tetrachloride was used to put out a fire in a container that contained calcium hypochlorite, this caused a very violent detonation.
- Fluorine gives rise to reactions that are particularly violent or even explosive with polychloroalkanes and iodoform.
- Dinitrogen tetroxide reacts very violently with trichloroethylene at 150°C.
- Nitric acid/dichloromethane mixture forms homogeneous solutions which detonate on impact, or with friction or heat.
- Vinyl chloride detonated when it was introduced into a container which had been rinsed with nitric acid.
- It has been reported that vinyl bromide reacts violently with oxidants (unspecified).
- Anhydrous perchloric acid gives rise to a very violent reaction with trichloroethylene. It could be due to the decomposition of an intermediate perchlorate.

6.5.4 Carbon chain reactions

Many of the dangerous reactions previously mentioned affect the carbon chain directly. However, they are listed in the previous paragraphs for classification

6 • Reactions of organic chemicals

6.5 Halogen derivatives

purposes. Below are all the dangerous reactions that affect the double ethylene bond, triple acetylene bond and aromatic ring specifically.

■ Ethylene halogen derivatives

- A mixture of gaseous bromine and gaseous chlorotrifluoroethylene detonated.
- The following reaction was mentioned as not being dangerous:

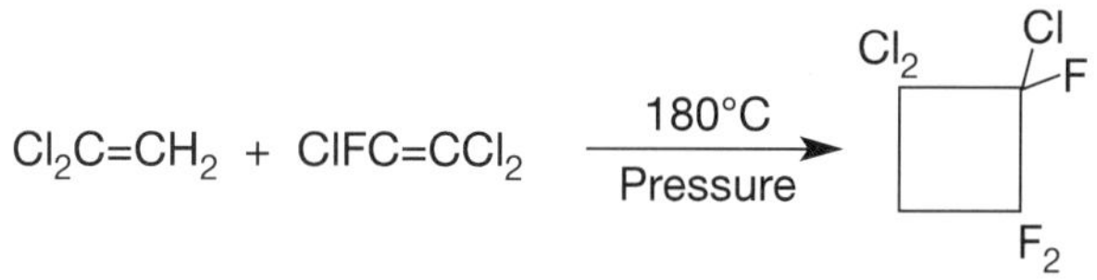

$$Cl_2C=CH_2 + ClFC=CCl_2 \xrightarrow[\text{Pressure}]{180°C} \text{cyclobutane with } Cl_2, Cl, F, F_2 \text{ substituents}$$

It was decided to increase the reagent quantities after carrying out the reaction three times without any incident. This time there was a detonation that was explained by the polymerisation of the halogen derivatives, whose exothermicity could not be controlled.

- Vinyl bromide polymerises violently when it is exposed to sunlight.
- Allyl chloride polymerises violently when an acid (H_2SO_4, BF_3, $AlCl_3$) is present.
- An ethylene/chlorotrifluoroethylene mixture polymerises violently under the effect of gamma radiation.
- The trichloroethylene/ozone mixture forms an explosive ozonide, which can be destroyed when it is in contact with an aqueous solution containing 5% of potassium iodide for twenty four hours.

The risks related to the polymerisation of ethylene halogen derivatives are mentioned in the 'reactivity' codes set by NFPA and the transport regulations of dangerous compounds (code 9). The available data is listed in the table below.

Ethylene halogen derivatives presenting a polymerisation risk

Compound	NFPA reactivity	Transportl
Allyl bromide	1	
1-Chloro-1-propene	2	
Chlorotrifluoroethylene	2	
Allyl chloride	1	
Vinyl chloride	2	339
Vinylidene chloride	2	
1,2-Dichloroethylene	2	
1',2'-Dichlorostyrene	1	
Fluoroethylene		239

■ Acetylene halogen derivatives

Their main danger is bonded to the instability of the triple bond and of the metal derivatives when the halogen derivative has a 'real' acetylene functional group.

- Propargyl bromide is an endothermic compound. It detonates on impact or when it is heated. When 70-80% of it is present in a solution with toluene it becomes insensitive to impact.

- Propargyl bromide forms explosive metal acetylides when copper or its alloys, silver or mercury are present.
- Chloroacetylene and dichloroacetylene are also endothermic and both inflammable when they are exposed to air. They need to be handled under inert gas.
- With diethyl ether, dichloroacetylene forms an azeotrope that contains 55.4% of dichloroacetylene, which is not explosive and is stable in air. If this solution is cooled excessively, the chlorinated derivative separates by forming a second liquid layer that can detonate when the compound is stirred or when the cap that seals the bottle is turned. The same thing happens when the homogeneous mixture is stirred with water. It is prepared safely by adding a catalytic quantity of methanol to a mixture of trichloroethylene and potassium hydride in tetrahydrofuran.

■ Aromatic halogen derivatives

The bonding of one or several halogen atoms with an aromatic ring creates a deactivation of the ring as with the electrophilic aromatic substitution and activates this ring as with the nucleophilic substitution due to the electron-withdrawing effect of halogens. Also, because of the electronic structure of halogen it will place the electrophilic substitutions in ortho-para position. If the halogen atom is in a benzyl position, there is a higher risk due to the halogen atom mobility, which will make it possible to form an electrophilic species. In this case the effect of halogen on the ring is confined to the induction effect.

- In one accident, benzyl bromide had been stored on zeolites. The bottle detonated after eight days because of the overpressure resulting from the formation of large quantities of hydrogen bromide. This accident was put down to the Friedel-Crafts reaction (see on p.256) of benzyl bromide, itself catalysed by zeolites. This is an identical behaviour to the one described with benzyl alcohol on p.256.
- Benzyl chloride 'decomposition' reactions have often been mentioned. In fact, this compound is stable (ΔH_d = -0.14 kJ/g between 290 and 370°C), but its propensity to give rise to Friedel-Crafts reactions with itself, has led to numerous accidents, caused by the overpressures created by the hydrogen chloride formed in massive quantities during this reaction. It can be stabilised by adding an agent sequestering it: tertiary amine or metal hydroxide. One of numerous accidents involved a metal container that contained benzyl chloride. It was thought that the electrophilic substitution reaction had been catalysed by iron trichloride traces formed by the interaction of hydrogen chloride traces with the rust present on the metal container.
- In a preparation reaction of benzyl acetate:

$$CH_3CO_2^- Na^+ + C_6H_5-CH_2-Cl \longrightarrow CH_3CO_2CH_2\ C_6H_5 + NaCl$$

a violent detonation of the mixture was observed after several operations without any incident. This reaction was carried out in the presence of pyridine traces as a sequestering agent of hydrogen chloride. The temperature was at 70°C to start with and then was brought up gradually to 115 then 135°C. It was thought that the explosion was due to the Friedel-Crafts reaction already

6 • Reactions of organic chemicals

6.5 Halogen derivatives

described caused by the accidental formation of iron trichloride from the reactor's rust and hydrogen chloride traces.

The nucleophilic substitution reaction was involved in the SEVESO (1976) 'disaster'. Another accident of the same nature might be mentioned before going into further details.

- The effect of sodium hydroxide on p-chloro-o-cresol according to the following equation:

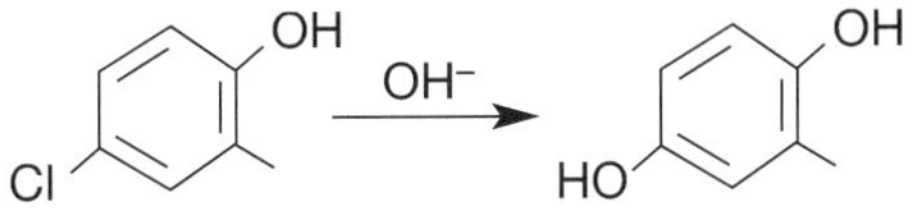

led to a violent detonation that was explained by bad monitoring of this reaction's exothermicity due to an increasing viscosity of the medium caused by the formation of diphenol.

- The SEVESO accident is the eighth accident of a series of identical accidents that happened between 1949 and 1976 during the treatment of 1,2,4,5-tetrachlorobenzene by sodium hydroxide, with the idea of preparing 2,4,5-trichlorophenol, which is the raw material of the 2,4,5-T defoliant synthesis (2,4,5-trichlorophenoxyacetic acid). Control of temperature is crucial to be able to carry out this reaction without any incident, the ideal being 125°C. If this limit is exceeded accidentally, there is a concurrent reaction that is very exothermic and forms a tetrachlorodibenzodioxin, which is regarded as being very toxic. The reaction exothermicity is such that it caused the reactor to detonate and the spreading of the product. The scheme below sums up the chemical aspect of this tragedy.

At a temperature that is lower than 125°C:

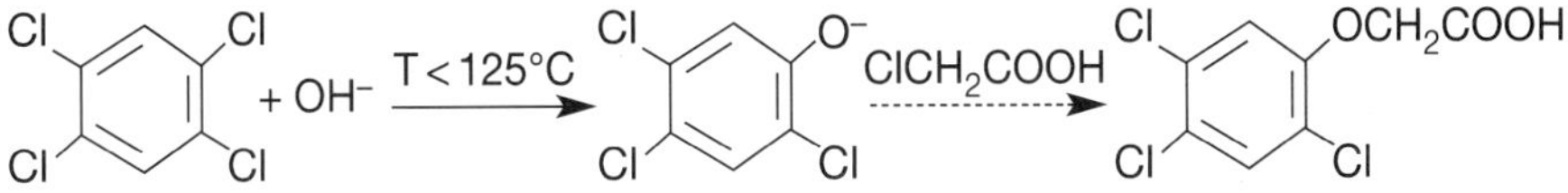

2,4,5-Trichlorophenoxyacetic acid or 2,4,5-T (defoliant)

If the temperature exceeds 125°C, it is surprising that so few lessons only have been learnt from the accidents that happened previously to the one in 1976.

6.5.5 Miscellaneous dangerous reactions

One common way of preparing aluminium chloride is by using the effect of a halogen derivative on alumina. The halogen derivative in the vapour state comes into contact with alumina when water and a catalyst are present. Hydrogen chloride and phosgene also form. If carbon tetrachloride or 1,1,1-trichloroethane at 20°C is used in the process and if the catalyst is a cobalt/molybdenum system, the reaction proves to be very difficult to control due to its exothermicity. This exothermicity is particularly high when the halogen derivative is adsorbed by the catalytic system. However, the danger can also come from the interaction between the polyhalogen derivative and the aluminium chloride formed.

6 • Reactions of organic chemicals

6.5 Halogen derivatives

Synopsis of dangerous reactions of halogen derivatives

Column	Substance
1	I-Bromo-2-propyne
2	Bromobenzene; chlorobenzene
3	Bromobutane; chlorobutane
4	Bromoethylene
5	Bromomethane
6	Bromotrichloromethane
7	Benzene bromide and chloride
8	2-Chloro-1,3-butadiene
9	Chloroacetylene; dichloroacetylene
10	Chlorocresol; tetrachlorobenzene
11	Chloroethane; chloromethane
12	Chlorotrifluoroethylene
13	Allyl chloride
14	Vinyl chloride
15	Di-, tri-, tetrachloroethylenes
16	Dichloromethane; dichloropropane
17	Epichlorhydrin
18	Iodomethane
19	Tetra-, trichloromethane; trichloroethane; tribromomethane
20	Tetrabromomethane
21	1,1,2,2-Tetrachloroethane

Reagents (rows):
- Hydrogen
- Group I (Li; Na; K...)
- Group II (Be; Mg; Ca...)
- Group III (B)
- Group IV (C; Si; Ge)
- Group V (N; P; As; Sb)
- Group VI (O_2) and O_2 liquid
- Group VII (Br_2)
- Transition metals
- Other metals (Al... Bi; Sn; Pb)
- Hydrides group III (B_xH_y)
- Hydrides group IV (SiH_4...)
- Hydrides group V (NH_3...)
- Water
- Hydrides group VI (H_2S...)
- Hydrogen halides
- Metal halides ($AlCl_3$; $FeCl_3$...)
- Non-metal halides ($SOCl_2$...)
- Metal oxides
- Non-metal oxides (O_3; N_2O_4)
- Peroxides (H_2O_2; Na_2O_2; KO_2)
- Acid oxidants and salts (HNO_3; MnO)
- Other acids and salts (H_2SO_4; $HClO_4$)
- ClO_x^-
- MnO_4^-, $Cr_2O_7^{2-}$
- Instability
- Polymerisation
- Bases
- Organometallics
- $(CH_3)_3CO^- K^+$
- Ethylene; Alcenes
- Benzene; Toluene
- Allyl alcohol

- Dichloromethane gives rise to very dangerous reactions with aluminium chloride:

$$AlCl_3 + CH_2Cl_2 \longrightarrow (Al_2Cl)_2CH_2$$

The aluminium bromide/dichloromethane mixtures are only stable at a low temperature (perhaps for the same reason).

- With allyl alcohol, carbon tetrachloride forms a complex mixture of epoxide compounds in C_4 that detonate during their distillation. One of these epoxides is thought to be the following:

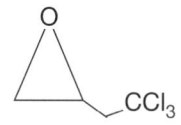

[epoxide structure with CCl$_3$ substituent]

The previous table sums up all reactions that have caused accidents. It complements the table on p.278 that lists the halogen derivatives/metal interactions.

6.6 Amines

Dangerous reactions can be classified into four main groups to which some reactions that are difficult to identify and dangerous reactions of hydrazines will be added:

1. reactions that are related to the basic property of the amine group, reactions that form unstable diazonium salts and the ones related to the instability of some amines,
2. oxidations,
3. halogenation of the amine group that leads to chloroamines (or bromoamines),
4. other dangerous reactions of hydrazines.

6.6.1 Reactions related to basicity of amines

■ Formation of unstable acid-base complexes

Amine basicity is related to the presence of a lone pair of electrons on the nitrogen atom. Usually the interaction of an amine with an acid is not dangerous. There can be an accident when the acid-base complex formed is unstable or the operating mode is not correct.

- With aluminium triacetylide, trimethylamine forms a complex which is extremely unstable. This is probably due to the presence of triple acetylene bonds.
- When bromine is present the same amine forms a complex that detonates if it is heated in a sealed tube:

$$Br_2 + (CH_3)_3N \longrightarrow [BrN(CH_3)_3]^+ Br^- \xrightarrow{\text{Sealed tube}} \text{Explosion}$$

With boron trichloride, phenylamine forms a complex. The reaction is very exothermic; if the medium is not cooled enough and if no solvent is used, nothing can be done to avoid detonation.

- Hexamethylenediamine forms unstable complexes with iodine or iodoform that detonate when they are heated at 138 and 178°C respectively

$$I_2 \cdot H_2N(CH_2)_6NH_2 \cdot I_2; \quad H_2N(CH_2)_6NH_2 \cdot HCl_3$$

The effect of chromium trioxide on pyridine forms a complex, which is very often used as an oxidant in organic synthesis:

$$\text{pyridine} + CrO_3 \longrightarrow [\text{pyridine} \rightarrow CrO_3]$$

- The synthesis of this complex is dangerous when the conditions are not right. Therefore, if the reaction is carried out between -18 and -15°C without using any solvent and when the medium is not stirred enough, there can be a detonation. Adding pyridine to oxide is an aggravating factor whereas adding oxide into pyridine will lower the risk. Ratcliffe[1] suggested a 'safe' method, which consists in dissolving chromium oxide in water first, then incorporating the resulting solution gradually into pyridine.
- A mixture of dichloromethane and ethylenediamine was distilled by water bath heating at 30°C. A very violent reaction caused a deflagration and the compounds to be dispersed. It was thought that the reaction responsible for the accident was the following:

$$H_2N-CH_2-CH_2-NH_2 + CH_2Cl_2 \longrightarrow [H_2N-CH_2-CH_2-\overset{\oplus}{N}H_2-CH_2Cl]\ Cl^{\ominus}$$

- A reaction of the same nature was probably responsible for a detonation that took place one hour after mixing carbon tetrachloride with tetraethylene pentamine. This reaction also occurs with monoamines, but it is not thought to be dangerous.
- When ethylenediamine comes into contact with silver perchlorate it detonates. This accident is explained by the formation of an amine/silver complex that is unstable.
- Similarly, another accident occurred when metallic silver came into contact with aziridine. According to the authors of the report, the accident was interpreted by the formation of an aziridine silver derivative. Comparing this behaviour with the one of ethylene oxide when silver is present, a danger which is of the same nature is demonstrated. The interpretation that had been given at the time was based on the presence of acetylene in ethylene oxide, whose silver derivatives are very sensitive explosives. It may be that acetylene traces were present in aziridine although none of the authors mentioned such as possibility as far as we know.

■ Effect of amines on epoxides

In the paragraph about ethers, it has already been mentioned that epoxides are very sensitive to bases. Given their basic behaviour, one can suspect that amines will be dangerous when peroxides are present.

- The accidental contamination by trimethylamine of a container that contained ethylene oxide caused the cylinder to detonate eighteen hours after adding the

---

1. Ratcliffe, *J. Org. Chem.*, 1970, 35, 4001.

oxide. This accident was put down to the polymerisation of ethy-lene oxide catalysed by amine.
- Isopropylamine reacts very violently with epichlorhydrin due to its very high exothermicity. If the mixture is made too quickly, the reaction cannot be controlled and the temperature reaches 350°C in six seconds even when it is cooled.

$$\text{epichlorhydrin} + (CH_3)_2CH-NH_2 \longrightarrow (CH_3)_2CH-NH-CH_2-CHOH-CH_2Cl$$

- A very similar explosive reaction was obtained with the same epoxide and with phenylamine, heterocyclic amines and N-substituted amines. In the last case, the reaction was described as 'not dangerous' below 60°C and dangerous above 70°C. The large number of accidents that bring this epoxide and the amines mentioned into play were all caused by the fact that the reaction exothermicity was badly controlled.

Accident during diethylamine neutralisation

- The following accident illustrates the accidental use of an unsuitable operating mode. Diethylamine was neutralised using sulphuric acid. The system taking in the acid broke down. A diethylammonium sulphate crust formed, which became soaked with diethylamine. When the system started working again the reaction of excessive quantities of amine with sulphuric acid and the sudden gas release caused the apparatus to detonate.

6.6.2 Diazotization reactions; diazonium salt and amine instability

A diazonium salt is obtained by the effect of sodium nitrite on an aromatic amine when an acid is present and at a low temperature:

$$Ar-NH_2 + NaNO_2 \xrightarrow{H\Sigma} Ar-N_2^+ \Sigma^-$$

The diazonium salt's stability depends on the structure of the aromatic group and above all, of the anion salt.

- Thus, in the first case, if the amine is o-aminobenzoic acid, an explosive diazonium salt with a zwitterion structure is obtained:

$$\text{o-aminobenzoic acid} \xrightarrow{NaNO_2} \text{diazonium zwitterion}$$

So far as the anionic nature is concerned, if $\Sigma- = N_3^-$; CrO_4^{2-}; NO_3^-; ClO_4^-; trinitrophenolate (or picrate); I_3^-; S^{2-}, or xanthate, the diazonium salt detonates on impact, friction, heating or when radiation is present.

If $\Sigma- = Cl^-$; ZnF_4^- or SO_4^{2-}, the diazonium salt is regarded as far more stable although accidents have been mentioned.

Finally, with the dianion benzene-1,3-dicarboxylate, the diazonium salts obtained have an excellent thermal stability:

benzene-1,3-dicarboxylate

The $C_6H_5-N_2^+$ Cl^- (benzenediazonium chloride) enthalpy of decomposition was calculated. Comparing its value (ΔH_d = -1.5 kJ/g) with the CHETAH criterion C_1 makes it moderately stable. Accidents have been observed during reactions that are schematised below and these illustrate what has been stated. They are all due to the explosive character of the diazonium sulphides formed. Here is one example:

o-, m-, ou p-Toluidine → (NaNO$_2$, H$_3$O$^+$) → diazonium → (S^{2-}) → [diazonium]$_2$ S^{2-}

According to many authors, amines are also relatively unstable. The following table gives the enthalpies of decomposition for a few amines.

Enthalpies of decomposition of some amines

Amine	ΔH_d (kJ/g)	Temperature (°C)
Aziridine	-2.020	130-390
o-Anisidine	-0.387	190-480
m-Anisidine	-0.518	410-500
p-Anisidine	-0.429	340-490
2-Methoxyethylamine	-0.230	140-200
2-Chloroaniline	-0.410	?
4-Chloroaniline	-0.630	210-400

Aziridine is the only one that has a serious instability risk according to CHETAH. It is due to the presence of a strained ring, which gives an endothermic character to the compound. As its stability should be more or less the same as ethylene oxide, so should be its propensity for polymerisation.

- When preparing aziridine according to the reaction below:

$ClCH_2-CH_2-NH_3^+$ Cl^- + NaOH$_{(aq. 33\%)}$ $\xrightarrow{T<50°C}$ aziridine

6 • Reactions of organic chemicals

6.6 Amines

there should be a vigorous stirring and the medium has to be diluted. If these conditions are not followed, a very violent reaction takes place that is often followed by a detonation. The interpretation of this reaction is ambiguous. It can be related to the exothermicity of either the preparation reaction or to the decomposition or polymerisation of aziridine.

- Accidents are often caused by the violent polymerisation of aziridine. To avoid it, this amine needs to be handled or stored in a diluted solution when it is cold, and alkaline hydroxide in the solid state is present.

The reactions that open this ring are extremely simple and very exothermic.

- The reaction below can only be controlled at 30-80°C when phenylamine and aluminium chloride are present; the reactants are diluted in an aromatic solvent and the order in which they are added is: phenylamine, aluminium chloride and then aziridine:

$$C_6H_5-NH_2 + \underset{\triangle}{\overset{H}{\underset{N}{\triangle}}} \xrightarrow{AlCl_3} C_6H_5-NH-CH_2-CH_2$$

6.6.3 Oxidation reactions

Like most organic compounds, amines are inflammable. When they are solid at ambient temperature, they can also form explosive air/dust mixtures. Accidents of this type have been mentioned with p-phenylenediamine.

■ Effect of ozone

This reaction is characteristic of aromatic and ethylene compounds. The amine group does not play a direct role in it.

- With ozone, phenylamine forms a triozonide, which separates in the form of a gelatinous mass that is extremely unstable.

■ Effect of peroxides

- Amine/hydrogen peroxide mixtures are explosive when they are prepared using specific proportions. Accidents have been reported with phenylamine and quinoline.
- 2-Picoline reacted very violently and suddenly with 30% hydrogen peroxide when ferrous sulphate was present. The apparatus detonated and the gases and liquids that spread caused a violent fire. The accident was put down to the fact that the medium was not stirred enough.
- When phenylamine comes into contact with sodium peroxide and water (formation of hydrogen peroxide) it combusts spontaneously.
- Performic acid at 90% gives rise to a very violent reaction with phenylamine.
- The same happens with peroxymonosulphuric acid (H_2SO_5).

Aromatic amines and in particular, tertiary amines, catalyse the decomposition of organic peroxides in very small quantities. They are used to start the radical polymerisations of numerous monomers.

6.6 Amines

- A drop of amine added to benzoyl peroxide is enough to cause a deflagration or a detonation of the peroxide, depending on to what extent the apparatus is confined. Phenylamine, N,N-dimethylaniline, N,N-dimethyl-p-toluidine react the same way.
- This is also the case when a drop of phenylamine, ethylenediamine or N,N-dimethylaniline is added to diisopropyl diperoxycarbonate:

$$(CH_3)_2HC-O-O-\underset{\underset{O}{\|}}{C}-O-\underset{\underset{O}{\|}}{C}-O-O-CH(CH_3)_2$$

■ Effect of fluorine

- When gaseous fluorine comes into contact with phenylamine, N,N-dimethylaniline or pyridine, there is incandescence.

■ Effect of nitric acid and nitrogen oxides

- Amines combust immediately when they are in contact with fuming nitric acid. The accidents described mainly involved aromatic amines (phenylamine, N-ethylaniline, o-toluidine, xylidines, benzidine) but also triethylamine. In the last case the ignition can take place at a temperature starting at -60°C.
- However, 96% nitric acid is not thought to ignite phenylamine except if sulphuric acid is present.
- Usually the nitric acid/amine interaction is more dangerous when impurities that can play a catalytic role are present. This goes for metal oxides such as copper oxides, iron (III) oxide and divanadium pentoxide. Salts such as sodium or ammonium metavanadates, iron trichloride, alkaline chromates and dichromates, cyanoferrates and alkaline or nitrosopentacyanoferrates can also act as catalysts.
- A cyclohexylamine/nitric acid mixture was used as a rocket propellant. On the other hand, cyclohexylammonium nitrate dissolves in fuming nitric acid 'without consequences'.

Amine/nitrogen oxide mixtures are not less dangerous.

- When a surplus of nitrogen tetroxide is added to triethylamine in the absence of a solvent, there is a detonation, even at a temperature that is lower than 0°C.
- With aziridine, nitrogen pentoxide forms a compound that is very unstable and is thought to have the following structure:

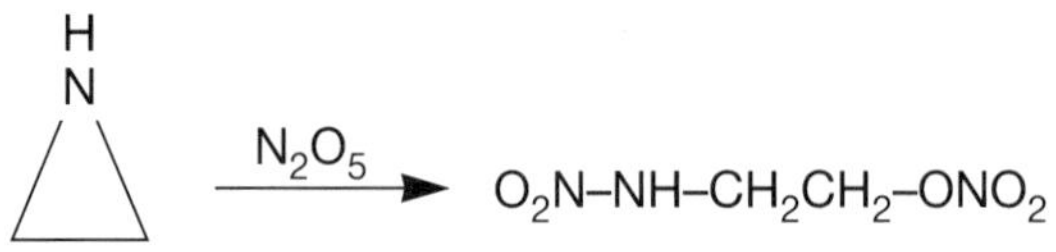

$$\text{aziridine} \xrightarrow{N_2O_5} O_2N-NH-CH_2CH_2-ONO_2$$

6.6.4 Halogenation of amines

Primary or secondary amines can form chloroamines when hypochlorites or analogues are present. This reaction is of the same nature as the one that affects alcohols and chloroamines and are unstable as organic hypochlorites. The halogenation agent can also be N-chloro- or N-bromosuccinimide:

$$Z-NH_2 + NaOCl \text{ or } Ca(OCl)_2 \longrightarrow Z-NH-Cl$$

- Thus, the following reaction led to a chloroamine, which was stored at 20°C and detonated three months after being prepared:

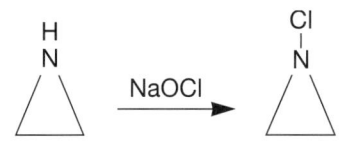

(aziridine + NaOCl → N-chloroaziridine)

- N-chloro- or N-bromosuccinimide gives rise to a halogenation reaction of phenylamine or benzylamine that is very violent:

(succinimide) NCl (Br) + $C_6H_5-NH_2$ or $C_6H_5-CH_2-NH_2$ $\longrightarrow$ Violent reaction

6.6.5 Miscellaneous dangerous reactions

- Preparing diphenylamine by using the equation below gives rise to a very violent reaction. The initial pressure of 7.6 bar suddenly reached 17 bar. A previous test did not cause any incident:

$$C_6H_5-NH_3^+Cl^- \xrightarrow[7.6 \text{ bar}]{240-260°C} C_6H_5-NH-C_6H_5$$

- Quinoline/linseed oil (I) mixture was used to purify thionyl chloride (II: $SOCl_2$). Incorporating (I) into (II) does not seem to represent any danger. On the other hand, incorporating (II) + (I) causes a very violent decomposition.
- Ethanolamine is used to absorb carbon dioxide. A tank that contained the absorption product combusted spontaneously in air, probably because of an unknown impurity and because an amine evaporation had stripped a heating coil of this compound. The fire went out later due to the carbon dioxide release caused by the compound's temperature rise.

6.6.6 Hydrazine reactions

The same dangerous reactions occur with these compounds as with amines, and are even more dangerous Hydrazines are very strong reducing agents and the fact that the weak N-N bond is present makes them rather unstable.

Instability

Hydrazine is hardly stable (see on p.166). Stability increases with the substitution degree and the hydrocarbon group size. The table below illustrates this typical property.

Hydrazine	ΔH_d (kJ/g)	Temperatures (°C)
1,1-Dimethylhydrazine	1.15	280-380
Phenylhydrazine	0.662	280-375

Oxidation reactions

- Methylhydrazine can combust spontaneously in the presence of air when in the form of a film at the surface of a porous substance. All oxidants give rise to dangerous reactions with this compound.
- Heating phenylhydrazine when hydroperoxides are present causes very violent reactions.
- The same is true for lead dioxide.

6.7 Nitrated derivatives

The main thing that the compounds listed here have in common is instability. The three active groups: nitro, nitrite or nitrate are called explosophoric groups because of the instability properties they give to the molecules in which they are present. The most dangerous reactions they give rise to can be explained by either the instability of the reactants involved at the beginning, the intermediates involved or the products obtained in the reaction. The active groups also have a property of attraction that makes the close hydrogen atom 'acid' and makes it possible to obtain 'salts' that are even more unstable than the compound used at the beginning. Finally, if another group is present, it will have a reactivity that is aggravated by the presence of nitro, nitrite and nitrate functional groups.

6.7.1 Instability

The NFPA 'reactivity' code, labour regulations, enthalpy and temperature of decomposition of the compound are all listed in the following table. They help to identify a compound's instability. There are a few compounds used in the explosives industry added to the list. These codes are described in Part I (see p. 119-123).

Note that no data in a group means that nothing was mentioned in the sources. The - sign in the labour regulations group means that no code was attributed to the compound, which means that the instability risk is not considered as being important.

There is not much correlation between the NFPA estimates and those of the EU. The record of accidents is a rather good illustration of the table.

6 • Reactions of organic chemicals

6.7 Nitrated derivatives

Synopsis of dangerous reactions of amines

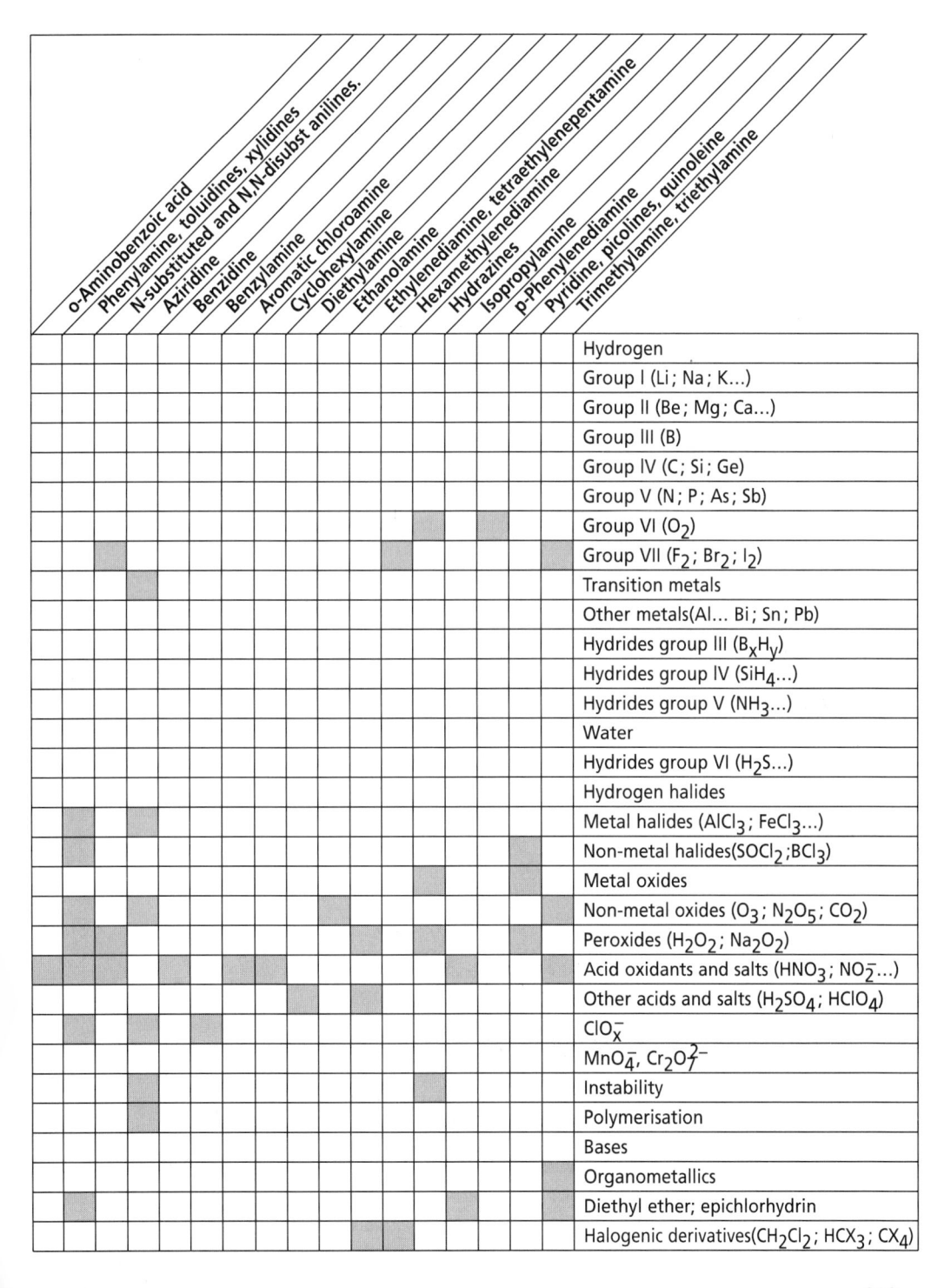

Amine (columns, left to right)
o-Aminobenzoic acid
Phenylamine, toluidines, xylidines
N-substituted and N,N-disubst anilines
Aziridine
Benzidine
Benzylamine
Aromatic chloroamine
Cyclohexylamine
Diethylamine
Ethanolamine
Ethylenediamine, tetraethylenepentamine
Hexamethylenediamine
Hydrazines
Isopropylamine
p-Phenylenediamine
Pyridine, picolines, quinoleine
Trimethylamine, triethylamine

Reagents (rows):

- Hydrogen
- Group I (Li; Na; K...)
- Group II (Be; Mg; Ca...)
- Group III (B)
- Group IV (C; Si; Ge)
- Group V (N; P; As; Sb)
- Group VI (O$_2$)
- Group VII (F$_2$; Br$_2$; I$_2$)
- Transition metals
- Other metals (Al... Bi; Sn; Pb)
- Hydrides group III (B$_x$H$_y$)
- Hydrides group IV (SiH$_4$...)
- Hydrides group V (NH$_3$...)
- Water
- Hydrides group VI (H$_2$S...)
- Hydrogen halides
- Metal halides (AlCl$_3$; FeCl$_3$...)
- Non-metal halides (SOCl$_2$; BCl$_3$)
- Metal oxides
- Non-metal oxides (O$_3$; N$_2$O$_5$; CO$_2$)
- Peroxides (H$_2$O$_2$; Na$_2$O$_2$)
- Acid oxidants and salts (HNO$_3$; NO$_2^-$...)
- Other acids and salts (H$_2$SO$_4$; HClO$_4$)
- ClO$_x^-$
- MnO$_4^-$, Cr$_2$O$_7^{2-}$
- Instability
- Polymerisation
- Bases
- Organometallics
- Diethyl ether; epichlorhydrin
- Halogenic derivatives (CH$_2$Cl$_2$; HCX$_3$; CX$_4$)

6 • Reactions of organic chemicals

6.7 Nitrated derivatives

Principal indicators of instability of nitrogen derivatives

Compound Reactivity	NFPA Code	Labour Code	ΔH_d [θ] (kJ/g)	T_d (°C)
Nitromethane	4	5	3.92 [?]	230 Explosion
Nitroethane	3	9		335-382
1-Nitropropane	3	–		
2-Nitropropane	3	–		
Tetranitromethane	3	8		
Ethyl nitrite	4	2		90 Explosion
Pentyl nitrite	2			250 Explosion
Methyl nitrate			4.42 [?]	65 Explosion
Ethyl nitrate		2		85 Explosion
Propyl nitrate	3			
Isopentyl nitrate	2			
Ethyleneglycol dinitrate	2			
Nitroglycerine		3		218 (starts at 50-60°)
Pentaerythritol tetranitrate	3			205-215 (starts at 150)
Cellulose nitrate	3	1-3		
Nitrobenzene	0	–	1.76 [360-490]	
1,2-Dinitrobenzene	4	–		
1,3-Dinitrobenzene	4	–	4.6 [?]	
1,4-Dinitrobenzene	4	–		
Trinitrobenzene	4	2		
Dinitrotoluenes (2,4- and 2,6-)	3	–		250
2,4,6-Trinitrotoluene	4	2	5.1 [?]	232-29
Trinitroxylene		2		h
1-Nitronaphthalene	0			
Tetranitronaphthalene		2		
2-Nitrophenol	–	–	2.15 [?]	
4-Nitrophenol	–	–	1.58 [240-440]	
2,4-Dinitrophenol	3	–		
2,4,6-Trinitrophenol	4	2-4		
2,4,6-Trinitrophenol (salts)		3		Na :250
4,6-Dinitro-o-cresol	1	1		
Trinitrocresol		2-4		
Trinitroresorcinol (styphnic acid)		2		
Lead styphnate		3		
2,4,6-Trinitroanisole		2		
Chloropicrin	1	–		
1,1-Dichloro-I-nitroethane	3	–		
1-Chloro-I-nitropropane	3	–		
1-Chloro-2-nitrobenzene	1		1.83 [350-450]	
1-Chloro-3-nitrobenzene	1			
1-Chloro-4-nitrobenzene	1	–	2.05 [300-450]	
1-Chloro-2,4-dinitrobenzene	4	–		
2-Nitroaniline	3	–	1.81 [280-380]	223
3-Nitroaniline	3	–	1.88 [280-380]	247
4-Nitroaniline	3	–	1.88 [280-380]	
Dinitroanilines (three isomers)	3	–		

6 • Reactions of organic chemicals
6.7 Nitrated derivatives

■ Aliphatic nitrated derivatives

- Nitromethane detonates on impact and under the effect of violent overpressure. Due to the lower toxicity of nitromethane it was often used instead of nitrobenzene, which is a solvent for the Friedel-Crafts reactions. Nitromethane boils at a lower temperature (101°C) than nitrobenzene (210°C), which forces the operation to proceed in an autoclave at the same temperature as with nitrobenzene (155°C) at a pressure of 10 bar. Under such conditions an exothermic reaction has caused nitromethane to detonate and the autoclave to be destroyed.

- The use of other compounds as joint solvents, for instance, often makes these unstable compounds sensitive. So, a nitromethane/acetone mixture is a very sensitive explosive.

- In the same way, the stability of nitromethane, nitroethane and 1-nitropropane are very much weakened by the presence of metal oxides. A study that was carried out on twenty four oxides showed that the most active are cobalt, nickel, copper and silver oxides and especially dichromium and diferric trioxides, which cause nitroethane to detonate at 245°C.

- Nitromethane had been stored with zeolites in order to dry it. More zeolites are added after several weeks. The resulting violent eruption was explained by the formation of sodium salt sediments.

- The enthalpy of absorption of 1- and 2-nitropropane on breathing mask cartridges made with carbon is such that the decomposition of the nitrated derivative can cause its ignition. This accident is aggravated when the cartridge also contains metal oxides such as copper (II) oxide or manganese dioxide.

- Tetranitromethane is thought to be relatively stable in the pure state. Mixing it with nitrobenzene (which is stable), 1- or 4-nitrotoluene, nitroxylene, 1,3-dinitro-benzene or 1-nitronaphthalene leads to mixtures that detonate with extreme violence.

- Tetranitromethane is obtained by the effect of nitric acid on acetic anhydride. A large number of very violent accidents arecaused by the fact that the temperature was not properly monitored. The detonation speeds are very high.

- Tetranitromethane detonates violently when ferrocene is present. This happens even when it is diluted in cyclohexane or methanol.

■ Aliphatic nitrates

- Methyl nitrate is prepared by the effect of nitric acid on methanol in the presence of sulphuric acid. Operating conditions are critical, given the exothermicity of the reaction and the compound thermal instability (see previous table). Moreover, methyl nitrate is very sensitive to impact. It combusts spontaneously at 250-316°C when it is in the vapour state and even when the vapour is diluted by an inert gas. The flame temperature reaches 2600°C.

- Propyl nitrate is very sensitive to impact. However, this sensitivity is lowered when adding 1 to 2% of propane, butane, chloroform, dimethyl ether or diethyl ether.

- Isopropyl nitrate combusts when it decomposes. It is a very good propellant for rockets (CHETAH factor C_3 = -68 g%). However, it is rather dangerous to store. This is the reason why it is no longer used very often.

6 • Reactions of organic chemicals

6.7 Nitrated derivatives

- Hexanitroethane is relatively insensitive to impact, friction or heat. It becomes sensitive when hydroxylated compounds are present.

■ Aromatic and complex nitrated derivatives

- 2,4- or 2,6-Dinitrotoluene decomposes at a temperature starting at 280°C. In the presence of traces of impurities, it decomposes at a much lower temperature. A mixture of both isomers had been kept at a temperature of 210 instead of 125°C for ten days in a 50mm pipe. It detonated. This accident cannot be avoided when the temperature is greater than 150°C.
- The temperature of explosive decomposition of trinitrotoluene very much depends on the compound purity. The table below shows the influence of various impurities on this compound.

Temperature of decomposition of trinitrotoluene

Impurity	None	Red HgO	Na_2CO_3	KOH/CH_3OH	KOH
T_d (°C)	297	192	218	162	192

- 2,4-Dinitrophenol, which is regarded as unstable is sold 'phlegmatised' with 15% of water.
- Trinitrophenol can only be stored safely in the form of a paste with water. Lead, mercury, copper, zinc, iron and nickel salts are sensitive to impact, friction and heat. Sodium, ammonium and amine salts give rise to explosions. When it was poured on to a cement floor, trinitrophenol formed a calcium salt that detonated when it came into contact with shoes. Trinitrophenol salts in the form of moist paste are stable. Aluminium salt is not explosive, but combusts spontaneously when in contact with water.
- 1-Chloro-2-nitrobenzene was distilled under reduced pressure by steam heating. The compound detonated at the end of the distillation.
- Chloropicrin detonates on impact above a certain critical volume (700 kg).
- During a decanting operation of chloropicrin/propargyl bromide insecticide mixture, a very violent detonation took place when it came into contact with a pump that heated up accidentally. Note that this risk seems rather obvious since propargyl bromide (see NFPA stability code: 4; see also 'halogen derivatives') is as unstable as chloropicrin.
- 2-Nitroaniline is stable when it is pure. When it is added to another compound it is dangerously destabilised. This is what happens every time it is involved in a chemical reaction.

6.7.2 Base action

With aliphatic derivatives that have a hydrogen atom bonded to the carbon atom which has the nitro, nitrite or nitrate functional group, this hydrogen atom has a mobility that makes it easy to form the corresponding anion due to the effect of a base. Unfortunately this anion is unstable and detonates immediately when dry. Even if these conditions are not fulfilled, the reaction that involves this type of intermediate is always dangerous. With aromatic nitrated derivatives the base-

/nitrated derivative interaction is not less dangerous. This can be interpreted by the destabilising effect of bases vis-à-vis nitrated derivatives. The most varied basic compounds have been involved in accidents: alkaline hydroxides, alkaline alkoxides, ammonia, amines etc.

- When bases are present, nitromethane gives rise to explosive mixtures that are sensitive. Potassium hydroxide, sodium carbonate, ammonia, phenylamine, 1,2-diaminoethane, morpholine and methylamine have been involved in accidents of this nature.
- In the presence of diethylenetriamine, the nitromethane/dichloromethane mixture is sensitive to impact even at -50°C.
- The nitromethane/methanol mixture becomes a very sensitive explosive when hydrazine is present.
- The effect of sodium hydride on nitromethane forms a salt that detonates if the reaction is carried out without using a solvent. When tetrahydrofuran is present, the reaction is violent if the temperature is greater than 40°C.

$$NaH + CH_3-NO_2 \xrightarrow{THF} H_2 + Na^+ \ {}^-CH_2-NO_2$$

- Organoaluminium or organozinc compounds (R_mMX_n; R = CH_3, C_2H_5; X = Br, I; M = Al, Zn; THF = tetrahydrofuran) cause the immediate ignition of nitromethane.
- Operations which involve any reaction between a base and nitroethane should be carried out in such a way that the formation of a salt in the 'dry' state is impossible, otherwise it will detonate.
- With amines and in the presence of heavy metal oxides (Ag, Hg), 1-nitropropane forms very unstable salts.
- 30 g of tetranitromethane were treated with sodium ethoxide. A violent detonation took place when the ethoxide was added. It was thought that the reaction corresponded to the equation below, which led to an extremely unstable compound:

$$C_2H_5O^- \ Na^+ + C(NO_2)_4 \longrightarrow (NO_2)_3C^- \ Na^+$$

- A mixture made of 80.5% of phenylamine and 19.5% of tetranitromethane combusts after a period of induction of 35 to 55 seconds and then detonates, 'if the liquid height exceeds a certain level in the container'. In the same way a mixture of pyridine and tetranitromethane sometimes detonates after a period of induction of several hours.
- When aliphatic amines come into contact with cellulose nitrate, this gives rise to charring without ignition. If these amines are replaced by polyamines or amino-alcohols, the mixture combusts spontaneously.
- When a reaction between potassium hydroxide with nitrobenzene was carried out, methanol was intended to be used as a solvent. The technician forgot the solvent and the reaction speeded up on its own, damaging the 6 m³ reactor in which it was carried out. An analysis showed that this reaction is not violent if methanol is either absent or present in large quantities, but becomes violent when methanol is present in the form of traces:

$$C_6H_5-NO_2 \xrightarrow{OH^-} (HO)C_6H_4-NO_2$$

6 • Reactions of organic chemicals

6.7 Nitrated derivatives

Reactions that involve bases and nitrobenzene as a solvent have caused numerous accidents.

- With a base, 2-nitrotoluene gives rise to a very exothermic reaction that is difficult to control. The reaction is as follows:

$$\text{2-nitrotoluene} \xrightarrow{OH^-} \text{dinitro-bibenzyl derivative}$$

- 2-Nitrotoluene was treated by hydrogen chloride in acetic acid. Then the compound was introduced into a container of sodium hydroxide to neutralise the medium. The mixture detonated immediately.
- 4-Nitrotoluene had been distilled under reduced pressure. The residue of the distillation detonated eight hours after its end. Amongst the possible causes mentioned was the accidental presence of a base in the medium. Another cause could have been the presence of a quantity of 2,4-dinitrotoluene (by-product of the preparation reaction of the distilled compound) that was larger than thought.
- Sodium oxide and 2,4-dinitrotoluene, both in the solid state, were added together to react. There was a very quick reaction that caused the nitrated derivative to combust and cause a fire in the workshop. The violence of this reaction was put down to the absence of any thinner.
- A trinitrobenzene/alkaline hydroxide mixture is explosive. It is thought that this mixture forms explosive salts.
- Trinitrotoluene leads to explosive mixtures with alkaline hydroxides. If this mixture is made in methanol, it is explosive, even at a temperature of -65°C (see table about trinitrotoluene decomposition temperatures on p.296).
- A mixture of 80% of nitromethane and 20% of 1,2-diaminoethane was added to 'tetryl' accidentally, which combusted immediately.

Tetryl

- Distilling 2-nitroethanol gave rise to a detonation, which was explained by the accidental presence of a base in the distillation apparatus.
- 2-Nitrophenol in the molten state gives rise to a very violent reaction with an aqueous solution of 85% of potassium hydroxide. 4-Nitrophenol and the same hydroxide in the molar proportion of 1/1.5 gives rise to a deflagration.
- In the presence of ammonia or sodium hydroxide, 2,4-dinitrophenol forms explosive phenates that cannot be heated in a closed container.
- Chloropicrin reacts violently with phenylamine in excess at 145°C. An accidental mixture of this nitrated derivative with sodium hydroxide in an alcoholic solution

gave rise to an extremely violent reaction. Introducing chloropicrin into sodium methoxide, in a methanolic solution, gives rise to a violent reaction above 50°C. Below this temperature the reaction does not happen until the reagent accumulation favours its start. Then it can not be controlled. Note that polyhalogen derivatives themselves give rise to dangerous reactions with bases. Thus, the presence of the nitro group is probably an aggravating factor.

- 1-Chloro-4-nitrobenzene mixed with potassium hydroxide in the molar proportion of 1/1.15 causes a deflagration. When adding the same compound to sodium methoxide in a 450 l container, this gives rise to an exothermic reaction, which destroyed the reactor and caused a fire. At the time the authors saw this behaviour as abnormal.

- Hexogen, that was submitted to the action of calcium hydroxide in the presence of water, started to decompose at 100°C, causing its ignition and then its detonation.

6.7.3 Acid action

All acids but especially Lewis acids (particularly aluminium chloride), give rise to dangerous interactions with nitrated derivatives and nitrates (there is not much information about nitrates). Aluminium chloride causes a large number of accidents due to nitrobenzene and sometimes nitromethane when used as a solvent in Friedel-Crafts reactions for which aluminium chloride is the common catalyst.

- With sulphuric, phosphoric, nitric and formic acids, nitromethane forms mixtures that are likely to detonate.

- The interaction between nitromethane, the etherate of boron trifluoride and silver oxide gives rise to an extremely dangerous reaction. However, it is difficult to interpret it. Can it be explained by the unstable property of silver tetrafluoroborate or the destabilising effect of boron trifluoride on nitromethane?

- The following reaction gives rise to a violent detonation:

$$CH_2=CH_2 + AlCl_3 + CH_3-NO_2 \xrightarrow[\text{[Ni]}]{\text{30-60 bar}} \text{Explosion}$$

this was attributed to the decomposition of the complex formed by the interaction of nitromethane with aluminium chloride. However, since the detonation took place when the mixture came into contact with the nickel catalyst, other interpretations might be suggested.

- A nitromethane/aluminium chloride complex had been prepared. A gaseous alkene was added to the complex; the pressure reached 5.6 bar and the temperature 2°C. The medium had been stirred at the beginning of the operation and then interrupted. A temperature rise caused the autoclave to detonate and the medium to carbonise entirely.

- The contact of a quaternary ammonium salt, with nitromethane increases its sensitivity to impact.

- An attempt to prepare salicylaldehyde was carried out according to the equation below:

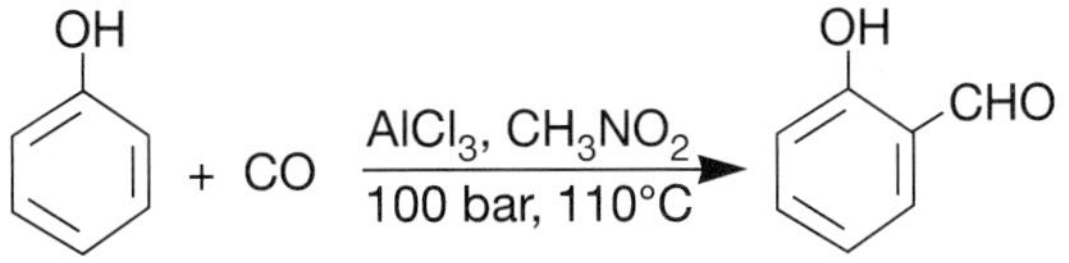

$$\text{C}_6\text{H}_5\text{OH} + \text{CO} \xrightarrow[\text{100 bar, 110°C}]{\text{AlCl}_3, \text{CH}_3\text{NO}_2} \text{2-HOC}_6\text{H}_4\text{CHO}$$

The reaction led to a violent explosion. The aluminium chloride/nitromethane interaction was regarded as being responsible for this accident.

- When alkyl nitrates and in particular methyl and ethyl nitrates are added to acids (H_2SO_4, $SnCl_4$, BF_3) they give rise to a very violent reaction with the formation of large quantities of gas. The presence of impurities, nitrogen oxides, transition metal oxides increases the sensitivity of these mixtures to detonation.

Nitrobenzene was also frequently involved in accidents due to its reactions with acids (above all Lewis acids).

- Nitrobenzene was 'washed' with an aqueous solution of 5% sulphuric acid before being distilled. Tars formed. The distillation took place in an apparatus made of iron. At the end of the distillation, hot tars were attacking the iron, producing hydrogen, leading to detonation. A post-accident analysis was carried out using differential thermal analysis. It was observed that when nitrobenzene is introduced into sulphuric acid, tars form immediately. If acid with a concentration of 85% in water is added to nitrobenzene, a highly exothermic transformation takes place at 200°C. If acid concentration is 70%, the transformation is hardly exothermic but starts at a temperature of 150-170°C.

- The preparation of quinolines using Skaup reaction uses the following scheme:

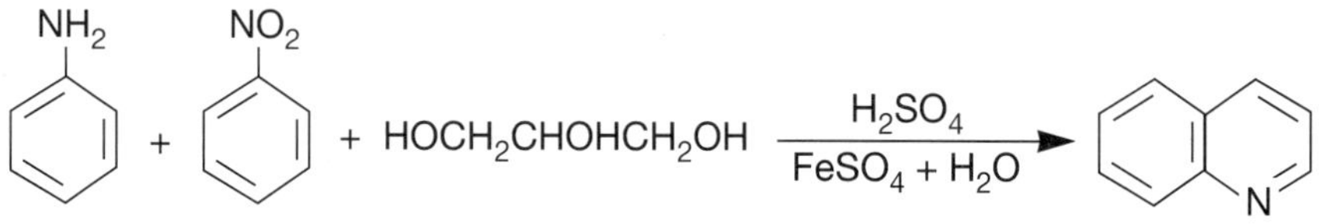

$$\text{PhNH}_2 + \text{PhNO}_2 + HOCH_2CHOHCH_2OH \xrightarrow[FeSO_4 + H_2O]{H_2SO_4} \text{quinoline}$$

In one accident, there was a series of mistakes during the operation: temperature of 32°C instead of 20°C; sulphuric acid in excess; too large quantities (450 l of reagents); etc. As a result, the reaction got out of control and caused the reactor to break.

- When nitrobenzene is in the presence of aluminium chloride at a temperature greater than 90°C the mixture obtained is thermally unstable . Its decomposition is explosive; pressure rises considerably and in a very short period of time. The following reaction that leads to the formation of very unstable compounds has been identified:

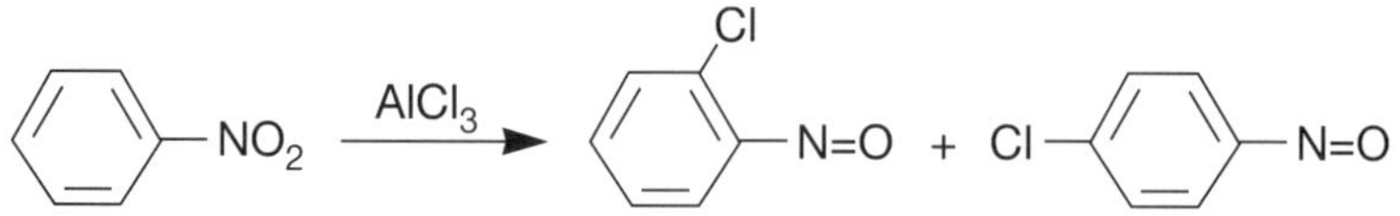

$$Ph\text{-}NO_2 \xrightarrow{AlCl_3} \text{(2-Cl-C}_6\text{H}_4\text{)-N=O} + Cl\text{-C}_6\text{H}_4\text{-N=O}$$

Such a reaction is possible every time nitrobenzene is used as a solvent in Friedel-Crafts reactions.

- Aluminium chloride was introduced into recovered nitrobenzene containing 5% phenol. This operation led to a violent detonation. A post-accident study showed that, at ambient temperature, the mixture of these three reactants makes the temperature rise and that at a temperature starting at 120°C, the reaction becomes extremely violent.

- Nitrobenzene/tin tetrachloride mixture decomposes violently at a temperature starting at 160°C.

6.7.4 Reactions involving side chain and complex groups

A large number of nitrated derivatives are aromatic. They can give rise to reactions of electrophilic aromatic substitution although the electron-attracting nitro group deactivates the aromatic nucleus. Besides, if a halogen atom is bonded to the aromatic ring, the presence of a nitro group will for the same reason activate the nucleophilic aromatic substitution of this halogen atom. Finally, other groups can also react 'on their behalf'. In these three cases, reactions that would not usually represent any danger could become dangerous due to the presence of one (or several) nitrated group(s), which would destabilise the reaction compounds. Numerous accidents illustrate these three situations.

- The compound that resulted from the reaction of oleum with nitrobenzene detonated after being stored at 150°C for several hours. The reaction was as follows:

$$\text{C}_6\text{H}_5\text{NO}_2 \xrightarrow{\text{H}_2\text{SO}_4,\ \text{SO}_3} \text{3-O}_2\text{N-C}_6\text{H}_4\text{-SO}_3\text{H}$$

- A differential analysis study showed that such a medium has an exothermic peak at a temperature starting at 145°C and that 3-nitrobenzenesulphonic acid decomposes violently at a temperature starting at 200°C. It is very interesting to mention an instability factor, which has not been discussed as far as we know. It is the simultaneous presence of two incompatible functional groups in a molecule. Indeed, as just seen in the previous paragraph, the acid groups destabilise the nitro group (see para 2.2.1).

- The situation that was described in the previous accident happened again and even twice, with o-nitrotoluene and p-nitrotoluene during their sulphonation. Oleum containing 24% sulphur trioxide had been added to o-nitrotoluene at 32°C. The reaction went out of control and caused the 2 l reactor to break and a very large volume of carbonised compound to be ejected (this was probably due to the decomposition of the sulphonic acid formed):

$$\text{o-CH}_3\text{-C}_6\text{H}_4\text{-NO}_2 \xrightarrow[32°C]{\text{H}_2\text{SO}_4,\ \text{SO}_3} \text{CH}_3\text{-C}_6\text{H}_3(\text{NO}_2)\text{-SO}_3\text{H}$$

- Sulphuric acid at 93% was added to p-nitrotoluene. The temperature reached 160°C due to a failure of the thermal control system. The sulphonic acid formed decomposed violently at this temperature. The post-accident investigation showed that the decomposition started between 160 and 190°C. In fourteen minutes the temperature rose to 190-224°C and in one minute and thirty seconds to 224-270°C. A large volume of gas was then released during the eruption. The phenomena caused by the decomposition of nitrated derivatives in the presence of sulphuric acid will be addressed several times. What these incidents have in common is the formation of large carbonised volumes. This phenomenon is common with sulphonic acids. The nitro group role is to destabilise intermediate compounds and final compounds and to generate large quantities of gas (nitrogen?) during the decomposition, hence the formation of a large amount of carbonised 'foams'.

- The same type of accident was also described as an effect of chlorosulphonic acid on o-nitrophenol:

o-nitrophenol + ClSO$_3$H at 4°C → 2-nitro-4-chlorosulphonylphenol (OH, NO$_2$, SO$_2$Cl)

A violent decomposition took place when the final compound was decanted. The technical analysis showed that the chlorosulphonic derivative obtained decomposes at a temperature starting at 24-27°C. Comparing the thermal behaviours of the four compounds just described could be interesting in terms of finding out more about stability factors of a molecule and is a good test to check the CHETAH method (see para 2.3).

The following are examples of accidents that are probably due to nucleophilic substitutions.

- When 1-chloro-4-nitrobenzene was treated with a solution of sodium methoxide in methanol it gives rise to an unusually violent explosion.
- The same chloronitrated derivative detonated after being heated at 100°C with water for an hour.
- When a mixture of 1-chloro-2-nitrobenzene and ammonia in an aqueous solution was exposed to a temperature of 160-180°C under 30-40 bar, the mixture went out of control and caused the reactor to detonate. Amongst the factors that contributed to this accident were an excess of the nitrated compound, bad cooling and failure of the reactor anti-explosion disk (which did not break).
- An accident that was perfectly identical involved 1-chloro-2,4-dinitrobenzene and aqueous ammonia that are heated at 170°C under 4 bar. The reaction was the following:

1-chloro-2,4-dinitrobenzene + NH$_4$OH → 2,4-dinitroaniline (NH$_2$, NO$_2$, NO$_2$)

This reaction had previously been carried out and details published, and had not created any incident.

- The exothermicity and violence of the reaction of hydrazine hydrate with 1-chloro-2,4-dinitrophenol caused destruction of the reactor in which the reaction was carried out. This reaction is of the same kind as the previous one (2,4-dinitrophenylhydrazine preparation).

Finally, there are a few reactions of nitroanilines which involve the amine group and for which the danger is entirely or partly related to the presence of nitrated functional groups.

6 • Reactions of organic chemicals

6.7 Nitrated derivatives

- When o-Nitroaniline is treated by sodium nitrite in acid medium and then by an inorganic sulphide, it forms an explosive diazonium sulphide. Note that even though the presence of a nitrated group does not help, it certainly is not a factor that is vital to cause the explosion, since this is a property that is common to all these diazonium salts whatever the nature of the substitution on the ring. The situation is exactly the same with p-nitroaniline.
- The problem is the same when attempting to carry out the following reaction in the thermal conditions mentioned below:

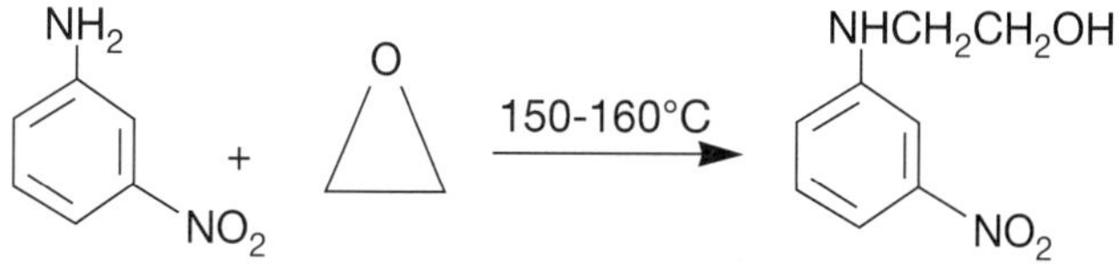

$$\text{m-NO}_2\text{-C}_6\text{H}_4\text{-NH}_2 + \text{ethylene oxide} \xrightarrow{150-160°C} \text{m-NO}_2\text{-C}_6\text{H}_4\text{-NHCH}_2\text{CH}_2\text{OH}$$

When the medium reached a temperature of 130°C, a detonation interrupted the process. As previously, it is not known with certainty whether the nitro group was the accident factor or not, or ethylene oxide for which the ring opening is extremely exothermic.

- When o-nitroaniline is heated in the presence of concentrated sulphuric acid above 200°C it gives rise to a violent reaction after a period of induction. There is also the formation of a black 'foam' that has a volume 150 times greater than the original volume. This is a dangerous variant of the spectacular 'black snake' experiment that is made with a nitrated derivative of a very similar structure and is described on p. 343. This is a phenomenon that can be compared with the ones described concerning p-nitrotoluene sulphonation on p.301.
- There are a few similarities between this accident and the previous one. When dinitroanilines (all isomers) are submitted to the effect of hydrogen chloride in the presence of chlorine, which seems to play a catalytic role, they give rise to a very violent reaction after a period of induction that can be very long if the temperature is low. Again a large volume of gas is also released. 2,3-Dinitroaniline is the most reactive, 3,5-the less reactive.

6.7.5 Oxidation and/or nitration reactions

This section is mainly concerned with reactions that involve nitric acid, of which behaviour is even more difficult to interpret. Indeed, nitric acid is not only a strong oxidant but also a nitrating agent, especially when it is combined with sulphuric acid. With nitrated derivatives it can also be destabilising due to its acidic property, which facilitates the compounds' decomposition. Since most reactions give rise to a detonation it is even more difficult to interpret the accidents.

- Nitric acid forms very sensitive explosive mixtures with nitromethane.
- Nitromethane gives rise to an extremely violent reaction with calcium hypochlorite.
- Alkyl nitrates become spontaneously explosive if they contain nitrogen oxides.
- A nitrobenzene/nitrogen tetroxide mixture used to be used as an explosive liquid. However, it is not still used due to its high sensitivity to thermal load.
- When water is present, a nitrobenzene/nitric acid mixture gives rise to detonations of the same intensity as trinitrotoluene. Danger is at its highest for

6 • Reactions of organic chemicals
6.7 Nitrated derivatives

the stoichiometric mixture (73% of nitric acid), but dangerous when nitric acid concentration lies between 50-80%.

- Nitrobenzene was submitted to the action of a nitric acid/sulphuric acid mixture. The mixture detonated. It was thought that this accident was caused by a lack of stirring, which caused the reagents to accumulate. A study showed that during this reaction the temperature could reach ~2000°C/s.
- A nitrobenzene/potassium chlorate mixture is highly explosive.
- Mixtures of nitric acid with dinitrobenzenes and trinitrobenzene are very strong and sensitive explosive combinations.
- Tetranitronaphthalene was prepared by adding nitric acid/sulphuric acid mixture to 1-nitronaphthalene. A solid residue formed after a valve had broken down. The medium was heated to try to melt the residue. This operation gives rise to a precipitate present in large quantities that blocked the stirrer and caused the reactor to detonate violently.
- Trinitrotoluene forms highly explosive combinations when nitric acid and metals such as lead or iron are present.
- The following reaction often leads to destructive accidents. To avoid this, potassium dichromate in the solid state should not come into contact with the nitrated derivative.

$$2,4,6\text{-trinitrobenzene} \xrightarrow{Na_2Cr_2O_7, H_2SO_4} 2,4,6\text{-trinitrobenzoic acid}$$

(Structure: 1,3,5-trinitrobenzene reacting with $Na_2Cr_2O_7$ / H_2SO_4 to give 2,4,6-trinitrobenzoic acid with CO_2H group)

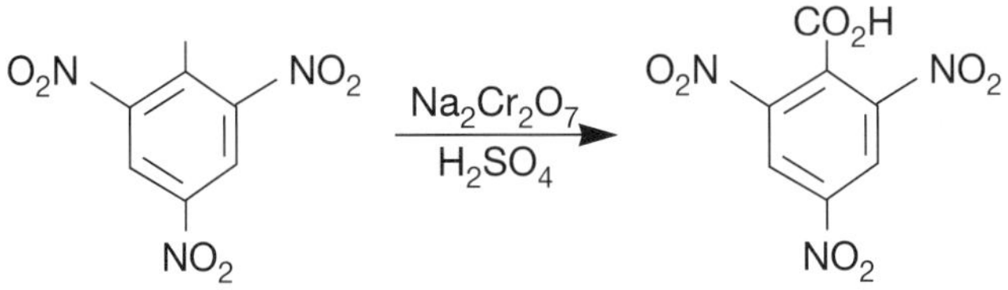

- A o-nitroaniline/nitric acid/magnesium mixture combusts spontaneously within a time limit of around fifty milliseconds depending on 2-nitroaniline proportion, which is around 25%.

6.7.6 Reactions caused by oxidising nature of nitrated derivatives

In this paragraph will be found the same difficulties as with the previous ones. All compounds more or less destabilise the nitrated derivatives. If these compounds are reducing agents, will it mean that this is why they caused a dangerous reaction with nitrated derivatives? For two of them, nitroethane and tetranitromethane, regulations suggest codes 9 and 8 respectively, which correspond to the codes of oxidising compounds. In this paragraph will be listed all reactions that involve metals as well as those that seem to involve nitrated derivatives as oxidants.

- Nitroalkanes are weak oxidants. They oxidise hydrocarbons, but it is difficult to create safe oxidation conditions since any excessive heating would cause nitroalkanes to decompose explosively.
- Nitromethane was treated with lithium aluminiumhydride in diethyl ether medium and at ambient temperature. This was followed by an explosion which pulverised the equipment. This accident can be explained by the fact that there was a redox reaction, but also by the formation of nitromethane lithium, unless

6 • Reactions of organic chemicals
6.7 Nitrated derivatives

it was due to the decomposition of lithium fulminate that could have been formed (see next paragraph).

- Nitromethane is very likely to detonate when aluminium powder is present. The same is true for a tetranitromethane/aluminium mixture. With aromatic nitrated derivatives, and in particular commercial explosives, the mixture with aluminium does not represent any danger. However, adding a drop of water causes spontaneous ignition that takes place within a time limit depending on quantities.
- Oxidation of hydrocarbons using tetranitromethane is extremely dangerous. For instance, when 10 g of this compound are mixed with toluene this led to a detonation that caused the death of ten people and twenty people to be seriously injured.
- Methyl and ethyl nitrates are promoters of the combustion of hydrocarbons.
- Nitrobenzene and aromatic nitrated explosives are mostly insensitive to impact. They become sensitive when potassium or Na-K alloy is present.
- 4-Nitrotoluene with sodium in diethyl ether medium forms a black residue that combusts spontaneously in air.
- 400 g of 2-Nitroanisole underwent catalytic hydrogenation at 250°C under 34 bar. The autoclave detonated. The lack of solvent and too drastic temperature and pressure conditions arementioned amongst the factors involved.
- The synthesis below had been carried out, but it was decided to distil the solvent before the end of the reaction; the reaction went out of control and the apparatus was pulverised. It is very likely that the very unstable compound that was obtained was responsible for this accident.

$$\text{o-}CH_3O\text{-}C_6H_4\text{-}NO_2 \xrightarrow{\text{Zn, NaOH, ethanol}} \text{o-}CH_3O\text{-}C_6H_4\text{-}N=N^+(O^-)\text{-}C_6H_4\text{-}OCH_3$$

6.7.7 Miscellaneous dangerous reactions

Following are a few reactions that are difficult to classify.

- Chloroform or bromoform/nitromethane mixtures can detonate very easily. This interaction has to be connected with the one described on p.297 that involved the CH_3NO_2/CH_2Cl_2 system, which by analogy could have been a factor in the accident described.
- When nitromethane comes into contact with silver nitrate, this can cause spontaneous explosions that are explained by the formation of explosive silver fulminate (see reaction below).
- Oxidising nitromethane electrochemically by using lithium perchlorate led to a violent detonation, which was explained by the formation of an unstable metal fulminate:

$$CH_3\text{-}NO_2 \xrightarrow{LiClO_4} Li^+NO_2\text{-}CH_2^- \xrightarrow{-H_2O} \bar{C}{\equiv}\overset{+}{N}\text{-}O^-Li^+$$

6 • Reactions of organic chemicals
6.7 Nitrated derivatives

Dangerous reactions of nitrated derivatives, nitrites and nitrates

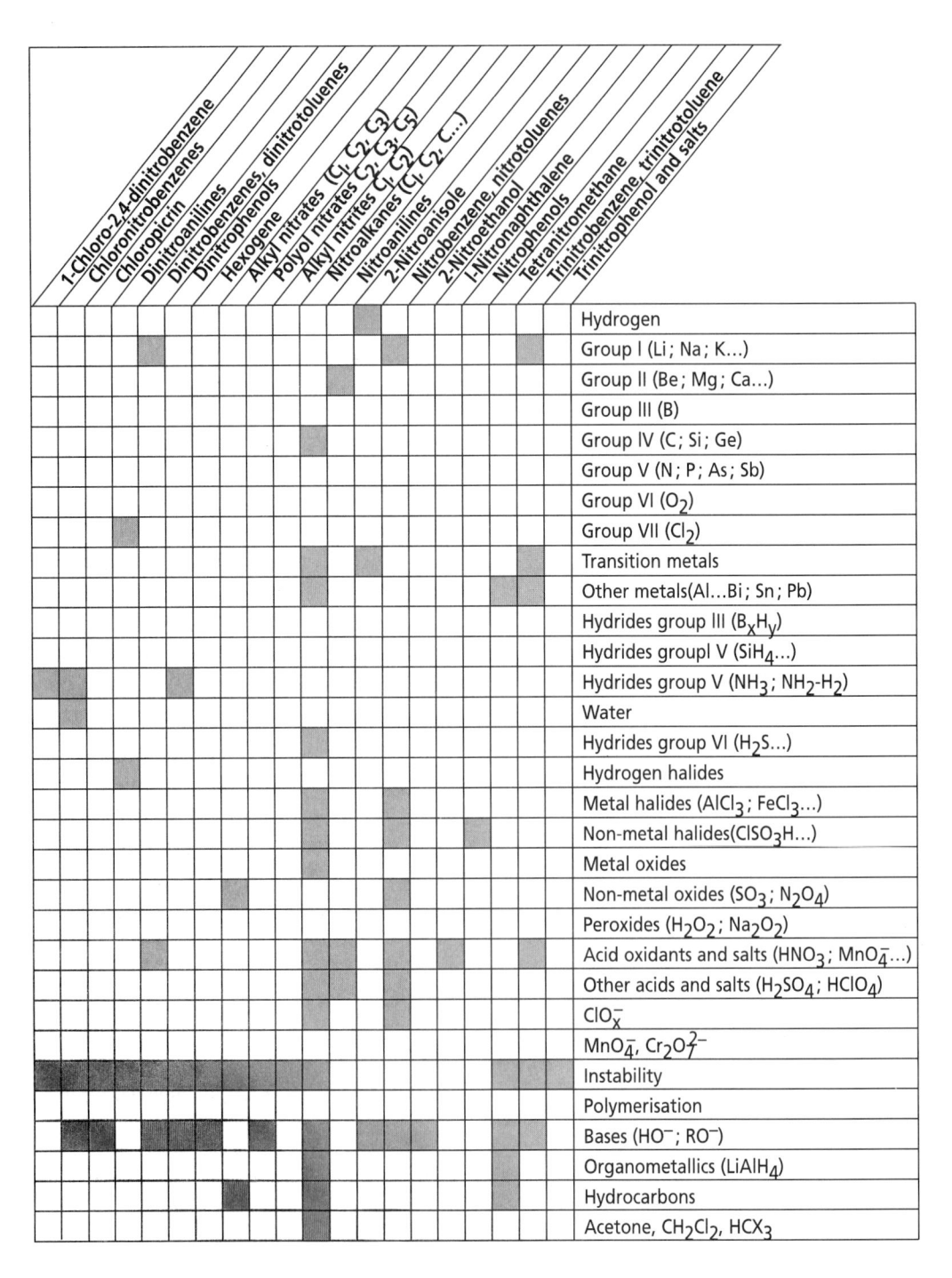

Columns (left to right):
1-Chloro-2,4-dinitrobenzene; Chloronitrobenzenes; Chloropicrin; Dinitroanilines; Dinitrobenzenes, dinitrotoluenes; Dinitrophenols; Hexogene; Alkyl nitrates (C₁, C₂, C₃); Polyol nitrates (C₂, C₃, C₅); Alkyl nitrites (C₁, C₂ ...); Nitroalkanes; Nitroanilines; Nitroanisole; 2-Nitroethanol; Nitrobenzene, nitrotoluenes; 1-Nitronaphthalene; Nitrophenols; Tetranitromethane; Trinitrobenzene, trinitrotoluene; Trinitrophenol and salts

Reagent rows:
- Hydrogen
- Group I (Li; Na; K...)
- Group II (Be; Mg; Ca...)
- Group III (B)
- Group IV (C; Si; Ge)
- Group V (N; P; As; Sb)
- Group VI (O₂)
- Group VII (Cl₂)
- Transition metals
- Other metals (Al...Bi; Sn; Pb)
- Hydrides group III (B$_x$H$_y$)
- Hydrides group IV (SiH₄...)
- Hydrides group V (NH₃; NH₂-H₂)
- Water
- Hydrides group VI (H₂S...)
- Hydrogen halides
- Metal halides (AlCl₃; FeCl₃...)
- Non-metal halides (ClSO₃H...)
- Metal oxides
- Non-metal oxides (SO₃; N₂O₄)
- Peroxides (H₂O₂; Na₂O₂)
- Acid oxidants and salts (HNO₃; MnO₄⁻...)
- Other acids and salts (H₂SO₄; HClO₄)
- ClO$_x^-$
- MnO₄⁻, Cr₂O₇²⁻
- Instability
- Polymerisation
- Bases (HO⁻; RO⁻)
- Organometallics (LiAlH₄)
- Hydrocarbons
- Acetone, CH₂Cl₂, HCX₃

306

- The nitrobenzene/phosphorus pentachloride mixture is stable up to 110°C. Above this temperature there is a decomposition that becomes more and more violent. Nitrogen oxides are also released in large quantities.

6.8 Aldehydes, ketones and acetals

The dangerous reactions of these are related to the propensity of the carbonyl group to react. Their reactivity depends on carbon chains' structure and in particular on their bonded groups. Carbonyl aldehydes are highly reactive; to such an extent that the first members of the series can give rise to polymerisation reactions. Besides, the carbonyl group activates hydrogen atoms, especially in the aldehyde group, but also the hydrogen atom which is bonded to the carbon of the group itself. Moreover, carbonyls activate double bonds of unsaturated compounds (polymerisations) and aromatic rings vis-à-vis nucleophilic substitution, because of its electron-withdrawing nature. Finally, these compounds are reducing, especially the aldehydes. Acetals and ketals are relatively inert; their reactivity is similar to ethers.

6.8.1 Oxidation reactions

■ Effect of oxygen

All these compounds, and especially the first members of the series, are highly inflammable, but aldehydes have the distinctive feature of being particularly inflammable, even in the absence of flames, merely by being in contact with objects that are heated moderately.

- A mixture of acetaldehyde/air of a concentration of 30-60% combusts when it is in contact with substances that are heated at 176°C. The mixture with oxygen of concentration of 60-80% combusts at 105°C. If the object is metallic and furthermore corroded, the ignition can become spontaneous even at ambient temperature. The same goes for corroded aluminium pipes where the AIT reaches 130°C, if vapour concentration is 55-57%. The AIT is also sensitive to the size and shape of the containers that contain acetaldehyde vapour.

- There also have been accidents that seemed to be the result of the mere contact of glyoxal with air.

This rather low AIT was explained by the quick formation of peroxides. With acetaldehyde the reaction is the following:

$$CH_3-C\overset{O}{\underset{H}{\diagup}} \xrightarrow{O_2} CH_3-C\overset{O}{\underset{O-OH}{\diagup}}$$

Peracetic acid lowers the AIT in this case. Besides, it has been demonstrated that the AIT depends on the partial pressure of peracetic acid formed and which settled on the container's walls that contains aldehyde. Metal oxides (rust, alumina) catalyse the formation of peroxidic compounds. This explains the effect of corroded metals that is described above. It is interesting to note that ketones, which peroxidise with more difficulty, have AITs that are much higher and that ethers, which behave in a similar way to aldehydes, also have a very low AIT.

- Oxidising acetaldehyde in air when cobalt acetate at -20°C is present gives rise to a detonation, if the medium is stirred. It has been put down to the formation of a very sensitive peroxidic compound. On the other hand, the presence of a halogen derivative inhibits this oxidation.
- An oxygen leak in an apparatus containing acetaldehyde that had been kept under nitrogen initially caused an oxidation which destroyed the apparatus.
- In some oxyacetylene welding equipment, there was a bottle of acetylene solution in acetone under pressure and a tube of compressed oxygen that was nearly empty, and in which the pressure had become lower than in the other bottle. Mixing both bottle contents was followed by the equipment's explosion. However, it is impossible to suggest that the danger is related to acetone, which seems to be a passive component, given the danger that is inherent to acetylene.
- $CH_3CH=CH-CH=O + [(CH_3)_2CH]_3Al$ + [solvent: isopropyl ether]

  The detonation was put down to the presence of crotonaldehyde peroxide or ether peroxide.

These peroxidations affect the aldehydic hydrogen atom, but also hydrocarbon positions in position α of the ketonic carbonyl as already seen (see alcohol group on p.253). Butanone is one of the key compounds that are involved in accidents of this type. The peroxidation is slow, but it seems that when other compounds that can also be moderately peroxidised are present the process is aggravated by their combination. We have already seen an example of such an interaction between 2-butanol and 2-butanone.

- The same goes for isopropanol/2-butanone. Isopropanol containing 0.5% of 2-butanone was distilled after being stored for four years. Distillation was interrupted by a violent explosion. It was thought that 2-butanone had played a catalytic role in the alcohol's peroxidation. A study showed that after a six months exposure to light (although a brown glass bottle was used), a quantity of 0.0015 mole of peroxide was present. If the bottle is kept in the dark, there is only 0.0009 mole of peroxide after five years. The authors of this study mention that acetone does not play a catalytic role in this peroxidation.
- This last comment forces one to reconsider the interpretation given to the following accident. A mixture of acetone and isoprene gives rise to the formation of peroxides that detonated spontaneously. One can ask oneself what role acetone plays since the presence of acetone is hardly necessary to the formation of explosive peroxides by isoprene in the presence of oxygen (see 'Hydrocarbons' on p.242).
- A similar problem arose in interpreting an accident due to the violent detonation of methylisobutylketone, which is well known for not being peroxidisable. This happened after a series of successive evaporations involving this compound. It was thought that peroxide that had accumulated caused this accident.

■ Effect of oxidising agents

Carbonyl compounds and especially aldehydes are very sensitive to the effect of strong oxidants.

6.8 Aldehydes, ketones and acetals

- When hydrogen peroxide is present, acetaldehyde forms polyethylideneperoxides, which are extremely sensitive explosives. Such peroxidation by hydrogen peroxide also has given rise to accidents with formol, propionaldehyde and acetone. The H_2O_2/reactant molar proportions play a crucial role in the detonation intensity, which is high when this ratio is greater than one. If proportions are stoichiometric, the detonation intensity is higher than for nitroglycerine. With acetone, peroxidic dimers and trimers apparently form, which detonate during evaporation operations; acetone has often been used (wrongly) as a solvent of oxidation reactions that are carried out with hydrogen peroxide.

- Benzaldehyde was oxidised by the action of a formic acid/hydrogen peroxide mixture on this compound. There was an extremely violent reaction. The reaction has to be carried out by adding the peroxide very slowly onto the benzaldehyde/formic acid mixture to allow the performic acid to react with benzaldehyde as it forms. A similar accident was observed when replacing benzaldehyde with formol.

- A mixture of ketene and hydrogen peroxide gives rise to a violent detonation that was put down to acetyl peroxide.

$$CH_2=C=O + H_2O_2 \longrightarrow CH_3-CO-OO-OC-CH_3$$

Nitric acid on its own or combined with sulphuric acid or/and with hydrogen peroxide has played an 'important' role in the accident history of carbonyl compounds.

- A mixture of fuming nitric acid and acetone caused the immediate ignition of both reactants.

- Oxidising cyclohexanone by using nitric acid (with the purpose of carrying out cyclohexane-1,2-dione synthesis) led to the mixture's detonation. The same reaction (that was made for the same purpose) with cyclohexanol was not apparently dangerous although the transitional formation of cyclohexanone also intervenes. It seems rather surprising that there is no danger in this reaction when using cyclohexanol (see alcohols on p.252).

- The effect of the nitric acid/hydrogen peroxide mixture on acetone when it is hot gives rise to an explosive oxidation, especially when the medium is confined. This situation also applies to a large number of ketones, and in particular, cyclic ketones. Cyclic di- and triperoxides form compounds that detonate, if there is no strict and very delicate thermal control. Accidents have been reported with butanone, 3-pentanone, cyclopentanone, cyclohexanone and methylcyclohexanones.

- Acetone was accidentally introduced into a reactor containing a sulphuric acid/nitric acid/hydrogen peroxide mixture. This led to an explosion due to acetone oxidation by the peroxymonosulphuric acid formed by the oxidising mixture.

- A formol (in an aqueous solution)/potassium permanganate mixture was used for disinfection work. This is a dangerous combination. Indeed, when mixing 100 cm^3 of aqueous formol with 50 g of permanganate in a plastic beaker, the beaker melted under the effect of the reaction's exothermicity. In the same way, a tower block was submitted to disinfection work by using 180 l of aqueous formol poured onto several kilograms of potassium permanganate. This gives rise to a huge fire that destroyed the tower block.

- Chromium trioxide has sometimes been used to destroy impurities present in acetone that was used as a solvent. This operation causes acetone to combust even at ambient temperature.
- An alkaline hypochlorite was used to carry out the following reaction:

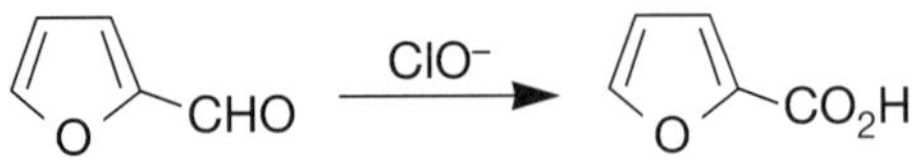

$$\text{furfural-CHO} \xrightarrow{ClO^-} \text{furfural-CO}_2\text{H}$$

Furfural was added drop by drop to a solution of 10% of hypochlorite at 20-25°C. At the end of this operation a violent detonation destroyed the equipment. The post-accident investigation showed that during the handling, pH decreases slowly down to 8.5, then suddenly to pH 2. At the same time the temperature rises suddenly to 70°C. The study showed that the same is true for benzaldehyde.

- Mercury chlorate or perchlorate with acetaldehyde forms compounds that detonate spontaneously.
- Acetone combusts when it comes into contact with chromyl dichloride.

6.8.2 Dangerous polymerisations

Polymerisation involves either aldehydic carbonyl or the double carbon-carbon bond in α position of the functional group.

With polymerisations that involve the aldehyde group:

- In the presence of an acid (sulphuric acid, acetic acid) or a metal (iron) that plays a catalytic role, acetaldehyde gives rise to polymerisation reactions that are often violent and cause the compound to overflow. An accident of this type has also been observed with anhydrous formol at -189°C.
- When it comes into contact with water, glyoxal can, under certain conditions, polymerise explosively.

With polymerisations that involve the non-saturation in α position of the carbonyl of aldehydes and ketones:

- When acid or base traces are present, acrolein gives rise to a very violent polymerisation after a period of induction that varies according to its purity. It decreases rapidly with the quantity of impurities that are present. Even when the acid is not very strong (NO, NO_2, SO_2, CO_2), the polymerisation is violent. Hydroquinone inhibits the polymerisation, but apparently not for very long.
- An old sample of acrolein (2 years) was stored in a fridge next to a bottle of dimethylamine. It detonated spontaneously. It was thought that aldehyde had polymerised under the effect of amine, which had diffused slowly through the caps.
- A store of 250 m³ of acrolein detonated. This was probably due to the spontaneous polymerisation of the aldehyde.
- Methylvinylketone, which is not inhibited, polymerises spontaneously in a violent fashion or even explosively. A serious accident involved the explosive polymerisation of this ketone under the effect of the heat generated by a fire close by.

6 • Reactions of organic chemicals
6.8 Aldehydes, ketones and acetals

- One can question whether the danger codes give an indication about the compounds that are likely to give rise to such dangerous polymerisations. For transport regulations it is code 9, for NFPA 1 to 4 in ascending order of danger is the 'reactivity' code. Unfortunately, in both cases, codes are ambiguous since they also refer to unstable compounds. However, code 9 in the transport regulations is attributed to methylvinylketone, but not to acrolein, which does not seem to make sense. By having a look at the acrolein enthalpy of decomposition, the technical literature gives the following value: ΔH_d = 0.864 kJ/g (70-380°C), which according to the CHETAH criteria, shows that it is a stable compound. For obvious reasons the same goes for methylvinylketone for which code 9 means that there is a risk of spontaneous polymerisation. The NFPA 'reactivity' code is more useful since it is more thorough. The following table lists all compounds that have a code greater than 0 (no risk).

Dangerous aldehydes, ketones and acetals according to NFPA reactivity Code

Compound	NFPA reactivity	Compound	NFPA reactivity
Acetaldehyde	2	Hexanal	1
Propionaldehyde	1	2-Methylpentanal	1
Acrolein	3	2-Ethylhexanal	1
Butyraldehyde	1	2-Ethyl-2-hexene-l-al	1
Crotonaldehyde	2	Methylvinylketone	2
Isobutyraldehyde	1	Hydroxybutanal	2
Methacrolein	2	Furfural	0

Comparing NFPA data for acrolein and methylvinylketone leads to the conclusion that transport regulations for aldehyde favour toxicity over the polymerisable character (336 instead of 339).

- A dangerous reaction that also affects the double bond in the α position of the carbonyl group is included in this section: the Diels-Alder reaction, which is responsible for a detonation that was explained by a wrong calculation of the liquid volumes involved:

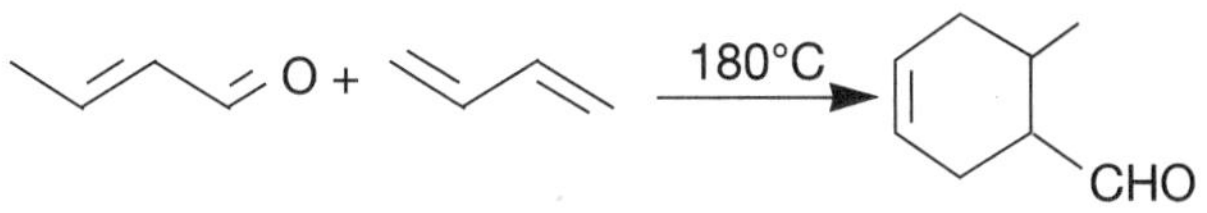

6.8.3 Reactions involving carbonyl group

- In acid or basic media and when phenol is present, formol forms resins that are 'phenolic'. To the author's knowledge, nine accidents have been mentioned that involved this reaction whose consequences were diminished by the detonation of the explosion vent disks of the reactors.
- A reaction of the same type is responsible for an accident that brought into play the effect of formol on phenylamine in the presence of perchloric acid:

$H_2C=O + C_6H_5NH_2 + HClO_4 \longrightarrow$ combustible and explosive resin

- $H_2C=O + CH_3NO_2$ [OH⁻] ⟶ $NO_2CH_2CH_2OH$ (I) + $HOCH(NO_2)CH_2OH$ (II) + $O_2N-C(CH_2OH)_3$ (III).
- After eliminating (I), by-products (II) and (III) combusted. However, note that the presence of nitromethane/base is a factor of the accident in itself as mentioned for nitrated derivatives on p.296.
- There was also a Cannizzaro reaction that gives rise to an accident. During a reaction that was carried out by using furfural this compound came into contact with sodium hydrogencarbonate that is used to check pHs. This compound catalysed a Cannizzaro reaction that could not be controlled and caused the compounds to combust:

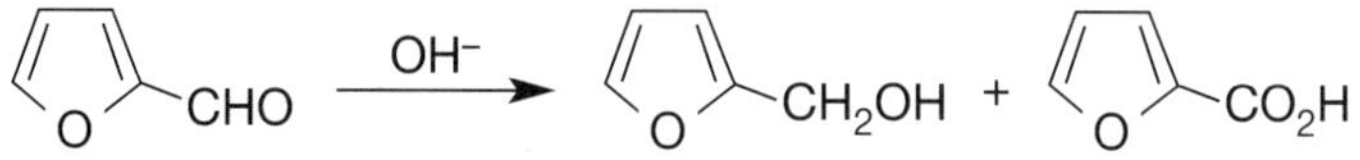

furfural-CHO —OH⁻→ furfural-CH$_2$OH + furfural-CO$_2$H

- There is also the haloform reaction (effect of a haloform on a ketone in a basic medium), already described with halogen derivatives on p.272 (the danger is more related to the interaction of the polyhalogen derivative with the base, according to the author). A large number of accidents involved the ketone as much as butanone. The accident below illustrates the danger of this reaction:
- $CH_3COCH_3 + CHCl_3$ + [KOH or $Ca(OH)_2$] ⟶ $(CH_3)_2C(OH)-CCl_3$. The reaction is very vigorous and exothermic. This accident was explained by the accidental presence of a base in the other compounds' mixture. The same accident is mentioned with $CHBr_3$ instead of $CHCl_3$.
- The reaction of dinitrogen pentoxide with acetaldehyde at -196°C is very dangerous when it is carried out without using any solvent. The danger comes from the instability of the compound formed:

$$CH_3-CHO + N_2O_5 \longrightarrow CH_3-CH(ONO_2)_2$$

6.8.4 Dangerous reactions related to the mobility of the α hydrogen atom.

■ Effect of bromine

- In the presence of a large quantity of bromine in excess, the following reaction was both violent and sudden:

$$CH_3-CO-CH_3 + Br_2 \longrightarrow BrCH_2-CO-CH_3$$

■ Effect of bases

- When acetone comes into contact with potassium tert-butylate, which is an extremely strong base, acetone combusts immediately. The same goes for butanone and methylisobutylketone.

■ Chloroacetone polycondensation

- Finally, two accidents are mentioned that brought into play the polycondensation of chloroacetone (in a glass bottle), which was exposed to sunlight. There was an extremely violent reaction. The first stage of it can be written down as following :

$$CH_3-CO-CH_2Cl + H_3C-CO-CH_2Cl \xrightarrow{h\nu} CH_3-CO-CH_2-CH_2-CO-CH_2Cl \longrightarrow ...$$

■ Sulphur dichloride destruction

- Acetone is used. However, if sulphur dichloride in excess is present in too large quantities, the reaction becomes particularly dangerous.

6.8.5 Miscellaneous dangerous reactions

- In the presence of phenylamine and hydroxylamine, chloral hydrate gives rise to a substantial release of hydrogen cyanide which is extremely toxic (see 'nitriles' on p.334). Reactions that would be classified as dangerous due to the toxicity of a compound formed are not being discussed often here, but an exception is being made because of the very high toxicity of hydrogen cyanide.

$$Cl_3C-CH(OH)_2 + C_6H_5-NH_2 + NH_2OH \longrightarrow HCN + ...$$

- While looking for the optimum operating conditions of the effect of dimethylamine on p-chloroacetophenone, the technicians heated the medium at 234°C; the reagents' proportion in weight being 1/4.22. The medium detonated not long after. It is likely that this was an aromatic nucleophilic substitution reaction as follows:

$$(CH_3)_2NH + Cl-C_6H_4-\overset{O}{\underset{\|}{C}}-CH_3 \longrightarrow (CH_3)_2N-C_6H_4-\overset{O}{\underset{\|}{C}}-CH_3$$

- Benzoquinone moistened by water gives rise to an important thermal load. A DTA showed that the exothermic peak is reached at 40-50°C and the compound decomposes at a temperature starting at 60-70°C. When a bottle of benzoquinone is exposed to sunlight, its temperature reaches 50-60°C rapidly.

6.8.6 Acetal reactions

These compounds have a certain similarity to ethers. Like the latter, their main danger comes from the easy peroxidation of all compounds of this class that have a hydrogen atom.

- This may be the reason why the NFPA 'reactivity' codes of some acetals reach 1 (paraldehyde) or even 2 (methylal and dioxolane).
- The detonation that took place during a distillation of 1,1-diethoxyethane was explained by the presence of peroxides.
- 2,2-Dimethoxypropane was being used to dry manganese and nickel perchlorates by heating above 65°C. A violent detonation interrupted the operation.

The small number of accidents found in the published sources might be explained by the fact that this type of compounds are hardly used.

6 • Reactions of organic chemicals

6.8 Aldehydes, ketones and acetals

Dangerous reactions of aldehydes, ketones and acetals

Acetaldehyde, propionaldehyde	Acetone, diethylketone	Acrolein	Benzaldehyde	Benzoquinone	Butanone, methylisobutylketone	Ketene	Cyclic ketones	Chloroacetone	4-Chloroacetophenone	Crotonaldehyde	1,1-Diethoxyethane, dimethoxymethane	2,2-Dimethoxypropane	Formol	Furfural	Glyoxal	Chloral hydrate	Methylvinylketone	
																		Hydrogen
																		Group I (Li; Na; K...)
																		Group II (Be; Mg; Ca...)
																		Group III (B)
																		Group IV (C; Si; Ge)
																		Group V (N; P; As; Sb)
■	■				■					■								Group VI (O_2)
■	■																	Group VII (Cl_2; Br_2)
■																		Transition metals
■																		Other metals (Al... Bi; Sn; Pb)
																		Hydrides group III (B_xH_y)
																		Hydrides groupl IV (SiH_4...)
													■					Hydrides group V (NH_3; NH_2OH)
		■										■						Water
																		Hydrides groupVI (H_2S...)
																		Hydrogen halides
																		Metal halides ($AlCl_3$; $FeCl_3$...)
■																		Non-metal halides (SCl_2...)
																		Metal oxides
	■																	Non-metal oxides (O_3; N_2O_5)
■	■		■							■								Peroxides (H_2O_2; Na_2O_2; ROOR...)
■	■	■								■								Acid oxidants and salts (HNO_3; MnO_4^-...)
																		Other acids and salts (H_2SO_4; $HClO_4$)
														■				ClO_x^-
																		MnO_4^-, $Cr_2O_7^{2-}$
																		Instability
■	■	■			■					■			■				■	Polymerisation
■	■																	Bases
																		Butadiene
		■																$CHCl_3$; $CHBr_3$
													■					Phenol
				■									■					Aniline, dimethylamine
									■									Nitromethane

6.9 Carboxylic acids

Compared with other groups, carboxylic acids appear to be relatively less dangerous. Like with all other organic compounds, most accidents are caused by oxidation reactions. Apart from those, dangers arise from formic acid that is hardly stable, and diacids and complex acids since the presence of a second acid grouping or another grouping destabilises the carboxyl group. Besides, their acidic properties can involve them in all dangerous reactions that take place in an acid medium. Nevertheless, they are not very strong acids and only those with a high acidity will behave dangerously. Amongst common acids, formic acid has a high acidity. Finally, if a non-saturation is present, the closeness of the active group boosts its reactivity and increases risk.

6.9.1 Oxidation reactions

■ Effect of oxygen

Analysing data in Part Three and comparing it with other compounds shows that acids are hardly inflammable. This is due to their low vapour pressure, which is linked to their propensity for hydrogen bonding in the liquid state, in particular in the case of the first series members that could have been the most inflammable ones. The only danger comes from solid acids at ambient temperature, which can easily give rise to dust 'clouds', which are explosive in the form of suspensions in air.

- This situation happens often with benzoic acid, which is used to heat blast calorimeters. Such accidents have sometimes caused the calorimeter to be destroyed. They can be explained by the fact that acid is used in the form of fine powder instead of flakes.

- Powdered aluminium had been added to oleic acid. The mixture detonated after being prepared. Such an accident could not be repeated and it was thought that it was caused by the presence of a peroxide formed by the effect of air on oleic acid. In fact, the acid functional group has obviously nothing to do with the peroxidation. It is more likely that the chain's double bond that activates β hydrogen atoms (allyl position) was involved in it. This is a well-known phenomenon since it is responsible for the rancidity of some oils and greases.

■ Effect of peroxides

It is possible to peroxidise the acid functional group with peroxides and especially hydrogen peroxide or sodium peroxide when traces of water are present.

- Formic acid forms performic acid when hydrogen peroxide is present:

$$H-CO_2H + H_2O_2 \longrightarrow H-CO-O-OH$$

The enthalpy of decomposition of this peracid is relatively high (ΔH_d = -1.83 kJ/g - average risk according to CHETAH criterion C_1). Its aqueous solutions are unstable. Solutions that contain 80% of peracid detonate when they are stirred (even at -10°C). The usual way of preparing this peracid involves the effect of hydrogen peroxide when metaboric acid is present; however, although this operating method was followed, serious accidents have occurred.

6 • Reactions of organic chemicals

6.9 Carboxylic acids

- Preparing peracetic acid by the action of hydrogen peroxide on acetic acid is as hazardous. If the temperature is too low, compounds accumulate and cause the medium to detonate. Using peracetic acid solution as an oxidant causes detonations when its concentration is too high or if evaporation is attempted. An accident happened during such an operation (see reaction below). The best way to eliminate this peracid at the end of the reaction is to heat it in a water bath at a temperature that should not exceed 50°C and under reduced pressure.

3-bromopyridine $\xrightarrow{CH_3CO_2H/H_2O_2}$ 3-bromopyridine N-oxide

- A mixture of 20 cm³ of acetic acid and 20 cm³ of hydrogen peroxide had been heated for four hours in the presence of 0.1 g of jute. A solution of 30% peroxide instead of 6% had been used. Peracetic acid formed quickly and its concentration reached 34%. Then, the medium detonated violently.

- Acetic acid that was added to sodium peroxide detonated violently. It could not be determined whether there was a direct oxidation reaction or if water could have transformed sodium peroxide into hydrogen peroxide.

- A mixture of formic acid and acetic acid detonates in the presence of a hydrogen peroxide/water mixture, if the peroxide concentration is greater than 50%.

- Treating an aromatic hydrocarbon with an aqueous solution that contains 30% hydrogen peroxide and in the presence of trifluoroacetic acid as a solvent has the advantage of oxidising the ring without touching the lateral chains. However, this method gives rise to a detonation if the solvent is eliminated before destroying hydrogen peroxide. It is possible to destroy it by using manganese dioxide.

■ Effect of other oxidising agents

- A very violent detonation occurred when a mixture of chromium trioxide and acetic acid was heated (this mixture is used as an oxidant). The investigation showed that this accident was caused by liquid acid being in contact with oxide particles. These particles glowed and ignited the air/acid vapour mixture, which caused the apparatus to detonate.

- Oxidising acetic acid by using chromium (VI) compounds is far less dangerous, if an aqueous potassium dichromate/sulphuric acid mixture is used. On the other hand, if hot evaporation of the medium is carried out and this gives rise to the formation of solid dichromate, the detonation due to the contact of solid salt with acetic acid cannot be avoided.

- The contact of solid chromium trioxide with butyric acid at 100°C causes the oxide to glow.

- A mixture of acetic acid with potassium permanganate heats up considerably. The mixture detonates if there is no efficient cooling.

- Nitric acid is used to treat some nuclear wastes. When trying to destroy it by heating it at 100°C with formic acid, the operation proves to be very risky. Large quantities of gas can be produced (CO_2, N_2O_4, NO, N_2, N_2O). Sulphuric acid catalyses this interaction and makes it more dangerous. Some salts are thought to inhibit it.

- An attempt to oxidise vanillin (3-methoxy-4-hydroxybenzaldehyde) was made using thallium trinitrate and when formic acid at 90°C as a solvent was present. The extremely violent redox reaction that followed was put down to the effect of salt on acid.

- A lactic acid/hydrogen fluoride/nitric acid mixture is used to polish metals. It is unstable and autocatalytic. After storing it for twelve hours the temperature rises to 90°C and there is significant gas release. Therefore such mixtures should not be kept.

- A mixture of alkaline hypochlorite/formic acid, which is used when treating certain industrial residues, detonates at 55°C.

- The effect of an aqueous solution of sodium chlorite on oxalic acid in the presence of sunlight gives rise to a violent detonation, which is supposedly due to the formation of chlorine dioxide that is unstable:

$$HO_2CCO_2H + NaClO_{2\ (aq)} \xrightarrow{h\nu,\ \Delta} ClO_2\ \text{(explosive)}$$

The enthalpy of this reaction was estimated experimentally at -1.88 kJ/g.

- The introduction of phthalic acid into sodium nitrite caused the medium to detonate. This was put down to the reaction below that leads to an unstable compound:

$$\text{benzene-1,2-(COOH)}_2 + NaNO_2 \xrightarrow{\Delta} \text{benzene-1,2-(COO-N=O)}_2$$

6.9.2 Instability

Generally speaking, carboxylic acids are stable. Instability affects either the first series member or acids that have another active group, including carboxyl group that is in close proximity to the acid group.

■ Formic acid

- The acid at 98-100% decomposes slowly. A study showed that if the acid is kept in a full bottle of 2.5 l at 25°C, a pressure of 7 bar is created in one year. With non-oxidised nickel, heated sulphuric acid catalyses the decomposition as follows:

$$H-CO_2H \longrightarrow CO + H_2O$$

- Diphosphorus pentoxide had been introduced into formic acid at 95% to obtain formic acid at 100%. An significant carbon monoxide release caused the compound to overflow. It can be assumed that the decomposition was catalysed by phosphoric acid that formed during the drying process.

- Formic acid at 98% had to be used as a solvent during a catalytic hydrogenation by using the palladium/carbon system. When the solvent came into contact with the catalyst there was a release of hydrogen. Does this accident result from the acid decomposition catalysed by palladium? In this case the decomposition would bring a different reaction into play from the one suggested previously. Is it more likely to be due to the acid reduction? Indeed, carboxylic acids and in particular the first series member which has an oxygen balance of -34.78g% have a relatively strong oxidising nature. It all depends on the reagents that are in the presence of this acid.

- The same ambiguity appears with the formic acid/aluminium mixture (the metal glows).

■ Diacids

- Oxalic acid is relatively stable by itself; it decomposes at a temperature starting at its melting point, which corresponds to 189.5°C, forming carbon dioxide, monoxide, formic acid and water.
- On the other hand, it forms a silver salt that is unstable and starts decomposing violently at 140°C. Under the same conditions monoacids give rise to slow decarboxylations.
- Malonic acid decomposes at a temperature starting at 135°C (its melting point). Superior diacids are more stable.
- The enthalpy of decomposition of fumaric acid was determined experimentally: -0.925 kJ/g. This value shows its unstable nature that is due to the second carboxyl group.

■ Pyruvic acid

- Pyruvic acid is not stable at ambient temperature when it is stored for a long period of time. It can only be stored in a refrigerated room. A bottle of this acid was stored in a laboratory at 25°C and detonated, probably because of the overpressure created by the formation of carbon dioxide. Indeed, with diacids and complex acids the decomposition is made by decarboxylation. In this particular case, this decomposition should give rise to acetaldehyde. It could be asked whether, in the exothermic conditions of this decomposition, a polymerisation of this aldehyde (see 'Aldehydes-ketones' on p.310) did not make the situation worse.

6.9.3 Reactions related to acidic characteristics

Carboxylic acids are weak acids. Formic acid and complex acids that have attracting groups not far from the carboxyl are exceptions.

- There was an attempt to neutralise formic acid by using magnesium hydrogencarbonate. By stirring the medium without care, a violent reaction of the acid with carbonate took place. Carbon dioxide was released in large quantities and caused the projection of the compounds into the surrounding area.
- Formic acid is acid enough to decompose nitromethane (see 'Nitrated derivatives' on p.298).
- Potassium tert-butylate, which is an extremely strong base, causes acetic acid to ignite immediately.

6.9.4 Polymerisation of unsaturated α, β acids

Four accidents that were the result of such a polymerisation reaction have been reported. Indeed, carboxyl group increases the reactivity of the double bond. The compound can be inhibited by using a phenolic inhibitor.

- Acrylic acid can polymerise dangerously even when it is inhibited, if the temperature is lower than its melting point (14°C). Indeed, its crystallisation gives rise to pure acid crystals that are not inhibited anymore.
- The violent polymerisation of acryl acid caused a violent fire on a boat that was transporting it. The investigation showed that this acid contained ethylidene norbornene, which is a very oxidisable compound that forms a peroxide in air, which caused the acid polymerisation.
- During a polymerisation operation of acryl acid in aqueous solution, in the presence of a primer and a moderator, the pump broke down and caused monomer to accumulate. Its polymerisation could not be controlled and the apparatus are destroyed.
- A container of inhibited methacryl acid had been stored outdoors. It crystallised (melting point: 16°C). The recrystallised acid polymerised violently not long after the container had been brought back into a heated room. The heat that was given off by the polymerisation caused the compound's vapourisation.

The propensity for polymerisation is given by codes 1 or 2 of the NFPA 'reactivity' code. The risk related to polymerisation is given in the transport regulations of dangerous compounds. The table below gives the available data. Crotonic acid is added, which has a surprising low NFPA code.

Compound	NFPA reactivity Code	Dangerous materials Code
Acrylic acid	2	89 SEE TEXT
Methacryl acid	2	89
Crotonic acid	0	–

6.9.5 Miscellaneous dangerous reactions

In this paragraph are listed the reactions that are difficult to classify or for which the interpretations given seem ambiguous.

- It is possible to make an acid chloride by the action of phosphorus trichloride on an acid. Three accidents have been described that involved the action of this compound on acetic, propionic and 2-furoic acids, respectively. This was explained by the formation of phosphine due to excessive heating that led to the decomposition of the phosphorus acid formed:

$$R-CO_2H + PCl_3 \longrightarrow R-COCl + P(OH)_3 \xrightarrow{\Delta} PH_3$$

In all cases the medium combusted spontaneously. This was put down to the spontaneous self-ignition of phosphine. (This interpretation does not convince the author.)

6 • Reactions of organic chemicals

6.9 Carboxylic acids

Synopsis of dangerous reactions of carboxylic acids

	Reactant
	Hydrogen
	Group I (Li; Na; K...)
	Group II (Be; Mg; Ca...)
	Group III (B)
	Group IV (C)
	Group V (N; P; As; Sb)
	Group VI (O_2)
	Group VII (Cl_2; Br_2)
	Transition metals (Pd)
	Other metals (Al)
	Hydrides group III (B_xH_y)
	Hydrides group IV (SiH_4...)
	Hydrides group V (NH_3; NH_2OH)
	Water
	Hydrides group VI (H_2S...)
	Hydrogen halides (HF)
	Metal halides ($AlCl_3$; $FeCl_3$...)
	Non-metal halides (PCl_3)
	Metal oxides (CrO_3)
	Non-metal oxides (P_2O_5)
	Peroxides (H_2O_2; Na_2O_2)
	Acid oxidants and salts (HNO_3; NO_2^-)
	Other acids and salts (H_2SO_4; $HClO_4$)
	ClO_x^- (ClO^-; ClO_2^-)
	MnO_4^-, $Cr_2O_7^{2-}$
	Instability
	Polymerisation
	Bases
	Metal hydrides (LiAlH$_4$)
	Nitromethane

Columns (acids): Acetic acid, Acrylic acid, Benzoic acid, Butyric acid, Formic acid, 2-Furoic acid, Lactic acid, Malonic acid, Methacryl acid, Oleic acid, Oxalic acid, Phthalic acid, Propionic acid, Pyruvic acid, Trifluoroacetic acid

6 • Reactions of organic chemicals

6.10 Esters

- Trifluoroacetic acid is supposed to form complexes that are very unstable with lithium tetrahydrogenaluminate. This could be the intermediate compound of this common reduction reaction:

$$R-CO_2H \xrightarrow[-H^2]{LiAlH_4} (R-CH_2O)_4Al^-Li^+ \xrightarrow{H_2O} RCH_2OH$$

whose hydrolysis gives rise to the corresponding alcohol.

- Finally, mention will be made of an accident that did not bring a chemical reaction into play, but a change of physical state. A bottle of formic acid was placed overnight in a refrigerator at -6°C. It detonated because the acid solidified (melting point: 8°C). The volume also increased on freezing (as with water).

6.10 Esters

Amongst the carbonyl derivatives, esters are the least reactive and the most stable. There are a few dangerous reactions that involve esters although they are often used in particular as solvents. Formates constitute a special case since they are less stable. Like acids, their intermediate position so far as functional carbon oxidation degree is concerned makes oxidation as well as reduction reactions possible. However, the latter are only possible with strong hence dangerous reducing agents. The esters' hydrolysis reaction, which is a balanced reaction, is not dangerous except in very specific situations. Finally, as for all carbonyl derivatives, there is the aggravating effect of the functional group on the reactivity of ethylene α-β double bond. With complex esters there is also the behaviour peculiar to the other groups. Lactones are notably more reactive (hence less stable) than linear esters especially when dealing with structures with tight rings (ketene, β-propiolactone), but these compounds are rarely used in the industry (apart from ε-caprolactone).

6.10.1 Instability

Even when ester has a double bond it still remains stable.

- The enthalpy of decomposition of allyl acetate (ΔH_d = -0.45 kJ/g at 170-470°C) makes it slightly dangerous according to CHETAH (see para 2.3).
- On the other hand, formates are less stable. This seems to be related to the H-CO-Σ atomic catenation in which Σ is an oxygen or chlorine atom. Nevertheless, with esters, the presence of a base as well as a high temperature and pressure are necessary. In the presence of sodium methoxide in a small quantity (0.5%), at 100°C and under 70 bar, methyl formate decomposes according to the reaction below; the violent release of carbon oxide creates a dangerous overpressure in the apparatus:

$$H-CO_2-CH_3 \xrightarrow[100°C \,;\, 70\, bar]{CH_3O^-} CH_3OH + CO$$

This is the reverse reaction of the industrial synthesis of methyl formate. This decomposition usually occurs at the end of the synthesis, if a quick treatment of the compound does not free it from the base in the medium.

- Diketene residues had been stored before being incinerated. They suddenly decomposed causing the container lid to fly off. Then the vapour combusted.

This instability of diketene or its decomposition in the presence of water explains why code 2 of the NFPA 'reactivity' code was attributed to it.

6.10.2 Oxidation reactions

Only strong oxidants can react dangerously with saturated esters. Unsaturated esters have often been involved, but the reaction mainly concerns non-saturation; the ester group can increase the risk by boosting the double bond reactivity. Therefore, reference to unsaturated esters will be concerned with this type of reaction.

- The only accident that involves a saturated ester is the result of an attempt to extract an organic residue containing hydrogen peroxide with ethyl acetate. The latter was mixed with methanol and refluxed with the residue and hydrogen peroxide in an aqueous solution. A second extraction was carried out with acetate and the liquid was then evaporated. The small quantity of the compound that remained after the evaporation detonated violently. It was thought that this detonation was the result of the violent decomposition of methyl hydroperoxide, peracetic acid and/or ethyl peracetate.

Besides, the formation of these three compounds can be explained by the following reactions:

$$CH_3-CO_2-C_2H_5 \underset{CH_3OH}{\overset{H_2O}{\rightleftharpoons}} CH_3-CO_2H \xrightarrow{H_2O_2} CH_3-CO_2-OH$$

$$CH_3-CO_2-OH \underset{}{\overset{CH_3OH}{\rightleftharpoons}} CH_3-CO_2-OC_2H_5$$

$$CH_3-OH \xrightarrow{H_2O_2} CH_3-OOH$$

These reactions do not bring a direct peroxidation of ester into play, but rather of the compounds present (methanol) and of the hydrolysis compounds.

6.10.3 Reduction reactions

- Ethyl acetate has sometimes been used to destroy lithium tetrahydrogen aluminate (the reaction is similar to the one that results from the effect of a carboxyl acid on this metal hydride described on p.321; the acid formed destroys the metal hydride). Such an attempt had been made for this purpose. It led to a very violent detonation.
- The reaction below, which is a common synthesis of aldehydes, is very violent and difficult to control:

$$R-C_6H_4-COOR + LiAlH_4 \longrightarrow R-C_6H_4-CH=O$$

- Diethyl succinate had been introduced into a mixture of sodium hydride and ethyl trifluoroacetate at 60°C. The medium detonated ten to twenty minutes after starting the addition of succinate. Two accidents that are identical to this one have been described.

6.10.4 Unsaturated ester reactions

These mainly affect two main types of esters $R-CO_2-R'$: those in which the non-saturation is in R' (vinyl acetate, essentially), those for which group R is unsaturated

6 • Reactions of organic chemicals
6.10 Esters

(acrylates, methacrylates). Finally, diketene has a particular reactivity due to the strong reactivity of the double bond at the top of the strained ring (a reaction on the double bond reduces the ring strain). The main danger comes from polymerisation, but other addition reactions on vinyl acetate double bond proved to be dangerous.

■ Vinyl acetate polymerisation

- This compound polymerises spontaneously and can only be stored at a low temperature and in the presence of an inhibitor (hydroquinone, for instance). Even in these conditions, storage can not exceed a period of six months.
- It is not recommended to use dehydrating agents such as silica gel since it catalyses this polymerisation. The same goes for alumina.
- Vinyl acetate had been heated at the reflux of toluene (111°C) in a 10 m^3 reactor to carry out a polymerisation in a solution of this compound. Because the compound mass was too critical, the exothermicity of the reaction could not be controlled; the reactor was destroyed and a toluene fire destroyed the site.
- In a moist air atmosphere, vinyl acetate polymerised spontaneously with extreme violence. The investigation led to the following interpretation. In the presence of moisture, the acetate hydrolysed forming acetaldehyde:

$$CH_3-CO_2-CH=CH_2 \xrightarrow{H_2O} CH_3-CO_2H + CH_3-CH=O$$
$$\text{Acetaldehyde}$$

Acetaldehyde peroxidised into peracetic acid when air was present:

$$CH_3-CH=O \xrightarrow{O_2} CH_3-CO_2-OH$$
$$\text{Peracetic acid}$$

The peracid catalysed the radical polymerisation of vinyl acetate.

- The reaction scheme below is carried out due to the effect of hydrogen peroxide on vinyl acetate when osmium tetroxide is present:

$$CH_3CO_2CH=CH_2 + H_2O_2 \xrightarrow{OsO_4} CH_3CO_2CHOHCH_2OH$$

A detonation took place while eliminating the excess of vinyl acetate by distillation, under reduced pressure, that was explained as before (formation of peracetic acid after acetate hydrolysis).

- Benzoyl peroxide had been introduced into vinyl acetate with the idea of polymerising it; ethyl acetate was used as a solvent. The polymerisation went out of control and vapourised the ester. Ester formed a vapour 'cloud', which detonated.

■ Polymerisation of acrylates

- A 4 l bottle of of methyl acrylate that had been stored for a long time detonated a few hours after being transported from the storage place to the laboratory. This explosion was explained by the formation of peroxides, which thanks to the stirring of the medium caused by the transport, gave rise to violent polymerisation. This type of ester has to be stored in the presence of an inhibitor, below 10°C and in an inert atmosphere. In this particular case, the compound had been inhibited, but the inhibitor may have been destroyed by the peroxides that had accumulated.

- Ethyl acrylate that was in a steel container was placed in a 4 l glass container. The climatic conditions were exceptionally hot and the fact that the container was exposed to light caused its polymerisation, which caused the container to detonate.

- Methyl methacrylate poses the same risks as the previous ones. This compound had been exposed to air for two months and polymerised partly. An attempt was made to recover the monomer. This monomer was evaporated by heating it to 60°C instead of 40°C. The medium detonated due to excessive heating of the peroxides that had formed. A study showed that in the presence of air or peroxide, methyl methacrylate gives rise to an auto-accelerated polymerisation during which the ester reaches a temperature of 90°C. Rust catalyses this polymerisation.

- Methyl methacrylate combusted in a beaker. This accident was the result of the compound being in contact with 'grains' of benzoyl peroxide that were on the beaker walls.

- Propionaldehyde was poured into a container that was intended for collecting residues from different chemical reactions and that already contained methyl methacrylate. The medium detonated not long after this operation and when closing the container. This could be explained (as has already been seen with vinyl acetate) by the fact that propionaldehyde was peroxidised and catalysed the methacrylate polymerisation that could not be controlled.

Diketene polymerisation

- Diketene polymerises very violently when bases, amines, acids and Lewis acids are present. Even sodium acetate at a concentration of 0.1% is enough to cause this polymerisation at a temperature starting at 60°C.

The danger codes give the risk that is related to the unsaturated esters polymerisation. The following table lists all available data. In these values as well as in the ones that are mentioned in this chapter, the correlation between both codes is not obvious.

Unsaturated esters: danger codes

Compound	NFPA reactivity	Transport Code (TR)
Vinyl acetate	2	339
Allyl acetate	–	not in code 9
Methyl acrylate	2	339
Ethyl acrylate	2	339
Butyl acrylate	2	39
Methyl methacrylate	2	339
Ethyl methacrylate	0	339
Butyl methacrylate	0	39
i-Butyl methacrylate	–	39
Ethyl crotonate	0	not in code 9

6 • Reactions of organic chemicals

6.10 Esters

Other dangerous reactions related to non-saturation
- In air at 50°C, vinyl acetate forms a polyperoxide, which detonates on separation.
- When ozone is present, vinyl acetate forms an ozonide that is explosive:

$$CH_3-CO_2-CH=CH_2 \xrightarrow{O_3} CH_3CO_2-CH(O-O)-CH_2-O$$

6.10.5 Miscellaneous dangerous reactions
- In the presence of potassium tert-butylate and as with most organic compounds, esters combust spontaneously after a period of induction that is generally very short. Such accidents have been reported with propyl formate, ethyl acetate and dimethyl carbonate.
- When dibutyl phthalate is treated with liquid chlorine at 118°C in a steel bomb, it gives rise to an explosive reaction.
- 2-Ethoxyethyl acetate detonated during distillation after air penetrated the distilling apparatus. The investigation could not provide any accurate explanation for this accident. This ester is well known for not being peroxidisable in the liquid phase. It was thought that the accident was due to the presence of ethyleneglycol in the medium that would have converted into dioxan, which is highly peroxidisable. This interpretation does not make sense and it is thought that this can be explained by a peroxidation in α position of the ether grouping of the ester that was made possible by the thermal conditions created during the distillation.
- Butyrolactone was introduced into a reactor containing 2,4-dichlorophenol, sodium hydroxide and n-butanol, that was used as a solvent, with the idea of making a herbicide following the reaction below:

[butyrolactone] + [2,4-dichlorophenol with OH] → (with C_4H_9OH / NaOH) → [dichlorophenyl ether with $OCH_2CH_2CH_2COOH$ substituent]

When the lactone was introduced, the temperature reached 165 and then 180°C very quickly even though an attempt was made to cool the medium. The reactor detonated and a fire broke out. It seems obvious that the temperature rise is due to the high reactivity of lactone, but the main factor in this accident seems to be related to the behaviour of dichlorophenol, which in such conditions gives rise to an aromatic nucleophilic substitution reaction that leads to the formation of a dichlorodioxin (see 'halogen derivatives' on p.283).

Saturated esters have a NFPA reactivity code of zero, but some carbonates are code 1. A possible interpretation for this code is the fact that these compounds are sensitive to hydrolysis. However, it is not dangerous.

Danger code for saturated esters

Compound	NFPA reactivity	Compound	NFPA reactivity
Dimethyl carbonate	1	Ethylene carbonate	1
Diethyl carbonate	1	Propylene carbonate	0
Diphenyl carbonate			

6 • Reactions of organic chemicals

6.10 Esters

Synopsis of dangerous reactions of esters

Ethyl acetate	2-Ethoxyethyl acetate	Vinyl acetate	Ethyl acetoacetate	Methyl and ethyl acrylates	γ-Butyrolactone	Dimethyl carbonate	Diketene	Methyl formate	Propyl formate	Methyl methacrylate	Butyl phthalate	Diethyl succinate	Methyl trichloroacetate	
														Hydrogen
														Group I (Li; Na; K...)
														Group II (Be; Mg; Ca...)
														Group III (B)
														Group IV (C)
														Group V (N; P; As; Sb)
	■							■						Group VI (O_2)
														Group VII (Cl_2)
														Transition metals (Zn)
														Other metals (Al)
														Hydrides group III (B_xH_y)
														Hydrides group IV (SiH_4...)
														Hydrides group V (NH_3; NH_2OH)
	■													Water
														Hydrides group VI (H_2S...)
														Hydrogen halides (HF)
■			■											Metal halides ($AlCl_3$; $FeCl_3$)
														Non-metal halides (PCl_3)
						■								Metal oxides (Fe_2O_3; Al_2O_3)
	■													Non-metal oxides (O_3)
■	■						■							Peroxides (H_2O_2; Na_2O_2)
														Acid oxidants and salts (HNO_3; NO_2^-)
				■										Other acids and salts (H_2SO_4; $HClO_4$)
														ClO_x^- (ClO^-; ClO_2^-)
														MnO_4^-, $Cr_2O_7^{2-}$
							■							Instability
		■		■						■				Polymerisation
														Bases
■	■			■										$(CH_3)_3CO^-$
■													■	Metal hydrates ($LiAlH_4$; NaH)
		■												Bromoalcohol
			■											2,4-Dichlorophenol
								■						Trimethylamine
				■										Propionaldehyde

6.10.6 Reactions involving complex esters

- The reaction of bromoalcohol with ethyl acetoacetate in the presence of metallic zinc has been used to make zinc chelates, the compound of cross-esterification formed according to:

$$CH_3-\overset{O}{\underset{||}{C}}-CH_2-CO_2C_2H_5 + (BrCH_2)_3C-CH_2OH \rightleftharpoons CH_3COCH_2CO_2CH_2C(CH_2Br)_3 + C_2H_5OH$$

To move the equilibrium, ethanol is distilled as it forms. After eliminating 80% of the ethanol, a sudden temperature rise was observed followed by a violent decomposition reaction. It was thought that this accident had resulted from the decomposition of an organobromozinc derivative formed by the contact of an excess of zinc with the brominated acetocetate formed.

- In an autoclave, methyl trichloroacetate had been mixed with trimethylamine. The medium was not cooled. A violent reaction caused a sudden pressure rise, which reached 400 bar. It was thought that an intramolecular dehydrohalogenation of chloroester had formed the unstable lactone according to the equation below. Its decomposition was responsible for the incident:

$$Cl_3C-CO_2CH_3 \xrightarrow{(CH_3)_3N} \text{[lactone with } Cl_2 \text{ and O]}$$

6.11 Acid anhydrides and chlorides

Acid chlorides and anhydrides, especially the former, are the most reactive carbonyl compounds, hence the most dangerous. The dangerous reaction that is involved in accidents usually corresponds to the equation below, in which Σ is a chlorine atom or a O_2C-R group and $H-\Sigma$ a compound with a 'mobile hydrogen':

$$R-\overset{O}{\underset{||}{C}}-\Sigma \xrightarrow{H-\Sigma'} R-\overset{O}{\underset{||}{C}}-\Sigma'$$

The CO-Σ bond breaking is the result of an electrophilic attack (on the carbonyl oxygen atom, hence the catalytic role of acids in these rupture reactions) or a nucleophilic one (on the carbonyl carbon atom whose positive property is due to the Σ electron-withdrawing property). The dangers of this type of reaction come from its speed and high exothermicity and/or instability of the products obtained in some cases. The accidents that are described below can make one believe that acid anhydrides in general and acetic anhydride in particular represent greater risks than acid chlorides since they constitute the accident factor of almost all accidents described. This is obviously related to their frequent use in synthesis rather than acid chlorides, that are rarely used.

6.11.1 Reactions involving breaking of CO-Σ bond

■ Hydrolysis

This reaction is particularly dangerous with the first series members of acid anhydrides and chlorides. Its consequences are particularly aggravated by the

6 • Reactions of organic chemicals

6.11 Acid anhydrides and chlorides

presence of an acid since it facilitates the nucleophilic attack of the carbonyl group carbon. With acetic anhydride the reaction is :

$$CH_3-CO-O-OC-CH_3 + H_2O \longrightarrow 2\ CH_3-CO_2H$$

- Adding aqueous acetic acid into acetic anhydride in error caused a particularly violent hydrolysis.
- It had been decided to purify N,N-dimethylaniline by mixing acetic anhydride, water and hydrochloric acid following a published operating method. However, a slight modification was made that consisted in using the double amount of reagents. The medium was cooled with ice. When hydrochloric acid was introduced, the anhydride hydrolysis was so violent that it caused the apparatus to detonate.
- A mixture of acetic anhydride/polyphosphoric acid/sodium acetate was used to carry out cyclodehydrations. The water treatment at the end of the reaction caused an extremely violent hydrolysis.
- Acetic anhydride was introduced into a container containing an aqueous solution of 48% tetrafluoroboric acid in order to obtain this acid in the anahydrous state. Although the operation was carried out at 0°C, it gave rise to a violent detonation.
- The same goes for treatments of aqueous solutions with 72% perchloric acid to obtain the anahydrous acid. Anhydride has to be added very slowly into the aqueous solution at a temperature that should not exceed 10°C otherwise the medium detonates. Generally speaking, it seems that the aqueous perchloric acid/acetic anhydride mixtures are explosive and detonate on impact.
- Acetyl chloride gives rise to hydrolysis reactions that are particularly violent. When a reactor was washed with water, this gave rise to a violent detonation that destroyed the equipment. It was caused by water being in contact with acetyl chloride residues that were in the pipes leading to the reactor.
- A hydrolysis of seven tons of phthaloyl chloride that was carried out in a herbicide synthesis workshop went out of control and detonated, causing the workshop to be destroyed.
- Chromium (III) oxide hydrate had been introduced into acetic anhydride causing a very violent hydrolysis of acetic anhydride and the spreading of the products.
- The reaction below had been carried out without causing any incident. At the end of the operation, an accident happened when water was introduced to destroy acetic anhydride. Its violent hydrolysis caused the mixture to boil.

Br—⟨⟩—⟨⟩—OH $(CH_3CO)_2O$ → Br—⟨⟩—⟨⟩—O_2CCH_3

The labour regulations, NFPA and transport regulations for dangerous substance codes enable one to spot the reactions that are likely to be dangerous when a compound comes into contact with water. The following table lists the compounds mentioned in the technical literature as well as the codes suggested that sometimes contradict each other depending on the source (both codes are given in this case). These codes are code 14 ('Reacts violently when it is in

| 6 • Reactions of organic chemicals | 6.11 Acid anhydrides and chlorides |

contact with water') for labour regulations, code 1 or 2 for NFPA and code X for the transport regulations. The '–' sign means that no code was suggested for the particular compound.

Danger codes for anhydrides and acid chlorides

Compound	Safety Code (Tr)	Transport Code (TR)	NFPA reactivity
Acetic anhydride	–	–	1
Propionic anhydride	–	–	1
Butyric anhydride	–	–	1
Maleic anhydride	–	–	1; 0
Phthalic anhydride	–	–	1; 0
Acetyl chloride	I4	X	2
Propionyl chloride	I4	–	1
Butyryl chloride			–
Hexonyl chloride			1
Benzoyl chloride	–	–	1
Oxalyl chloride	I4		–
Chloroacetyl chloride	–	X	0
Dichloroacetyl chloride	–	X	1
Trichloroacteyl chloride	I4	–	–

When analysing this table one can see that the authors of the different codes do not agree with each other.

■ Effect of alcohols

- A mixture of ethanol and acetic anhydride detonated and the compounds combusted causing a fire when sodium hydrogensulphate was introduced into the mixture by mistake. The acid nature of this salt obviously catalysed this alcoholysis.
- The reaction of acetic anhydride with glycerol in the presence of phosphoryl trichloride as a catalyst was carried out the following way:

$$HOH_2C-CHOH-CH_2OH \xrightarrow[POCl_3]{CH_3-COOCO-CH_3} CH_3CO_2CH_2-\underset{\underset{CH_3C}{|}}{\overset{O_2}{CH}}-CH_2O_2CCH_3$$

The extremely violent reaction was put down to the alcohol's high viscosity, which limited the dissipation of gas produced by the reaction. Refer to *alcohols and glycerol*.

6 • Reactions of organic chemicals
6.11 Acid anhydrides and chlorides

■ **Effect of acids**

- The following reaction was carried out:

$$CH_3COOCOCH_3 + H_3BO_3 \longrightarrow (CH_3COO)_3B$$

The method used came from a description in a publication that was taken from an old German publication. However, the publication did not say that the original German source mentioned the explosive nature of this reaction. A detonation took place when the temperature reached 60°C.

- A mixture of nitric acid/acetic anhydride is very sensitive and detonates very easily. This behaviour was explained by the following reaction:

$$(CH_3CO)_2O + HNO_3 \ (50\text{-}85\%) \longrightarrow CH_3\text{-}CO_2\text{-}ONO_2 \text{ and/or } C(NO_2)_4$$

The danger very much depends on the order in which the reagents are introduced and the acid concentration.

If the acid concentration is lower than 50%, acid has to be mixed into anhydride; if it is greater than 85%, anhydride has to be mixed into acid.

Below 50%, the danger is at its highest, since water, which is present in larger quantities, causes a reaction that cannot be controlled. Therefore, a mixture of 38% nitric acid/acetic anhydride detonated at ambient temperature a few hours after being prepared. Note that the commercial nitric acid that is most often handled is at 33 or 38%. Thus, danger is most of the time at its greatest.

- At 80-100°C a reaction of fuming nitric acid/sulphuric acid on phthalic anhydride had been carried out with the idea of nitrating of the aromatic nucleus. A violent decomposition interrupted the operation two hours after it had been started. It was thought that a reaction of the cleavage of the anhydride group had taken place and that the decomposition of a compound containing the -CO-O-NO_2 group was responsible for the accident. A study that was carried out later showed that aromatic nucleus nitration could be carried out by heating at 55-65°C without forming a nitrated compound of the anhydride group rupture. It seems that this kind of information should not be taken for granted.

- The effect of hypochlorous acid on acetic anhydride caused a violent detonation that was explained by the formation of a hypochlorite and/or dichlorine oxide, which are both explosive:

$$(CH_3CO)_2O \xrightarrow{HOC} CH_3\text{-}CO\text{-}O\text{-}Cl \text{ or } Cl_2O$$

■ **Other rupture reactions**

- An acetic anhydride/hydrogen peroxide mixture is used when bleaching textiles. This mixture has to be prepared in acid medium to enable the formation of peracetic acid, which is an active agent for bleaching. If the operation takes place in basic medium, acetyl peroxide forms; this is far less stable than the peracid and is spontaneously explosive. A large number of accidents arise from a lack of knowledge about this operating condition. The scheme below sums up the two transformations mentioned:

6 • Reactions of organic chemicals
6.11 Acid anhydrides and chlorides

$$(CH_3CO)_2O + H_2O_2 \xrightarrow[OH^-]{H^+} \begin{array}{l} CH_3-CO_2-OH \\ \\ CH_3-CO-O-O-OC-CH_3 \end{array}$$

- There was an attempt to carry out the reaction below by applying an operating method that is well known for being safe:

$$CH_3COOCOCH_3 + CrO_3 \longrightarrow CH_3-COCrO_2$$

However, the medium was not stirred or cooled. It detonated while it was being handled.

- The effect of sodium nitrite on phthalic anhydride when it was hot ended with a violent detonation of the medium. It was explained by the formation of a derivative containing the following unstable group: CO-ONO.

Note that in these three examples involving hydrogen peroxide, chromium trioxide and sodium nitrite, dangerous reactions have been described for carboxylic acids (see on p.316-317). They all referred to the three following systems: acetic acid-/hydrogen peroxide, acetic acid/chromium trioxide and o-phthalic acid/sodium nitrite. One can ask oneself whether the same reactions did not take place after the acetic and phthalic anhydride hydrolysis.

- The effect of hydrazine or hydroxylamine on phthalic anhydride is mentioned as being very dangerous.

6.11.2 Oxidation reactions

Only a few accidents have been mentioned involving oxidation of acid anhydrides and chlorides.

- It is dangerous to prepare phthalic anhydride because of the oxidation exothermicity and risks of accidental catalysis by rust. This reaction forms naphthoquinone as a by-product. This compound may have caused a large number of accidents (that caused the compounds to ignite spontaneously) causing the compounds to combust. These accidents may have been caused by the naphthoquinone oxidation catalysed by iron phthalates, which are present in this reaction. However, it will be seen later that phthalic anhydride can also decompose in certain conditions that may be combined here.

- Metal nitrates and acetic anhydride are particularly dangerous nitrating reagents if mixtures are made according to certain proportions. The danger also depends on the nature of the salt. Thus, with copper (II) nitrate or calcium nitrate, the mixture is always explosive whatever the proportions.

- Barium peroxide had been used instead of potassium permanganate to purify acetic anhydride. Several operations had been carried out and the technicians had realised that this medium gives rise to mild deflagrations. During the last test a very violent detonation took place. It was thought that acetyl peroxide had formed.
- One could logically think after what was mentioned before, using potassium permanganate is not dangerous. However, mixtures of acetic anhydride, acetic acid, potassium permanganate detonate if there is inefficient cooling during their preparation.

6.11.3 Instability

The functional groups that are analysed in this chapter are stable apart from the following specific structures:
- Formic anhydride and formyl chloride, which do not exist because of their instability.
- Anhydrides in which the group is part of a ring with five bonds (maleic, phthalic and succinic anhydrides, in particular).
- Maleic anhydride decomposes exothermically at a temperature starting at 150°C by releasing carbon dioxide, if it is in the presence of the following products:
- With aliphatic amines, the decomposition catalysis is moderate; with heterocyclic aromatic amines (pyridine, quinoline), 0.1% of amine is sufficient to cause maleic anhydride to decompose. An accident has also been mentioned with NaOH. This decomposition also takes place in the presence of sodium, lithium, ammonium, potassium, calcium, barium, magnesium and beryllium cations.
- Detonations sometimes occur during Diels-Alder reactions with maleic anhydride. They are explained by the decomposition of this anhydride.
- In the same way, phthalic anhydride gave rise to a violent detonation when it came into contact with copper (II) oxide.
- In the presence of bases, succinic anhydride gives rise to a very violent reaction after being heated for thirty hours. The temperature reached 550°C in some places of the apparatus when the glass container melted. No explanation was given. By analogy with the previous case, it can be assumed that the decomposition of the anhydride may have caused the accident.

6.11.4 Miscellaneous dangerous reactions

Here is a list of a few reactions that are difficult to classify or that involve the carbon chain and not the functional group itself, which plays an indirect role in this case. They concern acid chlorides.
- A container of acryloyl chloride containing 0.05% of phenothiazine used as an inhibitor, was exposed to very unfavourable climatic conditions that brought it to a temperature of 50°C for two days. After this time, it polymerised violently, shattering its container and forming a voluminous foam. Yet the container label mentioned that it could not be stored at temperatures greater than 4°C.

6 • Reactions of organic chemicals

6.11 Acid anhydrides and chlorides

Dangerous reactions of acid anhydrides and chlorides

Acetic anhydride	Maleic anhydride	Phthalic anhydride	Succinic anhydride	Acetyl chloride	Acryloyl chloride	Benzoyl chloride	Phthaloyl chloride	Propionyl chloride	
									Hydrogen
	■								Group I (Li; Na; K...)
									Group II (Be; Mg; Ca...)
									Group III (B)
									Group IV (C; Si; Ge)
									Group V (N; P; As; Sb)
									Group VI (O_2)
									Group VII (Cl_2; Br_2)
									Transition metals
									Other metals (Al; Bi; Sn; Pb)
									Hydrides group III (B_xH_y)
									Hydrides group IV (SiH_4...)
	■		■			■			Hydrides group V (NH_4^+; N_2H_4...)
■				■					Water
■									Hydrides group VI (H_2S...)
									Hydrogen halides
				■			■		Metal halides ($AlCl_3$; $FeCl_3$...)
									Non-metal halides ($POCl_3$...)
■									Metal oxides
■									Non-metal oxides (O_3; N_2O_5)
■									Peroxides (H_2O_2; Na_2O_2; BaO_2)
■									Acid oxidants (HNO_3)
■									Other acids and salts (H_2SO_4; $HClO_4$)
■									ClO_x^-
■									MnO_4^-, $Cr_2O_7^{2-}$
	■								NO_2^-; NO_3^-
									Instability
	■		■						Polymerisation
■									Bases
			■						Naphthalene
■									Ethanol, glycerol
				■					Ethers with few C atoms
■									Amines
■									Acetic acid and sodium acetate

6 • Reactions of organic chemicals — 6.12 Nitriles

- An accident involving propionyl chloride was not the result of a dangerous reaction, but of the physical properties of one of the reaction compounds. The reaction in question was the following:

$$(CH_3)_2CH-O-CH(CH_3)_2 + CH_3CH_2-\underset{}{\overset{O}{\underset{\|}{C}}}-Cl \xrightarrow[FeCl_3]{ZnCl_3} CH_3CH_2-CO_2CH(CH_3)_2 + (CH_3)_2CHCl$$

The thermal conditions used caused 2-chloropropane to vapourise (Eb: 35°C) creating an overpressure that damaged the reactor after twenty four hours.

■ Friedel-Crafts reaction with benzoyl chloride

- The following preparation is extremely dangerous if very precise operating conditions are not followed:

naphthalene + 2 C_6H_5-COCl $\xrightarrow{AlCl_3}$ 1,5-dibenzoylnaphthalene ($CO-C_6H_5$ and C_6H_5CO substituents)

Aluminium chloride has to be introduced gradually in the naphthalene/benzoyl chloride mixture and the mixture of these two reagents has to be brought to a temperature that is high enough so that it is in the molten state. Crystallised naphthalene residues are enough to cause the medium to detonate.

6.12 Nitriles

Nitrile reactivity and dangerous reactions are related to the following properties:
- unsaturated and endothermic nature of the carbon-nitrogen triple bond;
- active carbon atom sensitivity to nucleophilic attack due to the fact that the nitrogen atom is electronegative;
- nitrile group sensitivity to oxidation;
- activation of α,β-unsaturated nitrile double bonds especially in the presence of nucleophilic reagents;
- high mobility of hydrogen atom in α position of functional group.

When several properties are combined or when compounds belong to the lower members of the active series, the molecule becomes particularly dangerous.

6.12.1 Instability, 'functional polymerisation', additions to triple bond

Here is a list of all dangerous reactions that are related to nitrile functional group behaviour. By 'active polymerisation' is meant the polymerisation that affects the carbon-nitrogen triple bond. Polymerisations that are related to an ethylene double bond will be dealt with on p.336. So far as stability is concerned, it is difficult to say whether certain 'spontaneous' reactions of certain nitriles are

6 • Reactions of organic chemicals
6.12 Nitriles

caused by decomposition or polymerisation. Despite the nitrile group's endothermic nature, the known enthalpies of decomposition do not seem to show that this class of compounds is unstable, except possibly malonitrile. The table below lists all these values. They do not all appear to be coherent.

Enthalpy of decomposition of nitriles

Product	ΔH_d (kJ/g)	Temperatures (°C)
Malonitrile	−1.650	180-270
Succinonitrile	−0.700	?
Fumaronitrile	−0.454	340-380
Glutaronitrile	−0.098	200-340

- Succinonitrile that is kept at a temperature of 80°C for forty six hours caused a violent decomposition, after being heated to 195°C to eliminate some compound solidified in a pipe. Differential thermal analysis showed that the reaction was self-accelerated with a period of induction of thirty three hours at 200°C and an hour and a half at 280°C. No heat was given off during the period of induction. Once started, the decomposition is very fast and there is no inhibitor. The cyanide ion shortens the period of induction.
- At 130°C or at ambient temperature and in the presence of bases, malonitrile gives rise to a very violent polymerisation. The stability of this dinitrile in the molten state (mp: 32°C) decreases with the temperature rise and its fall in purity. It was assumed that decomposition never occurs below 100°C. However, when this compound was heated to 70-80°C for two months, this caused a very violent detonation.
- Hydrogen cyanide is highly endothermic. It polymerises violently when there is no inhibitor present. If the heat liberation brings the medium to 284°C, it causes the compound to detonate. The presence of a base catalyses the hydrogen cyanide polymerisation. Nevertheless, this reaction is inhibited by mineral acids. The most common polymerisation inhibitor is phosphoric acid.
- When an acetonitrile/sulphuric acid mixture is heated to 53°C, it gives rise to an exothermic reaction that could not be controlled, and brought the temperature in the reactor to 160°C. The same happens when adding sulphur trioxide to the same mixture, but in this case the reaction goes out of control at a temperature starting at 15°C. This behviour is explained by acetonitrile poly-merisation.
- An 'old' bottle (1 year) containing glycolonitrile with phosphoric acid used as a stabiliser showed the appearance of tars. It detonated during handling. The detonation was probably due to the polymerisation of nitrile that was made possible by the fact that phosphoric acid was isolated by tars. Besides, the cap was 'cemented' by tars that had already formed round the cap. A similar accident happened thirteen days after distilling the same nitrile.
- Preparing imidates is dangerous:

$$ROH + HC\equiv N \xrightarrow{HCl} H-C\overset{\oplus}{\underset{NH\ Cl^{\ominus}}{\diagup OR}}$$

The fact that hydrogen chloride is rapidly introduced into the hydrogen cyanide/alcohol mixture, with the purpose of preparing an imidate, leads to a detonation, even when the medium is cooled considerably. This danger can be avoided by proceeding to a slow addition of acid, in a medium that is properly stirred and cooled.

6.12.2 Oxidation reactions

- A reaction of dinitrogen tetroxide with acetonitrile in the presence of indium was carried out. The technician wanted to accelerate the process by stirring the medium, causing its detonation. This accident was put down to the violent oxidation of the acetonitrile accumulated, by nitrogen oxide catalysed by indium.
- An acetonitrile/nitric acid mixture is a highly sensitive explosive.
- The $CH_3CN/(CH_3)_2C(NO_2)CN$ mixture is used as a nitration reagent. It is potentially dangerous, possibly because of the nitro group's oxidising nature.
- Mixtures of acetonitrile with perchloric acid are potentially explosive.
- The action of anhydrous iron (III) perchlorate on acetonitrile gives rise to an extremely violent reaction. If the perchlorate is hydrated, it is not the same.

6.12.3 Reactions involving carbon chain and complex nitriles

■ **Polymerisation of α,β-unsaturated nitriles**

- Traces of mineral acids are sufficient to cause the very violent polymerisation of acrylonitrile.
- The same happens in the presence of bases, but usually at a temperature below 60°C.
- Products that are likely to form free radicals also start a very violent polymerisation if the mixture is made in uncontrolled conditions. Primers are usually tert-butyl or benzoyl peroxide or azobis-isobutyronitrile.
- Acrylonitrile came into contact with silver nitrate and was kept in this way for a long time. It gave rise to a violent detonation that was put down to nitrile polymerisation, which formed successive layers of polymer at the surface of the salt particles; the temperature rise that was caused accelerated the polymerisation gradually.

A large number of reactions from acrylonitrile or that lead to acrylonitrile, which are well known for not being dangerous, have actually given rise to accidents due to the accidental polymerisation of the compound.

As far as reactions from acrylonitrile are concerned:

- Bromine was added to acrylonitrile in small portions at 0°C and then by heating to 20°C between each portion. After adding half the amount of bromine, the temperature reached 70°C and the container detonated. The accident was explained by a violent polymerisation, catalysed by traces of hydrogen bromide that were the result of the following substitution reaction:

$$CH_2=CH-CN + Br_2 \longrightarrow CH_2=CBr-CN + HBr$$

6 • Reactions of organic chemicals
6.12 Nitriles

- The same probably happened with the following reaction:

$$R_2NH + CH_2=CH-CN + [C_6H_5N(CH_3)_3] + OH^- \longrightarrow R_2N-CH_2-CH_2-CN$$

The secondary amine that was used could be tetrahydrocarbazole or pyrol. The reaction was known and not mentioned as being dangerous. The authors of this new experiment used four times the amounts recommended in the method published. They also introduced the ionic compound at 0°C and stopped the cooling rapidly. These changes were sufficient to cause the medium to heat up and then detonate. It was considered to be due to the nitrile polymerisation caused by ammonium salt.

So far as reactions that form acrylonitrile are concerned:

- In the presence of bases, amines or mineral acids, ethylenecyanohydrin dehydrates into acrylonitrile, which then polymerises violently.
- The same goes for acetocyanohydrin with sulphuric acid.
- When propionitrile was heated at reflux during 24 hours and in the presence of N-bromosuccinimide, it gave rise to a detonation. The scheme below illustrates the interpretation that was given for this accident:

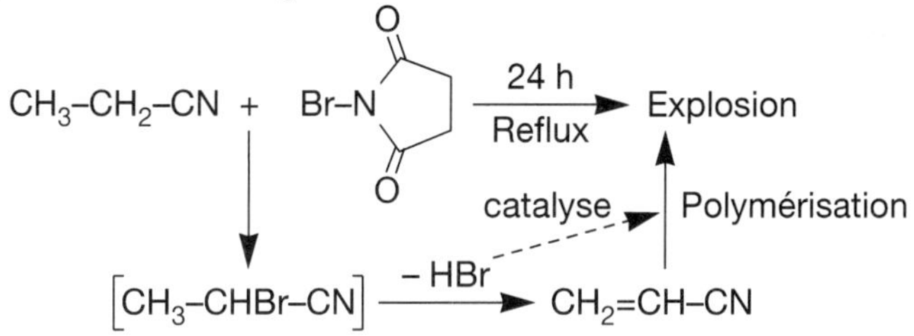

$$CH_3-CH_2-CN + Br-N(succinimide) \xrightarrow[\text{Reflux}]{24\ h} \text{Explosion}$$

$$[CH_3-CHBr-CN] \xrightarrow{-HBr} CH_2=CH-CN \xrightarrow{\text{catalyse}} \text{Polymérisation}$$

The first stage of this reaction illustrates the reactivity of hydrogen atom in α position of the nitrile group.

The NFPA reactivity codes indicate the risks that are related to the propensity for polymerisation and/or instability of the nitrile group and of the potential ethylene double bond. The table below gives the available data with some contradictions between different sources, as usual (for acetonitrile, code 1 seems to make more sense)

Code of danger for nitriles

Compound	NFPA reactivity	Compound	NFPA reactivitye
Hydrogen cyanide	2	Ethylenecyanohydrin	1
Acetonitrile	3; 0	Lactonitrile	–
Propionitrile	1	Acetocyanohydrin	2
Acrylonitrile	2		

■ Dangerous reactions related to the presence of another functional group

The previous examples regarding nitrile-alcohols corresponded to this classification. However, since their danger is related to acrylonitrile polymerisation they were classified in the previous paragraph. The three examples below exclude this type of interpretation.

- The preparation of tetramethylsuccinonitrile by decomposition at 90-92°C in azobis-isobutyronitrile (AIBN) heptane gives rise to a detonation. The medium was not stirred and the accident is related to the instability of this diazoic compound, which had probably accumulated (the nitrile groups are not responsible for it). If AIBN is introduced into heptane slowly, at 90°C and under vigorous stirring, the reaction seems safer.
- The dehydrohalogenation of 4-chlorobutyronitrile in cyclopropanecarbonitrile is dangerous if it is carried out in the presence of solid sodium hydroxide. The danger is caused by the formation of solid crusts on the reactor's walls that causes the reaction to be out of control. This dehydrohalogenation does not represent any danger in the presence of the base present in an aqueous solution.
- The electron-withdrawing nitrile group gives a strong acid character to cyanoacetic acid (pKa = 2.46). Like any such strong acid it gives rise to a detonation when it is put into contact with furfuryl alcohol, at the beginning of heating, due to the polymerisation of the alcohol present.

6.12.4 Miscellaneous dangerous reactions

- The electrolytic preparation of cyanuryl chloride from a chloride and hydrogen cyanide is dangerous.
- When ammonium chloride is used, there is a risk that nitrogen trichloride forms, which is explosive. However, this is not a dangerous reaction involving a nitrile group.
- The effect of sodium nitride on trichloroacetonitrile in the presence of ammonium chloride was used to make tetrazole. The operation was interrupted by the medium detonating.
- During an attempt at destroying benzyl cyanide residues with sodium hypochlorite, a detonation was caused that was probably due to the formation of nitrogen trichloride. However, it might be asked if it was not due to the nitrile group oxidation by the hypochlorite present.
- The preparation of ethyl cyanoacetate by the reaction of sodium cyanide on ethyl chloroacetate, which had not caused any incident after being carried out about twenty times, gives rise to a violent eruption of the medium during a further operation. No explanation could be provided.

6.13 Amides and analogues

Amides often give rise to accidents that are difficult to interpret because so many reagents are present and/or because of the complexity of the reactions that are brought into play. It is difficult to find a classification for this group. The first point is the fact that most accidents are due to dimethylformamide (DMF), which is much used as a polar aprotic solvent. When attempting to classify these types of dangerous reactions with this compound, as a model, it can be said that they are mainly due to:

- its basic nature that makes it very dangerous to handle in the presence of polyhalogen derivatives and inorganic and organic acid chlorides;
- its reducing character that makes it very sensitive to oxidants.

6 • Reactions of organic chemicals

6.13 Amides and analogues

Isocyanates react violently with water, alcohols and more generally compounds with a mobile hydrogen.

6.13.1 Violent interaction of amides and analogues with halogens, halogen derivatives and acid chlorides

■ Effect of halogens

- The effect of chlorine on dimethylformamide (DMF) gives rise to significant heat liberation that can cause the compounds to overflow. It is assumed that accidents are related to the reaction below that forms an unstable compound:

$$\underset{H}{\overset{O}{\|}}{-}C{-}N(CH_3)_2 \xrightarrow{Cl_2} \underset{H}{\overset{O}{\|}}{-}C{-}\overset{+}{N}(CH_3)_3\ Cl^- + CO\ (traces)$$

However, one can have some doubts about the interpretation given after comparing it with the following accident.

- Bromine had been mixed with DMF in an autoclave. A very exothermic reaction took place; the temperature reached 100°C and the pressure 135 bar, and then the autoclave detonated. The following interpretation was suggested: an unsaturated ammonium salt was formed and decomposed violently in N-bromoamine that is unstable:

$$H{-}\overset{O}{\overset{\|}{C}}{-}N(CH_3)_2 \xrightarrow{Br_2} (CH_3)_2\overset{\oplus}{N}{=}CH{-}OH\ \overset{\ominus}{Br} \longrightarrow BrN(CH_3)_2 + CO + HBr$$

■ Effect of inorganic halogen derivatives

- DMF gives rise to a very violent reaction with arsene trifluoride or thionyl chloride. In the latter case, an accident that happened had taken place a few hours after mixing a total weight of 200 kg of both compounds. It was assumed that the reaction responsible for this accident had been catalysed by powdered metallic zinc or iron. It was shown that this mixture remains the same for forty eight hours in the presence of 90 ppm of one of these metals (this is the exact quantity of iron that is contained in industrial thionyl chloride). However, adding 200 ppm of metal causes a very violent reaction when starting the stirring, twenty two hours after preparing the mixture.

The way urea behaves when inorganic halogen derivatives are present is as dangerous. It seems that the reasons for this are either related to the instability of the complex formed or the formation of N-halogen compounds.

- The first interpretation seems to be responsible for the violent reaction of titanium tetrachloride with urea after being heated at 80°C for six weeks.
- The interaction of a sodium or calcium hypochlorite or phosphorus pentachloride with urea causes a violent detonation, that has been put down to the decomposition of the nitrogen trichloride formed.
- When chromyl chloride (CrO_2Cl_2) is mixed with urea the mixture combusts immediately after being formed. The interpretation is rather problematic. It may be due to oxidation.

6 • Reactions of organic chemicals

6.13 Amides and analogues

Synopsis of dangerous reactions of nitriles

	Acetocyanohydrin	Acetonitrile	Cyanoacetic acid	Acrylonitrile	Azobis-isobutyronitrile	4-Chlorobutyronitrile	Hydrogen cyanide	Benzyl cyanide	Ethylene cyanohydrin	Glycolonitrile	Malononitrile	Propionitrile	Succinonitrile	Trichloroacetonitrile
Hydrogen														
Group I (Li; Na; K...)														
Group II (Be; Mg; Ca...)														
Group III (B)														
Group IV (C; Si; Ge)														
Group V (N; P; As; Sb)														
Group VI (O_2)														
Group VII (Br_2)			■											
Transition metals														
Other metals (Al... Bi; Sn; Pb; In)	■													
Hydrides group III (B_xH_y)														
Hydrides group IV (SiH_4...)														
Hydrides group V (NH_3...)														
Water														
Hydrides group VI (H_2S...)														
Hydrogen halides			■	■		■								
NH_4Cl					■									
Metal halides ($AlCl_3$; $FeCl_3$...)														
Non-metal halides ($SOCl_2$...)														
Metal oxides														
Non-metal oxides (SO_3; N_2O_4)		■												
Organic peroxides (H_2O_2; Na_2O_2)			■											
Acid oxidants (HNO_3; HNO_2)		■												
Other Acids and salts (H_2SO_4; $HClO_4$)	■		■					■						
ClO_x^-				■										
MnO_4^-, $Cr_2O_7^{2-}$														
N_3^-								■						
Instability				■	■									
Polymerisation		■		■	■		■							
Bases				■	■		■							
Alcohols					■									
Furfuryl alcohol														
Amines							■							
N-Bromosuccinimide									■					
$(CH_3)_2C(NO_2)CN$														

Effect of organic halogen derivatives

- Mixing DMF with carbon tetrachloride is potentially dangerous. The same is true for hexachloroethane and hexachlorocyclohexane, especially when iron is present. However, the interaction is not considered to be dangerous with dichloromethane or 1,2-dichloroethane. There is a greater danger when iron is present. When the DMF/CCl_4 mixture is analysed using DTA, it has a unique exotherm at around 100°C. When 3% of iron is present, there is a double exotherm at 56 and 94°C respectively. The temperature then increases and reaches 283°C.

- Dimethylacetamide (DMA) behaves the same way; the interaction exothermicity is higher than with DMF. A differential thermal analysis of the 1:1 mixture of DMA/CCl_4 shows two exothermic peaks at 91-97 and 97-172°C (maximum at 147°C), respectively. An experiment made in an autoclave shows that the final temperature reaches 450°C and a pressure of 12.8 bar that is bonded to a substantial gas release. If 1% of iron is present, the first exotherm occurs at 71°C.

- The following reaction that was made in order to produce a halogen maleimide and carried out with organic reagents in the molten state, gave rise to a violent detonation when the temperature reached 118°C.

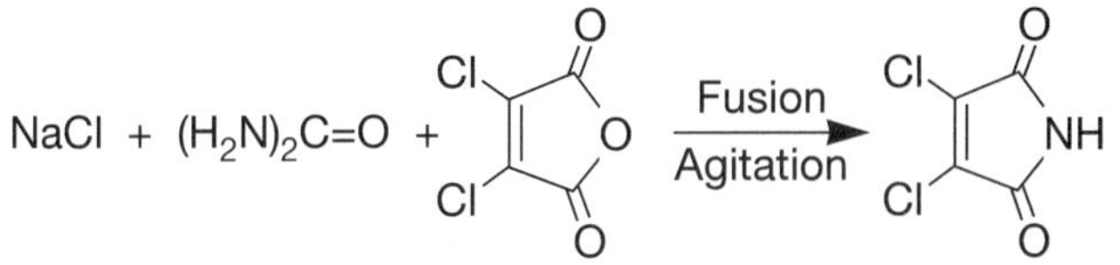

$$NaCl + (H_2N)_2C=O + \underset{Cl}{\underset{|}{\text{Cl}}}\text{-maleic anhydride} \xrightarrow[\text{Agitation}]{\text{Fusion}} \text{Cl-maleimide-Cl}$$

Note that for both examples, the metal/CCl_4 interaction is dangerous whether an amide is present or not. This makes it difficult to interpret the role played by amide.

Effect of acid chlorides

- Mixing phosgene with DMF is dangerous. Nevertheless, the danger is less than with thionyl chloride (described on p. 339).

- With cyanuryl chloride, DMF forms a 1:1 complex that decomposes violently at a temperature starting at 60°C by releasing carbon dioxide, and is also thought to form an 'unsaturated quaternary ammonium salt' (refer to the effect of bromine on DMF as a possible form of this ammonium in the effect of halogens):

$$\text{cyanuryl chloride} \xrightarrow{\text{DMF}} \xrightarrow{> 60°C} CO_2$$

6.13.2 Oxidation reactions

- A mixture of 27% of formamide, 51% of calcium nitrate, 12% of ammonium nitrate and 10% of water detonates at -20°C. On adding powdered aluminium, this mixture becomes more disruptive.

- DMF is often used as a solvent in oxidation reactions. Thus, in the reaction below, the medium combusts when chromium trioxide in the solid state comes into contact with the reagents:

$$R-CHOH-R' \xrightarrow[DMF]{CrO_3} R-CO-R'$$

The same happens when oxide is diluted in DMF in the following proportion (by weight): 2 g/18 cm^3.

This behaviour is related to the reducing nature of DMF. DMA, which is less reducing, is more suitable.

- A solution of 20% potassium permanganate in DMF detonated five minutes after making the mixture. An experiment that was carried out using micro-quantities showed an exotherm three to four minutes after the mixing that was followed by mild deflagration.
- A mixture of DMF and magnesium nitrate hexahydrate decomposed violently when it was heated above the salt's melting point. It was thought that in those conditions, nitric acid had formed and oxidised DMF.
- When one mole of sodium nitrite is mixed with three moles of urea, this can give rise to a detonation, if the mixture is heated until the urea melts. The reaction responsible for this is thought to be the following:

$$(H_2N)_2C=O \xrightarrow[Fusion]{NaNO_2} Na^+ \; N\equiv C-O^-$$

6.13.3 Reduction reactions

The problem that is posed here is more or less the same as previously. DMF is also used as a solvent in reduction reactions. When the reducing agent is highly reactive (alkali, organometallic metal, metal hydride), the danger can be significant.

- Heating a sodium suspension in DMF gives rise to a very violent reaction.
- Triethylaluminium was introduced into DMF and heated. This gave rise to a violent detonation.
- When sodium hydride was heated to 50°C with DMF, this gave rise to an exothermic reaction; the temperature then reached 75°C. An external cooling was then carried out, but was not successful. The reaction went out of control and the reactor's content overflowed. A test that was carried out on a small scale showed that the reaction started at 26-50°C, depending on the DMF's degree of drying and then accelerated. DMA's behaviour is the same; the reaction starts at a temperature of 40°C.
- A mixture containing 15.7% of sodium borohydride in DMF decomposed when it was heated by forming trimethylamine. The temperature of the solid residue formed rose to 310°C. This interaction occurs after a period of induction, which depends on temperature (45 hours at 62°C; 45 minutes at 90°C). This period is reduced, if formic acid is present because of the formation of the F320 salt (according to the authors).

Such an accident happened on an larger scale involving 83 kg of the mixture. There was a very substantial fire that was probably caused by trimethylamine self-ignition.

6.13.4 Instability of amides

These compounds are not very stable and the first series members decompose at low temperatures. However, these decompositions do not represent any danger. They become dangerous in very specific operating conditions or with nitrated aromatic amides.

- A Fischer reagent had been made with pyridine, iodine, sulphur trioxide and formamide, instead of methanol. The bottle detonated after being stored for a couple of months. The authors put it down to the decomposition of formamide into ammonia and carbon oxide, which created the overpressure that caused the bottle to detonate.
- When p-nitroacetanilide is heated to 200-250°C in the presence of small quantities of sulphuric acid, it gives rise to a very spectacular decomposition and the formation of a huge quantity of carbonised 'foam'. This extraordinary reaction is called 'black snake'.

6.13.5 Miscellaneous dangerous reactions

- Alkyl nitrate solutions in DMF are stable at 25°C, but detonate on impact at 200°C.
- Tetraphosphorus hexoxide gives rise to extremely violent reactions with DMF.
- An N-nitrose compound that is unstable is obtained by the effect of dinitrogen trioxide on caprolactam at a low temperature. The compound detonates when the medium is not cooled enough.

$$\text{caprolactam} \xrightarrow{N_2O_3} \text{N-nitrosocaprolactam}$$

- Urea was treated with oxalic acid and carbon. The operation was carried out in the presence of anhydrous copper sulphate in order to detect the water formed, and gases were expected to bubble through a barium hydroxide solution to be able to see carbon dioxide. Unfortunately, the apparatus was closed by mistake. It detonated due to the large quantity of gases formed in the reaction:

$$2(H_2N)_2C=O + 3\ HO_2C-CO_2H \xrightarrow{570°C} 5\ CO_2 + 3\ CO + 4\ NH_3 + H_2O$$

6.13.6 Dangerous isocyanate reactions; Bhopal accident

- In Bhopal, one of the three stores of 13 m^3 of methyl isocyanate gave rise to an accident during which there was a huge spreading of the highly toxic isocyanate. Two thousand people were killed. The reasons for this accident are still not entirely clear. It could be due to the transformation of the compound into trimethyl isocyanurate, under the effect of an impurity that would have played a catalytic role. It was also thought that there could have been an interaction with an amine. A violent hydrolysis reaction resulting from an accidental contact with water is more likely to be responsible for the accident:

$$CH_3-N=C=O + H_2O \longrightarrow CO_2 + CH_3-NH_2$$

6 • Reactions of organic chemicals
6.13 Amides and analogues

Synopsis of dangerous reactions of amides and analogues

	ε-Caprolactam	2,4 Toluene diisocyanate	Dimethylacetamide (DMA)	Dimethylformamide (DMF)	Formamide	Hexamethylenediisocyanate	Methyl isocyanate	p-Nitroacetanilide	Urea	
										Hydrogen
			■							Group I (Li; Na; K...)
										Group II (Be; Mg; Ca...)
										Group III (B)
										Group IV (C; Si; Ge)
										Group V (N; P; As; Sb)
										Group VI (O_2)
			■							Group VII (Cl_2; Br_2; I_2)
										Transition metals
										Other metals (Al... Bi; Sn; Pb)
										Hydrides group III (B_xH_y)
										Hydrides group IV (SiH_4...)
										Hydrides group V (NH_3...)
				■						Water
										Hydrides group VI (H_2S...)
										Hydrogen halides
			■							Metal halides ($AlCl_3$; $FeCl_3$...)
										CrO_2Cl_2
			■							Non-metal halides ($SOCl_2$...)
										Metal oxides
	■									Non-metal oxides (N_2O_3; P_2O_3; SO_3)
										Peroxides (H_2O_2; Na_2O_2)
										Acid oxidants (HNO_3)
				■						Other acids and salts (H_2SO_4; $HClO_4$)
										ClO_x^-
			■							MnO_4^-, $Cr_2O_7^{2-}$
										NO_3^-; NO_2^-
		■		■						Instability
										Polymerisation
		■								Bases
			■	■						Metal hydrates (NaH; $NaBH_4$)
			■	■						Organometallics ($(C_2H_5)_3Al$)
		■								Alcohols
			■							Halogen derivatives (CCl_4; C_2Cl_6...)
			■							Pyridine
			■							Alkyl nitrates
				■						Oxalic acid
				■						2,3-Dichloromaleic anhydride
			■							Acid chlorides ($COCl_2$...)

344

- The interaction of hexamethylenediisocyanate with alcohols is explosive when bases are present and there is no solvent present. The same is true for 2,4 toluylene diisocyanate. The reaction gives rise to an urethane formed through a similar process as the previous one (with water):

$$R-N=C=O + R'-OH \longrightarrow RNH-CO_2R'$$

- A solution of hexamethylenediisocyanate in DMF gives rise to a very violent reaction.

6.14 Miscellaneous substances

As far as reactivity is concerned, there is no link between the different classes of compounds this chapter is concerned with. Therefore, they will be analysed separately, excluding borates, since no accident involving them could be found in the sources (which does not mean that they are not dangerous, but that they are hardly used).

6.14.1 Sulphur derivatives

These are mainly thiols, dimethylsulphoxide (DMSO), which is used very often as a polar aprotic solvent, and sulphates. The first two classes will give rise to dangerous oxidation reactions. Sulphates can be easily hydrolysed. They will then form sulphuric acid and have very dangerous properties. Besides, they are alkylating agents (which explains the carcinogenic nature of the first series member), which enables them to give rise to very unstable alkylation compounds, under certain circumstances. DMSO is hardly stable especially when certain types of compounds are present, and easily forms salts and complexes that are unstable. Since it is used as a solvent throughout the world, it comes into contact with a whole range of reagents creating an enormous number of dangerous situations.

■ **Oxidation reactions**

- The effect of concentrated nitric acid on thiols is potentially dangerous. It forms a sulphonic acid:

$$R-SH + HNO_3 \longrightarrow R-SO_3H$$

The exothermicity of this reaction can be sufficient to cause thiols to combust. There is a safer method, which consists in operating in a nitrogen atmosphere at temperatures reaching 1-2°C. Impure butanethiol (containing 28% of propanethiol and 7% of pentanethiol) combusts immediately when it comes into contact with nitric acid at 96%. With pure pentanethiol, after two handlings that did not cause any incident, the third one ended with the spontaneous ignition of thiol. Dodecanethiol and hexadecanethiol detonate when they are put into contact with fuming nitric acid. However, the reaction is mentioned as being safe with nitric acid at 33%, if the temperature is lower than or equal to 35°C.

- Thiols that are treated by mercury (II) oxide give rise to a very violent detonation if no solvent is used.

- There was an attempt to treat spreadings of organic sulphides or thiols with calcium hypochlorite in the solid state. These treatments usually ended with a violent reaction followed by the compounds igniting. Nevertheless, this does not represent any danger when using sodium hypochlorite solutions at 15%.
- Dimethyl sulphide detonates when it is heated at 210°C with oxygen. This was predictable since this temperature is very close to the AIT of the sulphide present in air and thus *a fortiori* in oxygen.
- The effect of nitric acid on dimethyl sulphide in the presence of dioxan gives rise to a detonation, even at liquid nitrogen temperatures. There is no factor in this description that can prove whether the accident is related to sulphide or dioxan, which probably gives rise to dangerous reactions with nitric acid.
- Adding benzoyl peroxide to methyl sulphide without using any solvent, with the purpose of making the corresponding sulphoxide or sulphone ended with detonation of the peroxide.
- Xenon difluoride detonates in the presence of dimethyl sulphide if no solvent is used. This behaviour is not surprising at all.
- With fuming nitric acid, thiophene gives rise to a very violent reaction which can not be controlled and leads to complete oxidation. However, 2-nitrothiophene can be obtained by checking the temperature carefully and by using a solvent.
- Dinitrogen tetroxide gives rise to a violent reaction with dimethylsulphoxide. This reaction can lead to a detonation. The same happens when using nitric acid that contains less than 14% of water. In both cases, the reaction should form dimethylsulphone.
- Dimethylsulphoxide combusts when it comes into contact with potassium permanganate in the solid state.
- The treatment of thiourea using a nitric acid/hydrogen peroxide mixture is used to make a thiourea peroxide, which decomposes explosively by giving rise to a mixture of sulphur and sulphur dioxide, when attempting to carry out air drying.

■ Instability of dimethylsulphoxide and of its transformation compounds

- When dimethylsulphoxide was heated at 150°C for a prolonged period and before being distilled, it decomposed violently. Two accidents of this nature have been mentioned. The presence of halogen derivatives traces increases the risk of a decomposition which is highly exothermic. The enthalpy of decomposition corresponds to 0.85 kJ/g at 180°C. Adding zinc oxide or sodium carbonate delays the reaction and decreases the exothermicity. A study has shown that decomposition starts at the boiling point (189°C). The temperature of the decomposition is not modified by the inhibitors. If the decomposition is carried out in an autoclave, pressures of 60 bar can be reached very quickly.
- When there is no diluent, organic acid chlorides and metal halides react very violently with DMSO. This goes for acetyl chloride, benzenesulphonyl ($C_6H_5SO_2Cl$), cyanuryl chloride, phosphorus and phosphoryl trichlorides, tetrachlorosilane, sulphur, thionyl, and sulphuryl chlorides. With oxalyl chloride, the reaction is explosive at ambient temperature, but can be controlled at -60°C in a solution with dichloromethane. The dangerous reactions are thought to be

due to the decomposition of DMSO into formol and methanethiol, which is followed by the violent polymerisation of formol.

- The same is true for organic halogen derivatives. The effect of bromoethane, which was carried out in a sealed tube gives rise to a detonation, after heating that lasted for one hundred and twenty hours, at 65°C. The transformation that was thought to be brought into play during this accident is:

$$CH_3-\underset{\underset{O}{\|}}{S}-CH_3 \xrightarrow[65°C]{CH_3Br} [(CH_3)_2\overset{\oplus}{S}O]\,Br^{\ominus} \xrightarrow{180°C} CH_2=O + CH_3SO_3H$$

A study showed that, if the salt present decomposes at 180°C, its temperature of decomposition reaches 74-80°C when it is in a solution with DMSO. The compounds formed are the ones that are mentioned, but formol is in the form of polymer. There is a substantial gas release during the decomposition. This gas certainly caused the sealed tube to detonate. Another decomposition reaction has been suggested:

$$[(CH_3)_2\overset{\oplus}{S}O]\,Br^{\ominus} \longrightarrow CH_2=O + CH_3-S-CH_3 + HBr$$

Although methanesulphonic acid was found, it was assumed that the above decomposition reaction first occurred and that methanesulphonic acid was a compound of a later transformation.

- Acid anhydrides behave in a similar way to acid chlorides. In this paragraph is included an accident that is the result of a very violent interaction of trifluoroacetic anhydride with DMSO. This reaction can only be controlled by cooling at -40°C, in a solution with dichloromethane.
- With DMSO and aliphatic sulphoxides, dry perchloric acid also forms very unstable salts. If the acid concentration is 70% or more, when it comes into contact with DMSO, this gives rise to an immediate explosion. Aromatic sulphoxides give rise to far less dangerous interactions.
- The same goes for periodic acid. The 1.5M solutions of this acid in DMSO are explosive. Solutions of less than 0.15M are not dangerous. Between these two limits, the situation is intermediate.
- Diborane gives rise to explosive mixtures with DMSO.
- In the presence of potassium, DMSO forms the following salt, which detonates when there is no solvent by forming very large quantities of a charred compound. Tetrahydrofuran is a good solvent for this salt:

$$\begin{array}{c} CH_3 \\ \diagdown \\ S-O^-\ K^+ \\ \diagup \\ CH_3 \end{array}$$

- Salts with the same structure have been put forward to explain the detonations or ignitions that were observed during several accidents involving DMSO and sodium hydride, sodium isopropylate and potassium tert-butylate. However, it is known that the last-named base causes the ignition of nearly all organic compounds.

6 • Reactions of organic chemicals

6.14 Miscellaneous dangerous reactions

■ **Reactions that are related to the solvent properties of dimethylsulphoxide**

- The solvation property of the cations of this very polar aprotic solvent can make some salts more stable. Therefore, aluminium, sodium, mercury or silver perchlorate solutions are explosive. The same goes for iron (III) nitrate solutions.
- Dissolving sulphur trioxide in DMSO is very exothermic and has given rise to accidents, whenever the addition was not carried out slowly and under cooling.
- Trichloroacetic acid had been poured on to copper 'wool' then rinsed with DMSO. After being in contact for twenty seconds, the ballon flask's contents were violently ejected and the flask's neck was bent due to the heat that was given off. This accident was put down to the formation of a carbene:

$$Cu + Cl_3C-CO_2H \xrightarrow{DMSO} CuCl_2 + \left[HO_2C-\underset{..}{C} \diagup^{Cl} \right]$$

■ **Dialkyl sulphates**

- Moist diethyl sulphate was stored in an iron reservoir, which detonated after a little while. This accident was explained by the hydrolysis of the sulphate present that gives rise to the formation of sulphuric acid. By reacting with iron, sulphuric acid formed hydrogen that caused the overpressure responsible for the detonation.
- There was an attempt to destroy 1 l of dimethyl sulphate by using a concentrated aqueous solution of ammonia. This gives rise to an extremely violent reaction. This operation needs to be carried out by introducing the base gradually.
- Incorporating a tertiary amine into dimethyl sulphate without using any solvent caused a violent detonation.
- When diethyl sulphate comes into contact with potassium tert-butylate, this gives rise to immediate ignition.
- Barium chlorite had been added to dimethyl sulphate. The mixture combusted immediately. This behaviour was explained by the formation of methyl chlorite. This accident can also be explained by an oxidation of sulphate by chlorite, whose exothermicity would have caused the compound's self-ignition. The author believes that the instability of methyl chlorite would give rise to a violent detonation rather than an ignition.

■ **Miscellaneous dangerous reactions**

- Phosphorus (III) oxide reacts violently with DMSO by forming large quantities of charring compounds.
- When preparing anhydrous DMSO by using magnesium perchlorate, DMSO detonated during its distillation, although the published method followed did not mention any danger. This accident was explained by the presence of methanesulphonic acid in DMSO and the formation of perchloric acid under the effect of this sulphonic acid on magnesium salt. The risks related to the interaction of perchloric acid with DMSO have been mentioned on p. 247.

- The acidic properties of methanesulphonic acid that have just been mentioned have been responsible for two other accidents. When this acid is contact with methyl and vinyl oxide, this caused the latter to polymerise violently. The electrolysis of methanesulphonic acid with an aqueous solution of hydrogen fluoride gives rise to a violent detonation that was put down to the formation of oxygen difluoride that is explosive.

- A violent detonation took place when preparing this compound using a commonly-used method.

$$CH_2=CH-CH_2Cl + NaSCN \xrightarrow{5.5 \text{ bar}} CH_2=CH-CH_2-N=C=S$$

It could be due either to the decomposition of allyl isiothiocyanate or its polymerisation.

6.14.2 Phosphorus compounds

There are three main dangers with phosphorus compounds:
- the very strong oxidisable nature of alkyl phosphines and to a lesser extent of phosphites;
- the nucleophilic character of phosphines that can give rise to unstable complexes with some acceptors;
- the hydrolysable property of phosphites and phosphates that give rise to acids, which can prove to be dangerous under certain circumstances.

Phosphines

- In the presence of air, trimethyl, triethyl, and tributylphosphine combust spontaneously. In the presence of pure oxygen, even though it was at a low temperature, triethylphosphine detonated. In the same conditions, triphenyl-phosphine does not seem to be dangerous. Trimethylphosphine can be stored safely in the air in the form of a complex with silver iodide.

- With palladium (II) perchlorate, triethylphosphine gives rise to an explosive complex, if it is heated, even moderately:

$$(C_2H_5)_3P + Pd(ClO_4)_2 \longrightarrow \left[[(C_2H_5)_3P]_3PdClO_4\right]^{\oplus} ClO_4^{\ominus} \xrightarrow{\Delta} \text{Explosion}$$

Phosphites and phosphates

- At the end of a distillation under reduced pressure of dibutyl hydrogen-phosphite, an air inlet that was used without taking any precautions caused the spontaneous ignition of the product,

- The residue of a distillation of trimethyl phosphate that was carried out on a large scale detonated violently. No explanation could be given.

- When heating a mixture of diethyl hydrogenphosphite and p-nitrophenol in the absence of a solvent, this caused the medium to detonate. This accident could have been caused by the destabilisation of the nitrated derivative by acid.

- An attempt to dry trimethyl phosphite by using magnesium perchlorate ended with the compounds' detonation. This accident was put down to the formation of methyl perchlorate. However, by analogy with a similar accident described in the sulphur series (see on p. 345), one can ask oneself whether traces of phosphorus acid that resulted from the hydrolysis of phosphite would have formed perchloric acid.

6.14.3 Silicon compounds

Compared with carbon, which is above it in group IV of the periodic table, silicon has a particular affinity for oxygen with which it forms silicon-oxygen bonds that are particularly strong. This property gives silicons their thermal stability. Therefore, silicon compounds are particularly oxidisable. This goes for those compounds that have silicon-carbon bonds and especially silicon-hydrogen. In the latter case, hydrogensilanes have the reactivities of metal hydrides. Although these properties enable one to forecast the dangers that are related to the oxidation of these compounds, the accidents that are described below mention only a few reactions of this type. For reasons that are partly similar, compounds with silicon-chlorine bonds will have a particularly hydrolysable character, which as a result will find them classified under code 'X' in the transport regulations of dangerous compounds and '14' by safety regulations. For the same reason, they will have the NFPA reactivity code '1'. Their dangerous reactions are be due to the high mobility of chlorine bonded to silicon.

■ **Silicon-carbon bond reactions**

- The reaction below is explosive when it is carried out without any solvent and at 100°C. It can be controlled below 30°C:

$$(CH_3)_4Si + Cl_2 + SbCl_3 \xrightarrow{100°C} (CH_3)_3SiCl$$

■ **Silicon-hydrogen bond reactions**

- Methyldichlorosilane (CH_3SiHCl_2) combusts spontaneously in the presence of potassium permanganate, lead oxide and dioxide, copper (II) oxide and silver oxide, even when they are in an atmosphere of inert gas.
- Mixing trichlorosilane, acetonitrile and diphenylsulphoxide, carried out at 10°C, detonated. This accident was put down to the exothermic addition reaction of the silicon-hydrogen bond on the carbon-nitrogen triple bond of nitrile. Other interpretations are possible: for instance, the effect of traces of hydrogen chloride formed by the hydrolysis of chlorosilane on acetonitrile.

■ **Silicon-chlorine bond reactions**

- Trimethylchlorosilane reacts violently with water.
- Tetrachlorosilane was added to aqueous ethanol (the presence of water was accidental). There was no proper stirring during this operation, which led to the formation of two liquid layers of compounds that did not react. The very fast and exothermic reaction of the alcoholysis-hydrolysis of chlorosilane started violently and the large compoundion of hydrogen chloride caused the reactor to detonate.
- With tetrachlorosilane, sodium metal gives rise to mixtures that are explosive on impact. This situation is exactly similar to the one with tetrachloromethane.

6.14 Miscellaneous dangerous reactions

Synopsis of dangerous reactions of sulphur, phosphorus and silicon derivatives

Column headers (diagonal, left to right):
1. Methanesulphonic acid
2. Dimethylsulphoxide (DMSO)
3. Allyl isothiocyanate
4. Dimethyl, diethyl sulphate
5. Dimethyl sulphide
6. Thiols of C_3, C_4, C_5, C_2, C_6
7. Thiophene
8. Thiourea
9. Hydrogenphosphites of C_2H_5; C_4H_9
10. Trimethyl phosphite, phosphate
11. Trialkylphosphines
12. Methyldichlorosilane
13. Silicon tetrachloride
14. Tetramethylsilane
15. Trichlorosilane
16. Trimethylchlorosilane

1	2	3	4	5	6	7	8	9	10	11	12	13	14	15	16	Reagent
																Hydrogen
	▓									▓						Group I (Na; K...)
																Group II (Be; Mg; Ca...)
																Group III (B)
																Group IV (C; Si; Ge)
																Group V (N; P; As; Sb)
	▓				▓		▓			▓						Group VI (O_2)
												▓				Group VII (Cl_2; Br_2 XeF_2)
																Transition metals
																Other metals (Al... Bi; Sn; Pb)
▓																Hydrides group III (B_xH_y)
																Hydrides group IV (SiH_4...)
	▓															Hydrides group V (NH_3...)
	▓										▓					Water
																Hydrides group V (H_2S....)
▓																Hydrogen halides (HF)
										▓						Metal halides ($SbCl_3$)
▓																Non-metal halides ($SOCl_2$...)
			▓			▓										Metal oxides
▓																Non-metal oxides (N_2O_4; P_2O_3)
				▓												Peroxides (H_2O_2; Na_2O_2)
																Acid oxidants (HNO_3)
																Other acids (H_2SO_4; $HClO_4$; HIO_4)
▓																ClO_x^-
										▓						NO_3^-; NO_2^-
																MnO_4^-, $Cr_2O_7^{2-}$
																Instability
		▓														Polymerisation
																Bases
	▓	▓														Potassium tert-butylate
																Ethanol
																p-Nitrophenol
																Methyl and vinyl oxide
		▓														Tertiary amines
	▓															Methyl bromide
	▓															Trichloroacetic acid
												▓				Acetonitrile
	▓															Organic acid chlorides
																Trifluoroacetic anhydride

PART THREE
Tables

Introduction to Part Three

The tables contained in Part Three are derived from a number of sources, listed below. Ambiguities and contradictions occur between these sources which are presented in the tables in the form of abbreviations.

Published references

- *Material Safety Data Sheets* (MSDS), CD-ROM, Sigma-Aldrich-Fluka, 1995 **subscription.**
- *Catalogue de produits chimiques*, Fuka 1995-1996.
- *SAX's Dangerous Properties of Dangerous Materials*, CD-ROM version of the 8th edition, Van Nostrand Reinhold, 1995.
- *MSDS*, Canadian Centre for Work Hygiene and Safety, A1 CD-ROM version, 1995 **subscription.**
- *Bretherick's Reactive Chemical Hazards Database*, version 1, 4th edition CD-ROM, Butterworth-Heinemann, 1994.
- *Sécuridisque*, CD-ROM version of the Code du travail, Cabinet Beugnette publisher, 19, rue Poincaré, 88210 Senones, 1995 **subscription**, they distribute the Canadian Centre disk previously mentioned for France.
- *Guide de la chimie international*, Chimedit, 1994-1995.
- *The Merck Index*, Merck & Co.
- *Lange's Handbook of Chemistry*, Mc Graw Hill.
- *Handbook of Chemistry and Physics*, CRC Press.
- *Handbook of Laboratory Safety*, CRC Press.
- *Traité pratique de sécurité*, produits dangereux pour l'homme et l'environnement, Centre national de prévention et de protection.
- *Constantes*, Techniques de l'ingénieur, updated in 1989.
- *Health and Safety Manual*, Lawrence Livermore national laboratory, 1991-1996.
- *Guide for Safety in the Chemical Laboratory*, Manufacturing Chemists Association.
- *Fire Protection Handbook*, NFPA.
- *Comment utiliser en toute sécurité les produits chimiques dangereux*, edition by subscription, Weka.

Internet sites

Free access to MSDS. Some or all of these sites may not still exist due to the short life of some of them.

- http://hazard.com/msds/
- http://www.chemsafe.com

- http://www.ps.uga.edu/rtk/msds.htm
- http://odin.chemistry.uakron.edu/erd/
- gopher://ecosys.drdr.Virginia.EDU:70/11/library/gen/toxics
- http://www.wco.com/jray/pyro/safety/msds
- gopher://gaia.ucs.orst.edu:70/11/osu-i+s/osu-d+o/ehs/msds/Product
- http://www.chemexper.be/
- http://ace.orst.edu/info/extoxnet/ghindex.html

Abbreviations used

C	Cleveland (apparatus measuring flash point)
cc	Closed cup
LC50	Lethal Concentration 50
MLC	Minimum Lethal Concentration
oc	Open cup
dec	Decomposition
LD50	Lethal Dose 50
MLD	Minimum Lethal Dose
Eb	Boiling point
g	Gaseous state
IDLH	Immediately dangerous to life or health
INRS	Institut National de Recherche et de Sécurité
ip	Intraperitoneal
iv	Intravenous
r	Rabbit
l	Liquid state
LEL	Lower Explosive Limit
UEL	Upper Explosive Limit
MAK	*Maximale Arbeitsplatzkonzentration*
NFPA	National Fire Protection Association
o	Orally
p	Cutaneous
P_{fla}	Flash point
PM	Pensky-Martens (apparatus measuring flash point)
P_{vap}	Vapour pressure
r	Rat
m	Mouse
S	Setaflash (apparatus measuring flash point)
s	Solid state

scu	Subcutaneous
STEL	Short time exposure limit
subl	Sublimation
T	Tag (apparatus measuring flash point)
AIT	Autoinflammation Temperature
THF	Tetrahydrofurane
Tr	Code du travail (risk degree for inorganic data)
TR	Code du transport (risk degree of dangerous materials)
TWA	Abbreviation for TLV/TWA: Threshold limit value/Time weighted average
LVE	Limiting Value of Exposure
MVE	Mean Value of Exposure

Presentation of data in tables

The following tables are characterised by the fact that they provide all experimental values that are given by our sources for each chemical product. When several values exist for one product, they are separated by semicolons.

- Boiling points are in degrees Celsius. When the value stands on its own, the boiling point is under standard pressure conditions. If not, it is presented by two figures that are separated by a slash (/), the first figure being the boiling point, and the second the pressure in mmHg. When several boiling points are given under normal pressure, the ones thought to be more relevant are underlined.

- Vapour pressures are in mbar, the temperature in °C in square brackets. If there are several values at the same temperature, the latter is not repeated. For instance, 6,8; 7; 10.7[20] means that three different sources offered a vapour pressure at 20°C of 6.8 mbar, 7 mbar and 10.7 mbar.

When some values seem questionable a question mark in brackets is added next to these values. If two very different values are given for one substance the doubtful data source is noted. For instance, for o-cresol, two LD50 values for the rat orally are mentioned as follows: LD50 o-r: 121; 1350 (Merck). This means that the 1350 value that seems high and that applies to a phenol that is particularly corrosive and toxic (see Code du travail (ie Labour Code)) was suggested by the *Merck Index*.

- The Code du travail and NFPA hazard codes are explained in Part I in the references that are concerned with each risk. For inorganic products the inflammability, reactivity and toxicity codes are mentioned. For organic products the toxicity code is the only one that is given. Indeed, the inflammability code can be found easily (see para 1.5.1) and the reactivity code is included in chapter 6 under dangerous reactions of inorganic products.

Hydrogen/helium/lithium
beryllium

7 • TABLE OF INORGANIC COMPOUNDS

Name: Empirical formula • Molecular mass (g/mol) • Registry number (CAS).
(P) Physical data: ΔH_f^0 (kJ/mol) (physical state) • Eb (°C/mmHg) (volatile product) • P_{vap} (mBar) (°C) (volatile product).
(I) Inflammability data: LEL/UEL (%) • P_{fla}(°C) • AIT (°C) • NFPA (flame); NFPA (reactivity) •Safety code (Tr).
(T) Toxicity: Safety regulations (Tr) • NFPA • Transport code (TR) • LVE (MVE) (ppm) F, USA, D • STEL (ppm); IDLH (ppm).
(D) Lethal and toxic codes: LC (mg/l/duration of exposure); LD (mg/kg).

HYDROGEN

Hydrogen: H_2 • 2.02 • 12385-13-6
(P) Eb=-252.8
(I) LEL/UEL = 4.1/74.2 • AIT = 400; 571 • NFPA: 4 ; – • Tr: 12
(T) NFPA: 0

Water : H_2O • 18.02 • 7732-18-5
(P) $\Delta H_f^0 = -241.8$ (g); –285.8 (1) • Eb = 1002.9
(I) NFPA: 0.3.4 0 • Tr: –

Hydrogen peroxide: H_2O_2 • 34.01 • 7722-84-1
(P) $\Delta H_f^0 = -136.3$ (g); –187.8 (1) • Eb = 150.2
(I) NFPA: 0; 3 (90 %); 2 (<90 %) • Tr: 8
(T) Tr: 34 (c > 20 %) • TWA (USA): 1 • IDLH: 75
(D) LC50 r: 2/4h; MLC m : 0.321/?; LD50 o-r: 1520; LD50 o-m: 2000; LD50 cu-r: 4060; MLD cu-ra 500

HELIUM

Helium: He • 4 • 7440-59-7
(P) Eb = –268.9
(I) NFPA: –; – • Tr: –
(T) Asphyxiant Gas

LITHIUM

Lithium: Li • 6.94 • 7439-93-2
(I) NFPA: 1; 2 • Tr: 14/15
(T) Tr: 34 • NFPA; 3 • TR: –
(D) LD50 ip-m: 1000

Lithium bromide: LiBr • 86.85 • 7550-35-8
(P) $\Delta H_f^0 = -351.2$ (s)
(D) LD50 o-r: 1800; LD50 o-m: 1840; LD50 ip-m: 1160; LD50 scu-m: 1680

Lithium carbonate: Li_2CO_3 • 73.89 • 554-13-2
(P) $\Delta H_f^0 = -1215.9$ (s)
(D) LC50 r: 1.8/?; LD50 o-r: 525; 710; LD50 ip-r: 156; LD50 scu-r: 434; LD50 iv-r: 241; LD50 o-m: 530; LD50 ip-m: 236; LD50 scu-m: 413; LD50 iv-m: 497

Lithium chloride: LiCl • 42.39 • 7447-41-8.
(P) $\Delta H_f^0 = -408.6$ (s)
(D) LD50 o-r: 526; 757; LD50 o-m: 1165; LD50 o-ra: 800; 850; LD50 ip-r: 514; LD50 ip-m: 606; LD50 iv-m: 363; LD50 scu-r: 499; LD50 scu-m: 828

Lithium chromate: Li_2CrO_4 • 129.88 • 14307-35-8

Lithium fluoride: LiF • 25.94 • 7789-24-4
(P) $\Delta H_f^0 = -616.0$ (s)
(D) LD50 o-r: 143

Lithium hydroxide: LiOH • 24.95 • 1310-65-2
(P) $\Delta H_f^0 = -484,9$ (s)
(T) Tr: 35 • TR: 80
(D) LC50 r: 0.96/?; MLD o-m: 200; LD50 scu-m: 300

Lithium hydride: LiH • 133.85 • 10377-51-2
(P) $\Delta H_f^0 = -270.4$ (s)

Lithium sulphate; Li_2SO_4 • 109.94 • 10377-48-7
(P) $\Delta H_f^0 = -1436.5$ (s)
(D) LD50 o-m: 1190; LD50 scu-m: 953

Lithium tetraborate: $Li_2B_4O_7$ • 169.12 • 12007-60-2

BERYLLIUM

Beryllium: Be • 9.01 • 7440-41-7
(I) NFPA: 1; 1
(T) Tr: 49-25-26-36/37/38-43-48/23 • NFPA: 4 • TR: 64 (powder) • VLE (VME) (F): (0.002mg/m^3);   TWA (USA): 0.002mg/m^3(Be)
(D) LD50 iv-r: 0.496; DTM iv-ra: 20

Beryllium carbide: Be_2C • 30.04 • 506-66-1
(P) $\Delta H_f^0 = -117.15$(s)
(T) NFPA: 4

Beryllium chloride $BeCl_2$ • 79.92 • 7787-47-5
(P) $\Delta H_f^0 = -490.4$(s)
(I) NFPA: – ; 1
(T) NFPA: 4
(D) LD50 o-r: 86; LD50 o-m: 92; LD50 ip-r: 0.52; 44; LD50 ip-m: 106

7 • Table of inorganic compounds — Boron/carbon

- Name: Empirical formula • Molar mass (g/mol) • Registry number (CAS).
- **P** Physical data: ΔH_f^0 (kJ/mol) (physical state) • Eb (°C/mmHg) (if substance volatile) • P_{vap} (mBar) (°C) (if substance volatile).
- **I** Inflammability data: LEL/UEL (%) • P_{fla}(°C) • AIT (°C) • NFPA (flame); NFPA (reactivity) • Labour Code (L).
- **T** Toxicity: Labour Code (L) • NFPA • Transport Code (TR) • LVE (MVE) (ppm) F, USA, D • STEL (ppm); IDLH (ppm).
- **D** Toxic and lethal doses: LC (mg/l/exposure time); LD (mg/kg).

Beryllium fluoride : BeF_2 • 47.01 • 7787-49-7
- **P** $\Delta H_f^0 = -1026.8(s)$
- **I** NFPA: – ; 1
- **T** NFPA: 4
- **D** MTC r: 0.02/? 0.049/?; LD50 o-r: 98; LD50 o-m: 100; LD50 iv-m: 1.8; LD50 scu-m: 20

Beryllium hydroxide $Be(OH)_2$ • 43.03 • 13327-32-7
- **P** $\Delta H_f^0 = -902.5(s)$
- **T** NFPA: 4
- **D** MLD iv-r: 3.821

Beryllium nitrate: $Be(NO_3)_2$ • 133.02 • 13597-99-4
- **I** NFPA: – ; 1
- **T** NFPA: 4
- **D** LD50 ip-m: 500; MLD iv-r: 500

Beryllium oxide: BeO • 25.01 • 1304-56-9
- **P** $\Delta H_f^0 = -609.4(s)$
- **T** Tr: 49-25-26-36/ 37/38-43-48/23 • NFPA: 4 • LVE(MVE)(F): (0.002mg/m^3) ; TWA (USA): 0.002mg/m^3(Be)
- **D** MTC r: 0.017/?; 0.028/?; MTD iv-ra: 600

Beryllium potassium double sulphate: $BeK_2S_2O_8$ • 279.32 • 53684-48-3
- **I** NFPA: – ; 1
- **T** NFPA: 3
- **D** MTC r: 0.668/?; LD50 ip-m: 41; LD50 iv-m: 5; LD50 iv-r: 4

BORON

Boron: B • 10.81 • 7440-42-8
- **I** NFPA: 2
- **T** NFPA: 2
- **D** LD50 o-m: 2000

Boric acid: H_3BO_3 • 61.83 • 10043-35-3
- **P** $\Delta H_f^0 = -1094.3(s)$
- **I** NFPA: –; – • Tr –
- **T** NFPA: 2 • TR: – • LVE(MVE) (F): 10 mg/m^3
- **D** LC50 r: 0.016/?; MLC r: 0.028/? ; LD50 o-r: 2660; 3160; 3500-4100; 5140; LD50 o-m: 3450; LD50 iv-r: 1330; LD50 iv-m: 1240; LD50 scu-r:1400

Boron carbide : B_4C • 55.25 • 12069-32-8

Diborane: B_2H_6 • 27.67 • 19287-45-7
- **P** $\Delta H_f^0 = +31.3(g)$; +35.6(g) • Eb = 92.5
- **I** LEL/UEL = 0.9/98 • P_{fla} = 68 • AIT = 40-50 • NFPA: 4 ;3 • Tr: –
- **T** NFPA: 3 • TR: – ; LVE(MVE) (F): 0.1
- **D** LC50 r: 0.057/?; 0.046/4h; LC50 s: 0.033/?

Magnesium diborate: $Mg(BO_2)_2$ • 109.92 • 13703-82-7

Boron nitrate: BN • 24.82 • 10043-11-5
- **P** $\Delta H_f^0 = -254.4(s)$

Boron oxide: B_2O_3 • 69.62 • 1303-86-2
- **P** $\Delta H_f^0 = -1273.5(s)$
- **T** NFPA: 2 • TR: –
- **D** LD50 o-m: 3163; LD50 ip-m: 1868

Magnesium perborate: $Mg B_2O_6$ • 141.92 • 14635-87-1

Sodium perborate : $NaBO_3$ • 81.80 • 10486-00-7
- **P** Eb = 60 (dec $-O_2$)
- **D** Monohydrate: LD50 iv-ra: 78; Tetrahydrate: LD50 o-r: 1200; LD50 o-m: 1060; LD50 ip-m: 538

Sodium tetraborate $Na_2B_4O_7$ • 201.22 • 1330-43-4
- **T** Borax • NFPA: 2 • LVE(MVE) (F): anhy: 1mg/m^3; hydrate: 5 mg/m^3
- **D** Hydrate: LD50 o-r: 2650; 3200-3400; 4500-5000; 5660; Anhydrous: LD50 o-r: 2400-2660; 3400-3800; LD50 o-m: 2000; LD50 ip-m: 2711; LD50 iv-m: 13200

Boron trichloride: BCl_3 • 117.17 • 10294-34-5
- **P** $\Delta H_f^0 = -403.8(g); -427.2(l)$ • Eb = 12.5 • Pvap = 1300(20); 1900(30); 3200(50); 26/28-34
- **I** Tr: 14
- **T** NFPA: 2
- **D** MLC r: 0.097/?; MLC m : 0.097/?; LC50 r: 0.097/20h; 12.379/1h

Boron trifluoride; BF_3 • 67.81 • 7637-07-2
- **P** $\Delta H_f^0 = -1136.0(g)$ • Eb = 99.9; 100.3.
- **I** NFPA: 0 ; 1 • Tr: 14
- **T** Tr: 26-35 • NFPA: 2 • TR: 26 • LVE(MVE) (F): 1; TWA (USA): 1; MAK (D): 1
- **D** LC50 s: 3.46/2h; LC50 r: 1.18/?

CARBON

Carbon: C • 12.01 • 7440-44-0
- **I** NFPA: 1 • Tr: –
- **T** LVE(MVE) (F): 3.5 mg/m^3
- **D** LD50 iv-m: 440; MTD scu-r: 167

Carbon dioxide: CO_2 • 44.01 • 124-38-9
- **P** $\Delta H_f^0 = -393.5(g)$ • Eb = –78.5 (subl) • Pvap = 58400(20)
- **T** NFPA: – • TR: 20 • LVE(MVE) (F): (5000); TWA (USA): 5000; MAK (D): 5000 • STEL: 30000
- **D** MTC r: 109.85/?; MLC m : 36.615/?

7 • Table of inorganic compounds — Nitrogen

Name: Empirical formula • Molar mass (g/mol) • Registry number (CAS).
(P) Physical data: ΔH_f^0 (kJ/mol) (physical state) • Eb (°C/mmHg) (if substance volatile) • P_{vap} (mBar) (°C) (if substance volatile).
(I) Inflammability data: LEL/UEL (%) • P_{fla}(°C) • AIT (°C) • NFPA (flame); NFPA (reactivity) • Labour Code (Tr).
(T) Toxicity: Labour Code (Tr) • NFPA • Transport Code (TR) • LVE (MVE) (ppm) F, USA, D • STEL (ppm); IDLH (ppm).
(D) Toxic and lethal doses: LC (mg/l/exposure time); LD (mg/kg).

Carbon monoxide: CO • 28.01 • 630-08-0
- (P) $\Delta H_f^0 = -110.5(g)$ • Eb = 191.5
- (I) LEL/UEL = 12/75; 12.5/74 • AIT = 607; 652; 700; NFPA: 4; 0 • Tr: 12
- (T) Tr: 23 • NFPA: 2 • TR: 236 • LVE(MVE) (F): (50); TWA (USA): 25; MAK (D): 30 • IDLH: 1500
- (D) LC50 r: 2.104/4h; MTC r: 0.175/24h; LC50 m: 2.845/4h

Carbon suboxide C_3O_2 • 68.03 • 504-64-3
- (P) Eb = 7 • P_{vap} = 782-785(0)
- (I) LEL/UEL = 6/30

NITROGEN

Nitric acid: HNO_3 • 63.01 • 7597-37-2
- (P) $\Delta H_f^0 = -135.1(g); -173.22(l); -174.1(l)$ • Eb = 83; 86; 120 (azeotrope with H_2O) • P_{vap} = 4(20) ; 7.3 (30)
- (I) NFPA: 0 ; 1 • Tr: 8
- (T) Tr: 35 • NFPA: 2 • TR: 885 • LVE(MVE) (F): 4 (2); TWA (USA): 2; MAK (D): 10; STEL: 4; IDLH: 100

Sodium amide: $NaNH_2$ • 39.01 • 7782-92-5
- (P) $\Delta H_f^0 = -123.85$ (s)
- (I) NFPA: 2; – • Tr: 14/15-19
- (T) Tr: 34 • NFPA: 3 • TR: -

Ammonia: NH_3 • 17.03 • 7664-41-7
- (P) $\Delta H_f^0 = -46.11; -45.9$ • Eb = -33.35 • P_{vap} = 2000(18); 5000(5); 8700(20)
- (I) LEL/UEL = 15/28 • AIT = 630 • NFPA: 1 ; 0 • Tr: 10
- (T) Tr: 23 • NFPA: 3 • TR: 268 • LVE(MVE) (F): 50 (25); TWA (USA): 35; MAK (D): 35 • IDLH: 500
- (D) LC50 m: 2.94/?; 3.36/?; LC50 l: 4.87/?; LD50 o-r: 350 (aqueous solution); LD50 iv-m: 91 (aqueous solution)

Hydrogen azide: HN_3 • 43.03 • 7782-79-8
- (P) $\Delta H_f^0 = 294.14$ (g); 264.0 (l) • Eb = 37
- (I) NFPA: – ; 4 • Tr: –
- (T) NFPA: 3 • TR: – • TWA (USA): 0.1; MAK (D): 0.1
- (D) LC50 r: 1.93/1h; LD50 ip-m: 22

Sodium azide: NaN_3 • 65.01 • 26628-22-8
- (P) $\Delta H_f^0 = 21.7$ (s) • $\Delta Hd = -49.4$ kJ/mole (at 230-260°C)
- (I) NFPA: 2 ; 2 • Tr: –
- (T) Tr: 28-32 • NFPA: 3 • TR: – • LVE(MVE) (F): 0.1 (0.3 mg/m^3); TWA (USA): 0.11; MAK (D): 0.07
- (D) LD50 o-r: 27; LD50 o-m: 27; LD50 ip-r: 30; LD50 ip-m: 28; 18 ; LD50 iv-m: 19; LD50 scu-r: 35 ; LD50 scu-m: 23; 17; LD50 cu-ra: 20

Diammonium carbonate : $(NH_4)_2CO_3$ • 96.09 • 10361-29-2
- (P) $\Delta H_f^0 = -942.24$ (aqueous solution)
- (T) NFPA: – • TR: –
- (D) LD50 iv-m: 96

Ammonium hydrogen carbonate : NH_4HCO_3 • 79.06 • 1066-33-7
- (P) $\Delta H_f^0 = -849.35$ (s)
- (I) NFPA: –; – • Tr: –
- (T) NFPA: – • TR: –
- (D) LD50 iv-r: 245

Potassium cyanide; KCN • 65.12 • 151-50-8
- (P) $\Delta H_f^0 = -112.97$ (s)
- (I) NFPA: 0 ; 0 • Tr: –
- (T) 26/27/28-32 • NFPA: 3 • TR: 66 • LVE(MVE) (F): (5 mg/m^3 (CN); TWA (USA): 5 mg/m^3 (CN); MAK (D): 5 mg/m^3 (CN)
- (D) LD50 o-r: 5; 10; LD50 o-m: 8.5; LD50 o-ra: 5; LD50 ip-r: 4; LD50 ip-m: 6 ; LD50 iv-m: 26; LD50 scu-r: 9; LD50 scu-ra: 4

Sodium cyanide; NaCN • 49.01 • 143-33-9
- (P) $\Delta H_f^0 = -90.71$ (s)
- (I) NFPA: 0 ; 0; Tr: –
- (T) 26/27/28-32 • NFPA: 2 • TR: 66 • LVE(MVE) (F):(5 mg/m^3 (CN)); TWA (USA): 5 mg/m^3 (CN); MAK (D): 5 mg/m^3 (CN)
- (D) LD50 o-r: 15; 6.4; LD50 o-m: 5.8; LD50 ip-r: 59; 43; LD50 ip-m: 4.3; LD50 scu-m: 3.6; LD50 scu-ra: 2.2

Nitrogen dioxide : NO_2 • 46.01; 10102-44-0 (NO_2) • 10544-72-6 (N_2O_4)
- (P) $\Delta H_f^0 = 33.18$ (g); 33.8 (g); -19.58 (l, N_2O_4); 9.16 (g, N_2O_4); 9.7 (g, N_2O_4) • Eb = 21 • P_{vap} = 960
- (I) NFPA: 0 ; 1 • Tr: –
- (T) Tr: 26-37 • NFPA:3 • TR: 265 • LVE(MVE) (F): 3; TWA (USA): 1; MAK (D): 5 • STEL: 5
- (D) LC50 r: 0.107; 0.165/4h; LC50 m: 1.88/10'; LC50 l: 0.593/15'

Hydrazine: N_2H_4 32.05 • 302-01-2
- (P) $\Delta H_f^0 = 95.40$ (g); 50.63 (l); -242.73(l, hydrate) • Eb = 113.5 • Pvap = 10; 13.3; 21(20); 19.86(25); 33(30); 81(50)
- (I) LEL/UEL = 4.7/100 • P_{fla}=38cc; 52; 52oc• AIT = depending on contact, 23 (rust); 132 (iron); 156 (steel); 270 (glass) • NFPA: 3 ; 2 •Tr: –
- (T) Tr: 45-23/24/25-34-43 • NFPA: 3 • TR: 86 • LVE(MVE) (F): – (0.1); TWA (USA): 0.1; MAK (D): 0.1 • IDLH: 80
- (D) LC50 r: 0.746; LC50 m: 0.329 ; LD50 o-r: 60; LD50 o-m: 59; LD50 ip-r: 59; LD50 ip-m: 62; LD50 iv-r: 55; LD50 iv-m: 57; LD50 cu-ra: 91 ; LD50 iv-ra: 20

Hydrazine chlorohydrate : $N_2H_6Cl_2$ • 104.97 • 5341-61-7
- (P) $\Delta H_f^0 = -367.36$(s, 2HCl); -196.65 (s; HCl)
- (I) NFPA: – ; – • Tr: –
- (T) Tr: 45-23/24/25-43 • NFPA: – • TR: –
- (D) LD50 o-r: 128; LD50 o-m: 126; LD50 ip-r: 126; LD50 ip-m: 133; LD50 iv-r: 118 ; LD50 iv-m: 122

7 • Table of inorganic compounds

Oxygen/fluorine

Name: Empirical formula • Molar mass (g/mol) • Registry number (CAS).
(P) Physical data: ΔH_f^0 (kJ/mol) (physical state) • Eb (°C/mmHg) (if substance volatile) • P_{vap} (mBar) (°C) (if substance volatile).
(I) Inflammability data: LEL/UEL (%) • P_{fla}(°C) • AIT (°C) • NFPA (flame); NFPA (reactivity) • Labour Code (Tr).
(T) Toxicity: Labour Code (Tr) • NFPA • Transport Code (TR) • LVE (MVE) (ppm) F, USA, D • STEL (ppm); IDLH (ppm).
(D) Toxic and lethal doses: LC (mg/l/exposure time); LD (mg/kg).

Hydrazine sulphate : $N_2H_6SO_4$ • 130.12 • 10034-93-2
- (P) $\Delta H_f^0 = -958.97$(s)
- (I) NFPA – ; – • Tr: –
- (T) Tr: 45-23/24/25-43 • NFPA: – • TR: –
- (D) LD50 o-r: 601; 501; LD50 o-m: 740, 434; LD50 o-ra: 100; LD50 ip-r: 230; LD50 ip-m: 152; LD50 scu-m: 455

Hydroxylamine: NH_2OH • 33.03 • 7803-49-8
- (P) $\Delta H_f^0 = -114.2$ (s) • Eb = 56.5; 110; Exploded at 129 • Pvap = 13.3 (47)
- (I) NFPA: 3 ; 3 ; Tr: –
- (T) 34-40 (solution) • NFPA: 1 • TR: –
- (D) LD50 o-m: 408; LD50 ip-r: 59; LD50 ip-m: 60; 70; LD50 scu-r: 29

Hydroxylamine chlorohydrate: NH_2OHCl • 69.49 • 5470-11-1
- (P)
- (I) NFPA: 3 ; – • Tr: –
- (T) Tr: 20/22-36/38 • NFPA: 2 • TR: –
- (D) LD50 o-m: 408; LD50 ip-m: 10

Ammonium nitrate : NH_4NO_3 • 80.04 • 6484-52-2
- (P) $\Delta H_f^0 = -365.56$ (s)
- (I) NFPA: 1 ; 3 • Tr: 8
- (T) NFPA: 2 • TR: 589
- (D) LD50 o-r: 2220; 4820; 5300; LD50 o-m: 2085

Potassium nitrate ; KNO_3 • 101.10 • 7757-79-1
- (P) $\Delta H_f^0 = -494.63$ (s) • Eb = Decomposition at 400
- (I) NFPA: 0 ; 1 • Tr: 7
- (T) NFPA: 1 • TR: –
- (D) LD50 o-r: 3750; LD50 o-ra: 1901

Sodium nitrate $NaNO_3$ • 84.99 • 7631-99-4
- (P) $\Delta H_f^0 = -467.85$ (s) • Eb = dec at 380
- (I) NFPA: 0 ; 2 • Tr: 8
- (T) 36/37/38; NFPA: 1 • TR: –
- (D) LD50 o-r: 4300; 3236; 1267; LD50 o-ra: 2680; LD50 iv-m: 175

Potassium nitrite; KNO_2 • 85.10 • 7758-09-0
- (P) $\Delta H_f^0 = -369.82$ (s) • Eb = dec at 320
- (I) NFPA: – ; – • Tr: 8
- (T) Tr: 25; NFPA: – • TR: 50
- (D) LD50 o-ra: 108; 200; 235; LD50 scu-r: 85

Sodium nitrite ; $NaNO_2$ • 69.00 • 7632-00-0
- (P) $\Delta H_f^0 = -358.65$ (s)
- (I) NFPA: 2 ; – • Tr: 8
- (T) Tr: 25; NFPA: 3 • TR: 50
- (D) LC50 r: 1.45; 5.5 ; LD50 o-r: 85; LD50 o-m: 214; 175; LD50 o-ra: 186; LD50 ip-m: 158; LD50 iv-r: 65; LD50 scu-r: 10 ; LD50 scu-m: 150

Nitrous oxide; N_2O • 44.01 • 10024-97-2
- (P) $\Delta H_f^0 = 82.05$ (g) • Eb = – 88
- (I) NFPA: – ; – • Tr: –
- (T) NFPA: – • TR: –; TWA (USA): 50; MAK (D): 100
- (D) LC50 r: 1.068/4h; LC50 m: 2.7

Nitric oxide: NO • 30.01 • 10102-43-9.
- (P) $\Delta H_f^0 = 90.25$ (g) • Eb = – 151.7 • $P_{vap} = 50800(20)$
- (I) NFPA: – ; 3 • Tr: 44
- (T) Tr: 23/24/25-34 • NFPA: 3 • TR: – ; LVE(MVE) (F): – (25)
- (D) LC50 r: 1.068; LC50 m: 0.393; LC50 ra: 0.386

Nitrogen trichloride ; NCl_3 • 120.37 • 10025-85-1
- (P) $\Delta H_f^0 = 230.10$ (l) • 60 < Eb < 71 • Explosion at 93; < 60 • $P_{vap} = 200(20)$
- (I) NFPA: – ; – • Tr: –
- (T) NFPA: – • TR: –
- (D) LC50 r: 0.545/1h

OXYGEN

Oxygen: O_2 • 32 • 7782-44-7
- (P) Eb = –183
- (I) NFPA: 0 • Tr : 8 (liquid)
- (T) Tr: 34 (liquid) • NFPA: 3

Ozone: O_3 • 48 • 10028-15-6
- (P) $\Delta H_f^0 = +142.20$(g) • Eb = –111.9
- (I) NFPA: 3
- (T) NFPA: 3 • LVE(MVE) (F): 0.2 (0.1) • TWA (USA): 0.1; MAK (D): 0.1
- (D) LC50 r: 0.009/?; 9.58/4h; LC50 m: 0.025/3h

FLUORINE

Fluorine: F_2 • 38.00 • 7782-41-4
- (P) Eb = –188.14
- (I) NFPA: 0; 3 • Tr : 7
- (T) Tr: 26-35 • NFPA: 4 • TR: prohibited • LVE(MVE) (F): 1 (–); TWA (USA): 1; MAK (D): 0.1 • STEL: 2
- (D) LC50 r: 0.197/1h; LC50 m: 0.237/1h; LC50 ra: 0.326/30'

Fluoroboric acid : HBF_4 • 87.81 • 16872-11-0
- (P) $\Delta H_f^0 = -1571.09$(aq) • Eb = dec at 130
- (I) NFPA: –; –
- (T) NFPA: 3

Hydrofluoric acid : HF • 20.01 • 7664-39-3
- (P) Eb = 112 (38% in H_2O)
- (I) NFPA: 0; 0
- (T) Tr: 26/27/28-35 • NFPA: 4 • TR: 886 • LVE(MVE) (F): 3 (–); TWA (USA): 3; MAK (D): 3
- (D) LC50 r: 0.803/1h; LC50 m: 0.284/1h; MLD ip-r: 25 (sources inprecise if HF [aq] or HF (g agitated)].

7 • Table of inorganic compounds

Neon/sodium

Name: Empirical formula • Molar mass (g/mol) • Registry number (CAS).
(P) Physical data: ΔH_f^0 (kJ/mol) (physical state) • Eb (°C/mmHg) (if substance volatile) • P_{vap} (mBar) (°C) (if substance volatile).
(I) Inflammability data: LEL/UEL (%) • P_{elc} (°C) • AIT (°C) • NFPA (flame); NFPA (reactivity) • Labour Code (Tr).
(T) Toxicity: Labour Code (Tr) • NFPA • Transport Code (TR) • LVE (MVE) (ppm) F, USA, D • STEL (ppm); IDLH (ppm).
(D) Toxic and lethal doses: LC (mg/l/exposure time); LD (mg/kg).

Fluorosilicic acid: H_2SiF_6 • 144.09 • 16961-83-4

Ammonium difluoride: NH_4HF_2; • 57.04 • 1341-49-7
- (P) $\Delta H_f^0 = -802.91(s)$ • Eb = 239/750 • $P_{vap} = 1(20)$

Potassium difluoride: KHF_2 • 78.10 • 7789-29-9
- (P) $\Delta H_f^0 = -927.7(s)$
- (I) NFPA: 0; 0
- (T) Tr: 25-34• NFPA: 3 TR: 80 • LVE(MVE) (F): (2.5 mg/m³); TWA (USA): 2.5 mg/m³ (F–); MAK (D): 2.5 mg/m³ (F–)

Sodium difluoride: $NaHF_2$ • 61.99 • 1333-83-1
- (P) $\Delta H_f^0 = -920.3(s)$
- (T) Tr: 25-34 • TR: 80 • LVE(MVE) (F): (2.5 mg/m³); TWA (USA): 2.5 mg/m³ (F–); MAK (D): 2.5 mg/m³ (F–)

Ammonium fluoride NH_4F • 37.04 • 12125-01-8
- (P) $\Delta H_f^0 = -464.0(s)$
- (T) Tr: 23/24/25 • TR: 60 • LVE(MVE) (F): (2.5 mg/m³); TWA (USA): 2.5 mg/m³ (F–); MAK (D): 2.5 mg/m³ (F–)
- (D) LD50 ip-r: 31

Hydrogen fluoride: HF • 20.01 • 7664-39-3
- (P) $\Delta H_f^0 = -273.3(g); -299.8(l)$ • Eb = 19.51 • P_{vap}: 1100(20); 1500(30); 2800(50)
- (I) NFPA: 0; 1
- (T) Tr: 26/27/28-35 • NFPA: 4 • TR: 886 • LVE(MVE) (F): 3; TWA (USA): 3; MAK (D): 3
- (D) LC50 r: 1.062/?; 4.133/5'; LC50 m: 0.315/?; 5.195/5'

Potassium fluoride: KF • 58.10 • 7789-23-3
- (P) $\Delta H_f^0 = -567.3(s)$ • $E_b = 1505$
- (I) NFPA: 0; 0
- (T) Tr: 23/24/25 • NFPA: 3 • TR: 60 • LVE(MVE) (F): (2.5 mg/m³); TWA (USA): 2.5 mg/m³ (F–); MAK (D): 2.5 mg/m³ (F–)
- (D) LD50 o-r: 245; LD50 ip-r: 64; LD50 ip-m: 40

Sodium fluoride: NaF • 41.99 • 7681-49-7
- (P) $\Delta H_f^0 = -576.6(s)$
- (I) NFPA: –; –
- (T) Tr: 25-32-36/38 • NFPA: 3 • LVE(MVE) (F): (2.5 mg/m³); TWA (USA): 2.5 mg/m³ (F–); MAK (D): 2.5 mg/m³ (F–)
- (D) LD50 o-r: 52; 64; 80;180; LD50 o-m: 57; LD50 o-ra: 200; LD50 ip-r: 22; LD50 ip-m: 38; LD50 iv-r: 26; LD50 iv-m: 51; LD50 scu-r: 175; LD50 scu-m: 0.0115; 70

Ammonium hexafluorosilicate : $(NH_4)_2SiF_6$ • 178.15 • 16919-19-8
- (P) $\Delta H_f^0 = -2681.7(s)$
- (T) Tr: 23/24/25 • TR: 60 • LVE(MVE) (F): (2.5 mg/m³); TWA (USA): 2.5 mg/m³ (F–); MAK (D): 2.5 mg/m³ (F–)
- (D) MLD o-r: 100; LD50 o-m: 70

Potassium hexafluorosilicate : K_2SiF_6 • 220.27 • 16871-90-2
- (P) $\Delta H_f^0 = -2956.0(s)$
- (T) Tr: 23/24/25 • TR: 60 • LVE(MVE) (F): (2.5 mg/m³); TWA (USA): 2.5 mg/m³ (F–); MAK (D): 2.5 mg/m³ (F–)
- (D) LD50 o-r: 156; LD50 o-m: 70

Sodium hexafluorosilicate : Na_2SiF_6 • 188.06 • 16893-85-9
- (P) $\Delta H_f^0 = -2909.6(s)$
- (T) Tr: 23/24/25 • TR: 60 • LVE(MVE) (F): (2.5 mg/m³); TWA (USA): 2.5 mg/m³ (F–); MAK (D): 2.5 mg/m³ (F–)
- (D) LD50 o-r: 125; LD50 o-m: 70; MLD o-ra: 125; LD50 o-ra: 125

Potassium tetrafluoroborate : KBF_4 • 125.90 • 13755-29-8
- (P) $\Delta H_f^0 = -1881.96(s)$
- (I) NFPA: 0; 0
- (D) LD50 ip-r: 240; LD50 ip-m: 590; LD50 ip-ra: 380

NEON

Neon: Ne • 20.18 • 7440-09-1
- (I) Eb = 246
- (T) Asphyxiant gas.

SODIUM

Sodium: Na • 22.99 • 7440-23-5
- (P) Eb = 883
- (I) NFPA: 1; 2 • Tr 14/15
- (T) Tr: 34 • NFPA: 3 • TR: X423
- (D) LD50 ?-m: 4000

Sodium hydrogencarbonate : $NaHCO_3$ • 84.01 • 144-55-8
- (P) $\Delta H_f^0 = -950.8(s)$
- (I) NFPA: –;–
- (D) LD50 o-r: 4220; LD50 o-m: 3360

Sodium carbonate : Na_2CO_3 • 105.99 • 497-19-8
- (P) $\Delta H_f^0 = -1130.77(s); -1431.26(1.H_2O); -4081.32(10.H_2O)$
- (I) NFPA: –;–
- (T) Tr: 36 • NFPA: 2
- (D) LC50 r: 2.1-2.5/?; LC50 m:1.2/?; LD50 o-r: 4090; LD50 o-m: 6600; LD50 ip-m: 117; LD50 scu-m: 2210

Sodium hydroxide : NaOH • 40.00 • 1310-73-2
- (P) $\Delta H_f^0 = -425.6(s)$
- (I) NFPA: 0; 1 • Tr: –
- (T) Tr: c>5%: 35; 2% < c < 5%: 34; 0.5% < c < 2%: 36/38 • NFPA: 3 • TR: 80 • LVE(MVE) (F): (2 mg/m³); TWA (USA): 2 mg/m³; MAK (D): 2 mg/m³ • IDLH: 200 mg/m³
- (D) LD50 ip-m: 40; MLD o-ra: 500

7 • Table of inorganic compounds

Magnesium/aluminium

Name: Empirical formula • Molar mass (g/mol) • Registry number (CAS).
(P) **Physical data:** ΔH_f^0 (kJ/mol) (physical state) • Eb (°C/mmHg) (if substance volatile) • P_{vap} (mBar) (°C) (if substance volatile).
(I) **Inflammability data:** LEL/UEL (%) • P_{elc} (°C) • AIT (°C) • NFPA (flame); NFPA (reactivity) • Labour Code (Tr).
(T) **Toxicity:** Labour Code (Tr) • NFPA • Transport Code (TR) • LVE (MVE) (ppm) F, USA, D • STEL (ppm); IDLH (ppm).
(D) **Toxic and lethal doses:** LC (mg/l/exposure time); LD (mg/kg).

Sodium peroxide: Na_2O_2 • 77.98 • 1313-60-6
(P) $\Delta H_f^0 = -510.9(s)$ • Eb = 460 (dec)
(I) NFPA: 0; 2 • Tr: 8
(T) Tr: 35 • NFPA: 3

Sodium sesquicarbonate: $Na_3H(CO_3)_2$ • 190.00 • 533-96-0

MAGNESIUM

Magnesium: Mg • 24.31 • 7439-95-4
(I) NFPA: 1; 2 • Tr : 15-17 (pyrophoric powder); 11-15 (shavings or stabilised powder)
(T) NFPA: 0 • TR: 423

Magnesium carbonate: $MgCO_3$ • 84.31 • 82597-01-1; 546-93-0
(P) $\Delta H_f^0 = -1095.8(s)$
(T) NFPA: 1 • TWA (USA): 10 mg/m^3
(D) LD50 iv-m: 16

Magnesium dibromide: $MgBr_2$ • 184.11 • 7789-48-2
(P) $\Delta H_f^0 = -524.3(s)$

Magnesium dichloride: $MgCl_2$ • 95.21 • 7786-30-3
(P) $\Delta H_f^0 = -641.3(s)$
(I) NFPA: –;–
(T) NFPA: 1
(D) anhydrous: LD50 o-r: 2800; LD50 ip-m: 775; 1338; LD50 iv-m: 14; hydrate: LD50 o-r: 8100; LD50 o-m: 7600

Magnesium difluoride MgF_2 • 62.30 • 7783-40-6
(P) $\Delta H_f^0 = -1124.2(s)$
(T) Tr: 36/37/38 • LVE(MVE) (F): (2.5 mg/m^3); TWA (USA): 2.5 mg/m^3 (F); MAK (D): 2.5 mg/m^3 (F)

Bismagnesium(dihydrogenphosphate): $MgH_2(PO_4)_2$ • 218.28 • 13092-66-5

Magnesium dihydroxide: $Mg(OH)_2$ 58.32 • 1309-42-8
(P) $\Delta H_f^0 = -924.5(s)$
(D) (Brucite) LD50 o-r: 8500; LD50 o-m: 8500; LD50 ip-r: 2780 ; LD50 ip-m: 815

Magnesium dinitrate: $Mg(NO_3)_2$ • 148.31 • 10377-60-3
(P) $\Delta H_f^0 = -790.7(s)$
(I) NFPA: 0; 1
(T) NFPA: 1
(D) LD50 o-r: 5440

Magnesium fluorosilicate: $MgSiF_6$ • 166.38 • 18972-56-0

Magnesium oxide: MgO • 40.30 • 1309-48-4
(P) $\Delta H_f^0 = -601.6(s)$
(I) NFPA: –;–
(T) NFPA: 2; 1 • LVE(MVE) (F): (10 mg/m^3); TWA (USA): 10 mg/m^3; MAK (D): 10 mg/m^3
(D) LD50 o-m: 810

Magnesium perchlorate: $MgCl_2O_8$ • 223.21 • 10034-81-8
(P) $\Delta H_f^0 = -568.9(s)$
(I) NFPA: 0; 1
(T) NFPA: 1 • LD50 ip-m: 1500

Magnesium sulphate: $MgSO_4$ • 120.36 • 7487-88-9
(P) $\Delta H_f^0 = -1284.9(s)$ • Eb = 150 (6H$_2$O); 200(7H$_2$O)
(D) Anhydrous: MLD o-m: 5000; MLD o-ra: 3000; LD50 scu-m: 645; 980; LD50 scu-r: 1200

Magnesium sulphite $MgSO_3$ • 104.86 • 7757-88-2
(P) $\Delta H_f^0 = -1008.34(s)$

ALUMINIUM

Aluminium: Al • 26.98 • 7429-90-5
(I) NFPA: 1; 1 • Tr : 15-17 (pyrophoric); 10-17 (stabilised).
(T) TR: 40 (coated); 423 (uncoated) • LVE(MVE) (F): (5 mg/m^3); TWA (USA): 5 mg/m^3

Sodium aluminate: $NaAlO_2$ • 81.97 • 11138-49-1
(P) $\Delta H_f^0 = -1135.12(s)$

Aluminium borate : $Al_2B_2O_6$ • 171.58 • 11121-16-7
(T) LVE(MVE) (F): (2 mg/m^3)

Aluminium borohydride: $Al(BH_4)_3$ • 71.51 • 16962-07-5
(P) $\Delta H_f^0 = 13.0(g); -16.3(l)$ • Eb = 44.5
(I) LEL/UEL = 5/90 • NFPA: 4; 2
(T) NFPA: 3

Aluminium chlorate : $Al(ClO_3)_3$ • 277.34 • 15477-33-5
(T) LVE(MVE) (F): (2 mg/m^3)

Aluminium chloride : $AlCl_3$ • 133.34 • 7446-70-0
(P) $\Delta H_f^0 = -705.63(s)$
(I) NFPA: 0; 2 (anhydrous)
(T) Tr: 34 • NFPA: 3 • TR: 80 • LVE(MVE) (F): (2 mg/m^3)
(D) Anhydrous: LD50 o-r: 3730; LD50 o-m: 770; 3805; Hydrate: LD50 o-r: 3310; LD50 o-m: 1990; LD50 ip-r: 728; LD50 ip-m: 940

Aluminium fluoride: AlF_3 • 83.98 • 7784-18-1
(P) $\Delta H_f^0 = -1510.3(s)$
(T) LVE(MVE) (F): (2.5 mg/m^3); TWA (USA): 2.5 mg/m^3 (F); MAK (D): 2.5 mg/m^3 (F)
(D) LD50 o-m: 103

Aluminium fluorosilicate : $Al_2(SiF_6)_3$ 480.19 • 17099-70-6
(D) MLD o-guinea pig: 5000; MLD scu-guinea pig: 4000

Aluminium nitrate: $Al(NO_3)_3$ • 213 • 13473-90-0
(I) NFPA: 2; –
(T) LVE(MVE) (F): (2 mg/m^3)
(D) Anhydrous: LD50 o-r: 4280; Hydrate: LD50 o-r: 264; 3671; LD50 o-m: 3980; LD50 ip-r: 901; LD50 ip-m: 320; 1587

7 • Table of inorganic compounds — Silicon/phosphorus

Name: Empirical formula • Molar mass (g/mol) • Registry number (CAS).
(P) Physical data: ΔH_f^0 (kJ/mol) (physical state) • Eb (°C/mmHg) (if substance volatile) • P_{vap} (mBar) (°C) (if substance volatile).
(I) Inflammability data: LEL/UEL (%) • P_{elc} (°C) • AIT (°C) • NFPA (flame); NFPA (reactivity) • Labour Code (Tr).
(T) Toxicity: Labour Code (Tr) • NFPA • Transport Code (TR) • LVE (MVE) (ppm) F, USA, D • STEL (ppm); IDLH (ppm).
(D) Toxic and lethal doses: LC (mg/l/exposure time); LD (mg/kg).

Aluminium nitride: AlN • 40.99 • 24304-00-5
(P) $\Delta H_f^0 = -318.0(s)$
(T) Tr: 37 • TWA (USA): 10 mg/m^3

Aluminium oxychloride: Al$_2$ClH$_5$O$_5$ • 174.45 • 1327-41-9
(P) $\Delta H_f^0 = -793.29$ (AlOCl, s)

Aluminium oxide (alumina): Al$_2$O$_3$ • 101.96 • 1344-28-1
(P) $\Delta H_f^0 = -1675.7(s)$
(T) LVE(MVE) (F): (10 mg/m^3); TWA (USA): 10 mg/m^3; MAK (D): 6 mg/m^3 (vapour)

Aluminium phosphate : AlPO$_4$ • 121.95 • 7784-30-7
(P) $\Delta H_f^0 = -1733.8(s)$
(T) TWA (USA): 2 mg/m^3

Aluminium phosphide: AlP • 57.96 • 20859-73-8
(P) $\Delta H_f^0 = -166.5(s)$
(I) Tr : 15
(T) Tr: 15/29-28-32 • MLC mammal: 0.002/?

Aluminium silicate: Al$_2$SiO$_5$ • 162.05 • 12141-46-7
(P) $\Delta H_f^0 = -2596.17$ (s)

Magnesium aluminium silicate : Al$_2$MgSi$_2$O$_8$ • 262.43 • 12511-31-8

Aluminium sulphate : Al$_2$(SO$_4$)$_3$ • 342.14 • 10043-01-3
(P) $\Delta H_f^0 = -3435.06(s)$
(T) TWA (USA): 2 mg/m^3
(D) Anhydrous: LD50 o-r: 1930; LD50 o-m: 6207; LD50 ip-m: 270; 1735; Hydrate: LD50 ip-r: 61 ; LD50 ip-m: 997

Aluminium ammonium sulphate: AlNH$_4$(SO$_4$)$_2$ • 237.14 • 7784-25-0
(T) LVE(MVE) (F): 2 mg/m^3

Aluminium potassium sulphate AlK(SO$_4$)$_2$ • 258.20 • 10043-67-1

Aluminium sodium sulphate: AlNa(SO$_4$)$_2$ • 242.09 • 10102-71-3

Ammonium tetrachloroaluminate: NH$_4$AlCl$_4$ • 186.83 • 7784-14-7

Aluminium trioxide: Al(OH)$_3$ • 78.00 • 21645-51-2
(P) $\Delta H_f^0 = -1284.49(s)$
(D) MLD ip-m: 150

SILICON

Silicon: Si • 28.09 • 440-21-3
(T) Tr: 36/37/38 • TR: 40 • LVE(MVE) (F): (10 mg/m^3); TWA (USA): 10 mg/m^3 (dust)
(D) LD50 o-r: 3160

Silicic acid: H$_4$SiO$_4$ • 96.11 • 10193-36-9
(P) $\Delta H_f^0 = -1481.1(s)$
(T) NFPA: 1

Asbestos: – • – • 1332-21-4
(T) Tr: 45-48/23 • NFPA: 2 • Amosite: VLE(F): 1 fibre/cm^3; TWA (USA): 0.5 fibre/cm^3; Crocidolite: VLE(F): 0.5 fibre/cm^3; TWA (USA): 0.2 fibre/cm^3; Other fibres: VLE(F): 1 fibre/cm^3; TWA (USA): 2 fibres/cm^3

Silicion carbide: SiC • 40.10 • 409-21-2
(P) $\Delta H_f^0 = \alpha$: 71.55(s); ß: 73.22(s)

Silicon nitride: Si$_3$N$_4$ • 140.28 • 12033-89-5
(P) $\Delta H_f^0 = 743.50(s)$

Silane: SiH$_4$ • 32.12 • 7803-62-5
(P) $\Delta H_f^0 = +34.3(g)$ • Eb = –112
(I) AIT = burns spontaneously in air • NFPA: 3 • Tr : 12
(T) NFPA: 3 • LVE(MVE) (F): (5)
(D) LC50 r: 12.825/4h; MLC m : 12.825/4h

Potassium silicate : various formulae; 1312-76-1

Sodium silicate : various formulae; 6834-92-0;
(T) Tr: 38-41 • NFPA: 1
(D) LD50 o-r: 1153; LD50 o-m: 770

Silica: SiO$_2$ • 60.08 • 14808–60–7
(P) $\Delta H_f^0 = -910.94$ (s, quartz)
(T) NFPA: 2 • Amorphous: 10 mg/m^3 • Quartz: 0.1 mg/m^3 (3.106 f/m^3) • Tridymite: 50 µg/m^3
(D) LD50 o-r: 3160; LD50 iv-r: 15

PHOSPHORUS

Phosphorus: P • 30.97; 7723-14-0 (P) • 12185-10-3(P$_4$)
(P) Eb = 280 • P_{vap}: 0.033(20); 0.076(30); 0.32(50) • AIT = 30; 30-45 (white) ; 300 (red)
(I) NFPA: 3; 1 (white); 1; 1 (red) • Tr : 17 (white); 11-16 (red)
(T) (white) Tr: 26/28-35 • NFPA: 3 • TR:446 (molten); 46 (solid) • LVE(MVE) (F): 0.3 mg/m^3 ; (0.1 mg/m^3); TWA (USA): 0.1 mg/m^3 ; MAK (D): 0.1 mg/m^3
(D) LD50 o-r: 12
(T) (red) NFPA: 0 • TR: 40

Metaphosphoric acid :(HPO$_3$)$_n$ • n.79.98 • 10343-62-1
(P) $\Delta H_f^0 = -948.5(s)$

Orthophosphoric acid: H$_3$PO$_4$ • 98.00 • 7664-38-2
(P) $\Delta H_f^0 = -1284.4(s); -1271.7(l);$ Eb = 213 (H$_2$O)
(I) NFPA: 0; 0 • Tr : –
(T) Tr: c > 25%: 34; 10% < c < 25: 36/38 • NFPA: 2 • TR: 80 • LVE(MVE) (F): 3 mg/m^3 ; (1mg/m^3); TWA (USA): 1 mg/m^3
(D) LD50 o-r: 1530; LD50 cu-ra: 2740

Potassium dihydrogenphosphate : KH$_2$PO$_4$ • 136.09 • 7778-77-0
(P) $\Delta H_f^0 = -1568.3(s)$

Sodium dihydrogenphosphate : NaH$_2$NaPO$_4$ • 119.98 • 7558-80-7
(P) $\Delta H_f^0 = -1536.78(s)$

7 • Table of inorganic compounds

Sulphur

Name: Empirical formula • Molar mass (g/mol) • Registry number (CAS).
(P) **Physical data:** ΔH_f^0 (kJ/mol) (physical state) • Eb (°C/mmHg) (if substance volatile) • P_{vap} (mBar) (°C) (if substance volatile).
(I) **Inflammability data:** LEL/UEL (%) • P_{elc} (°C) • AIT (°C) • NFPA (flame); NFPA (reactivity) • Labour Code (Tr).
(T) **Toxicity:** Labour Code (Tr) • NFPA • Transport Code (TR) • LVE (MVE) (ppm) F, USA, D • STEL (ppm); IDLH (ppm).
(D) **Toxic and lethal doses:** LC (mg/l/exposure time); LD (mg/kg).

(D) LD50 o-r: 8290

Sodium hexametaphosphate(Calgon): $Na_6P_6O_{18}$ • 611.77 • 10124-56-8

(D) LD50 o-m: 7250; MLD o-ra: 140; LD50 ip-m: 870; LD50 iv-m: 62; LD50 scu-m: 1300

Dipotassium hydrogenphosphate K_2HPO_4 • 174.18 • 7758-11-4

Disodium hydrogenphosphate : Na_2HO_4P • 141.96 • 7558-79-4

(P) $\Delta H_f^0 = -1748.1(s)$ • Eb = 95 (H_2O)

(D) LD50 o-r: 12930; 17000; MLD ip-r: 1000; MLD scu-r: 1000

Sodium ammonium hydrogenphosphate $NaNH_4PO_4$ • 137.01 • 13011-54-6

Phosphorus pentachloride: PCl_5 • 208.24 • 10026-13-8

(P) $\Delta H_f^0 = -443.5(s)$ • Eb = 160; 166.8 (dec) • P_{vap} = 0.11(30); 0.8(50)

(I) NFPA: 0; 2 Tr : –

(T) Tr: 34-37 • NFPA: 3 • TR: 80 • LVE(MVE) (F): (0.1); TWA (USA): 0.1; MAK(D): 0.1

(D) LC50 r: 0.205/?; LD50 o-r: 660

Diphosphorus pentasulphide : P_2S_5 (P_4S_{10}) • 222.25 • 1314-80-3

(P) Eb = 514 • AIT = 142

(I) NFPA: 1; 2

(T) NFPA: 3 • TWA (USA): 1 mg/m³

(D) LD50 o-r: 389

Diphosphorus pentoxide: P_2O_5 • 141.94 • 1314-56-3

(P) $\Delta H_f^0 = -2984.03(s, P_4O_{10}); -365.83(s)$

(I) NFPA: 0; 2

(T) Tr: 35 • NFPA: 3 • TR: 80 • LVE(MVE) (F): (1 mg/m³); MAK (D): 1 mg/m³

(D) LC50 r: 1.217/?; LC50 m: 0.271/?; LC50 ra: 1.69/?

Ammonium phosphate: $NH_4H_2PO_4$ • 115.03 • 7722-76-1

(P) $\Delta H_f^0 = -1445.07(s)$

(D) MLC r: 2.773/1h

Diammonium phosphate $(NH_4)_2HPO_4$ • 132.06 • 7783-28-0

(P) $\Delta H_f^0 = -1566.91(s)$ • Eb = 155 (dec)

Sodium phosphate : Na_3PO_4 • 163.94 • 7601-54-9

(P) $\Delta H_f^0 = -1917.40$

(D) 12 H_2O: LD50 o-r: 7400; LD50 ip-m: 430; MLD iv-ra: 1580 ; LD50 cu-ra: > 7940

Phosphine: PH_3 • 34.00 • 7803-51-2

(P) $\Delta H_f^0 = +5.4(g); +22.8(g)$ (contradictory information) • Eb = 87.7 • P_{vap} = 36000(20); 43000(30); 64000(50)

(I) AIT = 100 • NFPA: 4

(T) NFPA: 3 • TR: unauthorised • LVE(MVE) (F): 0.3 (0.1); TWA (USA): 0.3; MAK (D): 0.1; STEL: 1

(D) LC50 r: 0.015/4h ; MLC m : 0.380/?; MLC ra 3.53/20'

Ammonium phosphite: $(NH_4)_2HPO_3$ • 116.06 • 51503-61-8

Sodium polymetaphosphate : $(NaPO_3)_n$ • n. 101.96 • 50813-16-6

Potassium pyrophosphate: $K_4P_2O_7$ • 330.34 • 7320-34-5

Phosphorus trichloride: PCl_3 • 137.33 • 7719-12-2

(P) $\Delta H_f^0 = -287.0(g); -319.7(l)$ • Eb = 75.5-76 • P_{vap} = 127(20); 194(30); 415(50)

(I) NFPA: 0; 2

(T) Tr: 34-37 • NFPA: 3 • TR: 80 • LVE(MVE) (F): (0.2); TWA (USA): 0.2; MAK (D): 0.5; STEL: 0.5

(D) LC50 r: 0.59 /4h; LD50 o-r: 550; 18

Phosphoryl trichloride: $POCl_3$ • 153.33 • 10025-87-3

(P) $\Delta H_f^0 = -558.5(g); -597.1(l)$ • Eb = 105.3-105.8; 107 • P_{vap} = 36(20); 60(30); 150(50)

(I) NFPA: –;–

(T) Tr: 34-37 • NFPA: 3 • TR: 80 • LVE(MVE) (F): (0.1); TWA (USA): 0.1; MAK (D): 0.2

(D) LC50 r: 0.2/?; 0.302/?; LD50 o-r: 380

Sodium trimetaphosphate: $Na_3P_3O_9$ • 305.89 • 7785-84-4

(D) LD50 ip-r: 3650; MLD iv-ra: 240

Sodium tripolyphosphate: $Na_5P_3O_{10}$ • 367.86 • 7758-29-4

(P) $\Delta H_f^0 = -4409.94(s)$

(D) LD50 o-r: 3900; 6500; 4100; 5400; LD50 o-m: 3100; 3210; LD50 ip-r: 525; LD50 ip-m: 700; LD50 iv-m: 71; 74; LD50 scu-r: 2090; LD50 scu-m: 900

SULPHUR

Sulphur: S • 32.06 • 7704-34-9

(P) Eb = 444.7

(I) AIT = 232 • NFPA: 1;–

(T) NFPA: 2

(D) MLD o-ra: 10; MLD iv-r: 8; MLD iv-ra: 5

Chlorosulphonic acid: $HOS(Cl)O_2$ • 116.52 • 7790-94-5

(P) $\Delta H_f^0 = -601.7(l)$ • Eb = 151-152/755 • P_{vap} = 1.33(25); 1.33(32); 4.4(37.7)

(I) LEL/UEL = 3.3/37.7 (This substance is not inflammable and this data comes from the Aldrich data bank) • NFPA: 0 ; 2 • Tr : 14

(T) Tr: 35-37 • NFPA: 3 • TR: 88

7 • Table of inorganic compounds

Sulphur

Name: Empirical formula • Molar mass (g/mol) • Registry number (CAS).
(P) Physical data: ΔH_f^0 (kJ/mol) (physical state) • Eb (°C/mmHg) (if substance volatile) • P_{vap} (mBar) (°C) (if substance volatile).
(I) Inflammability data: LEL/UEL (%) • P_{elc} (°C) • AIT (°C) • NFPA (flame); NFPA (reactivity) • Labour Code (Tr).
(T) Toxicity: Labour Code (Tr) • NFPA • Transport Code (TR) • LVE (MVE) (ppm) F, USA, D • STEL (ppm); IDLH (ppm).
(D) Toxic and lethal doses: LC (mg/l/exposure time); LD (mg/kg).

Sulphamic or amidosulphuric acid: H_3NSO_3 • 97.09 • 5329-14-6
- (I) NFPA: –; 3
- (T) Tr: 36/38 • NFPA: 2
- (D) MLD o-r: 3160; LD50 o-m: 1312

Sulphurous acid: H_2SO_3 • 82.03 • 7782-99-2
- (P) $\Delta H_f^0 = -608.81$(aq, 1 mol/l)

Sulphuric acid: H_2SO_4 • 98.07 • 7664-93-9
- (P) $\Delta H_f^0 = -740.57$(g); –814.0(l)
- (I) NFPA: 0 (inflammability); 1; 2
- (T) Tr: > 15%: 35; 5%15%: 36/38 • NFPA: 3 • TR: 80 • LVE(MVE) (F): 3 mg/m³; (1mg/m³); TWA (USA): 1 mg/m³; MAK (D): 1 mg/m³ • STEL: 3 mg/m³
- (D) LC50 r: 0.51/2h; LC50 m: 0.32/2h; LD50 o-r: 2140

Sulphuryl chloride: SO_2Cl_2 • 134.96 • 7791-25-5
- (P) $\Delta H_f^0 = -364.0$(g); –394.1(l) • Eb = 69.1 • $P_{vap} = 147(20)$; 227(30); 500(50)
- (I) NFPA: –;– • Tr : 14
- (T) Tr: 34-37 • NFPA: 3 • TR: X88
- (D) LC50 r: 0.735/?; 1.358/?

Thionyl chloride: $SOCl_2$ • 118.97 • 7719-09-7
- (P) $\Delta H_f^0 = -212.55$(g); –245.6(l) • Eb = 76; 78.8/746 • $P_{vap} = 124(20)$; 133(21.4); 188(30); 435(50)
- (I) NFPA: –;– • Tr : 14
- (T) Tr: 34-37 • NFPA: 3 • TR: X88 • TWA (USA): 1
- (D) LC50 r: 2.473/1h

Disulphur dichloride: S_2Cl_2 • 135.03 • 10025-67-9
- (P) $\Delta H_f^0 = -19.5$(g); –59.4(l); –51.88(l) • Eb = 135.6; 138 • $P_{vap} = 9.06$; 9.2(20); 16(30); 44(50)
- (I) AIT = 230 • NFPA: 1; 1 • Tr : 14
- (T) Tr: 34-37 • NFPA: 2 • TR: 88 • TWA (USA): 1; MAK (D): 1
- (D) LC50 m: 0.642/?; LC50 r: 1.4/1h ; 2.985/1h

Sulphur dichloride: SCl_2 • 102.97 • 10545-99-0
- (P) $\Delta H_f^0 = -19.66$(g); –50.0(l); –50.21(l) • Eb = 59 • $P_{vap} = 227(20)$; 240(22); 845(55)
- (I) NFPA: 2;– • Tr : 14
- (T) Tr: 34-37 • NFPA: 3 • TR: X88

Sulphur dioxide: SO_2 • 64.06 • 7446-09-5
- (P) $\Delta H_f^0 = -296.83$(g); –320.39(l) • Eb = 10 • $P_{vap} = 2400(20)$; 4600(30); 8400(50)
- (I) NFPA: 0; 0 • Tr : –
- (T) Tr: 23-36/37 • NFPA: 3 • TR: 26 • LVE(MVE) (F): 5 (2); TWA (USA): 2; MAK (D): 2; STEL: 5
- (D) MLC r: 2,66/?; MLC: 15.966/?; LC50 r: 6.706/1h; LC50 m: 7.983/30'

Carbon disulphide: CS_2 • 76.13 • 75-15-0
- (P) $\Delta H_f^0 = +116.6$(g); +89.0(l) • Eb = 46.3-46.5 • $P_{vap} = 400(20)$; 560(30); 1200(50)
- (I) LEL/UEL = 0.6/60; 1.25/50; 1.3/50; $P_{fla} = 30$; AIT = 102; 125; NFPA: 3; 0 • Tr : 11
- (T) Tr: 36/38-48/23-62-63; 26 • NFPA: 2 • TR: 336 • LVE(MVE) (F): 25 (10); TWA (USA): 10; MAK (D): 10 • IDLH: 500
- (D) LC50 r: 0.079/?; LC50 m: 0.031/?; LD50 o-r: 3188; LD50 o-m: 2780

Sulphuryl fluoride: F_2SO_2 • 102.06 • 2699-79-8
- (P) $\Delta H_f^0 = -758.56$(g) • Eb = –54.4
- (I) NFPA: –;–
- (T) NFPA: 3 • TWA (USA): 5
- (D) LC50 r: 12.815/?; 4.205/4h; MLC m : 52.473/?; 5.092/1h; MLC ra: 21.218/1h; LD50 o-r: 100

Sulphur hexafluorine : SF_6 • 146.05 • 2551-62-4
- (P) $\Delta H_f^0 = -1220.5$(g) • Eb = –63.8 (subl)
- (I) NFPA: –; –
- (T) NFPA: – • TWA (USA): 1000
- (D) LD50 iv-ra: 5790

Potassium hydrogensulphate: $KHSO_4$ • 136.16 • 7646-93-7
- (P) $\Delta H_f^0 = -1160.6$(s)
- (I) Tr: –
- (T) Tr: 34-37 • TR: 80
- (D) LD50 o-r: 2340

Sodium hydrogensulphite : $NaHSO_3$ • 104.06 • 7631-90-5
- (I) NFPA: –;– • Tr : –
- (T) Tr: 22 • NFPA: – • TR: 80 • TWA (USA): 5 mg/m³
- (D) LD50 o-r: 2000; LD50 ip-r: 475; 650; LD50 ip-m: 675; 750; LD50 ip-ra: 300; LD50 iv-r: 115; LD50 iv-m: 130; LD50 iv-ra: 65

Ammonium hydrogensulphide : NH_4HS 51.11 • 12124-99-1
- (P) $\Delta H_f^0 = -156.9$(s) • Eb = 88.4 (under 19 atm)
- (D) LD50 o-r: 168; MLD o-m: 80; MLD iv-m: 2; LD50 scu-m: 132; MLD scu-ra: 8; MLD cu-m: 2457

Sodium hydrogensulphide NaHS • 56.06 • 16721-80-5
- (P) $\Delta H_f^0 = -237.23$(s)
- (D) LD50 ip-r: 14.5; 30; LD50 scu-m: 200; LD50 cu-m: 18

Sodium hydrosulphite : $Na_2S_2O_4$ • 174.10 • 7775-14-6
- (P) $\Delta H_f^0 = -1232.19$(s)
- (T) Tr: 22-31 • TWA (USA): 10 mg/m³ (hydrate)

Potassium metabisulphite : $K_2S_2O_5$ • 222.31 • 16731-55-8

Diammonium peroxodisulphate $(NH_4)_2S_2O_8$ • 228.19 • 7727-54-0
- (P) $\Delta H_f^0 = -1648.08$(s) • Eb = 120 (dec)
- (I) NFPA: 2;–
- (T) NFPA: 1 • TWA (USA): 2 mg/m³
- (D) LD50 o-r: 689; 750; 820; LD50 ip-r: 226; LD50 iv-ra: 178

367

7 • Table of inorganic compounds — Chlorine

Name: Empirical formula • Molar mass (g/mol) • Registry number (CAS).
(P) Physical data: ΔH_f^0 (kJ/mol) (physical state) • Eb (°C/mmHg) (if substance volatile) • P_{vap} (mBar) (°C) (if substance volatile).
(I) Inflammability data: LEL/UEL (%) • $\bullet_{Pelc}$ (°C) • AIT (°C) • NFPA (flame); NFPA (reactivity) • Labour Code (Tr).
(T) Toxicity: Labour Code (Tr) • NFPA • Transport Code (TR) • LVE (MVE) (ppm) F, USA, D • STEL (ppm); IDLH (ppm).
(D) Toxic and lethal doses: LC (mg/l/exposure time); LD (mg/kg).

Potassium peroxodisulphate : $K_2(SO_4)_2$ • 270.31 • 7727-21-1
(P) $\Delta H_f^0 = -1916.1(s)$
(I) NFPA: 0; 1
(T) NFPA: 1 • TWA (USA): 2 mg/m^3
(D) LD50 o-r: 802

Sodium peroxodisulphate : $Na_2S_2O_8$; 238.09; 7775-27-1
(P) $\Delta H_f^0 = -1825.06$ (aq)
(D) LD50 ip-m: 226; MLD iv-ra: 178

Ammonium sulphamate $NH_4NH_2SO_3$ • 114.12 • 7773-06-0
(P) Eb = 160 (dec)
(I) NFPA: 1;–
(T) NFPA: 1 • TWA (USA): 10 mg/m^3
(D) LD50 o-r: 1600; 2000; LD50 o-m: 3100; MLD ip-r: 800

Diammonium sulphate : $(NH_4)_2SO_4$ • 132.13 • 7783-20-2
(P) $\Delta H_f^0 = -1180.85(s)$
(D) LD50 o-r: 2840; 3000; LD50 o-m: 640; LD50 ip-m: 610

Potassium sulphate: K_2SO_4 • 174.25 • 7778-80-5
(P) $\Delta H_f^0 = -1437.8(s)$ • Eb = 1689
(D) LD50 o-r: 6600; LD50 scu-ra: 3000

Sodium sulphate: Na_2SO_4 • 142.04 • 7757-82-6
(P) $\Delta H_f^0 = -1387.1$
(D) LD50 o-m: 5989; MLD iv-m: 1220; LD50 iv-ra: 1220

Diammonium sulphite : $(NH_4)_2SO_3$ • 116.14 • 10196-04-0
(P) $\Delta H_f^0 = -885.33(s)$ • Eb = 60-70 (hydrate: →H_2O); 150 (subl)

Potassium sulphite: K_2SO_3 • 158.25 • 10117-38-1
(P) $\Delta H_f^0 = -1125.5(s)$

Sodium sulphite : Na_2SO_3 • 126.04 • 7757-83-7
(P) $\Delta H_f^0 = -1100.8(s)$
(I) NFPA: –;–
(T) NFPA: 2
(D) LD50 o-m: 820; MLD o-ra: 2825; LD50 ip-m: 950; LD50 iv-r: 115; LD50 iv-m: 130; 175; MLD iv-m: 130; MLD iv-ra: 65; MLD scu-ra: 300

Hydrogen sulphide: H_2S • 34.08 • 7783-06-4
(P) $\Delta H_f^0 = -20.16(g); -20.6(g)$ • Eb = -60.33; -60.7 • P_{vap} = 18300(20); 23100(30); 36500(50)
(I) LEL/UEL = 4/46; 4.3/45.5 • AIT = 260; 270 • NFPA: 4; 0 • Tr : 12; 13
(T) Tr: 26 • NFPA: 3 • TR: 236 • LVE(MVE) (F): 10 (5) (at 2 ppm); TWA (USA): 10; MAK (D): 10 • STEL: 15; IDLH: 300
(D) LC50 r: 0.627/?; LC50 m: 0.896/1h; 0.951/1h

Diammonium sulphide: $(NH_4)_2S$ • 68.14 • 12135-76-1
(P) $\Delta H_f^0 = -231.79$ (aq, 1 mole/l)
(I) NFPA: 2; –
(T) NFPA: – • VLE: id H_2S
(D) MLD : o-r: 80; LD50 ip-m: 190; LD50 iv-m: 25; LD50 scu-m: 132; MLD cu-m: 2460; MLD cu-ra: 119

Sodium sulphide: Na_2S • 78.04 • 1313-82-2
(P) $\Delta H_f^0 = -364.8(s)$
(I) NFPA: 1; 0
(T) Tr: 31-34 • NFPA: 2; 3 • TR: 40 (anhydrous or < 30% H_2O)
(D) LD50 o-r: 208; LD50 o-m: 205; LD50 ip-r: 147

Sodium thiocarbonate : Na_2CS_3 • 154.17 • 534-18-9
(P) Eb = 75 (dec)

Ammonium thiocyanate: NH_4CNS • 76.12 • 1762-95-4
(P) Eb = 170 (dec)
(I) NFPA: –;–
(T) NFPA: 1
(D) LD50 o-r: 760; LD50 o-m: 500; MLD o-m: 330; MLD ip-m: 500

Ammonium thiosulphate : $(NH_4)_2S_2O_3$ • 148.20 • 7783-18-8
(P) $\Delta H_f^0 = -917.13$(aq) • Eb = 150 (dec)
(D) LD50 o-r: 2890

Sodium thiosulphate : $Na_2S_2O_3$ • 158.10 • 7772-98-7
(P) $\Delta H_f^0 = -1122.99(s); -2607.93(s,5H_2O)$
(I) NFPA: –; –
(T) NFPA: 1
(D) Anhydride: MLD scu-ra: 400

Sulphur trioxide: SO_3 • 80.06 • 7446-11-9
(P) $\Delta H_f^0 = -395.7(g); -441.0(l); -454.7(s)$ • Eb = 44.8 (α ; ß) • P_{vap} = 373(25); 680(38)
(I) Tr : 14
(T) Tr: 34, 35-37 • NFPA: 3 • TR: X88
(D) LC50 r: 1.154/?; LD50 o-r: 2140

CHLORINE

Chlorine: Cl_2 • 70.91 • 7782-50-5
(P) Eb = -34.05; -34.6 • P_{vap} = 6900(20)
(I) NFPA: 0; 1 • Tr : 5-8
(T) Tr: 23-36/37/38 • NFPA: 3 • TR: 266 • LVE(MVE) (F): 1 (–); TWA (USA): 0.5; MAK (D): 0.5 • STEL: 1; IDLH: 30
(D) LC50 r: 0.865/1h; LC50 m: 0.304/1h

7 • Table of inorganic compounds

Argon/potassium

Name: Empirical formula • Molar mass (g/mol) • Registry number (CAS).
(P) Physical data: ΔH_f^0 (kJ/mol) (physical state) • Eb (°C/mmHg) (if substance volatile) • P_{vap} (mBar) (°C) (if substance volatile).
(I) Inflammability data: LEL/UEL (%) • P_{fla}(°C) • AIT (°C) • NFPA (flame); NFPA (reactivity) • Labour Code (Tr).
(T) Toxicity: Labour Code (Tr) • NFPA • Transport Code (TR) • LVE (MVE) (ppm) F, USA, D • STEL (ppm); IDLH (ppm).
(D) Toxic and lethal doses: LC (mg/l/exposure time); LD (mg/kg).

Hydrochloric acid : HCl • 36.46 • 7647-01-0
- (P) $\Delta H_f^0 = -92.3$(g); -167.16 (aq, 1 mol/l) • Eb = 108.08 (azeotrope at 20.22% HCl)
- (I) NFPA: 0; 0 • Tr : –
- (T) Tr: c > 25%: 34-37; 10 c 25%: 36/37/38; 22-35; 36/38 • NFPA: 3 • TR: 80 • LVE(MVE) (F): 5 (–); TWA (USA): 5; MAK (D): 5 • IDLH: 100

Perchloric acid: HClO$_4$ • 10046 • 7601-90-3
- (P) $\Delta H_f^0 = -40.6$(l)
- (I) NFPA: 0; 3 • Tr : 5; 8 (c > 50%)
- (T) Tr: > 50% : 35; 10% < c < 50 %: 34 • NFPA: 3 • TR: 85 (c < 50%); 558 (c > 50%)
- (D) LD50 o-r: 1100; LD50 scu-m: 250

Potassium chlorate : KClO$_3$ • 122.55 • 3811-04-9
- (P) $\Delta H_f^0 = -397.7$(s)
- (I) NFPA: 0; 1 • Tr : 9
- (T) Tr: 20/22 • NFPA: 1 • TR: 50
- (D) LD50 o-r: 1870; MLD o-ra: 2000

Sodium chlorate: NaClO$_3$ • 106.44 • 7775-09-9
- (P) $\Delta H_f^0 = -365.8$(s)
- (I) NFPA: 0; 2 • Tr : 9
- (T) Tr: 20/22; 22 (UE) • NFPA: 1 • TR: 50
- (D) LD50 o-r: 1200; MLD o-ra: 8000

Sodium chlorite: NaClO$_2$ • 90.44 • 7758-19-2
- (P) $\Delta H_f^0 = -307.0$(s)
- (I) NFPA: 1(inflammability); 2; 1 (reactivity)
- (T) NFPA: 1
- (D) LD50 o-r: 165; 350; LD50 o-m: 350

Ammonium chloride: NH$_4$Cl • 53.49 • 12125-02-9
- (P) $\Delta H_f^0 = -314.4$(s)
- (I) Tr : –
- (T) Tr: 22-36 • LVE(MVE) (F): –(10mg/m^3); TWA (USA): 10 mg/m^3 • STEL: 20 mg/m^3
- (D) LD50 o-r: 1650; LD50 o-m: 1300; MLD o-ra: 1000; LD50 ip-m: 1439; LD50 iv-m: 358

Hydrogen chloride : HCl • 36.46 • 7647-01-0
- (P) $\Delta H_f^0 = -92.3$(g) • Eb = –85.05 • P_{vap} = 42600(20); 53200(30); 80600(50)
- (I) NFPA: 0; 0 • Tr : –
- (T) Tr: 35-37 • NFPA: 3 • TR: 286 • LVE(MVE) (F): 5 (–); TWA (USA): 5; MAK (D): 5 • IDLH: 100
- (D) LC50 r: 4.741/1h; 8.6/30'; 7.134/30'; 48.262/5'; LC50 m: 1.681/1h; 3.25/30'; 4.012/30'; 17.056/5'; LD o-ra: 900; LD50 ip-m: 1449

Potassium chloride: KCl • 74.55 • 7447-40-7
- (P) $\Delta H_f^0 = -436.5$(s)
- (D) LD50 o-r: 2600; 3020; LD50 o-m: 383; LD50 ip-r: 660; LD50 ip-m: 1181; LD50 iv-r: 39; 142; LD50 iv-m: 117

Sodium chloride: NaCl • 58.44 • 7647-14-5
- (P) $\Delta H_f^0 = -411.2$;
- (I) NFPA: –; –
- (T) NFPA: 0
- (D) LD50 o-r: 3000; LD50 o-m: 4000; MLD o-ra: 8000; LD50 ip-m: 2602; 6614; LD50 iv-m: 645; LD50 scu-m: 3000; 3150

Chlorine dioxide: ClO$_2$ • 67.45 • 10049-04-4
- (P) $\Delta H_f^0 = +102.5$(g) • Eb = 9.9; 11 (explosion) • P_{vap} = 1400(20)
- (I) NFPA: 3; – • Tr : –
- (T) NFPA: 3 • LVE(MVE) (F): 0.3 (0.1); TWA (USA): 0.1; MAK (D): 0.1 • STEL: 0.3; IDLH: 10/1h
- (D) MLC r: 1.808/15'

Sodium hypochlorite: NaClO • 74.44 • 7681-52-9
- (P) $\Delta H_f^0 = -347.27$(aq)
- (I) NFPA: –; –
- (T) Tr: c > 10% Cl$_2$: 31-34; c 10% Cl$_2$: 31-36/38 • NFPA: 2
- (D) LD50 o-r: 8910; LD50 o-m: 5800; LD50 cu-ra: > 10500

Ammonium perchlorate: NH$_4$ClO$_4$ • 117.49 • 7790-98-9
- (P) $\Delta H_f^0 = -295.3$(s)
- (I) NFPA: 0; 4 • Tr : 9 – 44
- (T) NFPA: 1; 2
- (D) LD50 o-r: 3500; 4200; LD50 o-m: 1900; LD50 ?-ra: 1900

Potassium perchlorate KClO$_4$ • 138.55 • 7778-74-7
- (P) $\Delta H_f^0 = -432.8$(s)
- (I) NFPA: 0; 2 • Tr : 9
- (T) Tr: 22 • NFPA: 1 • TR: 50

Sodium perchlorate : NaClO$_4$ • 122.44 • 7601-89-0
- (P) $\Delta H_f^0 = -383.3$(s)
- (I) NFPA: 0; 2 • Tr : 9
- (T) Tr: 22 • NFPA: 2 • TR: 50
- (D) LD50 o-r: 2100; LD50 ip-m: 551

ARGON

Argon: Ar • 39.95 • 7440-37-1
- (P) Eb = 185.86
- (I) NFPA: –; –

POTASSIUM

Potassium: K • 39.10 • 7440-09-7
- (I) NFPA: 1; 2 • Tr : 14/15
- (T) Tr: 34 • NFPA: 3

Potassium carbonate: K$_2$CO$_3$ • 105.99 • 584-08-7
- (P) $\Delta H_f^0 = -1130.7$(s)
- (T) NFPA: 3
- (D) LD50 o-r: 1870

Name: Empirical formula • Molar mass (g/mol) • Registry number (CAS).
(P) Physical data: ΔH_f^0 (kJ/mol) (physical state) • Eb (°C/mmHg) (if substance volatile) • P_{vap} (mBar) (°C) (if substance volatile).
(I) Inflammability data: LEL/UEL (%) • P_{elc} (°C) • AIT (°C) • NFPA (flame); NFPA (reactivity) • Labour Code (Tr).
(T) Toxicity: Labour Code (Tr) • NFPA • Transport Code (TR) • LVE (MVE) (ppm) F, USA, D • STEL (ppm); IDLH (ppm).
(D) Toxic and lethal doses: LC (mg/l/exposure time); LD (mg/kg).

Potassium hydrogencarbonate : $KHCO_3$ • 84.01 • 298-14-6
(P) $\Delta H_f^0 = -963.2(s)$
(T) NFPA: 0

Potassium hydroxide : KOH • 56.11 • 1310-58-3
(P) $\Delta H_f^0 = -424.8(s)$
(I) NFPA: 0; 1 • Tr : –
(T) Tr: Anh.: 35; c > 5%: 34; 1% < c < 5%: 36/38 • NFPA: 3 • TR: 80 • LVE(MVE) (F): (2 mg/m^3); TWA (USA): 2 mg/m^3
(D) LD50 o-r: 273; 365

CALCIUM

Calcium: Ca • 40.08 • 7440-70-2
(I) NFPA: 1; 2 • Tr : 15
(T) NFPA: 1 • TR: 423 • LVE(MVE) (F): - (2 mg/m^3); TWA (USA): 2 mg/m^3

Calcium bis(dihydrogenphosphate) $Ca(H_2PO_4)_2$ • 234.05 • 7758-23-8
(D) LD50 o-r: 17500; LD50 o-m: 15250

Calcium borate : CaB_4O_7 • 195.32 • 12007-56-6
(P) $\Delta H_f^0 = -3360.25(s)$
(D) LD50 o-r: 5600; LD50 o-m: 5900; LD50 ip-m: 3900

Calcium bromide : $CaBr_2$ • 199.90 • 7789-41-5
(P) $\Delta H_f^0 = -683.25(s); -663.0(l); -2506,22(s, 6H_2O)$
(D) LD50 ip-r: 437; LD50 ip-m: 740; LD50 scu-m: 1580

Calcium carbonate : $CaCO_3$ • 100.09 • 471-34-1; 1317-65-3
(P) $\Delta H_f^0 = -1207.13(s)$
(T) NFPA: 1 • TWA (USA): 10 mg/m^3
(D) LD50 o-r: 6450

Calcium carbide: CaC_2 • 64.10 • 75-20-7
(P) $\Delta H_f^0 = -59.83(s)$
(I) NFPA: 4; 2 • Tr : 15
(T) NFPA: 1 • TR: 423

Calcium chlorate $Ca(ClO_3)_2$ • 206.98 • 10137-74-3
(I) NFPA: 2; –
(D) MLD o-r: 4500; MLD ip-r: 625

Calcium chloride: $CaCl_2$ • 110.98 • 10043-52-4
(P) $\Delta H_f^0 = -795.4(s);$
(I) NFPA: –; –
(T) Tr: 36
(D) LD50 o-r: 1000; >2500; LD50 o-m: 1940; MLD o-ra: 1380; LD50 ip-r: 264; LD50 ip-m: 210; 245; LD50 iv-r: 161; LD50 iv-m: 42; LD50 scu-r: 2630; LD50 scu-m: 823

Calcium chromate $CaCrO_4$ • 156.08 • 13765-19-0
(P) $\Delta H_f^0 = -329.6(s)$
(I) Tr : –
(T) Tr: 45-22 • LVE(MVE) (F): (0.05 mg/m^3); TWA (USA): 0.001 mg/m^3
(D) LD50 o-r: 327

Calcium cyanamide : $CaCN_2$ • 80.11 • 156-62-7
(P) $\Delta H_f^0 = -83.8(s)$ • Eb = 1150-1200 (subl)
(T) Tr: 22-37-41 • TR: 423 • LVE(MVE) (F): (0.5 mg/m^3); TWA (USA): 0.5 mg/m^3; MAK (D): 1 mg/m^3
(D) MLC r: 0.086/4h; LC50 r: > 0.150/4h; LD50 o-r: 158; LD50 o-m: 334; LD50 o-ra: 1400; LD50 ip-m: 100; LD50 iv-r: 125; LD50 iv-m: 282; LD50 cu-r: 84; LD50 cu-ra: 590

Calcium cyanide: $Ca(CN)_2$ • 92.12 • 592-01-8
(P) $\Delta H_f^0 = -44.1(s)$
(I) NFPA: 0; 0 • Tr : –
(T) Tr: 28-32 • NFPA: 2 • TR: prohibited • LVE(MVE) (F): (5 mg/m^3); TWA (USA): 5 mg/m^3 (CN); MAK (D): 5 mg/m^3 (CN)
(D) LD50 o-r: 39

Calcium fluoride : CaF_2 • 78.08 • 7789-75-5
(P) $\Delta H_f^0 = -1228.0(s)$
(T) TWA (USA): 2.5 mg/m^3 (F)
(D) LD50 o-r: 4250; LD50 ip-m: 2638; LD50 cu-m: 2775

Calcium hexafluorosilicate : $Ca SiF_6$ • 182.17 • 16925-39-6
(T) LVE(MVE) (F): (2.5 mg/m^3); TWA (USA): 2.5 mg/m^3 (F); MAK (D): 2.5 mg/m^3 (F)

Calcium hydrogenphosphate: $CaHPO_4$ • 136.06 • 7757-93-9
(P) $\Delta H_f^0 = -3104.70$
(D) LD50 o-r: > 4640

Calcium hydroxide $Ca(OH)_2$ • 74.10 • 1305-62-0
(P) $\Delta H_f^0 = -985.2(s)$
(I) NFPA: –; –
(T) NFPA: 2 • TWA (USA): 5 mg/m^3
(D) LC50 ?: 0.005/?; LD50 o-r: 7340; LD50 o-m: 7300; 7340

Calcium hypochlorite : $CaCl_2O_2$ • 142.98 • 7778-54-3
(P) $\Delta H_f^0 = -754.38(s)$
(I) NFPA: 1; 2 • Tr : 8
(T) Tr: 31-34 • NFPA: 1
(D) LD50 o-r: 850

Calcium hypophosphite : $CaH_4P_2O_4$ • 170.06 • 7789-79-9
(P) $\Delta H_f^0 = -1752,68(s)$

Calcium nitrate $Ca(NO_3)_2$ • 164.10 • 10124-37-5
(P) $\Delta H_f^0 = -938.39(s); -2132.33(s; 4H_2O)$
(I) NFPA: 0; 0
(T) NFPA: 1
(D) LD50 o-r: 3900 (4H$_2$O); LD50 cu-ra: 500

7 • Table of inorganic compounds

Scandium/titanium/vanadium

Name: Empirical formula • Molar mass (g/mol) • Registry number (CAS).
- (P) **Physical data**: ΔH_f^0 (kJ/mol) (physical state) • Eb (°C/mmHg) (if substance volatile) • P_{vap} (mBar) (°C) (if substance volatile).
- (I) **Inflammability data**: LEL/UEL (%) • P_{elc} (°C) • AIT (°C) • NFPA (flame); NFPA (reactivity) • Labour Code (Tr).
- (T) **Toxicity**: Labour Code (Tr) • NFPA • Transport Code (TR) • LVE (MVE) (ppm) F, USA, D • STEL (ppm); IDLH (ppm).
- (D) **Toxic and lethal doses**: LC (mg/l/exposure time); LD (mg/kg).

Calcium oxide: CaO • 56.08 • 1305-78-8
- (P) $\Delta H_f^0 = -634.9(s)$
- (I) NFPA: 0; 1
- (T) NFPA: 1 • TWA (USA): 5 mg/m^3

Calcium permanganate: CaMn$_2$O$_8$ • 277.96 • 10118-76-0
- (T) TWA (USA): 5 mg/m^3
- (D) LD50 iv-ra: 50

Calcium phosphate : Ca$_3$(PO$_4$)$_2$ • 310.18 • 12167-74-7
- (P) $\Delta H_f^0 = -4120.82(s)$

Calcium phosphide: Ca$_3$P$_2$ • 182.18 • 1305-99-3
- (P) $\Delta H_f^0 = -506.26(s)$
- (I) Tr : 15/29
- (T) Tr: 15/29-28

Calcium pyrophosphate: Ca$_2$P$_2$O$_7$ • 254.10 • 7790-76-3
- (P) $\Delta H_f^0 = -3338.83(s)$

Calcium sulphate: CaSO$_4$ • 136.14 • 7778-18-9
- (P) $\Delta H_f^0 = -1434.11(s)$
- (T) TWA (USA): 10 mg/m^3

Calcium sulphide: CaS • 72.14 • 20548-54-3
- (P) $\Delta H_f^0 = -482.4(s)$
- (T) Tr: 31-36/37/38

Calcium thiocyanate: Ca(CNS)$_2$ • 156.24 • 2092-16-2
- (D) MLD o-m.: 120; MLD iv-ra: 250

SCANDIUM

Scandium: Sc • 44.96 • 7440-20-2

TITANIUM

Titanium: Ti • 47.90 • 7440-32-6
- (I) AIT = 250; 329 (powder); 1200 (compact) • NFPA: –; –
- (T) NFPA: 0 • TWA (USA): 10 mg/m^3

Titanium carbide : TiC • 59.91 • 12070-08-5
- (P) $\Delta H_f^0 = -184.1(s)$

Titanium diboride: TiB$_2$ • 69.50 • 12045-63-5
- (P) $\Delta H_f^0 = -280.33(s)$
- (T) TWA (USA): 10 mg/m^3

Titanium dihydride: TiH$_2$ • 49.90 • 7704-98-5
- (P) $\Delta H_f^0 = -144.35(s)$

Titanium dioxide: TiO$_2$ • 79.90 • 13463-67-7
- (P) $\Delta H_f^0 = -944.0(s)$
- (I) NFPA: –; –
- (T) NFPA: 1 • LVE(MVE) (F): (10 mg/m^3); TWA (USA): 10 mg/m^3; MAK (D): 10 mg/m^3
- (D) LC50 r: 6.82/4h; LD50 o-r: > 5000; > 20000; 24000; LD50 cu-ra: >10000

Titanium nitride: TiN • 61.89 • 25583-20-4
- (P) $\Delta H_f^0 = -338.07(s)$

Titanium sulphate: Ti(SO$_4$)$_2$ • 240.0 • 13825-74-6

Titanium tetrachloride: TiCl$_4$ • 189.70 • 7550-45-0
- (P) $\Delta H_f^0 = -763.2(g); -804.2(l)$ • Eb = 136.4 • P_{vap} = 12.7; 13(20); 21(30); 56(50)
- (I) NFPA: 0; 1
- (T) Tr: 34-36/37 • NFPA: 3 • TR: 80
- (D) LC50 o-r: 0.300/?; 0.360/4h; LC50 o-m: 0.100/2h

Titanium trichloride: TiCl$_3$ • 154.25 • 7705-07-9
- (P) $\Delta H_f^0 = -720.9(s)$

VANADIUM

Vanadium: V • 50.94 • 7440-62-2
- (D) LD50 scu-ra: 59

Vanadium carbide : VC • 62.95

Vanadyl chloride: VOCl$_3$ • 173.29 • 7727-18-6
- (P) $\Delta H_f^0 = -695.6(g); -734.7(l)$
- (I) NFPA: 2; –
- (T) TWA (USA): 0.05 mg/m^3
- (D) LD50 o-r: 140

Divanadium pentoxide : V$_2$O$_5$ • 181.88 • 1314-62-1
- (P) $\Delta H_f^0 = -1550.6(s)$
- (I) NFPA: 2; – • Tr : –
- (T) Tr: 20 • NFPA: – • TR: 60 • LVE(MVE) (F): (0.05 mg/m^3); TWA (USA): 0.05 mg/m^3; MAK (D): 0.05 mg/m^3
- (D) MLC r: 0.070/2h; >2/1h; LD50 o-r: 10; > 50; LD50 o-m: 23; LD50 ip-r: 12; LD75 ip-r: 4.5; MLD iv-ra: 10; LD50 scu-r: 14; LD50 scu-m: 10; MLD scu-ra: 20

Vanadyl sulphate: VOSO$_4$ • 163.00 • 27774-13-6
- (P) $\Delta H_f^0 = -1309.17(s)$
- (T) TWA (USA): 0.05 mg/m^3
- (D) LD50 o-r: 448 (5H$_2$O); LD50 o-m: 467 (5H$_2$O); LD50 ip-r: 74 (5H$_2$O); LD50 ip-m: 113 (5H$_2$O); LD50 ip-m: 144; LD50 ip-ra: 16; MLD iv-ra: 16; MLD scu-r: 140; LD50 scu-m: 560; LD50 cu-ra: 4450

7 • Table of inorganic compounds — Chromium

Name: Empirical formula • Molar mass (g/mol) • Registry number (CAS).
(P) Physical data: ΔH_f^0 (kJ/mol) (physical state) • Eb (°C/mmHg) (if substance volatile) • P_{vap} (mBar) (°C) (if substance volatile).
(I) Inflammability data: LEL/UEL (%) • P_{elc} (°C) • AIT (°C) • NFPA (flame); NFPA (reactivity) • Labour Code (Tr).
(T) Toxicity: Labour Code (Tr) • NFPA • Transport Code (TR) • LVE (MVE) (ppm) F, USA, D • STEL (ppm); IDLH (ppm).
(D) Toxic and lethal doses: LC (mg/l/exposure time); LD (mg/kg).

Divanadium trioxide: V_2O_3 • 149.88 • 1314-34-7
- (P) $\Delta H_f^0 = -1218.8(s)$
- (I) AIT = spontaneous inflammation
- (T) TWA (USA): 0.05 mg/m³
- (D) LD50 o-m: 382; LD50 o-m: 130; LD50 scu-m: 130

Sodium vanadate: $NaVO_3$ • 121.93 • 13718-26-8
- (P) $\Delta H_f^0 = -1145.8(s)$
- (I) NFPA: 0; 0 (by analogy with the ammonium salt)
- (D) MLD o-r: 98; 200; LD50 o-m: 74.6; MLD o-ra: 200; LD50 ip-m: 36; LD50 ip-r: 12; MLD iv-ra: 17

CHROMIUM

Chromium: Cr • 52.00 • 7440-47-3
- (I) AIT = 580 (suspension); 400 (layer) • NFPA: 2; –
- (T) NFPA: 0 • TWA (USA): 0.5 mg/m³ (Cr)
- (D) LD50 ?-r: 27.5

Ammonium chromate: $(NH_4)_2CrO_4$ • 152.1 • 7788-98-9
- (P) $\Delta H_f^0 = -1167.34(s)$
- (I) Tr : 8
- (T) TWA (USA): 0.05 mg/m³

Potassium chromate: K_2CrO_4 • 194.20 • 7789-00-6
- (P) $\Delta H_f^0 = -1403.73(s)$
- (T) TWA (USA): 0.05 mg/m³
- (D) LD50 o-r: 180; LD50 ip-m: 32; MLD scu-ra: 12

Sodium chromate: Na_2CrO_4 • 161.98 • 7775-11-3
- (P) $\Delta H_f^0 = -1342.23(s)$
- (I) NFPA: –; –
- (T) Tr: 36/37/38-43 • NFPA: 3 • TWA (USA): 0.05 mg/m³
- (D) LD50 o-r: 52; LD50 ip-r: 57; LD50 ip-m: 32; MLD iv-ra: 32; MLD scu-ra: 243; LD50 cu-ra: 1690

Chromyl dichloride: $CrCl_2O_2$ • 154.90 • 14977-61-8
- (P) $\Delta H_f^0 = -538.06(g); -579.48(l)$ • Eb = 115.7; 117 • P_{vap} = 20; 26.7(20); 35(30); 90(50)
- (I) NFPA: –;– • Tr : 8
- (T) Tr: 35 • NFPA: 3 • TR: 88 • LVE(MVE) (F): (0.05 mg/m³); TWA (USA): 0.025
- (D) LD50 scu-m: 5.5

Ammonium dichromate: $(NH_4)_2Cr_2O_7$ • 252.10 • 7789-09-5
- (P) $\Delta H_f^0 = -1806.65(s)$; dec = 170
- (I) NFPA: 2; 1(inflammability); 1 (reactivity) • Tr : 1-8
- (T) Tr: 36/37/38-43 • NFPA: 3; 1 • TR: 50 • LVE(MVE) (F): (0.05 mg/m³); TWA (USA): 0.05 mg/m³
- (D) LD50 o-r: 67.5; LD50 iv-r: 30

Potassaium dichromate: $K_2Cr_2O_7$ • 294.20 • 7778-50-9
- (P) $\Delta H_f^0 = -2061.46(s)$
- (I) NFPA: 0; 1 • Tr : 8
- (T) Tr: 36/37/38-43 • NFPA: 3; 1 • TWA (USA): 0.05 mg/m³
- (D) LC50 r: 0.094/?; LD50 o-r: 57; LD50 o-m: 190; LD50 ip-m: 37; LD50 scu-m: 100; MLD scu-ra: 10; LD50 cu-ra: 1640

Sodium dichromate: $Na_2Cr_2O_7$ • 261.98 • 10588-01-9
- (P) $\Delta H_f^0 = -1978.61(s)$
- (I) NFPA: –;– • Tr : 8
- (T) Tr: 36/37/38-43 • NFPA: 3 • TWA (USA): 0.05 mg/m³
- (D) LC50 r: 0.124/?; LD50 o-r: 50; 105; LD50 cu-ra: 1000; MLD iv-m: 26; MLD iv-ra: 26

Chromium(III) oxide: Cr_2O_3 • 152.00 • 1308-38-9
- (P) $\Delta H_f^0 = -1139.7(s)$
- (I) Tr : 8
- (T) Tr: 35-40 • TWA (USA): 0.5 mg/m³ (Cr)
- (D) LD50 o-r: > 5000; > 10000

Chromium(VI) oxide: CrO_3; • 100.00 • 1333-82-0
- (P) Eb = 250 (dec: Cr_2O_3)
- (I) NFPA: 0; 1 • Tr : 8
- (T) Tr: 49-25-35-43 • NFPA: 3 • TR: 58 (anhydrous); 80 (solution) • LVE(MVE) (F): 0.1 (0.5); TWA (USA): 0.5
- (D) LC50 r: 0.217/?; LD50 o-r: 52; 80; LD50 o-m: 127; LD50 ip-r: 58.4; LD50 ip-m: 14; 29; LD50 iv-r: 9.26; LD50 iv-m: 17.1; MLD scu-m: 20; LD50 cu-ra: 57

Chromium(III) sulphate: $Cr_2(SO_4)_3$ • 392.16 • 15005-90-0
- (P) $\Delta H_f^0 = -609.61$(s, anhydrous)
- (T) TWA (USA): 0.5 mg/m³ MLD iv-m: 247; MLD iv-r: 144 (7H_2O)

Chromium potassium sulphate: $CrK(SO_4)_2$ • 283.21 • 7788-99-0
- (D) LD50 iv-r: 112 (12H_2O)

Chromium trichloride: $CrCl_3$ • 158.35 • 10025-73-7
- (P) $\Delta H_f^0 = -556.5(s)$
- (T) Tr: 22-36/37/38 • TWA (USA): 0.5 mg/m³
- (D) LC50 m: 0.0315/2h; LD50 o- r: 1870; 1790 (hydrate) ;LD50 ip-m: 140; 434; LD50 ip-m: 520 (hydrate); MLD iv-m: 801 (hydrate); MLD iv-m: 400; MLD iv-ra: 288

Chromium trinitrate: $Cr(NO_3)_3$ • 238.03 • 13548-38-4
- (T) TWA (USA): 0.5 mg/m³
- (D) LD50 o-r: 1790; LD50 o-r; 3250 (9H_2O); LD50 o-m: 2976; LD50 ip-m: 110; LD50 scu-m: 3232

7 • Table of inorganic compounds

Manganese/iron

Name: Empirical formula • Molar mass (g/mol) • Registry number (CAS).
(P) Physical data: ΔH_f^0 (kJ/mol) (physical state) • Eb (°C/mmHg) (if substance volatile) • P_{vap} (mBar) (°C) (if substance volatile).
(I) Inflammability data: LEL/UEL (%) • P_{elc} (°C) • AIT (°C) • NFPA (flame); NFPA (reactivity) • Labour Code (Tr).
(T) Toxicity: Labour Code (Tr) • NFPA • Transport Code (TR) • LVE (MVE) (ppm) F, USA, D • STEL (ppm); IDLH (ppm).
(D) Toxic and lethal doses: LC (mg/l/exposure time); LD (mg/kg).

MANGANESE

Manganese: Mn • 54.94 • 7439-96-5
(I) NFPA: 2; – Tr :
(T) Tr: 48 (vapour) • NFPA: 2 • LVE(MVE) (F): 3mg/m³ (vapour) (5; 1 mg/m³) (layered; vapour); TWA (USA): idem; MAK (D): 0.5 mg/m³ (expressed by reference to Mn)
(P) LD50 o-r: 9000

Manganese dichloride : $MnCl_2$ • 125.84 • 7773-01-5
(P) $\Delta H_f^0 = -481.3(s)$
(T) TWA (USA): 5 mg/m³
(D) Anhydrous: LD50 o-r: 250; LD50 o-m: 1031; 1720; LD50 ip-r: 147; LD50 ip-m: 121; LD50 iv-r: 92.6; LD50 iv-m: 171; MLD iv-ra: 65; LD50 scu-m: 180; MLD scu-m: 210; MLD scu-ra: 180; Hydrate: LD50 o-r: 1480; 1484; LD50 ip-r: 138; 1484; LD50 ip-m: 144; 190; LD50 iv-m: 710

Manganese dioxide : MnO_2 • 86.94 • 1313-13-9
(P) $\Delta H_f^0 = -520.0(s)$
(T) Tr: 20/22 • TWA (USA): 5 mg/m³ (Mn)
(D) MLD iv-ra: 45; LD50 scu-m: 422

Manganese (II) oxide: MnO • 70.94 • 1344-43-0
(P) $\Delta H_f^0 = -385.2(s)$ • Eb = $(+O_2 \to Mn_3O_4)$
(T) TWA (USA): 5 mg/m³ (Mn)
(D) LD50 scu-m: 1000

Manganese(III) oxide: Mn_2O_3 • 157.88 • 1317-34-6
(P) $\Delta H_f^0 = -959.0(s)$ • Eb = 1080 (dec $\to O_2$)
(T) TWA (USA): 5 mg/m³ (Mn)
(D) LD50 scu-m: 616

Potassium permanganate : $KMnO_4$ • 158.04 • 7722-64-7
(P) $\Delta H_f^0 = -837.2(s)$
(I) NFPA: 0;(1; 0(reactivity) • Tr : 8
(T) Tr: 22 • NFPA: 1 • TR: 50 • TWA (USA): 5 mg/m³ (Mn)
(D) LD50 o-r: 1090; LD50 o-m: 2157; MLD o-ra: 600; MLD iv-ra: 70; LD50 scu-m: 500

Manganese sulphate : $MnSO_4$ • 151.00 • 7785-87-7; 10101-68-5 ($4H_2O$)
(P) $\Delta H_f^0 = -1065.25(s)$
(I) NFPA: –;–
(T) NFPA: 2 • TWA (USA): 5 mg/m³ (Mn)
(D) LD50 o-r: 50 (anhydrous); LD50 ip-m: 332 (anhydrous); LD50 ip-m: 120 (dihydrate); LD50 ip-m: 534 (tetrahydrate)

Trimanganese tetroxide: Mn_3O_4 • 228.82 • 1317-35-7
(P) $\Delta H_f^0 = -1387.8(s)$
(T) TWA (USA): 1 mg/m³ (Mn)

IRON

Iron: Fe • 55.85 • 7439-89-6
(D) LD50 o-r: > 5000; 30000; MLD ip-ra: 20

Ferrous(II) chloride: $FeCl_2$ • 126.75 • 7758-94-3
(P) $\Delta H_f^0 = -341.8(s)$
(T) Tr: 22-36/38 • NFPA: 1 • TWA (USA): 1 mg/m³ (Fe)
(D) Anhydrous: LD50 o-r: 450; LD50 ip-m: 59; Hydrate: LD50 o-r: 984; MLD o-ra: 890; LD50 ip-m: 93; MLD scu-ra: 189

Ferric chloride : $FeCl_3$ • 162.20 • 7705-08-0
(P) $\Delta H_f^0 = -399.5(s)$
(T) Tr: 34 • NFPA: 1 • TR: 80 • TWA (USA): 1 mg/m³ (Fe)
(D) Anhydrous: LD50 o-r: 900; 1160; 1872; LD50 o-m: 895; 1280; LD50 ip-m: 68; 142; LD50 iv-m: 58; Hexahydrate: MLD o-r: 900; LD50 ip-m: 260; LD50 iv-m: 49; MTD iv-r: 2580; MLD iv-ra: 7200

Ferric(II) hexacyanoferrate: $Fe_7C_8N_{18}$ • 859.25 • 14038-43-8
(T) TWA (USA): 5 mg/m³ (CN)
(D) LD50 ip-r: 2100; LD50 ip-m: 2000

Potassium(III) hexacyanoferrate : $FeK_3C_6N_6$ • 329.27 • 13746-66-2
(T) NFPA: 1 • TWA (USA): 5 mg/m³ (CN)
(D) LD50 o-r: 1600; MLD o-r: 1600; LD50 o-m: 2970

Ferric nitrate: $Fe(NO_3)_3$ • 241.86 • 7782-61-8
(T) (non-hydrate) TWA (USA): 1 mg/m³ (Fe)
(D) LD50 o-r: 3250

Iron(II) oxide: FeO • 71.85 • 1345-25-1
(P) $\Delta H_f^0 = -272.0(s)$
(T) MAK (D): 6 mg/m³ (dust)

Ferric oxide : Fe_2O_3 • 159.70 • 1309-37-1
(P) $\Delta H_f^0 = -824.2(s)$
(T) LVE(MVE) (F): (5 mg/m³) (Fe); TWA (USA): 5 mg/m³ (vapour) (Fe); MAK (D): 6 mg/m³ (dust) (Fe)
(D) LD50 ip-r: 5500; LD50 ip-m: 5400

Iron (II) ammonium sulphate: $Fe(NH_4)_2(SO_4)_2$ • 284.04 • 7783-85-9
(T) TWA (USA): 1 mg/m³ (Fe)
(D) LD50 o-r: 3250

Ferrous sulphate: $FeSO_4$ • 151.91; 7720-78-7 • 7782-63-0 ($7H_2O$)
(P) $\Delta H_f^0 = -928.4(s)$
(T) TWA (USA): 1 mg/m³ (Fe)
(D) Anhydrous: LD50 o-r: 319; LD50 o-m: 680; 979; LD50 ip-m: 64; 289; LD50 iv-m: 112; 115; LD50 scu-r: 155; LD50 scu-m: 60.3; Hydrate: MLD o-r: 1390; LD50 o-m: 1520; LD50 ip-m: 245; LD50 iv-m: 51; 65

7 • Table of inorganic compounds — Cobalt/nickel/copper

Name: Empirical formula • Molar mass (g/mol) • Registry number (CAS).
(P) Physical data: ΔH_f^0 (kJ/mol) (physical state) • Eb (°C/mmHg) (if substance volatile) • P_{vap} (mBar) (°C) (if substance volatile).
(I) Inflammability data: LEL/UEL (%) • P_{elc} (°C) • AIT (°C) • NFPA (flame); NFPA (reactivity) • Labour Code (Tr).
(T) Toxicity: Labour Code (Tr) • NFPA • Transport Code (TR) • LVE (MVE) (ppm) F, USA, D • STEL (ppm); IDLH (ppm).
(D) Toxic and lethal doses: LC (mg/l/exposure time); LD (mg/kg).

Ferric sulphate: $Fe_2(SO_4)_3$ • 399.88 • 10028-22-5
- (P) $\Delta H_f^0 = -2581.53(s)$
- (T) Hydrate: LD50 ip-m: 601

Tricobalt tetroxide: Co_3O_4 • 240.80 • 1308-06-1
- (P) $\Delta H_f^0 = -891.0(s)$
- (T) TWA (USA): 0.05 mg/m³ (Co)
- (D) LD50 o-r: > 5000

COBALT

Cobalt: Co • 58.93 • 7440-48-4
- (P) Eb = 2870; ~3100
- (I) NFPA: –;– • Tr : –
- (T) Tr: 42/43 • NFPA: 2 • TWA (USA): 0.05 mg/m³ (Co); TWA in draft: 0.02 mg/m³
- (D) LC50 r: > 0.010/? ; LD50 o-r: >5000; 6171; MLD o-r: 1500; MLD o-ra: 750; LD50 ip-r: 100; MLD ip-r: 250; MLD ip-m: 100; MLD iv-r: 100; MLD iv-ra: 100

Cobalt bromide: $CoBr_2$ • 218.75 • 7789-43-7
- (P) $\Delta H_f^0 = -220.9(s)$
- (D) LD50 o-r: 406

Cobalt carbonate: $CoCO_3$ • 118.94 • 513-79-1
- (P) $\Delta H_f^0 = -712.95(s)$
- (D) LD50 o-r: 640

Cobalt chloride: $CoCl_2$ • 129.83 • 7646-79-9
- (P) $\Delta H_f^0 = -312.5(s)$
- (T) TWA (USA): 0.1 mg/m³ (Co)
- (D) Anhydrous: LD50 o-r: 80; LD50 o-m: 80; MLD o-ra: 1272; LD50 ip-r: 17.4; LD50 ip-m: 49; LD50 iv-r: 20; LD50 iv-m: 23.3; MLD scu-m: 100; Hydrate: LD50 o-r: 766; LD50 ip-r: 35; LD50 ip-m: 90; MLD iv-ra: 25.4; MLD scu-r: 121; MLD scu-m: 100; MLD scu-ra: 200

Cobalt dihydroxide $Co(OH)_2$ • 92.95 • 21041-93-0
- (P) $\Delta H_f^0 = -539.7(s)$

Potassium(III) hexanitrocobaltate: $K_3N_6CoO_{12}$ • 452.28 • 13782-01-9

Cobalt nitrate: $Co(NO_3)_2$ • 182.95 • 10141-05-6
- (P) $\Delta H_f^0 = -420.5(s)$
- (I) NFPA: 0; 1
- (T) NFPA: 1
- (D) Anhydrous: LD50 o-r: 434; MLD o-ra: 80; 250; MLD scu-ra: 75; Hydrate: LD50 o-r: 691

Cobalt oxide : CoO • 74.93 • 1307-96-6
- (P) $\Delta H_f^0 = -237.9(s)$
- (T) Tr: 22-43
- (D) LD50 o-r: 202; LD50 scu-m: 125

Cobalt sulphate: $CoSO_4$ • 154.99 • 10124-43-3
- (P) $\Delta H_f^0 = -888.3(s)$
- (D) Anhydrous: LD50 o-r: 424; MLD o-ra: 1800; LD50 ip-m: 143; Hydrate: LD50 o-r: 582; 768

NICKEL

Nickel: Ni • 58.71 • 7440-02-0
- (I) NFPA: 2
- (T) Tr: 40-43 • TWA (USA): Soluble: 0.1 mg/m³ (Ni), in draft 0.05 mg/m³; others 1 mg/m³ (Ni)
- (D) LC50 r: > 0.01/?; MLD o-r: 5000; LD50 ip-r: 250; MLD ip-ra: 7; MLD iv-m: 50; MLD scu-r: 12.5; MLD scu-ra: 7.5

Nickel carbonate : $NiCO_3$ • 118.72 • 3333-67-3
- (T) Tr: 22-40-43 • TWA (USA): 1 mg/m³ (Ni)

Nickel chloride : $NiCl_2$ • 129.61 • 7718-54-9
- (P) $\Delta H_f^0 = -305.31(s); -2103.55(s, 6H_2O)$
- (I) NFPA: 0; 0
- (T) Tr: 45-25-43 • NFPA: 1
- (D) Anhydrous: LD50 o-r: 105; LD50 o-m: 369; LD50 ip-r: 20.597; LD50 ip-m: 11; 26; LD50 iv-r: 68.1; LD50 iv-m: 20; Hydrate: LD50 o-r: 175; LD50 ip-m: 48

Nickel cyanide: $Ni(CN)_2$ • 110.75 • 557-19-7
- (P) $\Delta H_f^0 = +127.61(s)$
- (T) TWA (USA): 0.1 mg/m³ (Ni)

Nickel nitrate: $Ni(NO_3)_2$ • 182.73 • 13138-45-9
- (P) $\Delta H_f^0 = -415.05(s)$
- (I) NFPA: 0; 1
- (T) NFPA: 1 • TWA (USA): 0.1 mg/m³ (Ni)
- (D) Anhydrous: MLD iv-m: 9 • Hydrate: LD50 o-r: 1620

Nickel oxide : NiO • 74.71 • 1313-99-1
- (P) $\Delta H_f^0 = -239.74(s)$
- (T) Tr: 49-43 • TWA (USA): 1 mg/m³ (Ni)
- (D) MLD scu-r: 25; MLD scu-m: 50; MLD scu-ra: 9

Nickel sulphate: $NiSO_4$ • 154.77 • 7786-81-4
- (P) $\Delta H_f^0 = -872.91(s)$
- (I) NFPA: 0
- (T) Tr: 22-40-42/43 • NFPA: 1 • LVE(MVE) (F): 0.1 mg/m³ (Ni)
- (D) Anhydrous: LD50 ip-r: 500; LD50 ip-m: 20; 55; MLD iv-m: 7.64; 8; MLD iv-ra: 33; MLD scu-ra: 33; Hydrate: LD50 o-r: 264; 275; MLD scu-ra: 500; MLD iv-ra: 36

COPPER

Copper: Cu • 63.54 • 7440-50-8
- (I) NFPA: –; –
- (T) NFPA: 2 • TWA (USA): 0.2 mg/m³ (Cu) (vapour); 1 mg/m³ (Cu) (dust)
- (D) LD50 ip-m: 3.5

7 • Table of inorganic compounds — Zinc

- **P** **Physical data:** ΔH_f^0 (kJ/mol) (physical state) • Eb (°C/mmHg) (if substance volatile) • P_{vap} (mBar) (°C) (if substance volatile).
- **I** **Inflammability data:** LEL/UEL (%) • P_{elc} (°C) • AIT (°C) • NFPA (flame); NFPA (reactivity) • Labour Code (Tr).
- **T** **Toxicity:** Labour Code (Tr) • NFPA • Transport Code (TR) • LVE (MVE) (ppm) F, USA, D • STEL (ppm); IDLH (ppm).
- **D** **Toxic and lethal doses:** LC (mg/l/exposure time); LD (mg/kg).

Cupric bromide: $CuBr_2$ • 223.35 • 7789-45-9
- **P** $\Delta H_f^0 = -141.8(s)$
- **T** TWA (USA): 1 mg/m^3

Basic copper carbonate: $Cu_2(OH)_2CO_3$ • 221.11 • 12069-69-1
- **P** $\Delta H_f^0 = -1051.44(s)$
- **D** LD50 o-r: 1350; LD50 o-ra: 159

Cupric chloride: $CuCl_2$ • 134.44 • 1344-67-8
- **P** $\Delta H_f^0 = -220.1$
- **D** LD50 o-r: 140; 584; LD50 o-m: 190; 233; LD50 ip-r: 14.7; LD50 ip-m: 7.4; LD50 iv-r: 5; LD50 iv-m: 17.5

Cupric nitrate: $Cu(NO_3)_2$ • 187.55 • 3251-23-8
- **P** $\Delta H_f^0 = -302.92(s)$
- **I** NFPA: 0; 1
- **T** NFPA: 1
- **D** Hydrate: LD50 o-r: 794; 940; LD50 o-m: 430; LD50 ip-r: 20; LD50 ip-m: 17.1; LD50 iv-r: 14.7; LD50 iv-m: 23.3

Cupric oxide: CuO • 79.54 • 1317-38-0
- **P** $\Delta H_f^0 = -157.3(s)$

Cupric perchlorate: $CuCl_2O_8$ • 262.45 • 10294-46-9 • 17031-32-2 (2H$_2$O)
- **D** Hydrate: LD50 o-r: 29

Copper sulphate: $CuSO_4$ • 159.60 • 7758-98-7
- **P** $\Delta H_f^0 = -771.4$
- **T** Tr: 22-36/38 • NFPA: 1 • TWA (USA): 1 mg/m^3 (Cu)
- **D** Anhydrous: LD50 o-r: 300; LD50 o-m: 369; LD50 ip-r: 7; 20; LD50 ip-m: 7; 18; LD50 iv-r: 48.9; LD50 iv-m: 23.3; 50 ; LD50 iv-ra: 10; MLD iv-ra: 4; LD50 scu-r: 43; LD50 scu-m: 500; Hydrate: LD50 o-r: 300; 960; LD50 ip-r: 18.7; LD50 ip-m: 33

Cupric sulphide: CuS • 95.60 • 1317-40-4
- **P** $\Delta H_f^0 = -53.1(s)$

Cuprous chloride: CuCl • 98.99 • 7758-89-6
- **P** $\Delta H_f^0 = -137.2(s)$
- **I** NFPA: -;-
- **T** Tr: 22 • NFPA: 3 • TWA (USA): 1 mg/m^3 (Cu)
- **D** LC50 m: 1.008/?; LD50 o-r: 140; 265; LD50 o-m: 347

Cuprous cyanide: CuCN • 89.56 • 544-92-3
- **P** $\Delta H_f^0 = +96.2(s)$
- **I** NFPA: -;-
- **T** Tr: 26/27/28-32 • NFPA: 3 • TR: 60 • LVE(MVE) (F): –(5 mg/m^3); TWA (USA): 5 mg/m^3 (CN); MAK (D): 5 mg/m^3
- **D** LD50 o-r: 500

Cuprous oxide: Cu_2O • 143.08 • 1317-39-1
- **P** $\Delta H_f^0 = -168.6(s)$ • Eb = 1800 ($\to O_2$)
- **T** Tr: 22
- **D** LD50 o-r: 470

ZINC

Zinc: Zn • 65.37 • 7440-66-6
- **I** NFPA: 1; 1 • Tr: 10-15(stabilised); 15-17(pyrophoric)
- **T** NFPA: 0 • TWA (USA): 1 mg/m^3 (Zn)
- **D** LD50 ip-m: 15

Zinc carbonate: $ZnCO_3$ • 125.39 • 3486-35-9
- **P** $\Delta H_f^0 = -812.8(s)$

Zinc chloride: $ZnCl_2$ • 136.27 • 7646-85-7
- **P** $\Delta H_f^0 = -415.1(s)$
- **I** NFPA: 0; 2 Tr: –
- **T** Tr: 34 • NFPA: 2 • TR: 80 • LVE(MVE) (F): – (1 mg/m^3) (Zn); TWA (USA): 1 mg/m^3 (Zn) • STEL: 2 mg/m^3 (Zn)
- **D** LC50 r: 1.2–2/?; 1.96/10'; LD50 o-r: 350; LD50 o-m: 329; 350; LD50 ip-r: 58; LD50 ip-m: 24; 31; LD50 iv-r: 3.69; MLD iv-r: 30; 60-90; LD50 iv-m: 9.09; MLD iv-ra: 11; LD50 scu-m: 330

Zinc chromate: $ZnCrO_4$ • 181.37 • 13530-65-9
- **T** Tr: 45-22-43; LVE(MVE) (F): (0.05 mg/m^3) (Zn); TWA (USA): 0.01 mg/m^3 (Cr)
- **D** MLD iv-m: 30

Zinc cyanide: $Zn(CN)_2$ • 117.41 • 557-21-1
- **P** $\Delta H_f^0 = +95.81(s)$
- **D** MLD ip-r: 100

Zinc fluoride: ZnF_2 • 103.37 • 7783-49-5
- **P** $\Delta H_f^0 = -764.4(s)$
- **T** Tr: 23/24/25-34 • TWA (USA): 2.5 mg/m^3 (F)

Zinc hexafluorosilicate: $ZnSiF_6$ • 207.46 • 16871-71-9
- **T** TWA (USA): 2.5 mg/m^3 (F)
- **D** MLD o-r: 100 • MLD scu-m: 280

Zinc nitrate: $Zn(NO_3)_2$ • 189.39 • 7779-88-6
- **P** $\Delta H_f^0 = -483.7(s)$
- **D** Hydrate: LD50 o-r: 1190; LD50 o-m: 926; LD50 ip-r: 133; LD50 ip-m: 110

Zinc oxide: ZnO• 81.37 • 1314-13-2
- **P** $\Delta H_f^0 = -350.5(s)$
- **I** NFPA: -;-
- **T** NFPA: 3 • LVE(MVE) (F): id USA; TWA (USA): 5 mg/m^3 (Zn) (vapour); 10 mg/m^3 (Zn) (dust)
- **D** LC50 m: 2.5/?; LD50 o-r: 630; 7950; LD50 o-m: 7950; LD50 ip-r: 240

Zinc peroxide: ZnO_2 • 97.37 • 1314-22-3

Zinc phosphide: Zn_3P_2; 258.05; 1314-84-7
- **P** $\Delta H_f^0 = -472.79(s)$
- **I** Tr: 15/29
- **T** Tr: 15/29-28-32
- **D** LD50 o-r: 12; 40; 41-47; LD50 o-m: 40; MLD o-ra: 40

7 • Table of inorganic compounds — Gallium/germanium/arsenic

Name: Empirical formula • Molar mass (g/mol) • Registry number (CAS).
(P) Physical data: ΔH_f^0 (kJ/mol) (physical state) • Eb (°C/mmHg) (if substance volatile) • P_{vap} (mBar) (°C) (if substance volatile).
(I) Inflammability data: LEL/UEL (%) • P_{elc} (°C) • AIT (°C) • NFPA (flame); NFPA (reactivity) • Labour Code (Tr).
(T) Toxicity: Labour Code (Tr) • NFPA • Transport Code (TR) • LVE (MVE) (ppm) F, USA, D • STEL (ppm); IDLH (ppm).
(D) Toxic and lethal doses: LC (mg/l/exposure time); LD (mg/kg).

Zinc sulphate $ZnSO_4$ • 161.43 • 7733-02-0 • 7446-20-0 ($7H_2O$)
- (P) $\Delta H_f^0 = -982.8(s)$; $-3077.75(s, 7H_2O)$
- (T) Tr: 36/38
- (D) Anhydrous: LD50 o-r: 2949; MLD o-r: 2200; LD50 o-m: 57; 1890; MLD o-ra: 2000; LD50 ip-r: 260; LD50 ip-m: 29; 71.75; MLD scu-r: 33; 330; MLD iv-r: 50; 500; MLD iv-ra: 44; MLD scu-m: 15; 1500; MLD scu-ra: 300
- (T) (heptahydrate) Tr: 36/38
- (D) MLD o-r: 2150; 2200; LD50 o-m: 2200; MLD o-m: 2200; MLD o-ra: 1910; 2200; LD50 ip-r: 200; LD50 ip-m: 75; 260; MLD iv-r: 49; MLD iv-ra: 44; MLD scu-r: 33; 330

Zinc sulphide: ZnS • 97.44 • 1314-98-3
- (P) $\Delta H_f^0 = -192.6(s)$
- (D) LC50 r: > 5.04/4h; LD50 o-r: > 2000; LD50 cu-r: > 2000

GALLIUM

Gallium: Ga • 69.72 • 7440-55-3
- (P) $\Delta H_f^0 = 5.6$ (l)
- (T) NFPA: 1

Gallium nitrate : $Ga(NO_3)_3$ • 255.75 • 13494-90-1
- (D) LD50 o-m: 4360; LD50 ip-r: 67.5; LD50 ip-m: 80; LD50 iv-m: 55; MLD scu-r: 72; LD50 scu-m: 600

Gallium trichloride: $GaCl_3$ • 176.07 • 13450-90-3
- (P) $\Delta H_f^0 = -524.7(s)$
- (D) MLC r: 0.316/3h; LD50 ip-m: 93.4; LD50 iv-r: 47; LD50 iv-ra: 43; LD50 scu-r: 306; LD50 scu-ra: 245

Digallium trioxide: Ga_2O_3 • 187.44 • 12024-21-4
- (P) $\Delta H_f^0 = -1089.1(s)$
- (D) LD50 o-m: 10000

GERMANIUM

Germanium: Ge • 72.59 • 7440-56-4
- (I) NFPA: 2
- (T) NFPA: 2

Germanium dioxide: GeO_2 • 104.59 • 1310-53-8
- (P) $\Delta H_f^0 = -580.0(s)$
- (T) NFPA: 2
- (D) LD50 o-r: 1250; LD50 o-m: 1250; LD50 ip-r: 750; LD50 ip-m: 1550; LD50 scu-r: 1910; LD50 scu-m: 2550; LD50 scu-ra: 845

Germanium tetrachloride: $GeCl_4$ • 214.39 • 10038-98-9
- (P) $\Delta H_f^0 = -495.8(g)$; $-531.8(l)$ • Eb = 83.1
- (T) NFPA: 3
- (D) LC50 m: 44/2h; LD50 iv-m: 56

ARSENIC

Arsenic: As • 74.92 • 7440-38-2
- (P) $\Delta H_f^0 = +14.6$ (s, As yellow) • $P_{vap} = 1.33(372)$
- (I) NFPA: 2
- (T) Tr: 23/25 • NFPA: 3 • TWA (USA): 0.2 mg/m^3 (As)
- (D) LD50 o-r: 763; LD50 o-m: 145; LD50 ip-r: 13.39; LD50 ip-m: 46.2; MLD scu-ra: 300

Arsenic acid : H_3AsO_4 • 141.95 • 7778-39-4
- (P) $\Delta H_f^0 = -906.3$ (s, $1/2H_2O$)
- (I) NFPA: – ; –
- (T) Tr: 45-23/25 • NFPA: 3 • TR: 66(pure); 60(solution) • TWA (USA): 0.01 mg/m^3 (As)
- (D) LD50 o-r: 8; 48; LD50 o-m: 55; MLD o-ra: 5

Calcium arsenate $Ca_3As_2O_8$ • 398.08 • 7778-44-1
- (P) $\Delta H_f^0 = -3297.83(s)$
- (I) NFPA: – ; –
- (T) Tr: 23/25 • NFPA: 3 • TWA (USA): 0.2 mg/m^3 (As)
- (D) LD50 o-r: 20; LD50 o-m: 794; MLD o-ra: 50; LD50 cu-r: 2400

Potassium arsenate KH_2AsO_4 • 180.04 • 7784-41-0
- (P) $\Delta H_f^0 = -1180.72(s)$
- (T) TWA (USA): 0.1 mg/m^3 (As)

Sodium arsenate : NaH_2AsO_4 • 185.91 • 7778-43-0 • 10048-95-0 ($7H_2O$)
- (P) $\Delta H_f^0 = -1386.78(s, 7H_2O)$
- (T) TWA (USA): 0.1 mg/m^3 (As)
- (D) MLD o-ra: 51; MLD ip-r: 49; MLD iv-r: 85; MLD iv-ra: 28

Sodium arsenite : $NaAsO_2$ • 129.91 • 7784-46-5
- (P) $\Delta H_f^0 = -660.53(s)$
- (T) TWA (USA): 0.2 mg/m^3 (As)
- (D) LD50 o-r: 41 • MLD ip-r: 9

Gallium arsenide: GaAs • 144.64 • 1303-00-0
- (P) $\Delta H_f^0 = -71.0(s)$

Arsine : AsH_3 • 77.95 • 7784-42-1
- (P) $\Delta H_f^0 = +66.4(g)$ • $P_{vap} = 15.1(20)$ (unrealistic figure)
- (T) Tr: 23/25 • TR: prohibited • LVE(MVE) (F): 0.2 (0.05) • TWA (USA): 0.05 ; MAK (D): 0.05
- (D) MLC r: 0.300/15'; 0.35/10'; MLC m : 0.070/3h; LC50 r: 0.032/?; MLC ra: 0.5/15'

Arsenic pentasulphide : As_2S_5 • 310.20 • 1303-34-0

Arsenic pentoxide : As_2O_5 • 229.84 • 1303-28-2
- (P) $\Delta H_f^0 = -924.9(s)$
- (T) Tr: 45-23/25 • TR: 60 • TWA (USA): 0.01 mg/m^3 (As)
- (D) LD50 o-r: 8 • LD50 o-m: 55 • MLD iv-ra: 6

7 • Table of inorganic compounds

Selenium/bromine/krypton

Name: Empirical formula • Molar mass (g/mol) • Registry number (CAS).
- **(P) Physical data:** ΔH_f^0 (kJ/mol) (physical state) • Eb (°C/mmHg) (if substance volatile) • P_{vap} (mBar) (°C) (if substance volatile).
- **(I) Inflammability data:** LEL/UEL (%) • P_{elc} (°C) • AIT (°C) • NFPA (flame); NFPA (reactivity) • Labour Code (Tr).
- **(T) Toxicity:** Labour Code (Tr) • NFPA • Transport Code (TR) • LVE (MVE) (ppm) F, USA, D • STEL (ppm); IDLH (ppm).
- **(D) Toxic and lethal doses:** LC (mg/l/exposure time); LD (mg/kg).

Arsenic trichloride : $AsCl_3$ • 181.27 • 7784-34-1
- **(P)** $\Delta H_f^0 = -305.01(l)$ • Eb = 130.2 • P_{vap} = 10.8(20); 19(30); 53(50)
- **(I)** NFPA: –; –
- **(T)** Tr: 23/25 • NFPA: 3 • TR: 66 • TWA (USA): 0.01 mg/m³ (As)
- **(D)** MLC m : 2.547/?

Arsenic trioxide : As_2O_3 • 197.84 • 1327-53-3
- **(P)** $\Delta H_f^0 = -793.79(s)$
- **(I)** NFPA: –; –
- **(T)** Tr: 45-28-34 • NFPA: 3 • TR: 60 • LVE(MVE) (F): (0.2 mg/m³); TWA (USA): 0.1 mg/m³ (As)
- **(D)** LD50 o-r: 14.6; 15.1; 20; LD50 o-m: 31.5; 39.4; 45; LD50 o-ra: 20.19; MLD o-ra: 4; LD50 ip-r: 871; LD50 iv-m: 10.7; MLD iv-ra: 10.56; LD50 scu-m: 9.8

Arsenic trisulphide: As_2S_3 • 246.04 • 1303-33-9
- **(P)** $\Delta H_f^0 = -169.00(s)$
- **(T)** TWA (USA): 0.2 mg/m³ (As)
- **(D)** LD50 o-r: 185; LD50 o-m: 254; LD50 ip-m: 215; LD50 cu-r: 936

SELENIUM

Selenium: Se • 78.96 • 7782-49-2
- **(P)** $\Delta H_f^0 = 0$(grey); +5.02(black); +6.70(red); P_{vap} = 1.33(356)
- **(I)** NFPA: 2; –
- **(T)** Tr: 23/25-33 • NFPA: 3 • TWA (USA): 0.2 mg/m³ (Se)
- **(D)** MLC r: 0.033/8h; LD50 o-r: 6700; LD50 iv-r: 6; LD50 iv-ra: 2500

Selenium dioxide: SeO_2 • 110.96 • 7446-08-4
- **(P)** $\Delta H_f^0 = -225.4(s)$ • Eb = 315 (subl) • P_{vap} = 1.33(157)
- **(I)** NFPA: –; –
- **(T)** Tr: 23/25-33 • NFPA: 3 • TR: 60 • TWA (USA): 0.2 mg/m³ (Se)
- **(D)** MLC ra: 5.89/20'; LD50 o-r: 68.1; LD50 o-m: 23.3; LD50 ip-r: 3.6; LD50 ip-m: 4.3; LD50 iv-r: 2.71; LD50 iv-m: 9.2; LD50 scu-ra: 4

Arsenic hemiselenide: As_2Se • 228.78 • 1303-35-1
- **(T)** Tr: 23/25-33 • TWA (USA): 0.2 mg/m³ (Se)

Selenium hexafluoride: SeF_6 • 192.96 • 7783-79-1
- **(P)** $\Delta H_f^0 = -1117.0(g)$ • Eb = 34 • P_{vap} = 868(48.7)
- **(I)** NFPA: –; –
- **(T)** Tr: 23/25-33 • NFPA: 3 • TWA (USA): 0.05 (Se); MAK (D): 0.1 mg/m³ (Se)
- **(D)** MLC r: 0.080/1h; MLC m : 0.080/3h; MLC ra: 0.080/?

Sodium selenite: Na_2SeO_3 • 172.94 • 10102-18-8
- **(P)** $\Delta H_f^0 = -958.55(s)$
- **(T)** Tr: 23/25-33 • TWA (USA): 0.2 mg/m³ (Se); MAK (D): 0.1 mg/m³ (Se)
- **(D)** Hydrate: LD50 o-r: 7; LD50 o-m: 7.08; LD50 o-ra: 2.25; 5; LD50 ip-r: 3; MLD ip-r: 5.476; LD50 iv-r: 3; LD50 iv-m: 5; LD50 scu-m: 13

BROMINE

Bromine: Br_2 • 159.82 • 7726-95-6
- **(P)** Eb = 58.2 • P_{vap} = 231(20); 233(21)
- **(I)** NFPA: 0; 1
- **(T)** Tr: 26-35 • NFPA: 4 • TR: 886 • LVE(MVE) (F): 0.1 (–) • TWA (USA): 0.1; MAK (D): 0.1 • STEL: 0.3 IDLH: 10
- **(D)** LC50 r: 2.7/?; 5/?; LC50 m: 4.989/9'; MLC ra: 1.197/6.5h; LD50 o-r: 2600; LD50 o-m: 3100; LD50 o-ra: 4160

Hydrobromic acid HBr (aqueous solution) • 80.92 • 10035-10-6
- **(P)** Eb = 126
- **(I)** NFPA: –; –
- **(T)** Tr: 34-37 • NFPA: 3 • TR: 80 • TWA (USA): 3; MAK (D): 5
- **(D)** LC50 r: 9.265/1h; LC50 m: 2.741/1h; LD50 ip-r: 76

Potassium bromate: $KBrO_3$ • 167.01 • 7758-01-2
- **(P)** $\Delta H_f^0 = -360.2(s)$ • Eb = 470 (dec)
- **(I)** Tr : 9
- **(D)** LD50 o-r: 321; LD50 o-m: 289; MLD o-ra: 250; LD50 ip-m: 140; 177; MLD iv-ra: 360

Sodium bromate : $NaBrO_3$ • 150.90 • 7789-38-0
- **(P)** $\Delta H_f^0 = -334.1(s)$
- **(T)** Tr: 45-25 • NFPA: 1 • TR: 50
- **(D)** LD50 o-r: 400; MLD o-ra: 250; LD50 ip-m: 140; LD50 cu-ra: 2000

Ammonium bromide : NH_4Br • 97.96 • 12124-97-9
- **(P)** $\Delta H_f^0 = -270.8(s)$ • Eb = 235 (under vacuum); 452 (subl) • P_{vap} = 1.33 (198.3)
- **(D)** LD50 o-r: 2700; LD50 o-m: 2860; LD50 ip-m: 559

Hydrogen bromide: HBr • 80.92 • 10035-10-6
- **(P)** $\Delta H_f^0 = -36.3(g)$ • Eb = 66.8 • P_{vap} = 21300(20)
- **(I)** NFPA: –; –
- **(T)** Tr: 35-37 • NFPA: 3 • TR: 286 • TWA (USA): 3 • MAK (D): 5

Potassium bromide: HBr • 119.01 • 7758-02-3
- **(P)** $\Delta H_f^0 = -393.8(s)$
- **(D)** LD50 o-r: 3070; LD50 o-m: 3120; LD50 ip-m: 1030

Sodium bromide: NaBr • 102.90 • 7647-15-6
- **(P)** $\Delta H_f^0 = -361.1(s)$
- **(D)** LD50 o-r: 3500; LD50 o-m: 7000; MLD o-ra: 580; LD50 ip-m: 5000; LD50 scu-r: 2900; LD50 scu-m: 5020

KRYPTON

Krypton: Kr • 83.80 • 7439-90-9
- **(P)** Eb = –153.35

7 • Table of inorganic compounds
Rubidium/strontium/zirconium niobium/molybdenum

Name: Empirical formula • Molar mass (g/mol) • Registry number (CAS).
(P) Physical data: ΔH_f^0 (kJ/mol) (physical state) • Eb (°C/mmHg) (if substance volatile) • P_{vap} (mBar) (°C) (if substance volatile).
(I) Inflammability data: LEL/UEL (%) • P_{elc} (°C) • AIT (°C) • NFPA (flame); NFPA (reactivity) • Labour Code (Tr).
(T) Toxicity: Labour Code (Tr) • NFPA • Transport Code (TR) • LVE (MVE) (ppm) F, USA, D • STEL (ppm); IDLH (ppm).
(D) Toxic and lethal doses: LC (mg/l/exposure time); LD (mg/kg).

RUBIDIUM

Rubidium: Rb • 85.47 • 7440-17-7
(I) NFPA: 3; –
(T) NFPA: 2
(D) LD50 ip-m: 1200

Rubidium carbonate: Rb_2CO_3 • 230.95 • 584-09-8
(P) $\Delta H_f^0 = -1135.96$(s)
(D) LD50 o-r: 2625; LD50 ip-r: 450

Rubidium chloride: RbCl • 120.92 • 7791-11-9
(P) $\Delta H_f^0 = -435.35$(s)
(D) LD50 o-m: 3800; LD50 ip-m: 1625

STRONTIUM

Strontium: Sr • 87.62 • 7440-24-6
(P) $\Delta H_f^0 = +7.61$(l)
(I) NFPA: 2; –
(T) NFPA: 2

Strontium carbonate: $SrCO_3$ • 147.62 • 1633-05-2
(I) NFPA: – ; –
(D) LD50 o-r: 5000

Strontium chlorate: $Sr(ClO_3)_2$ • 327.74 • 7791-10-8

Strontium chloride $SrCl_2$ • 158.52 • 10476-85-4 • 10025-70-4 ($6H_2O$)
(P) $\Delta H_f^0 = -828.85$(s)
(D) Anhydrous: LD50 o-r: 2250; MLD o-ra: 7500; LD50 ip-r: 222; LD50 ip-m: 1643; LD50 iv-m: 148; Hydrate: LD50 ip-m: 1253; MLD iv-r: 400

Strontium chromate : $SrCrO_4$ • 203.62 • 7789-06-2
(T) Tr: 45-22 • LVE(MVE) (F): –[0.05mg/m³ (CrO_3)]; TWA(USA): 0.0005 mg/m³
(D) LD50 o-r: 312

Strontium hydroxide: $Sr(OH)_2$ • 121.63 • 18480-07-4
(P) $\Delta H_f^0 = -969.01$(s)

Strontium nitrate: $Sr(NO_3)_2$ • 211.64 • 10042-76-9
(P) $\Delta H_f^0 = -978.22$(s)
(I) NFPA: 0; 1
(T) NFPA: 1 • TWA (USA): 0.05 mg/m³ (Sr)
(D) LD50 o-r: 2750; LD50 o-m: 1826; LD50 o-ra: 3865; LD50 ip-r: 500; 540

Strontium oxide : SrO • 103.62 • 1314-11-0
(P) $\Delta H_f^0 = -592.04$(s)

Strontium sulphate : $SrSO_4$ • 183.68 • 7759-02-6
(P) $\Delta H_f^0 = -1453.10$(s)

Strontium sulphide: SrS• 119.68 • 1314-96-1
(P) $\Delta H_f^0 = -430.95$(s)

ZIRCONIUM

Zirconium: Zr • 91.224 • 7440-67-7
(P) $\Delta H_f^0 = 0$ (α); +7.15(ß)
(I) NFPA: 4; 1
(T) NFPA: 1 • TWA (USA): 5 mg/m³ (Zr) • STEL: 10 mg/m³ (Zr)

Zirconyl chloride: $ZrOCl_2$ • 178.12 • 7699-43-6 • 15461-27-5 (H_2O)
(T) NFPA: 1 • TWA (USA): 5 mg/m³ (Zr) • STEL: 10 mg/m³ (Zr)
(D) LD50 o-r: 3500; LD50 ip-r: 400; LD50 ip-m: 335; LD50 scu-r: 1227; MLD scu-r: 500

Zirconium hydroxide : $Zr(OH)_4$ • 159.25 • 14475-63-9
(T) NFPA: 1 • TWA (USA): 5 mg/m³ (Zr) • STEL: 10 mg/m³ (Zr)

Zirconium oxide: ZrO_2 • 123.22 • 1314-23-4
(P) $\Delta H_f^0 = -1097.46$(s)
(T) NFPA: 1 • TWA (USA): 5 mg/m³ (Zr) • STEL: 10 mg/m³(Zr)

Zirconium silicate: $ZrSiO_4$ • 183.31 • 14940-68-2 • 10101-52-7
(P) $\Delta H_f^0 = -2023.80$(s)
(T) NFPA: 1 • TWA (USA): 5 mg/m³ (Zr)

Zirconium sulphate: $Zr(SO_4)_2$ • 283.34 • 14644-61-2 • 34806-73-0
(T) NFPA: 1 • TWA (USA): 5 mg/m³ (Zr) • STEL: 10 mg/m³ (Zr)
(D) LD50 o-r: 3500; LD50 ip-r: 175; MLD scu-r: 500

Zirconium tetrachloride: $ZrCl_4$ • 233.02 • 10026-11-6
(P) $\Delta H_f^0 = -980.52$(s) • Eb = 331 (subl)
(T) NFPA: 1 • TWA (USA): 5 mg/m³ (Zr) • STEL: 10 mg/m³ (Zr)
(D) LD50 o-r: 1688; LD50 o-m: 489; 665

NIOBIUM

Niobium: Nb • 92.906 • 7440-03-1
(I) NFPA: 2
(T) NFPA: – • TWA (USA): 6 mg/m³ (Nb)

Niobium pentachloride : $NbCl_5$ • 270.16 • 10026-12-7
(P) $\Delta H_f^0 = -797.47$(s) • Eb = ~250
(D) LD50 o-r: 1400; LD50 o-m: 829; LD50 ip-r: 40; LD50 ip-m: 61

Niobium pentoxide : Nb_2O_5 • 265.81 • 1313-96-8
(P) $\Delta H_f^0 = -1899.54$(s)
(D) LD50 o-m: > 4000

MOLYBDENUM

Molybdenum: Mo • 95.94 • 7439-98-7
(I) NFPA: 2
(T) TWA (USA): 10 mg/m³ (Mo) • MAK (D): 15 mg/m³

7 • Table of inorganic compounds
Technetium/ruthenium/rhodium palladium/silver

Name: Empirical formula • Molar mass (g/mol) • Registry number (CAS).
(P) **Physical data:** ΔH_f^0 (kJ/mol) (physical state) • Eb (°C/mmHg) (if substance volatile) • P_{vap} (mBar) (°C) (if substance volatile).
(I) **Inflammability data:** LEL/UEL (%) • P_{elc} (°C) • AIT (°C) • NFPA (flame); NFPA (reactivity) • Labour Code (Tr).
(T) **Toxicity:** Labour Code (Tr) • NFPA • Transport Code (TR) • LVE (MVE) (ppm) F, USA, D • STEL (ppm); IDLH (ppm).
(D) **Toxic and lethal doses:** LC (mg/l/exposure time); LD (mg/kg).

Ammonium molybdate : $(NH_4)_2MoO_4$ • 196.04 • 13106-76-8
• 12054-85-2(H_2O)
(T) NFPA: 1 • TWA (USA): 5mg/m^3 (Mo)
(D) LD50 o-r: 330; MLD o-ra: 1870

Sodium molybdate : Na_2MoO_4 • 205.92 • 7631-95-0 • 10102-40-6
(T) NFPA: 1 • TWA (USA): 5 mg/m^3 (Mo)
(D) Anhydrous: LD50 scu-m: 570; Hydrate: LD50 ip-r: 520;
LD50 ip-m: 257

Molybdenum sulphide : MoS_2 • 160.07 • 1317-33-5
(P) $\Delta H_f^0 = -235.14$(s) • Eb = 450 (subl) dec in air
(T) NFPA: 1 • TWA (USA): 10 mg/m^3 (Mo)

Molybdenum trioxide : MoO_3 • 143.94 • 1313-27-5
(P) $\Delta H_f^0 = -748.10$(s)
(T) NFPA: 1 • TWA (USA): 10 mg/m^3 (Mo)
(D) LD50 o-r: 101; 125; MLD ip-ra: 10; LD50 scu-m: 94

TECHNETIUM

No accrurate data available.

RUTHENIUM

Ruthenium: Ru • 101.07 • 7440-18-8
(I) NFPA: 2

Ruthenium oxide: RuO_4 • 165.07 • 20427-56-9
(P) $\Delta H_f^0 = -239.32$(s) • Eb = 40° (F: 25)

Ruthenium trichloride: $RuCl_3$ • 207.42 • 10049-08-8
(P) $\Delta H_f^0 = -205.02$(s)
(D) LD50 ip-r: 360; LD50 ip-m: 108

RHODIUM

Rhodium: Rh • 102.91 • 7440-16-6
(I) NFPA: 2
(T) TWA (USA): 0.001 mg/m^3 (Rh)
(D) LD50 o-r: 1302; LD50 ip-r: 280; LD50 iv-r: 198;
LD50 iv-ra: 215

Rhodium sulphate : $Rh_2(SO_4)_3$ • 493.98 • 10489-46-0
(T) TWA (USA): 0.001 mg/m^3 (Rh)

Rhodium trichloride : $RhCl_3$ • 209.26 • 10049-07-7
(P) $\Delta H_f^0 = -299.156$

PALLADIUM

Palladium: Pd • 106.3 • 7440-05-3
(P) Eb = 2970; 3167
(I) Tr : 8 (Attribution by Weka code)

Palladium chloride : $PdCl_2$ • 177.30 • 7647-10-1
(P) $\Delta H_f^0 = -171.54$(s)
(T) Tr: 23/24/25
(D) Anhydrous: LD50 o-r: 31; 200; 2704; LD50 o-m: > 1000;
LD50 ip-r: 70; LD50 ip-m: 87; 174; LD50 iv-r: 3;
MLD iv-ra: 18.6; Hydrate: LD50 o-r: 576; LD50 ip-r: 85;
MLD iv-ra: 18 (too high?)

SILVER

Silver: Ag • 107.868 • 7440-22-4
(I) NFPA: 2
(T) NFPA: 1 • LVE(MVE) (F): [0.1 mg/m^3 (Ag)] ;
MAK (D): 0.01 mg/m^3

Silver bromide : AgBr • 187.77 • 7785-23-1
(P) $\Delta H_f^0 = -100.3$(s)

Silver carbonate : Ag_2CO_3 • 275.75 • 534-16-7
(P) $\Delta H_f^0 = -505.85$(s)
(D) LD50 o-r: 3731; LD50 o-m: 2168; LD50 ip-r: 156;
LD50 ip-m: 39

Silver chloride : AgCl • 143.32 • 7783-90-6
(P) $\Delta H_f^0 = -127.00$(s)

Silver cyanide : AgCN • 133.89 • 506-64-9
(P) $\Delta H_f^0 = +146.00$(s)
(T) TWA (USA): 5 mg/m^3 (CN);
(D) LD50 o-r: 123

Silver fluoride : AgF • 126.88 • 7775-41-9
(P) $\Delta H_f^0 = -204.6$(s)

Silver iodide : AgI • 234.77 • 7783-96-2
(P) $\Delta H_f^0 = -61.8$(s)

Silver nitrate : $AgNO_3$ • 169.88 • 7761-88-8
(P) $\Delta H_f^0 = -124.4$(s) • NF: 0; 1
(T) Tr: 34 • NFPA: 1 • TR: 50 • TWA(USA): 0.01 mg/m^3 (Ag)
(D) LD50 o-r: > 500; 1173; LD50 o-m: 50; MLD o-ra: 800;
LD50 ip-r: 83; LD50 ip-m: 17; 22; 34.5; MLD iv-ra: 8.8

Silver oxide : Ag_2O • 231.74 • 20667-12-3
(P) $\Delta H_f^0 = -31.1$(s)
(D) MLD o-r: 2820; LD50 o-m: 1027; LD50 ip-r: 70; LD50 ip-m: 37

Silver perchlorate : $AgClO_4$ • 207.32 • 7783-93-9 • 14242-05-8
(P) $\Delta H_f^0 = -31.13$(s)
(T) TWA(USA): 0.01 mg/m^3 (Ag)

Silver permanganate : $AgMnO_4$ • 226.80 • 7783-98-4

Silver phosphate : Ag_3PO_4 • 418.58 • 7784-09-0
(T) TWA(USA): 0.01 mg/m^3 (Ag)

7 • Table of inorganic compounds
Cadmium/indium

Name: Empirical formula • Molar mass (g/mol) • Registry number (CAS).
(P) **Physical data:** ΔH_f^0 (kJ/mol) (physical state) • Eb (°C/mmHg) (if substance volatile) • P_{vap} (mBar) (°C) (if substance volatile).
(I) **Inflammability data:** LEL/UEL (%) • P_{elc} (°C) • AIT (°C) • NFPA (flame); NFPA (reactivity) • Labour Code (Tr).
(T) **Toxicity:** Labour Code (Tr) • NFPA • Transport Code (TR) • LVE (MVE) (ppm) F, USA, D • STEL (ppm); IDLH (ppm).
(D) **Toxic and lethal doses:** LC (mg/l/exposure time); LD (mg/kg).

Silver sulphate : Ag_2SO_4 • 311.79 • 10294-26-5
(P) $\Delta H_f^0 = -715.88(s)$
(T) TWA(USA): 0.01 mg/m^3 (Ag)ph

Silver sulphide : Ag_2S • 247.80 • 21548-73-2
(P) $\Delta H_f^0 = -32,6(s)$

CADMIUM

Cadmium: Cd • 112.40 • 7440-43-9
(I) NFPA: 2
(T) NFPA: 3 • TWA (USA): 0.01 mg/m^3 (Cd)
(D) LC50 r: 0.025/30'; MLC m : 0.17/?; LD50 o-r: 225; 2330; LD50 o-m: 890; MLD o-ra: 70; LD50 ip-r: 4; LD50 ip-m: 5.7; LD50 iv-r: 1.8; 3; MLD iv-ra: 5; LD50 scu-m: 9; MLD scu-ra: 6

Cadmium bromide : $CdBr_2$ • 272.22 • 7789-42-6
(P) $\Delta H_f^0 = -316.2(s)$
(D) LD50 o-r: 322

Cadmium carbonate : $CdCO_3$ • 172.41 • 513-78-0
(P) $\Delta H_f^0 = -750.61(s)$
(T) TWA (USA): 0.05 mg/m^3 (Cd)
(D) LD50 o-r: 438; LD50 o-m: 310

Cadmium chloride: $CdCl_2$ • 183.30 • 10108-64-2
(P) $\Delta H_f^0 = -391.50(s)$
(I) NFPA: –; –
(T) Tr: 45-48/23/25 • NFPA: 3 • TR: 60 • TWA (USA): 0.01 mg/m^3 (Cd)
(D) Anhydrous: LC50 m: 2.3/?; LD50 o-r: 88; LD50 o-m: 60; 175; MLD o-ra: 70; LD50 ip-r: 1.8; LD50 ip-m: 9; LD50 iv-r: 5.463; LD50 iv-m: 3.5; LD50 iv-ra: 2.5; LD50 scu-r: 15.17; LD50 scu-m: 3.2; Hydrate: LD50 o-m: 194; LD50 ip-m: 4567

Cadmium cyanide : $Cd(CN)_2$ • 164.44 • 542-83-6
(P) $\Delta H_f^0 = +162.34(s)$
(T) Tr: 26/27/28-32-33-40 • TR: 66 • TWA (USA): 0.01 mg/m^3 (Cd)Cadmium

Cadmium fluoride : CdF_2 • 150.30 • 7790-79-6
(P) $\Delta H_f^0 = -700.30(s)$
(T) Tr: 23/25-33-40 • TR: prohibited • LVE(MVE)(F): [2.5mg/m^3 (F)]; TWA (USA): 0.01 mg/m^3 (Cd)

Cadmium hydroxide : $Cd(OH)_2$ • 146.42 • 21041-95-2
(P) $\Delta H_f^0 = -560.70(s)$

Cadmium iodide : CdI_2 • 366.22 • 7790-80-9
(P) $\Delta H_f^0 = -203.3(s)$
(T) Tr: 23/25-33-40 • LVE(MVE) (F): (0.05 mg/m^3 (Cd))
(D) LD50 o-r: 222; LD50 o-m: 166

Cadmium nitrate: $Cd(NO_3)_2$ • 236.42 • 10325-94-7 • 10022-68-1 (4 H_2O)
(P) $\Delta H_f^0 = -456.31(s)$
(T) Tr: 20/21/22 • TR: 60 • LVE(MVE) (F): [0.05 mg/m^3 (Cd)] ; TWA (USA): 0.01 mg/m^3 (Cd)
(D) Anhydrous: LC50 m: 3.5/?; LD50 o-m: 100; Hydrate: LD50 o-r: 0.300

Cadmium oxide: CdO • 128.40 • 1306-19-0
(P) $\Delta H_f^0 = -258.4(s)$
(I) NFPA: –; –
(T) Tr: 49-22-48/23/ 25 • NFPA: 3 • TR: 60 • LVE(MVE) (F): (0.05 mg/m^3 (Cd)); TWA (USA): 0.01 mg/m^3 (Cd)
(D) MLC r: 0.01/?; 0.5/10'(vapour); LC50 m: 0.34/10'; 0.25/2h; MLC m : 0.7/10'(vapour); LC50 ra: 2.5/10'; 3/15'; LD50 o-r: 72; LD50 ip-r: 12; LD50 iv-r: 25; LD50 scu-m: 94

Cadmium sulphate : $CdSO_4$ • 208.46 • 10124-36-4 • 13477-21-9 (4 H_2O)
(P) $\Delta H_f^0 = -933.3(s)$
(T) Tr: 49-22-48/23/ 25 • TR: 60 • TWA (USA): 0.01 mg/m^3 (Cd)
(D) LD50 o-r: 280; LD50 o-m: 88; LD50 ip-m: 7; 12.76

Cadmium sulphide : CdS • 144.46 • 1306-23-6
(P) $\Delta H_f^0 = -161.92(s)$
(D) LC50 m: 1.35/?; LD50 o-r: 7080; LD50 o-m: 1166

INDIUM

Indium: In • 114.82 • 7440-74-6
(I) NFPA: 2
(T) NFPA: 3 • TWA (USA): 0.1 mg/m^3 (In)
(D) MLD scu-m.: 10

Indium antimonide: InSb • 236.57 • 1312-41-0
(P) $\Delta H_f^0 = -30.5(s)$
(T) TWA (USA): 0.1 mg/m^3 (In)

Indium oxide: In_2O_3 • 277.64 • 1312-43-2
(P) $\Delta H_f^0 = -925.8(s)$
(T) TWA (USA): 0.1 mg/m^3 (In)

Indium sulphate : $In_2(SO_4)_3$ • 517.82 • 13464-82-9
(P) $\Delta H_f^0 = -2786.54(s)$
(T) TWA (USA): 0.1 mg/m^3 (In)
(D) LD50 o-r: 1200; MLD o-ra: 1300; 1800; LD50 iv-r: 5.63; LD50 iv-ra: 670; LD50 scu-r: 22.5; LD50 scu-ra: 2600

Indium trichloride: $InCl_3$ • 221.17 • 10025-82-8
(P) $\Delta H_f^0 = -537.2(s)$
(T) TWA (USA): 0.1 mg/m^3 (In)
(D) LD50 ip-m: 5; 9.5; MLD iv-ra: 0.64; MLD scu-r: 10; MLD scu-m: 60; MLD scu-ra: 2.35

7 • Table of inorganic compounds
Tin/antimony/tellurium

Name: Empirical formula • Molar mass (g/mol) • Registry number (CAS).
(P) Physical data: ΔH_f^0 (kJ/mol) (physical state) • Eb (°C/mmHg) (if substance volatile) • P_{vap} (mBar) (°C) (if substance volatile).
(I) Inflammability data: LEL/UEL (%) • P_{elc} (°C) • AIT (°C) • NFPA (flame); NFPA (reactivity) • Labour Code (Tr).
(T) Toxicity: Labour Code (Tr) • NFPA • Transport Code (TR) • LVE (MVE) (ppm) F, USA, D • STEL (ppm); IDLH (ppm).
(D) Toxic and lethal doses: LC (mg/l/exposure time); LD (mg/kg).

TIN

Tin : Sn • 118.71 • 7440-31-5
(P) $\Delta H_f^0 = 0$ (s, white); -2.1(s, grey)
(I) NFPA: 1
(T) NFPA: 0 • TWA (USA): 2 mg/m^3 (Sn); MAK (D): 2 mg/m^3 (Sn)

Stannous chloride : SnCl$_2$ • 189.59 • 7772-99-8
(P) $\Delta H_f^0 = -325.1$(s)
(T) TWA (USA): 2 mg/m^3 (Sn)
(D) Anhydrous: LD50 o-r: 700; LD50 o-m: 250; 1200; MLD o-ra: 40; 10000; LD50 ip-r: 316; LD50 ip-m: 105; LD50 iv-r: 43; LD50 iv-m: 17.8; Hydrate: LD50 iv-r: 7830

Stannous fluoride : SnF$_2$ • 156.69 • 7783-47-3
(T) TWA (USA): 2 mg/m^3 (Sn)
(D) LD50 o-r: 377; LD50 o-m: 184; 210; LD50 ip-m: 16.15

Stannic bromide : SnBr$_4$ • 438.33 • 7789-65-5
(P) $\Delta H_f^0 = -377.4$(s)
(T) TWA (USA): 2 mg/m^3 (Sn)
(D) LD50 iv-m: 18

Stannic chloride : SnCl$_4$ • 260.39 • 7646-78-8
(P) $\Delta H_f^0 = -471.5$(g); -511.3(l) • Eb = 114.1 • NFPA: 0; 1
(T) Tr: 34-37 • NFPA: 3 • TWA (USA): 2 mg/m^3 (Sn)
(D) LC50 r: 2.3 /10'; LD50 ip-m: 41; 99; 101; LD50 iv-m: 32

Tin disulphide: SnS$_2$ • 182.81 • 1315-01-1
(P) $\Delta H_f^0 = -167.36$(s)
(T) TWA (USA): 2 mg/m^3 (Sn)

Tin(IV) oxide: SnO$_2$ • 150.69 • 1332-29-2
(P) $\Delta H_f^0 = -577.6$(s)
(T) TWA (USA): 2 mg/m^3 (Sn)

Potassium stannate (IV): K$_2$SnO$_3$ • 244.88 • 12125-03-0 (3H$_2$O)
(T) TWA (USA): 2 mg/m^3 (Sn)

Sodium stannate (IV) : Na$_2$SnO$_3$ • 212,67 • 12058-66-1 (3H$_2$O)
(T) TWA (USA): 2 mg/m^3 (Sn)
(D) LD50 o-r: 3457; LD50 o-m: 2132

ANTIMONY

Antimony: Sb • 121.75 • 7440-36-0
(P) $\Delta H_f^0 = 0$; 10.63 (Sb "explosive")
(I) NFPA: 2
(T) Tr: 20/22 • NFPA: 3 • LVE(MVE) (F): [0.5 mg/m^3 (Sb)] ; TWA (USA): 0.5 mg/m^3 (Sb); MAK (D): 0.5 mg/m^3 (Sb)
(D) LD50 o-r: 100; 7000; LD50 ip-m: 90

Antinomy pentasulphide: Sb$_2$S$_5$ • 403.80 • 1315-04-4
(I) NFPA: 1; 1
(T) NFPA: 3 • TWA (USA): 0.5 mg/m^3 (Sb)
(D) LD50 ip-r: 1500; LD50 ip-m: 458

Antinomy pentoxide : Sb$_2$O$_5$ • 323.50 • 1314-60-9
(P) $\Delta H_f^0 = -971.94$(s)
(D) MLD ip-r: 4000; LD50 ip-r: 4000

Stibine: SbH$_3$ • 124.78 • 7803-52-3
(P) $\Delta H_f^0 = +145.1$(g) • Eb = -18.4; -17.1
(I) NFPA: 2; –
(T) Tr: 20/22 • NFPA: 3 • LVE(MVE) (F): -(0.1); TWA (USA): 0.1 • MAK (D): 0.1
(D) MLC m : 0.519/?; LD50 iv-ra: 8

Antinomy trichloride: SbCl$_3$ • 228.10 • 10025-91-9
(P) $\Delta H_f^0 = -382.2$(l) • Eb = 220; 223; 283 • P_{vap} = 0.16(20); 0.3(30); 1.33(49.2); 1.9(50)
(I) NFPA: –; –
(T) Tr: 34-37 • NFPA: 3 • TR: 80 • LVE(MVE) (F): id TWA; TWA (USA): 0.5 mg/m^3 (Sb)
(D) LD50 o-r: 525; LD50 ip-m: 13; 978

Antinomy trifluoride : SbF$_3$ • 178.75 • 7783-56-4
(P) $\Delta H_f^0 = -915.5$(s) • Eb = 319 (subl)
(T) Tr: 23/24/25 • LVE(MVE) (F): id TWA; TWA (USA): 0.5 mg/m^3 (Sb); 2.5 mg/m^3 (F)
(D) LD50 o-m: 804

Antimony trioxide : Sb$_2$O$_3$ • 291.50 • 1309-64-4
(P) $\Delta H_f^0 = -689.94$(s)
(I) NFPA:–; –
(T) Tr: 40 • NFPA: 3 • TR: 60 • TWA (USA): 0.5 mg/m^3 (Sb)
(D) LD50 o-r: 7000 (Baker); LD50 o-r; 100 (Caledon lab.); LD50 o-r: > 20000 (Aldrich); LD50 ip-r: 3250; LD50 ip-m: 172; MLD scu-ra: 2.5

Antimony trisulphide: Sb$_2$S$_3$ • 339.78 • 1345-04-6
(P) $\Delta H_f^0 = -174.89$ (black); -147.28(orange)
(T) Tr: 20 • LVE(MVE) (F): id TWA; TWA (USA): 0.5 mg/m^3 (Sb)
(D) MLD ip-r: 1000; 1390; LD50 ip-m: 209

TELLURIUM

Tellurium: Te • 127.6 • 13494-80-9
(I) NFPA: 2
(T) Tr: 25 • NFPA: 2 • TR: 60 • LVE(MVE) (F):-(0.1 mg/m^3); TWA (USA): 0.1 mg/m^3
(D) LC50 r: > 2.42/4h; LD50 o-r: 83; > 5000; LD50 o-m: 20; MLD o-ra: 67

7 • Table of inorganic compounds

Iodine/xenon/barium

Name: Empirical formula • Molar mass (g/mol) • Registry number (CAS).
- (P) **Physical data:** ΔH_f^0 (kJ/mol) (physical state) • Eb (°C/mmHg) (if substance volatile) • P_{vap} (mBar) (°C) (if substance volatile).
- (I) **Inflammability data:** LEL/UEL (%) • P_{elc} (°C) • AIT (°C) • NFPA (flame); NFPA (reactivity) • Labour Code (Tr).
- (T) **Toxicity:** Labour Code (Tr) • NFPA • Transport Code (TR) • LVE (MVE) (ppm) F, USA, D • STEL (ppm); IDLH (ppm).
- (D) **Toxic and lethal doses:** LC (mg/l/exposure time); LD (mg/kg).

Potassium tellurate (IV): K_2TeO_3 • 253.80 • 15571-91-2
- (P) $\Delta H_f^0 = -1012.53(s)$
- (T) TWA (USA): 0.1 mg/m^3 (Te)

XENON

Xenon: Xe; • M = 131.29; CAS N°: 7440-63-3;• Toxic Code : NFPA: 0

IODINE

Iodine: I_2 • 253.80 • 7553-56-2
- (P) $\Delta H_f^0 = 0(s)$; +62.4(g) • Eb = 184; 185.24 • P_{vap} = 0.04(0); 0.35(20); 0.72(30); 2.8(50)
- (I) NFPA: –; –
- (T) Tr: 20/21 • NFPA: 3 • TR: 80 • LVE(MVE) (F): - (0.1); TWA (USA): 0.1; MAK (D): 0.1 • IDLH: 10
- (D) MLC r: 0.8/1h; LD50 o-r: 14000; LD50 o-m: 22000; LD50 o-ra: 10000; LD50 scu-r: 10500; MLD scu-ra: 175

Iodic acid : See hydrogen iodide
- (P) Eb = 127 (57% solution)
- (I) NFPA: 0; 0 • Tr : –
- (T) Tr: c > 25%: 34 • NFPA: 3 • TR: 80

Potasssium iodate: KIO_3 • 214.00 • 7758-05-6
- (P) $\Delta H_f^0 = -501.4(s)$
- (T) NFPA: 2
- (D) MLD o-m: 531; LD50 ip-m: 136

Sodium iodate : $NaIO_3$ • 197.89 • 7681-55-2
- (P) $\Delta H_f^0 = -481.8(s)$
- (I) Tr : 8
- (T) NFPA: 2
- (D) LD50 ip-m:119; LD50 iv-m:108; MLD iv-ra: 75

Ammonium iodide: NH_4I • 145.000 • 12027-06-4
- (P) $\Delta_f^0 = -204(s)$
- (T) TWA (USA): 25 (salt of ammoniac)

Hydrogen iodide: HI • 127.91 • 10034-85-2
- (P) $\Delta H_f^0 = +24.9(g)$; +26.5(g) • P_{vap} = 7100(20); 9400(30); 15000(50)
- (I) NFPA: –; –
- (T) Tr: 35-37 • NFPA: 3

Potassium iodide: KI• 166.00 • 7681-11-0
- (P) $\Delta H_f^0 = -327.9(s)$
- (T) NFPA: 2
- (D) MLD o-m: 1862 ; LD50 ip-m: 1117; LD50 iv-r: 120; 167; 285

Sodium iodide: NaI• 149.89 • 7681-82-5
- (P) $\Delta H_f^0 = -287.8(s)$ • Eb = 1300; 1304
- (I) NFPA: –; –
- (T) NFPA: 2
- (D) LD50 o-r: 4340; LD50 o-m: 1000; MLD o-m: 1650; MLD ip-m: 430; 869; LD50 iv-r: 1060

BARIUM

Barium: Ba • 137.36 • 7440-39-3
- (I) NFPA: 2
- (T) NFPA: 2 • TWA (USA): 0.5 mg/m^3 (Ba)

Barium bromide: $BaBr_2$ • 297.14 • 10553-31-8
- (P) $\Delta H_f^0 = -757.3(s)$
- (T) TWA (USA): 0.5 mg/m^3 (Ba)

Barium carbonate : $BaCO_3$ • 197.35 • 513-77-9
- (P) $\Delta H_f^0 = -1216.29(s)$
- (I) NFPA: 0; 0
- (T) Tr: 22 • NFPA: 1 • TR: 60 • TWA (USA): 0.5 mg/m^3 (Ba)
- (D) LD50 o-r: 200; 418; 630; 800; MLD o-m: 200; LD50 ip-m: 50; LD50 iv-r: 20

Barium chlorate $Ba(ClO_3)_2$ • 304.24 • 13477-00-4
- (P) $\Delta H_f^0 = -762.74(s)$; Eb = 250 (dec O_2)
- (I) NFPA: 0; 1 • Tr: 9
- (T) Tr: 20/22 • NFPA: 1 • TR: 56 • TWA (USA): 0.5 mg/m^3 (Ba) • MAK (D): 0.5 mg/m^3 (Ba)

Barium chloride : $BaCl_2$ • 208.24 • 10361-37-2
- (P) $\Delta H_f^0 = -858.6(s)$
- (I) NFPA: 0; –
- (T) Tr: 20/22 • NFPA: 2 • TR: 60 • TWA (USA): 0.5 mg/m^3 (Ba); MAK (D): 0.5 mg/m^3 (Ba)
- (D) LD50 o-r: 118; MLD o-m: 70; MLD o-ra: 170; LD50 ip-m: 54; 184; MLD iv-r: 20; LD50 iv-m: 12; LD50 scu-r: 178; MLD scu-m: 10

Barium chromate : $BaCrO_4$ • 255.36 • 10294-40-3
- (P) $\Delta H_f^0 = -1445.99(s)$
- (T) TWA (USA): 0.5 mg/m^3 (Ba); 0.05 mg/m^3 (Cr)
- (D) LD50 o-r: > 2000

Barium cyanide: $Ba(CN)_2$ • 189.38 • 542-62-1
- (P) $\Delta H_f^0 = -218.40(s)$
  • Eb = Decomposes slowly in air.
- (T) Tr: 26/27/28-32 • TR: prohibited LVE(MVE)(F):id USA; TWA (USA): 5 mg/m^3 (CN); 0.5 mg/m^3 (Ba); MAK (D): id USA

Barium fluoride : BaF_2 • 175.34 • 7787-32-8
- (P) $\Delta H_f^0 = -1207.1(s)$
- (T) Tr: 20/22 • TR: 60 • LVE(MVE) (F): (id USA); TWA (USA): 2.5 mg/m^3 (F); MAK (D): id USA
- (D) LD50 o-r: 250; LD50 ip-m: 29.91

Barium fluorosilicate : $BaSiF_6$ • 279.43 • 17125-80-3
- (P) $\Delta H_f^0 = -2952.23(s)$
- (D) LD50 o-r: 175; MLD o-ra: 175

7 • Table of inorganic compounds

Mercury/lead

Name: Empirical formula • Molar mass (g/mol) • Registry number (CAS).
(P) Physical data: ΔH_f^0 (kJ/mol) (physical state) • Eb (°C/mmHg) (if substance volatile) • P_{vap} (mBar) (°C) (if substance volatile).
(I) Inflammability data: LEL/UEL (%) • P_{elc} (°C) • AIT (°C) • NFPA (flame); NFPA (reactivity) • Labour Code (L).
(T) Toxicity: Labour Code (Tr) • NFPA • Transport Code (TR) • LVE (MVE) (ppm) F, USA, D • STEL (ppm); IDLH (ppm).
(D) Toxic and lethal doses: LC (mg/l/exposure time); LD (mg/kg).

Barium hydroxide : $Ba(OH)_2$ • 171.35 • 17194-00-2
- (P) $\Delta H_f^0 = -944.70$(s, anhydrous)
- (I) NFPA: –; –
- (T) NFPA: 2 • TWA (USA): 0.5 mg/m^3 (Ba)
- (D) Hydrate: LD50 o-r: 333; LD50 ip-m: 255

Barium nitrate : $Ba(NO_3)_2$ • 261.36 • 10022-31-8
- (P) $\Delta H_f^0 = -992.1$(s)
- (I) NFPA: 0; 1
- (T) NFPA: 1 • TR: 56 • TWA (USA): 0.5 mg/m^3 (Ba); MAK (D): 0.5 mg/m^3 (Ba)
- (D) LD50 o-r: 355; MLD o-ra: 150; LD50 iv-m: 8.5; 16

Barium oxide : BaO • 153.34 • 1304-28-5
- (P) $\Delta H_f^0 = -553.5$(s)
- (T) TWA (USA): 0.5 mg/m^3 (Ba)
- (D) LD50 scu-m: 50

Barium peroxide : BaO_2 • 169.34 • 1304-29-6
- (P) $\Delta H_f^0 = -634.29$(s); Eb = 800 (dec)
- (I) NFPA: 0; 1 • Tr: 8
- (T) Tr: 20/22 • NFPA: 1 • TR: 56 • TWA (USA): 0.5 mg/m^3 (Ba)
- (D) LD50 scu-m: 50

Barium sulphate : $BaSO_4$ • 233.40 • 7727-43-7
- (P) $\Delta H_f^0 = -1473.2$(s)
- (T) TWA (USA): 10 mg/m^3 (Ba)

Barium sulphide: BaS • 169.40 • 21109-95-5
- (P) $\Delta H_f^0 = -460.0$(s)
- (T) Tr: 20/22-31 • TR: 60 • TWA (USA): 0.5 mg/m^3 (Ba)

MERCURY

Mercury: Hg • 200.59 • 7439-97-6
- (P) $\Delta H_f^0 = +61.4$(g) • Eb = 356.58
- (I) NFPA: –; –
- (T) Tr: 23-33 • NFPA: 3 • TWA (USA): 0.05 mg/m^3; MLC ra: 0.029/?

Mercuric chloride : $HgCl_2$ • 271.50 • 7487-94-7
- (P) $\Delta H_f^0 = -224.3$(s)
- (I) NFPA: –; –
- (T) Tr: 28-34-48/23/ 25 • NFPA: 3 • TR: 60 • LVE(MVE) (F): id USA; TWA (USA): 0.1 mg/m^3 (Hg)
- (D) MLC m: 0.300/10'; LD50 o-r: 1; LD50 o-m: 6; 10; MLD o-ra: 40; LD50 ip-r: 3.31; LD50 ip-m: 5; 6; LD50 iv-r: 1.27; LD50 iv-m: 4.99; LD50 scu-r: 14; LD50 scu-m: 4.5; LD50 cu-r: 41

Mercuric cyanide : $Hg(CN)_2$ • 252.63 • 592-04-1
- (P) $\Delta H_f^0 = +253.59$(s); +261.5(s)
- (T) TWA (USA): 0.1 mg/m^3 (Hg)
- (D) LD50 o-r: 18; MLD o-r: 25; 26; LD50 o-m: 33; MLD ip-r: 7.5; MLD iv-ra: 2

Mercuric iodide : HgI_2 • 454.39 • 7774-29-0
- (P) $\Delta H_f^0 = -105.44$(red); –102.93(yellow)
- (T) TWA (USA): 0.1 mg/m^3 (Hg)
- (D) LD o-r: 18; 40; LD50 o-m: 17; LD50 ip-m: 4.2; 60; LD50 cu-r: 75

Mercuric nitrate: $Hg(NO_3)_2$ • 324.61 • 10045-94-0
- (T) TWA (USA): 0.1 mg/m^3 (Hg)
- (D) LD50 o-r: 26; 51.4; LD50 o-m: 25; 29; LD50 ip-m: 7.2; 8; MLD scu-m: 20; LD50 cu-r: 75

Mercuric oxide : HgO • 216.59 • 21908-53-2
- (P) $\Delta H_f^0 = -90.8$(red); –90.5(yellow) • Eb = 500 (decO_2)
- (T) Tr: 26/27/28-33 • TR: 60 • LVE(MVE) (F): (id USA); TWA (USA): 0.1 mg/m^3 (Hg)
- (D) LD50 o-r: 18; LD50 o-m: 16; LD50 ip-m: 4.5; LD50 cu-r: 315

Mercuric sulphate : $HgSO_4$ • 296.65 • 7783-35-9
- (P) $\Delta H_f^0 = -707.5$(s)
- (T) TWA (USA): 0.1 mg/m^3 (Hg)
- (D) LD50 o-r: 57; LD50 o-m: 25; 40; LD50 ip-m: 6.3; LD50 cu-r: 625

Mercuric sulphide : HgS • 232.65 • 1344-48-5
- (P) $\Delta H_f^0 = -53.56$(black); –58.16(red)

Mercury thiocyanate : $Hg(CNS)_2$ • 316.79 • 592-85-8
- (P) $\Delta H_f^0 = +200.8$(s); dec = 165
- (T) TWA (USA): 0.1 mg/m^3 (Hg)
- (D) LD50 o-r: 46; LD50 o-m: 24.5; LD50 ip-m: 3.5; LD50 cu-r: 685

Mercurous chloride : Hg_2Cl_2 • 472.09 • 10112-91-1
- (P) $\Delta H_f^0 = -265.4$(s)
- (T) Tr: 22-36/37/38 • TR: – • LVE(MVE) (F): (id USA); TWA (USA): 0.1 mg/m^3 (Hg)
- (D) LD50 o-r: 166; 210; LD50 o-m: 180; LD50 ip-m: 10; LD50 cu-r: 1500

Mercurous nitrate : $HgNO_3$ • 262.60 • 10415-75-5
- (T) TWA (USA): 0.1 mg/m^3 (Hg)
- (D) LD50 o-r: 170; 182; 297; LD50 o-m: 49.3; 388; LD50 ip-m: 5; LD50 cu-r: 2330

Mercurous sulphate : Hg_2SO_4 • 497.24 • 7783-36-0
- (P) $\Delta H_f^0 = -743.1$(s)
- (T) TWA (USA): 0.1 mg/m^3 (Hg)
- (D) LD50 o-r: 205; LD50 o-m: 152; LD50 ip-m: 11.5; LD50 iv-m: 5600; LD50 cu-r: 1175

LEAD

Lead: Pb • 207.19 • 7439-92-1
- (I) NFPA: 2
- (T) NFPA: 3 • TWA (USA): 0.15 mg/m^3 (Pb); MAK (D): 0.10 mg/m^3 (Pb)
- (D) MLD ip-r: 1000

7 • Table of inorganic compounds — Bismuth

Name: Empirical formula • Molar mass (g/mol) • Registry number (CAS).
(P) Physical data: ΔH_f^0 (kJ/mol) (physical state) • Eb (°C/mmHg) (if substance volatile) • P_{vap} (mBar) (°C) (if substance volatile).
(I) Inflammability data: LEL/UEL (%) • P_{fla} (°C) • AIT (°C) • NFPA (flame); NFPA (reactivity) •Labour Code (Tr).
(T) Toxicity: Labour Code (L) • NFPA • Transport Code (TR) • LVE (MVE) (ppm) F, USA, D • STEL (ppm); IDLH (ppm).
(D) Toxic and lethal doses: LC (mg/l/exposure time); LD (mg/kg).

Lead nitride: PbN_6 • 291.25 • 13424-46-9
(P) ΔH_f^0 = +476.24(s); +478.23(s)
(I) Tr : 3
(T) Tr: 61-20/22-33 • TR: prohibited

Lead carbonate : $PbCO_3$ • 267.20 • 598-63-0
(P) ΔH_f^0 = –699.15(s) • Eb = 400 (dec → CO_2)
(I) NFPA :- ;-
(T) NFPA: 3

Lead chloride : $PbCl_2$ • 278.09 • 7758-95-4
(P) ΔH_f^0 = –359.4(s)
(T) TWA (USA): 0.15 mg/m^3 (Pb)

Lead chromate: $PbCrO_4$ • 323.19 • 7758-97-6
(P) ΔH_f^0 = –930.9(s)
(T) Tr: 61-33-40 • TR: 60 • LVE(MVE) (F): (0.05 mg/m^3) (Cr); TWA (USA): 0.05 mg/m^3 (Pb); 0.012 mg/m^3 (Cr)
(D) LD50 o-m: 12000

Lead hexafluorosilicate : $PbSiF_6$ • 349.28 • 25808-74-6
(T) Tr: 20/22-33
(D) MLD o-r: 250

Lead iodide : PbI_2 • 461.01 • 10101-63-0
(P) ΔH_f^0 = –175.5(s)
(T) Tr: 20/22-33

Lead nitrate: $Pb(NO_3)_2$ • 331.21 • 10099-74-8
(I) NFPA: 0; 1
(T) Tr: 20/22-33 • NFPA: 3 • TWA (USA): 0.15 mg/m^3 (Pb)
(D) MLD ip-r: 270; LD50 ip-m: 74; LD50 iv-r: 93

Lead oxide : PbO • 223.19 • 1317-36-8
(P) ΔH_f^0 = –219 (red); –217.3 (yellow)
(I) NFPA: 1
(T) Tr: 61-20/22-33 • NFPA: 3 • TR: 60 • TWA (USA): 0.15 mg/m^3 (Pb)
(D) LD50 o-r: 4300; MLD ip-r: 430; LD50 ip-m: 17700

Lead peroxide: PbO_2 • 239.19 • 1309-60-0
(P) ΔH_f^0 = –277.4
(T) TWA (USA): 0.15 mg/m^3 (Pb)

Lead sulphate: $PbSO_4$ • 303.25 • 7446-14-2
(P) ΔH_f^0 = –920.0(s)
(T) TWA (USA): 0.15 mg/m^3 (Pb)

Lead sulphide : PbS • 239.25 • 1314-87-0
(P) ΔH_f^0 = –100.3(s)
(T) Tr: 20/22-33 • TWA (USA): 0.15 mg/m^3 (Pb)
(D) MLD ip-r: 1847

Lead tetroxide (or minimum) **:** Pb_3O_4 • 685.60 • 1314-41-6
(P) ΔH_f^0 = –718.4(s)
(T) TWA (USA): 0.15 mg/m^3 (Pb)
(D) LD50 ip-r: 630

BISMUTH

Bismuth: Bi • 208.98 • 7440-69-9
(D) LD50 o-r: 500; 5000 ; LD50 o-m: 10000

Bismuth nitrate : $Bi(NO_3)_3$ • 395.08 • 10361-44-1
(D) LD50 o-r: 4042; LD50 ip-m: 2500; MLD iv-m: 21

Bismuth oxychloride: BiOCl • 260.33 • 7787-59-9
(P) ΔH_f^0 = –366.9(s)
(D) LD50 o-r: 22000

Bismuth (III) oxide: Bi_2O_3 • 465.96 • 1304-76-3
(P) ΔH_f^0 = –573.9(s)
(D) LD50 o-r: 5000; LD50 o-m: 10000;The similarity of the LD proposed for bismuth and its oxide appears to arise from a confusion of attribution of these values to these compounds in the sources.

Bismuth trichloride : $BiCl_3$ • 315.34 • 7787-60-2
(P) ΔH_f^0 = –379.1(s)
(I) NFPA: –; –
(T) NFPA: 2
(D) LD50 o-r: 3334; LD50 o-m: 2250

Hydrocarbons

8 • TABLE OF ORGANIC COMPOUNDS

Name: Empirical formula • Molar mass(g/mol)• Registry number (CAS).
(P) **Physical data:** ΔH_f^0 (kJ/mol) (physical state) • Eb (°C/mmHg) (if substance volatile) • P_{vap} (mBar) (°C) (if substance volatile).
(I) **Inflammability data:** LEL/UEL (%) • P_{fla} (°C) • AIT (°C).
(T) **Toxicity:** Safety Code (Tr) • NFPA • Transport Code (TR) • LVE (VME) (ppm) F, USA, D • STEL (ppm); IDLH (ppm).
(D) **Lethal and toxic doses:** LC (mg/l/duration of exposure); LD (mg/kg).

HYDROCARBONS
• Alkanes, cycloalkanes

Methane: CH_4 • 16.05 • 74-82-8
(P) $\Delta H_f^0 = -74.82$ (g) • Eb = –161
(I) LEL/UEL = 5/15; 5.3/15 • P_{fla} = –223; –188; –183; AIT = 537; 595; 650
(T) NFPA: 1 • TR: 223

Ethane: C_2H_6 • 30.08 • 74-84-0
(P) $\Delta H_f^0 = -84.68$ (g) • Eb = –88 • P_{vap} = 38500(20); 46900(30)
(I) LEL/UEL = 2.9/13; 3/12.5; 3.2/12.5 • P_{fla} = –183; –135; –130; –93 • AIT = 471; 510; 515; 530
(T) NFPA: 1 • TR: 223

Propane: C_3H_8 • 44.11 • 74-98-6
(P) $\Delta H_f^0 = -103.85$ (g) • Eb = –42 • P_{vap} = 7700(20); 8500(20); 9064(20); 8531(21); 10800(30); 17200(50)
(I) LEL/UEL = 1.7/9.5; 2.1/9.5; 2.2/9.5; 2.3/9.5 • P_{fla} = –104 • AIT = 470; 450; 432; 467
(T) NFPA: 1 • TR: 23 • TWA (USA): 1000; MAK (D): 1000

Cyclopropane: C_3H_6; 42.09; 75-19-4
(P) $\Delta H_f^0 = 53.30$ (g) • Eb = –33 • P_{vap} = 5300
(I) LEL/UEL = 2.4/10.4; 2.4/10.3 • P_{fla} = • AIT = 497; 495; 500
(T) NFPA: 1 • TR: 23

Butane: C_4H_{10} • 58.14 • 106-97-8
(P) $\Delta H_f^0 = -126.15$ (g); –147.75 (l) • Eb = –0.5 • P_{vap} = 1150(20); 2046(19); 2100(20); 2027(19)
(I) LEL/UEL = 1.5/8.5; 1.8/8.4; 1.9/8.5 • P_{fla} = –60; –138 • AIT = 365; 405; 429
(T) NFPA: 1 • TR: 23 • LVE(MVE) (F) –(800); TWA (USA): 800; MAK (D): 1000
(D) LC50 r: 658/4h; m: 680/2h

2-Methylpropane: C_4H_{10} • 58.14 • 75-28-5
(P) $\Delta H_f^0 = -134.52$ (g) • Eb = –12 • P_{vap} = 2047(21); 2400(16); 3000(20); 4100(30); 6900(50)
(I) LEL/UEL = 1.4/8.3; 1.8/8.4; 1.8/8.5; 1.9/8.5 • P_{fla} = –83; –81 • AIT = 460; 462
(T) NFPA: 1 • TR: 23; MAK (D): 1000
(D) LC 50 r: 1352/15'

Pentane: C_5H_{12}; 72.17; 109-66-0
(P) $\Delta H_f^0 = -146.44$ (g) • Eb = 35-36.1; 18.5/400; 1.9/200; –12.6/100; 3.6 • P_{vap} = 573(20); 568(20); 532(20); 666(25); 682(30); 1600(50); 1865(55)
(I) LEL/UEL = 1.4/7.8; 1.45/7.5; 1.4/~8.3; 1.5/7.8; 1.4/8 • P_{fla} = –49; < –40; –40cc • AIT = 260; 285; 309
(T) NFPA: 1 • TR: 33 • LVE(MVE) (F) –(600); TWA (USA): 600; MAK (D): 1000 • STEL: 750; IDHL: 16000
(D) LC50 r: 364/4h; MLC m: 384/?; LD50 iv-m: 446

2-Methylbutane: C_5H_{12}; 72.17; 78-78-4
(P) $\Delta H_f^0 = -154.47$ (g) • Eb = 28-30 • P_{vap} = 791(20); 780(20); 793(21); 791(21); 1100(30); 2100(50); 2356(55)
(I) LEL/UEL = 1.4/7.6; 1.4/8.3 • P_{fla} = 56; < –51; –51 • AIT = 420
(T) NFPA: 1 • TR: 33 • TWA (USA): 120 (alkanes); MAK (D): 1000
(D) MLC m: 419/?; LD50 iv-m: 4015

2,2-Dimethylpropane: C_5H_{12} • 72.17 • 463-82-1
(P) $\Delta H_f^0 = -165.98$ (g) • Eb = 9.5 • P_{vap} = 1500(20); 2100(30); 3700(50)
(I) LEL/UEL = 1.4/7.5 • P_{fla} = –65; < –7 • AIT = 450
(T) NFPA: - • TR: 23 • MAK(D): 1000
(D) LD50 ip-m: 100

Cyclopentane: C_5H_{10} • 70.15 • 287-92-3
(P) $\Delta H_f^0 = -77.24$(g); –105.77 (l) • Eb = 48-49.5; P_{vap} = 350(20); 520(30); 532(31); 1040(50)
(I) LEL/UEL = 1.1/8.7; 1.5/8.7 • P_{fla} = –42; –37cc; –25; –20; < –7; –7 • AIT = 360; 380
(T) NFPA: 1 • TR: 33 • LVE(MVE)(F): – (600); TWA (USA): 600
(D) MLC m: 110/?

Hexane: C_6H_{14} • 86.20 • 110-54-3
(P) $\Delta H_f^0 = -167.19$ (g); –198.82 (l) • Eb = 68-70 • P_{vap} = 133(16); 160(20); 176(20); 240(25); 248(30); 540(50)
(I) LEL/UEL = 1.1/7.4; 1.2/7.4; 1.2/7.5; 1.2/7.7; 1/7.8.; 1.1/7.5 • P_{fla} = –26; –23; –13; –11Tag cc • AIT = 225; 233; 240; 261
(T) Tr: 48/20 • NFPA: 1 • TR: 33 • LVE(MVE)(F): – (50); TWA (USA): 50; MAK(D): 50 • IDLH: 5000
(D) MLC m: 95/?; 120/?; 143/?; LD50 o-r: 28710; 32340; 50 ip-r: 9100

385

8 • Table of organic compounds

Hydrocarbons

Name: Empirical formula • Molar mass(g/mol)• Registry number (CAS).
(P) Physical data: ΔH_f^0 (kJ/mol) (physical state) • Eb (°C/mmHg) (if substance volatile) • P_{vap} (mBar) (°C) (if substance volatile).
(I) Inflammability data: LEL/UEL (%) • P_{fla} (°C) • AIT (°C).
(T) Toxicity: Safety Code (Tr) • NFPA • Transport Code (TR) • LVE (VME) (ppm) F, USA, D • STEL (ppm); IDLH (ppm).
(D) Lethal and toxic doses: LC (mg/l/duration of exposure); LD (mg/kg).

2-Methylpentane: C_6H_{14} • 86.20 • 107-83-5
- **(P)** $\Delta H_f^0 = -174.47$ (g) • Eb = 59-60.3 • $P_{vap} = 133(20)$; 240(21); 290(38)
- **(I)** LEL/UEL = 1/7.4; 1/7; 1.2/7 • $P_{fla} = -40; -31; -23; -7$ • AIT = 260; 300; 306
- **(T)** NFPA: 1 • TR: 33 • LVE(MVE) (F): – (500); TWA (USA): 500; MAK(D): 200 • STEL: 1000

3-Methylpentane: C_6H_{14} • 86.20 • 94-14-0
- **(P)** $\Delta H_f^0 = -171.63$ (g) • Eb = 63-64 • $P_{vap} = 490(38); 133(10.5)$
- **(I)** LEL/UEL = ~1.2/~7.7; 1.2/7 • $P_{fla} = -7; \sim -7$ • AIT = 300
- **(T)** NFPA: 1 • TR: 33 • LVE(MVE) (F): – (500); TWA (USA): 500; MAK(D): 200 • STEL: 1000

2,2-Dimethylbutane: C_6H_{14} • 86.20 • 75-83-2
- **(P)** $\Delta H_f^0 = -185.56$ (g) • Eb = 49-50 • $P_{vap} = 533(31)$
- **(I)** LEL/UEL = 1.2/7.0; 1.2/7.7; 1.2/7(100) • $P_{fla} = -48; < -34; -29$ • AIT = 425
- **(T)** NFPA: 0; 1 • TR: 33

Cyclohexane: C_6H_{12} • 84.18 • 110-82-7
- **(P)** $\Delta H_f^0 = -123.14$ (g); -156.23 (l) • Eb = 80-81; 60.8/400; 42/200; 25.5/100; 14.7/60; 6.7/40 • $P_{vap} = 104(20); 127(20); 160(30); 275(50); 133(61)$
- **(I)** LEL/UEL = 1.2/8.3; 1.3/8.4; 1.1/8.7; 1/9 • $P_{fla} = -26; -20; -18$Tag cc • AIT = 245; 260
- **(T)** NFPA: 1 • TR: 33 • LVE(MVE) (F): 375 (300); TWA (USA): 300; MAK(D): 300
- **(D)** LC50 : 70/?; (Mammals); MLC m: 11.92/?; LD50 o-r: 12700; 29820; o-m: 813; ip-m: 1297; LDM iv-ra: 77 o-ra: 5500

Methylcyclopentane: C_6H_{12} • 84.18 • 96-37-7
- **(P)** $\Delta H_f^0 = -106.69$ (g) • Eb = 72 • $P_{vap} = 133(18); 150(20); 230(30); 480(50)$
- **(I)** LEL/UEL = 1.0/8.3 • $P_{fla} = -29; -17; < -7; -7$ • AIT = 258; 315
- **(T)** TR: 33

Heptane: C_7H_{16} • 100.20 • 142-82-5
- **(P)** $\Delta H_f^0 = -187.78; -224.39$ (l) • Eb = 98-99; -2.1/10 • $P_{vap} = 48(20); 66.5(25); 110.4(37.)$
- **(I)** LEL/UEL = 1.1/6.7 • $P_{fla} = -4$ cc; -1 oc • AIT = 220; 223
- **(T)** NFPA: 1 • TR: 33 • LVE(MVE) (F): – (400); TWA (USA): 400; MAK(D): 500 • STEL: 500
- **(D)** LC50 r: 103/4h; MLC m: 64.915/?; LD50 iv-m: 222

Cycloheptane: C_7H_{14} • 98.19 • 291-64-5
- **(P)** $\Delta H_f^0 = -156.77$ (l) • Eb = 116-119 • $P_{vap} = 58.7(38)$
- **(I)** LEL/UEL = 1.1/ • $P_{fla} = 6; 15; 20$
- **(T)** TR: 33

Methylcyclohexane: C_7H_{14} • 98.19 • 108-87-2
- **(P)** $\Delta H_f^0 = -154.77$ (g); -190.16 (l) • Eb = 101 • $P_{vap} = 48(20); 49.3(20); 53(22); 111(38); 133(42)$
- **(I)** LEL/UEL = 1.1/–; 1.1/6.7; 1.2/6.7 • $P_{fla} = -4$ cc; 4 • AIT = 250; 260; 265; 285
- **(T)** NFPA: 2 • TR: 33 • LVE(MVE) (F): 400; TWA (USA): 400; MAK(D): 500
- **(D)** LC50 m: 41.5/2h; ra: 61/?; LD50 o-m: 2250; o-ra: 4000

Octane: C_8H_{18} • 114.26 • 111-65-9
- **(P)** $\Delta H_f^0 = -208.45$ (g); -295.95 (l) • Eb = 125-126; 66/100; 19.2/10; $-14/1$ • $P_{vap} = 13.3(19); 15(20); 19(25); 67(50)$
- **(I)** LEL/UEL = 0.8/6; 1/3.2; 1/4.7; 1/6 • $P_{fla} = 8$; 13 cc; 15; 22co • AIT = 190; 210; 220; 226; 240
- **(T)** NFPA: 0 • TR: 33 • LVE(MVE) (F): – (300); TWA (USA): 300; MAK(D): 500 • STEL: 375
- **(D)** LC50 r: 118/4h

2,2,4-Trimethylpentane: C_8H_{18} • 114.26 • 540-84-1
- **(P)** $\Delta H_f^0 = -224.24$ (g); -259.28 (l) • Eb = 99.3 • $P_{vap} = 54.7(20); 54.12(21); 117.3(38)$
- **(I)** LEL/UEL = 1/6; 1.1/6 • $P_{fla} = -12$ cc; -7; 4 • AIT = 410; 415; 418
- **(T)** NFPA: 2 • TR: 33 • MAK(D): 500 • STEL: 400

Cyclooctane: C_8H_{16} • 112.22 • 292-64-8
- **(P)** $\Delta H_f^0 = -169,,79$ (l) • Eb = 150-152 • $P_{vap} = 21.3(38)$
- **(I)** LEL/UEL = 0.95/– • $P_{fla} = 28$ • AIT = 290

Ethylcyclohexane: C_8H_{16} • 112.22 • 1678-91-7
- **(P)** $\Delta H_f^0 = -212.21$ (l) • Eb = 130-132 • $P_{vap} = 13(20); 33(38); 61.5(50)$
- **(I)** LEL/UEL = 0.9/6.6 • $P_{fla} = 19; 35$ • AIT = 260; 262

Nonane: C_9H_{20} • 128.29 • 111-84-2
- **(P)** $\Delta H_f^0 = -229.14$ (g); -275.47 (l) • Eb = 150.8 • $P_{vap} = 5(20); 4.7(20); 4.26(20); 13.3(38); 8.7(30); 24(50)$
- **(I)** LEL/UEL = 0.7/5.6; 0.8/2.9; 0.9/2.9 • $P_{fla} = 31$ • AIT = 190; 204; 206
- **(T)** NFPA: 0 • TR: 30
- **(D)** LC50 r: 16.5/4h; LD50 iv-m: 218

Decane: $C_{10}H_{22}$ • 142.29 • 124-18-5
- **(P)** $\Delta H_f^0 = -301.04$ (l) • Eb = 173-174.1; 58/10 • $P_{vap} = 1.33(16.5); 1.33(20); 1.9(20); 2.9(30); 10(50)$
- **(I)** LEL/UEL = 0.7/5.4; 0.8/2.6; 0.8/5.4 • $P_{fla} = 44$ Tag oc; 46 cc • AIT = 205; 208; 210; 250
- **(T)** NFPA: 0 • TR: 30
- **(D)** LC50 m: 72.3/2h

Decahydronaphthalene: (Z+E); $C_{10}H_{18}$ • 138.28 • 91-17-8
- **(P)** Eb = 186-190
- **(I)** LEL/UEL = 0.7(100)/4.9(100) • $P_{fla} = 50; 52; 54; 58$ • AIT = 255; 262
- **(T)** Tr: 21-36/37/38 • NFPA: 2 • TR: 30
- **(D)** LC50 r: 4/4h; LD50 o-r: 4170; cu-ra: 5900

8 • Table of organic compounds — Hydrocarbons

Name: Empirical formula • Molar mass(g/mol)• Registry number (CAS).
(P) Physical data: ΔH_f^0 (kJ/mol) (physical state) • Eb (°C/mmHg) (if substance volatile) • P_{vap} (mBar) (°C) (if substance volatile).
(I) Inflammability data: LEL/UEL (%) • P_{fla} (°C) • AIT (°C).
(T) Toxicity: Safety Code (Tr) • NFPA • Transport Code (TR) • LVE (VME) (ppm) F, USA, D • STEL (ppm); IDLH (ppm).
(D) Lethal and toxic doses: LC (mg/l/duration of exposure); LD (mg/kg).

Undecane: $C_{11}H_{24}$ • 156.31 • 1120-21-4
- (P) ΔH_f^0 = –270.42 (g); –326.56 (l) • Eb = 195-196 • P_{vap} = < 0.53; 1.4(20)
- (I) LEL/UEL = 0.6/6.5 • P_{fla} = 59; 65; 66 oc• AIT = 240
- (T) NFPA: 0
- (D) LD50 iv-m: 517

Dodecane: $C_{12}H_{26}$ • 170.38 • 112-40-3
- (P) ΔH_f^0 = –352.13 (l) • Eb = 213-216.3 • P_{vap} = 1.33(20); 1.33(47)
- (I) LEL/UEL = 0.6/– • P_{fla} = 71; 74; 83 • AIT = 200; 202; 204
- (T) NFPA: 0
- (D) MLD iv-m: 2672

Cyclododecane: $C_{12}H_{24}$ • 168.32 • 294-62-2
- (P) Eb = 60/1
- (I) LEL/UEL = • P_{fla} = 89

HYDROCARBONS
• Alkenes, cylcoalkenes, alkynes

Ethylene: C_2H_4 • 28.06 • 74-85-1
- (P) ΔH_f^0 = 52.3 (g) • Eb = –104; –102; –102/700 • P_{vap} = 2026(–90.8); 10130(–52.8); 30397(–14.2); 50663(8.9); 51680(10)
- (I) LEL/UEL = 2.7/34; 2.7/36; 3/34; 3/36 • P_{fla} = –135; –100 • AIT = 425; 450; 490; 543
- (T) Tr: 40 • NFPA: 1 • TR: 23 ou 223;if liquefied
- (D) MLC m: 1108/?

Acetylene: C_2H_2 • 26.04 • 74-86-2
- (P) ΔH_f^0 = 226.73 (g) • Eb = –75; –84 • P_{vap} = 40530(17); 17237(21); >40000(20)
- (I) LEL/UEL = 1.5/82; 2.2/80; 2.3/100; 2.5/82; 2.5/80; 3/65 • P_{fla} = –18 Tag cc • AIT = 300; 305; 335
- (T) NFPA: 1
- (D) MLC r: 974/?

Propylene: C_3H_6 • 42.09 • 115-07-1
- (P) ΔH_f^0 = 20.42 (g) • Eb = –48 • P_{vap} = 10130(19.8); 9600 (20); 10500 (20); 13100(30); 15400(38); 20700(50)
- (I) LEL/UEL = 1.4/7.1; 2/11; 2/11.1; 2/11.7; 2.4/10.1; 2.4/11.1; 2.7/36 • P_{fla} = –108; –100; AIT = 455; 459; 497
- (T) NFPA: 1 • TR: 23

Allene: C_3H_4 • 40.07 • 463-49-0
- (P) ΔH_f^0 = 192.13 (g) • Eb = –34.5 • P_{vap} = 7600(20); 9059(21); 7380(21)
- (I) LEL/UEL = 1.5/11.5; 1.7/12; 2.1/11.5; 2.1/13

Propyne: C_3H_4 • 40.07 • 74-99-7
- (P) ΔH_f^0 = 185.43 (g) • Eb = –23 • P_{vap} = 5166(20)
- (I) LEL/UEL = 1.7/11.7; 2.15/12.5; 2.4/11.7; 2.5/80 • P_{fla} = –51
- (T) NFPA: 2 • TWA (USA): 1000; MAK(D): 1000

1-Butene: C_4H_8 • 56.11 • 106-98-9 (25167-67-3)
- (P) ΔH_f^0 = –0.13 (g) • Eb = –6.3 • P_{vap} = 1650(20); 2500(20); 2585(21); 2550(21.1); 4638 (21); 3400(30); 6000(50)
- (I) LEL/UEL = 1.6/9.3; 1.6/10 • P_{fla} = –80; –12 • AIT = ~371; 384; 440
- (T) NFPA: 1 • TR: 23

2-Butene (E): C_4H_8 • 56.11 • 624-64-6
- (P) ΔH_f^0 = –11.17 (g) • Eb = 1; 3 • P_{vap} = 2122(21)
- (I) LEL/UEL = 1.8/9.7 • P_{fla} = –73; –20; < –6 • AIT = 324
- (T) NFPA: 1
- (D) LC50 m: 1.216/?

2-Butene (Z): C_4H_8 • 56.11 • 590-18-1
- (P) ΔH_f^0 = –6.99 (g) • Eb = 3.7; 2.5; 4 • P_{vap} = 1880(21)
- (I) LEL/UEL = 1.7/9 • P_{fla} = –73; –12 • AIT = 324

Isobutene: C_4H_8 • 56.11 • 115-11-7
- (P) ΔH_f^0 = –16.9 (g) • Eb = –7; 49.3/100; –67.9/30; –82/10; –105/1 • P_{vap} = 1700(20); 1820(10); 1675(21); 2600(20); 2650 (21); 3500(30); 6100(50)
- (I) LEL/UEL = 1.8/8.8; 1.8/9.6; 1.6/–; 1.6/10 • P_{fla} = –80; –76; < –10 • AIT = 465
- (T) NFPA: 1 • TR: 23
- (D) LC50 r: 620/4h; m: 415/2h

1,3-Butadiene: C_4H_8 • 56.11 • 106-99-0
- (P) ΔH_f^0 = 110.16 (g) • Eb = – 4 • P_{vap} = 2026(15.3); 1500(20); 2460(21); 2453(21); 5066(47); 10130(76)
- (I) LEL/UEL = 1.1/12.5; 2/11.5; 2/12; 2/11; 1.1/11.5 • P_{fla} = –116; –85; –76; –60; <–17; AIT = 152; 415; 420; 429; 450
- (T) Tr: 45 • NFPA: 2 • TR: 239 • TWA (USA): 10; MAK(D): 15 ppm after polymerisation. Otherwise: 5 ppm
- (D) LC50 r: 285/4h ; m: 259/2h; 270/2h; l: 552/?; MLC ra: 583/?; LD50 o-r: 5480

1-Butyne: C_4H_6 • 54 • 107-00-6
- (P) ΔH_f^0 = 165.18 (g) • Eb = 8.3 • P_{vap} = 1650(21); 2688(37)
- (I) P_{fla} = < –34; <7; < –1 Tag co

2-Butyne: C_4H_6 • 54 • 503-17-3
- (P) ΔH_f^0 = 146.31 (g) • Eb = 24-27 • P_{vap} = 791(20)
- (I) P_{fla} = < –34; –31; –25

1-Pentene: C_5H_{10} • 70.14 • 109-76-1
- (P) ΔH_f^0 = –20.92 (g) • Eb = 30-31 • P_{vap} = 665(20); 708(20); 720(20); 1012(30)
- (I) LEL/UEL = 1.4/8.7; 1.5/8.7 • P_{fla} = –28 cc; –18 cc • AIT = 271; 290
- (T) NFPA: 1 • TR: 33

8 • Table of organic compounds — Hydrocarbons

Name: Empirical formula • Molar mass(g/mol)• Registry number (CAS).
Ⓟ Physical data: ΔH_f^0 (kJ/mol) (physical state) • Eb (°C/mmHg) (if substance volatile) • P_{vap} (mBar) (°C) (if substance volatile).
Ⓘ Inflammability data: LEL/UEL (%) • P_{fla} (°C) • AIT (°C).
Ⓣ Toxicity: Safety Code (Tr) • NFPA • Transport Code (TR) • LVE (VME) (ppm) F, USA, D • STEL (ppm); IDLH (ppm).
Ⓓ Lethal and toxic doses: LC (mg/l/duration of exposure); LD (mg/kg).

2-Pentene (E): C_5H_{10} • 70.14 • 646-04-8
Ⓟ $\Delta H_f^0 = -31.76$ (g) • Eb = 35-36; 37
Ⓘ $P_{fla} = -45; -18$
Ⓣ NFPA: 0 • TR: 33

2-Pentene (Z): C_5H_{10} • 70.14 • 627-20-3
Ⓟ $\Delta H_f^0 = -28.07$ (g) • Eb = 36-37 • $P_{fla} = -27; -18$;
Ⓣ NFPA: 2

2-Methyl-1-butene: C_5H_{10} • 70.14 • 563-46-2
Ⓟ $\Delta H_f^0 = -36.32$ (g) • Eb = 38-39
Ⓘ $P_{fla} = -48; \sim -28; -20; < -7$

3-Methyl-1-butene: C_5H_{10} • 70.14 • 563-45-1
Ⓟ $\Delta H_f^0 = -28.95$ (g) • Eb = 20-20.6; 31.11 • $P_{vap} = 1032(20); 1150(20); 1821(38)$
Ⓘ LEL/UEL = 1.4/–; 1.4/8.7; 1.5/9.1; 1.6/9.1 • $P_{fla} = -58; < -7; -6$ • AIT = 365
Ⓣ Tr: 36/37/38 • NFPA: 2

2-Methyl-2-butene: C_5H_{10} • 70.14 • 513-35-9 (26760-64-5)
Ⓟ $\Delta H_f^0 = -42.55$ (g) • Eb = 30; 37-39 • $P_{vap} = 528$
Ⓘ LEL/UEL = 1.5/9; 1.6/8.7 • $P_{fla} = -45; -20; -18; -18$co • AIT = 273
Ⓣ NFPA: 2

Isoprene: C_5H_8 • 68.13 • 78-79-5
Ⓟ $\Delta H_f^0 = 75.73$ (g) • Eb = 33-34 • $P_{vap} = 405(10); 533(15.4); 640(20);$
Ⓘ LEL/UEL = 1/9.7; 1.5/10; 2/9 • $P_{fla} = -54; -48$ • AIT = 220; 427; 440
Ⓣ NFPA: 2 • TR: 339
Ⓓ LC50 r: 180/4h; m: 139/2h

1.3-Pentadiene (E): C_5H_8 • 68.13 • 2004-70-8
Ⓟ $\Delta H_f^0 = 78.07$ (g) • Eb = 42 • $P_{vap} = 827(38)$
Ⓘ LEL/UEL = 2/8.3; 1/7 • $P_{fla} = -43$cc ; -28cc
Ⓓ LC50 r: 140/2h; m: 1.1/2h; LD50 iv-m: 18

Cyclopentene: C_5H_8 • 68.13 • 142-29-0
Ⓟ $\Delta H_f^0 = 32.93$ (g); 4.27 (l) • Eb = 44-46; 44.24 • $P_{vap} = 421(20); 1190(50); 1895(65);$
Ⓘ LEL/UEL = 1.5/– • $P_{fla} = -34; -29$ • AIT = 394
Ⓣ Tr: 21/22-(36/37/38) • NFPA: 1 • TR: 33
Ⓓ LC50 m: 2.76/? ; LD50 o-r: 1660; 2140; cu-ra: 1230; 1590

Cyclopentadiene: C_5H_6 • 66.11 • 542-92-7
Ⓟ $\Delta H_f^0 = 133.89$ (g) • Eb = 40-43 • $P_{vap} = 330(12)$
Ⓘ $P_{fla} = -3; 25$ • AIT = 640
Ⓣ LVE(MVE) (F): –75 • TWA (USA): 75; MAK(D): 75

1-Pentyne: C_5H_8 • 68.12 • 627-19-0
Ⓟ $\Delta H_f^0 = 144.35$ (g) • Eb = 39-41 • $P_{vap} = 469(20)$
Ⓘ $P_{fla} = < -34; -34; -20$
Ⓣ NFPA: 1 • TR: 33

2-Pentyne: C_5H_8 • 68.12 • 627-21-4
Ⓟ $\Delta H_f^0 = 128.87$ (g) • Eb = 56.1 • $P_{vap} = 524(38)$
Ⓘ $P_{fla} = -30; -20$
Ⓣ NFPA: 1

1-Hexene: C_6H_{12} • 84.16 • 592-41-6
Ⓟ $\Delta H_f^0 = -41.67$ (g); –72.38 (l) • Eb = 62-64.5 • $P_{vap} = 180(20); 207(21); 280(30); 413(38); 580(50)$
Ⓘ LEL/UEL = 1.2/6.9; 1.2/– • $P_{fla} = -26$cc; –20; –9 • AIT = 252; 265
Ⓣ NFPA: 1 • TR: 33

2-Hexene: C_6H_{12} • 84.16 • 592-43-8
Ⓟ $\Delta H_f^0 = -51.34$ (l) [E]; –52.51 (l) [Z] • Eb = 67-69
Ⓘ $P_{fla} = -20; < -7$
Ⓣ NFPA: 1

2,3-Dimethyl-1-butene: C_6H_{12} • 84.16 • 563-78-0
Ⓟ $\Delta H_f^0 = -55.73$ (g) • Eb = 56 • $P_{vap} = 550(38)$
Ⓘ LEL/UEL = 1.2/– • $P_{fla} = -29; -8$ • AIT =359

3,3-Dimethyl-1-butene: C_6H_{12} • 84.16 • 558-37-2
Ⓟ $\Delta H_f^0 = -43.14$ (g) • Eb = 40-42;
Ⓘ LEL/UEL = 1.2/– • $P_{fla} = -29; -8$

2,3-Dimethyl-2-butene: C_6H_{12} • 84.16 • 563-79-1
Ⓟ $\Delta H_f^0 = -59.2$ (g) • Eb = 73 • $P_{vap} = 480(20)$
Ⓘ $P_{fla} = -18; -8$ • AIT = 549

Cyclohexene: C_6H_{10} • 82.15 • 110-83-8
Ⓟ $\Delta H_f^0 = -3.36$ (g); –38.83 (l) • Eb = 82-84 • $P_{vap} = 90(20); 94(20); 140(30); 213(38); 340(50)$
Ⓘ LEL/UEL = 1.2/–; 1/5; 1.3/7.7; 1.2(100)/– • $P_{fla} = -30; -20; -12; < -6; -6$ • AIT = 310
Ⓣ NFPA: 1 • TR: 33 • LVE(MVE) (F): –(300); TWA (USA): 300; MAK(D): 300
Ⓓ LD50 o-r: 1940

4-Methylcyclohexene: C_7H_{12} • 96.17 • 591-47-9
Ⓟ Eb = 101-103 • $P_{vap} = 34(38)$
Ⓘ $P_{fla} = -7; -1$ oc ; –1
Ⓣ NFPA: 1

4-Vinylcyclohexene: C_8H_{12} • 108.2 • 100-40-3
Ⓟ Eb = 126-131 • $P_{vap} = 13.6(20); 15(20); 34(38)$
Ⓘ $P_{fla} = 14; 16$cc; 16oc; 19; 21Tag oc • AIT = 270; 280
Ⓣ Tr: 38 (40) • TWA (USA): 0.1
Ⓓ LC50 r: 36/4h ; m: 27/?; LD50 o-r: 2563; cu-ra: 16640

Essence of turpentine: $C_{10}H_{16}$ • 136.24 • 8006-64-2
Ⓟ Eb = 154-175 • $P_{vap} = 5.33(20); 6.67(25)$
Ⓘ LEL/UEL = 0.8/–; 0.8/6 • $P_{fla} = 31-46; 35$cc • AIT = 250; 253
Ⓣ NFPA: 1 • TR: 30 • LVE(MVE) (F): –100; TWA (USA): 100; MAK(D): 100
Ⓓ LC50 m: 19/2h; LD50 iv-m: 1180; o-r: 5760

8 • Table of organic compounds

Hydrocarbons

Name: Empirical formula • Molar mass(g/mol)• Registry number (CAS).
- **(P) Physical data:** ΔH_f^0 (kJ/mol) (physical state) • Eb (°C/mmHg) (if substance volatile) • P_{vap} (mBar) (°C) (if substance volatile)
- **(I) Inflammability data:** LEL/UEL (%) • P_{fla} (°C) • AIT (°C).
- **(T) Toxicity:** Safety Code (Tr) • NFPA • Transport Code (TR) • LVE (VME) (ppm) F, USA, D • STEL (ppm); IDLH (ppm).
- **(D) Lethal and toxic doses:** LC (mg/l/duration of exposure); LD (mg/kg).

α–Pinene:
$C_{10}H_{16}$ • 136.24 • 7785-70-8
- (P) $\Delta H_f^0 = -16.32(l)$ • Eb = 154-156; 52.5/20 • P_{vap} = 13(38)
- (I) P_{fla} = 33 • AIT = 255
- (T) Tr: 36/37/38

β–Pinene:
$C_{10}H_{16}$ • 136.24 • 80-56-8 (19902-08-0)
- (P) $\Delta H_f^0 = -7.53$ • Eb = 162-165 • P_{vap} = 2.7(20)
- (I) P_{fla} = 32; 36
- (T) Tr: 36/37/38 • NFPA: 1
- (D) LD50 o-r: 4700

Limonene: $C_{10}H_{16}$ • 136.24 • 138-86-3
- (P) $\Delta H_f^0 = -54.39$ (l) • Eb = 170-180; 175-177; 110/100; 71/20 • P_{vap} = 1.33(20); 2.1(20); 3.6(30); 11.2(50)
- (I) LEL/UEL = 0.7/6.1(150) • P_{fla} = 42; 43; 44; 45; 46 • AIT = 237; 255
- (T) NFPA: 2 • TR: 30
- (D) LD50 o-r: 4400; o-m: 5550; 5600; ip-r: 3600; iv-r: 110; ip-m: 600; ivm: 1010

1-Heptene: C_7H_{14} • 98.19 • 592-76-7
- (P) $\Delta H_f^0 = -62.30$ (g); –97.41 (l) • Eb = 94 • P_{vap} = 160(20); 135(38); 225(50); 395(65);
- (I) LEL/UEL = 2.7/34; 2.7/36 • $P_{fla} = -8; <-1; -1$ • AIT = 260; 375
- (T) Tr: 20/22 • TR: 33

2-Heptene (E): C_7H_{14} • 98.19 • 14686-13-6
- (P) Eb = 95-98 • P_{vap} = 117(38)
- (I) LEL/UEL = 1/– • $P_{fla} = < -34; -1$

3-Heptene (E): C_7H_{14} • 98.19 • 14686-14-7
- (P) Eb = 95-97
- (I) $P_{fla} = < -7; -7; -1$

Cycloheptene: C_7H_{12} • 96.17 • 628-92-2
- (P) Eb = 112-116
- (I) $P_{fla} = < 23; -6; 15$
- (T) TR: 33

1-Octene: C_8H_{16} • 112.2 • 111-66-0
- (P) $\Delta H_f^0 = -82.97$ (g); –123.51 (l); • Eb = 121-122; 61.6/100 • P_{vap} = 47(20); 47(38)
- (I) LEL/UEL = 0.7/3.9 • P_{fla} = 8; 21 oc • AIT = 250
- (T) NFPA: 1 • TR: 33
- (D) LD50 o-r: > 10000; cu-ra: > 10000

2-Octene (E): C_8H_{16} • 112.2 • 13389-42-9
- (P) Eb = 122-124 • P_{vap} = 41.3(38)
- (I) LEL/UEL = 0.9/– • P_{fla} = 14; 21
- (T) NFPA: 1 • TR: 33

3-Octene (E): C_8H_{16} • 112.2 • 14919-01-8
- (P) Eb = 121-122 • P_{vap} = 50(38)
- (I) LEL/UEL = 0.9/– • P_{fla} = 14

2.4.4-Trimethyl-1-pentene: C_8H_{16} • 112.2 • 107-39-1 (25167-70-8)
- (P) $\Delta H_f^0 = -147.32$ (l) • Eb = 100-102; 118 • P_{vap} = 103(38); 110(38)
- (I) LEL/UEL = 0.8/4.8; 0.9/– • P_{fla} = 29; –7; –4cc; 2co • AIT = 305; 390; 415
- (T) NFPA: 1 • TR: 33

2.4.4-Trimethyl-2-pentene: C_8H_{16} • 112.2 • 107-40-4
- (P) $\Delta H_f^0 = -144.10$ (l) • Eb = 102-105 • P_{vap} = 111(38)
- (I) LEL/UEL = 0.9/– • $P_{fla} = -1; 2$ • AIT = 307

Dicyclopentadiene: $C_{10}H_{12}$ • 132.2 • 77-73-6
- (P) $\Delta H_f^0 = 116.73$ (s) • Eb = 164-170; 172; 64-5/15 • P_{vap} =2(20); 5(30); 13(38); 13.3(48); 11(50)
- (I) LEL/UEL = 1/10 • P_{fla} = 26; 32co; 39 • AIT = 503; 509
- (T) Tr: 20/22-36/37/38 • NFPA: 1 • TR: 30 • LVE(MVE) (F): –(5); TWA (USA): 5; MAK(D): 0.5
- (D) LC50 r: 2/4h ; s: 0.77/4h ; l: 4.17/4h; LD50 o-r: 353; 820; o-m: 190; ip-r: 200; ip-m: 200; cu-ra: 5080

HYDROCARBONS
• Aromatic compounds

Benzene: C_6H_6 • 78.12 • 71-43-2
- (P) $\Delta H_f^0 = 82.93$ (g); 48.99 (l) • Eb = 80-81; 80.093-80.094 • P_{vap} = 99.5(20); 101(20); 133(26); 155(30); 221(38); 365(50); 400(50)
- (I) LEL/UEL = 1.2/8; 1.3/8; 1.3/7.1; 1.4/7.1; 1.4/8.0 • $P_{fla} = -11; -11$ Tagcc • AIT = 498; 555; 560; 562; 580
- (T) Tr: 45-48/23/24/25 • NFPA: 3 • TR: 33 • LVE(MVE) (F): – (5); TWA (USA): 10 (project: 0.1) • STEL: 5; ILDH: 2000
- (D) LC50 r: 31.9/7h ; s: 31.84/? ; LD50 o-r: 930; 3306; 4894; o-m: 4700; ip-r: 2890; ip-m: 340; cu-ra: > 9400

Toluene: C_7H_8 • 92.15 • 108-88-3
- (P) $\Delta H_f^0 = 50.02$ (g); 12.01 (l) • Eb = 110-112; 89.5/400; 52/100; 32/40; 6.4/10 • P_{vap} = 13.3(6.4); 29(20); 29.53(20); 34.7(25); 49(30); 54(32); 70(38)
- (I) LEL/UEL = 1/7; 1.27/7; 1.3/7.0; 1.4/6.7 • P_{fla} = 4cc; 2 Tagcc; 5 Tagcc; 6; 7 Tagcc; 9 Tagcc • AIT = 480; 536; 552
- (T) Tr: 20 • NFPA: 2 • TR : 33 • LVE(MVE) (F): 150 (100); TWA (USA): 50; MAK(D): 100 • STEL: 150; ILDH: 2000
- (D) LC50 r: 49/4h; m: 20/8h; ra: 207/40'; LD50 o-r: 636; 5000; 7530; ip-r: 1332; iv-r: 1960; ip-m: 59; 640; 1120; scu-m: 2250; cu-ra: 12124; 14100

o-Xylene: C_8H_{10} • 106.18 • 95-47-6
- (P) $\Delta H_f^0 = 19.00$ (g); –24.43 (l) • Eb = 143-144.4; 122/400; 81.3/100; 59.5/40; 32.1/10; 20.2/5 • P_{vap} = 6.7(20); 6.9(20); 8(20); 8.8(25)
- (I) LEL/UEL = 1/6; 1/6.4; 1/7.6; 1.1/7 • P_{fla} = 17cc; 28; 30; 31; 32; 46Tag oc • AIT = 465
- (T) Tr: 20/21-38 • NFPA: 2 • TR: 30 • LVE(MVE) (F): 150 (100); TWA (USA): 100; MAK(D): 100 • STEL: 100; 150
- (D) LC50 r: 19.91/4h ; 19.72/4h; LD50 o-r: 4300; 5000; 7710; ip-r: 2459; 3810; ip-m: 1364; 1739; 2110; cu-ra: 14000

8 • Table of organic compounds — Hydrocarbons

Name: Empirical formula • Molar mass(g/mol)• Registry number (CAS).
(P) **Physical data:** ΔH_f^0 (kJ/mol) (physical state) • Eb (°C/mmHg) (if substance volatile) • P_{vap} (mBar) (°C) (if substance volatile).
(I) **Inflammability data:** LEL/UEL (%) • P_{fla} (°C) • AIT (°C).
(T) **Toxicity:** Safety Code (Tr) • NFPA • Transport Code (TR) • LVE (VME) (ppm) F, USA, D • STEL (ppm); IDLH (ppm).
(D) **Lethal and toxic doses:** LC (mg/l/duration of exposure); LD (mg/kg).

m-Xylene: C_8H_{10} • 106.18 • 108-38-3
(P) $\Delta H_f^0 = 17.25$ (g); −25.44 (l) • Eb = 137-139; 117/400; 77/100; 55.3/40; 28.3/10; 6.9/11 • P_{vap} = 8(20); 8.8(20); 11.06(20); 11(25); 13.3(28.3); 21(38)
(I) LEL/UEL = 1/7.6; 1.1/7.0 • P_{fla} = 25cc; 27; 29; 35 • AIT = 465; 525; 528; 530
(T) Tr: 20/21-38 • NFPA: 2 • TR: 30 • LVE(MVE) (F): 150 (100); TWA (USA): 100; MAK(D): 100 • STEL: 100; 150
(D) LC50 r: 19.91/4h ; 19.72/4h; LD50 o-r: 4300; 5000; 7710; ip-r: 2459; 3810; ip-m: 1364; 1739; 2110; cu-ra: 14000

p-Xylene: C_8H_{10} • 106.18 • 106-42-3
(P) $\Delta H_f^0 = 17.96$ (g); −24.43 (l) • Eb = 137-138.5; 116/400; 76/100; 54/40; 27/10; −8.1/1 • P_{vap} = 8.2(20); 11.5(20); 12(20); 13.3(27.3); 16(30); 45(50)
(I) LEL/UEL = 1/7.6; 1.1/7; 1.1/7.6 • P_{fla} = 25cc; 27 Tag co; 36 • AIT = 525; 528
(T) Tr: 20/21-38 • NFPA: 2 • TR: 30 • LVE(MVE) (F): 150 (100); TWA (USA): 100; MAK(D): 100 • STEL: 100; 150
(D) LC50 r: 19.91/4h ; 19.72/4h; LD50 o-r: 4300; 5000; 7710; ip-r: 2459; 3810; ip-m: 1364; 1739; 2110; cu-ra: 14000

Ethylbenzene: C_8H_{10} • 106.18 • 100-41-4
(P) $\Delta H_f^0 = 29.79$ (g); −12.47 (l) • Eb = 136-137; 114/400; 74/100; 53/40; 26/10; −9.8/1 • P_{vap} = 9.3(20); 9.46(20); 13.3(26); 80(62)
(I) LEL/UEL = 1/6.7; 1/7.8; 1.2/6.8 • P_{fla} = 15; 18cc; 21 Tagcc • AIT = 430; 432; 466
(T) Tr: 20 • NFPA: 2 • TR: 30 (pur) • LVE(MVE) (F): − (100); TWA (USA): 100; MAK(D): 100 • STEL: 125
(D) LC50 r: 13.4/4h; m: 50/2h; LD50 o-r: 3500; 5460; ip-m: 2270; cu-ra: 17800

Styrene: C_8H_8 • 104.16 • 100-42-5
(P) $\Delta H_f^0 = 147.36$ (g); 103.89 (l) • Eb = 142-146; 46/20; 33.6/10; 31/10 • P_{vap} = 6(20); 6.52(20); 6.7(20); 11.7(30); 12.7(30); 16.5(38); 35(50)
(I) LEL/UEL = 1.1/6.1; 1.1/8; 1/6 • P_{fla} = 32; 31 Tagcc • AIT = 490
(T) NFPA: 2 • TR: 39 • LVE(MVE) (F):− (50); TWA (USA): 50; MAK(D): 20 • STEL: 100
(D) LC50 r: 25/4h ; m: 9.5/4h; LD50 o-r: 2650; 5000; o-m: 316; ip-r: 898; ip-m: 660 44.5; iv-m: 90 5.2

Phenylacetylene: C_8H_6 • 102.14 • 536-74-3
(P) Eb = 140-143 • P_{vap} = 23.5(38)
(I) P_{fla} = 27; 30
(D) LD50 o-r: 5000; iv-m: 100

Diphenylacetylene: $C_{14}H_{10}$ • 178.24 • 01-65-5
(P) $\Delta H_f^0 = 312.38$ (s) • Eb = 300; 170/19

Propylbenzene: C_9H_{12} • 120.21 • 103-65-1
(P) Eb = 157-160; 135.6/400; 113.5/200; 94/100; 71.6/40; 56.8/20; 43.4/10; 31.3/5; 6.3/1 • P_{vap} = 2.67(20); 13.3(43)
(I) LEL/UEL = 0.8/6 • P_{fla} = 30; 42; 45; 47 • AIT = 450
(T) Tr: 37 • NFPA: 2
(D) LC50 r: 292/1h ; s: 20/? ; LD50 o-r: 6040

Cumene: C_9H_{12} • 120.21 • 98-82-8
(P) $\Delta H_f^0 = 3.93$ (g); −41.21 (l) • Eb = 152-153; 152.7 • P_{vap} = 5(20); 5.3(20); 10.66(20); 13.3(30); 12.9(38); 13.3(38); 26.7(52)
(I) LEL/UEL = 0.8/6.5; 0.9/6.5 • P_{fla} = 31; 33; 34; 39cc; 44; 46 • AIT = 420; 425
(T) Tr: 37 • NFPA:1; 0 • TR: 30 • LVE(MVE) (F): − (50); TWA (USA): 50; MAK(D): 50
(D) LC50 r: 39.26/4h; 27/4h ; s: 24.7/2h; 10/7h ; LD50 o-r: 1400; 2910; o-m: 12750; cu-ra: 12300

Allylbenzene: C_9H_{10} • 118.19 • 300-57-2
(P) Eb = 156; 50/12
(I) P_{fla} = 33; 37; 40
(D) LD50 o-r: 2900; o-m: 2900

p-Cymene: $C_{10}H_{14}$ • 134.24 • 99-87-6
(P) $\Delta H_f^0 = -78.24$ (l) • Eb = 176-177 • P_{vap} = 1.33(17); 2(20); 10.7(20); 25(65);
(I) LEL/UEL = 0.7(100)/5.6; 0.7/5.6 • P_{fla} = 47cc; 51; 53 (impure product) • AIT = 436
(T) NFPA: 2 • TR: 30
(D) LD50 o-r: 4750

2-Ethyltoluene: C_9H_{12} • 120.21 • 611-14-3
(P) Eb = 164-166
(I) P_{fla} = 39; 43; 53 • AIT = 440
(D) LC50 m: 54/4h; LD50 o-r: 5000

3-Ethyltoluene: C_9H_{12} • 120.21 • 620-14-4
(P) Eb = 159-161
(I) P_{fla} = 39; 44 • AIT = 479

4-Ethyltoluene: C_9H_{12} • 120.21 • 622-96-8
(P) Eb = 161-163;
(I) LEL/UEL = 1.2/7 • P_{fla} = 36; 39; 43 • AIT = 475; 535
(D) LD50 o-r: 5000

Mesitylene: C_9H_{12} • 120.21 • 108-67-8
(P) $\Delta H_f^0 = -63.51$ (l) • Eb = 162-165; 141/400; 99/100; 76/40; 61/20; 47.4/10; 9.6/1 • P_{vap} = 2(20); 2.48(20); 2.8(20); 3.31(25); 5.2(30); 16(50)
(I) LEL/UEL = 0.8(100)/6.1(100) • P_{fla} = 44; 46; 54 • AIT = 550; 558
(T) Tr: 37 • NFPA: 0 • TR: 30 • LVE(MVE) (F): − (25); TWA (USA): 25
(D) LC50 r: 24/4h

8 • Table of organic compounds — Hydrocarbons

Name: Empirical formula • Molar mass(g/mol)• Registry number (CAS).
(P) Physical data: ΔH_f^0 (kJ/mol) ( ) • Eb (°C/mmHg) (if substance volatile) • P_{vap} (mBar) (°C) (if substance volatile).
(I) Inflammability data: LEL/UEL (%) • P_{fla} (°C) • AIT (°C).
(T) Toxicity: Safety Code (Tr) • NFPA • Transport Code (TR) • LVE (VME) (ppm) F, USA, D • STEL (ppm); IDLH (ppm).
(D) Lethal and toxic doses: LC (mg/l/duration of exposure); LD (mg/kg).

α-Methylstyrene: C_9H_{10} • 118.19 • 98-83-9
- (P) ΔH_f^0 = –113.02 (g); 70.29 (l) • Eb = 152.4; 165-166; 47/60; 7.4/1
  • P_{vap} = 2.53(20); 2.67(20)
- (I) LEL/UEL = 0.9/6.1; 0.9/6.6; 1.9/6.1; 0.7/3.4; 0.7/6.1
  • P_{fla} = 43; 46; 54Tagcc; 56; 58co • AIT = 420; 494; 574
- (T) Tr: 36/37 • NFPA: 1
- (D) LD50 o-r: 4900; o-m: 4500

2-Methylstyrene: C_9H_{10} • 118.19 • 611-15-4
- (P) Eb = 169-171 • P_{vap} = 2.66(20)
- (I) P_{fla} = 51; 58
- (T) Tr: 20 (36/37/38) • NFPA: 2 • TR: 39 • LVE(MVE) (F): – (50); TWA (USA): 50; MAK(D): 100 • STEL: 100

3-Methylstyrene: C_9H_{10} • 118.19 • 100-80-1
- (P) Eb = 170-171 • P_{vap} = 3.5(20); 5.5(30); 14(50)
- (I) P_{fla} = 51
- (T) Tr: 20 (36/37/38) • NFPA: 2 • TR: 39 • LVE(MVE) (F): – (50); TWA (USA): 50; MAK(D): 100 • STEL: 100

4-Methylstyrene: C_9H_{10} • 118.19 • 622-97-9
- (P) Eb = 175; 60/12 • P_{vap} = < 1.33(20); 1.7(20); 3.2(30); 10(50)
- (I) LEL/UEL = 1.1/5.3 • P_{fla} = 45 • AIT = 514
- (T) Tr: 20 (36/37/38) • NFPA: 2 • TR: 39 • LVE(MVE) (F): – (50); TWA (USA): 50; MAK(D): 100 • STEL: 100
- (D) LC50 m: 3.02/? ; LD50 o-m: 1072; o-r: 2255; 4000; 3160; ip-r: 2324; ip-m: 581; iv-m: 280; cu-ra: > 5000

n-Butylbenzene: $C_{10}H_{14}$ • 134.24 • 104-51-8
- (P) ΔH_f^0 = –13.81 (g); –78.12 (l) • Eb = 180-183.2; 159.2/400; 136.9/200; 116.2/100; 102.6/60; 92.4/40; 76.3/20; 62/10; 22.7/1
  • P_{vap} = 1.33(20); 1.33(23); 1.37(23); 3.2(38)
- (I) LEL/UEL = 0.8/5.8 • P_{fla} = 54; 59; 71 oc • AIT = 410; 412
- (T) NFPA: 2 • TR: 30
- (D) LD50 o-r: 5000

s-Butylbenzene: $C_{10}H_{14}$ • 134.24 • 135-98-8
- (P) ΔH_f^0 = –66.40 (l) • Eb = 172-173.5; 150.3/400; 128.8/200; 109.5/100; 96/60; 86.2/40; 70.6/20; 57/10; 44.2/5; 18.6/1
  • P_{vap} = 1.33(18.6); 1.33(20); 5.33(38); 7(44); 20(60); 26(70); 80(96)
- (I) LEL/UEL = 0.8/6.9 • P_{fla} = 45; 52cc; 52Tag co; 63co
  • AIT = 415; 420; 417
- (T) NFPA: 2 • TR: 30
- (D) LD50 o-r: 2240; LD50 cu-ra: > 16000

t-Butylbenzene: $C_{10}H_{14}$ • 134.24 • 98-06-6
- (P) ΔH_f^0 = –70.71 (l) • Eb = 167-169.1; 145.8/400; 123.7/200; 103.8/100; 90.6/60; 80.8/40; 65.6/20; 51.7/10; 39/5; 13/1
  • P_{vap} = 1.33(13); 1.33(20); 6.4(38)
- (I) LEL/UEL = 0.7(100)/5.7(100); 0.8/5.6 • P_{fla} = 34; 44; 60Tag co
  • AIT = 445; 450
- (T) Tr: 20/21-33 • NFPA: 2 • TR: 30
- (D) MLDo-r: 10000

Isobutylbenzene: $C_{10}H_{14}$ • 134.24 • 538-93-2
- (P) ΔH_f^0 = –69.79 (l) • Eb = 173; 145.2/400; 120.7/200; 99/100; 84.1/60; 73.2/40; 54.7/20; 37.3/10; 21.1/5; 9.8/1
  • P_{vap} = 1.33(14); 5.6(38)
- (I) LEL/UEL = 0.8/6 • P_{fla} = 45; 55cc; 60 • AIT = 425; 430; 427
- (T) NFPA: 2

1.2-Diethylbenzene: $C_{10}H_{14}$ • 134.24 • 135-01-3 (25340-17-4)
- (P) ΔH_f^0 = –18.95 (g) • Eb = 183-184 • P_{vap} = 1.33(21)
- (I) P_{fla} = 49; 56 • AIT = 395
- (T) Tr: 36/37/38 • NFPA: 2
- (D) LD50 o-r: 5000

1.3-Diethylbenzene: $C_{10}H_{14}$ • 134.24 • 141-93-5
- (P) ΔH_f^0 = –21.84 (g) • Eb = 181-182
- (I) P_{fla} = 50; 55 • AIT = 449
- (T) Tr: 36/37/38 • NFPA: 2
- (D) LD50 o-r: 5000

1.4-Diethylbenzene: $C_{10}H_{14}$ • 134.24 • 105-05-5
- (P) ΔH_f^0 = –22.26 (g) • Eb = 182; 184 • P_{vap} = 1.33(20)
- (I) P_{fla} = 55 • AIT = 429
- (T) Tr: 36/37/38 • NFPA: 2
- (D) LD50 o-r: 5000

Divinylbenzene: $C_{10}H_{10}$ • 130.20 • 1321-74-0
- (P) Eb = 195-198; 200 • P_{vap} = 1(20); 1.13(30); 1.2(30)
- (I) LEL/UEL = 0.3/-; 1.1/6.2 • P_{fla} = 64; 69; 74

n-Pentylbenzene: $C_{11}H_{16}$ • 148.25 • 538-68-1
- (P) Eb = 202; 205; 81/10
- (I) P_{fla} = 65

Indene: C_9H_8 • 116.17 • 95-13-6
- (P) ΔH_f^0 = 110.42 (l) • Eb = 182; 157./400; 135.6/200; 115/110; 114.7/100; 100.8/60; 90.7/40; 74/20; 58.5/10; 44.3/5; 16.4/1
  • P_{vap} = 1.7(20); 2.9(30); 8.7(50)
- (I) P_{fla} = 51; 58; 61; 78
- (D) LC50 r: 14/?

Naphthalene: $C_{10}H_8$ • 128.18 • 91-20-3
- (P) ΔH_f^0 = 75.31 (s) • Eb = 210; 218; 193.2/400; 167.7/200; 145.5/100; 130.2/60; 119.3/40; 101.7/20; 85.8/10
  • P_{vap} = 0.04(20); 0.04(25); 0.066(20); 0.13(30)1.1(50); 1.33(53)
- (I) LEL/UEL = 0.9/5.9; 0.88/5.9 • P_{fla} = 79 Tagcc; 79co; 88Tagcc
  • AIT = 525; 540; 526; 567
- (T) Tr: 22-20/21/22 • NFPA: 2 • LVE(MVE) (F): – (10); TWA (USA): 10; MAK(D): 10 • STEL: 15; ILDH: 500
- (D) LC50 r: > 0.34/1h ; LD50 o-r: 490; 1780; o-m: 533; o-ra: 3000; ip-m: 150; iv-m: 100; scu-m: 964; cu-ra:> 20000

8 • Table of organic compounds

Alcohols-Polyols

Name: Empirical formula • Molar mass(g/mol)• Registry number (CAS).
(P) Physical data: ΔH_f^0 (kJ/mol) (physical state) • Eb (°C/mmHg) (if substance volatile) • P_{vap} (mBar) (°C) (if substance volatile).
(I) Inflammability data: LEL/UEL (%) • P_{fla} (°C) • AIT (°C).
(T) Toxicity: Safety Code (Tr) • NFPA • Transport Code (TR) • LVE (VME) (ppm) F, USA, D • STEL (ppm); IDLH (ppm).
(D) Lethal and toxic doses: LC (mg/l/duration of exposure); LD (mg/kg).

Tetrahydronaphthalene: $C_{10}H_{12}$ • 132.22 • 119-64-2
(P) $\Delta H_f^0 = -25.52$ (l) • Eb = 204-207.2; 181.8/400; 157.2/200; 135.3/100; 121.3/60; 110.4/40; 93.8/20; 79/10; 65.3/5; 38/1
• P_{vap} = 0.24(20); 0.27(20); 0.8(30); 1.33(38); 2.8(50)
(I) LEL/UEL = 0.8/5; 0.5(150)/ 5(150); 0.8(100)/5.0(150)
• P_{fla} = 71; 75; 77oc; 78; 82cc • AIT = 384; 425
(T) Tr: 36/38 • NFPA: 1 • TR: 30
(D) LD50 o-r: 1620; 2860; cu-ra: 17000

1-Methylnaphthalene: $C_{11}H_{10}$ • 142.21 • 90-12-0
(P) $\Delta H_f^0 = 56.19$ (l) • Eb = 241-245
(T) Tr: 22 • NFPA: 2
(D) LD50 o-r: 1840

2-Methylnaphthalene: $C_{11}H_{10}$ • 142.21 • 91-57-6
(P) $\Delta H_f^0 = 44.85$ (l) • Eb = 242; 245; 241
(I) P_{fla} = 98

1.1-Diphenylethylene: $C_{14}H_{12}$ • 180.26 • 530-48-3
(P) $\Delta H_f^0 = 172.42$ (l) • Eb = 270-277; 250/400; 222.8/200; 198.6/100; 183.4/60; 170.8/40; 151.8/20; 135/10; 119.6/5; 87.4/1

1.2-Diphenylethylene (E): $C_{14}H_{12}$ • 180.26 • 103-30-0
(P) Eb = 305; 307
(D) LD50 o-m: 920; ip-r: 6500; iv-m: 34; ip-m: 1150

Diphenyl: $C_{12}H_{10}$ • 154.22 • 92-52-4
(P) $\Delta H_f^0 = 100.50$ (s); 119.24 (s) • Eb = 254-256
• P_{vap} = 0.007(20); 0.07(20); 0.017(30); 0.15(50)
(I) LEL/UEL = 0.6/5.8; 0.7/3.4; 0.6(100)/5.8(166)
• P_{fla} = 110; 112cc • AIT = 540; 570
(T) Tr: 36/37/38 • NFPA: 2 • LVE(MVE) (F): – (0.2); TWA (USA): 0.2; MAK(D): 0.2
(D) LC50 r: 0.2/?; LD50 o-r: 2400; 3280; o-m: 1900; iv-m: 56; cu-ra: >5010; 5010

Diphenylmethane: $C_{13}H_{12}$ • 168.24 • 101-81-5
(P) $\Delta H_f^0 = 88.91$ (l) • Eb = 262-266; 237.5/400; 210.7/200; 186.3/100; 170.2/60; 157./40; 139.8/20; 122.8/10; 107.4/5; 76/1
• P_{vap} = 1.33(76)
(I) P_{fla} = > 109; 130 • AIT = 436; 485
(T) NFPA: 1

Cyclohexylbenzene: $C_{12}H_{16}$ • 160.28 • 827-52-1
(P) Eb = 237-239; 240.0 • P_{vap} = 1.33(68)
(I) P_{fla} = 81; 98oc
(T) NFPA: 2
(D) LD50 o-r: 5000; ip-m: 248; iv-m: 67

Anthracene: $C_{14}H_{10}$ • 178.24 • 120-12-7
(P) $\Delta H_f^0 = 121.34$ (s) • Eb = 340-355 (subl) • P_{vap} = 1.33(145)
(I) LEL/UEL = 0.6/– • P_{fla} = 120cc • AIT = 540
(T) Tr: 36/37/38-42/43 • NFPA: 0; 1
(D) LD50 ip-m: 430

9-Methylanthracene: $C_{15}H_{12}$ • 192.27 • 779-02-2
(P) Eb = 196
(D) LD50 o-m: 700

Phenanthrene: $C_{14}H_{10}$ • 178.24 • 85-01-8
(P) $\Delta H_f^0 = 114.22$ (s) • Eb = 338-340 • P_{vap} = 0.00012(20); 1.33(118)
(T) Tr: 22-43
(D) LD50 o-m: 700; ip-m: 700 ; iv-m: 56

Fluorene: $C_{13}H_{10}$ • 166.23 • 86-73-7
(P) Eb = 295; 298
(I) P_{fla} = 151
(D) LD50 ip-m: 2000

o-Terphenyl: $C_{18}H_{14}$ • 230.32 • 84-15-1
(P) Eb = 332; 337
(I) P_{fla} = > 110; 163oc
(D) LD50 o-r: 1900; o-m: 13200

p-Terphenyl: $C_{18}H_{14}$ • 230.32 • 92-94-4
(P) Eb = 383; 389; 405
(I) P_{fla} = 207oc
(D) LD50 o-m: 13200

ALCOHOLS-POLYOLS
•Saturated alcohols

Methanol: CH_4O • 32.05 • 67-56-1
(P) $\Delta H_f^0 = -201.08$(g); –239.03(l) • Eb = 65; 49.9/400; 34.8/200; 21.2/100; 12.1/60; 5/40; –6/20; –16.2/10; –25.3/5; –44/1
• P_{vap} = 123; 128; 130(20); 133(21.2); 167(25); 350(40); 535(50); 546.5(50)
(I) LEL/UEL = 5.5/44; 6/36.5; 6/36; 6.7/36.5 • P_{fla} = 11; 12Tagcc; 16Tag oc• AIT = 384; 455; 464; 470
(T) Tr: 23/25 • NFPA: 1 • TR: 336 • LVE(MVE) (F): 1000 (200); TWA (USA): 200; MAK (D): 200 • STEL: 250; IDLH: 25000
(D) LC50 r: 83/4h; **LD50** o-r: 5628; LD50 o-m: 7300; 870; LD50 o-ra: 14200; LD50 ip-r: 7529; 9540; LD50 ip-m: 10765; LD50 ip-ra: 1826; LD50 iv-r: 2131; LD50 iv-m: 4710; LD50 iv-ra: 8907; LD50 scu-m: 9800; LD50 cu-ra: 15800; 20000

Ethanol: C_2H_6O • 46.08 • 64-17-5
(P) $\Delta H_f^0 = -234.43$(g); –276.98(l) • Eb = 78.32; 78.5 • P_{vap} = 53(19); 53(20); 59.5(20); 80(25); 500(60)
(I) LEL/UEL = 3.2/19; 3.3/19; 3.3/19(60); 3.3/24.5; 3.5/15; 4.3/19 • P_{fla} = 9; 12; 13Tagcc; 19oc
• AIT = 361; 363; 363 at 425; 371 at 427
(T) NFPA: 0 • TR: 33 • LVE(MVE) (F): 5000 (1000); TWA (USA): 1000
(D) LC50 r: 37.63/10h; LC50 m: 39/4h; **LD50** o-r: 7060; 10600; LD50 o-m: 3450; LD50 o-ra: 6300; LD50 ip-r: 3600; 3750; LD50 ip-m: 933; LD50 ip-ra: 963; LD50 iv-r: 1440; LD50 iv-m: 1973; LD50 iv-ra: 2374; LD50 scu-m: 8285

8 • Table of organic compounds — Alcohols-Polyols

Name: Empirical formula • Molar mass(g/mol) • Registry number (CAS).
(P) Physical data: ΔH_f^0 (kJ/mol) (physical state) • Eb (°C/mmHg) (if substance volatile) • P_{vap} (mBar) (°C) (if substance volatile).
(I) Inflammability data: LEL/UEL (%) • P_{fla} (°C) • AIT (°C).
(T) Toxicity: Safety Code (Tr) • NFPA • Transport Code (TR) • LVE (VME) (ppm) F, USA, D • STEL (ppm); (ppm).
(D) Lethal and toxic doses: LC (mg/l/duration of exposure); LD (mg/kg).

1-Propanol: C_3H_8O • 60.11 • 71-23-8
- **(P)** $\Delta H_f^0 = -256.52(g); -304.01(l)$ • Eb = 97-98 • P_{vap} = 13.3(14.7); 17.3; 18.7; 20(20); 26.7(30); 133.5(53)
- **(I)** LEL/UEL = 2.0/13.5; 2.1/13.5; 2.1/13.7 • P_{fla} = 15cc; 22Tagcc; 24Tagcc; 25; 28Tag oc• AIT = 370; 405; 413; 433
- **(T)** NFPA: 1 • TR: 30 • LVE(MVE) (F): (200); TWA (USA): 200 • STEL: 250
- **(D)** LC50 r: 59/?; LC50 m: 48 /?; LD50 o-r: 1870; 1900; 5400; LD50 o-m: 6800; LD50 o-ra: 2825; LD50 ip-r: 2164; LD50 ip-m: 3125; LD50 ip-ra: 515; LD50 iv-r: 590; LD50 iv-m: 697; LD50 iv-ra: 483; LD50 scu-m: 4700; LD50 cu-ra: 4060; 5000; 5040

2-Propanol: C_3H_8O • 60.11 • 67-63-0
- **(P)** $\Delta H_f^0 = -272.55(g); -317.6(l)$ • Eb = 81-83; 67./400; 53/200; 39.5/100; 30.5/60; 23.8/40; 12.7/20; 2.4/10; -7/5; -26.1/1 • P_{vap} = 11.9(0); 22.7(10); 41.5; 44(20); 59(25)
- **(I)** LEL/UEL = 2/12; 2/12.7; 2.1/ 13.5; 2.3/12.7; 2.5/12; 2.5/12(93.3) • P_{fla} = 11cc; 12Tagcc; 17oc; 21oc • AIT = 399; 425; 440; 456; 460; 485
- **(T)** NFPA: 1 • TR: 33 • LVE(MVE) (F): 400; TWA (USA): 400; MAK (D): 400 • STEL: 500; IDLH: 20000
- **(D)** LC50 r: 39.26/8h; LC50 m: 3141/3h ; LC50 l: 31.41/?; LD50 o-r: 4420; 5045; 5111; 5800; 5840; LD50 o-m: 3600; 4800; LD50 o-ra: 6410; 7900; LD50 ip-r: 933; 2735; LD50 ip-m: 4477; LD50 ip-ra: 667; LD50 iv-r: 1088; 1099; LD50 iv-m: 1509; LD50 iv-ra: 1184; LD50 scu-r:; LD50 scu-m: 6000; LD50 cu-ra: 6291; 13000

1-Butanol: $C_4H_{10}O$ • 74.14 • 71-36-3
- **(P)** $\Delta H_f^0 = -274.68(g); -327.11(l)$ • Eb = 116-118 • P_{vap} = 5; 5.3; 5.7; 6.7; 7.3(20); – 8.6(25); 42(50)
- **(I)** LEL/UEL = 1.4/11.2; 1.4/11.3 • P_{fla} = 29Tagcc; 30; 34 to 38; 34cc; 37 • AIT = 330; 340; 342; 365; 367
- **(T)** Tr: 20 • NFPA: 1 • TR: 30 • LVE(MVE) (F): 50 (-); TWA (USA): 50; MAK (D): 100 • IDLH: 8000
- **(D)** LC50 r: 24.21/4h; **LD50** o-r: 790; 4360; LD50 o-ra: 4250; LD50 ip-m: 603; LD50 iv-r: 310; LD50 iv-m: 377; LD50 cu-ra: 3400; 4200

2-Butanol: $C_4H_{10}O$ • 74.14 • 78-92-2 • 15892-23-6
- **(P)** $\Delta H_f^0 = -292.94(g); -342.59(l)$ • Eb = 99-100 • P_{vap} = 13.3; 16.1; 16.7; 17.3 (20)
- **(I)** LEL/UEL = 1.7(100)/9.8(100); 1.7/9.8 • P_{fla} = 14; 22; 24Tagcc; 26; 31 oc • AIT = 390; 404; 406; 427
- **(T)** Tr: 20 • NFPA: 1 • TR: 30 • TWA (USA): 100; MAK (D): 100 • IDLH: 8000
- **(D)** LC50 r: 48.42/4h; **LD50** o-r: 6480; LD50 o-ra: 4893; LD50 ip-r: 1193; LD50 ip-m: 771; LD50 iv-r: 138; LD50 iv-m: 764

Isobutanol: $C_4H_{10}O$ • 74.14 • 78-83-1
- **(P)** $\Delta H_f^0 = -283.35(g)$ • Eb = 106-109 • P_{vap} = 9.5; 10.7; 12 (20); 13.3(21.7)
- **(I)** LEL/UEL = 1.2/10.9(100); 1.6/10.9; 1.7/10.6; 1.7/10.9(100); 1.9/10.2 • P_{fla} = 28Tagcc; 30; 38Tag oc• AIT = 415; 417; 426; 430; 433
- **(T)** NFPA: 1 • TWA (USA): 50; MAK (D): 100 • IDLH: 8000
- **(D)** LC50 r: 24.21/4h; **LD50** o-r: 2460; LD50 o-ra: 3750; LD50 ip-r: 720; LD50 ip-m: 1801; LD50 ip-ra: 323; LD50 iv-r: 340; LD50 iv-m: 417; LD50 cu-ra: 3400; 4240

t-Butanol: $C_4H_{10}O$ • 74.14 • 75-65-0
- **(P)** $\Delta H_f^0 = -312.57(g); -359.24(l)$ • Eb = 82-83 • P_{vap} = 40; 41.2 (20); 53(24.5); 56(25); 220(50)
- **(I)** LEL/UEL = 2.3/8; 2.4/8 • P_{fla} = 10; 11cc • AIT = 470; 477; 479
- **(T)** NFPA: 1 • TR: 33 • TWA (USA): 100; MAK (D): 100 • STEL: 150; IDLH: 8000
- **(D)** LD50 o-r: 3500; LD50 o-ra: 3559; LD50 ip-m: 933; LD50 iv-m: 1538

1-Pentanol: $C_5H_{12}O$ • 88.10 • 71-41-0
- **(P)** $\Delta H_f^0 = -355.64$ (l) • Eb = 136-138; 74/50; 48/10 • P_{vap} = 1.33(13.6); 1.33; 2 (20); 13.3(45)
- **(I)** LEL/UEL = 1.2/10(100); 1.2/- • P_{fla} = 33; 38cc; 48; 57; AIT = 300
- **(T)** NFPA: 1
- **(D)** LC50 r: 14/6h; LC50 m: 14/6h; **LD50** o-ra: 200; LD50 o-r: 2200; 3030; LD50 ip-r: 579; LD50 ip-m: 970; LD50 iv-r: 196; LD50 iv-m: 184; LD50 cu-ra: 4490

2-Pentanol: $C_5H_{12}O$ • 88.10 • 6032-29-7
- **(P)** $\Delta H_f^0 = -366.94(l)$ • Eb = 116-119.3; 62/60 • P_{vap} = 5.3(20); 11.2(30); 40(50)
- **(I)** LEL/UEL = 1.2/9 • P_{fla} = 33; 39; 40; 41 oc • AIT = 340; 343 to 385; 347
- **(T)** Tr: 20 • NFPA: 1 • TR: 30
- **(D)** LD50 o-r: 1470; LD50 o-ra: 3500; LD50 ip-r: 2130; LD50 cu-ra: 2821

3-Pentanol: $C_5H_{12}O$ • 88.10 • 584-02-1
- **(P)** $\Delta H_f^0 = -370.28(l)$ • Eb = 114-116; 114-115/749; 113/738 • P_{vap} = 9.3(20)
- **(I)** LEL/UEL = 1.2/8; ~1.2/~9; 1.2/9 • P_{fla} = 19; 30; 33cc; 34; 40 • AIT = 360; 434
- **(T)** Tr: 20 • NFPA: 1 • TR: 30
- **(D)** LD50 o-r: 1870; LD50 cu-ra: 2520

Isoamylic alcohol: $C_5H_{12}O$ • 88.10 • 123-51-3
- **(P)** $\Delta H_f^0 = -356.46(l)$ • Eb = 131-132 • P_{vap} = 2.7; 3; 3.72(20); 6.4(30); 24(50)
- **(I)** LEL/UEL = 1.2/8; 1.2/9; 1.2/10; 1.2/9(100) • P_{fla} = 43cc; 45cc; 46; 55co • AIT = 339; 350; 353
- **(T)** Tr: 20 • NFPA: 1 • TWA (USA): 100; MAK (D): 100 • STEL: 150
- **(D)** LD50 o-r: 1300; LD50 o-ra: 3438; 4250; 5748; LD50 ip-m: 233; LD50 iv-m: 234; LD50 iv-ra: 1570; LD50 scu-m: 7480; LD50 cu-ra: 3212; 3970

8 • Table of organic compounds

Alcohols-Polyols

Name: Empirical formula • Molar mass(g/mol)• Registry number (CAS).
(P) **Physical data:** ΔH_f^0 (kJ/mol) (physical state) • Eb (°C/mmHg) (if substance volatile) • P_{vap} (mBar) (°C) (if substance volatile).
(I) **Inflammability data:** LEL/UEL (%) • P_{fla} (°C) • AIT (°C).
(T) **Toxicity:** Safety Code (Tr) • NFPA • Transport Code (TR) • LVE (MVE) (ppm) F, USA, D • STEL (ppm); IDLH (ppm).
(D) **Lethal and toxic doses:** LC (mg/l/duration of exposure); LD (mg/kg).

t-Amylic alcohol: $C_5H_{12}O$ • 88.10 • 75-85-4
- (P) $\Delta H_f^0 = -329.86(g); -379.49(l)$ • Eb = 100-103
  • $P_{vap} = 13.3(17.2); 16(20); 27(30); 80(50)$
- (I) LEL/UEL = 1.2/9; ~1.2/~8 • P_{fla} = 19cc; 21 oc 24; 40
  • AIT = 435; 437
- (T) Tr: 20 • NFPA: 1 • TR: 33
- (D) LD50 o-r: 1000; LD50 o-ra: 2028; LD50 ip-r: 1530; LD50 iv-m: 610; LD50 scu-r: 1400; LD50 scu-m: 2100

2-Methyl-1-butanol: $C_5H_{12}O$ • 88.10 • 137-32-6 • 34713-94-5
- (P) $\Delta H_f^0 = -356.48(l)$ • Eb = 126-130 • P_{vap} = 4; 25(20)
- (I) LEL/UEL = 1.2/8; 1.4/9; 1.9/10 • P_{fla} = 43; 50co • AIT = 384; 400
- (T) Tr: 20 • NFPA: 2 • TR: 30
- (D) LD50 o-r: 1000; 4920; LD50 ip-r: 1900; LD50 cu-ra: 3540

3-Methyl-2-butanol: $C_5H_{12}O$ • 88.10 • 598-75-4
- (P) $\Delta H_f^0 = -366.10(l)$ • Eb = 112-114; 110/742 • P_{vap} = 3(20)
- (I) LEL/UEL = 1.2/8 • P_{fla} = 26; 35co; 39cc; 43 • AIT = 347
- (T) Tr: 20

2.2-Dimethyl-1-propanol: $C_5H_{12}O$ • 88.10 • 75-84-3
- (P) Eb = 113-114 • P_{vap} = 21.3(20); 16.6(30); 67.3(46); 711(100)
- (I) LEL/UEL = 1.2/8; 1.7/9.7 • P_{fla} = 28; 37 • AIT = 400; 430
- (D) LD50 o-r: 3600

Cyclopentanol: $C_5H_{10}O$ • 86.14 • 96-41-3
- (P) $\Delta H_f^0 = -300.16(l)$ • Eb = 139-141; 53/10 • P_{vap} = 1.3(20); 15(50); 30(65)
- (I) P_{fla} = 51; 60; 62 • AIT = 375
- (T) TR: 30

1-Hexanol: $C_6H_{14}O$ • 102.20 • 111-27-3
- (P) $\Delta H_f^0 = -317.57(g); -379.49(l)$ • Eb = 154-158; 138/400; 119.6/200; 102.8/100; 92/60; 83.7/40; 70.3/20; 58.2/10; 47.2/5; 24.4/1 • P_{vap} = 0.93; 1(20); 1.33(24.4); 1(25); 1.33(25.6); 8(50)
- (I) LEL/UEL = 1.2/7.7; 1.3/- • P_{fla} = 59; 63 • AIT = 292
- (T) Tr: 22 • NFPA: 1 • TR: 30
- (D) LD50 o-r: 720; 4590; LD50 o-m: 1950; LD50 iv-m: 103; LD50 cu-ra: 3100

2-Hexanol: $C_6H_{14}O$ • 102.20 • 626-93-7
- (P) Eb = 136-140
- (I) P_{fla} = 41; 45; 58
- (T) Tr: 22 • TR: 30
- (D) LD50 o-r: 1410; LD50 cu-ra: 3560

3-Hexanol: $C_6H_{14}O$ • 102.20 • 623-37-0
- (P) Eb = 134-135
- (I) P_{fla} = 41; 45
- (T) Tr: 22 • TR: 30

4-Methyl-2-pentanol: $C_6H_{14}O$ • 102.20 • 108-11-2
- (P) Eb = 129-132; 58.2/40; 33.3/10
  • P_{vap} = 3.73; 4.65; 4.9; 7(20); 11(30); 33(50)
- (I) LEL/UEL = 1/5.5; 0.9(100)/5.8 • P_{fla} = 41cc; 54co • AIT = 349
- (T) Tr: 37 • NFPA: 2 • TR: 30 • TWA (USA): 25; MAK (D): 25
  • STEL: 40
- (D) LC50 r: 8.34/4h; LD50 o-r: 2590; LD50 o-m: 1000; LD50 iv-m: 812; LD50 cu-ra: 3560

3.3-Dimethyl-2-butanol: $C_6H_{14}O$ • 102.20 • 464-07-3
- (P) Eb = 119-121
- (I) P_{fla} = 26; 28

Cyclohexanol: $C_6H_{12}O$ • 100.16 • 108-93-0
- (P) $\Delta H_f^0 = -348.19(l)$ • Eb = 162; 141.4/400; 104/100; 56/10
  • P_{vap} = 0.56; 1.2(20); 1.33(20); 1.33(21); 1.31(25); 3(30); 10.5(50)
- (I) LEL/UEL = 1.25/12.25; 2.0/- ; 2.4/11.2 • P_{fla} = 68Tagcc
  • AIT = ~290; 300
- (T) Tr: 20/22-37/38 • NFPA: 1 • LVE(MVE) (F): 75 (50); TWA (USA): 50; MAK (D): 50
- (D) LD50 o-r: 2060; LD50 o-ra: 2200; LD50 ip-m: 1352; LD50 ip-ra: 1420; LD50 iv-r: 272; LD50 scu-m: 2480; LD50 cu-ra: 12000; > 12000

1-Heptanol: $C_7H_{16}O$ • 116.23 • 111-70
- (P) $\Delta H_f^0 = -331.79(g); -400.83(l)$ • Eb = 176; 155.6/400; 136.6/200; 119.5/100; 99.8/40; 85.8/20; 74.7/10; 64.3/5; 42.4/1
  • P_{vap} = 0.15; 0.7(20); 0.45(30); 2.5(50)
- (I) P_{fla} = 71; 73; 77 • AIT = 349
- (T) Tr: 22 • TR: 30
- (D) LC50 m: 6.6/2h; LD50 o-r: 500; LD50 o-m: 1500; LD50 o-ra: 750; LD50 cu-ra: 2000

2-Heptanol: $C_7H_{16}O$ • 116.23 • 543-49-7
- (P) Eb = 158-162 • P_{vap} = 1.33(14); 1.33(14.6)
- (I) P_{fla} = 59; 64; 71oc
- (T) Tr: 21/22; 21-36 • NFPA: 0 • TR: 30
- (D) LD50 o-r: 2580; LD50 cu-ra: 1780

3-Heptanol: $C_7H_{16}O$ • 116.23 • 589-82-2
- (P) Eb = 156.2; 66/20 • P_{vap} = 0.67; 1 (20)
- (I) P_{fla} = 54; 60cco
- (T) Tr: 22-36/38 • NFPA: 0 • TR: 30
- (D) LD50 o-r: 1870; LD50 cu-ra: 4360

4-Heptanol: $C_7H_{16}O$ • 116.23 • 589-55-9
- (P) Eb = 156
- (I) P_{fla} = 48; 54
- (T) Tr: 36 • TR: 30

2.4-Dimethyl-3-pentanol: $C_7H_{16}O$ • 116.23 • 600-36-2
- (P) Eb = 139-140
- (I) P_{fla} = 37; 49 • AIT = 370; 398

8 • Table of organic compounds

Alcohols-Polyols

Name: Empirical formula • Molar mass(g/mol)• Registry number (CAS).
- (P) **Physical data:** ΔH_f^0 (kJ/mol) (physical state) • Eb (°C/mmHg) (if substance volatile) • P_{vap} (mBar) (°C) (if substance volatile).
- (I) **Inflammability data:** LEL/UEL (%) • P_{fla} (°C) • AIT (°C).
- (T) **Toxicity:** Safety Code (Tr) • NFPA • Transport Code (TR) • LVE (MVE) (ppm) F, USA, D • STEL (ppm); IDLH (ppm).
- (D) **Lethal and toxic doses:** LC (mg/l/duration of exposure); LD (mg/kg).

2-Methylcyclohexanol: $C_7H_{14}O$ • 114.21 • 583-59-5
- (P) $\Delta H_f^0 = -415.89$ (l) isomer E • Eb = 163-167
- (I) LEL/UEL = 1.1/- ; 1.3/- • P_{fla} = 58; 64; 68cc • AIT = 296
- (T) Tr: 20 • NFPA: 3; – • TR: 30 • LVE(MVE) (F): – (50); TWA (USA): 50; MAK (D): 50

3-Methylcyclohexanol: $C_7H_{14}O$ • 114.21 • 591-23-1
- (P) $\Delta H_f^0 = -416.31$(l) isomer Z • Eb = 163; 170-172
- (I) LEL/UEL = 1.3/- • P_{fla} = 62; 67
- (T) Tr: 20

4-Methylcyclohexanol: $C_7H_{14}O$ • 114.21 • 589-91-3
- (P) $\Delta H_f^0 = -433.46$(l) isomer E • Eb = 170-175
- (I) LEL/UEL = 1.3/- • P_{fla} = 39 (?); 68
- (T) Tr: 20

Cycloheptanol: $C_7H_{14}O$ • 114.21 • 502-41-0
- (P) Eb = 184-185; 77-81/11
- (I) P_{fla} = 60; 71; 75

1-Octanol: $C_8H_{18}O$ • 130.28 • 111-87-5
- (P) $\Delta H_f^0 = -425.09$(l) • Eb = 195-196; 103/16 • P_{vap} = 0.032; 0.13; 0.3 (20); 0.19(25); 0.11(30); 0.9(50)
- (I) LEL/UEL = 0.8/- • P_{fla} = 81cc; 84 • AIT = 270; 272
- (T) Tr: 22-36/38 • NFPA: 1 • TR: 30
- (D) LD50 o-m: 1790; LD50 iv-m: 69

2-Octanol: $C_8H_{18}O$ • 130.28 • 123-96-6 (4128-31-8)
- (P) Eb = 179-180; 107.4/60; 83.3/20; 70/10; 57.6/5; 32.8/1
- (I) P_{fla} = 61; 71; 76; 88
- (T) NFPA: 1 • TR: 30

3-Octanol: $C_8H_{18}O$ • 130.28 • 20296-29-1 (589-98-0)
- (P) Eb = 175-179; 76/21 • P_{vap} = ~1.33(20)
- (I) P_{fla} = 65; 68
- (T) TR: 30
- (D) LD50 o-r: > 5000; LD50 cu-ra: > 5000

2-Ethyl-1-hexanol: $C_8H_{18}O$ • 130.28 • 104-76-7
- (P) Eb = 180-186; 84/15 • P_{vap} = 0.07; 0.144; 0.27; 0.4; 0.5(20); 13.3(79); 53.3(105)
- (I) LEL/UEL = 0.88/9.7; 1.1/7.4; 0.88(104)/9.7(113); 1.8/12.7 • P_{fla} = 72cc; 77; 81 • AIT = 190; 250; 270; 287
- (T) Tr: 21-36 • NFPA: 2
- (D) LD50 o-r: 2049; 2460; 4000; 5000; 10397; LD50 o-m: 2500; LD50 o-ra: 1180; LD50 ip-r: 500; 650; LD50 ip-m: 759; LD50 scu-r: 650; LD50 cu-ra: 1970; 2390; > 3000

ALCOHOLS-POLYOLS
• Unsaturated alcohols

Allylic alcohol: C_3H_6O • 58.09 • 107-18-6
- (P) $\Delta H_f^0 = -132.01$ (g) • Eb = 95-98 • P_{vap} = 13.3(10.5); 22.7; 24(20); 32(20); 32(25); 183(55)
- (I) LEL/UEL = 2.5/18; 3/18 • P_{fla} = 21cc; 21oc; 24cc • AIT = 375; 378
- (T) Tr: 23/24/25-36/37/38 • NFPA: 3 • TR: 663 • LVE(MVE) (F): 4 (2); TWA (USA): 2; MAK (D): 2 • STEL: 4
- (D) LC50 r: 0.18/8h; LC50 m: 0.5/2h; **LD50** o-r: 64; LD50 o-m: 96; LD50 o-ra: 71; LD50 ip-r: 37; 42; LD50 ip-m: 60; LD50 iv-m: 78; LD50 cu-ra: 45

2-Buten-1-ol (Z+E): C_4H_8O • 72.12 • 627-27-0 • 6117-91-5
- (P) Eb = 118-122
- (I) P_{fla} = 33; 38; 56
- (T) Tr: 21/22
- (D) LC50 r: 5.89/4h; **LD50** o-r: 793; LD50 cu-ra: 1270

3-Buten-2-ol: C_4H_8O • 72.12 • 598-32-3
- (P) Eb = 96-98
- (I) P_{fla} = 16
- (T) Tr: 20 • TR: 30

2-Methyl-2-propen-1-ol: C_4H_8O • 72.12 • 513-42-8
- (P) Eb = 112-115 • P_{vap} = 10(20)
- (I) P_{fla} = 32 • AIT = 320
- (T) Tr: 36/37/38

Propargyl alcohol: C_3H_4O • 56.07 • 107-19-7
- (P) Eb = 114-115; 70/147.6; 30/20.6 • P_{vap} = 15.5; 18(20)
- (I) LEL/UEL = 1.9/86.2; 3.1/71; 3.4/- • P_{fla} = 33co; 36co; 39 • AIT = 365
- (T) Tr: 23/24/25-34 • NFPA: 3 • TR: 63 • TWA (USA): 1 (skin); MAK(D): 2
- (D) LC50 r: 2/2h; LC50 m: 2/2h; **LD50** o-r: 56; 70; LD50 o-m: 50; LD50 scu-ra: 25; LD50 cu-ra: 16; 88

3-Butyn-2-ol: C_4H_6O • 70.10 • 2028-63-9
- (P) Eb = 108
- (I) P_{fla} = 27; 44
- (T) Tr: 23/24/25-36/38
- (D) LD50 o-m: 30

2-Methyl-3-butyn-2-ol: C_5H_8O • 84.13 • 115-19-5
- (P) Eb = 103-105; 52/80; 20/12 • P_{vap} = 20(20)
- (I) LEL/UEL = 1.8/16.6 • P_{fla} = 20; 22; 24; 25co; < 21 • AIT = 349
- (T) Tr: 22-41
- (D) LC50 m: 2/?; **LD50** o-r: 1950; 1420; LD50 o-m: 500; 1800; LD50 ip-m: 3600; LD50 iv-m: 2340; LD50 scu-m: 2340

8 • Table of organic compounds

Alcohols-Polyols

Name: Empirical formula • Molar mass(g/mol)• Registry number (CAS).
(P) Physical data: ΔH_f^0 (kJ/mol) (physical state) • Eb (°C/mmHg) (if substance volatile) • P_{vap} (mBar) (°C) (if substance volatile).
(I) Inflammability data: LEL/UEL (%) • P_{fla} (°C) • AIT (°C).
(T) Toxicity: Safety Code (Tr) • NFPA • Transport Code (TR) • LVE (MVE) (ppm) F, USA, D • STEL (ppm); IDLH (ppm).
(D) Lethal and toxic doses: LC (mg/l/duration of exposure); LD (mg/kg).

ALCOHOLS-POLYOLS
• Aromatic alcohols

Benzyl alcohol: C_7H_8O • 108.15 • 100-51-6
- (P) $\Delta H_f^0 = -161.04(l)$ • Eb = 203-205.7; 183/400; 160/200; 141.7/100; 129.3/60; 119.8/40; 105.8/20; 92.6/10; 80.8/5; 58/1 • $P_{vap} = 0.027$; 0.13; 0.2(20); 0.13(25); 1.33(58); 5(77); 17.7(100)
- (I) LEL/UEL = 1.3/13 • P_{fla} = 94; 96cc; 101cc; 104oc • AIT = 436
- (T) Tr: 20/22 • NFPA: 2
- (D) LC50 r: 8.83/4h; LD50 o-r: 1230; LD50 o-m: 1580; LD50 ip-r: 400; LD50 iv-r: 53; 64; LD50 iv-m: 324; LD50 scu-r: 1700; LD50 cu-ra: 2000

1-Phenyl ethanol: $C_8H_{10}O$ • 122.18 • 98-85-1 (13323-81-4)
- (P) Eb = 204-205; 204/745 • P_{vap} = 0.09; 0.13; 1(20)
- (I) LEL/UEL = 1.8/4.9 • P_{fla} = 84; 96co; 99 • AIT = 480
- (D) LD50 o-r: 400; LD50 o-m: 558; LD50 scu-m: 250; LD50 cu-ra: 2500

2-Phenyl ethanol: $C_8H_{10}O$ • 122.18 • 60-12-8
- (P) Eb = 219-221/750; 154/100; 94-96/15; 104/14; 100/12; 100/10 • P_{vap} = 1.33(58)
- (I) P_{fla} = 102co; 102
- (T) Tr: 22-36/38 • NFPA: 1
- (D) LD50 o-r: 1790; LD50 o-m: 800; LD50 ip-m: 200; LD50 cu-ra: 790

1-Phenyl-1-propanol: $C_9H_{12}O$ • 136.21 • 93-54-9
- (P) Eb = 219; 103/19; 107/15; 103/14; 78/3
- (I) P_{fla} = 86; 90
- (T) Tr: 22
- (D) LD50 o-r: 1586; LD50 o-m: 500; LD50 scu-m: 700

2-Phenyl-1-propanol: $C_9H_{12}O$ • 136.21 • 98103-87-8
- (P) Eb = 217-219; 112/10; 114/14; 112/16
- (I) P_{fla} = 90; 93
- (D) LD50 o-r: 2300; LD50 cu-ra: > 5000

3-Phenyl-1-propanol: $C_9H_{12}O$ • 136.21 • 122-97-4
- (P) Eb = 235-238; 119-121/12 • P_{vap} = 0.5(20)
- (I) P_{fla} = 106; 109; 120 • AIT = 430
- (D) LD50 o-r: 2300; LD50 cu-ra: 5000

3-Phenyl-2-propen-1-ol: $C_9H_{10}O$ • 134.18 • 104-54-1
- (P) Eb = 250; 224.6/400; 199.8/200; 177./100; 162/60; 151/40; 133.7/20; 117./10; 102.5/5; 72.6/1 • P_{vap} = 1.33(20)
- (I) P_{fla} = >109; ~126
- (T) Tr: 22-36-43
- (D) LD50 o-r: 2000; LD50 o-m: 2675; LD50 cu-ra: > 5000

1-Phenyl-2-propyn-1-ol: C_9H_8O • 132.16 • 64599-56-0
- (P) Eb = 231-233; 135/13; 135/17; 135-136/13
- (I) P_{fla} = 99
- (T) Tr: 22
- (D) LD50 ip-m: 240

ALCOHOLS-POLYOLS
• Polyols

Ethylene glycol: $C_2H_6O_2$ • 62.08 • 107-21-1
- (P) $\Delta H_f^0 = -389.32(g); -454.80(l)$ • Eb = 195-198; 140/97; 100/18; 100/13; 70/3; 20/0.06 • P_{vap} = 0.05; 0.07; 0.08; 0.1; 0.16(20); 1.3(54); 6.7(80)
- (I) LEL/UEL = 1.8/12.8; 3.2/15.3; 3.2/53 • P_{fla} = >109; 111cc; 115co; 116Tagcc; 116Coc • AIT = 399; 410; 413
- (T) Tr: 22 • NFPA: 1 • LVE(MVE) (F): 50 (-); TWA (USA): 50; MAK (D): 10
- (D) LC50 r: 10.876; LD50 o-r: 4700; 8540; LD50 o-m: 5500; 7500; LD50 ip-r: 5010; LD50 ip-m: 5614; LD50 iv-r: 3260; LD50 iv-m: 3000; LD50 scu-r: 2800; LD50 scu-m: 2700; LD50 cu-ra: 9530; 19500

Propane-1,2-diol: $C_3H_8O_2$ • 76.11 • 57-55-6
- (P) $\Delta H_f^0 = -500.41(l)$ • Eb = 188.2; 168.1/400; 149.7/200; 132/100; 119.9/60; 111.2/40; 96.4/20; 83.2/10; 70.8/5; 45.5/1 • P_{vap} = 0.1; 0.105; 0.13; 0.26(20); 0.17(25)
- (I) LEL/UEL = 2.6/12.6; 2.6/12.5 • P_{fla} = 99oc; 103; 107 • AIT = 371; 414; 420
- (T) NFPA: 0
- (D) LD50 o-r: 20000; 25900; LD50 o-m: 22000; LD50 o-ra: 18500; LD50 ip-r: 6660; 13000; LD50 iv-r: 6423; LD50 ip-m: 9718; LD50 iv-m: 6630; LD50 scu-r: 22500; LD50 scu-m: 17370; LD50 cu-ra: 20800

Propane-1,3-diol: $C_3H_8O_2$ • 76.11 • 504-63-2
- (P) $\Delta H_f^0 = -520.49(l)$ • Eb = 210-212; 214(dec) • P_{vap} = 0.11; 0.13(20); 13.3 (100)
- (I) P_{fla} = 79cc; 131 • AIT = 400
- (T) NFPA: 1
- (D) LD50 o-r: 10000; LD50 o-m: 4773

Butane-1,2-diol: $C_4H_{10}O_2$ • 90.14 • 584-03-2 (26171-83-5)
- (P) Eb = 190; 194; 196-197; 191-192/747
- (I) P_{fla} = 90; 93
- (T) NFPA: 1
- (D) LD50 o-r: 16000; LD50 o-m: 3720

Butane-1,3-diol: $C_4H_{10}O_2$ • 90.14 • 107-88-0
- (P) Eb = 204-208; 132/50; 90/10 • P_{vap} = 0.08(20)
- (I) P_{fla} = 108; 121Toc; 121 • AIT = 375; 394
- (T) NFPA: 1
- (D) LD50 o-r: 18610; 22800; LD50 o-m: 12980; LD50 scu-r: 20000 LD50 scu-m:; LD50 cu-ra: > 20000

Butane-1,4-diol: $C_4H_{10}O_2$ • 90.14 • 110-63-4
- (P) Eb = 228-230; 120-122/10
- (I) P_{fla} = >109; 121oc; 134 • AIT = 369
- (T) Tr: 20/21/22; 22
- (D) LD50 o-r: 1525; LD50 o-m: 2062; LD50 o-ra: 2531; LD50 ip-r: 1070; 1370; LD50 ip-m: 500

8 • Table of organic compounds — Phenols

Name: Empirical formula • Molar mass(g/mol)• Registry number (CAS).
- (P) **Physical data:** ΔH_f^0 (kJ/mol) (physical state) • Eb (°C/mmHg) (if substance volatile) • P_{vap} (mBar) (°C) (if substance volatile).
- (I) **Inflammability data:** LEL/UEL (%) • P_{fla} (°C) • AIT (°C).
- (T) **Toxicity:** Safety Code (Tr) • NFPA • Transport Code (TR) • LVE (MVE) (ppm) F, USA, D • STEL (ppm); IDLH (ppm).
- (D) **Lethal and toxic doses:** LC (mg/l/duration of exposure); LD (mg/kg).

Butane-2,3-diol: $C_4H_{10}O_2$ • 90.14 • 513-85-9
- (P) Eb = 181.7; 89/16 (erythro); 180/745; 172.7/742; 86/16; 78/10 (threo) • P_{vap} = 0.23(20)
- (I) P_{fla} = 85Tag oc• AIT = 402

Pentane-1,5-diol: $C_5H_{12}O_2$ • 104.17 • 111-29-5
- (P) Eb = 239-244; 260 • P_{vap} = < 0.013(20)
- (I) LEL/UEL = 1.3/13.1; 1.4/13.2 • P_{fla} = 129; 129co; 136Tagcc • AIT = 330; 334
- (T) Tr: 22
- (D) LD50 o-r: 2000; 5890; 10000; LD50 o-m: 6300; LD50 o-ra: 6300; LD50 cu-ra: > 20000

2,2-Dimethylpropane-1,3-diol: $C_5H_{12}O_2$ • 104.17 • 126-30-7
- (P) Eb = 207-212; 121/25; 93-94/3 • P_{vap} = 234.6(70)
- (I) LEL/UEL = 1.37(149)/18.8(177) • P_{fla} = 103; 107Tagcc; 152co • AIT = 387; 399
- (D) LD50 o-r: ~7000

Hexane-1,6-diol: $C_6H_{14}O_2$ • 118.20 • 629-11-8
- (P) Eb = 250; 253-260; 134/10 • P_{vap} = 0.71(20)
- (I) LEL/UEL = 6.6/16 • P_{fla} = 101; 130; 135Tagcc; 147 • AIT = 319
- (D) LD50 o-r: ~3000; 3730; LD50 cu-ra: > 2500; > 10000

2,3-Dimethylbutane-2,3-diol: $C_5H_{12}O_2$ • 104.17 • 76-09-5
- (P) Eb = 171-172/739; 172.8
- (I) P_{fla} = 71cc; 77
- (D) LD50 o-m: 3380

Glycerol: $C_3H_8O_3$ • 92.11 • 56-81-5
- (P) ΔH_f^0 = –668.44(l) • Eb = 290(dec); 263/400; 240/200; 220/100; 208/60; 182/20; 167.2/10; 153.8/5; 125.5/1; 138-140/0.1 • P_{vap} = ~0.002(20); ~0.015(50); 0.0033(50)
- (I) LEL/UEL = 0.9/- • P_{fla} = 160; 176oc; 198cc • AIT = 370; 393; 405
- (T) Tr: 36 • NFPA: 1 • LVE(MVE) (F): – (10 mg/m³); TWA (USA): 10 mg/m³
- (D) LC50 r: 0.57/1h; LD50 o-r: 12600; > 25000; LD50 o-m: 4090; LD50 o-ra: 27000; LD50 ip-r: 4420; LD50 ip-m: 8700; 8982; LD50 iv-r: 5566; LD50 iv-m: 4250; LD50 iv-ra: 53000; LD50 scu-r: 100; LD50 scu-m: 91; LD50 cu-ra: > 10000

Pentaerythritol: $C_5H_{12}O_4$ • 136.17 • 115-77-5
- (P) ΔH_f^0 = –920.48(s) • Eb = 276/30 • P_{vap} = 2.4(200); 80(250)
- (I) P_{fla} = 260co • AIT = 450
- (T) NFPA: 0 • LVE(MVE) (F): 15 mg/m³; TWA (USA): 10 mg/m³
- (D) LD50 o-r: 19500; LD50 o-m: 25500; LD50 o-ra: 18500

2-Butene-1,4-diol (Z): $C_4H_8O_2$ • 88.11 • 6117-80-2
- (P) Eb = 235; 131.5/12; 132/16
- (I) P_{fla} = >109; 127
- (D) LD50 o-r: 1250

2-Butyne-1,4-diol: $C_4H_6O_2$ • 86.10 • 110-65-6
- (P) Eb = 238; 194/100; 140/10; 125-127/2 • P_{vap} = < 0.13(20); 20(145)
- (I) P_{fla} = 152
- (T) Tr: 25-34 • TR: 60
- (D) LC50 r: 0.15/2h; LC50 m: 0.15/2h; LD50 o-r: 105; 100-136; LD50 o-m: 105; LD50 o-ra: 150; LD50 ip-r: 52.4; LD50 cu-ra: 659

PHENOLS

Phenol: C_6H_6O • 94.11 • 108-95-2
- (P) ΔH_f^0 = –165.02(s); –158.16(l) • Eb = 180-182; 86/20 • P_{vap} = 0.3; 0.48(20); 1.33(40); 3.5(50); 6.21(55); 54(100); 500(160)
- (I) LEL/UEL = 1.3/9.5; 1.5/9; 1.5/8.6; 1.7/8.6 • P_{fla} = 79cc; 855 oc• AIT = 605; 714
- (T) Tr: 24/25-34 • NFPA: 3 • TR: 68 • LVE(MVE) (F): 5 (5); TWA (USA): 5 (skin); MAK (D): 5 • IDLH: 100
- (D) LC50 r: 0.316; LC50 m: 0.177; LD50 o-r: 317; 530; LD50 o-m: 270; LD50 ip-r: 127; LD50 ip-m: 180; LD50 iv-m: 112; LD50 scu-r: 460; LD50 scu-m: 344; LD50 cu-ra: 630; 850; LD50 cu-r: 669

o-Cresol: C_7H_8O • 108.14 • 95-48-7
- (P) ΔH_f^0 = –128.62(g) • Eb = 191-192; 75/10 • P_{vap} = 0.4; 0.35(20); 0.75(30); 1.33(38.2); 2.19(50)
- (I) LEL/UEL = 1.3/-; 1.35/-; 1.47(148)/- • P_{fla} = 73; 81; 81-83 • AIT = 555; 558; 599
- (T) Tr: 24/25-34 • NFPA: 3;2 • TR: 60 • LVE(MVE) (F): – (5); TWA (USA): 5 (skin); MAK (D): 5 • IDLH: 250
- (D) LC50 m: 0.179; LD50 o-r: 121; 1350 (Merck); LD50 o-m: 344; LD50 ip-r: 200; LD50 scu-r: 65; LD50 scu-m: 410; LD50 cu-ra: 890; LD50 cu-r: 620

m-Cresol: C_7H_8O • 108.14 • 108-39-4
- (P) ΔH_f^0 = –132.34(g) • Eb = 202-203; 86/10 • P_{vap} = 0.05; 0.13(20); 0.19(30); 1.1(50); 1.33(52)
- (I) LEL/UEL = 1/-; 1.1/1.4; 1.06(150)/1.35(150); 1.1(150)/1.4 • P_{fla} = 73; 86cc; 94 • AIT = 555; 557; 559
- (T) Tr: 24/25-34 • NFPA: 3;2 • TR: 60 • LVE(MVE) (F): – (5); TWA (USA): 5 (skin); MAK (D): 5 • IDLH: 250
- (D) LD50 o-r: 242; 2020 (Merck); LD50 o-m: 828; LD50 o-ra: 1400; LD50 ip-m: 168; LD50 iv-ra: 280; LD50 scu-r: 900; LD50 scu-m: 450; LD50 cu-ra: 2050; LD50 cu-r: 1100

p-Cresol: C_7H_8O • 108.14 • 106-44-5
- (P) ΔH_f^0 = –152.32(g) • Eb = 202; 179.4/400; 140/100; 117.7/40; 102.3/20; 88.6/10; 76.5/5; 53/1 • P_{vap} = 0.05; 1.33(20); 0.18(30); 0.9(50); 1.33(53)
- (I) LEL/UEL = 1/-; 1.1/1.4; 1.1(150)/- • P_{fla} = 86cc; 89; 94 • AIT = 555; 559
- (T) Tr: 24/25-34 • NFPA: 3;2 • TR: 60 • LVE(MVE) (F): – (5); TWA (USA): 5 (skin); MAK (D): 5 • IDLH: 250
- (D) LD50 o-r: 207; 1800 (Merck); LD50 o-m: 344; LD50 o-ra: 620; LD50 ip-r: 25; LD50 iv-ra: 180; LD50 scu-r: 500; LD50 scu-m: 150; LD50 scu-m: 300; LD50 cu-ra: 301; LD50 cu-r: 750

8 • Table of organic compounds — Phenols

Name: Empirical formula • Molar mass(g/mol) • Registry number (CAS).
(P) Physical data: ΔH_f^0 (kJ/mol) (physical state) • Eb (°C/mmHg) (if substance volatile) • P_{vap} (mBar) (°C) (if substance volatile).
(I) Inflammability data: LEL/UEL (%) • P_{fla} (°C) • AIT (°C).
(T) Toxicity: Safety Code (Tr) • NFPA • Transport Code (TR) • LVE (MVE) (ppm) F, USA, D • STEL (ppm); IDLH (ppm).
(D) Lethal and toxic doses: LC (mg/l/duration of exposure); LD (mg/kg).

2-Ethylphenol: $C_8H_{10}O$ • 122.17 • 90-00-6
- (P) $\Delta H_f^0 = -208.82(s)$ • Eb = 195-197; 204-206
- (I) $P_{fla} = 78$
- (T) Tr: 22
- (D) LD50 o-m: 600; LD50 ip-m: 172

3-Ethylphenol: $C_8H_{10}O$ • 122.17 • 620-17-7
- (P) $\Delta H_f^0 = -214.26(s)$ • Eb = 212/750; 108-110/15; 108/20
- (I) $P_{fla} = 94$
- (T) Tr: 20/21/22-36/ 37/38
- (D) LD50 ip-m: 138

4-Ethylphenol: $C_8H_{10}O$ • 122.17 • 123-07-9
- (P) $\Delta H_f^0 = -224.39(s)$ • Eb = 218-219; 101/13 • $P_{vap} = 0.17(20)$
- (I) $P_{fla} = 100; 104$
- (T) Tr: 20/21/22-38

2,3-Xylenol: $C_8H_{10}O$ • 122.17 • 526-75-0
- (P) $\Delta H_f^0 = -157.19(g)$ • Eb = 217-218
- (I) $P_{fla} = 95$
- (T) Tr: 24/25-34 • TR: 60

2,4-Xylenol: $C_8H_{10}O$ • 122.17 • 105-67-9
- (P) $\Delta H_f^0 = -162.96(g)$ • Eb = 209-213 • $P_{vap} = 0.13(25); 1.3(52)$
- (I) $P_{fla} = 94; 96$
- (T) Tr: 24/25-34 • NFPA: 3 • TR: 60
- (D) LD50 o-r: 3200; LD50 o-m: 809; LD50 ip-m: 183; LD50 iv-m: 100; LD50 cu-m: 1040; LD50 cu-r: 1040

2,5-Xylenol: $C_8H_{10}O$ • 122.17 • 95-87-4
- (P) $\Delta H_f^0 = -161.71(g)$ • Eb = 212-213.5 • $P_{vap} = 1.3(91.2)$
- (I) $P_{fla} = 78; 86$ • AIT = 598
- (T) Tr: 24/25-34 • NFPA: 3 • TR: 60
- (D) LD50 o-r: 444; LD50 o-m: 383; LD50 cu-ra: 938

2,6-Xylenol: $C_8H_{10}O$ • 122.17 • 576-26-1
- (P) $\Delta H_f^0 = -161.83(g)$ • Eb = 203 • $P_{vap} = 1.3(91.2)$
- (I) $P_{fla} = 73$
- (T) Tr: 24/25-34 • NFPA: 3 • TR: 60
- (D) LD50 o-r: 296; LD50 o-m: 450; 980; LD50 o-ra: 700; LD50 ip-m: 150; LD50 iv-m: 80; LD50 su-m: 920; LD50 cu-r: 2325

3,4-Xylenol: $C_8H_{10}O$ • 122.17 • 95-65-8
- (P) $\Delta H_f^0 = -156.64(g)$; Eb = 225-227; $P_{vap} = 1.3(91.2)$
- (I) $P_{fla} = 110$
- (T) Tr: 24/25-34 • NFPA: 3 • TR: 60
- (D) LD50 cu-ra: 1000

2-Allylphenol: $C_9H_{10}O$ • 134.18 • 1745-81-9
- (P) Eb = 220; 230; 94/8
- (I) $P_{fla} = 88; 99$
- (T) Tr: 20/21
- (D) LD50 ip-m: 256

4-Nonylphenol: $C_{15}H_{24}O$ • 220.36 • 104-40-5
- (P) Eb = 293-297
- (I) $P_{fla} = 141; 150$
- (T) Tr: 22-34 • NFPA: 2
- (D) LD50 o-r: 1620; LD50 o-m: 1231; LD50 cu-ra: 2140

o-Phenylphenol: $C_{12}H_{10}O$ • 170.21 • 90-43-7
- (P) Eb = 280-284; 152/15; 145/14 • $P_{vap} = 0.4(20); 0.7(100); 9.3(140)$
- (I) $P_{fla} = 124$
- (T) Tr: 36/38 • NFPA: 2
- (D) LD50 o-r: 2000; 2480; LD50 o-m: 1050; LD50 ip-m: 50

1-Naphthol: $C_{10}H_8O$ • 144.18 • 90-15-3
- (P) $\Delta H_f^0 = -21.34(g)$ • Eb = 278-280; 283; 288; 184/40 • $P_{vap} = 1.33(94)$
- (I) LEL/UEL = 0.8/5 • $P_{fla} = 125; 153co$ • AIT = 541
- (T) Tr: 21/22-36/38-41 • NFPA: 2
- (D) LD50 o-r: 1870; 2400; 2590; LD50 o-m: 275; LD50 o-ra: 9000; LD50 ip-r: 250; LD50 cu-ra: 880

2-Naphthol: $C_{10}H_8O$ • 144.18 • 135-19-3
- (P) $\Delta H_f^0 = -42.26(g)$ • Eb = 285-286; 295 • $P_{vap} = 1(50); 13.3(145.5)$
- (I) $P_{fla} = 153; 160cc$
- (T) Tr: 20/22 • NFPA: 2
- (D) LD50 o-r: 1960; LD50 o-m: 100; LD50 ip-m: 97.5; LD50 scu-r: 2940; LD50 scu-m: 100; LD50 scu-m: 3000

Pyrocatechol: $C_6H_6O_2$ • 110.11 • 120-80-9
- (P) $\Delta H_f^0 = -361.08; -361.68$ (s) • Eb = 245-246; 221.5/400; 197.7/200; 176/100; 161.7/60; 150.6/40; 134/20; 118.3/10; 104/5 • $P_{vap} = 0.2(20); 0.35(30); 0.75(50); 1.33(75); 13.3(118.3)$
- (I) LEL/UEL = 1.97/- • $P_{fla} = 127cc; 137$
- (T) Tr: 21/22-36/38 • LVE(MVE) (F): – (5); TWA (USA): 5 (skin)
- (D) LD50 o-r: 260; LD50 o-m: 260; LD50 ip-m: 68; 190; LD50 scu-r: 110; LD50 scu-m: 247; LD50 cu-ra: 800

Resorcinol: $C_6H_6O_2$ • 110.11 • 108-46-3
- (P) $\Delta H_f^0 = -367.98(s)$ • Eb = 277; 281; 178/16 • $P_{vap} = 0.01(20); 0.02(30); 0.06(50); 1.33(108.4)$
- (I) LEL/UEL = 1.4/- • $P_{fla} = 127cc; 170$ • AIT = 605; 608
- (T) Tr: 22-36/38 • NFPA: –; 0; 2 • TR: 60 • LVE(MVE) (F): – (10); TWA (USA): 10 • STEL: 20
- (D) LD50 o-r: 301; LD50 o-m: 200; LD50 ip-m: 215; 240; LD50 scu-r: 400; LD50 scu-m: 213; LD50 cu-ra: 3360

Hydroquinone: $C_6H_6O_2$ • 110.11 • 123-31-9
- (P) $\Delta H_f^0 = -364.34(s)$ • Eb = 285-287; 286.2 • $P_{vap} = 0.00015(20); 1.3(130); 1.33(132.4)$
- (I) $P_{fla} = 165cc$ • AIT = 498; 515
- (T) Tr: 45-20/22; 20/22 • NFPA: 2 • TR: 60 • LVE(MVE) (F): – (2 mg/m^3); TWA (USA): 2 mg/m^3 • MAK (D): 2 mg/m^3
- (D) LD50 o-r: 320; LD50 o-m: 245; LD50 o-ra: 200; LD50 ip-r: 170; LD50 iv-r: 115; LD50 ip-ra: 125; LD50 scu-r: 300; LD50 scu-m: 182; 190

8 • Table of organic compounds — Ethers

Name: Empirical formula • Molar mass(g/mol)• Registry number (CAS).
(P) Physical data: ΔH_f^0 (kJ/mol) (physical state) • Eb (°C/mmHg) (if substance volatile) • P_{vap} (mBar) (°C) (if substance volatile).
(I) Inflammability data: LEL/UEL (%) • P_{fla} (°C) • AIT (°C).
(T) Toxicity: Safety Code (Tr) • NFPA • Transport Code (TR) • LVE (MVE) (ppm) F, USA, D • STEL (ppm); IDLH (ppm).
(D) Lethal and toxic doses: LC (mg/l/duration of exposure); LD (mg/kg).

Pyrogallol: $C_6H_6O_3$ • 126.11 • 87-66-1
(P) Eb = 309 • P_{vap} = 13.3(167.7)
(T) Tr: 20/21/22 • NFPA: 3
(D) LD50 o-m: 300; LD50 o-ra: 1600; LD50 ip-m: 400; LD50 scu-r: 650; LD50 scu-m: 566

ETHERS

• Aliphatic ethers

Dimethyl ether: C_2H_6O • 46.07 • 115-10-6
(P) ΔH_f^0 = –184.05(g) • Eb = –25 at –24 • P_{vap} = 2400(0); 5065; 5100; 5333(20); 6900(30); 11400(50)
(I) LEL/UEL = 2.7/18.6; 3.4/18; 3.4/27.0 • P_{fla} = –41 • AIT = 235; 240; 350
(T) NFPA: 3 • TR: – • MAK (D): 1000
(D) LC50 r: 0.308/?; LC50 m: 726/30'

Methyl ethyl ether: C_3H_8O • 60.11 • 540-67-0
(P) ΔH_f^0 = –216.44(g) • Eb = 10-11 • P_{vap} = 1600(20); 2300(30); 4100(50)
(I) LEL/UEL = 2.0/10.1 • P_{fla} = –37 • AIT = 190
(T) NFPA: 2

Diethyl ether: $C_4H_{10}O$ • 74.12 • 60-29-7
(P) ΔH_f^0 = –252.13(g); –273.22(l) • Eb = 34.6; 35 • P_{vap} = 580; 587; 589(20); 712(25)
(I) LEL/UEL = 1.7/36; 1.8/48; 1.85/36; 1.9/36 • P_{fla} = –45cc; –39 • AIT = 160; 170; 180
(T) NFPA: 2 • TR: 33 • LVE(MVE) (F): 500 (400); TWA (USA): 400; MAK (D): 400 • STEL: 500
(D) LC50 r: 137/2h; LC50 m: 58.3/30'; LC50 l: 199/?; LD50 o-r: 1215; LD50 ip-m: 2420; LD50 iv-m: 996

Ethyl vinyl ether: C_4H_8O • 72.11 • 109-92-2
(P) ΔH_f^0 = –140.71(g) • Eb = 33-36; 17.9/400; 2.2/200; –11.5/100; –48.1/10; –74.3/1 • P_{vap} = 246(0); 388(10); 560; 567; 571(20); 1919(55); 3071(70)
(I) LEL/UEL = 1.3/12; 1.4/28; 1.7/28 • P_{fla} = < –46; –45cc; –18 • AIT = 178; 190; 200; 202
(T) NFPA: 2 • TR: 339
(D) LC50 m: 324; **LD50** o-r: 6153; 8160; LD50 cu-ra: > 20000

Divinyl ether: C_4H_6O • 70.10 • 109-93-3
(P) ΔH_f^0 = –39.87(g) • Eb = 28.4-29; 39
(I) LEL/UEL = 1.2/27; 1.7/27; 1.7/36.5 • P_{fla} = < –30; –30; < –20 • AIT = 360
(T) NFPA: 2 • TR: 339
(D) MLC m: 149/?

Butyl vinyl ether: $C_6H_{12}O$ • 100.16 • 111-34-2
(P) Eb = 92-94.4 • P_{vap} = 56(20)
(P) P_{fla} = –12; –9; –9 oc; –6; –1 • AIT = 224
(T) NFPA: 2
(D) LC50 m: 62; **LD50** o-r: 10000; LD50 cu-ra: 4240

Dipropyl ether : $C_6H_{14}O$ • 102.18 • 111-43-3
(P) ΔH_f^0 = –292.88(g) • Eb = 88-91 • P_{vap} = 25.7; 73(20); 80(25); 115(30); 255(50)
(I) LEL/UEL = 1.3/7; 1.7/- • P_{fla} = –28; –20co; 4; 7; < 21 • AIT = 215
(T) TR: 33
(D) LC50 m: 163/15'; **LD50** iv-m: 204

Diisopropyl ether $C_6H_{14}O$ • 102.18 • 108-20-3
(P) ΔH_f^0 = –318.82(g); –351.20(l) • Eb = 65-69 • P_{vap} = 160; 175; 180(20); 200; 227(25); 270(30); 560(50)
(I) LEL/UEL = 1(100)/21(100); 1.1/4.5; 1.4/7.9 • P_{fla} = –28cc; –22; –12; –9 oc • AIT = 405; 441; 443
(T) NFPA: 2 • TR: 33 • TWA (USA): 250; 500; MAK (D): 500
(D) LC50 r: 162/?; LC50 m: 131/?; LC50 l: 121/?; LD50 o-r: 8470; LD50 ip-m: 812; LD50 cu-ra: 20000

Butyl methyl ether: $C_5H_{12}O$ • 88.15 • 628-28-4
(P) Eb = 69-71 • P_{vap} = 268(20); 213(25)
(I) LEL/UEL = 1.7/8.4 • P_{fla} = –30; –10 • AIT = 460
(D) LC50 m: 176/?

Methyltert-butyl ether (MTBE): $C_5H_{12}O$ • 88.15 • 1634-04-4
(P) Eb = 54-56 • P_{vap} = 279.24; 319(20); 327(25); 1019(55)
(I) LEL/UEL = 1.6/15.1; 1.6/8.4; 2.5/15.1 • P_{fla} = –27cc; –25; –9 • AIT = 191; 193; 224; 460
(T) Tr: 36/37/38 • TR: 33
(D) LC50 r: 85/4h; LC50 m: 141/15'; LD50 o-r: 4000; LD50 o-m: 5960; 5970; LD50 ip-m: 1700

Isobutyl vinyl ether: $C_6H_{12}O$ • 100.16 • 109-53-5
(P) Eb = 83-83.2 • P_{vap} = 90(20)
(I) LEL/UEL = 2.6/14 • P_{fla} = –7; –9; –13; –15; –18
(T) NFPA: 2
(D) LC50 r: 59/4h; LD50 o-r: 17000; LD50 cu-ra: 20000

Dibutyl ether: $C_8H_{18}O$ • 130.23 • 142-96-1
(P) ΔH_f^0 = –653.12(l) • Eb = 140-143; 47/31 • P_{vap} = 6.4(20)
(I) LEL/UEL = 0.9/8.5; 1.5/7.6; 0.85/7.2 • P_{fla} = 25; 33cc; 37cc • AIT = 184; 194
(T) Tr: 36/37/38 • NFPA: 2 • TR: 30
(D) LC50 r: 21.27/4h; LC50 m: 169/15'; **LD50** o-r: 7400; 11000; LD50 cu-ra: 10000

Diisobutyl ether: $C_8H_{18}O$ • 130.23 • 628-55-7
(P) Eb = 120-125 • P_{vap} = 27(20); 20(25); 24-30(29); 50-56(43); 77-83(52)
(I) LEL/UEL = 0.9/6.4 • P_{fla} = 3; 8; 16 • AIT = 330
(D) LC50 m: 156/15'

Diamyl ether : $C_{10}H_{22}O$ • 158.30 • 693-65-2
(P) Eb = 187-188; 190; 70/12 • P_{vap} = 0.7(20)
(I) P_{fla} = 57cc; 57oc • AIT = 171
(T) NFPA: 1
(D) LD50 iv-m: 164

8 • Table of organic compounds

Ethers

Name: Empirical formula • Molar mass(g/mol)• Registry number (CAS).
- (P) **Physical data:** ΔH_f^0 (kJ/mol) (physical state) • Eb (°C/mmHg) (if substance volatile) • P_{vap} (mBar) (°C) (if substance volatile).
- (I) **Inflammability data:** LEL/UEL (%) • P_{fla} (°C) • AIT (°C).
- (T) **Toxicity:** Safety Code (Tr) • NFPA • Transport Code (TR) • LVE (MVE) (ppm) F, USA, D • STEL (ppm); IDLH (ppm).
- (D) **Lethal and toxic doses:** LC (mg/l/duration of exposure); LD (mg/kg).

Diisoamyl ether : $C_{10}H_{22}O$ • 158.30 • –
- (P) Eb = 172-173; 60/10 • P_{vap} = 4.8(40)
- (I) P_{fla} = 46; 57.5

Hexyl ether $C_{12}H_{26}O$ • 186.38 • 112-58-3
- (P) Eb = 228-229 • P_{vap} = 0.1(20)
- (T) NFPA: 2
- (D) LD50 o-r: 30900; LD50 cu-ra: 6000

ETHERS
• Aliphatic ethers of glycols and ether-alcohols

1,2-Dimethoxyethane: $C_4H_{10}O_2$ • 90.12 • 110-71-4
- (P) ΔH_f^0 = –376.64(l) • Eb = 82- 83.5; 20/61.5; 16/50; –14/10 • P_{vap} = 64; 80; 81.6(20); 125(30); 270(50)
- (I) LEL/UEL = 1.6/10.4 • P_{fla} = –1cc; 1; 5 • AIT = 202; 205; 245
- (T) Tr: 20 • NFPA: 2 • TR: 33

1,2-Diethoxyethane: $C_6H_{14}O_2$ • 118.18 • 629-14-1
- (P) Eb = 115-116; 121.4; 122 • P_{vap} = 12.2; 12.5(20)
- (P) P_{fla} = 20; 22; 35oc • AIT = 205; 208

2-Methoxyethanol: $C_3H_8O_2$ • 76.10 • 109-86-4
- (P) Eb = 124.5; 56/50; 34-41/20; 27/10 • P_{vap} = 8.22; 8.3; 11; ~11(20); 13(25); 32; 49(50)
- (I) LEL/UEL = 1.8/14; 1.8/19.8; 2.3/19.8; 2.5/14; 2.5/20.0; 2.5/24.5; 2.5/19.8 • P_{fla} = 38cc; 39Tcc; 43.3Tag oc; 46oc • AIT = 285; 288
- (T) Tr: 60-61-20/21/ 22 • NFPA: 2 • TR: 30
  • LVE(MVE) (F): – (5); TWA (USA): 5; 25 (skin); 2 (female); MAK (D): 5 • STEL: 5 (female)
- (D) LC50 r: 4.7/7h; LC50 m: 4.6/7h; LC50 l: 1.38/?;
  LD50 o-r: 2370; 2460; LD50 o-m: 2560; LD50 o-ra: 890;
  LD50 ip-r: 2500; LD50 ip-m: 2147; LD50 iv-r: 2068; 2140;
  LD50 cu-ra: 1280

2-Ethoxyethanol: $C_4H_{10}O_2$ • 90.12 • 110-80-5
- (P) Eb = 135; 64/50 • P_{vap} = 5.1(20); 9.3(30); 27(50)
- (I) LEL/UEL = 1.7/15.6; 1.8/14; 1.8/15.7; 2.6/15.7 • P_{fla} = 40cc; 42cc; 44; 49oc; 57oc • AIT = 235; 238
- (T) Tr: 60-61-20/21/ 22 • NFPA: 2 • TR: 30
  • LVE(MVE) (F): – (5); TWA (USA): 5; MAK (D): 20
- (D) LC50 r: 7.36/7h; LC50 m: 6.7/7h; **LD50** o-r: 2125; 3000;
  LD50 o-m: 2451; LD50 o-ra: 1275; LD50 ip-r: 2800;
  LD50 ip-m: 1707; LD50 iv-r: 2400; LD50 iv-m: 3900;
  LD50 iv-ra: 900; LD50 scu-r: 3400; LD50 cu-ra: 3300; 3500

2-Isopropoxyethanol: $C_5H_{12}O_2$ • 104.15 • 109-59-1
- (P) Eb = 142-144; 44/13 • P_{vap} = 3.47; 30.6(20); 6.9(25)
- (I) LEL/UEL = 1.6/13.0 • P_{fla} = 43; 45oc; 47 • AIT = 345
- (T) Tr: 20/21-36 • TR: 30 • LVE(MVE) (F): – (25); TWA (USA): 25; MAK (D): 5
- (D) LC50 r: 3.1/4h; LC50 m: 8.21/7h; **LD50** o-r: 5660;
  LD50 o-m: 4900; LD50 ip-r: 800; LD50 ip-m: 1860;
  LD50 cu-ra: 1600

2-Butoxyethanol: $C_6H_{14}O_2$ • 118.18 • 111-76-2
- (P) Eb = 168.4-172; 94/50; 62/10 • P_{vap} = 0.8; 0.9; < 1.33(20);
  1.1(25); 8(50); 400(140)
- (I) LEL/UEL = 1/10.6; 1.1/10.6; 1.1/10.7; 1.1/12.7; 1.9/10.3
  • P_{fla} = 60cc; 65.6Tcc; 71.1PMcc; 71; 74oc; 85Coc
  • AIT = 230; 240; 244
- (T) Tr: 20/21/22-37 • NFPA: 2 • LVE(MVE) (F): – (25);
  TWA (USA): 25; MAK (D): 20
- (D) LC50 r: 2.15/4h; 2.9/4h; 3.34/7h; LC50 m: 3.35/7h;
  LD50 o-r: 470; 1480; 2616; LD50 o-m: 1230;
  LD50 o-ra: 300; LD50 ip-r: 220; LD50 ip-m: 536;
  LD50 ip-ra: 220; LD50 iv-r: 307; 340; LD50 iv-m: 1130;
  LD50 iv-ra: 252; LD50 scu-m: 500; LD50 cu-ra: 220; 631

Diethyleneglycol: $C_4H_{10}O_3$ • 106.12 • 111-46-0
- (P) ΔH_f^0 = –628.44(l) • Eb = 245.8; 125-126/11 • P_{vap} = 0.0067;
  0.013(20); 1.33(91.8); 6.7(120)
- (I) LEL/UEL = 0.7/22; 1.7/10.6; 2/12.3; 2/- • P_{fla} = 124; 137cc; 143oc;
  151.7PMcc; 151.7Coc • AIT = 229; 370
- (T) Tr: 22 • NFPA: 1
- (D) LC50 m: 0.13/2h; **LD50** o-r: 4700; 12565; 14800; 20760;
  LD50 o-m: 23700; LD50 o-ra: 4400; LD50 ip-r: 7700;
  LD50 ip-m: 9719; LD50 iv-r: 6565; LD50 iv-ra: 2000;
  LD50 scu-r: 18800; LD50 scu-m: 5000; LD50 cu-ra: 9530; 11890

Diethyleneglycolmonomethylether: $C_5H_{12}O_3$ • 120.15 • 111-77-3
- (P) Eb = 189-194.2; 110/50 • P_{vap} = 0.08; 0.13; 0.27(20)
- (I) LEL/UEL = 1.38(167)/22.7(135); 1.5/9.5; 1.6/16.1
  • P_{fla} = 77; 83; 85; 85oc; 87; 87Tcc; 93Coc • AIT = 215; 220
- (T) NFPA: 1
- (D) LD50 o-r: 5500; 9210; LD50 o-m: 8220; LD50 o-ra: 7190;
  LD50 ip-r: 2722; 3000; LD50 ip-m: 2611; LD50 cu-ra: 6540; 6690

Diethyleneglycolmonoethylether: $C_6H_{14}O_3$ • 134.18 • 111-90-0
- (P) Eb = 195; 200; 202; 120/50; 87/10 • P_{vap} = 0.10; 0.13; 0.27(20);
  0.17(25)
- (I) LEL/UEL = 1.2/11.6; 1.2/23.5 • P_{fla} = 90cc; 94oc; 96Too
  • AIT = 190; 204
- (T) NFPA: 1
- (D) LD50 o-r: 5500; 6500; 8690; LD50 o-m: 6600; LD50 ip-r: 6310;
  LD50 ip-m: 2300; LD50 iv-ra: 2500; LD50 scu-r: 6000;
  LD50 scu-m: 5500; LD50 cu-ra: 8500

Triethyleneglycol: $C_6H_{14}O_4$ • 150.18 • 112-27-6
- (P) ΔH_f^0 = –804.16(l) • Eb = 276; 285; 288; 165/14; 125-127/0.1
  • P_{vap} = < 0.013(20); 1.33(114); 1.3(117)
- (I) LEL/UEL = 0.9/9; 0.9/9.2 • P_{fla} = 166Coc; 172cc; 172PMcc; 177
  • AIT = 371
- (T) NFPA: 1
- (D) LD50 o-r: 17000; 15000-22000; LD50 o-m: 18500; 21000;
  LD50 o-ra: 8400; LD50 ip-m: 8141; 8941; LD50 iv-r: 11700;
  LD50 iv-m: 6500; 7300-9500; LD50 iv-ra: 1900; LD50 scu-m: 8750

400

8 • Table of organic compounds — Ethers

Name: Empirical formula • Molar mass(g/mol) • Registry number (CAS).
- (P) **Physical data:** ΔH_f^0 (kJ/mol) (physical state) • Eb (°C/mmHg) (if substance volatile) • P_{vap} (mBar) (°C) (if substance volatile).
- (I) **Inflammability data:** LEL/UEL (%) • P_{fla} (°C) • AIT (°C).
- (T) **Toxicity:** Safety Code (Tr) • NFPA • Transport Code (TR) • LVE (MVE) (ppm) F, USA, D • STEL (ppm); IDLH (ppm).
- (D) **Lethal and toxic doses:** LC (mg/l/duration of exposure); LD (mg/kg).

ETHERS
• Aromatic ethers and ether-phenols

Anisole: C_7H_8O • 108.14 • 100-66-3
- (P) $\Delta H_f^0 = -72.38(g)$ • Eb = 154-155.5; 93/100; 70.7/40; 55.8/20; 42.2/10; 30/5; 5.4/1 • P_{vap} = 3.6(20); 6.7(30); 13.3(42.2); 20(50)
- (I) LEL/UEL = 1.2/- • P_{fla} = 41; 44; 50; 51cc; 52Coc • AIT = 475
- (T) Tr: 36/37/38 • NFPA: 1 • TR: 30
- (D) LC50 m: 3; **LD50** o-r: 3700; LD50 o-m: 2800

2-Methylanisole: $C_8H_{10}O$ • 122.17 • 578-58-5
- (P) Eb = 170-172
- (I) P_{fla} = 51; 56

3-Methylanisole: $C_8H_{10}O$ • 122.17 • 100-84-5
- (P) Eb = 175-180
- (I) P_{fla} = 54; 59
- (T) Tr: 20

4-Methylanisole: $C_8H_{10}O$ • 122.17; 104-93-8
- (P) Eb = 173-177
- (I) P_{fla} = 53; 59
- (T) Tr: 22
- (D) LD50 o-r: 1920

Diphenyl ether : $C_{12}H_{10}O$ • 170.21 • 101-84-8
- (P) $\Delta H_f^0 = -14.56(l)$ • Eb = 252; 257-259 • P_{vap} = 0.08(20); 0.16(30); 0.52(50)
- (I) LEL/UEL = 0.8/1.5 • P_{fla} = 96oc; 113cc; 115; 125 • AIT = 610; 615; 620; 646
- (T) Tr: 36/37/38
- (D) LD50 o-r: 3370

Dibenzyl ether : $C_{14}H_{14}O$ • 198.27 • 103-50-4
- (P) Eb = 295-298; 173-174/21; 158-160/10; 126/2 • P_{vap} = 0.03(20)
- (I) P_{fla} = 135cc
- (T) Tr: 23/24/25
- (D) LD50 o-r: 2500

1.2-Dimethoxybenzene: $C_8H_{10}O_2$ • 138.17 • 91-16-7
- (P) $\Delta H_f^0 = -290.37(l)$ • Eb = 205-207
- (I) P_{fla} = 72; 87
- (T) Tr: 22
- (D) LD50 o-r: 890; 1360; LD50 o-m: 700; 2020

1.3-Dimethoxybenzene: $C_8H_{10}O_2$ • 138.17 • 151-10-3
- (P) Eb = 217; 85-87/7
- (I) P_{fla} = 88
- (T) Tr: 36/38
- (D) LD50 ip-m: 900

1.4-Dimethoxybenzene: $C_8H_{10}O_2$ • 138.17 • 150-78-7
- (P) Eb = 212-213; 109/20 • P_{vap} = < 1.33(20)
- (I) AIT = 422
- (T) Tr: 36/38
- (D) LD50 o-r: 3600; LD50 ip-r: 1100

Gaiacol: $C_7H_8O_2$ • 124.14 • 90-05-1
- (P) Eb = 204-206; 53-55/4 • P_{vap} = 0.1(20); 0.15(25); 0.3(30); 1.2(50)
- (I) P_{fla} = 82cc
- (T) Tr: 22-36/38 • NFPA: 2 • TR: 60
- (D) LC50 m: 7.6; **LD50** o-r: 520; 725; LD50 o-m: 621; 890; LD50 iv-m: 170; LD50 scu-r: 900; LD50 cu-ra: 4600

4-Methoxyphenol: $C_7H_8O_2$ • 124.14 • 150-76-5
- (P) Eb = 243-246; 144/10 • P_{vap} = < 0.013(20)
- (I) P_{fla} = > 109; 124cc; 131oc • AIT = 420
- (T) Tr: 22-37 • NFPA: 2 • TWA (USA): 5mg/m^3
- (D) LD50 o-r: 1600; LD50 ip-r: 725; LD50 ip-m: 250

ETHERS
• Heterocyclic ethers and ether-alcohols

Ethylene oxide : C_2H_4O • 44.05 • 75-21-8
- (P) $\Delta H_f^0 = -52.63(g)$ • Eb = 10-11 • P_{vap} = 657.3(0); 1455; 1460(20); 3955.7(50)
- (I) LEL/UEL = 2.6/99; 3/100 • P_{fla} = -30; -20; -18oc • AIT = 429-440; 571
- (T) Tr: 45-46-23-36/ 37/38 • NFPA: 2 • LVE(MVE) (F): 10 (5); TWA (USA): 1 • STEL: 10; IDLH: 800
- (D) LC50 r: 1.44/4h; LC50 m: 1.5/4h; **LD50** o-r: 72; LD50 ip-m: 175; LD50 iv-m: 290; LD50 iv-ra: 175; LD50 scu-r: 187

Propylene oxide: C_3H_6O • 58.08 • 75-56-9
- (P) $\Delta H_f^0 = -92.76(g)$ • Eb = 34-35 • P_{vap} = 258(0); 400(10); 532(17.); 589; 592; 600; 661; 980(20); 827(30)
- (I) LEL/UEL = 1.9/24; 1.9/15; 2/22; 2/39; 2.8/37 • P_{fla} = -35cc; -37cc; -37oc; -37Tag oc• AIT = 430; 448; 747
- (T) Tr: 45-20/21/22-36/37/38 • NFPA: 2 • TR: 33 • LVE(MVE) (F): – (20); TWA (USA): 20; MAK (D): 20
- (D) LC50 r: 9.16/4h; LC50 m: 4/4h; **LD50** o-r: 380; 930; 1140; LD50 o-m: 440; LD50 ip-r: 150; 364; LD50 ip-m: 175; LD50 cu-ra: 1500; 1245

8 • Table of organic compounds — Ethers

Name: Empirical formula • Molar mass(g/mol)• Registry number (CAS).
- (P) **Physical data:** ΔH_f^0 (kJ/mol) (physical state) • Eb (°C/mmHg) (if substance volatile) • P_{vap} (mBar) (°C) (if substance volatile).
- (I) **Inflammability data:** LEL/UEL (%) • P_{fla} (°C) • AIT (°C).
- (T) **Toxicity:** Safety Code (Tr) • NFPA • Transport Code (TR) • LVE (MVE) (ppm) F, USA, D • STEL (ppm); IDLH (ppm).
- (D) **Lethal and toxic doses:** LC (mg/l/duration of exposure); LD (mg/kg).

1,2-Epoxybutane: C_4H_8O • 72.11 • 106-88-7
- (P) Eb = 62-65; 63.3 • P_{vap} = 187(20)
- (I) LEL/UEL = 1.5/18.3; 1.7/19; 3.1/25.1 • P_{fla} = –26cc; –15; –12 • AIT = 369
- (T) Tr: 45-20/21/22-36/37/38; 22-36/37 • TR: 339 • TWA (USA): 400
- (D) LC50 r: 11.45/4h; **LD50** o-r: 500; 920; LD50 cu-ra: 2100; ~1760

Propylglycidylether: $C_6H_{12}O_2$ • 116.16 • –
- (P) Eb = 154 • P_{vap} = 6.2(20)
- (I) P_{fla} = 45; 57

Isopropylglycidylether: $C_6H_{12}O_2$ • 116.16 • 4016-14-2
- (P) Eb = 131-137 • P_{vap} = 12.5(25)
- (I) P_{fla} = 33; 35
- (T) Tr: 20/21/22-40-42/43; 42/43 • NFPA: 3 • LVE(MVE) (F): – (50); TWA (USA): 50; MAK (D): 50 • STEL: 75
- (D) LC50 r: 5.26/8h; LC50 m: 7.18/4h; **LD50** o-r: 4200; LD50 o-m: 1300; LD50 cu-ra: 9650

Butylglycidylether: $C_7H_{14}O_2$ • 130.19 • 2426-08-6
- (P) Eb = 164-8 • P_{vap} = 3.5(20); 5.5(30); 12(50)
- (I) P_{fla} = 22; 54 • AIT = 215
- (T) Tr: 40-43; 20-43 • LVE(MVE) (F): – (25); TWA (USA): 25; MAK (D): 25
- (D) LC50 r: 5.52/8h; LC50 m: 0.26/?; **LD50** o-r: 1660; 2050; LD50 o-m: 1530; LD50 ip-r: 1140; LD50 ip-m: 700; LD50 cu-ra: 2520

Phenylglycidylether: $C_9H_{10}O_2$ • 150.19 • 122-60-1
- (P) Eb = 245-247 • P_{vap} = 0.04(20); 0.013(25); 10.7(112); 11.4(120)
- (I) LEL/UEL = 1.1/- • P_{fla} = > 109; 114
- (T) Tr: 45-21-43; 21-43 • LVE(MVE) (F): – (1); TWA (USA): 1
- (D) LD50 o-r: 3850; LD50 o-m: 1400; LD50 scu-m: 760; LD50 cu-ra: 1500

Diglycidylether: $C_6H_{10}O_3$ • 130.16 • 2238-07-5
- (P) Eb = 260; 98-99/11 • P_{vap} = 0.12(25)
- (T) LVE(MVE) (F): – (0.1); TWA (USA): 0.1; MAK (D): 0.1

Furan: C_4H_4O • 68.08 • 110-00-9
- (P) ΔH_f^0 = –34.43(g) • Eb = 31-33 • P_{vap} = 636; 670(20); 1000(30); ~2026(50); 2183(55)
- (I) LEL/UEL = 2.3/14.3 • P_{fla} = –36cc; 0 • AIT = 390
- (T) Tr: 26; - • NFPA: 1 • TR: 33
- (D) LC50 r: 9.45/1h; MLC r: 86.06/?; LC50 m: 0.12/1h; **LD50** ip-r: 5200; LD50 ip-m: 7

2-Methylfuran: C_5H_6O • 82.10 • 534-22-5
- (P) Eb = 61-64 • P_{vap} = 185(20)
- (I) P_{fla} = –30; –22; <–7
- (T) Tr: 20/21/22 • NFPA: 2
- (D) LC50 r: 1.68/4h; **LD50** o-r: 167

Benzofuran: C_8H_6O • 118.14 • 271-89-6
- (P) Eb = 170-175; 62-63/15
- (I) P_{fla} = 50; 56
- (T) Tr: 23

2,5-Dihydrofuran: C_4H_6O • 70.09 • 1708-29-8
- (P) Eb = 84-88; 66-67 (?)
- (I) P_{fla} = –18; –15

Tetrahydrofuran: C_4H_8O • 72.11 • 109-99-9
- (P) ΔH_f^0 = –216.19(l) • Eb = 64-67; 25/176 • P_{vap} = 70(0); 152(15); 180; 191; 200(20); 263(25); 625(50)
- (I) LEL/UEL = 1.5/12; 1.8/11.8; 2/11.8; 2.3/11.8 • P_{fla} = –20; –17Tcc; –15 • AIT = 205; 230; 260; 321
- (T) Tr: 36/37 • NFPA: 2 • TR: 33 • LVE(MVE) (F): – (200); TWA (USA): 200; MAK (D): 200 • STEL: 250
- (D) LC50 r: 61.84/3h; LC50 m: 24/2h; **LD50** o-r: 1650; LD50 ip-r: 2900; LD50 ip-m: 1900

2,5-Dimethyltetrahydrofuran: $C_6H_{12}O$ • 100.16 • 1003-38-9
- (P) Eb = 90-93
- (I) P_{fla} = –12; 26
- (D) LD50 o-r: 4000

Tetrahydropyran: $C_5H_{10}O$ • 86.14 • 142-68-7
- (P) ΔH_f^0 = –255.64(l) • Eb = 87-89
- (I) P_{fla} = –20; –15; –7; –4

Dihydropyran: C_5H_8O • 84.12 • 110-87-2
- (P) ΔH_f^0 = –156.90(l) • Eb = 84-88
- (I) P_{fla} = –18; –15

1,4-Dioxan: $C_4H_8O_2$ • 88.11 • 123-91-1
- (P) ΔH_f^0 = –315.06(g); –353.42(l) • Eb = 101.1-102.5; 81.8/400; 62.3/200; 45.1/100; 33.8/60; 25.2/40; 12/20 • P_{vap} = 36; 39; 41(20); 53(25); 53.3(25.2); 67(30); 160(50)
- (I) LEL/UEL = 1.7/25.2; 1.9/-; 1.9/22.5; 1.97/22.5; 1.97/22.25; 2.0/22; 2.0/22.2 • P_{fla} = 5-18; 12Tcc; 23Tco • AIT = 180; 300; 375
- (T) Tr: 36/37-40 • NFPA: 2 • TR: 33 • LVE(MVE) (F): 40 (10); TWA (USA): 25; MAK (D): 50
- (D) LC50 r: 46/2h; LC50 m: 37/2h; **LD50** o-r: 2900; 4200; 5200; 7120; LD50 o-m: 5700; LD50 o-ra: 2000; LD50 ip-r: 800; 5600; LD50 ip-m: 790; LD50 iv-ra: 1500; LD50 ip-ra: 7600; 7858

Glycidol: $C_3H_6O_2$ • 74.08 • 556-52-5
- (P) Eb = 161; 167dec; 62/15; 66/2.5; 25/0.9 • P_{vap} = 1.2(20); 1.2(25); 3.3(65)
- (I) P_{fla} = 66; 71; 81 • AIT = 415
- (T) Tr: 23-21/22-36/37/38-40-42/43; 21/22-23-36/37/38-42/43 • NFPA: 3 • LVE(MVE) (F): – (25); TWA (USA): 25; MAK (D): 50
- (D) LC50 r: 1.76/8h; LC50 m: 1.36/4h; **LD50** o-r: 420; 850; LD50 o-m: 431; LD50 ip-r: 200; LD50 ip-m: 500; LD50 cu-ra: 1980

8 • Table of organic compounds — Halogenated derivatives

Name: Empirical formula • Molar mass(g/mol)• Registry number (CAS).
(P) Physical data: ΔH_f^0 (kJ/mol) (physical state) • Eb (°C/mmHg) (if substance volatile) • P_{vap} (mBar) (°C) (if substance volatile).
(I) Inflammability data: : LEL/UEL (%) • P_{fla} (°C) • AIT (°C).
(T) Toxicity: Safety Code (Tr) • NFPA • Transport Code (TR) • LVE (MVE) (ppm) F, USA, D • STEL (ppm); IDLH (ppm).
(D) Lethal and toxic doses: LC (mg/l/duration of exposure); LD (mg/kg).

Furfurylic alcohol: $C_5H_6O_2$ • 98.10 • 98-00-0
- (P) $\Delta H_f^0 = -276.35(l)$ • Eb = 170-171/750; 151.8/400; 133.1/200; 115.9/100; 104/60; 95.7/40; 81/20; 71-72/11; 68/10; 56/5; 31.8/1 • $P_{vap} = 0.53$; 0.56; 0.67(20); 1.1(30); 1.3(31.8); 4.5(50); 7.3(55)
- (I) LEL/UEL = 1.8/16.3(72-122); 1.8/16.3 • P_{fla} = 64; 75cc; 75oc; 79cc; 87 • AIT = 390; 491
- (T) Tr: 20/21/22 • TR: 60 • LVE(MVE) (F): – (10); TWA (USA): 10; MAK (D): 10; 50 • STEL: 15
- (D) LC50 r: 0.934/4h; LC50 m: 2.4/6h; **LD50** o-r: 88.3; 177; 275; LD50 o-m: 160; LD50 ip-r: 650; LD50 iv-ra: 650; LD50 scu-r: 85; LD50 cu-r: 3825; LD50 cu-ra: 400

Tetrahydrofurfurylic alcohol: $C_5H_{10}O_2$ • 102.13 • 97-99-4
- (P) $\Delta H_f^0 = -435.55(l)$ • Eb = 173-178; 82/20 • P_{vap} = 0.33; 2.3(20); 1(25); 2.3(40); 3.1(39)b
- (I) LEL/UEL = 1.5/9.7(72-122); 1.5/9.7 • P_{fla} = 74; 77oc; 84oc • AIT = 280; 282
- (T) Tr: 36
- (D) LD50 o-r: 1600; 2500; LD50 o-m: 2300; LD50 ip-r: 400; LD50 iv-ra: 725

HALOGENATED DERIVATIVES

• Alkyl halides

•*Fluorine derivatives*

Trifluoromethane: CHF_3 • 70.02 • 75-46-7
- (P) $\Delta H_f^0 = -693.33(g)$ • Eb = –84 • P_{vap} = 44640(20); 43782(21)
- (T) TR: 20

Tetrafluoromethane: CF_4 • 88.01 • 75-73-0
- (P) $\Delta H_f^0 = -933.03(g)$ • Eb = –127

Fluoroethane: C_2H_5F • 48.06 • 353-36-6
- (P) $\Delta H_f^0 = -263.17(g)$ • Eb = –38

.*Chlorine derivatives*

Chloromethane: CH_3Cl • 50.49 • 74-87-3
- (P) $\Delta H_f^0 = -81.96(g)$ • Eb = –23.7 • P_{vap} = 5060(21); 5100(21.1)
- (I) LEL/UEL = 7/19; 7.1/18.5; 8.1/17; 8.1/17.2; 10.7/17.4 • P_{fla} = < 0; ~0; < –24 • AIT = 625; 632
- (T) Tr: 40-48/20 • NFPA: 2 • TR: 236
  • LVE(MVE) (F): 100 (50); TWA (USA): 50; MAK (D): 50
  • STEL: 100
- (D) LC50 r: 17.14/1h; 5.3/4h; LC50 m: 4.54/6h; **LD50** o-r: 1800

Chloroethane: C_2H_5Cl • 64.52 • 75-00-3
- (P) $\Delta H_f^0 = -112.26(g)$ • Eb = 12.3 • P_{vap} = 622(0); 1324; 1333; 1400(20); 2177(25); 3977(55); 10130(92.6)
- (I) LEL/UEL = 3.6/14.8; 3.8/15.4 • P_{fla} = –50cc; –43oc • AIT = 510; 517; 519
- (T) NFPA: 2 • TR: 236 • LVE(MVE) (F): – (1000); TWA (USA): 1000
- (D) LC50 r: 152/2H; 160/2h; LC50 m: 146/2h

Dichloromethane: CH_2Cl_2 • 84.93 • 75-09-2
- (P) $\Delta H_f^0 = -95.39(g)$; –124.26(l) • Eb = 40.1 • P_{vap} = 196(0); 453; 464; 467; 471; 475(20); 505(22); 581(25); 1688(55)
- (I) LEL/UEL = 12.0/19.0; 14/22; 15.5/66.4 (O_2 pure)
  • P_{fla} = None; (–9 calculated)
  • AIT = 556; 605; 615; 624; 662; 665
- (T) Tr: 40 • NFPA: 2 • TR: 60 • LVE(MVE) (F): 500 (100); TWA (USA): 50; MAK (D): 100
- (D) LC50 r: 52/?; 88/30'; LC50 m: 50.61/7h; LC50 l: 34.76/7h; **LD50** o-r: 1600; 2121; 2136; LD50 ip-r: 916; LD50 ip-m: 437; LD50 scu-m: 6460

Trichloromethane: $CHCl_3$ • 119.38 • 67-66-3
- (P) $\Delta H_f^0 = -102.93(g)$; –132.21(l) • Eb = 60-62; 10/100 • P_{vap} = 133(10.4); 210; 213(20); 260(25); 705(50)
- (I) AIT = 982
- (T) Tr: 22-38-40-48/20/22 • NFPA: 2 • TR: 60 • LVE(MVE) (F): 50 (5); TWA (USA): 10; 2; MAK (D): 10 • STEL: 50
- (D) LC50 r: 47.7/4h; LC50 m: 28/?; **LD50** o-r: 908; LD50 o-m: 36; LD50 o-ra: 500; LD50 ip-r: 894; LD50 ip-m: 623; 1000; LD50 scu-m: 704

Tetrachloromethane: CCl_4 • 153.82 • 56-23-5
- (P) $\Delta H_f^0 = -95.86(g)$; –132.84(l) • Eb = 75-78; 23/100 • P_{vap} = 120; 122(20); 133(23); 153(25); 191(30); 415(50)
- (T) Tr: 23/24/25-40-48/23-59 • NFPA: 3 • TR: 60 • LVE(MVE) (F): 10 (-) TWA (USA): 5; MAK (D): 10 • STEL: 10; IDLH: 300
- (D) LC50 r: 50.39/4h; LC50 m: 58.11/8h; 60.94/?; **LD50** o-r: 2350; 2800; LD50 o-m: 8263; 12800; LD50 o-ra: 5760; LD50 ip-r: 1500; LD50 ip-m: 572; LD50 iv-ra: 5840; LD50 scu-r: 5070; LD50 scu-m: 31

1,1-Dichloroethane: $C_2H_4Cl_2$ • 98.96 • 75-34-3
- (P) $\Delta H_f^0 = -130.12(g)$; –160.25(l) • Eb = 57.3; 57-59 • P_{vap} = 233; 240(20); 306(25); 790(50)
- (I) LEL/UEL = 5.6/11.4; 5.9/15.9; 5.6/16 • P_{fla} = –10; –8; –6oc
  • AIT = 458
- (T) Tr: 22-36/37 • NFPA: 2 • TR: 33 • LVE(MVE) (F): – (200); TWA (USA): 100; MAK (D): 100
- (D) LC50 r: 53.17/4h; LD50 o-r: 725

1,2-Dichloroethane: $C_2H_4Cl_2$ • 98.96 • 107-06-2
- (P) $\Delta H_f^0 = -129.7(g)$; –165.23(l) • Eb = 82-84 • P_{vap} = 82; 87; 111; 116(20); 133(29.4); 137(30); 320(50)
- (I) LEL/UEL = 6.2/15.6; 6.2/15.9; 6.2/16 • P_{fla} = 13cc; 15; 18oc
  • AIT = 413; 440
- (T) Tr: 45-22-36/37/ 38 • NFPA: 2 • TR: 336
  • LVE(MVE) (F): – (10); TWA (USA): 10
- (D) LC50 r: 4.09/7h; LC50 m: 5/2h; LC50 l: 12.27/7h; **LD50** o-r: 670; 770; LD50 o-m: 413; 489; LD50 o-ra: 860; LD50 ip-r: 807; LD50 iv-r: 1000; LD50 ip-m: 470; LD50 scu-m: 380; LD50 cu-ra: 2800; 4886

8 • Table of organic compounds — Halogenated derivatives

Name: Empirical formula • Molar mass(g/mol)• Registry number (CAS).
(P) Physical data: ΔH_f^0 (kJ/mol) (physical state) • Eb (°C/mmHg) (if substance volatile) • P_{vap} (mBar) (°C) (if substance volatile).
(I) Inflammability data: LEL/UEL (%) • P_{fla} (°C) • AIT (°C).
(T) Toxicity: Safety Code (Tr) • NFPA • Transport Code (TR) • LVE (MVE) (ppm) F, USA, D • STEL (ppm); IDLH (ppm).
(D) Lethal and toxic doses: LC (mg/l/duration of exposure); LD (mg/kg).

1,1,1-Trichloroethane: $C_2H_3Cl_3$ • 133.40 • 71-55-6
(P) $\Delta H_f^0 = -142.67$(g) • Eb = 72-75; 20/100
 • P_{vap} = 133(20); 161(25); 445(50)
(I) LEL/UEL = 7.5/15.0; 7.5/15.5; 8.0/15.5; 8/10.5; 8(10)/14.2(22)
 • P_{fla} = Self inflammatory at 25°C in presence of strong illumination
 • AIT = 500; 537
(T) Tr: 20-59 • NFPA: 2 • TR: 60 • LVE(MVE) (F): 450 (300); TWA (USA): 350; MAK (D): 200 • STEL: 450
(D) LC50 r: 99/4h; 108.6/4h; LC50 m: 23.59/2h; **LD50** o-r: 9600; 10300; 15000; LD50 o-m: 6000; 11240; LD50 o-ra: 5660; LD50 ip-r: 3593; 5100; LD50 ip-m: 2568; 3636; LD50 scu-m: 16000

1,1,2,2-Tetrachloroethane: $C_2H_2Cl_4$ • 167.4 • 79-34-5
(P) $\Delta H_f^0 = -152.79$(g); -196.65(l) • Eb = 146.4-147; 83/100
 • P_{vap} = 6.8; 7; 10.7(20); 11.3(30); 13.3(33); 30(50)
(T) Tr: 26/27-40; 26/27 • NFPA: 3 • TR: 60
 • LVE(MVE) (F): 5 (1); TWA (USA): 1; MAK (D): 1
(D) LC50 r: 7.36/4h; LC50 m: 4.5/2h; **LD50** o-r: 250; 317; 800; LD50 o-m: 821; LD50 ip-m: 821; LD50 scu-m: 1108; LD50 scu-m: 500

1,1,1,2-Tetrachloroethane: $C_2H_2Cl_4$ • 167.4 • 630-20-6
(P) $\Delta H_f^0 = -149.37$(g) • Eb = 129-130; 138 • P_{vap} = 10.7
(T) Tr: 20/22-36/37/ 38
(D) LC50 r: 14.6/4h; LC50 l: 19.5/4h; **LD50** o-r: 670; LD50 o-m: 1500; LD50 ip-m: 1275; LD50 cu-ra: 20000

1-Chloropropane: C_3H_7Cl • 78.54 • 540-54-5
(P) $\Delta H_f^0 = -130.12$(g) • Eb = 45-47.2 • P_{vap} = 373; 380; 467(20); 1342(55)
(I) LEL/UEL = 2.6/10.7; 2.6/11.1; 2.6/11 • P_{fla} = < –18; –18cc; 18
 • AIT = 520
(T) Tr: 20/21/22 • NFPA: 2 • TR: 33
(D) LD50 o-r: 2000

2-Chloropropane: C_3H_7Cl • 78.54 • 75-29-6
(P) $\Delta H_f^0 = -146.44$(g) • Eb = 35-36 • P_{vap} = 592; 695(20); 1938(55)
(I) LEL/UEL = 2.8/10.7 • P_{fla} = –21; –32; –35; –42 • AIT = 590; 593
(T) Tr: 20/21/22 • NFPA: 2 • TR: 33

1,1-Dichloropropane: $C_3H_6Cl_2$ • 112.99 • 78-99-9
(P) Eb = 88
(I) LEL/UEL = 3.1/- • P_{fla} = 7; 21
(D) LC50 r: 18.32/4h; **LD50** o-r: 6500

1,2-Dichloropropane: $C_3H_6Cl_2$ • 112.99 • 78-87-5
(P) $\Delta H_f^0 = -165.69$(g) • Eb = 95-96.8 • P_{vap} = 53.3(19.4); 56(20); 88(30); 200(50)
(I) LEL/UEL = 3.4/14.5 • P_{fla} = 4; 16; 21oc; 38 (fire point)
 • AIT = 555; 557
(T) Tr: 20/22-40 • NFPA: 2 • TR: 33 • TWA (USA): 75; MAK (D): 75 • STEL: 110
(D) LC50 r: 14/8h; LC50 m: 4.62/2h; **LD50** o-r: 1380; 1947; 2196; LD50 o-m: 860; LD50 cu-ra: 8750

1,3-Dichloropropane: $C_3H_6Cl_2$ • 112.99 • 142-28-9
(P) $\Delta H_f^0 = -161.50$(g) • Eb = 118-122; 125 • P_{vap} = 20(20); 35(30); 90(50)
(I) P_{fla} = 21; 30; 32
(T) Tr: 20 • NFPA: 3 • TR: 30
(D) LD50 o-m: 3600

2,2-Dichloropropane: $C_3H_6Cl_2$ • 112.99 • 594-20-7
(P) Eb = 68-69
(I) P_{fla} = –5; 20
(T) Tr: 20

1,1,1-Trichloropropane: $C_3H_5Cl_3$ • 147.43 • 7789-89-1
(P) Eb = 140
(D) LC50 r: 48.26/4h; **LD50** o-r: 7460

1,2,3-Trichloropropane: $C_3H_5Cl_3$ • 147.43 • 96-18-4
(P) $\Delta H_f^0 = -185.86$(g) • Eb = 142; 152-156 • P_{vap} = 2.8; 2.9(20); 5.3(30); 16(50)
(I) LEL/UEL = 3.2/12.6 • P_{fla} = 74; 82oc • AIT = 304
(T) Tr: 20/21/22-40; 20/21/22 • NFPA: 3 • TR: 60
 • TWA (USA): 10; MAK (D): 50
(D) LC50 r: 6/4h; LC50 m: 3.4/2h; **LD50** o-r: 505; 320; LD50 o-m: 369; LD50 o-ra: 380; LD50 cu-ra: 1770

1-Chlorobutane: C_4H_9Cl • 92.58 • 109-69-3
(P) $\Delta H_f^0 = -147.28$(g) • Eb = 77-79; 10/50 • P_{vap} = 106; 108; 110(20); 170(30); 395(50); 107(78.4)
(I) LEL/UEL = 1/10.1; 1.8/10.1; 1.9/10.1 • P_{fla} = –12; –9cc; –9oc; –7
 • AIT = 239; 460; 471
(T) NFPA: 2 • TR: 33
(D) LC50 r: 30.27/4h; **LD50** o-r: 2670

2-Chlorobutane: C_4H_9Cl • 92.58 • 78-86-4
(P) $\Delta H_f^0 = -161.50$(g) • Eb = 65-69 • P_{vap} = 145; 160(20); 250(30); 560(50)
(I) LEL/UEL = 1.7/- • P_{fla} = –14; –10; 0 • AIT = 460
(T) NFPA: 2 • TR: 33
(D) LC50 r: 30.26/4h; **LD50** o-r: 17470; LD50 scu-m: 20000

1-Chloro-2-methylpropane: C_4H_9Cl • 92.58 • 513-36-0
(P) $\Delta H_f^0 = -159.41$(g) • Eb = 67-69
(I) LEL/UEL = 2.0/8.7 • P_{fla} = –10; –6; 21
(T) Tr: 37 • NFPA: 2

2-Chloro-2-methylpropane: C_4H_9Cl • 92.58 • 507-20-0
(P) $\Delta H_f^0 = -183.26$(g) • Eb = 50-52; 32.6/400; 14.6/200; –1/100; –11.4/60; –19/40 • P_{vap} = 332; 530(20); 1167(55)
(I) LEL/UEL = 1.8/10.1 • P_{fla} = –10; 0; 18 • AIT = 570
(T) NFPA: 2

1,4-Dichlorobutane: $C_4H_8Cl_2$ • 127.00 • 110-56-5
(P) Eb = 153-155; 161-163 • P_{vap} = 5(20)
(I) P_{fla} = 39; 52; 54 • AIT = 220
(T) NFPA: 2 • TR: –

8 • Table of organic compounds — Halogenated derivatives

Name: Empirical formula • Molar mass(g/mol) • Registry number (CAS).
(P) Physical data: ΔH_f^0 (kJ/mol) (physical state) • Eb (°C/mmHg) (if substance volatile) • P_{vap} (mBar) (°C) (if substance volatile).
(I) Inflammability data: LEL/UEL (%) • P_{fla} (°C) • AIT (°C).
(T) Toxicity: Safety Code (Tr) • NFPA • Transport Code (TR) • LVE (MVE) (ppm) F, USA, D • STEL (ppm); IDLH (ppm).
(D) Lethal and toxic doses: LC (mg/l/duration of exposure); LD (mg/kg).

1-Chloropentane: $C_5H_{11}Cl$ • 106.61 • 543-59-9
- (P) $\Delta H_f^0 = -174.89(g)$ • Eb = 108–109 • $P_{vap} = 27(20); 42(30); 140(50)$
- (I) LEL/UEL = 1.4/8.6 • $P_{fla} = 1; 3; 12oc; 13cc$ • AIT = 255; 260
- (T) Tr: 20/21/22 • NFPA: – • TR: 33

1-Chloro-3-methylbutane: $C_5H_{11}Cl$ • 106.61 • 107-84-6
- (P) Eb = ~100
- (I) LEL/UEL = 1.5/7.4 • $P_{fla} = < 21$
- (T) Tr: 20/21/22 • NFPA: 1 • TR: –

2-Chloro-2-methylbutane: $C_5H_{11}Cl$ • 106.61 • 594-36-5
- (P) Eb = 85–86; 88
- (I) LEL/UEL = 1.5/7.4 • $P_{fla} = -9cc; 3; 12$
- (T) Tr: 36/37/38 • NFPA: 2

Bromine derivatives

Bromomethane: CH_3Br • 94.95 • 74-83-9
- (P) $\Delta H_f^0 = -37.74(g)$ • Eb = 4 • $P_{vap} = 1666; 1893; 1900; 2426(20); 2431(25); 5259(55)$
- (I) LEL/UEL = 8.6/20 (under pressure); ~10/~16; 13.5/14.5 • AIT = 535; 538
- (T) Tr: 45-23-36/37/38; 23-36/37/38 • NFPA: 3 • TR: 26 • LVE(MVE) (F): – (5); TWA (USA): 5; MAK (D): 5
- (D) LC50 r: 2.029/6h; 1.173/8h; LC50 m: 1.54/2h; **LD50** o-r: 214

Dibromomethane: CH_2Br_2 • 173.85 • 74-95-3
- (P) $\Delta H_f^0 = -14.77(g)$ • Eb = 96–98 • $P_{vap} = 45; 47(20); 73(30); 180(50)$
- (T) Tr: 20 • NFPA: – • TR: 60
- (D) LC50 r: 40/2h; **LD50** o-r: 108; LD50 o-ra: 1000; LD50 scu-m: 3738

Tribromomethane: $CHBr_3$ • 252.75 • 75-25-2
- (P) $\Delta H_f^0 = 16.74(g)$ • Eb = 149–150; 46/15 • $P_{vap} = 5.9; 6.67(20); 10.8(30); 28(50)$
- (T) Tr: 23-36/38 • NFPA: 2 • TR: 60 • LVE(MVE) (F): – (0.5); TWA (USA): 0.5; MAK (D): 0.5
- (D) LC50 r: 45/4h; **LD50** o-r: 933; 1147; LD50 o-m: 1072; 1400; LD50 ip-r: 414; LD50 ip-m: 1274; LD50 scu-m: 820; 1820

Tetrabromomethane: CBr_4 • 331.65 • 558-13-4
- (P) Eb = 190 • $P_{vap} = 0.75(20); 1.45(30); 4.8(50); 53(96)$
- (T) Tr: 20/21/22; 26-36 • NFPA: 3 • TR: 60 • LVE(MVE) (F): – (0.1); TWA (USA): 0.1; MAK (D): 0.1 • STEL: 0.3
- (D) LD50 o-r: 1000; LD50 iv-m: 56; LD50 scu-m: 298

Bromoethane: C_2H_5Br • 108.98 • 74-96-4
- (P) $\Delta H_f^0 = -64.02(g); -92.01(l)$ • Eb = 37–40 • $P_{vap} = 140(0); 400; 507; 515; 520(20); 533(21); 624(25); 1560(50); 1745(55)$
- (I) LEL/UEL = 6.7/11.3; ~6.75/11.25 • $P_{fla} = -23; < -20; -20; 26$ • AIT = 496; 510
- (T) Tr: 20/21/22-40; 20/21/22 • NFPA: 2 • TR: 60 • LVE(MVE) (F): – (200); TWA (USA): 200; MAK(D): 200
- (D) LC50 r: 120.28/1h; 122.35/?; LC50 m: 72.35/1h; 73.41/?; **LD50** o-r: 1350; LD50 ip-r: 1750; LD50 ip-m: 2850; LD50 scu-m: 1820

1,2-Dibromoethane: $C_2H_4Br_2$ • 187.8 • 106-93-4
- (P) $\Delta H_f^0 = -81.17(l)$ • Eb = 131–133 • $P_{vap} = 8(10); 12.2; 14.3(20); 14.7; 15.6(25); 23(30); 59(50); 72(55)$
- (T) Tr: 45-23/24/25-36/37/38 • NFPA: 3 • TR: 60
- (D) LC50 r: 14.3/30'; LD50 o-r: 108; LD50 o-m: 250; LD50 o-ra: 55; LD50 ip-m: 220; LD50 scu-m: 300

1,1,2,2-Tetrabromoethane: $C_2H_2Br_4$ • 345.68 • 79-27-6
- (P) Eb = 223–229; 239dec; 151/54; 119/15 • $P_{vap} = 0.1; 0.13; 1.33(20); 0.16(30); 0.3(50)$
- (I) AIT = 335
- (T) Tr: 26-36 • NFPA: 3 • TR: 60 • LVE(MVE) (F): – (1); TWA (USA): 1; MAK (D): 1
- (D) LC50 r: 549/4h; **LD50** o-r: 1200; LD50 o-m: 269; LD50 o-ra: 400; LD50 ip-m: 443; LD50 cu-ra: 400

1-Bromopropane: C_3H_7Br • 123.01 • 106-94-5
- (P) $\Delta H_f^0 = -87.6(g)$ • Eb = 70–71 • $P_{vap} = 140; 191; 194(20); 500(50)$
- (I) LEL/UEL = 4.6/- • $P_{fla} = < -6; -1; 14; 21cc; 25$ • AIT = 490
- (T) Tr: 20 • NFPA: 2 • TR: 30
- (D) LC50 r: 250/30'; **LD50** o-r: 4000; LD50 ip-r: 2950; LD50 ip-m: 2530

2-Bromopropane: C_3H_7Br • 123.01 • 75-26-3
- (P) $\Delta H_f^0 = -97.07(g)$ • Eb = 58–60 • $P_{vap} = 224(20); 340(30); 735(50)$
- (I) LEL/UEL = 4.6/- • $P_{fla} = < -10; 1; 19; 21$
- (T) Tr: 20/21/22-36/37/38; 26/27/28 • NFPA: – • TR: 33
- (D) LD50 ip-r: 4837; LD50 ip-m: 4837

1,2-Dibromopropane: $C_3H_6Br_2$ • 201.91 • 78-75-1
- (P) $\Delta H_f^0 = -72.80(g)$ • Eb = 140–143; 79/100
- (I) $P_{fla} = 35$; None
- (T) Tr: 20/22
- (D) LC50 r: 12/4h; **LD50** o-r: 741; 1373; LD50 o-m: 676; LD50 ip-m: 75

1,3-Dibromopropane: $C_3H_6Br_2$ • 201.91 • 109-64-8
- (P) Eb = 167; 57/10 • $P_{vap} = 2.93(20)$
- (I) $P_{fla} = 47; 54cc; 56$
- (T) Tr: 20/22
- (D) LD50 ip-m: 473

1-Bromobutane: C_4H_9Br • 137.04 • 109-65-9
- (P) $\Delta H_f^0 = -107.32(g)$ • Eb = 99–104 • $P_{vap} = 43(20); 53(25); 68(30); 160; 200(50)$
- (I) LEL/UEL = 2.6/6.6(100); 2.6/6.6; 2.8(100)/6.6(100) • $P_{fla} = 18cc; 18oc; 22; 24$ • AIT = 265
- (T) Tr: 36/38; 37 • NFPA: 2 • TR: 33
- (D) LC50 r: 237/30'; **LD50** ip-r: 4450; LD50 ip-m: 1424; 6680

2-Bromobutane: C_4H_9Br • 137.04 • 78-76-2
- (P) $\Delta H_f^0 = -120.08(g); -155.64(l)$ • Eb = 90–92
- (I) LEL/UEL = 2.6/6.6 • $P_{fla} = 21cc$ • AIT = 265
- (T) Tr: 20/21-37 • NFPA: 2 • TR: 33

8 • Table of organic compounds — Halogenated derivatives

Name: Empirical formula • Molar mass(g/mol)• Registry number (CAS).
Ⓟ **Physical data:** ΔH_f^0 (kJ/mol) (physical state) • Eb (°C/mmHg) (if substance volatile) • P_{vap} (mBar) (°C) (if substance volatile).
Ⓘ **Inflammability data:** LEL/UEL (%) • P_{fla} (°C) • AIT (°C).
Ⓣ **Toxicity:** Safety Code (Tr) • NFPA • Transport Code (TR) • LVE (MVE) (ppm) F, USA, D • STEL (ppm); IDLH (ppm).
Ⓓ **Lethal and toxic doses:** LC (mg/l/duration of exposure); LD (mg/kg).

1-Bromo-2-methylpropane: C_4H_9Br • 137.04 • 78-77-3
Ⓟ Eb = 90-92
Ⓘ P_{fla} = ⩽0; –5; 18; 22
Ⓣ Tr: 37
Ⓓ LD50 ip-m: 1660

2-Bromo-2-methylpropane: C_4H_9Br • 137.04 • 507-19-7
Ⓟ ΔH_f^0 = –133.89(g); –164.43(l) • Eb = 71-74 • P_{vap} = 57(20)
Ⓘ P_{fla} = –18; 16cc; 18
Ⓣ Tr: 22
Ⓓ LD50 ip-r: 1250; LD50 ip-m: 4400

1,4-Dibromobutane: $C_4H_8Br_2$ • 215.94 • 110-52-1
Ⓟ Eb = 196-198; 78/13; 63-65/6
Ⓘ P_{fla} = 80; > 109
Ⓣ Tr: 37/38
Ⓓ LD50 ip-m: 300

1-Bromopentane: $C_5H_{11}Br$ • 151.07 • 110-53-2
Ⓟ ΔH_f^0 = –170.21(l) • Eb = 127-130; 129.7/740 • P_{vap} = 13(21)
Ⓘ P_{fla} = 30; 32
Ⓣ Tr: 36/37/38 • NFPA: 1
Ⓓ LC50 m: 26.8; LD50 ip-m: 1250

2-Bromopentane: $C_5H_{11}Br$ • 151.07 • 107-81-3
Ⓟ Eb = 116-120; 58/100 • P_{vap} = 105(50); 190(65)
Ⓘ P_{fla} = 20; 32
Ⓣ Tr: 36/37/38 • NFPA: – • TR: 33
Ⓓ LC50 m: 33; LD50 ip-m: 150

1-Bromo-3-methylbutane: $C_5H_{11}Br$ • 151.07 • 107-82-4
Ⓟ Eb = 118-121
Ⓘ P_{fla} = 22; 32
Ⓓ LD50 ip-r: 6150; LD50 ip-m: 420

1-Bromohexane: $C_6H_{13}Br$ • 165.10 • 111-25-1
Ⓟ ΔH_f^0 = –194.22(l) • Eb = 154-169; 154-158; 41/10 • P_{vap} = < 13.3(20)
Ⓘ P_{fla} = 47; 57
Ⓣ Tr: 23-36/38-40
Ⓓ LC50 r: 550/30'; 2.013/8h; LC50 m: 1.54/2h; LC50 l: 28.9/30'; LD50 o-r: 214; LD50 ip-r: 1226; LD50 ip-m: 1226; LD50 scu-r: 135

2-Bromohexane: $C_6H_{13}Br$ • 165.10 • –
Ⓟ Eb = 142
Ⓘ P_{fla} = 47

• *Iodine derivatives*

Iodoethane: C_2H_5I • 155.97 • 75-03-6
Ⓟ ΔH_f^0 = –8.37(g); –40.17(l) • Eb = 69-73 • P_{vap} = 133(18); 133(20)
Ⓘ P_{fla} = 53; None
Ⓣ Tr: 23/25-37/38 • NFPA: 2
Ⓓ LC50 r: 65/30'; LD50 ip-r: 330; LD50 ip-m: 560; LD50 scu-m: 1000

Iodomethane: CH_3I • 141.94 • 74-88-4
Ⓟ ΔH_f^0 = 13.77(g); –13.77(l) • Eb = 41-43; (dec: 270) • P_{vap} = 438; 533; 544(20); 533(25.3); 640(30); 1300(50); 1661(55)
Ⓣ Tr: 21-23/25-37/38-40 • NFPA: 3 • TR: 60 • TWA (USA): 2 (skin)
Ⓓ LC50 r: 1.3/4h; LC50 m: 5/1h; LD50 o-r: 150; LD50 ip-r: 101; LD50 ip-m: 172; LD50 scu-r: 110; LD50 scu-m: 110; 122

1-Iodobutane: C_4H_9I • 184.03 • 542-69-8
Ⓟ Eb = 129-131
Ⓘ P_{fla} = –10; 34
Ⓣ Tr: 20-36/37/38
Ⓓ LC50 r: 6.1/4h; LD50 ip-r: 692; LD50 ip-m: 101

2-Iodobutane: C_4H_9I • 184.03 • 513-48-4
Ⓟ Eb = 118-120
Ⓘ P_{fla} = –10; 23; 34
Ⓣ Tr: 36/37/38 • NFPA: – • TR: 33

• *Mixed F, Cl derivatives*

Fluorotrichloromethane: CCl_3F • 137.37 • 75-69-4
Ⓟ ΔH_f^0 = –284.93(g) • Eb = 24.1; 6.8/400; –9.1/200; –23/100; –32.3/60; –39/40; –49.7/20; –59/10; –67.6/5; –84.3/1 • P_{vap} = 886; 889(20); 924(21); 1290(30); 2400(50); 2700(55)
Ⓣ Tr: 20-59; – • NFPA: 1 • TR: – • LVE(MVE) (F): – (1000); TWA (USA): 1000; MAK (D): 1000
Ⓓ LC50 r: 720/15'; 554/20'; LC50 m: 554/30'; LC50 ra: 1385/?; LD50 ip-m: 1743

Dichlorodifluoromethane: CCl_2F_2 • 120.91 • 75-71-8
Ⓟ ΔH_f^0 = –284.93(g) • Eb = –30 to –29 • P_{vap} = 2026(–12.2); 5066(16.1); 5700(20); 1590(21); 2026(28); 7500(30); 10133(42.4); 12200(50)
Ⓘ AIT = 552
Ⓣ NFPA: 1 • TR: 20 • LVE(MVE) (F): – (1000); TWA (USA): 1000; MAK (D): 1000
Ⓓ LC50 m: 3.348/3h; LC50 ra: 3860/30'

1,2-Dichloro-1,1-difluoroethane: $C_2H_2Cl_2F_2$ • 134.94 • 1649-08-7
Ⓟ Eb = –29 • P_{vap} = 5847(21)

1-Chloro-1,1-difluoroethane: $C_2H_3ClF_2$ • 100.50 • 75-68-3
Ⓟ Eb = –9.2 • P_{vap} = 2927(20); 3900(30); 6800(50)
Ⓘ LEL/UEL = 4.4/18.5; 6.2/18; 6.2/17.9; 9/14.8 • AIT = 632
Ⓣ Tr: 36/37/38 • MAK(D): 1000
Ⓓ LC50 r: 2050/4h; LC50 m: 1758/2h

1,1,1,2-Tetrachlorodifluoroethane: $C_2Cl_4F_2$ • 203.82 • 76-11-9
Ⓟ ΔH_f^0 = –489.95(g) • Eb = 92 • P_{vap} = 53(20)
Ⓣ NFPA: 2 • TR: – • LVE(MVE) (F): – (500); TWA (USA): 500; MAK (D): 1000
Ⓓ LC50 r: 160/15'

8 • Table of organic compounds

Halogenated derivatives

Name: Empirical formula • Molar mass(g/mol)• Registry number (CAS).
(P) Physical data: ΔH_f^0 (kJ/mol) (physical state) • Eb (°C/mmHg) (if substance volatile) • P_{vap} (mBar) (°C) (if substance volatile).
(I) Inflammability data: LEL/UEL (%) • P_{fla} (°C) • AIT (°C).
(T) Toxicity: Safety Code (Tr) • NFPA • Transport Code (TR) • LVE (MVE) (ppm) F, USA, D • STEL (ppm); IDLH (ppm).
(D) Lethal and toxic doses: LC (mg/l/duration of exposure); LD (mg/kg).

1,1,2,2-Tetrachlorodifluoroethane; $C_2Cl_4F_2$ • 203.82 • 76-12-0
(P) Eb = 93 • P_{vap} = 80(20)
(T) TWA (USA): 1000
(D) LC50 r: 120/4h; LC50 m: 123/2h; **LD50** o-m: 800

1,1,2-Trichlorotrifluoroethane: $C_2Cl_3F_3$ • 187.37 • 76-13-1
(P) Eb = 47-48 • P_{vap} = 362; 379(20); 539(30); 1280(55)
(I) AIT = 680
(T) Tr: 36/37/38 • LVE(MVE) (F): 1250 (1000); TWA (USA): 1000; MAK (D): 500 • STEL: 1250
(D) LC50 r: 523/6h; LC50 m: 260/2h; 1503/90'; **LD50** o-r: 43000

1,2-Dichlorotetrafluoroethane: $C_2Cl_2F_4$ • 170.92 • 76-14-2
(P) Eb = 3.6-4.1 • P_{vap} = 1902(20)
(T) NFPA: 1 • TWA (USA): 1000; MAK (D): 1000
(D) LC50 r: 4800/30'; LC50 m: 4670/30'; LC50 ra: 4800/30'

Chloropentafluoroethane: C_2ClF_5; 154.47; 76-15-3
(P) Eb = –39.3 to –38 • P_{vap} = 2270(25)
(T) NFPA: – • TR: 20 • LVE(MVE) (F): – (1000); TWA (USA): 1000

•**Mixed F, Br derivatives**

Dibromodifluoromethane: CBr_2F_2 • 209.83 • 75-61-6
(P) ΔH_f^0 = –429.70(g) • Eb = 23-25 • P_{vap} = 830; 882; 920(20); 2723(55)
(T) TWA (USA): 100; MAK (D): 100
(D) LC50 r: 460/15'; LC50 m: 140/2h; 67/15'

Bromotrifluoromethane: $CBrF_3$ • 148.92 • 75-63-8
(P) ΔH_f^0 == –648.94(g) • Eb = –58 to 57 • P_{vap} = 14210; 15200(2); 13031(21); 14110(21.1); 18000(30); 27900(50)
(T) NFPA: 1 • TR: 20 • LVE(MVE) (F): – (1000); TWA (USA): 1000; MAK (D): 1000
(D) LC50 r: 416/1h; 491/15'; 4679/ 4h; LC50 m: 381

1,2-Dibromotetrafluoroethane: $C_2Br_2F_4$ • 259.84 • 124-73-2
(P) Eb = 24.5; 47 • P_{vap} = 380(20); 1296(55)
(D) LC50 r: 869/2h; LC50 m: 300/2h

• **Mixed Cl, Br derivatives**

Bromochloromethane: CH_2BrCl • 129.39 • 74-97-5
(P) ΔH_f^0 = –50.21(g) • Eb = 66-69 • P_{vap} = 147; 156(20); 250(30); 560(70)
(T) Tr: 36/37/38; 20 • NFPA: 2 • TR: 60 • LVE(MVE) (F): – (200); TWA (USA): 200; MAK (D): 200
(D) LC50 r: 158/15'; LC50 m: 12.03; 15.85/8h; **LD50** o-r: 5000; LD50 o-m: 4300; LD50 cu-ra: > 20000

Bromotrichloromethane: $CBrCl_3$ • 198.27 • 75-62-7
(P) ΔH_f^0 = –37.24(g) • Eb = 100-105.1; 0.6/10 • P_{vap} = 51(25)
(T) Tr: 20/21/22
(D) LD50 o-r: 100; LD50 ip-r: 119

1-Bromo-2-chloroethane: C_2H_4BrCl • 143.42 • 107-04-0
(P) Eb = 104-107 • P_{vap} = 4(20); 53(29.7)
(T) Tr: 23/25
(D) LD50 o-r: 64

1-Bromo-3-chloropropane: C_3H_6BrCl • 157.45 • 109-70-6
(P) Eb = 138-142; 142-145 • P_{vap} = 7.5(20); 12(30); 29(50)
(I) P_{fla} = 45; 81; none
(T) Tr: 20/22; 36/37/38 • NFPA: – • TR: 60
(D) LC50 r: 5.7; LC50 m: 7.27/2h; **LD50** o-r: 930; LD50 o-m: 1290

1-Bromo-3-chloro-2-methylpropane: C_4H_8BrCl • 171.47 • –
(P) Eb = 152-154
(I) P_{fla} = 47; > 109

HALOGENATED DERIVATIVES
• **Halogenated alkenes and alkynes**

Fluoroethylene: C_2H_3F • 46 • 75-02-5
(P) Eb = –72; –51 • P_{vap} = 23900(25); 30200(30); 46500(50)
(I) LEL/UEL = 2.6/21.7; 2.6/22; 2.9/28.9
(T) NFPA: 2 • TR: 239 • LVE(MVE) (F): – (2.5 mg/m^3 fluor); TWA (USA): 2.5 mg/m^3 fluor; MAK (D): 2.5 mg/m^3 fluor

Chloroethylene: C_2H_3Cl • 62.50 • 75-01-4
(P) ΔH_f^0 = 35.15(g) • Eb = –14 • P_{vap} = 3372(20); 3465; 3500(25)
(I) LEL/UEL = 3.8/22; 3.8/29.3; 4/22; 4/33
  • P_{fla} = –78cc; –61; –28Coc; –8 • AIT = 472
(T) Tr: 45 • NFPA: 2 • TR: 339 • LVE(MVE) (F): – (1); TWA (USA): 5
(D) LC50 r: 534/15'; **LD50** o-r: 500

1,1-Dichloroethylene: $C_2H_2Cl_2$ • 96.94 • 75-35-4
(P) ΔH_f^0 = 1.26(g); –24.27(l) • Eb = 30-32; 37
  • P_{vap} = 663; 667(20); 2138(55)
(I) LEL/UEL = 5.6/11.4; 5.6/13; 6.5/15.5; 7.3/16; 5.6/16
  • P_{fla} = –25; –22; –18oc; –15oc; –10oc • AIT = 440; 458; 519; 570
(T) Tr: 20-40 • NFPA: 2 • TR: – • LVE(MVE) (F): – (5); TWA (USA): 5; MAK (D): 2 • STEL: 20
(D) LC50 r: 26.23/4h; **LD50** o-r: 200; LD50 o-m: 194; LD50 scu-m: 3700

1,2-Dichloroethylene (E): $C_2H_2Cl_2$ • 96.94 • 156-60-5
(P) ΔH_f^0 = 4.18(g) • Eb = 47-49; 47.2/745
  • P_{vap} = 220; 347; 350(20); 520(30); 533(30.8); 1100(50)
(I) LEL/UEL = 6.2/16; 9.7/12.8 • P_{fla} = –15; 2cc; 4oc; 6 • AIT = 460
(T) Tr: 20 • NFPA: 2 • TR: – • LVE(MVE) (F): – (200); TWA (USA): 200; MAK (D): 200
(D) LD50 o-r: 770; 1280; LD50 ip-r: 76800; LD50 ip-m: 2000; ~2150; 4096

1,2-Dichloroethylene (Z): $C_2H_2Cl_2$ • 96.94 • 156-59-2
(P) ΔH_f^0 = 1.88(g); –27.61(l) • Eb = 60-61 • P_{vap} = 215; 223(20); 350(30); 533(41); 730(50)
(I) LEL/UEL = 3.3/15; 9.7/12.8; 6.2/16 • P_{fla} = 4cc; 6oc • AIT = 460

8 • Table of organic compounds — Halogenated derivatives

Name: Empirical formula • Molar mass(g/mol) • Registry number (CAS).
(P) **Physical data:** ΔH_f^0 (kJ/mol) (physical state) • Eb (°C/mmHg) (if substance volatile) • P_{vap} (mBar) (°C) (if substance volatile).
(I) **Inflammability data:** LEL/UEL (%) • P_{fla} (°C) • AIT (°C).
(T) **Toxicity:** Safety Code (Tr) • NFPA • Transport Code (TR) • LVE (MVE) (ppm) F, USA, D • STEL (ppm); IDLH (ppm).
(D) **Lethal and toxic doses:** LC (mg/l/duration of exposure); LD (mg/kg).

Trichloroethylene: C_2HCl_3 • 131.38 • 79-01-6
(P) $\Delta H_f^0 = -5.86$(g) • Eb = 86-88; 67/400; 48/200; 31.4/100; 20/60; −1/20; −12.4/10; −22.8/5; −43.8/1 • P_{vap} = 77; 80; 86(20); 124(30); 133(32); 280(50)
(I) LEL/UEL = 7.9/-; 7.9/90; 8/10.5; 12.5/90; 8.5(25)/14.5(35); 6.0(150)/47(150) • P_{fla} = 32cc; none • AIT = 410; 420
(T) Tr: 40 • NFPA: 2 • TR: 60 • LVE(MVE) (F): 200 (75); TWA (USA): 50; MAK (D): 50 • STEL: 100
(D) LC50 r: 155/1h; MLC r:45.45/4h; LC50 m: 52.4/4h; LD50 o-r: 3670; 4920; 5650; LD50 o-m: 2402; LD50 o-ra: 7330; LD50 ip-r: 1282; LD50 ip-m: 1831; LD50 iv-m: 34; LD50 scu-m: 16000; LD50 scu-m: 1800; LD50 cu-ra: > 20000

Tetrachloroethylene: C_2Cl_4 • 165.82 • 127-18-4
(P) $\Delta H_f^0 = -14.23$(g) • Eb = 121-122; 14/10 • P_{vap} = 19; 17; 22(20); 21.1(22); 32(30); 84(50)
(T) Tr: 40 • NFPA: 3 • TR: 60 • LVE(MVE) (F): − (50); TWA (USA): 25; MAK (D): 50 • STEL: 100
(D) LC50 r: 34/8h; LC50 m: 35.3/4h; 40.85/?; **LD50** o-r: 2629; 3005; 8850; LD50 o-m: 8100; LD50 ip-r: 4678

3-Chloro-1-propene: C_3H_5Cl • 76.53 • 107-05-1
(P) $\Delta H_f^0 = -0.63$(g) • Eb = 43-45 • P_{vap} = 391; 393; 396; 485(20); 1419(55)
(I) LEL/UEL = 2.9/11.1; 2.9/11.2; 3/11; 3.2/11.2; 3.3/11.2 • P_{fla} = −32cc; −28; −26; 4 • AIT = 390; 485
(T) Tr: 26-40; 26 • NFPA: 3 • TR: 336 • TWA (USA): 1
(D) LC50 r: 0.938/1h; 11/2h; LC50 m: 11.5/2h; **LD50** o-r: 64; 460; 700; LD50 o-m: 425; LD50 ip-m: 155; LD50 cu-ra: 2066

3-Chloro-1-propyne: C_3H_3Cl • 74.51 • 624-65-7
(P) Eb = 57-63
(I) P_{fla} = −16; 16; 18; 50(?)
(T) Tr: 23/24/25-34; 34
(D) LD50 o-r: 53

1,3-Dichloropropene: $C_3H_4Cl_2$ • 110.97 • 542-75-6
(P) Eb = 97-112; 103-110; 108; 104(Z); 112(E)
(I) P_{fla} = 27; 36
(T) Tr: 45-20/21-36/37/38-43; 20/21-25-36/37/38-43 • NFPA: 3 • TR: 30 • TWA (USA): 1
(D) LC50 m: 4.65/2h; **LD50** o-r: 250; 470; 713; LD50 o-m: 640; LD50 ip-r: 175; LD50 cu-ra: 504

2,3-Dichloropropene: $C_3H_4Cl_2$ • 110.97 • 78-88-6
(P) Eb = 93-95 • P_{vap} = 59(20)
(I) LEL/UEL = 2.6/7. • P_{fla} = 10; 23
(T) Tr: 22
(D) LC50 r: 2.27/4h; LC50 m: 3.1/2h; **LD50** o-r: 320; LD50 cu-ra: 1580

1-Chloro-2-butene: C_4H_7Cl • 90.56 • 591-97-9
(P) Eb = 85-86; (Z:84.1; E:84.8) • P_{vap} = 469(20)
(I) P_{fla} = −15; −10; AIT = 469
(T) Tr: 36/37

1-Chloro-3-butene: C_4H_7Cl; 90.56; 563-52-0
(P) Eb = 63-65
(I) LEL/UEL = 2.2/- • P_{fla} = −27; −19

3-Chloro-2-methyl-1-propene: C_4H_7Cl • 90.56 • 563-47-3
(P) Eb = 71-74 • P_{vap} = 136(20)
(I) LEL/UEL = 2.3/8.1; 2.3/9.3 • P_{fla} = −23; −18; −12; −10 • AIT = 481
(T) Tr: 20-40
(D) LC50 r: 34/30'

1-Chloro-2-methyl-2-propene: C_4H_7Cl • 90.56 • 513-37-1
(P) Eb = 68-71
(I) P_{fla} = −1

2-Chloro-1,3-butadiene: C_4H_5Cl • 88.54 • 126-99-8
(P) Eb = 59.4 • P_{vap} = 267(20); 400(30); 790(50)
(I) LEL/UEL = 2.5/20; 4.0/20 • P_{fla} = −20 • AIT = 440
(T) Tr: 20/22-36 • NFPA: 3 • TR: 336 • LVE(MVE) (F): − (10); TWA (USA): 10; MAK (D): 10
(D) LC50 r: 11.8/4h; LC50 m: 2.3/?; **LD50** o-r: 450; LD50 o-m: 146; LD50 iv-ra: 96; LD50 scu-r: 500; LD50 scu-m: 1000

1,4-Dichloro-2-butene: $C_4H_6Cl_2$ • 125.00 • 110-57-6
(P) Eb = 151-153; 156; 72/40; 74-75/40 • P_{vap} = 3; 3.5; 13.3(20); 6.5(30); 20(50)
(I) P_{fla} = 27; 55; 59
(T) Tr: 45-24/25-26-34 • TWA (USA): 0.005
(D) LC50 r: 0.316/4h; **LD50** o-r: 89 • LC50 m: 0.920/?; **LD50** o-m: 190; LD50 iv-m: 56; LD50 scu-m: 620

1,3-Dichloro-2-butene: $C_4H_6Cl_2$ • 125.00 • 926-57-8
(P) Eb = 123; 125-129
(I) P_{fla} = 27; 33
(T) NFPA: 2
(D) LC50 r: 3.93/?; LC50 m: 4.4/?

Bromoethylene: C_2H_3Br • 106.96 • 593-60-2
(P) $\Delta H_f^0 = 78.37$(g) • Eb = 16; −32/100 • P_{vap} = 1200(20); 1375(25); 1700(30); 2067(37.); 3300(50)
(I) LEL/UEL = 5.6/13.5; 6/15; 6/16 • P_{fla} = −18; <−8; 5; None • AIT = 529
(T) Tr: 40; − • NFPA: 2 • TR: 236 • TWA (USA): 5
(D) LD50 o-r: 500

1,2-Dibromoethylene: $C_2H_2Br_2$ • 185.86 • 540-49-8
(P) Eb = 107-112
(T) Tr: 20/21-36/37/38
(D) LD50 o-r: 117

8 • Table of organic compounds — Halogenated derivatives

Name: Empirical formula • Molar mass(g/mol)• Registry number (CAS).
(P) Physical data: ΔH_f^0 (kJ/mol) (physical state) • Eb (°C/mmHg) (if substance volatile) • P_{vap} (mBar) (°C) (if substance volatile).
(I) Inflammability data: LEL/UEL (%) • P_{fla} (°C) • AIT (°C).
(T) Toxicity: Safety Code (Tr) • NFPA • Transport Code (TR) • LVE (MVE) (ppm) F, USA, D • STEL (ppm); IDLH (ppm).
(D) Lethal and toxic doses: LC (mg/l/duration of exposure); LD (mg/kg).

1-Bromo-2-propene: C_3H_5Br • 120.99 • 106-95-6
(P) Eb = 70-71.3
(I) LEL/UEL = 4.3/7.3; 4.4/7.3 • P_{fla} = –5; –1cc • AIT = 295
(T) Tr: 23/24/25-36/ 37/38; 23/24/25 • NFPA: 3 • TR: 336
(D) LC50 r: 10/30'; LD50 o-r: 120; LD50 ip-r: 48; LD50 ip-m: 108

1-Bromo-2-propyne: C_3H_3Br • 118.97 • 106-96-7
(P) ΔH_f^0 = 230-270 (estimated value) • Eb = 82-85; 88-90; 33/130
(I) LEL/UEL = 3.0/- • P_{fla} = 3; 10; 18Coc • AIT = 324
(T) Tr: 28-36/37/38; 23/24/25 • NFPA: 3 • TR: 33

Allyl iodide: C_3H_5I • 167.98 • 556-56-9
(P) ΔH_f^0 = 57.32(l) • Eb = 101-103.1
(I) P_{fla} = 5; 16; 18
(T) Tr: 34 • NFPA: 3 • TR: 338

Chlorotrifluoroethylene: C_2ClF_3 • 116.47 • 79-38-9
(P) Eb = –28 • P_{vap} = 5600(20); 7500(30); 12600(50)
(I) LEL/UEL = 4.6/64.3; 8.4/36.7; 8.4/38.7; 8.4/39; 24/40.3 • P_{fla} = –28; none
(T) Tr: 20 • NFPA: 3 • TR: 236
(D) LC50 r: 4.11/4h; LC50 m: 12.33/7h; LD50 o-m: 628

HALOGENATED DERIVATIVES
• Aromatic halogenated derivatives

Fluorobenzene: C_6H_5F • 96.11 • 462-06-6
(P) ΔH_f^0 = –116.57(g) • Eb = 84-85.1 • P_{vap} = 80(20); 130(30); 305(50)
(I) P_{fla} = –15; –12; 1
(T) Tr: 20 • TR: 33
(D) LC50 r: 26.91/?; LC50 m: 45/2h; LD50 o-r: 4399

2-Fluorotoluene: C_7H_7F • 110.14 • 95-52-3
(P) Eb = 113-115 • P_{vap} = 25; 28(20); 41(30); 110(50)
(I) LEL/UEL = 1.3/- • P_{fla} = 8; 12
(T) Tr: 20 • TR: 33

3-Fluorotoluene: C_7H_7F • 110.14 • 352-70-5
(P) Eb = 114-116 • P_{vap} = 23(20); 39(30); 100(50)
(I) P_{fla} = 9; 12
(T) Tr: 20 • TR: 33

4-Fluorotoluene: C_7H_7F • 110.14 • 352-32-9
(P) ΔH_f^0 = –187.44(l) • Eb = 115-117 • P_{vap} = 21(20); 36(30); 100(50)
(I) P_{fla} = 10; 17
(T) Tr: 20/21/22; 20 • TR: 33

Trifluoromethylbenzene: $C_7H_5F_3$ • 146.12 • 98-08-8
(P) ΔH_f^0 = –600.36(g); –637.64(l) • Eb = 100-104 • P_{vap} = 14.7(0); 26.6(12); 40.7(20); 67(30); 165(50)
(I) P_{fla} = 12cc • AIT = 620
(T) TR: 33
(D) LC50 r: 70.89/4h; LD50 o-r: 15000 • LC50 m: 92.24/2h; LD50 ip-m: 100

Chlorobenzene: C_6H_5Cl • 112.56 • 108-90-7
(P) ΔH_f^0 = 10.79(l) • Eb = 131-132 • P_{vap} = 11.65; 15.7(20); 12(22); 13.3(22.2); 15.7(25)
(I) LEL/UEL = 1.3/7.1; 1.3/11; 1.3/7.1(150) • P_{fla} = 23; 28cc; 29Tcc • AIT = 590; 592; 636
(T) Tr: 20 • NFPA: 2 • TR: 30 • LVE(MVE) (F): – (75); TWA (USA): 10; MAK (D): 50
(D) LD50 o-r: 2910; LD50 o-m: 2300; LD50 o-ra: 2250; LD50 ip-r: 1655; LD50 ip-m: 515

1,2-Dichlorobenzene: $C_6H_4Cl_2$ • 147.00 • 95-50-1
(P) ΔH_f^0 = 29.96(g) • Eb = 172-179; 179-183; 86/18 • P_{vap} = 1.3; 1.33; 1.6(20); 1.7(25); 2.1(35); 8.2(50)
(I) LEL/UEL = 2/9; 2.2/9.2; 2.2/12.0 • P_{fla} = 65cc; 66Tcc; 68-78oc • AIT = 640; 645; 648
(T) Tr: 22-36/37/38 • NFPA: 2 • TR: 60 • LVE(MVE) (F): 50; TWA (USA): 25; MAK (D): 50 • STEL: 50
(D) LC50 r: 4.936/7h; LD50 o-r: 500; LD50 o-m: 4386; LD50 o-ra: 500; LD50 ip-r: 840; LD50 ip-m: 1228; LD50 iv-m: 400; LD50 iv-ra: 250; LD50 scu-r: 5000

1,3-Dichlorobenzene: $C_6H_4Cl_2$ • 147.00 • 541-73-1
(P) ΔH_f^0 = 26.44(g) • Eb = 173; 60-62/11; 53/10 • P_{vap} = 2; 2.4(20); 4.3(30); 6.67(39); 12(50)
(I) P_{fla} = 63; 65; 67
(T) Tr: 22
(D) LD50 ip-m: 1062

1,4-Dichlorobenzene: $C_6H_4Cl_2$ • 147.00 • 106-46-7
(P) ΔH_f^0 = 23.01(g) • Eb = 174.12; 55/10 • P_{vap} = 0.53; 0.85; 1.37; 1.7(20); 11/50; 13.3(54.8); 100(100)
(I) LEL/UEL = 2.2/12.0; 2.5/- • P_{fla} = 66cc • AIT = 640
(T) Tr: 22-36/38-40; 22-36/38 • NFPA: 2 • TR: –
• LVE(MVE) (F): 110 (75); TWA (USA): 75 (proposed at 10); MAK (D): 50; 75 • STEL: 110
(D) LD50 o-r: 500; LD50 o-m: 2950; LD50 o-ra: 2830; LD50 ip-r: 2562; LD50 ip-m: 2000; LD50 scu-m: 5145; LD50 cu-ra: > 2000

1,2,3-Trichlorobenzene: $C_6H_3Cl_3$ • 181.44 • 87-61-6
(P) Eb = 219-221 • P_{vap} = 0.1(25)
(I) LEL/UEL = 2.5/6.6 • P_{fla} = 99; 113cc; 126 • AIT = 569
(T) Tr: 20/21/22 • TR: 60 • MAK(D): 5
(D) LD50 o-r: 1830; LD50 ip-m: 1390

8 • Table of organic compounds — Halogenated derivatives

Name: Empirical formula • Molar mass(g/mol) • Registry number (CAS).
(P) Physical data: ΔH_f^0 (kJ/mol) (physical state) • Eb (°C/mmHg) (if substance volatile) • P_{vap} (mBar) (°C) (if substance volatile).
(I) Inflammability data: LEL/UEL (%) • P_{fla} (°C) • AIT (°C).
(T) Toxicity: Safety Code (Tr) • NFPA • Transport Code (TR) • LVE (MVE) (ppm) F, USA, D • STEL (ppm); IDLH (ppm).
(D) Lethal and toxic doses: LC (mg/l/duration of exposure); LD (mg/kg).

1,2,4-Trichlorobenzene: $C_6H_3Cl_3$ • 181.44 • 120-82-1
- (P) Eb = 212-214 • P_{vap} = 0.27(20); 1.3(38); 1.33(38.4); 2.7(50)
- (I) LEL/UEL = 2.5/6.6(150); 2.5/6.6 • P_{fla} = 99; 110cc • AIT = 571
- (T) Tr: 22-36/37/38; 20/21/22 • NFPA: 2 • TR: 60 • LVE(MVE) (F): – (5); TWA (USA): 5; MAK (D): 5
- (D) LD50 o-r: 756; LD50 o-m: 300; LD50 ip-m: 1223

1,3,5-Trichlorobenzene: $C_6H_3Cl_3$ • 181.44 • 108-70-3
- (P) Eb = 209
- (I) P_{fla} = 107; 126
- (T) Tr: 20/21/22-36/37/38; 20/21/22 • TR: 60 • MAK (D): 5
- (D) LD50 o-r: 800; LD50 o-m: 3350; LD50 ip-m: 2260

1,2,4,5-Tetrachlorobenzene: $C_6H_2Cl_4$ • 215.88 • 95-94-3
- (P) Eb = 240-246 • P_{vap} = < 0.13(25)
- (I) P_{fla} = 155cc; 160
- (T) Tr: 22-36/37/38
- (D) LD50 o-r: 1500; LD50 o-m: 1035; LD50 o-ra: 1500

Hexachlorobenzene: C_6Cl_6 • 284.76 • 118-74-1
- (P) ΔH_f^0 = –33.89(g); –130.96(s) • Eb = 322-326 • P_{vap} = 1.45x10^{-5} (20); 1.33(114.4)
- (I) P_{fla} = 242
- (T) Tr: 45-48/25 • NFPA: 1 • TR: 60 • TWA (USA): 0.025 mg/m^3
- (D) LC50 r: 3.6/?; LC50 m: 4/?; LC50 ra: 1.8/?; **LD50** o-r: 10000; LD50 o-m: 4000; LD50 o-ra: 2600

2-Chlorotoluene: C_7H_7Cl • 126.59 • 95-49-8
- (P) Eb = 156-160 • P_{vap} = 3.6(20); 13.3(43)
- (I) P_{fla} = 42; 48
- (T) Tr: 20 • TR: 30 • LVE(MVE) (F): – (50) ; TWA (USA): 50
- (D) LC50 r: 90.54/?; **LD50** o-r: 5700; LD50 o-m: 4400

3-Chlorotoluene: C_7H_7Cl • 126.59 • 108-41-8
- (P) Eb = 160-162
- (I) P_{fla} = 48; 50; 52
- (T) Tr: 20 • TR: 30 • LVE(MVE) (F): – (50) ; TWA (USA): 50

4-Chlorotoluene: C_7H_7Cl • 126.59 • 106-43-4
- (P) Eb = 159-162.4; 162-166 • P_{vap} = 13(44); 13.3(45)
- (I) P_{fla} = 50; 53
- (T) Tr: 20 • TR: 30 • LVE(MVE) (F): – (50) ; TWA (USA): 50
- (D) LC50 m: 34/2h; **LD50** o-r: 2100; 3600; LD50 o-m: 1900

2,4-Dichlorotoluene: $C_7H_6Cl_2$ • 161.03 • 95-73-8
- (P) Eb = 200; 113/50; 82-85/15; 77/10 • P_{vap} = 4(20)
- (I) P_{fla} = 79; 87
- (D) LD50 o-r: 2400; 4600; LD50 o-m: 2400; 2900

2,6-Dichlorotoluene: $C_7H_6Cl_2$ • 161.03 • –
- (P) Eb = 196-202; 73-76/11
- (I) P_{fla} = 79; 82

Benzyl chloride: C_7H_7Cl • 126.59 • 100-44-7
- (P) ΔH_f^0 = –32.64(l) • Eb = 179; 66/11 • P_{vap} = 1; 1.2; 1.33(20); 9.33(55); 13.7(60)
- (I) LEL/UEL = 1.1/14; 1.1/- • P_{fla} = 52; 60cc; 67cc; 67oc; 71oc; 73 • AIT = 525; 585
- (T) Tr: 22-23-37/38-40-41 • NFPA: 2 • TR: 68 • LVE(MVE) (F): 2 (1); TWA (USA): 1; MAK (D): 1
- (D) LC50 r: 0.782/2h; LC50 m: 0.417/2h; **LD50** o-r: 1231; LD50 o-m: 1500; LD50 scu-r: 1000

Benzylidene chloride: $C_7H_6Cl_2$ • 161.03 • 98-87-3
- (P) Eb = 205; 214; 82-84/11; 82/10 • P_{vap} = 0.4(20); 0.9(30); 3(50); 40(55)
- (I) LEL/UEL = 1.1/7.0 • P_{fla} = 92 • AIT = 585
- (T) Tr: 22-23/37/38-40-41 • TR: 60
- (D) LC50 r: 0.040/2h; LC50 m: 0.212/2h; **LD50** o-r: 3249; LD50 o-m: 2462

Trichloromethylbenzene: $C_7H_5Cl_3$ • 195.47 • 98-07-7
- (P) Eb = 220.8; 129/60; 105/25; 90-91/15; 89/10 • P_{vap} = 0.25(20); 0.49(30); 1.8(50)
- (I) LEL/UEL = 2.1/6.5 • P_{fla} = 97; 109 • AIT = 420
- (T) Tr: 45-22-23-37/ 38-41 • TR: 80
- (D) LC50 r: 0.153/2h; LC50 m: 0.064/2h; **LD50** o-r: 6000; LD50 o-m: 702; LD50 cu-ra: 4000

4-Chlorostyrene: C_8H_7Cl • 138.60 • 1073-67-2
- (P) Eb = 187-190; 192; 53/4 • P_{vap} = 0.91(20) • P_{fla} = 59; 68

1-Chloronaphthalene: $C_{10}H_7Cl$ • 162.62 • 90-13-1
- (P) ΔH_f^0 = 54.39(l) • Eb = 259.3; 230.8/400; 204.2/100; 180.4/100; 165.6/60; 153.2/40; 134.4/20; 117-120/12; 118.6/10; 104.8/5; 80.6/1 • P_{vap} = 0.013(20); 1.3(80.6)
- (I) P_{fla} = 121cc; 132 • AIT = 558
- (T) Tr: 22-37 • NFPA: 3
- (D) LD50 o-r: 1540; LD50 o-m: 1091; 1591

2-Chloronaphthalene: $C_{10}H_7Cl$ • 162.62 • 91-58-7
- (P) ΔH_f^0 = 55.23(s) • Eb = 256-264; 161/60; 132.6/20; 119.7/11
- (T) NFPA: 3
- (D) LD50 o-r: 2078; LD50 o-m: 886

Bromobenzene: C_6H_5Br • 157.02 • 108-86-1
- (P) ΔH_f^0 = 60.67(l) • Eb = 154-156.2; 132.3/400; 110.1/200; 68.8/40; 53.8/20; 40/10; 27./5; 49/1 • P_{vap} = 4(20); 5.33(25); 13.3(40)
- (I) LEL/UEL = 0.5/2.5; 6/36.5 • P_{fla} = 51cc; 65; fire point : 155 • AIT = 565
- (T) Tr: 38 • NFPA: 2 • TR: 30
- (D) LC50 r: 20.41/?; LC50 m: 21/2h; **LD50** o-r: 2699; LD50 o-m: 2700; LD50 o-ra: 3300; LD50 ip-r: 3882; LD50 ip-m: 817; 1000; LD50 scu-m: 2000

8 • Table of organic compounds — Halogenated derivatives

Name: Empirical formula • Molar mass(g/mol)• Registry number (CAS).
(P) **Physical data:** ΔH_f^0 (kJ/mol) (physical state) • Eb (°C/mmHg) (if substance volatile) • P_{vap} (mBar) (°C) (if substance volatile).
(I) **Inflammability data:** LEL/UEL (%) • P_{fla} (°C) • AIT (°C).
(T) **Toxicity:** Safety Code (Tr) • NFPA • Transport Code (TR) • LVE (MVE) (ppm) F, USA, D • STEL (ppm); IDLH (ppm).
(D) **Lethal and toxic doses:** LC (mg/l/duration of exposure); LD (mg/kg).

2-Bromotoluene: C_7H_7Br • 171.05 • 95-46-5
(P) Eb = 181; 58/13; 58/10
(I) P_{fla} = 79
(T) Tr: 20/22 • NFPA: 2
(D) LD50 o-m: 1864; LD50 ip-m: 1358

3-Bromotoluene: C_7H_7Br • 171.05 • 591-17-3
(P) Eb = 184; 160/400; 138/200; 117./100; 104.1/60; 93.9/40; 78.1/20; 64/10; 60-62/10; 50.8/5; 14.8/1
(I) P_{fla} = 59; 81
(T) Tr: 20/22
(D) LD50 o-m: 1436; LD50 ip-m: 1215

4-Bromotoluene: C_7H_7Br • 171.05 • 106-38-7
(P) Eb = 184.5; 116.4/400; 102.3/60; 91.8/40; 75.2/20; 71-72/15; 61.1/10; 47.5/5; 10.3/1
(I) P_{fla} = 84
(T) Tr: 20/22-36/37/ 38
(D) LD50 ip-m: 1741

Benzyl bromide: C_7H_7Br • 171.05 • 100-39-0
(P) ΔH_f^0 = 23.43(l) • Eb = 196-199; 127/80 • P_{vap} = 0.5(20)
(I) P_{fla} = 86; 92
(T) Tr: 36/37/38-40; 36/37/38 • NFPA: 3 • TR: 60

Iodobenzene: C_6H_5I • 204.01 • 591-50-4
(P) ΔH_f^0 = 162.55(g) • Eb = 188; 63-65/10
(I) P_{fla} = 74; 77
(T) Tr: 20/22
(D) LC50 r: 16.32/?; **LD50** o-r: 1749

HALOGENATED DERIVATIVES
• Halogenated alcohols

2-Fluoroethanol: C_2H_5FO • 64.07 • 371-62-0
(P) ΔH_f^0 = –464.97(s) • Eb = 100-104 • P_{vap} = 21.3(20)
(I) P_{fla} = 31; 34
(T) Tr: 26/27/28
(D) LC50 r: 0.200/10'; LC50 m: 1.10/10'; LC50 ra: 0.025/10'; **LD50** o-r: 5; LD50 ip-r: 2; LD50 iv-m: 10; LD50 scu-m: 15

2,2,2-Trifluoroethanol: $C_2H_3F_3O$ • 100.05 • 75-89-8
(P) ΔH_f^0 = –868.18(l) • Eb = 73-75; 77-80; 78 • P_{vap} = 69; 72(20); 93(25)
(I) LEL/UEL = 5.5/42; 8.4/28.8 • P_{fla} = 27cc; 29 • AIT = 479
(T) Tr: 20/21/22-36/38
(D) LC50 m: 2.9/2h; **LD50** o-r: 240; LD50 o-m: 366; LD50 ip-r: 210; LD50 ip-m: 158; LD50 iv-m: 250

2-Chloroethanol: C_2H_5ClO • 80.52 • 107-07-3
(P) ΔH_f^0 = –295.39(l) • Eb = 127-130; 44/20 • P_{vap} = 6.52; 6.7; 7.3(20); 13.3(30.3)
(I) LEL/UEL = 4.9/15.9; 5/16; 5.6/16 • P_{fla} = 40oc; 55Tcc; 60oc • AIT = 425
(T) Tr: 26/27/28 • NFPA: 3 • TR: 60 • LVE(MVE) (F): 1 (-); TWA (USA): 1; MAK (D): 1
(D) LC50 r: 0.11/4h; 0.29/?; LC50 m: 0.385/?; **LD50** o-r: 71; 76; 95; LD50 o-m: 81; LD50 ip-r: 58; LD50 ip-m: 97; LD50 iv-ra: 80; LD50 scu-r: 84; LD50 scu-m: 98; LD50 scu-m: 100; LD50 cu-ra: 67

2,2,2-Trichloroethanol: $C_2H_3Cl_3O$ • 149.40 • 115-20-8
(P) Eb = 150-154; 54/10 • P_{vap} = 1.33(20)
(I) P_{fla} = > 109
(T) Tr: 22
(D) LD50 o-r: 500; 600; LD50 o-m: 500; LD50 o-ra: 50; LD50 ip-r: 300; LD50 iv-m: 201

3-Chloro-1-propanol: C_3H_7ClO • 94.55 • 627-30-5
(P) Eb = 160-162; 48-50/0.2
(I) P_{fla} = 65; 93
(T) Tr: 36/37/38
(D) LD50 o-m: 2300

2-Bromoethanol: C_2H_5BrO • 124.98 • 540-51-2
(P) Eb = 149-150/750 (dec); 57/20; 48.5/17; 48/13; 47/10 • P_{vap} = 3.2(20)
(I) P_{fla} = 40; > 109
(T) Tr: 26/27/28-34-40
(D) LD50 ip-m: 80

2,2,2-Tribromoethanol: $C_2H_3Br_3O$ • 282.78 • 75-80-9
(P) Eb = 92/3; 92/10
(T) Tr: 22
(D) LD50 o-r: 1000; 1090; LD50 o-m: 930

3-Bromo-1-propanol: C_3H_7BrO • 139.01 • 627-18-9
(P) Eb = 70/10; 62/7
(I) P_{fla} = 65; 93

HALOGENATED DERIVATIVES
• Halogenated phenols

2-Chlorophenol: C_6H_5ClO • 128.56 • 95-57-8
(P) Eb = 174-176; 62-63/11 • P_{vap} = 1(12); 1.33(12.1)
(I) P_{fla} = 64; 74; 85 • AIT = 550
(T) Tr: 20/21/22 • NFPA: 3 • TR: 68
(D) LD50 o-r: 670; LD50 o-m: 345; LD50 ip-r: 230; LD50 ip-m: 235; LD50 scu-r: 950

3-Chlorophenol: C_6H_5ClO • 128.56 • 108-43-0
(P) ΔH_f^0 = –206.69(s) • Eb = 214 • P_{vap} = 1(44); 1.33(44.2)
(I) P_{fla} = > 44; > 110; 120
(T) Tr: 20/21/22 • NFPA: 3 • TR: 60
(D) LD50 o-r: 570; LD50 o-m: 521; LD50 ip-r: 355; LD50 scu-r: 1390

8 • Table of organic compounds — Amines

Name: Empirical formula • Molar mass(g/mol) • Registry number (CAS).
(P) Physical data: ΔH_f^0 (kJ/mol) (physical state) • Eb (°C/mmHg) (if substance volatile) • P_{vap} (mBar) (°C) (if substance volatile).
(I) Inflammability data: LEL/UEL (%) • P_{fla} (°C) • AIT (°C).
(T) Toxicity: Safety Code (Tr) • NFPA • Transport Code (TR) • LVE (MVE) (ppm) F, USA, D • STEL (ppm) • IDLH (ppm).
(D) Lethal and toxic doses: LC (mg/l/duration of exposure); LD (mg/kg).

4-Chlorophenol: C_6H_5ClO • 128.56 • 106-48-9
- (P) $\Delta H_f^0 = -197.90(s)$ • Eb = 217-220 • P_{vap} = 0.13(20); 1.33(49.8); 1(50)
- (I) P_{fla} = 115; 121
- (T) Tr: 20/21/22 • NFPA: 3 • TR: 60
- (D) LC50 r: 0.011/?; **LD50** o-r: 261; 670; LD50 o-m: 367; 1373; LD50 ip-r: 281; LD50 ip-m: 332; LD50 scu-r: 1030

2,4-Dichlorophenol: $C_6H_4Cl_2O$ • 163.00 • 120-83-2
- (P) Eb = 209-211 • P_{vap} = 0.1(20); 1.1(50); 1.33(53)
- (I) P_{fla} = 93; 114
- (T) Tr: 22-36/38 • NFPA: – • TR: 60
- (D) LD50 o-r: 580; LD50 o-m: 1276; LD50 ip-r: 430; LD50 ip-m: 153; LD50 scu-r: 1730

2,4,5-Trichlorophenol: $C_6H_3Cl_3O$ • 197.44 • 95-95-4
- (P) Eb = 253/760; 248/740 • P_{vap} = 0.02(30); 0.2(50)
- (I) P_{fla} = 133
- (T) Tr: 22-36/38-50/53 • TR: 60
- (D) LD50 o-r: 820; LD50 o-m: 600; LD50 ip-r: 355; LD50 iv-m: 56; LD50 scu-r: 2260

Pentachlorophenol: C_6HCl_5O • 266.32 • 87-86-5
- (P) $\Delta H_f^0 = -295.39(s)$ • Eb = 310dec • P_{vap} = $1.51 \times 10^{-4}(20)$; $3.0 \times 10^{-2}(53)$; 53.3(211)
- (T) Tr: 24/25-26-36/37/38-40-50/53 • NFPA: 3 • TR: 60 • LVE(MVE) (F): – (0.5 mg/m^3); TWA (USA): 0.5 mg/m^3; MAK (D): 0.5 mg/m^3
- (D) LC50 r: 0, 355/?; LC50 m: 0.225/?; **LD50** o-r: 27; 146; 175; LD50 o-m: 117; LD50 ip-r: 56; LD50 ip-m: 58; LD50 scu-r: 100

p-Chloro-m-cresol: C_7H_7ClO • 142.59 • 9-50-7
- (P) Eb = 235
- (I) P_{fla} = 118
- (T) Tr: 21/22-38 • NFPA: 3 • TR: 60
- (D) LD50 o-r: 1830; LD50 o-m: 600; LD50 ip-m: 30; LD50 iv-m: 70; LD50 scu-r: 400; LD50 scu-m: 360

HALOGENATED DERIVATIVES
• **Halogenated ethers**

Chloromethyl ether: C_2H_5ClO • 80.52 • 107-30-2
- (P) Eb = 59 • P_{vap} = 213; 245(20)
- (I) P_{fla} = 15; < 23
- (T) Tr: 45-20/21/22 • NFPA: – • TR: 336
- (D) LC50 r: 0.181/7h; LC50 m: 1.03/2h; LD50 o-r: 817

Methyl dichloromethyl ether: $C_2H_4Cl_2O$ • 114.96 • 4885-02-3
- (P) Eb = 83-85 • P_{vap} = 84(20)
- (I) P_{fla} = 11; 42
- (T) Tr: 20/21/22-36/37/38-40

2-Methoxychloroethane: C_3H_7ClO • 94.54 • 627-42-9
- (P) Eb = 88-92
- (I) P_{fla} = –7; 10; 14
- (T) Tr: 20/22-37/38

Bis(2-chloroethyl)ether: $C_4H_8Cl_2O$ • 143.02 • 111-44-4
- (P) Eb = 177-178.5; 67/15; 63/10 • P_{vap} = 0.53; 0.93; 1.07; 1.1(20); 2.1(30); 7.2(50)
- (I) P_{fla} = 55cc; 63cc; 85 • AIT = 365; 369
- (T) Tr: 26/27/28-40 • TR: 63 • LVE(MVE) (F): – (5); TWA (USA): 5; MAK (D): 10 • STEL: 10
- (D) LC50 r: 0.33/4h; LC50 m: 0.65/2h; **LD50** o-r: 75; LD50 o-m: 112; 209; LD50 o-ra: 126; LD50 cu-ra: 90; 720

Epichlorhydrin: C_3H_5ClO • 92.53 • 106-89-8
- (P) Eb = 115-7; 98/400; 79.3/200; 62/100; 42/40; 29/20; 16.6/10; –16.5/1 • P_{vap} = 4.8(0); 13.3(16); 16.7; 17.3(20); 18.4(21.1); 30.5(30)
- (I) LEL/UEL = 2.3/34; 2.3/34.4; 3.8/21.0 • P_{fla} = 21oc; 28; 33cc; 41oc • AIT = 385; 410; 414
- (T) Tr: 45-23/24/25-34-43 • NFPA: 3 • TR: 63 • LVE(MVE) (F): 2; TWA (USA): 2
- (D) LC50 r: 1.36/6h; 0.946/8h; **LD50** o-r: 90; LD50 o-m: 195; LD50 o-ra: 345; LD50 ip-r: 133; LD50 iv-r: 154; LD50 ip-ra: 118; LD50 scu-r: 150; LD50 cu-ra: 515

4-Chloroanisole: C_7H_7ClO • 142.59 • 623-12-1
- (P) Eb = 195-198; 202
- (I) P_{fla} = 74; 78

AMINES
• **Primary amines**

Methylamine: CH_5N • 31.06 • 74-89-5
- (P) $\Delta H_f^0 = -23.01(g)$ • Eb = –6; –6.3; –19.7/400; –32.4/200; –43.7/100; –73.8/10 • P_{vap} = 1862; 3100(20); 4200(30); 7800(50)
- (I) LEL/UEL = 4.9/20.7; 4.95/20.75; 4.9/20.8; 4.5/21 • P_{fla} = –60; –18; –10; 0oc; 1Tco • AIT = 430
- (T) Tr: 36/37 • NFPA: 3 • TR: 236 • LVE(MVE) (F): 10 (–); TWA (USA): 5; MAK (D): 10 • STEL: 15
- (D) LC50 r: 0.57/4h; LC50 m: 2.4/2h; **LD50** scu-r: 200; MLDscu-m: 2500

Ethylamine: C_2H_7N • 45.08 • 75-04-7
- (P) $\Delta H_f^0 = -46.02(g)$ • Eb = 17; –36/50 • P_{vap} = 1165; 1200(20); 5066(65)
- (I) LEL/UEL = 3.5/14; 4.95/21 • P_{fla} = –49; –37; –18oc; –16; –10 • AIT = 335; 382; 385
- (T) Tr: 36/37 • NFPA: 3 • TR: 236 • LVE(MVE) (F): 15 (10); TWA (USA): 5; MAK(D): 10 • STEL: 15; IDLH: 4000
- (D) LC50 r: 5.533/4h; **LD50** o-r: 400; 540; LD50 cu-ra: 390

8 • Table of organic compounds — Amines

Name: Empirical formula • Molar mass(g/mol)• Registry number (CAS).
(P) Physical data: ΔH_f^0 (kJ/mol) (physical state) • Eb (°C/mmHg) (if substance volatile) • P_{vap} (mBar) (°C) (if substance volatile).
(I) Inflammability data: LEL/UEL (%) • P_{fla} (°C) • AIT (°C).
(T) Toxicity: Safety Code (Tr) • NFPA • Transport Code (TR) • LVE (MVE) (ppm) F, USA, D • STEL (ppm); IDLH (ppm).
(D) Lethal and toxic doses: LC (mg/l/duration of exposure); LD (mg/kg).

Propylamine: C_3H_9N • 59.11 • 107-10-8
- (P) $\Delta H_f^0 = -72.38(g)$ • Eb = 47-49; 0.5/100; -37/10 • P_{vap} = 327; 330; 333; 338(20); 409(25); 1050(50); 1150; 1287(55)
- (I) LEL/UEL = 2.0/10.4; 2.1/13.6 • P_{fla} = -44; -37; -30; -26; < -17; -12cc • AIT = 318; 325; 340
- (T) Tr: 20/21/22-34 • NFPA: 3 • TR: 338
- (D) LC50 r: 5.584/4h; 7.06/4h; LC50 m: 2.5/2h; **LD50** o-r: 370; 570; MLDo-r: 570; LD50 cu-ra: 560

Isopropylamine: C_3H_9N • 59.11 • 75-31-0
- (P) $\Delta H_f^0 = -112.26(l)$ • Eb = 32-34; -24/50; -48/10 • P_{vap} = 605; 634; 637; 665(20); 766(25); 1300(40); 2223(55)
- (I) LEL/UEL = 2.3/10.4; 2/10.4 • P_{fla} = -50; -37cc; -37oc; -32; -26oc • AIT = 400-403
- (T) Tr: 36/37/38 • NFPA: 3 • TR: 338 • LVE(MVE) (F): – (5); TWA (USA): 5; MAK (D): 5 • STEL: 10
- (D) LC50 r: 9.67/4h; **LD50** o-r: 820; LD50 o-m: 2200; LD50 o-ra: 3200; LD50 scu-m: 550; LD50 cu-ra: 380; 550

Allylamine: C_3H_7N • 57.10 • 107-11-9
- (P) Eb = 52-55; 55-58 • P_{vap} = 282; 533(20); 533(35); 1127(55)
- (I) LEL/UEL = 2/22; 2.2/22 • P_{fla} = -29; -20; -12cc; -7 • AIT = 370; 374
- (T) Tr: 23/24/25 • NFPA: 3 • TR: 336
- (D) LC50 r: 0.668/4h; 0.413/8h; **LD50** o-r: 102; 106; LD50 o-m: 57; LD50 ip-m: 49; LD50 cu-ra: 35

Butylamine: $C_4H_{11}N$ • 73.14 • 109-73-9
- (P) $\Delta H_f^0 = -92.05(g)$; -127.7(l) • Eb = 76-78; 13/50; -14/10 • P_{vap} = 76; 91; 93; 102; 109(20); 116(25); 155(30); 350(50); 798(70)
- (I) LEL/UEL = 1.7/9.8; 1.7/10 • P_{fla} = -12cc; -12oc; -9cc; -7; -3cc; -1oc; 1; < 7 • AIT = 310; 312
- (T) Tr: 36/37/38 • NFPA: 2; 3 • TR: 338 • LVE(MVE) (F): 5 (–); TWA (USA): 5; MAK (D): 5
- (D) LC50 r: 4.2/4h; LC50 m: 0.8/2h; **LD50** o-r: 366; 500; LD50 o-m: 430; LD50 ip-m: 629; LD50 iv-m: 198; LD50 cu-ra: 850

s-Butylamine: $C_4H_{11}N$ • 73.14 • 33966-50-6 • 13952-84-6
- (P) $\Delta H_f^0 = -104.18(g)$ • Eb = 62-65; 63-68 • P_{vap} = 5.3(4); 180(20); 43(29)
- (I) P_{fla} = -25; -19; -9.5cc
- (T) Tr: 20/22-35 • MAK(D): 5
- (D) LD50 o-r: 152; LD50 cu-ra: 2500

t-Butylamine: $C_4H_{11}N$ • 73.14 • 75-64-9
- (P) $\Delta H_f^0 = -119.87(g)$; -150.50(l) • Eb = 44-46 • P_{vap} = 300; 393(20); 483(25); 1442(55)
- (I) LEL/UEL = 1.5/9.2; 1.7/9.8; 1.7(100)/ 9.8(100); 1.7/8.9 • P_{fla} = -38cc; -9; -7; 10 • AIT = 375; 380
- (T) Tr: 25-34; 36/37/ 38 • NFPA: 3 • MAK(D): 5
- (D) LC50 r: 3.8/4h; **LD50** o-r: 78; 464; LD50 o-m: 378; 900; LD50 o-ra: 375

Isobutylamine: $C_4H_{11}N$ • 73.14 • 78-81-9
- (P) $\Delta H_f^0 = -132.55(l)$ • Eb = 64-71; 67-69 • P_{vap} = 133(18.8); 266(32)
- (I) LEL/UEL = 1.9/10.8 • P_{fla} = -20; -13; -9 • AIT = 360; 370; 378
- (T) Tr: 22-35; 22-34 • NFPA: 3 • TR: 338 • MAK(D): 5
- (D) LD50 o-r: 224; 228; 232

n-Pentylamine: $C_5H_{13}N$ • 87.19 • 110-58-7
- (P) Eb = 103-104
- (I) LEL/UEL = 2.2/22 • P_{fla} = -1; 4; 7oc; 15
- (T) Tr: 36/37/38 • NFPA: 3 • TR: 338
- (D) MLDip-r: 3750

Isopentylamine: $C_5H_{13}N$ • 87.19 • 107-85-7 (598-74-3)
- (P) Eb = 95-97
- (I) LEL/UEL = 2.3/22 • P_{fla} = -1; 6cc; 18
- (T) Tr: 36/37/38
- (D) LD50 ip-m: 279

Neopentylamine: $C_5H_{13}N$ • 87.19 • 5813-64-9
- (P) Eb = 81; 81-82/741
- (I) P_{fla} = -1; -13; 25
- (D) LD50 o-r: 470; LD50 scu-m: 1120

Cyclopentylamine: $C_5H_{11}N$ • 85.15 • 1003-03-8
- (P) Eb = 106-108
- (I) P_{fla} = 13; 17
- (T) Tr: 37/38

Hexylamine: $C_6H_{15}N$ • 101.19 • 111-26-2
- (P) Eb = 131-133 • P_{vap} = 10.6; 12; 24(20)
- (I) LEL/UEL = 2.1/9.3 • P_{fla} = 8; 26cc; 29oc; 34oc • AIT = 270
- (T) Tr: 21/22-34 • NFPA: 2 • TR: –
- (D) LC50 r: 2.07/4h; **LD50** o-r: ~240; 670; LD50 cu-ra: 420

Cyclohexylamine: $C_6H_{13}N$ • 99.18 • 108-91-8
- (P) Eb = 134; 72/100; 56/50; 36/20; Azeotrope with water Eb = 96.4 • P_{vap} = 13; 14(20); 13.3(22); 20(30); 31(37.7); 60(50)
- (I) LEL/UEL = 1.6/9.4 • P_{fla} = 21; 27cc; 32oc • AIT = 265; 290; 293
- (T) Tr: 21/22-34 • NFPA: 2 • TR: 83 • LVE(MVE) (F): – (10); TWA (USA): 10; MAK (D): 10
- (D) LC50 m: 1.07/?; **LD50** o-r: 156; 300; 614; LD50 o-m: 224; LD50 ip-r: 200; 300; LD50 ip-m: 129; LD50 iv-m: 200; LD50 scu-m: 1150; LD50 cu-ra: 277; 280; 320

Heptylamine: $C_7H_{17}N$ • 115.25 • 111-68-2
- (P) Eb = 153-156 • P_{vap} = 11(20); 16(30); 32(50)
- (I) P_{fla} = 34; 44; 54oc
- (T) Tr: 34 • NFPA: 2 • TR: 83
- (D) LD50 ip-r: 75; LD50 ip-m: 100

8 • Table of organic compounds — Amines

Name: Empirical formula • Molar mass(g/mol)• Registry number (CAS).
(P) Physical data: ΔH_f^0 (kJ/mol) (physical state) • Eb (°C/mmHg) (if substance volatile) • P_{vap} (mBar) (°C) (if substance volatile).
(I) Inflammability data: LEL/UEL (%) • P_{fla} (°C) • AIT (°C).
(T) Toxicity: Safety Code (Tr) • NFPA • Transport Code (TR) • LVE (MVE) (ppm) F, USA, D • STEL (ppm); IDLH (ppm).
(D) Lethal and toxic doses: LC (mg/l/duration of exposure); LD (mg/kg).

Aniline: C_6H_7N • 93.14 • 62-53-3
- **(P)** ΔH_f^0 = +31.59(l) • Eb = 181-186; 69/10; 70-71/10
  • P_{vap} = 0.4; 0.8; 3(20); 0.93(25); 1.33(34.8); 5.33(55)
- **(I)** LEL/UEL = 1.3/-; 1.3/11; 1.3/20-25 • P_{fla} = 70cc; 70oc; 76 • AIT = 530; 540; 615; 770
- **(T)** Tr: 20/21/22-40-48/23/24/25 • NFPA: 3 • TR: 60 • LVE(MVE) (F): – (2); TWA (USA): 2 • MAK (D): 2 • IDLH: 100
- **(D)** LC50 r: 0.67/7h; **LD50** o-r: 250; 440; LD50 o-m: 464; LD50 ip-r: 420; LD50 ip-m: 492; LD50 iv-ra: 64; LD50 scu-m: 200; LD50 cu-ra: 820

Benzylamine: C_7H_9N • 107.17 • 100-46-9
- **(P)** Eb = 185; 90/12; 70-71/10 • P_{vap} = 0.6(20)
- **(I)** P_{fla} = 59; 63; 65cc; 72 • AIT = 390
- **(T)** Tr: 34 • NFPA: 3 • TR: 80
- **(D)** LD50 ip-m: 600

o-Toluidine: C_7H_9N • 107.17 • 95-53-4
- **(P)** Eb = 200-202; 89-90/11; 81/10 • P_{vap} = 0.13; 0.35(20); 0.35(25); 0.43(30); 0.88(38); 1.33(44); 2(50)
- **(I)** LEL/UEL = 1.5/- • P_{fla} = 85cc • AIT = 480; 482
- **(T)** Tr: 45-23/25-34; 45-23/25-36 • NFPA: 3 • TR: 60 • LVE(MVE) (F): – (2); TWA (USA): 2
- **(D)** LD50 o-r: 670; 940; LD50 o-m: 520; LD50 o-ra: 840; LD50 cu-ra: 3250

m-Toluidine: C_7H_9N; 107.17; 108-44-1
- **(P)** Eb = 200-204 • P_{vap} = 0.2(20); 0.53(30); 1.33(41); 2.4(50); 66.7(68)
- **(I)** LEL/UEL = 1.1/6.1; 1.1/6.6 • P_{fla} = 86; 90 • AIT = 482
- **(T)** Tr: 23/24/25-33 • NFPA: 3 • TR: 60 • TWA (USA): 2
- **(D)** LD50 o-r: 450; LD50 o-m: 740; LD50 o-ra: 750; LD50 ip-m: 116

p-Toluidine: C_7H_9N • 107.17 • 106-49-0
- **(P)** Eb = 200-201 • P_{vap} = 0.2(20); 0.35(25); 0.52(30); 0.88(38); 1.33(42); 2.1(50)
- **(I)** LEL/UEL = 1.1/6.6 • P_{fla} = 87cc; 89 • AIT = 482
- **(T)** Tr: 23/24/25-33; 23/24/25-33-40 • NFPA: 3 • TR: 60 • TWA (USA): 2
- **(D)** LD50 o-r: 656; LD50 o-m: 330; LD50 o-ra: 270; LD50 cu-ra: 890

2-Ethylaniline: $C_8H_{11}N$ • 121.20 • 578-54-1
- **(P)** Eb = 210-215 • P_{vap} = 0.15(20)
- **(I)** P_{fla} = 85oc; 91; 97
- **(T)** Tr: 23/24/25-33 • TR: 60 • TWA (USA): 2
- **(D)** LD50 o-r: 1260

4-Ethylaniline: $C_8H_{11}N$ • 121.20 • 589-16-2
- **(P)** Eb = 215-216 • P_{vap} = ~0.09(20); ~0.25(30); 1.1(50)
- **(I)** P_{fla} = 84
- **(T)** Tr: 23/24/25-33 • TR: 60 • TWA (USA): 2

2-Diphenylamine: $C_{12}H_{11}N$ • 169.23 • 90-41-5
- **(P)** ΔH_f^0 = +112.13(s) • Eb = 299
- **(I)** P_{fla} = > 109; 153 • AIT = 449; 452
- **(T)** Tr: 40 • NFPA: 2
- **(D)** LD50 o-r: 2340; LD50 o-ra: 1020

1-Naphthylamine: $C_{10}H_9N$ • 143.20 • 134-32-7
- **(P)** ΔH_f^0 = +67.78(s) • Eb = 301 • P_{vap} = 0.01(20); 0.02(30); 0.07(50); 1.33(104.3)
- **(I)** P_{fla} = > 109; 157
- **(T)** Tr: 45-22; 22 • NFPA: 2 • TR: 60
- **(D)** LD50 o-r: 779; LD50 ip-m: 96; LD50 scu-m: 300

2-Naphthylamine: $C_{10}H_9N$ • 143.20 • 91-59-8
- **(P)** ΔH_f^0 = +60.25(s) • Eb = 294; 306 • P_{vap} = 0.007(20); 0.015(30); 0.055(50); 1.33(108)
- **(I)** P_{fla} = 109
- **(T)** Tr: 45-22 • TR: 60 • LVE(MVE) (F): – (0.001)
- **(D)** LD50 o-r: 727; LD50 ip-m: 200

AMINES
• Secondary amines

Dimethylamine: C_2H_7N • 45.10 • 124-40-3
- **(P)** ΔH_f^0 = –18.83(g) • Eb = 6.9; 7.4 • P_{vap} = 1703; 2000(20); 2500(30); 4600(50); 4596(52)
- **(I)** LEL/UEL = 2.8/14.4; 2.8/14 • P_{fla} = [–17; –15] (aqueous 40 % sol.,) ; [–50; –55] (anhydrous); 6 (aqueous 25 %sol.); 1 • AIT = 400; 402; 430
- **(T)** Tr: 36/37 • NFPA: 3 • TR: 236 • LVE(MVE) (F): 10 (-); TWA (USA): 5; MAK (D): 10 • STEL: 15; IDLH: 2000
- **(D)** LC50 r: 8.374/6h; LC50 m: 8.715/2h; **LD50** o-r: 698; LD50 o-m: 316; LD50 o-ra: 240; LD50 ip-m: 736; LD50 iv-ra: 4000; MLDiv-ra: 4000

Diethylamine: $C_4H_{11}N$ • 73.16 • 109-89-7
- **(P)** ΔH_f^0 = –72.38(g) • Eb = 55-57; –7/50; –33/10 • P_{vap} = 242; 253; 260(20); 311(25); 533(38); 500(40); 975(55)
- **(I)** LEL/UEL = 1.7/10.1; 1.8/10.1; 2.0/11.8 • P_{fla} = –39; –36; –28cc; < –26; –25; ~ –23; –18Tcc • AIT = 290; 310; 312
- **(T)** Tr: 36/37 • NFPA: 2; 3 • TR: 338 • LVE(MVE) (F): 10 (-); TWA (USA): 5; MAK (D): 10 • STEL: 15; IDLH: 2000
- **(D)** LC50 r: 11.97/4h; **LD50** o-r: ~540; 540; LD50 o-m: 500; LD50 ip-m: 585; LD50 cu-ra: 820

N-Methylbutylamine: $C_5H_{13}N$ • 87.19 • 110-68-9
- **(P)** Eb = 87-91 • P_{vap} = 57(20)
- **(I)** P_{fla} = –1cc; 2; 13Toc
- **(T)** Tr: 34-20/21/22 • NFPA: 3
- **(D)** LC50 r: 7.132; **LD50** o-r: 420; LD50 ip-m: 471; LD50 iv-m: 122; LD50 cu-ra: 1260

8 • Table of organic compounds — Amines

Name: Empirical formula • Molar mass(g/mol) • Registry number (CAS).
- **(P) Physical data:** ΔH_f^0 (kJ/mol) (physical state) • Eb (°C/mmHg) (if substance volatile) • P_{vap} (mBar) (°C) (if substance volatile).
- **(I) Inflammability data:** LEL/UEL (%) • P_{fla} (°C) • AIT (°C).
- **(T) Toxicity:** Safety Code (Tr) • NFPA • Transport Code (TR) • LVE (MVE) (ppm) F, USA, D • STEL (ppm); IDLH (ppm).
- **(D) Lethal and toxic doses:** LC (mg/l/duration of exposure); LD (mg/kg).

Dipropylamine: $C_6H_{15}N$ • 101.22 • 142-84-7
- (P) Eb = 105-110; 37/50; 7/10 • P_{vap} = 24; 28; 38; 42(20); 37(30); 92(50)
- (I) LEL/UEL = 1.8/9.3 • P_{fla} = 3; 7cc; 17Toc • AIT = 260
- (T) Tr: 36/37/38 • TR: 338
- (D) LC50 r: 4.4/4h; LC50 m: 3.07/2h; **LD50** o-r: 460; 495; 930; LD50 ip-r: 75; LD50 cu-ra: 925; 1250

Diisopropylamine: $C_6H_{15}N$ • 101.22 • 108-18-9
- (P) Eb = 83-84.1; 16/50; −33/10 • P_{vap} = 66.7; 82(20); 112(25); 140(30); 300(50)
- (I) LEL/UEL = 1.1/8.5; 1.5/8.5 • P_{fla} = −17; −14cc; −6oc; −1oc • AIT = 285; 314
- (T) Tr: 36/37/38 • NFPA: 3 • TR: 338 • LVE(MVE) (F): − (5); TWA (USA): 5
- (D) LC50 r: 4.8/4h; LC50 m: 4.2/4h; LC50 ra: 9.136/3'; **LD50** o-r: 770; LD50 o-m: 2120; LD50 o-ra: 4700; LD50 cu-ra: > 10000

Diallylamine: $C_6H_{11}N$ • 97.18 • 124-02-7
- (P) Eb = 107-112 • P_{vap} = 24.2(20)
- (I) P_{fla} = 7; 15; 21
- (T) Tr: 22-36/37/38; 36/3 • TR: 338
- (D) LC50 r: 3.16/8h; 7.95/4h; **LD50** o-r: 578; 650; LD50 o-m: 355; 516; LD50 ip-m: 187; LD50 cu-ra: 280

N-Methylcyclohexylamine: $C_7H_{15}N$ • 113.20 • 100-60-7
- (P) Eb = 149; 76/18 • P_{vap} = 14.7(40)
- (I) P_{fla} = 29; 39; 46(Luchaire) • AIT = 170
- (T) Tr: 20/21/22-34-37; 34-20/21/22 • TR: 83
- (D) LD50 o-r: 400; MLDcu-ra: 2000

Dibutylamine: $C_8H_{19}N$ • 129.25 • 111-92-2
- (P) Eb = 159-161; 80/50; 47/10 • P_{vap} = 2; 2.5; 2.7(20); 3.1(25)
- (I) LEL/UEL = ~1.1/10; 1.1/- • P_{fla} = 40; 42cc; 51cc; 52oc; 57oc • AIT = 260; 311
- (T) Tr: 20/21/22 • NFPA: 3 • TR: 83
- (D) LC50 r: 2.644/4h; **LD50** o-r: 189; 220; 360; 550; LD50 o-m: 290; LD50 ip-r: 110; LD50 ip-m: 200; LD50 scu-r: 330; 494; LD50 cu-ra: 1010

Di-s-butylamine: $C_8H_{19}N$ • 129.25 • 4444-67-1
- (P) Eb = 134; 135/765
- (I) P_{fla} = 20; 24oc
- (T) Tr: 20/21/22

N-Isopropylcyclohexylamine: $C_9H_{19}N$ • 141.26 • 1195-42-2
- (P) Eb = 175; 60/16; 60-65/12
- (I) P_{fla} = 33; 42; 46
- (T) Tr: 34-37
- (D) LD50 ip-m: 550; LD50 iv-m: 40

Dihexylamine: $C_{12}H_{27}N$ • 185.40 • 143-16-8
- (P) Eb = 236; 240-245; 233-243
- (I) P_{fla} = 94oc; 104oc
- (T) Tr: 22-24-34
- (D) LD50 o-r: 380; LD50 iv-m: 10; LD50 cu-ra: 170

Dicyclohexylamine: $C_{12}H_{23}N$ • 181.36 • 101-83-7
- (P) Eb = 256; 214.5/300; 199/200; 174.4/100; 154.3/50; 135.5/25; 121/11; 119-120/10; 99.3/4; 83/1 • P_{vap} = < 0.1(20); 16(38)
- (I) LEL/UEL = 0.83/4.63 • P_{fla} = 96; 98cc; 99oc; > 99; 105cc; 110 • AIT = 240
- (T) Tr: 22-34 • NFPA: 3 • TR: 80
- (D) LD50 o-r: 240; 373; LD50 o-m: 500; LD50 scu-m: 135; LD50 scu-m: 500

N-Methylaniline: C_7H_9N • 107.17 • 100-61-8
- (P) ΔH_f^0 = +32.22(l) • Eb = 192-197 • P_{vap} = 0.4(20); 0.85(30); 3.3(50)
- (I) P_{fla} = 78; 85
- (T) Tr: 23/24/25-33 • TR: 60 • LVE(MVE) (F): − (0.5); TWA (USA): 0.5; MAK (D): 0.5
- (D) MLDo-ra: 280; MLDiv-ra: 24

N-Ethylaniline: $C_8H_{11}N$; 121.20 • 103-69-5
- (P) ΔH_f^0 = +3.77(l) • Eb = 205-206; 82-85/10 • P_{vap} = 0.4(20); 0.27(25); 0.75(30); 0.67(38); 1.33(38.5); 2.7(50)
- (I) LEL/UEL = 1.6/9.5 • P_{fla} = 85oc • AIT = 479
- (T) Tr: 23/24/25-33 • NFPA: 3 • TR: 60
- (D) LC50 r: > 1.13/4h; **LD50** o-r: 334; 1100; LD50 ip-r: 180; LD50 ip-m: 242; LD50 cu-ra: 4700

Diphenylamine: $C_{12}H_{11}N$ • 169.24 • 122-39-4
- (P) ΔH_f^0 = +130.00(s) • Eb = 302 • P_{vap} = 3.3x10^{-4}; 0.03(20); 0.0012(30); 1.33(108.3)
- (I) P_{fla} = 153cc • AIT = 630; 634
- (T) Tr: 23/24/25-33 • NFPA: 3 • TR: − • LVE(MVE) (F): − (10 mg/m^3); TWA (USA): 10 mg/m^3; MAK (D): 10 mg/m^3
- (D) LD50 o-r: 2000; LD50 o-m: 1750; LD50 scu-r: 3000

Dibenzylamine: $C_{14}H_{15}N$ • 197.28 • 103-49-1
- (P) Eb = 300dec; 270/250; 170/10
- (I) P_{fla} = 143
- (T) NFPA: 3

AMINES
• **Polyamines**

Ethylene diamine: $C_2H_8N_2$ • 60.10 • 107-15-3
- (P) ΔH_f^0 = −65.27(l) • Eb = 116-118; 49/50; 19/10 • P_{vap} = 12; 13.3; 14.3(20)
- (I) LEL/UEL = 2.7/16; 2.7/16.6; 3/18; 4.2/14.4 • P_{fla} = 34cc; 38oc; 43cc • AIT = 379; 385; 405
- (T) Tr: 21/22-34-43 • NFPA: 3 • TR: 83 • LVE(MVE) (F): 15 (10); TWA (USA): 10; MAK (D): 10 • IDLH: 2000
- (D) LC50 m: 0.3/?; **LD50** o-r: 500; 1160; LD50 ip-r: 76; LD50 ip-m: 200; LD50 scu-r: 300; LD50 scu-m: 424; LD50 scu-m: 500; LD50 cu-ra: 730

8 • Table of organic compounds — Amines

Name: Empirical formula • Molar mass(g/mol)• Registry number (CAS).
(P) Physical data: ΔH_f^0 (kJ/mol) (physical state) • Eb (°C/mmHg) (if substance volatile) • P_{vap} (mBar) (°C) (if substance volatile).
(I) Inflammability data: LEL/UEL (%) • P_{fla} (°C) • AIT (°C).
(T) Toxicity: Safety Code (Tr) • NFPA • Transport Code (TR) • LVE (MVE) (ppm) F, USA, D • STEL (ppm); IDLH (ppm).
(D) Lethal and toxic doses: LC (mg/l/duration of exposure); LD (mg/kg).

1,2-Diaminopropane: $C_3H_{10}N_2$ • 74.15 • 78-90-0
- (P) Eb = 119-123; 52/50; 23/10 • P_{vap} = 3.6; 10.6; 14; 18.7(20)
- (I) LEL/UEL = 1.9/11.1; 2.2/11.1 • P_{fla} = 24; 33oc; 49 • AIT = 360
- (T) Tr: 21/22-35; 22-35 • TR: 83
- (D) LD50 o-r: ~1300; 2230; LD50 ip-r: 593; LD50 scu-r: 2250; LD50 cu-ra: ~500; 500

1,3-Diaminopropane: $C_3H_{10}N_2$ • 74.15 • 109-76-2
- (P) Eb = 136-140 • P_{vap} = < 10.7(20)
- (I) LEL/UEL = 2.8/15.2 • P_{fla} = 24oc; 48Toc; 54 • AIT = 349; 380
- (T) Tr: 22-24-35; 22-24-34 • NFPA: 2 • TR: 83
- (D) LD50 o-r: ~700; 350; LD50 ip-r: 296; LD50 cu-ra: ~200; 200

1,4-Diaminobutane: $C_4H_{12}N_2$ • 88.18 • 110-60-1
- (P) Eb = 158; 60-61/11
- (I) LEL/UEL = 0.98/9.08 • P_{fla} = 51; 63
- (T) Tr: 34-37
- (D) LD50 o-m: 1600; LD50 o-ra: 1600; LD50 ip-m: 1750; LD50 iv-ra: 80; LD50 scu-r: 300; LD50 scu-m: 200

1,6-Diaminohexane: $C_6H_{16}N_2$ • 116.24 • 124-09-4
- (P) Eb = 199; 204-205; 100/20 • P_{vap} = 1.47(20); 2(50); 27(100)
- (I) LEL/UEL = 0.7/6.3 • P_{fla} = 81; 85; 93cc
- (T) Tr: 21/22-34-37 • NFPA: 2 • TR: 80 • TWA (USA): 0.5
- (D) LD50 o-r: 750; LD50 ip-m: 320; LD50 iv-m: 180; LD50 scu-m: 1300; LD50 cu-ra: 1110

p-Phenylenediamine: $C_6H_8N_2$ • 108.16 • 106-50-3
- (P) ΔH_f^0 = +3.05(s) • Eb = 267; P_{vap} = 1.44(100)
- (I) P_{fla} = 68(?); > 110; 156
- (T) Tr: 23/24/25-40-43; 23/24/25-43 • NFPA: 2 • TR: 60 • LVE(MVE) (F): – (0.1 mg/m³); TWA (USA): 0.1 mg/m³; MAK (D): 0.1 mg/m³
- (D) LD50 o-r: 80; MLDo-ra: 250; LD50 ip-r: 37; MLDiv-r: 50; LD50 ip-m: 50; MLDip-ra: 150; MLDiv-m: 300; MLDscu-r: 170; MLDscu-m: 140; MLDscu-m: 200; MLDcu-ra: 5000

Benzidine: $C_{12}H_{12}N_2$ • 184.26 • 92-87-5
- (P) ΔH_f^0 = +70.71(s) • Eb = 400-402
- (T) Tr: 45-22 • NFPA: 3 • LVE(MVE) (F): – (0.001)
- (D) LD50 o-r: 309; LD50 o-m: 214; LD50 o-ra: 200; LD50 ip-m: 110

Diethylenetriamine: $C_4H_{13}N_3$ • 103.20 • 111-40-0
- (P) ΔH_f^0 = –13.39(l) • Eb = 199-209 • P_{vap} = 0.11; < 0.13; 0.29; 0.37; 0.49(20)
- (I) LEL/UEL = 1/10; 1.4/9.2; 2/6.7; 4.4/16.1 • P_{fla} = 94; 97cc; 98PMcc; 102oc; 107Coc • AIT = 357; 395, 399; 360-400
- (T) Tr: 21/22-34-43 • NFPA: 3 • TR: 80 • LVE(MVE) (F): – (1); TWA (USA): 1
- (D) LD50 o-r: 1080; 2330; LD50 ip-r: 74; LD50 ip-m: 71; LD50 cu-ra: 1038; 1090

Triethylenetetramine: $C_6H_{18}N_4$ • 146.24 • 112-24-3
- (P) Eb = 267; 277; 284-287; 183/50; 144/10 • P_{vap} = < 0.013; 0.013(20); 15(50)
- (I) P_{fla} = 129; 135cc; 143oc; 149PMcc; 154Coc • AIT = 335; 338
- (T) Tr: 21-34-43 • NFPA: 3 • TR: 80
- (D) LD50 o-r: 2500; 2773; LD50 o-m: 1600; LD50 o-ra: 5500; LD50 ip-m: 468; LD50 iv-m: 350; LD50 cu-ra: 550; 805; 820

AMINES

• **Tertiary amines**

Trimethylamine: C_3H_9N • 59.13 • 75-50-3
- (P) ΔH_f^0 = –23.85(g) • Eb = 3-4 • P_{vap} = 1900(20); 1222(21); 2500(30); 4500(50)
- (I) LEL/UEL = 2/11.6 • P_{fla} = –65; –19oc; –13 to –8; –6; 3.3Toc (. aqueous 25% sol.) • AIT = 190
- (T) Tr: 36/37 • NFPA: 3 • TR: 236 • LVE(MVE) (F): 10 (–); TWA (USA): 5 • STEL: 15
- (D) LC50 r: 8.464/4h; LC50 m: 19/?; **LD50** o-r: 535; LD50 ip-m: 946; LD50 iv-m: 90; MLDiv-ra: 400; MLDscu-r: 1000; MLDscu-m: 800

N,N-Dimethylethylamine: $C_4H_{11}N$ • 73.14 • 598-56-1
- (P) Eb = 34-38 • P_{vap} = 558(20); 1818(55)
- (I) P_{fla} = –46; –36; –28
- (T) Tr: 20/22-34 • TR: 338 • LVE(MVE) (F): 25 (5); MAK (D): 25

N,N-Dimethylallylamine: $C_5H_{11}N$ • 85.15 • 2155-94-4
- (P) Eb = 63
- (I) P_{fla} = –20; 8
- (T) Tr: 20-36/37/38

Triethylamine: $C_6H_{15}N$ • 101.22 • 121-44-8
- (P) ΔH_f^0 = –99.58(g) • Eb = 86-91; 19/50; –10/10 • P_{vap} = 61.6; 69; 72(20); 76(25)
- (I) LEL/UEL = 1.2/8; 1.6/9.3 • P_{fla} = –17; –15; –11cc; –8Tcc; < –7; –6cc; –6Toc; 1; 7oc • AIT = 215; 230; 248; 311
- (T) Tr: 36/37 • NFPA: 2 • TR: 338 • LVE(MVE) (F): 10 (–); TWA (USA): 10; MAK (D): 10 • STEL: 15; IDLH: 1000
- (D) LC50 r: 4.138/4h, 10.76/4h; LC50 m: 6.0/2h; **LD50** o-r: 460; 730; LD50 o-m: 546; LD50 ip-m: 405; LD50 cu-ra: 416; 570

N-Ethyldiisopropylamine: $C_8H_{19}N$ • 129.25 • 7087-68-5
- (P) Eb = 126-128 • P_{vap} = 41.33(37.7)
- (I) P_{fla} = 3; 10
- (T) Tr: 20/21/22-34

Triallylamine: $C_9H_{15}N$ • 137.25 • 102-70-5
- (P) Eb = 150-151 • P_{vap} = 119.97(80)
- (I) P_{fla} = 30; 39Toc
- (D) LC50 r: 2.8/4h; 3.109/8h; **LD50** o-r: 1030; LD50 o-m: 492; LD50 ip-m: 187; LD50 cu-ra: 400

8 • Table of organic compounds — Amines

Name: Empirical formula • Molar mass(g/mol) • Registry number (CAS).
- (P) **Physical data:** ΔH_f^0 (kJ/mol) (physical state) • Eb (°C/mmHg) (if substance volatile) • P_{vap} (mBar) (°C) (if substance volatile).
- (I) **Inflammability data:** LEL/UEL (%) • P_{fla} (°C) • AIT (°C).
- (T) **Toxicity:** Safety Code (Tr) • NFPA • Transport Code (TR) • LVE (MVE) (ppm) F, USA, D • STEL (ppm); IDLH (ppm).
- (D) **Lethal and toxic doses:** LC (mg/l/duration of exposure); LD (mg/kg).

Tripropylamine: $C_9H_{21}N$ • 143.31 • 102-69-2
- (P) $\Delta H_f^0 = -207.15(l)$ • Eb = 155-158; 40-42/11 • $P_{vap} = 3.87(20)$
- (I) LEL/UEL = 0.7/5.6 • $P_{fla} = 34; 36; 41oc$ • AIT = 179; 190
- (T) Tr: 20/21-25-34 • NFPA: 2
- (D) LC50 r: 1.465/4h; LC50 m: 3.8/2h; **LD50** o-r: 72; LD50 cu-ra: 429

Tributylamine: $C_{12}H_{27}N$ • 185.40 • 102-82-9
- (P) $\Delta H_f^0 = -281.67(l)$ • Eb = 213-216; 147/100; 115/30; 88-90/10 • $P_{vap} = 0.40(20); 3.2(55)$
- (I) LEL/UEL = 1.4/6; 0.6/4.6 • $P_{fla} = 63; 68cc; 70; 80oc; 82; 86oc$ • AIT = 209
- (T) Tr: 23/24/25-34 • NFPA: 2
- (D) MLC r: 0.568/4h; **LD50** o-r: 114; 540; LD50 o-m: 114; LD50 o-ra: 615; MLDscu-r: 380; LD50 cu-ra: 250

Triisobutylamine: $C_{12}H_{27}N$ • 185.40 • 1116-40-1
- (P) Eb = 189-191; 192-193/770; 84/15
- (I) $P_{fla} = 50; 57; 62$
- (T) Tr: 36/37/38

Triamylamine: $C_{15}H_{33}N$ • 227.46 • 621-77-2
- (P) Eb = 232; 81-83/0.2
- (I) $P_{fla} = 88; 97; 102$
- (T) Tr: 36/37/38 • NFPA: 2

N,N-Diethylaniline: $C_{10}H_{15}N$ • 149.26 • 91-66-7
- (P) Eb = 215-217; 62-66/3 • $P_{vap} = 0.2; 1(20); 0.4(30); 1.33(49.7); 1.4(50); 13.3(92)$
- (I) $P_{fla} = 85cc; 88; 97$ • AIT = 332; 629
- (T) Tr: 23/24/25-33 • NFPA: 3 • TR: 60
- (D) LC50 r: 1.92/4h; **LD50** o-r: 782; LD50 ip-r: 420

N,N-Dimethylaniline: $C_8H_{11}N$ • 121.20 • 121-69-7
- (P) $\Delta H_f^0 = +34.31(l)$ • Eb = 192-194; 75-78/10 • $P_{vap} = 0.53; 0.67; 1.33(20); 1.33(29.5); 1.3(30); 4.5(50); 13.3(70)$
- (I) LEL/UEL = 1.2/7; 1/7 • $P_{fla} = 61; 63cc; 75cc$ • AIT = 316; 370
- (T) Tr: 23/24/25-33; 23/24/25-33-40-51-53 • NFPA: 3 • TR: 60 • LVE(MVE) (F): – (5); TWA (USA): 5; MAK (D): 5 • STEL: 10
- (D) LC50 r: 0.250/4h; **LD50** o-r: ~1100; 1348; 1410; LD50 cu-ra: ~1700; 1770

N,N-Dimethyl-p-toluidine: $C_9H_{13}N$ • 135.23 • 99-97-8
- (P) Eb = 207-211; 215; 90-92/10
- (I) LEL/UEL = 1.2/7 • $P_{fla} = 76; 83cc$
- (T) Tr: 23/24/25-33
- (D) LD50 ip-m: 212

AMINES

• Heterocyclic amines

Aziridine: C_2H_5N • 43.08 • 151-56-4
- (P) $\Delta H_f^0 = +123.43(g)$ • Eb = 55-56 • $P_{vap} = 84(0); 213(20); 285(25); 333(30); 780(50)$
- (I) LEL/UEL = 3.3/54.8; 3.6/46 • $P_{fla} = -24; -13$ • AIT = 320; 322; 325
- (T) Tr: 45-46-26/27/28-34 • NFPA: 3 • TR: 336 • LVE(MVE) (F): – (0.5); TWA (USA): 0.5 (skin)
- (D) LC50 r: 0.100/2h; LC50 m: 0.400/2h; MLC ra: 0.100/2h; **LD50** o-r: 15; LD50 ip-r: 3500; MLDcu-ra: 10

Pyrrolidine: C_4H_9N • 71.14 • 123-75-1
- (P) $\Delta H_f^0 = -3.60(g); -41.17(l)$ • Eb = 86-89 • $P_{vap} = 65.32(20); 170(39); 260(50)$
- (I) LEL/UEL = 1.6/10.6; 2.9/13.0 • $P_{fla} = 2; 3Tcc$ • AIT = 269; 320; 344
- (T) Tr: 20/22-35; 20/22-34 • NFPA: 2 • TR: 338
- (D) LC50 m: 1.3/2h; **LD50** o-r: 300; LD50 o-m: 450; LD50 ip-m: 420; LD50 iv-m: 56

Pyrrole: C_4H_5N • 67.10 • 109-97-7
- (P) $\Delta H_f^0 = +63.09(l)$ • Eb = 129-131
- (I) $P_{fla} = 33; 36; 39Tcc$
- (T) NFPA: 2
- (D) LD50 o-ra: 147; LD50 ip-m: 98; MLDip-ra: 150; LD50 scu-m: 61; MLDscu-m: 250

N-Methylpyrrole: C_5H_5N • 81.12 • 96-54-8
- (P) Eb = 108-112; 115 • $P_{vap} = 20(20)$
- (I) LEL/UEL = 1/6 • $P_{fla} = 0; 3; 14; 16$
- (T) Tr: 20 • NFPA: 3

Pyrazole: $C_3H_4N_2$ • 68.09 • 288-13-1
- (P) $\Delta H_f^0 = +118.41(s)$ • Eb = 186-188
- (D) LD50 o-r: 1100; 1430; LD50 o-m: 409; 1456; 1498; LD50 ip-r: 900; LD50 iv-r: 1021; 1294; LD50 ip-m: 538; LD50 iv-m: 374; 1430

Piperidine: $C_5H_{11}N$ • 85.17 • 110-89-4
- (P) $\Delta H_f^0 = -88.07(l)$ • Eb = 104-106; 18/20 • $P_{vap} = 30.65; 33(20); 53(29.2); 56(30); 85(37.); 140(50)$
- (I) LEL/UEL = 1.3/10.3 • $P_{fla} = 3; 16$ • AIT = 320
- (T) Tr: 23/24-34 • NFPA: 2 • TR: 338
- (D) LC50 r: 13.93/4h; LC50 m: 6/2h; **LD50** o-r: 400; 448; LD50 o-m: 30; LD50 o-ra: 145; LD50 ip-m: 50; LD50 cu-ra: 320

N-Methylpiperidine: $C_6H_{13}N$ • 99.18 • 626-67-5
- (P) Eb = 105-107
- (I) $P_{fla} = 0; 3; < 23$
- (T) Tr: 36/37/38
- (D) LD50 ip-m: 400

8 • Table of organic compounds — Amines

Name: Empirical formula • Molar mass(g/mol)• Registry number (CAS).
(P) Physical data: ΔH_f^0 (kJ/mol) (physical state) • Eb (°C/mmHg) (if substance volatile) • P_{vap} (mBar) (°C) (if substance volatile).
(I) Inflammability data: LEL/UEL (%) • P_{fla} (°C) • AIT (°C).
(T) Toxicity: Safety Code (Tr) • NFPA • Transport Code (TR) • LVE (MVE) (ppm) F, USA, D • STEL (ppm); IDLH (ppm).
(D) Lethal and toxic doses: LC (mg/l/duration of exposure); LD (mg/kg).

2-Methylpiperidine: $C_6H_{13}N$ • 99.18 • 109-05-7
- (P) Eb = 117-119
- (I) P_{fla} = 8; 10
- (T) Tr: 36/37/38
- (D) MLD scu-m: 300

3-Methylpiperidine: $C_6H_{13}N$ • 99.18 • 626-56-2
- (P) Eb = 122-126
- (I) P_{fla} = 17; 21
- (T) Tr: 36/37/38

4-Methylpiperidine: $C_6H_{13}N$ • 99.18 • 626-58-4
- (P) Eb = 122-126
- (I) P_{fla} = 7; 9; 14
- (T) Tr: 36/37/38

N-Ethylpiperidine: $C_7H_{15}N$ • 113.23 • 766-09-6
- (P) Eb = 126-131
- (I) P_{fla} = 16; 19 • AIT = 230
- (T) Tr: 20/22-36/37/38
- (D) LD50 iv-m: 56

2-Ethylpiperidine: $C_7H_{15}N$ • 113.23 • 1484-80-6
- (P) Eb = 140-142
- (I) P_{fla} = 17; 28; 31
- (T) Tr: 37/38

Piperazine: $C_4H_{10}N_2$ • 86.16 • 110-85-0
- (P) ΔH_f^0 = −45.61(s) • Eb = 145-146; 149 • P_{vap} = 1.07(20); 15(50); 303(111)
- (I) LEL/UEL = 3.9/11.9; 4/14 • P_{fla} = 88oc; 107; 109 • AIT = 340
- (T) Tr: 34 • TR: 80
- (D) LC50 m: 5.4/2h; LD50 o-r: 1900; LD50 o-m: 600; LD50 ip-r: 3700; LD50 iv-r: 1340; LD50 scu-r: 3700; LD50 cu-ra: 4000

Pyridine: C_5H_5N • 79.11 • 110-86-1
- (P) ΔH_f^0 = +140.69(g); +100.25(l) • Eb = 115-116 • P_{vap} = 13.3(13.2); 20.5; 21.33; 24(20); 27(25); 53(38); 93(50); 500(100)
- (I) LEL/UEL = 1.7/10.6; 1.8/12.4; 1.8/12.5 • P_{fla} = 17-20; 18cc; 19Tcc; 20Tcc • AIT = 482; 550
- (T) Tr: 20/21/22 • NFPA: 2 • TR: 336 • LVE(MVE) (F): 10 (5); TWA (USA): 5; MAK (D): 5 • IDLH: 3600
- (D) LC50 r: 29.168/1h; 12.94/4h; LD50 o-r: 891; 1580; LD50 o-m: 1500; LD50 ip-r: 866; LD50 ip-m: 950; 1108; MLDip-ra: 15; LD50 iv-r: 360; LD50 iv-m: 420; 538; LD50 scu-r: 1000; LD50 scu-m: 1250; LD50 cu-ra: 1121

2-Ethylpyridine: C_7H_9N • 107.18 • 100-71-0
- (P) ΔH_f^0 = −5.02(l) • Eb = 147-149
- (I) P_{fla} = 23; 29; 37cc

3-Ethylpyridine: C_7H_9N • 107.18 • 536-78-7
- (P) Eb = 165-166
- (I) P_{fla} = 48; 51

4-Ethylpyridine; C_7H_9N • 107.18 • 536-75-4
- (P) Eb = 164-166; 168; 169.6-170
- (I) P_{fla} = 39; 47; 50; 54

2-Methylpyridine: C_6H_7N • 93.14 • 109-06-8
- (P) ΔH_f^0 = +100.67(g); +57.6(l) • Eb = 127-129 • P_{vap} = 10; 12(20); 13.3(24.4); 20(30); 51(50)
- (I) LEL/UEL = 1.4/8.6 • P_{fla} = 26; 29; 39oc • AIT = 534; 538
- (T) Tr: 20/21/22-36/37 • NFPA: 2 • TR: 30
- (D) MLC r: 15.24/4h; LC50 m: 9/?; LD50 o-r: 790; 1410; LD50 o-m: 674; LD50 ip-r: 200; LD50 ip-m: 529; LD50 cu-ra: 410

3-Methylpyridine: C_6H_7N • 93.14 • 108-99-6
- (P) ΔH_f^0 = +65.14(l) • Eb = 141-144 • P_{vap} = 5.87(20)
- (I) LEL/UEL = 1.3/8.7 • P_{fla} = 36; 39 • AIT = 537
- (T) Tr: 20/21/22-36/37/38; 22-36/37/38 • TR: 30
- (D) LD50 o-r: 400; LD50 ip-r: 150; LD50 ip-m: 596; LD50 iv-m: 298

4-Methylpyridine: C_6H_7N • 93.14 • 108-89-4
- (P) ΔH_f^0 = +56.82(l) • Eb = 142-145 • P_{vap} = 5.33; 5.8(20)
- (I) LEL/UEL = 1.3/8.7 • P_{fla} = 39; 50; 57oc • AIT = 500; 537
- (T) Tr: 20/22-24-36/37/38 • NFPA: 2 • TR: 30
- (D) MLC r: 3.809/4h; LC50 m: 4/?; LD50 o-r: 440; 1290; LD50 o-m: 350; LD50 ip-r: 163; LD50 ip-m: 335; LD50 cu-ra: 270

2-Vinylpyridine: C_7H_7N • 105.15 • 100-69-6
- (P) ΔH_f^0 = +155.64(l) • Eb = 158-160; 79-82/29; 60/15; 49/15; 48-51/12; 52/4 • P_{vap} = 13.3(44.5); 5.5(50); 96(100)
- (I) P_{fla} = 32; 42; 46
- (T) Tr: 25-36/37/38; 23/24/25-34-42/43 • TR: 639
- (D) LC50 r: 0.61/?; LC50 m: 0.46/?; LD50 o-r: 100; LD50 o-m: 420

4-Vinylpyridine: C_7H_7N • 105.15 • 100-43-6
- (P) Eb = 62/20; 62-65/15; 58-61/12
- (I) LEL/UEL = 1.3/10.7 • P_{fla} = 48; 51; 61 • AIT = 440
- (T) Tr: 25-36/37/38; 23/25-34-42/43 • TR: 639
- (D) LC50 r: 0.17/?; MLC r: 8.60/2h; LC50 m: 0.38/2h ; LD50 o-r: 100; LD50 o-m: 161

2-Aminopyridine: $C_5H_6N_2$ • 94.13 • 504-29-0
- (P) Eb = 204-211; 105/20
- (I) P_{fla} = 92; 102; 110
- (T) Tr: 23/24/25-36/37/38; 23/24/25-33 • NFPA: 3 • LVE(MVE) (F): − (0.5); TWA (USA): 0.5; MAK (D): 0.5
- (D) LD50 o-r: 200; LD50 o-m: 145; LD50 ip-r: 48; LD50 ip-m: 25; 35; LD50 iv-r: 29; LD50 iv-m: 23; LD50 scu-m: 70

8 • Table of organic compounds

Amines

Name: Empirical formula • Molar mass(g/mol) • Registry number (CAS).
- **(P)** Physical data: ΔH_f^0 (kJ/mol) (physical state) • Eb (°C/mmHg) (if substance volatile) • P_{vap} (mBar) (°C) (if substance volatile).
- **(I)** Inflammability data: LEL/UEL (%) • P_{fla} (°C) • AIT (°C).
- **(T)** Toxicity: Safety Code (Tr) • NFPA • Transport Code (TR) • LVE (MVE) (ppm) F, USA, D • STEL (ppm); IDLH (ppm).
- **(D)** Lethal and toxic doses: LC (mg/l/duration of exposure); LD (mg/kg).

3-Aminopyridine: $C_5H_6N_2$ • 94.13 • 62-08-8
- **(P)** Eb = 248; 250-252; 131/12
- **(I)** P_{fla} = 88
- **(T)** Tr: 23/24/25-33

4-Aminopyridine: $C_5H_6N_2$ • 94.13 • 504-24-5
- **(P)** Eb = 260-270; 273; 180/13
- **(T)** Tr: 28-36/37/38

Quinoline: C_9H_7N • 129.17 • 91-22-5
- **(P)** ΔH_f^0 = +156.19(l) • Eb = 235-238; 163.2/100; 136.7/40; 119.8/20; 103.8/10; 89.6/5; 59.7/1 • P_{vap} = 0.093(20); 1.33(59.7)
- **(I)** LEL/UEL = 1.2/7; 1.2/- • P_{fla} = 92; 99; 101cc • AIT = 480
- **(T)** Tr: 21/22 • NFPA: 2
- **(D)** LD50 o-r: 331; 460; MLD ip-m: 64; MLD scu-m: 200; LD50 cu-ra: 540

Isoquinoline: C_9H_7N • 129.17 • 119-65-3
- **(P)** ΔH_f^0 = +158.57(s) • Eb = 242-243
- **(I)** P_{fla} = 102; 107
- **(T)** Tr: 20/21/22-36/38
- **(D)** LD50 o-r: 360; MLD ip-m: 128; LD50 cu-ra: 590

Quinaldine: $C_{10}H_9N$ • 143.20 • 91-63-4
- **(P)** ΔH_f^0 = +164.43(l) • Eb = 246-248; 105-110/10
- **(I)** P_{fla} = 79; 84cc
- **(T)** Tr: 20/21/22-36 • NFPA: 3
- **(D)** LD50 o-r: 1230; LD50 cu-ra: 1870

Carbazole: $C_{12}H_9N$ • 167.22 • 86-74-8
- **(P)** ΔH_f^0 = +126.78(s) • Eb = 352-355; 200/147 • P_{vap} = 533(323)
- **(I)** P_{fla} = 220
- **(T)** Tr: 22 • NFPA: 1
- **(D)** MLD o-r: 500; LD50 ip-m: 200

Nicotine: $C_{10}H_{14}N_2$ • 162.26 • 22083-74-5 (54-11-5)
- **(P)** ΔH_f^0 = +39.31(l) • Eb = 243-248; 247dec; 123-125/17 • P_{vap} = 0.056(20); 0.15(30); 0.36(50); 1.33(61.8)
- **(I)** LEL/UEL = 0.7/4.0; 0.75/4.0 • P_{fla} = 101 • AIT = 244
- **(T)** Tr: 25-27 • NFPA: 4 • TR: 60 • LVE(MVE) (F): – (0.05); TWA (USA): 0.05; MAK (D): 0.05
- **(D)** LD50 o-r: 50; LD50 o-m: 3; 230 (Merck); LD50 ip-r: 15; LD50 ip-m: 6; 9.5; LD50 ipra: 14; LD50 iv-r: 1; 7; LD50 iv-m: 0.3; LD50 iv-ra: 6; LD50 scu-r: 25; LD50 scu-m: 16; LD50 cu-ra: 50

Hexamethylenetetramine: $C_6H_{12}N_4$ • 140.22 • 100-97-0
- **(P)** ΔH_f^0 = +125.52(s) • Eb = 280; 285(subl; partial dec) • P_{vap} = < 0.0133(20)
- **(I)** P_{fla} = 250 • AIT = 409
- **(T)** Tr: 42/43 • NFPA: 2 • TR: 40
- **(D)** LD50 o-m: 569; MLD o-m: 512; LD50 iv-r: 9200; LD50 iv-m: 18; MLD scu-r: 200; LD50 scu-m: 215

AMINES
• Hydrazines

Methylhydrazine: CH_6N_2 • 46.09 • 60-34-4
- **(P)** ΔH_f^0 = +94.35(g); +53.97(l) • Eb = 87-90 • P_{vap} = 50(20); 66(25); 88(50); 275(55)
- **(I)** LEL/UEL = 2.5/97±2 • P_{fla} = 14; 21Coc; 23; 27; < 27 • AIT = 195
- **(T)** Tr: 24/25-26-40; 45-23/24/25 • TR: 338 • LVE(MVE) (F): – (0.2); TWA (USA): 0.2 (skin) (based on study)
- **(D)** LC50 r: 0.064/4h; LC50 m: 0.107/4h; **LD50** o-r: 32.5; 70.7; LD50 o-m: 29; 33; LD50 ip-r: 21; LD50 ip-m: 15; LD50 iv-r: 17; LD50 iv-m: 33; LD50 iv-ra: 12; LD50 scu-r: 35; LD50 scu-m: 25; LD50 cu-ra: 95

1,1-Dimethylhydrazine: $C_2H_8N_2$ • 60.12 • 57-14-7
- **(P)** ΔH_f^0 = +49.37(l) • Eb = 60-64; 62-64/753 • P_{vap} = 137.3(20); 209.3(25); 250(30); 590(50); 754.5(55)
- **(I)** LEL/UEL = 2/95; 2.4/20 • P_{fla} = ~ –15; –15; –10; 1 • AIT = 247; 249
- **(T)** Tr: 45-23/25-34 • NFPA: 3 • TR: 338 • LVE(MVE) (F): – (0.1); TWA (USA): 0.5; (project: 0.01)
- **(D)** LC50 r: 0.620/4h; LC50 m: 0.423/4h; **LD50** o-r: 122; LD50 o-m: 265; LD50 ip-r: 102; LD50 ip-m: 113; LD50 iv-r: 119; LD50 iv-m: 250; LD50 cu-ra: 1060

1,2-Dimethylhydrazine: $C_2H_8N_2$ • 60.12 • 540-73-8
- **(P)** ΔH_f^0 = +55.65(l) • Eb = 81 • P_{vap} = 75(20); 112(30); 260(50)
- **(I)** P_{fla} = < 23; 23
- **(T)** Tr: 45-23/24/25 • TR: 336
- **(D)** LC50 r: 0.688/4h; **LD50** o-r: 100; 160; LD50 o-m: 36; LD50 ip-r: 29; 163; LD50 iv-r: 176; LD50 ip-m: 35; LD50 iv-m: 29; 100; LD50 scu-r: 220; LD50 scu-m: 24

Phenylhydrazine: $C_6H_8N_2$ • 108.16 • 100-63-0
- **(P)** ΔH_f^0 = +142.39(l); Eb = 238-241; 243.5(dec); 173.5/100; 148.2/40; 131.5/20; 115.8/10; 101.6/5; 71.8/1; 52/0.06 • P_{vap} = 0.13(25); 1.2(30); 4(50); 1.33(71.8); 130(173)
- **(I)** P_{fla} = 88 • AIT = 173; 195
- **(T)** Tr: 23/24/25-36; 23/24/25-36-40 • NFPA: 3 • TR: 60 • LVE(MVE) (F): – (5); TWA (USA): 0.1; MAK (D): 5 • IDLH: 250
- **(D)** LC50 r: 2.61/?; LC50 m: 0.04/?; **LD50** o-r: 188; MLD o-m: 175; LD50 o-ra: 80; MLD ip-m: 170; MLD scu-r: 40; MLD scu-m: 80

AMINES
• Amino-alcohols

Ethanolamine: C_2H_7NO • 61.10 • 141-43-5
- **(P)** Eb = 170-172; 100/50; 70-72/12; 68/10; 69-70/10 • P_{vap} = 0.27; 0.5; 0.64; 0.7; 5.2(20); 8(60); 20(80); 55(100)
- **(I)** LEL/UEL = 2.5/17; 5.5/17 • P_{fla} = 85cc; 91; 93oc; 96PMcc; 104Coc • AIT = 385; 779
- **(T)** Tr: 36/37/38 • NFPA: 2 • TR: 80 • LVE(MVE) (F): – (3); TWA (USA): 3; MAK (D): 3 • STEL: 6; IDLH 1000
- **(D)** LD50 o-r: 1720; 10200 (Merck); LD50 o-m: 700; LD50 o-ra: 1000; LD50 ip-r: 67; 981; LD50 ip-m: 50; LD50 iv-r: 225; 800; LD50 scu-r: 1500; LD50 cu-r: 1500; LD50 cu-ra: 1000

8 • Table of organic compounds — Amines

Name: Empirical formula • Molar mass(g/mol)• Registry number (CAS).
(P) Physical data: ΔH_f^0 (kJ/mol) (physical state) • Eb (°C/mmHg) (if substance volatile) • P_{vap} (mBar) (°C) (if substance volatile).
(I) Inflammability data: LEL/UEL (%) • P_{fla} (°C) • AIT (°C).
(T) Toxicity: Safety Code (Tr) • NFPA • Transport Code (TR) • LVE (MVE) (ppm) F, USA, D • STEL (ppm); IDLH (ppm).
(D) Lethal and toxic doses: LC (mg/l/duration of exposure); LD (mg/kg).

2-Dimethylaminoethanol: $C_4H_{11}NO$ • 89.16 • 108-01-0
(P) Eb = 132-136; 135/758 • P_{vap} = 5.33; 5.6; 6.12; 10.67(20); 133(81.5)
(I) LEL/UEL = 1.9/10; 1.4/12.2 • P_{fla} = 31; 39cc; 41oc • AIT = 220; 225; ~245
(T) Tr: 36/37/38 • TR: 30
(D) LC50 r: 4.5/4h; 5.984/4h; LC50 m: 3.25/?; **LD50** o-r: 2000; 2130; LD50 ip-r: 1080; LD50 ip-m: 234; LD50 scu-m: 961; LD50 cu-ra: 1220; 1370

2-Ethylaminoethanol: $C_4H_{11}NO$ • 89.16 • 110-73-6
(P) Eb = 160-161; 165; 167; 169-170 • P_{vap} = < 1.33(20)
(I) P_{fla} = 71oc; 81
(T) Tr: 22; 22-34
(D) LD50 o-r: 1000; LD50 ip-r: 1170; LD50 cu-ra: 360

2-Diethylaminoethanol: $C_6H_{15}NO$ • 117.22 • 100-37-8
(P) Eb = 158-163.5; 100/80; 100/20; 55/10 • P_{vap} = 1.33; 1.87; 2; 2.5; 28(20); 4.5(30); 14(50)
(I) LEL/UEL = 0.7/10.1; 1.4/11.7; 1.8/28; 6.7/11.7 • P_{fla} = 48; 49Tcc; 50; 52cc; 54Toc; 57; 60oc; 67oc • AIT = 260; 270
(T) Tr: 36/37/38 • NFPA: 3 • TR: 30 • LVE(MVE) (F): – (10); TWA (USA): 10; MAK (D): 10
(D) MLC r: 4.5/4h; LC50 m: 5/?; **LD50** o-r: 1300; 1320; 2460; LD50 ip-r: 1220; LD50 ip-m: 192; LD50 iv-m: 188; LD50 scu-m: 650; 1361; LD50 cu-ra: 1110; 1116; 1260

2-Anilinoethanol: $C_8H_{11}NO$ • 137.20 • 122-98-5
(P) Eb = 268; 285; 209/100; 157-160/17; 147-152/11; 150-152/10; 104/1 • P_{vap} = < 0.013; 0.01
(I) P_{fla} = > 109; 148; 152oc • AIT = 410
(T) Tr: 24-36
(D) LD50 o-r: 2230; LD50 ip-m: 137; MLD ip-m: 176; MLD ip-ra: 44; LD50 cu-ra: 63

1-Amino-2-propanol: C_3H_9NO • 75.13 • 78-96-6
(P) Eb = 158-162 • P_{vap} = 0.8; < 1.33(20); 3.5(40); 36.7(80); 240(120)
(I) P_{fla} = 71oc; 73; 77 • AIT = 335; 410
(T) Tr: 34
(D) LD50 o-r: 1715; 4260; MLD ip-m: 250; LD50 cu-ra: 1640

3-Amino-1-propanol: C_3H_9NO • 75.13 • 156-87-6
(P) Eb = 184-187; 168/500 • P_{vap} = 2.8; 3(60); 16(80)
(I) P_{fla} = > 79Toc; 79; 89; 98; 101
(T) Tr: 21-34
(D) MLD o-r: 2830

1-Dimethylamino-2-propanol: $C_5H_{13}NO$ • 103.17 • 108-16-7
(P) Eb = 121-127 • P_{vap} = 10.7; 17.5; 23.3(20); 464(100)
(I) LEL/UEL = 2.7/11.1 • P_{fla} = 26; 32; 34 • AIT = 225
(T) Tr: 22-34
(D) LD50 o-r: 1890

3-Dimethylamino-1-propanol: $C_5H_{13}NO$ • 103.17 • 3179-63-3
(P) Eb = 161-164
(I) P_{fla} = 36; 41; 58
(T) Tr: 22-37/38

2-Amino-1-butanol: $C_4H_{11}NO$ • 89.16 • 13054-87-0 (96-20-8)
(P) Eb = 172; 176-183; 178; 79-80/10
(I) P_{fla} = 74oc; 84; 91; 95
(T) Tr: 22-34-37
(D) LD50 o-m: 2300; MLD ip-m: 250; LD50 iv-m: 316

Diethanolamine: $C_4H_{11}NO_2$ • 105.16 • 111-42-2
(P) Eb = 268-269dec; 217/150; 189/50; 152/10; 115-120/0.1; 20/0.01 • P_{vap} = < 0.013(20); 6.7(138)
(I) LEL/UEL = 1.6/-; 1.6/9.8; 1.6/10.6 • P_{fla} = 137oc; 152oc; 166Coc; 168PMcc • AIT = 660; 662
(T) Tr: 36/38 • NFPA: 1 • LVE(MVE) (F): – (3); TWA (USA): 3
(D) LD50 o-r: 710; 1410; 12760(Merck); LD50 o-m: 3300; LD50 o-ra: 2200; LD50 ip-r: 120; 2300; LD50 iv-r: 778; LD50 scu-r: 2200; LD50 scu-m: 3553; LD50 cu-ra: 1298; 12200

Triethanolamine: $C_6H_{15}NO_3$ • 149.22 • 102-71-6
(P) Eb = 335(estimee); 355dec; 360; 245/50; 206-207/15; 203/10; 190-193/5 • P_{vap} = < 0.013; 0.013; 0.02(20); 2.7(100); 6.3(120); 30(160); 13.3(205); 200(270)
(I) LEL/UEL = 1.3/8.5 • P_{fla} = 179cc; 184cc; 184oc; 191oc; 204Coc; 208PMcc • AIT = 305; 315
(T) Tr: 36/37/38; 36/38 • NFPA: 1 • TWA (USA): 5 mg/m³
(D) LD50 o-r: 8000; 8680; 11700; LD50 o-m: 5846; 7400; LD50 o-ra: 2200; LD50 ip-m: 1450; LD50 cu-ra: > 18000; > 20000

AMINES
• **Amino phenols**

2-Aminophenol: C_6H_7NO • 109.14 • 95-55-6
(P) Eb = > 170(subl.)
(I) P_{fla} = 168
(T) Tr: 20/21/22 • TR: 60
(D) LD50 o-r: 1300; 3670; LD50 o-m: 1250; MLD ip-r: 300; LD50 ip-m: 200; LD50 scu-r: 37

3-Aminophenol: C_6H_7NO • 109.14 • 591-27-5
(P) Eb = ~250; 164/11
(T) Tr: 20/21/22 • TR: 60
(D) LC50 r: 1.162/?; **LD50** o-r: 924; 1450; LD50 o-m: 401; MLD ip-r: 1000; LD50 ip-m: 150; 225; MLD scu-m: 70

4-Aminophenol: C_6H_7NO • 109.14 • 123-30-8
(P) Eb = 284
(T) Tr: 20/21/22 • TR: 60
(D) LD50 o-r: 375; 2400; LD50 o-m: 420; LD50 o-ra: 10000; MLD ip-m: 100; MLD scu-r: 37; MLD scu-m: 470; LD50 cu-ra: > 10000

8 • Table of organic compounds — Amines

Name: Empirical formula • Molar mass(g/mol)• Registry number (CAS).
(P) Physical data: ΔH_f^0 (kJ/mol) (physical state) • Eb (°C/mmHg) (if substance volatile) • P_{vap} (mBar) (°C) (if substance volatile).
(I) Inflammability data: LEL/UEL (%) • P_{fla} (°C) • AIT (°C).
(T) Toxicity: Safety Code (Tr) • NFPA • Transport Code (TR) • LVE (MVE) (ppm) F, USA, D • STEL (ppm); IDLH (ppm).
(D) Lethal and toxic doses: LC (mg/l/duration of exposure); LD (mg/kg).

8-Quinolinol: C_9H_7NO • 145.17 • 148-24-3
- (P) Eb = ~267; 122/0.1
- (T) Tr: 20/21/22 • NFPA: 2
- (D) LC50 r: > 1.2l/6h; **LD50** o-r: 1200; LD50 o-m: 20000; LD50 ip-m: 43; 48; LD50 scu-m: 84

AMINES
• Amino ethers

2-Methoxyethylamine: C_3H_9NO • 75.13 • 109-85-3
- (P) Eb = 87-90; 90-95
- (I) P_{fla} = 4; 9; 12
- (T) Tr: 34
- (D) LD50 ip-m: 400

o-Anisidine: C_7H_9NO • 123.17 • 90-04-0
- (P) Eb = 225; 90/4 • P_{vap} = < 0.13(30)
- (I) P_{fla} = 98; 100
- (T) Tr: 45-26/27/28-33 • NFPA: 2 • TR: 60
  • LVE(MVE) (F): – (0.1); TWA (USA): 0.1; MAK (D): 0.1
- (D) LD50 o-r: 1150; 2000; LD50 o-m: 1400; LD50 o-ra: 870

m-Anisidine: C_7H_9NO • 123.17 • 536-90-3
- (P) Eb = 251; 132/7; 81-86/2
- (I) P_{fla} = > 109; 126
- (T) Tr: 26/27/28-33

p-Anisidine: C_7H_9NO • 123.17 • 104-94-9
- (P) Eb = 240-243; 246; 115/13
- (I) P_{fla} = 122 • AIT = 514
- (T) Tr: 26/27/28-33; 45-26/27/28-33 • NFPA: 2 • TR: 60
  • LVE(MVE) (F): – (0.1); TWA (USA): 0.1; MAK (D): 0.1
- (D) LD50 o-r: 1320; 1400; LD50 o-m: 1410; LD50 o-ra: 2900; LD50 ip-r: 1400; LD50 ip-m: 806; LD50 cu-r: 3200

3,3'-Dimethoxybenzidine: $C_{14}H_{16}N_2O_2$ • 244.32 • 119-90-4
- (I) P_{fla} = 206
- (T) Tr: 45-22 • TR: 60 • TWA (USA): 0.03 mg/m³
- (D) LD50 o-r: 1920

Morpholine: C_4H_9NO • 87.14 • 110-91-8
- (P) Eb = 127-129; 20/6 • P_{vap} = 8.8; 9.33; 10.7(20); 13.3(23); 13.5(25); 41.3(38); 33(50)
- (I) LEL/UEL = 1.8/10.8; 1.8/15.2 • P_{fla} = 31; 35; 38oc
  • AIT = 255; 310
- (T) Tr: 20/21/22-34 • NFPA: 2 • TR: 30
  • LVE(MVE) (F): 30 (20); TWA (USA): 20; MAK (D): 20
- (D) LC50 r: 28.5l2/8h; LC50 m: 1.32/2h; **LD50** o-r: 1050; 1450; LD50 o-m: 525; LD50 ip-m: 413; LD50 cu-ra: 500

N-Methylmorpholine: $C_5H_{11}NO$ • 101.17 • 109-02-4
- (P) Eb = 115.4; 117/764
- (I) P_{fla} = 14
- (T) Tr: 20/21/22-34

N-ethylmorpholine: $C_6H_{13}NO$ • 115.20 • 100-74-3
- (P) Eb = 136-139 • P_{vap} = 6.65; 8; 20(20); 12(30); 28; 40(50)
- (I) LEL/UEL = 2/10.6; 1.5/- ; 1.9/11.8 • P_{fla} = 15; 27; 30cc; 32oc
  • AIT = 160; 190; 240
- (T) Tr: 34; 20/22-36 • NFPA: 2 • LVE(MVE) (F): – (5); TWA (USA): 5
- (D) LC50 m: 18/2h; **LD50** o-r: 1640; 1780; LD50 o-m: 1200; LD50 iv-m: 180

AMINES
• Halogenated amines

2-Chloroaniline: C_6H_6ClN • 127.58 • 95-51-2
- (P) Eb = 207-210; 94-95/11 • P_{vap} = 0.13; 0.5(20); 0.41(25); 0.36(30); 0.96(38); 1.3(46.3)
- (I) LEL/UEL = 2.4/14.2 • P_{fla} = 97
- (T) Tr: 23/24/25-33 • NFPA: 3
- (D) LD50 o-r: 256

3-Chloroaniline: C_6H_6ClN • 127.58 • 108-42-9
- (P) Eb = 228-230.5; 95-96/11 • P_{vap} = 0.031(20); 0.08(30); 1.33(63.5)
- (I) P_{fla} = 118; 123 • AIT = 705
- (T) Tr: 23/24/25-33 • NFPA: 3
- (D) LC50 m: 0.550/4h; **LD50** o-r: 256; LD50 o-m: 334; LD50 ip-r: 200; LD50 ip-m: 200; LD50 cu-r: 250

4-Chloroaniline: C_6H_6ClN • 127.58 • 106-47-8
- (P) Eb = 229-232; 113-114/11 • P_{vap} = 0.02; 0.065; 0.2(20); 0.15(30); 0.40(38); 1.3(59.3)
- (I) P_{fla} = > 120 • AIT = 685; 200
- (T) Tr: 23/24/25-33; 45-23/24/25-33 • NFPA: 3
- (D) LD50 o-r: 300; 310; LD50 o-m: 100; LD50 ip-r: 420; LD50 ip-m: 200; LD50 cu-r: 3200; LD50 cu-ra: 360

2,3-Dichloroaniline: $C_6H_5Cl_2N$ • 162.02 • 608-27-5
- (P) Eb = 252; 120-124/10
- (I) P_{fla} = 115
- (T) Tr: 23/24/25-33

2,4-Dichloroaniline: $C_6H_5Cl_2N$ • 162.02 • 554-00-7
- (P) Eb = 245
- (I) P_{fla} = 115 • AIT = 211
- (T) Tr: 23/24/25-33
- (D) LD50 o-r: 1600; LD50 o-m: 400; LD50 ip-r: 400; LD50 ip-m: 400

2,5-Dichloroaniline: $C_6H_5Cl_2N$ • 162.02 • 95-82-9
- (P) Eb = 251
- (I) P_{fla} = > 110; 139; 166 • AIT = > 189
- (T) Tr: 23/24/25-33 • NFPA: 3
- (D) LD50 o-r: 1600; 2900; LD50 o-m: 1600; LD50 o-ra: 3750; LD50 ip-r: 400; LD50 ip-m: 400; LD50 iv-m: 56

8 • Table of organic compounds
Nitrated, nitrite, nitrate derivatives

Name: Empirical formula • Molar mass(g/mol)• Registry number (CAS).
(P) Physical data: ΔH_f^0 (kJ/mol) (physical state) • Eb (°C/mmHg) (if substance volatile) • P_{vap} (mBar) (°C) (if substance volatile).
(I) Inflammability data: LEL/UEL (%) • P_{fla} (°C) • AIT (°C).
(T) Toxicity: Safety Code (Tr) • NFPA • Transport Code (TR) • LVE (MVE) (ppm) F, USA, D • STEL (ppm); IDLH (ppm).
(D) Lethal and toxic doses: LC (mg/l/duration of exposure); LD (mg/kg).

2,6-Dichloroaniline: $C_6H_5Cl_2N$ • 162.02 • 608-31-1
- (I) P_{fla} = 118 • AIT = 217
- (T) Tr: 23/24/25-33

3,4-Dichloroaniline: $C_6H_5Cl_2N$ • 162.02 • 95-76-1
- (P) Eb = 272 • P_{vap} = 1.33(80.5)
- (I) LEL/UEL = 2.8/7.2 • P_{fla} = 135; 166 • AIT = 187
- (T) Tr: 23/24/25-33 • TR: 60
- (D) LD50 o-r: 648; LD50 o-m: 740; LD50 ip-r: 280; LD50 ip-m: 310

3,5-Dichloroaniline: $C_6H_5Cl_2N$ • 162.02 • 626-43-7
- (P) Eb = 259-260/741
- (I) P_{fla} = > 109; 133
- (T) Tr: 23/24/25-33

3,3'-Dichlorobenzidine: $C_{12}H_{10}Cl_2N_2$ • 253.14 • 91-94-1
- (P) Eb = dec
- (T) Tr: 45-21-43 • TWA (USA): 0.1 mg/m³

NITRATED, NITRITE, NITRATE DERIVATIVES
Nitrated aliphatic derivatives

Nitromethane: CH_3NO_2 • 61.05 • 75-52-5
- (P) ΔH_f^0 = –74.76(g); –113.09(l) • Eb = 100-102; 101.2; 46.6/100; 27.5/40; 14.1/20; 2.8/10; –7.9/5; –29/1 • P_{vap} = 36.4; 37.1(20); 48.8(25); 150(50)
- (I) LEL/UEL = 4/-; 7.3/-; 7.3/63(33) • P_{fla} = 35cc; 44 • AIT = 415; 418
- (T) Tr: 22 • NFPA: 1 • TR: – • LVE(MVE) (F): – (100); TWA (USA): 100; MAK (D): 100 • STEL: 150
- (D) LD50 o-r: 940; 1210; LD50 o-m: 950; 1440; LD50 ip-m: 110; MLD iv-ra: 750

Nitroethane: $C_2H_5NO_2$ • 75.08 • 79-24-3
- (P) ΔH_f^0 = –102.09(g) • Eb = 112-116 • P_{vap} = 20.8(20); 35(30); 92(50)
- (I) LEL/UEL = 4.0/-; 3.4/-. • P_{fla} = 28; 30; 41oc • AIT = 410; 414
- (T) Tr: 20/22 • NFPA: 1 • TR: 30 • LVE(MVE) (F): – (100); TWA (USA): 100; MAK (D): 100 • STEL: 150
- (D) LD50 o-r: 1100; LD50 o-m: 860; MLD o-ra: 500; LD50 ip-m: 310

1-Nitropropane: $C_3H_7NO_2$ • 89.11 • 108-03-2
- (P) ΔH_f^0 = –125.52(g); –167.53(l) • Eb = 131.6; 130/728; 110/401; 100/400; 70/94; 20/7.5; 10/4 • P_{vap} = 10(20); 17(30); 48(50)
- (I) LEL/UEL = 2.2/-; 2.6/- • P_{fla} = 25; 34Tcc; 34Toc; 49oc • AIT = 421
- (T) Tr: 20/21/22 • NFPA: 1 • TR: 30 • LVE(MVE) (F): – (25); TWA (USA): 25; MAK (D): 25 • STEL: 35
- (D) LC50 r: 11.3/8h; LC50 ra: 18.2/3h; **LD50** o-r: 455; LD50 o-m: 800; MLD o-ra: 250; LD50 ip-m: 250

2-Nitropropane: $C_3H_7NO_2$ • 89.11 • 79-46-9
- (P) Eb = 120.3; 110/564; 90/300; 60/95; 10/7 • P_{vap} = 13.3(15.8); ~17.3(20); 24(25); 76(50)
- (I) LEL/UEL = 2.2/-; 2.6/- • P_{fla} = 24Toc; 26; 28Tcc; 34; 37; 39oc • AIT = 425; 428
- (T) Tr: 45-20/22 • NFPA: 1 • TR: 30 • LVE(MVE) (F): – (10); TWA (USA): 10
- (D) LC50 r: 1.458/6h; LC50 m: 10/2h; LC50 ra: 8.68/5h; **LD50** o-r: 720; MLD o-ra: 500; LD50 ip-m: 75

1-Nitrobutane: $C_4H_9NO_2$ • 103.12 • 627-05-4
- (P) ΔH_f^0 = –144.0(g) • Eb = 148-15
- (I) P_{fla} = 44; 47
- (T) TR: 30

2-Nitrobutane: $C_4H_9NO_2$ • 103.12
- (P) ΔH_f^0 = –163.67(g)

1-Nitropentane: $C_5H_{11}NO_2$ • 117.17 • 628-05-7
- (P) Eb = 70-72/11; 75-76/23
- (I) P_{fla} = 59; 61
- (T) Tr: 36/37/38

1-Nitrohexane: $C_6H_{13}NO_2$ • 131.18 • 646-14-0
- (P) Eb = 78-80/11; 91-92/24
- (I) P_{fla} = 72
- (T) Tr: 36/37/38

Tetranitromethane: CN_4O_8 • 196.05 • 509-14-8
- (P) ΔH_f^0 = +37.24(l) • Eb = 125.7; 40/25.8; 30/14.9; 20/8.4; 13.8/5.7; 0/1.9 • P_{vap} = 11.2(20); 13.3(22.7); 20(30); 57(50); 73(55)
- (I) P_{fla} = > 109
- (T) Tr: 23/24/25-36/37/38 • NFPA: 1 • TR: 559 (not permitted) • LVE(MVE) (F): – (1); TWA (USA): 0.005; MAK (D): 1
- (D) LC50 r: 0.144/4h; LC50 m: 0.433/4h; **LD50** o-r: 130; LD50 o-m: 375; LD50 ip-m: 53; LD50 iv-r: 13; LD50 iv-m: 63

NITRATED, NITRITE, NITRATED DERIVATIVES
• Aliphatic nitrites

Methyl nitrite: CH_3NO_2 • 61.05 • 624-91-9
- (P) ΔH_f^0 = –64.02(g); Eb = –12
- (D) LC50 r: 0.624/4h

Ethyl nitrite: $C_2H_5NO_2$ • 75.08 • 109-95-5
- (P) ΔH_f^0 = –104.18(g) • Eb = 16.4-17
- (I) LEL/UEL = 3.0/50; 3.1/50; 4.1/50 • P_{fla} = –35cc; 14(?) • AIT = Explosive at 90°C
- (T) Tr: 20/21/22 • NFPA: 2

Butyl nitrite: $C_4H_9NO_2$ • 103.14 • 544-16-1
- (P) Eb = 65; 75; 78.2 (slight dec)
- (I) P_{fla} = –13; 10
- (T) Tr: 23/25 • TR: 33
- (D) LC50 r: 1.771/4h; 3.872/1h; LC50 m: 2.392/1h; **LD50** o-r: 83; LD50 o-m: 171; LD50 ip-m: 158; 169

8 • Table of organic compounds

Nitrated, nitrite, nitrate derivatives

Name: Empirical formula • Molar mass(g/mol)• Registry number (CAS).
(P) Physical data: ΔH_f^0 (kJ/mol) (physical state) • Eb (°C/mmHg) (if substance volatile) • P_{vap} (mBar) (°C) (if substance volatile).
(I) Inflammability data: LEL/UEL (%) • P_{fla} (°C) • AIT (°C).
(T) Toxicity: Safety Code (Tr) • NFPA • Transport Code (TR) • LVE (MVE) (ppm) F, USA, D • STEL (ppm); IDLH (ppm).
(D) Lethal and toxic doses: LC (mg/l/duration of exposure); LD (mg/kg).

s-Butyl nitrite: $C_4H_9NO_2$ • 103.14 • 924-43-6
- (P) Eb = 68
- (T) Tr: 20/22 • TR: 33
- (D) LC50 m: 7.39/1h; LD50 o-m: 423; LD50 ip-m: 496

Isobutyl nitrite: $C_4H_9NO_2$ • 103.14 • 542-56-3
- (P) Eb = 66-68
- (I) P_{fla} = –21
- (T) Tr: 20/22 • TR: 33
- (D) LC50 r: 3.277/4h, 4.05/1h; LC50 m: 4.358/1h; LD50 o-r: 410; LD50 o-m: 205; LD50 ip-m: 169

t-Butyl nitrite: $C_4H_9NO_2$ • 103.14 • 540-80-7
- (P) Eb = 61-63; 34/250
- (I) P_{fla} = –13; –10
- (T) Tr: 20/22 • TR: 33
- (D) LC50 r: 45.78/1h; LD50 o-r: 308; LD50 ip-m: 496

Pentyl nitrite: $C_5H_{11}NO_2$ • 117.17 • 463-04-7
- (P) Eb = 96-99; 104.5
- (I) P_{fla} = –40; 10 • AIT = 205; 209; Explosive above 250°C
- (T) Tr: 20/22 • NFPA: 1 • TR: 33

Isopentyl nitrite: $C_5H_{11}NO_2$ • 117.17 • 110-46-3
- (P) Eb = 97-99; 95-100 • P_{vap} = 705(50)
- (I) P_{fla} = –20; 9; < 23 • AIT = 209
- (T) Tr: 20/21/22-36/37/38; 20/22 • TR: 33
- (D) LC50 r: 3.43/4h; 6.05/1h; LC50 m: 6.85/30'; LD50 o-r: 505; LD50 ip-m: 130; LD50 iv-m: 51

NITRATED, NITRITE, NITRATE DERIVATIVES
• Aliphatic nitrates

Methyl nitrate: CH_3NO_3 • 77.05 • 598-58-3
- (P) ΔH_f^0 = –124.68(g); –158.99(l) • Eb = explosive at 65°C
- (D) LC50 r: 4.018/4h; LC50 m: 18.73/4h; LD50 o-r: 344; LD50 o-m: 1820

Ethyl nitrate: $C_2H_5NO_3$ • 91.08 • 625-58-1
- (P) ΔH_f^0 = –153.97(g) • Eb = 88.7; explosive at 85°C
- (I) LEL/UEL = < 3.8/-; 3.8/- • P_{fla} = 10cc

Propyl nitrate: $C_3H_7NO_3$ • 105.11 • 627-13-4
- (P) ΔH_f^0 = –174.05(g) • Eb = 110-111 • P_{vap} = 24(20); 40(30); 100(50)
- (I) LEL/UEL = 2/100 • P_{fla} = 20; 23 • AIT = 175; 177; 190
- (T) NFPA: 2 • LVE(MVE) (F): – (25); TWA (USA): 25; MAK (D): 25 • STEL: 40
- (D) LD50 iv-ra: 200

Isopropyl nitrate: $C_3H_7NO_3$ • 105.11 • 1712-64-7
- (P) Eb = 100-103
- (I) LEL/UEL = -/100 • P_{fla} = 11; 13
- (D) LC50 m: 65/2h

Pentyl nitrate: $C_5H_{11}NO_3$ • 133.17 • 1002-16-0
- (P) Eb = 145
- (I) P_{fla} = 52oc

Isopentyl nitrate: $C_5H_{11}NO_3$ • 133.17 • 543-87-3
- (P) Eb = 147-148; 152-157
- (I) P_{fla} = 52
- (D) LD50 ip-m: 480

Ethyleneglycol dinitrate: $C_2H_4N_2O_6$ • 152.08 • 628-96-6
- (P) Eb = Explosive at 114°C
- (T) Tr: 26/27/28-33 • NFPA: 2 • TWA (USA): 0.05; MAK (D): 0.05
- (D) LD50 o-r: 616; MLD scu-m: 300

Nitroglycerine: $C_3H_5N_3O_9$ • 227.11 • 55-63-0
- (P) ΔH_f^0 = –370.70(l) • P_{vap} = 3.31x10^{-4}; 3.47x10^{-4}(20); 1.33(127)
- (I) AIT = 270
- (T) Tr: 26/27/28-33 • LVE(MVE) (F): – (0.15); TWA (USA): 0.05; MAK (D): 0.05

Pentaerythritol tetranitrate $C_5H_8N_4O_{12}$ • 316.17 • 78-11-5
- (P) ΔH_f^0 = –538.90(s)
- (T) TR: Unauthorised

Cellulose nitrate: – • – • 9004-70-0
- (I) P_{fla} = 13
- (T) NFPA: 2
- (D) LD50 o-r: > 5000; LD50 o-m: > 5000

NITRATED, NITRITE, NITRATE DERIVATIVES
• Nitrated aromatic derivatives

Nitrobenzene: $C_6H_5NO_2$ • 123.12 • 98-95-3
- (P) ΔH_f^0 = +15.90(l) • Eb = 210-211; 83-84/10 • P_{vap} = 0.2(20); 0.4(25); 1.33(44.4); 1.9(50); 66.7(120); 500(185)
- (I) LEL/UEL = 1.8(93)/-; 1.8/40 • P_{fla} = 85; 88cc • AIT = 480; 482
- (T) Tr: 26/27/28-33 • NFPA: 3 • TR: 60 • LVE(MVE) (F): – (1); TWA (USA): 1; MAK (D): 1 • STEL: 2; IDLH: 200
- (D) LD50 o-r: 640; 780; LD50 o-m: 590; MLD o-ra: 700; LD50 ip-r: 640; MLD scu-r: 800; MLD scu-m: 286; LD50 cu-r: 2100; MLD su-m: 480; MLDcu-ra: 600

1,2-Dinitrobenzene: $C_6H_4N_2O_4$ • 168.12 • 528-29-0
- (P) ΔH_f^0 = +8.62(s) • Eb = 302; 319/773 • P_{vap} = 1(100)
- (I) P_{fla} = 150cc
- (T) Tr: 26/27/28-33-40 • NFPA: 3 • TR: 60
  • LVE(MVE) (F): – (0.15); TWA (USA): 0.15; MAK (D): 0.15

1,3-Dinitrobenzene: $C_6H_4N_2O_4$ • 168.12 • 99-65-0
- (P) ΔH_f^0 = –16.90 • Eb = 291; 297; 300-303
- (I) P_{fla} = 150
- (T) Tr: 26/27/28-33-40 • NFPA: 3; 4 • TR: 60
  • LVE(MVE) (F): – (0.15); TWA (USA): 0.15; MAK (D): 0.15
- (D) LD50 o-r: 60; 83; 85; LD50 o-m: 75; MLD o-ra: 400; LD50 ip-r: 28; LD50 cu-ra: 1900

8 • Table of organic compounds

Nitrated, nitrite, nitrate derivatives

Name: Empirical formula • Molar mass(g/mol) • Registry number (CAS).
(P) **Physical data:** ΔH_f^0 (kJ/mol) (physical state) • Eb (°C/mmHg) (if substance volatile) • P_{vap} (mBar) (°C) (if substance volatile).
(I) **Inflammability data:** LEL/UEL (%) • P_{fla} (°C) • AIT (°C).
(T) **Toxicity:** Safety Code (Tr) • NFPA • Transport Code (TR) • LVE (MVE) (ppm) F, USA, D • STEL (ppm); IDLH (ppm).
(D) **Lethal and toxic doses:** LC (mg/l/duration of exposure); LD (mg/kg).

1,4-Dinitrobenzene: $C_6H_4N_2O_4$ • 168.12 • 100-25-4
(P) Eb = 299; 183.4/34
(I) P_{fla} = 150
(T) Tr: 26/27/28-33-40 • NFPA: 3 • TR: 60
 • LVE(MVE) (F): – (0.15); TWA (USA): 0.15; MAK (D): 0.15
(D) LD50 o-cat: 29.4

1,3,5-Trinitrobenzene: $C_6H_3N_3O_6$ • 213.12 • 99-35-4
(P) ΔH_f^0 = –43.51(s) • Eb = dec; 175/1.5
(I) P_{fla} = (Code NFPA: 4)
(T) Tr: 26/27/28-33 • NFPA: 2 • TR: Unauthorised
(D) LD50 o-r: 450; LD50 o-m: 572; LD50 iv-m: 32

2-Nitrotoluene: $C_7H_7NO_2$ • 137.15 • 88-72-2
(P) Eb = 220-222.3; 225; 120/31 • P_{vap} = 0.13(20); 0.27(25); 0.35(30); 0.67(38); 1.33(50)
(I) LEL/UEL = 2.2/- • P_{fla} = 95; 106cc • AIT = 304
(T) Tr: 23/24/25-33; 45-23/24/25-33 • NFPA: 2 • TR: –
 • TWA (USA): 2; MAK (D): 5
(D) LC50 r: 0.79/?; LC50 m: 0.328/?; LD50 o-r: 891; LD50 o-m: 970; LD50 o-ra: 1750

3-Nitrotoluene: $C_7H_7NO_2$ • 137.15 • 99-08-1
(P) Eb = 230-232.6; 231.9; 156.9/100; 130.7/40; 112.8/20; 96/10; 81/5; 50.2/1 • P_{vap} = 0.13(20); 1.33(50.2)
(I) LEL/UEL = 2.2/- • P_{fla} = 102; 106; 112cc
(T) Tr: 23/24/25-33; 22-33 • NFPA: 2 • TR: –
 • LVE(MVE) (F): – (2); TWA (USA): 2; MAK (D): 5
(D) LC50 r: 0.693/?; LC50 m: 0.425/?; LD50 o-r: 1072; LD50 o-m: 330; LD50 o-ra: 1750; 2400

4-Nitrotoluene: $C_7H_7NO_2$ • 137.15 • 99-99-0
(P) Eb = 238.3 • P_{vap} = 0.16(20); 0.27(30); 1(50); 1.33(53.7); 6.7(85); 53.3(136)
(I) LEL/UEL = 1.6/- • P_{fla} = 90; 106cc • AIT = 389
(T) Tr: 23/24/25-33 • NFPA: 2; 3 • TR: – • TWA (USA): 2; MAK (D): 5
(D) LC50 r: 0.975/?; LC50 m: 0.419/?; LD50 o-r: 1960; 2144; LD50 o-m: 1231; LD50 o-ra: 1750; LD50 ip-r: 940

2,4-Dinitrotoluene: $C_7H_6N_2O_4$ • 182.15 • 121-14-2
(P) ΔH_f^0 = –71.55(s) • Eb = 300 (dec) • P_{vap} = ~0.13(50); 1.33(102.7); 133(157.7)
(I) P_{fla} = 155; 207; 212
(T) Tr: 23/24/25-33; 45-23/24/25-33 • NFPA: 3 • TR: 60
 • TWA (USA): 1.5 mg/m³
(D) LC50 r: LD50 o-r: 268; LC50 m: LD50 o-m: 790

2,6-Dinitrotoluene: $C_7H_6N_2O_4$ • 182.15 • 606-20-2
(T) Tr: 23/24/25-33; 45-23/24/25-33 • TR: 60
 • TWA (USA): 1.5 mg/m³
(D) LC50 r: LD50 o-r: 177; LC50 m: LD50 o-m: 621

2,4,6-Trinitrotoluene: $C_7H_5N_3O_6$ • 227.15 • 118-96-7
(P) ΔH_f^0 = –66.94(s) • P_{vap} = 0.065(85)
(I) P_{fla} = Code NFPA: 4 • AIT = 295 (unconfined)
(T) Tr: 23/24/25-33 • NFPA: 2 • Unauthorised
 • LVE(MVE) (F): – (0.05); TWA (USA): 0.05; MAK (D): 0.01

o-Nitrodiphenyl: $C_{12}H_9NO_2$ • 199.22 • 86-00-0
(P) Eb = 325; 330; 1205/30; 70/13; 166/4 • P_{vap} = 2.67(140)
(I) P_{fla} = > 109; 143; 179 • AIT = 180
(T) NFPA: 2 • TR: –
(D) LD50 o-r: 1230; LD50 o-ra: 1580

p-Nitrodiphenyl: $C_{12}H_9NO_2$ • 199.22 • 92-93-3
(P) Eb = 340; 224/30 • P_{vap} = 40(224)
(T) Tr: 45

1-Nitronaphthalene: $C_{10}H_7NO_2$ • 173.18 • 86-57-7
(P) ΔH_f^0 = +42.68(s) • Eb = 304 • P_{vap} = 0.013(35)
(I) P_{fla} = 164cc; 167; 170
(T) Tr: 23/24/24-40; 25-40 • NFPA: 1 • TR: 40
(D) LD50 o-r: 120; LD50 ip-r: 86

2-Nitronaphthalene: $C_{10}H_7NO_2$ • 173.18 • 581-89-5
(P) Eb = 312-313; 165/15
(T) Tr: 45 • TR: 40
(D) LD50 o-r: 4400; LD50 o-ra: 2650; LD50 ip-m: 1300

NITRATED, NITRITE, NITRATE DERIVATIVES
• Nitro alcohols

2-Nitroethanol: $C_2H_5NO_3$ • 91.07 • 625-48-9
(P) ΔH_f^0 = –350.79(l) • Eb = 203-206; 194/765
(I) P_{fla} = > 109
(T) Tr: 36/37/38

2-Nitro-1-propanol: $C_3H_7NO_3$ • 105.10 • 2902-96-7
(P) Eb = 72-74/1; 99/10
(I) P_{fla} = 99; 101

NITRATED, NITRITE, NITRATE DERIVATIVES
• Nitro phenols

2-Nitrophenol: $C_6H_5NO_3$ • 139.12 • 88-75-5
(P) Eb = 214.5-216 • P_{vap} = 1.33(49.3); 1.5(50)
(I) P_{fla} = 102
(T) Tr: 20/21/22-33 • NFPA: 3 • TR: 60
(D) LD50 o-r: 334; 2828 (Merck); LD50 o-m: 1297; 1300; LD50 ip-m: 378; MLD scu-r: 1100; MLD scu-m: 1700; LD50 cu-ra: > 7940

8 • Table of organic compounds

Nitrated, nitrite, nitrate derivatives

Name: Empirical formula • Molar mass(g/mol) • Registry number (CAS).
(P) Physical data: ΔH_f^0 (kJ/mol) (physical state) • Eb (°C/mmHg) (if substance volatile) • P_{vap} (mBar) (°C) (if substance volatile).
(I) Inflammability data: LEL/UEL (%) • P_{fla} (°C) • AIT (°C).
(T) Toxicity: Safety Code (Tr) • NFPA • Transport Code (TR) • LVE (MVE) (ppm) F, USA, D • STEL (ppm); IDLH (ppm).
(D) Lethal and toxic doses: LC (mg/l/duration of exposure); LD (mg/kg).

3-Nitrophenol: $C_6H_5NO_3$ • 139.12 • 554-84-7
- (P) Eb = 194/70
- (T) Tr: 20/21/22-33 • NFPA: 3 • TR: 60
- (D) LD50 o-r: 328; 447; LD50 o-m: 1070; MLD ip-m: 70; MLD scu-r: 500

4-Nitrophenol: $C_6H_5NO_3$ • 139.12 • 100-02-7
- (P) Eb = 279 (with decomposition) • P_{vap} = 0.8(120); 9.2(65)
- (I) P_{fla} = 169 • AIT = 282
- (T) Tr: 20/21/22-33 • NFPA: 3 • TR: 60
- (D) LD50 o-r: 250; 350; 616; LD50 o-m: 380; 467; LD50 ip-m: 75; MLD scu-r: 200

2,4-Dinitrophenol: $C_6H_4N_2O_5$ • 184.12 • 51-28-5
- (P) ΔH_f^0 = −232.63(s)
- (T) Tr: 23/24/25-33 • NFPA: 3 • TR: –
- (D) LD50 o-r: 30; LD50 o-m: 45; LD50 o-ra: 30; LD50 ip-r: 20; LD50 iv-r: 72; LD50 ip-m: 26; LD50 iv-m: 56; LD50 scu-r: 25; LD50 scu-m: 58

2,4,6-Trinitrophenol: $C_6H_3N_3O_7$ • 229.12 • 88-89-1
- (P) ΔH_f^0 = −214.35(s) • Eb = Explosion • P_{vap} = 0.065(85); 1.33(195)
- (I) P_{fla} = 150 • AIT = 300
- (T) Tr: 23/24/25 • NFPA: 2 • TR: –
  • LVE(MVE) (F): (0.1 mg/m³); TWA (USA): 0.1 mg/m³; MAK (D): 0.1 mg/m³
- (D) LD50 o-r: 200; MLD o-ra: 120; LD50 ip-m: 56.3

4,6-Dinitro-o-cresol: $C_7H_6N_2O_5$ • 198.15 • 534-52-1
- (P) P_{vap} = 1.051×10⁻⁴(20)
- (T) Tr: 27/28-33-36-40-44 • NFPA: 3 • TR: 60
  • LVE(MVE) (F): – (0.2 mg/m³); TWA (USA): 0.2 mg/m³; MAK (D): 0.2 mg/m³
- (D) LD50 o-r: 7; 10; LD50 o-m: 75; MLD o-ra: 400; MLD ip-r: 28; LD50 cu-r: 200; LD50 cu-ra: 1900

NITRATED, NITRITE, NITRATE DERIVATIVES
• Nitro ethers

2-Nitroanisole: $C_7H_7NO_3$ • 153.15 • 91-23-6
- (P) Eb = 268-273; 277
- (I) P_{fla} = > 109; > 112; 200 • AIT = 464
- (T) Tr: 45-22; 22 • TR: 60
- (D) LD50 o-r: 740; 1980; LD50 o-m: 1300; 1450

3-Nitroanisole: $C_7H_7NO_3$ • 153.15 • 555-03-3
- (P) Eb = 258; 265; 121-123/8
- (I) P_{fla} = > 109
- (T) Tr: 22 • TR: 60

4-Nitroanisole: $C_7H_7NO_3$ • 153.15 • 100-17-4
- (P) Eb = 260
- (I) P_{fla} = 130
- (T) Tr: 22-33 • TR: 60
- (D) LD50 o-r: 2300; 2600; LD50 o-m: 1710; LD50 ip-r: 1400; LD50 ip-m: 698

NITRATED, NITRITE, NITRATE DERIVATIVES
• Nitrated halogenated derivatives

Chloropicrin: CCl_3NO_2 • 164.37 • 76-06-2
- (P) Eb = 112.3/766; 112/757 • P_{vap} = 25.3(20); 40(34); 53.3(33.8)
- (T) Tr: 22-26-36/37/38 • NFPA: 4 • TR: 66
  • LVE(MVE) (F): – (0.1); TWA (USA): 0.1; MAK (D): 0.1
- (D) LC50 r: 0.094/4h; LC50 m: 0.06/4h; 1.6/10*; LC50 ra: 0.8/20*; LD50 o-r: 250; LD50 ip-m: 25

1,1-Dichloro-1-nitroethane: $C_2H_3Cl_2NO_2$ • 143.96 • 594-72-9
- (P) Eb = 124 • P_{vap} = 17(20); 27(30); 55(50)
- (I) P_{fla} = 76oc
- (T) Tr: 23/24/25 • NFPA: 3 • TR: 60 • LVE(MVE) (F): – (2); TWA (USA): 2; MAK (D): 10
- (D) MLC ra: 0.580/6h; LD50 o-r: 410; MLD o-ra: 150; LD50 ip-m: 240

1-Chloro-1-nitropropane: $C_3H_6ClNO_2$ • 123.54 • 600-25-9
- (P) Eb = 142; 170.6/745
- (I) P_{fla} = 62
- (T) Tr: 20/22 • NFPA: 3
- (D) LC50 m: 66/3h; MLC ra: 2/6h; LD50 o-r: 197; LD50 o-m: 510; MLD o-ra: 50; LD50 scu-m: 165

1-Chloro-2-nitrobenzene: $C_6H_4ClNO_2$ • 157.56 • 88-73-3
- (P) Eb = 245-246 • P_{vap} = 0.053(25); 0.15(37.7)
- (I) LEL/UEL = 1.4/8.7 • P_{fla} = > 109; 123; 127
- (T) Tr: 22; 23/24/25-33-36/37/38 • NFPA: 3 • TR: 60
- (D) LD50 o-r: 268; 288; LD50 o-m: 135; LD50 o-ra: 280; LD50 cu-ra: 400

1-Chloro-3-nitrobenzene: $C_6H_4ClNO_2$ • 157.56 • 121-73-3
- (P) Eb = 236; 117/112
- (I) P_{fla} = 103; 115
- (T) Tr: 22; 23/24/25-33-36/37/38 • NFPA: 3 • TR: 60
- (D) LD50 o-r: 420;470; LD50 o-m: 380

1-Chloro-4-nitrobenzene: $C_6H_4ClNO_2$ • 157.56 • 100-00-5
- (P) Eb = 242 • P_{vap} = 0.019(20); 0.12(25); 0.31(38)
- (I) P_{fla} = > 110; 110; 124; 127
- (T) Tr: 23/24/25-33-40; 23/24/25 • NFPA: 3 • TR: 60
  • TWA (USA): 0.1
- (D) LD50 o-r: 420; LD50 o-m: 440; 650; LD50 ip-r: 420; LD50 cu-ra: 3400

8 • Table of organic compounds

Aldehydes - ketones

Name: Empirical formula • Molar mass(g/mol)• Registry number (CAS).
- **(P)** Physical data: ΔH_f^0 (kJ/mol) (physical state) • Eb (°C/mmHg) (if substance volatile) • P_{vap} (mBar) (°C) (if substance volatile).
- **(I)** Inflammability data: LEL/UEL (%) • P_{fla} (°C) • AIT (°C).
- **(T)** Toxicity: Safety Code (Tr) • NFPA • Transport Code (TR) • LVE (MVE) (ppm) F, USA, D • STEL (ppm); IDLH (ppm).
- **(D)** Lethal and toxic doses: LC (mg/l/duration of exposure); LD (mg/kg).

1-Chloro-2,4-dinitrobenzene: $C_6H_3ClN_2O_4$ • 202.56 • 97-00-7
- **(P)** Eb = 315 • P_{vap} = 53(34)
- **(I)** LEL/UEL = 2.0/22 • P_{fla} = 185; 194 • AIT = 432
- **(T)** Tr: 23/24/25-33 • NFPA: 3
- **(D)** LD50 o-r: 780; 1070; LD50 ip-r: 280; LD50 cu-ra: 130

NITRATED, NITRITE, NITRATE DERIVATIVES
• Nitro amines and other complex nitrogenous substances

2-Nitroaniline: $C_6H_6N_2O_2$ • 138.14 • 88-74-4
- **(P)** ΔH_f^0 = −14.43(s) • Eb = 284.5 (calc, Merck)
  • P_{vap} = 0.0004(20); 0.013(30); 0.11(50); 1; 1.33(104)
- **(I)** P_{fla} = 168oc • AIT = 521
- **(T)** Tr: 23/24/25-33 • NFPA: 3 • TR: −
- **(D)** LD50 o-r: 1600; LD50 o-m: 1070

3-Nitroaniline: $C_6H_6N_2O_2$ • 138.14 • 99-09-2
- **(P)** ΔH_f^0 = −18.66(s) • Eb = 286; 306.4 (calc, Merck)
  • P_{vap} = 0.008(30)1.33(119)
- **(I)** P_{fla} = 196 • AIT = 521
- **(T)** Tr: 23/24/25-33 • NFPA: 3 • TR: −
- **(D)** LD50 o-r: 535; LD50 o-m: 308

4-Nitroaniline: $C_6H_6N_2O_2$ • 138.14 • 100-01-6
- **(P)** ΔH_f^0 = −41.46(s) • Eb = 332 (calc, Merck); 336; 260/100
  • P_{vap} = 1.33(142.4)
- **(I)** P_{fla} = 164; 199; 213
- **(T)** Tr: 23/24/25-33 • NFPA: 3 • TR: − • TWA (USA): 0.5
  • STEL: 1
- **(D)** LD50 o-r: 750; LD50 o-m: 810; MLD ip-r: 600; LD50 ip-m: 250

2,4-Dinitroaniline: $C_6H_5N_3O_4$ • 183.14 • 97-02-9
- **(P)** ΔH_f^0 = −68.2(s) • P_{vap} = 13(20)
- **(I)** P_{fla} = 224; 244
- **(T)** Tr: 26/27/28-33 • NFPA: 3 • TR: 60
- **(D)** LD50 o-m: 370; MLD ip-r: 250; MLD ip-m: 400

Hexogene: $C_3H_6N_6O_6$ • 222.15 • 121-82-4
- **(I)** P_{fla} = Code NFPA 3
- **(T)** NFPA: 2 • TWA (USA): 1.5 mg/m³
- **(D)** LD50 o-r: 100; LD50 o-m: 59; MLD o-ra: 500; MLD ip-r: 10; MLD iv-r: 18; LD50 iv-m: 19

Nitrosodiethylamine: $C_4H_{10}N_2O$ • 102.16 • 55-18-5
- **(P)** Eb = 175-177; 47/5
- **(T)** Tr: 45-23/24/25
- **(D)** LD50 ip-r: 75; LD50 iv-r: 280; LD50 ip-m: 132; LD50 scu-r: 195

ALDEHYDES - KETONES
• Aliphatic aldehydes

Formol (anhydrous): CH_2O • 30.03 • 50-00-0
- **(P)** ΔH_f^0 = −115.9(g) • Eb = −16; −19; −19.5; −33/400; −46/200; −57.3/100; −65/60; −70.6/40; −79.6/20; −88/10
- **(I)** LEL/UEL = 7.0/73.0 • AIT = 430
- **(T)** Tr: 23/24/25-34-40-43 • NFPA: 2 • TR: −
- **(D)** LC50 r: 0.203/?; 0.590/?; LC50 m: 0.400/2h

Formol (aqueous solution):
- **(P)** Eb = 96-101 • P_{vap} = 1.73(20); 70(37)
- **(I)** LEL/UEL = 7.0/73.0. 7.0/73(>70) • P_{fla} = 50 (sol.15%); 56; 60cc; 64; 68; 50-80; 85 (sol.37%) • AIT = 300; 423; 430
- **(T)** Tr: 23/24/25-34-40-43 • NFPA: 2 • TR: −
  • LVE(MVE) (F): 2 (−); TWA (USA): 0.3; 0.75; MAK(D): 0.5 • STEL: 2; 3
- **(D)** LD50 o-r: 100; 800; LD50 o-m: 42; LD50 iv-r: 87; MLD ip-m: 16; LD50 scu-r: 420; LD50 scu-m: 300; MLD scu-m: 240; LD50 cu-ra: 270

Acetaldehyde: C_2H_4O • 44.05 • 75-07-0
- **(P)** ΔH_f^0 = −166.36(g); −192.30(l) • Eb = 20-21; 2.5/100; −26/20
  • P_{vap} = 980; 986; 1000; 1009(20); 1450(30). 2800(50); 3221(55); 10000(100)
- **(I)** LEL/UEL = 4.0/57; 4/60; 4.1/57 • P_{fla} = −50Toc; −40oc; −38Tcc; −27 • AIT = 140; 165; 175; 185; 204; 365
- **(T)** Tr: 36/37-40; 36/37/38-40 • NFPA: 2 • TR: 33 • LVE(MVE) (F): −(100); TWA (USA): 25; MAK(D): 50 • STEL: 150
- **(D)** LC50 r: 2.70; 15.6/4h; 37/30`; LC50 m: 2.70/4h; 30/2h; LD50 o-r: 661; 1930; LD50 o-m: 900; MLD ip-r: 500; LD50 iv-m: 212; LD50 scu-r: 640; LD50 scu-m: 560; LD50 cu-ra: 3540

Glyoxal: $C_2H_2O_2$ • 58.04 • 107-22-2
- **(P)** ΔH_f^0 = −211.96(g) • Eb = 51/776; 50; 104 (solution 40%)
  • P_{vap} = 21.3; 24(20)
- **(I)** P_{fla} = nonexistent (solution 40%); > 115 • AIT = 284
- **(T)** Tr: 36/38 • NFPA: 2 • TR: −;
- **(D)** LD50 o-r: 1100; ~4000; 7070; MLD ip-r: 100; LD50 ip-m: 200

Propionaldehyde: C_3H_6O • 58.08 • 123-38-6
- **(P)** ΔH_f^0 = −192.05(g) • Eb = 46-50; 49; 47/740; 45/687 • P_{vap} = 337; 343(20); 1294(55)
- **(I)** LEL/UEL = 2.3/21; 2.6(31)/ 17(26); 2.9/17; 3.0/16 • P_{fla} = −40; −29cc; −26; −9oc; −7oc; 15 • AIT = 190; 207
- **(T)** Tr: 36/37/38 • NFPA: 2 • TR: 33
- **(D)** LC50 r: 19.319/4h; LC50 m: 21.8/2h; LD50 o-r: 1410; MLD o-m: 800; LD50 scu-r: 820; LD50 scu-m: 680; LD50 cu-ra: 5040

8 • Table of organic compounds

Aldehydes - ketones

Name: Empirical formula • Molar mass(g/mol) • Registry number (CAS).
- (P) **Physical data:** ΔH_f^0 (kJ/mol) (physical state) • Eb (°C/mmHg) (if substance volatile) • P_{vap} (mBar) (°C) (if substance volatile).
- (I) **Inflammability data:** LEL/UEL (%) • P_{fla} (°C) • AIT (°C).
- (T) **Toxicity:** Safety Code (Tr) • NFPA • Transport Code (TR) • LVE (MVE) (ppm) F, USA, D • STEL (ppm); IDLH (ppm).
- (D) **Lethal and toxic doses:** LC (mg/l/duration of exposure); LD (mg/kg).

Acrolein: C_3H_4O • 56.06 • 107-02-8
- (P) $\Delta H_f^0 = -85.77(g); -117.03(l)$ • Eb = 51-53; 52.5; 17.5/200; 2.5/100; -7.5/60; -64.5/1 • P_{vap} = 280; 286; 293; 300(20); 1066(55)
- (I) LEL/UEL = 2.8/31. 3/31 • P_{fla} = −29; −26cc; −18oc • AIT = 219; 235 (unstable); 278
- (T) Tr: 25-26-34 • NFPA: 3 • TR: 336 • LVE(MVE) (F): 0.1 (–); TWA (USA): 0.1; MAK(D): 0.1 • STEL: 0.3; 0.5
- (D) LC50 r: 0.3/30'; 0.151/6h; LC50 m: 0.151/6h; LD50 o-r: 25.9; 46; LD50 o-m: 40; LD50 o-ra: 7; LD50 ip-r: 4; LD50 ip-m: 9; LD50 scu-r: 50; LD50 scu-m: 30; LD50 cu-ra: 200; 562

Butyraldehyde: C_4H_8O • 72.11 • 123-72-8
- (P) $\Delta H_f^0 = -205.02(g)$ • Eb = 68; 74-76; 74.8 • P_{vap} = 113; 120; 122(20)
- (I) LEL/UEL = 1.4/12.5; 2/11.8; 2.4/12.5; 2.5/12.5; 2.5/– • P_{fla} = −11; −7cc; 5 • AIT = 190; 198; 210; 230
- (T) NFPA: 2 • TR: 33
- (D) MLC r: 23/4h; LC50 : 174/30'; LC50 m: 44.61/2h; LD50 o-r: 2490; 5890; LD50 ip-r: 800; LD50 ip-m: 1140; MLD scu-r: 10000; LD50 scu-m: 2700; LD50 cu-ra: 3560

Isobutyraldehyde: C_4H_8O • 72.11 • 78-84-2
- (P) $\Delta H_f^0 = -218.61(g)$ • Eb = 60-64 • P_{vap} = 88(4.4); 153; 187(20); 281(30); 600(50)
- (I) LEL/UEL = 1.6/10.6; 2(32)/ 10(25) • P_{fla} = −40cc; −25oc; 18 • AIT = 195; 223; 250; 254
- (T) Tr: 20/22-36/37/38 • NFPA: 2 • TR: 33
- (D) LC50 r: 23.594/4h; LC50 m: 39.5/2h; LD50 o-r: 960; 2810; 3700; LD50 cu-ra: 7130

Crotonaldehyde: C_4H_6O; • 70.09 • 4170-30-3 • 123-73-9
- (P) Eb = 99-104 • P_{vap} = 40; 43(20); 31(25); 193(55)
- (I) LEL/UEL = 2.1/15.5; 2.15/19.5 • P_{fla} = 8; 13 • AIT = 207; 232
- (T) Tr: 23-36/37/38; 23-36/37/38-40 • NFPA: 3 • TR: 33 • LVE(MVE) (F): (2)
- (D) LC50 r: 0.2/2h; 4/30'; LC50 m: 0.580/2h; 1.51/2h; LD50 o-r: 206; 300; LD50 o-m: 104; 240; LD50 ip-m: 160; LD50 scu-r: 140; LD50 scu-m: 160; LD50 cu-ra: 380

Methacrolein: C_4H_6O • 70.09 • 78-85-3
- (P) Eb = 68-70; 73.5; P_{vap} = 160; 369(20); 927(55)
- (I) P_{fla} = −15; 2oc
- (T) NFPA: 3

Pentanal: $C_5H_{10}O$ • 86.13 • 110-62-3
- (P) $\Delta H_f^0 = -227.2(g)$ • Eb = 100-103 • P_{vap} = 26(20)
- (I) LEL/UEL = ~2.1/-; 1.4/- • P_{fla} = 1; 4; 12cc • AIT = 210; 219
- (T) Tr: 20-36/37/38; 36/37/38 • NFPA: – • TR: 33 • LVE(MVE); (F); (50); TWA (USA): 50
- (D) MLC r: 14.09/4h; LD50 o-r: 3200; 4581; LD50 o-m: 6400; LD50 cu-ra: 4857

2-Methylbutyraldehyde: $C_5H_{10}O$ • 86.13 • 96-17-3
- (P) Eb = 90-92
- (I) P_{fla} = 4
- (T) Tr: 36
- (D) LC50 r: 49.32/4h; LD50 o-r: 6400; LD50 cu-ra: 5730

3-Methylbutyraldehyde: $C_5H_{10}O$ • 86.13 • 590-86-3
- (P) Eb = 91-93 • P_{vap} = 40(20)
- (I) P_{fla} = −5; −1 • AIT = 239
- (T) Tr: 36/37/38; 36/37
- (D) LC50 r: 56.36/4h; 90.86/?; LC50 m: 50.77/?; LD50 o-r: 5600; 8910; LD50 o-m: 4750; LD50 cu-ra: 3180

3-Methylcrotonaldehyde: C_5H_8O • 84.12 • 107-86-8
- (P) Eb = 133-135; 153 • P_{vap} = 9.33(20)
- (I) P_{fla} = 33
- (T) Tr: 23-36/37/38; 36/37/38

Glutaraldehyde(anhydrous): $C_5H_8O_2$ • 100.12 • 111-30-8
- (P) Eb = 188 (dec); 106-108/50; 71-72/10 • P_{vap} = 23(20); 13(70.5)
- (I) P_{fla} = 56
- (T) Tr: 23/25-34-41-43; 33-23/24/25-42/43 • NFPA: 1 • TR: 80 • LVE(MVE) (F): 0.2; TWA (USA): 0.2; MAK(D): 0.2
- (D) LC50 r: 0.48/4h; 4/4h; LD50 o-r: 134; 320; 595; LD50 o-m: 100; LD50 ip-r: 18; LD50 iv-r: 10; 15; LD50 ip-m: 14; LD50 iv-m: 15; LD50 scu-r: 2390; LD50 scu-m: 1430; LD50 cu-ra: 560; 640; > 2000; 2500; 2560

Hexanal: $C_6H_{12}O$ • 100.16 • 66-25-1
- (P) $\Delta H_f^0 = -248.40(g)$ • Eb = 119-124; 128.7; 130-131; 28/12 • P_{vap} = 11.5; 12; 14(20); 13.3(25)
- (I) P_{fla} = 25; 30; 32oc • AIT = 220
- (T) Tr: –; 36/37/38; 37 • NFPA: 2 • TR: 30
- (D) LC50 r: 8.196/4h; LD50 o-r: 4890; LD50 o-m: 8292

2-Methylpentanal: $C_6H_{12}O$ • 100.16 • 123-15-9
- (P) Eb = 115-117
- (I) P_{fla} = 16
- (T) Tr: 36/37/38

2-Ethylbutyraldehyde: $C_6H_{12}O$ • 100.16 • 97-96-1
- (P) Eb = 117-119 • P_{vap} = 18.3(20)
- (I) LEL/UEL = 1.2/7.7 • P_{fla} = 12; 21oc
- (T) Tr: 36/37/38
- (D) MLC r: 32.77/4h; LD50 o-r: 3980; LD50 cu-ra: 5990

Heptanal: $C_7H_{14}O$ • 114.19 • 111-71-7
- (P) $\Delta H_f^0 = -264.01(g)$ • Eb = 152.8; 40-42/11; 42.5/10 • P_{vap} = 2.6(20); 5.6(30); 20(50)
- (I) P_{fla} = 35; 38; 40
- (T) Tr: 36/37/38 • NFPA: 2 • TR: 30
- (D) LD50 o-r: 14000; LD50 o-m: 20000

8 • Table of organic compounds — Aldehydes - ketones

Name: Empirical formula • Molar mass(g/mol)• Registry number (CAS).
- **(P) Physical data:** ΔH_f^0 (kJ/mol) (physical state) • Eb (°C/mmHg) (if substance volatile) • P_{vap} (mBar) (°C) (if substance volatile).
- **(I) Inflammability data:** LEL/UEL (%) • P_{fla} (°C) • AIT (°C).
- **(T) Toxicity:** Safety Code (Tr) • NFPA • Transport Code (TR) • LVE (MVE) (ppm) F, USA, D • STEL (ppm); IDLH (ppm).
- **(D) Lethal and toxic doses:** LC (mg/l/duration of exposure); LD (mg/kg).

Cyclohexanecarboxaldehyde: $C_7H_{12}O$ • 112.17 • 2043-61-0
- (P) Eb = 162-164
- (I) P_{fla} = 41
- (T) Tr: 37

Octanal: $C_8H_{16}O$ • 128.21 • 124-13-0
- (P) ΔH_f^0 = –286.66(g) • Eb = 163.4; 168; 171; 66-68/10; 63/10
  • P_{vap} = 2.67(20); 66(90); 142(110); 278(130)
- (I) P_{fla} = 52cc; 55 • AIT = 210
- (T) Tr: –; 36/37/38 • NFPA: 2 • TR: 30
- (D) LD50 o-r: 5630; LD50 cu-ra: 6350

2-Ethyl-2-hexen-1-al: $C_8H_{14}O$ • 126.20 • 645-62-5 (64344-45-2)
- (P) Eb = 175; 55/13.5; 60/13; 58-60/13 • P_{vap} = 1.33; 13.3(20)
- (I) P_{fla} = 53; 68oc
- (T) Tr: 36/37/38 • NFPA: 2
- (D) LD50 o-r: 3000

Nonanal: $C_9H_{18}O$ • 142.24 • 124-19-6
- (P) ΔH_f^0 = –310.43(g) • Eb = 79-81/12 • P_{vap} = ~0.35(20)
- (I) P_{fla} = 63; 72
- (T) Tr: 36/38

Decanal: $C_{10}H_{20}O$ • 156.27 • 112-31-2
- (P) Eb = 207-209; 93-95/10; 92/10 • P_{vap} = ~0.2(20)
- (I) P_{fla} = 85
- (D) LC50 r: LD50 o-r: 3730; LC50 m: LD50 o-m: < 41750; LD50 cu-ra: 5040

Isodecanal: $C_{10}H_{20}O$ • 156.27 • –
- (P) Eb = 197; 207-209 • P_{vap} = 0.2(20)
- (I) P_{fla} = 85
- (T) Tr: 36/37/38

ALDEHYDES - KETONES
• Aromatic aldehydes and complex groups

Benzaldehyde: C_7H_6O • 106.12 • 100-52-7
- (P) ΔH_f^0 = –40.04(g); –89.24(l); –88.82(l) • Eb = 178-179; 62/10
  • P_{vap} = 0.8; 1; 1.33(20); 1.33(26.2); 1.7(30); 5.33(45); 6.7(50)
- (I) LEL/UEL = 1.4/-; 2.15/13.5 • P_{fla} = 62cc; 64 • AIT = 190; 192
- (T) Tr: 22 • NFPA: 2 • TR: 30
- (D) LC50 r: 47/?; LD50 o-r: 1300; LD50 o-m: 28; LD50 ip-m: 9; MLD scu-r: 5000; LD50 scu-m: 5000

2-Methylbenzaldehyde: C_8H_8O • 120.15 • 529-20-4
- (P) Eb = 196-200; 200-202; 94-96/15; 68-72/6
- (I) P_{fla} = 67; 77
- (T) Tr: 23/24/25
- (D) LD50 o-r: 2250 (o-+m-+p-)

3-Methybenzaldehyde: C_8H_8O • 120.15 • 620-23-5
- (P) Eb = 199; 80-82/11
- (I) P_{fla} = 78; 83

4-Methybenzaldehyde: C_8H_8O • 120.15 • 104-87-0
- (P) Eb = 204-205; 221; 82-85/11; 96/10
- (I) P_{fla} = 76; 79; 85 • AIT =435

2,4-Dimethylbenzaldehyde: $C_9H_{10}O$ • 134.18 • 15764-16-6
- (P) Eb = 215; 102-103/14 • P_{vap} = 1(20)
- (I) P_{fla} = 88

Cinnamaldehyde: C_9H_8O • 132.16 • 104-55-2 (14371-10-9)
- (P) Eb = 246 (light dec); 250-253; 222.4/400; 199.3/200; 177.7/100; 163.7/60; 152.2/40; 135.7/20; 125-128/11; 120/10; 105.8/5; 76.1/1 • P_{vap} = 1.33(20)
- (I) P_{fla} = 49; 71; 120
- (T) Tr: 36/37/38
- (D) LD50 o-r: 2220; LD50 o-m: 2225; LD50 ip-m: 610; LD50 iv-m: 75

3-Hydroxybutyraldehyde: $C_4H_8O_2$ • 88.11 • 107-89-1
- (P) Eb = 83/20; 79/12 (dec at 85)
- (I) P_{fla} = 66oc; 83 • AIT = 245; 250

Salicylaldehyde: $C_7H_6O_2$ • 122.12 • 90-02-8;
- (P) ΔH_f^0 = –279.91(l) • Eb = 196-197 • P_{vap} = < 1.3(20); 1.33(33)
- (I) P_{fla} = 77cc
- (T) Tr: 22-36/37/38
- (D) LD50 o-r: 520; LD50 ip-m: 231; LD50 scu-r: 900; MLD scu-r: 1000; LD50 cu-r: 600; LD50 cu-ra: 3000

2-Anisaldehyde: $C_8H_8O_2$ • 136.15 • 135-02-4
- (P) Eb = 238; 248; 250; 118-121/12; 122/10
- (I) P_{fla} = 118; 120
- (T) NFPA: 2
- (D) LD50 o-r: 2500

3-Anisaldehyde: $C_8H_8O_2$ • 136.15 • 591-31-1
- (P) Eb = 230; 143/50; 100-103/10; 62/1
- (I) P_{fla} = > 109; 110; 121

4-Anisaldehyde: $C_8H_8O_2$ • 136.15 • 123-11-5
- (P) ΔH_f^0 = –267.36(l) • Eb = 247-248; 89-90/1.5
- (I) P_{fla} = 108; 116; 118oc; 121
- (T) Tr: 22
- (D) LD50 o-r: 1510

Furfural: $C_5H_4O_2$ • 96.09 • 98-01-1
- (P) ΔH_f^0 = –200.0(l) • Eb = 161.8; 103/100; 72/25; 67./20; 18.5/1
  • P_{vap} = 1.5; 2.3; 2.67(20); 3.3(25); 10(50); 18(55); 500(140)
- (I) LEL/UEL = 2.1/19.3; 2.1/- • P_{fla} = 60cc; 68oc; 73 • AIT = 316; 392
- (T) Tr: 23/25-40; 23/25; 21-25-36/37/38-43 • NFPA: 1; 2 • TR: 30
  • LVE(MVE) (F): 2 (-); TWA (USA): 2; MAK(D): 5
  • IDLH: 250
- (D) LC50 r: 0.687/4h; MLC m: 1.454/6h; LD50 o-r: 50; 65; 127; LD50 o-m: 400; LD50 ip-r: 20; LD50 ip-m: 102; LD50 iv-m: 152; LD50 scu-r: 148; LD50 scu-m: 119; MLD scu-m: 240; MLDcu-ra: 620

428

8 • Table of organic compounds

Aldehydes - ketones

Name: Empirical formula • Molar mass(g/mol) • Registry number (CAS).
(P) Physical data: ΔH_f^0 (kJ/mol) (physical state) • Eb (°C/mmHg) (if substance volatile) • P_{vap} (mBar) (°C) (if substance volatile).
(I) Inflammability data: LEL/UEL (%) • P_{fla} (°C) • AIT (°C).
(T) Toxicity: Safety Code (Tr) • NFPA • Transport Code (TR) • LVE (MVE) (ppm) F, USA, D • STEL (ppm); IDLH (ppm).
(D) Lethal and toxic doses: LC (mg/l/duration of exposure); LD (mg/kg).

Chloral (anhydrous): C_2HClO • 147.39 • 75-87-6
(P) Eb = 97-98 • P_{vap} = 46.67; 52(20); 84(30); 200(50)
(T) Tr: 25-36/38 • TR: 60
(D) LD50 ip-m: 600

Chloral hydrate: $C_2H_3ClO_2$ • 165.40 • 302-17-0
(P) Eb = 97.5 • P_{vap} = 13(20); 29(30); 100(50)
(T) Tr: 25-36/38 • NFPA: 2 • TR: 60
(D) LD50 o-r: 479; LD50 o-m: 1100; LD50 ip-r: 472; LD50 ip-m: 580; MLD scu-m: 800; LD50 cu-r: 3030

Chloroacetaldehyde (anhydrous): C_2HCl
(P) Eb = 85-86; 157 • P_{vap} = < 0.027(20)
(I) P_{fla} = 29 • AIT =204
(T) LVE(MVE) (F): 1 (-); TWA (USA): 1; MAK(D): 1
(D) LC50 r: 0.650/1h; LD50 o-r: 75; 89; LD50 o-m: 82; LD50 ip-r: 7; LD50 ip-m: 7; LD50 iv-ra: 4640; 5522; LD50 cu-ra: 224; 267

Chloroacetaldehyde hydrate: C_2H_3ClO (40% soln) • 78.50 • 107-20-0
(P) Eb = 90-100 • P_{vap} = 139(25); 133(45)
(I) P_{fla} = 53; 70; 88
(T) Tr: 23/24/25-36/37/38 • NFPA: 3 • TR: 60 • LVE(MVE) (F): 1 (-); TWA (USA): 1; MAK(D): 1

2-Chlorobenzaldehyde: C_7H_5ClO • 140.57 • 89-98-5
(P) ΔH_f^0 = –118.83(l) • Eb = 209-215; 211.9; 84/10 • P_{vap} = 1.7(50); ~13(84)
(I) P_{fla} = 87 • AIT = 396
(T) Tr: 34 • TR: 60
(D) LD50 ip-m: 10

3-Chlorobenzaldehyde: C_7H_5ClO • 140.57 • 587-04-2
(P) ΔH_f^0 = –126.36(l) • Eb = 213-215
(I) P_{fla} = 88
(T) Tr: 36/37/38; TR: 60

4-Chlorobenzaldehyde: C_7H_5ClO • 140.57 • 104-88-1
(P) ΔH_f^0 = –146.86(s) • Eb = 214;79-80/11; 72/3
(I) P_{fla} = 87
(T) Tr: 22-36/37/38; 22-38 • TR: 60
(D) LD50 o-r: 1575; LD50 o-m: 1400

ALDEHYDES - KETONES

• **Aliphatic ketones**

Ketene: C_2H_2O • 42.04 • 463-51-4
(P) ΔH_f^0 = –61.09(g); Eb = –56
(T) NFPA: 3 • TWA (USA): 0.5; MAK(D): 0.5 • STEL: 1.5
(D) MLC r: 0.091/2h; MLC m: 0.040/30'; MLC ra: 0.091/2h; LD50 o-r: 1300

Acetone: C_3H_6O • 58.08 • 67-64-1
(P) ΔH_f^0 = –216.65(g); –247.61(l) • Eb = 56-57 • P_{vap} = 233; 240; 245; 256(20); 533(39.5); 2200(80)
(I) LEL/UEL = 2/13; 2.15/13; 2.2/12.8; 2.2/13; 2.5/13; 2.5/13; 2.6/12.8; 3/13 • P_{fla} = –20Tcc; –18Tcc; –9oc (10% water 27 °C) • AIT = 465; 538; 540; 465-560
(T) NFPA: 1 • TR: 33 • LVE(MVE) (F): –(750); TWA (USA): 750; MAK(D): 1000 • STEL: 1000
(D) LC50 r: 50.1/8h; MLC m: 110/1h; LD50 o-r: 5800; 8432; 9750; LD50 o-m: 3000; LD50 o-ra: 5300; 5340; MLD ip-r: 500; LD50 iv-r: 5500; LD50 ip-m: 1297; MLD iv-m: 4000; LD50 cu-ra: 20000

Butanone: C_4H_8O • 72.11 • 78-93-3
(P) ΔH_f^0 = –235.39(g); –273.17(l) • Eb = 79-80; 25/100; 6/40 • P_{vap} = 32.4(0); 56.6(10); 94.64; 95; 99; 103; 104; 105(20)
(I) LEL/UEL = 1.8/9.5; 1.8/10.1; 1.8/11.5; 1.8/12.5; 2/10 • P_{fla} = –9Tcc; –6oc; –7Tcc; –6Tcc; –4Toc; –3; –1Toc • AIT = 403; 505; 514; 516; 550-615
(T) Tr: 36/37 • NFPA: 1 • TR: 33 • LVE(MVE) (F): –(200); TWA (USA): 200; MAK(D): 200 • STEL: 300
(D) LC50 r: 23.5/8h; 28.343/4h; LC50 m: 40/2h; LD50 o-r: 2737; 4800; 5522; LD50 o-m: 4050; LD50 ip-r: 607; LD50 ip-m: 616; LD50 cu-ra: 6480; 13000

Methylvinylketone: C_4H_6O • 70.09 • 78-94-4
(P) Eb = 79-81.4; 37/145; 34/120; 32-34/60 • P_{vap} = 95(20); 160(30); 335(50); 413(55)
(I) LEL/UEL = 2.14/15.64 • P_{fla} = –7cc; –2 • AIT = 491
(T) Tr: 25-26-36/38; 26/27/28-36/37/38 • NFPA: 2 • TR: 339
(D) LC50 r: 0.007/4h; LC50 m: 0.008/2h; LD50 o-r: 31; 35; LD50 o-m: 33; 35; LD50 ip-m: 76

Butane-2,3-dione: $C_4H_6O_2$ • 86.09 • 431-03-8
(P) ΔH_f^0 = –365.85(l) • Eb = 88-89 • P_{vap} = 69.6(20)
(I) P_{fla} = –26; –10; 3; 7; 27 • AIT = 284; 365
(T) Tr: 36/37/38
(D) LD50 o-r: 1580; LD50 o-m: 250; LD50 ip-r: 400; LD50 ip-m: 249; LD50 cu-ra: < 5000

2-Pentanone: $C_5H_{10}O$ • 86.13 • 107-87-9
(P) ΔH_f^0 = –296.52(l) • Eb = 100-105 • P_{vap} =16; 36; 37(20); 29(30); 92(50)
(I) LEL/UEL = 1.5/8; 1.5/8.2; 1.6/-1.6/7.2; 1.6/8.2; 1.55/8.15 • P_{fla} = 7Tcc; 10; 13; 70(?) • AIT = 449; 452; 505
(T) Tr: 22 • NFPA: 2 • TR: 33 • LVE(MVE) (F): –(200); TWA (USA): 200; MAK(D): 200 • STEL: 250
(D) MLC r: 5.284/4h; LD50 o-r: 1600; 3730; LD50 o-m: 1600; LD50 ip-r: 1250; 800; LD50 ip-m: 1600; LD50 cu-ra: 6500

8 • Table of organic compounds

Aldehydes - ketones

- **Name:** Empirical formula • Molar mass(g/mol)• Registry number (CAS).
- (P) **Physical data:** ΔH_f^0 (kJ/mol) (physical state) • Eb (°C/mmHg) (if substance volatile) • P_{vap} (mBar) (°C) (if substance volatile).
- (I) **Inflammability data:** LEL/UEL (%) • P_{fla} (°C) • AIT (°C).
- (T) **Toxicity:** Safety Code (Tr) • NFPA • Transport Code (TR) • LVE (MVE) (ppm) F, USA, D • STEL (ppm); IDLH (ppm).
- (D) **Lethal and toxic doses:** LC (mg/l/duration of exposure); LD (mg/kg).

3-Pentanone: $C_5H_{10}O$ • 86.13 • 96-22-0
- (P) $\Delta H_f^0 = -258.65(g)$ • Eb = 101-103 • P_{vap} = 13.2(17); 20; 35.8; 37.5(20); 29(30); 92(50)
- (I) LEL/UEL = 1.6/-; 1.6/7.7; 1.6(25)/7.7 • P_{fla} = 5; 13cc; 13oc • AIT = 445; 452
- (T) NFPA: 1 • TR: 33 • LVE(MVE) (F): –(200); TWA (USA): 200
- (D) MLC r: 28.18/4h; LD50 o-r: 2100; 2140; MLD ip-r: 1250; LD50 iv-m: 513; LD50 cu-ra: 20000

Methylisopropylketone: $C_5H_{10}O$ • 86.13 • 563-80-4
- (P) Eb = 93-94 • P_{vap} = 29(20); 53(30); 160(50)
- (I) LEL/UEL = 1.2/8; 1.4/7.5; 1.8/9.0 • P_{fla} = –3; –1; 6 • AIT = 430; 475; 505
- (T) TR: 33 • LVE(MVE) (F): – (200); TWA (USA): 200
- (D) LC50 r: 20.079/4h; LD50 o-r: 148; LD50 o-m: 2572; LD50 ip-r: 800; LD50 ip-m: 200; LD50 cu-ra: 6350

Acetylacetone: $C_5H_8O_2$ • 100.12 • 123-54-6;
- (P) $\Delta H_f^0 = -378.71(g); -423.96(l)$ • Eb = 136-140; 139/746 • P_{vap} = 8; 8.5; 9.3(20)
- (I) LEL/UEL = 1.7/11.4; 2.4/11.6 • P_{fla} = 30; 34cc; 38; 40cc; 41Toc • AIT = 340; 349
- (T) Tr: 22 • NFPA: 2 • TR: 30
- (D) MLC r: 4.163/4h; LD50 o-r: 1000; LD50 o-m: 951; MLD ip-r: 400; LD50 ip-m: 750; LD50 cu-ra: 810; 5000

Pentane-2,3-dione: $C_5H_8O_2$ • 100.12 • 600-14-6
- (P) Eb = 110-112
- (I) P_{fla} = 19
- (D) LD50 o-r: 3000; LD50 cu-ra: > 2500

Cyclopentanone: C_5H_8O • 84.12 • 120-92-3
- (P) $\Delta H_f^0 = -192.59(g); -235.31(l)$ • Eb = 127-131; 23/10 • P_{vap} = 45(50); 95(65)
- (I) P_{fla} = 26; 30cc • AIT = 430
- (T) Tr: 36/38 • NFPA: 2 • TR: 30
- (D) LD50 ip-m: 1950; MLD scu-m: 2600

2-Cyclopenten-1-one: C_5H_6O • 82.10 • 930-30-3
- (P) Eb = 151-154; 64-65/19
- (I) P_{fla} = 42; 63
- (T) Tr: 37

2-Hexanone: $C_6H_{12}O$ • 100.16 • 591-78-6
- (P) Eb = 126-128 • P_{vap} = 3.5; 3.6; 13.3; 14.66(20); 13.3(38.8); 29(50)
- (I) LEL/UEL = 1.22/8.0; 1.3/8.1 • P_{fla} = 19cc; 23; 35oc • AIT = 423; 530; 533
- (T) Tr: 48/23 • NFPA: 2 • TR: – • LVE(MVE) (F): 8 (5); TWA (USA): 5; MAK(D): 5
- (D) LC50 r: 32.77/4h; LD50 o-r: 2590; LD50 o-m: 2430; MLD ip-r: 914; LD50 cu-ra: 4800

3-Hexanone: $C_6H_{12}O$ • 100.16 • 589-38-8
- (P) Eb = 120-124
- (I) LEL/UEL = 1/8 • P_{fla} = 14oc; 34oc
- (T) Tr: –; 21 • NFPA: • TR: 33
- (D) MLC r: 16.386/4h; LD50 o-r: 3360; LD50 cu-ra: 3170

Methylisobutylketone: $C_6H_{12}O$ • 100.16 • 108-10-1
- (P) Eb = 115-118; 30/10 • P_{vap} = 6.7; 8; 20; 20.93; 21.33(20); 13.3(30); 44(50)
- (I) LEL/UEL = 1.35/7.6; 1.4/7.5; 1.2(93)/8(93) • P_{fla} = 13; 15cc; 16Tcc; 17cc; 23cc; 23Toc • AIT = 448; 459
- (T) NFPA: 2 • TR: 33 • LVE(MVE) (F): – (50); TWA (USA): 50; MAK(D): 100 • STEL: 75
- (D) LC50 r: 32.77/4h; LC50 m: 23.3/?; LD50 o-r: 2000; 2080; LD50 o-m: 2671; LD50 ip-r: 400; LD50 ip-m: 268; LD50 cu-ra: > 16000; > 20000

Ethylisopropylketone: $C_6H_{12}O$ • 100.16 • 565-69-5
- (P) Eb = 112-116
- (I) P_{fla} = 11; 13

3,3-Dimethyl-2-butanone: $C_6H_{12}O$ • 100.16 • 75-97-8
- (P) Eb = 103-107
- (I) P_{fla} = 3; 12; 17; 23
- (D) LC50 m: 5.7/?; LD50 o-r: 610; LD50 o-m: 1625; LD50 o-ra: 900

Hexane-2,3-dione: $C_6H_{10}O_2$ • 114.14 • 3848-24-6
- (P) Eb = 128-130
- (I) P_{fla} = 28
- (T) Tr: 36/37/38
- (D) LD50 o-r: > 5000; LD50 cu-ra: > 5000

Hexane-2,5-dione: $C_6H_{10}O_2$ • 114.14 • 110-13-4
- (P) Eb = 185-193; 188; 68-70/11 • P_{vap} = 0.57; 1.73(20)
- (I) P_{fla} = 65; 78cc; 80; 85 • AIT = 490; 493
- (T) Tr: 36/38 • NFPA: 1
- (D) MLC r: 9.336/4h; LD50 o-r: 2076; 2700; LD50 o-m: 2386

Hexane-3,4-dione: $C_6H_{10}O_2$ • 114.14 • 4437-51-8
- (P) Eb = 123-125;129
- (I) P_{fla} = 27; 29; 30; 32

Mesityl oxide: $C_6H_{10}O$ • 98.14 • 141-79-7
- (P) Eb = 128-131; 130; 72.1/100; 26/20;–8.7/1 • P_{vap} = 10; 11.6(20); 13(25); 13.3(26); 48(50)
- (I) LEL/UEL = 1.6/- • P_{fla} = 3; 26; 31cc • AIT = 340; 344
- (T) Tr: 20/21/22 • NFPA: 3 • TR: 30 • LVE(MVE) (F): – (15); TWA (USA): 15; MAK(D): 25 • STEL: 25
- (D) LC50 r: 9/4h; MLC r: 10.2/?; LC50 m: 10/2h; LD50 o-r: 1120; LD50 o-m: 710; LD50 o-ra: 1000; LD50 ip-m: 354; MLD scu-ra: 840; LD50 cu-ra: 5150

8 • Table of organic compounds

Aldehydes - ketones

Name: Empirical formula • Molar mass(g/mol)• Registry number (CAS).
(P) Physical data: ΔH_f^0 (kJ/mol) (physical state) • Eb (°C/mmHg) (if substance volatile) • P_{vap} (mBar) (°C) (if substance volatile).
(I) Inflammability data: LEL/UEL (%) • P_{fla} (°C) • AIT (°C).
(T) Toxicity: Safety Code (Tr) • NFPA • Transport Code (TR) • LVE (MVE) (ppm) F, USA, D • STEL (ppm); IDLH (ppm).
(D) Lethal and toxic doses: LC (mg/l/duration of exposure); LD (mg/kg).

Cyclohexanone: $C_6H_{10}O$ • 98.14 • 108-94-1
- **(P)** $\Delta H_f^0 = -230.12(g)$ • Eb = 154-156; 90.4/100; 77.5/60; 67./40; 52.5/20; 38.7/10; 26.4/5; 1.4/1 • P_{vap} = 2.67; 4.53; 4.7(20); 6.4(25); 13.3(38.7); 180(100)
- **(I)** LEL/UEL = 1.1(100)/9.4; 1.1/8.1; 1.3/9.4 • P_{fla} = 44 (Luchaire); 44Tcc; 44PMcc; 47Toc; 63cc • AIT = 420; 430
- **(T)** Tr: 20-40; 20 • NFPA: 1 • TR: 30 • LVE(MVE) (F): –(25); TWA (USA): 25; MAK(D): 50 • IDLH: 5000
- **(D)** LC50 r: 32.11/4h; LD50 o-r: 1535; 1620; 3460; LD50 o-m: 1400; LD50 o-ra: 1600; LD50 ip-r: 1130; LD50 ip-m: 1230; 1350; LD50 ipra: 1540; LD50 scu-r: 2170; MLD scu-m: 1300; LD50 cu-ra: 948; 1000; > 1000

2-Cyclohexen-1-one: C_6H_8O • 96.13 • 930-68-7
- **(P)** Eb = 168; 171-173; 61-62/10
- **(I)** P_{fla} = 34; 56; 58
- **(T)** Tr: 20/22
- **(D)** LC50 r: 0.983/4h; LD50 o-r: 220; LD50 o-m: 170; LD50 cu-ra: 70

Benzoquinone: $C_6H_4O_2$ • 108.10 • 106-51-4;
- **(P)** $\Delta H_f^0 = -185.48(s)$ • Eb = sublimation • P_{vap} = 0.12(20); 0.133(25)
- **(I)** P_{fla} = 77cc; 293 • AIT = 434
- **(T)** Tr: 23/25-36/37/38 • NFPA: 3 • TR: 60 • LVE(MVE) (F): 0.3 (0.1); TWA (USA): 0.1; MAK(D): 0.1
- **(D)** LD50 o-r: 130; LD50 ip-r: 30; LD50 iv-r: 25; LD50 ip-m: 8.5; 9; LD50 scu-m: 94

2-Heptanone: $C_7H_{14}O$ • 114.19 • 110-12-3
- **(P)** Eb = 144; 149-152; 151.5; 111/21 • P_{vap} = 2.7; 2.85; 3.47(20); 22(50); 61(70)
- **(I)** LEL/UEL = 1.11(65)/7.9(121); 1.0/5.5 • P_{fla} = 38cc; 41; 43oc; 47; 49oc • AIT = 393; 420; 533
- **(T)** Tr: 22 • NFPA: 1 • TR: 30 • LVE(MVE) (F): – (50); TWA (USA): 50
- **(D)** MLC r: 18.681/4h; LD50 o-r: 800; 1670; LD50 o-m: 730; LD50 ip-r: 800; LD50 ip-m: 400; LD50 cu-ra: 12600

3-Heptanone: $C_7H_{14}O$ • 114.19 • 106-35-4
- **(P)** Eb = 145-149 • P_{vap} = 1.5; 5.05(20); 2.4(30); 6.8(50)
- **(I)** P_{fla} = 38; 41; 46oc • AIT = 390
- **(T)** Tr: 20-36 • NFPA: 1 • TR: 30 • LVE(MVE) (F): – (50); TWA (USA): 50
- **(D)** MLC r: 9.34/4h; LD50 o-r: 2760; LD50 cu-ra: > 20000

4-Heptanone: $C_7H_{14}O$ • 114.19 • 123-19-3
- **(P)** Eb = 141-145; 144 • P_{vap} = 1; 6.93; 7(20); 2.3(30); 9.5(50)
- **(I)** P_{fla} = 49cc • AIT = 430
- **(T)** TR: 30 • LVE(MVE) (F): – (50); TWA (USA): 50
- **(D)** LC50 r: 12.563/6h; MLC r: 18.681/4h; LD50 o-r: 3730; LD50 cu-ra: 5660

Diisopropylketone: $C_7H_{14}O$ • 114.19 • 565-80-0
- **(P)** $\Delta H_f^0 = -311.29(g)$ • Eb = 122-125 • P_{vap} = 16(20); 27(31); 133(65); 533(103)
- **(I)** P_{fla} = 9; 15; 19; 24oc • AIT = 478
- **(D)** LC50 r: > 16.12/6h; LD50 o-r: 3536

5-Methyl-2-hexanone: $C_7H_{14}O$ • 114.19 • 110-12-3
- **(P)** Eb = 142-145 • P_{vap} = 5.3; 6(20); 1.33(28)
- **(I)** LEL/UEL = 1.35(93)/8.2(93); 1.2/8.0 • P_{fla} = 17; 35cc; 36Tcc; 40; 43oc • AIT = 191; 424; 455
- **(T)** TR: 30 • LVE(MVE) (F): – (50); TWA (USA): 50
- **(D)** MLC r: 18.681/4h; LD50 o-r: 3200; 4760; LD50 o-m: 2542; 3200; LD50 ip-r: 400; LD50 ip-m: 800; LD50 cu-ra: 10000

Cycloheptanone: $C_7H_{12}O$ • 112.17 • 502-42-1
- **(P)** $\Delta H_f^0 = -299.16(l)$ • Eb = 178-181; 66-70/16
- **(I)** P_{fla} = 55; 57
- **(D)** LD50 ip-m: 750; MLD scu-m: 930

2-Methylcyclohexanone: $C_7H_{12}O$ • 112.17 • 583-60-8
- **(P)** $\Delta H_f^0 = -287.6(l)$ • Eb = 162-166 • P_{vap} = 3.3(20); 4.7(30); 8.7(50)
- **(I)** LEL/UEL = 1.2/- • P_{fla} = 46; 48; 59
- **(T)** Tr: 20 • NFPA: 3 • TR: 30 • LVE(MVE) (F): –(50); TWA (USA): 50; MAK(D): 50 • STEL: 75
- **(D)** LD50 o-r: 1980; 2140; LD50 o-ra: 1000; LD50 ip-m: 200; MLD iv-m: 270; LD50 cu-ra: 1635

3-Methylcyclohexanone: $C_7H_{12}O$ • 112.17 • 591-24-2
- **(P)** Eb = 167-170; 174
- **(I)** P_{fla} = 48; 51
- **(T)** Tr: 20/21/22

4-Methylcyclohexanone: $C_7H_{12}O$ • 112.17 • 589-92-4
- **(P)** Eb = 169-171; 60-62/10
- **(I)** P_{fla} = 40; 48
- **(T)** Tr: 20
- **(D)** LC50 r: LD50 o-r: 800; LC50 m: LD50 o-m: 1600

5-Methyl-3-heptanone: $C_8H_{16}O$ • 128.21 • 541-85-5
- **(P)** Eb = 157-162 • P_{vap} = 2.4; 2.67(20); 2.7(25)
- **(I)** LEL/UEL = < 0.9/- • P_{fla} = 43; 57oc; 59
- **(T)** Tr: 36/37 • NFPA: – • TR: 30 • LVE(MVE) (F): – (25); TWA (USA): 25
- **(D)** LD50 o-r: 3500; LD50 o-m: 3800; LD50 cu-ra: > 16000

Methylcyclohexylketone: $C_8H_{14}O$ • 126.20 • 823-76-7
- **(P)** Eb = 180-183
- **(I)** P_{fla} = 57; 66
- **(T)** Tr: 36/38

8 • Table of organic compounds — Aldehydes - ketones

Name: Empirical formula • Molar mass(g/mol)• Registry number (CAS).
(P) **Physical data:** ΔH_f^0 (kJ/mol) (physical state) • Eb (°C/mmHg) (if substance volatile) • P_{vap} (mBar) (°C) (if substance volatile).
(I) **Inflammability data:** LEL/UEL (%) • P_{fla} (°C) • AIT (°C).
(T) **Toxicity:** Safety Code (Tr) • NFPA • Transport Code (TR) • LVE (MVE) (ppm) F, USA, D • STEL (ppm); IDLH (ppm).
(D) **Lethal and toxic doses:** LC (mg/l/duration of exposure); LD (mg/kg).

Diisobutyketone: $C_9H_{18}O$ • 142.24 • 108-83-8
(P) Eb = 165-170 • P_{vap} = 1.6; 2.27; 2.6; 5(20); 7.5(30); 16(50)
(I) LEL/UEL = 0.8(100)/6.2(100); 0.8/7.1 • P_{fla} = 44; 48cc; 60 • AIT = 395
(T) Tr: 37 • NFPA: 1 • TR: 30 • LVE(MVE) (F): – (25); TWA (USA): 25; MAK(D): 50
(D) MLC r: 11.635/4h; LD50 o-r: 5750; LD50 cu-ra: 16000; 17000

Phorone: $C_9H_{14}O$ • 138.21 • 504-20-1
(P) Eb = 198-199; 88/17 • P_{vap} = 0.51(20); 1.33(42)
(I) P_{fla} = 79; 85oc
(T) NFPA: 2

Isophorone: $C_9H_{14}O$ • 138.21 • 78-59-1
(P) Eb = 211-216; 141/100; 87/10 • P_{vap} = 0.16; 0.33; 1(20); 0.73(30); 1.33(38); 2.7(50); 20(100)
(I) LEL/UEL = 0.8/3.8; 0.84/- • P_{fla} = 84oc; 86oc; 88PMcc; 93oc; 96oc; 104Coc • AIT = 205; 460; 462
(T) Tr: 36/37/38 • NFPA: 2 • TR: 30 • LVE(MVE) (F): 5 (–); TWA (USA): 5; MAK(D): 5 • IDLH: 800
(D) MLC r: 10.40/4h; LD50 o-r: 1870, 2330; 2582; LD50 o-m: 2690; LD50 cu-ra: 1199; 1500

Camphor (natural isomer): $C_{10}H_{16}O$ • 152.24 • 464-49-3
(P) Eb = 204-209 • P_{vap} = 0.52(18.2); 5.2(65.1); 5.33(70); 520(180.3)
(I) LEL/UEL = 0.6/3.5 • P_{fla} = 65 • AIT = 466
(T) Tr: 22/36 • NFPA: 2 • TWA (USA): 2; MAK(D): 2 • STEL: 3
(D) LC50 r: LD50 o-r: ; MLC m: 0.400/3h; LD50 o-m: 1310; MLD o-ra: 2000; MLD ip-r: 900; 3500; LD50 ip-m: 3000; LD 90 iv-m: 525; LD50 scu-r: 70; MLD scu-r: 1700; MLD scu-m: 200; 2200

ALDEHYDES - KETONES
• Aromatic ketones and complex groups

Acetophenone: C_8H_8O • 120.15 • 98-86-2
(P) ΔH_f^0 = –86.86(g); –142.55(l) • Eb = 202; 84/12; 83-85/11 • P_{vap} = 1.33(15); 0.4(20); 0.6(25)
(I) P_{fla} = 76cc; 81; 82Coc; 105cc • AIT = 535; 569; 571
(T) Tr: 22-36 • NFPA: 1 • TR: 30 • TWA (USA): 10
(D) LD50 o-r: 815; 900; LD50 o-m: 740; LD50 ip-m: 200; MLD scum-: 330; LD50 cu-ra: 15900

Anthraquinone: $C_{14}H_8O_2$ • 208.22 • 84-65-1
(P) ΔH_f^0 = –207.53(s) • Eb = 376-377; 379-381 • P_{vap} = 1.33(190)
(I) P_{fla} = 185cc • AIT = 650
(T) Tr: 36/37/38; 43 • NFPA: 1
(D) LC50 r: > 1.3/4h; LD50 o-m: 3500; > 5000; LD50 ip-r: 3500

1-Hydroxy-2-butanone: $C_4H_8O_2$ • 88.11 • 5077-67-8
(P) Eb = 148; 78/60
(I) P_{fla} = 57; 59

3-Hydroxy-2-butanone: $C_4H_8O_2$ • 88.11 • 513-86-0
(P) Eb = 140-148; 147-148
(I) P_{fla} = 41; 47; 51
(T) Tr: 36/38
(D) MLD scu-r: 14000

Diacetone alcohol: $C_6H_{12}O_2$ • 116.16 • 123-42-2
(P) Eb = 167-168; 108.2/100; 72/20; 58.8/10; 22/1 • P_{vap} = 1; 1.07; 1.1; 1.3; 1.47(20); 2.3(25); 26.7(72)
(I) LEL/UEL = 1.4/8.1; 1.8/6.9 • P_{fla} = 8cc (commercial prod) ; 13oc ((Commercial prod.); 23-38(depending on purity); 45; 51cc; 56; 58; 59Scc; 61oc; 64 • AIT = 600; 603; 640
(T) Tr: 36 • NFPA: 1 • TR: 30 • LVE(MVE) (F): –(50); TWA (USA): 50; MAK(D): 50 • IDLH: 2100
(D) LD50 o-r: 4000; LD50 o-m: 3950; LD50 ip-m: 933; LD50 cu-ra: 13000; 13500

Chloroacetone: C_3H_5ClO • 92.53 • 78-95-5
(P) Eb = 119-120; 61/50; 20/12 • P_{vap} = 16; 27(20); 105(50)
(I) P_{fla} = 27; 35; 40 • AIT = 610
(T) Tr: 23/24/25-36/37/38 • TR: 60 • TWA (USA): 1
(D) LC50 r: 1.008/1h; LD50 o-r: 100; LD50 o-m: 127; LD50 ip-r: 80; LD50 ip-m: 92; MLDcu-r: 100; LD50 cu-ra: 141

1,1-Dichloroacetone: $C_3H_4Cl_2O$ • 126.97 • 513-88-2
(P) Eb = 117-120
(I) P_{fla} = 24; 46; 85
(T) Tr: 34
(D) LD50 o-m: 250

3-Chloro-2-butanone: C_4H_7ClO • 106.55 • 4091-39-8
(P) Eb = 114-117 • P_{vap} = 27(20)
(I) LEL/UEL = 2.3/- • P_{fla} = 21; 23; 28; 36 • AIT = 460
(T) Tr: 36/37/38

1-Chloro-3-pentanone: C_5H_9ClO • 120.58 • 32830-97-0
(P) Eb = 68/20
(I) P_{fla} = 51
(T) Tr: 36/37/38

5-Chloro-2-pentanone: C_5H_9ClO • 120.58 • 5891-21-4
(P) Eb = 172-175; 70/27; 71-72/20
(I) P_{fla} = 35; 62; 67

2-Chlorocyclopentanone: C_5H_7ClO • 118.56 • 694-28-0
(P) Eb = 72-74/12
(I) P_{fla} = 77
(T) Tr: 36/37/38

2-Chlorocyclohexanone: C_6H_9ClO • 132.59 • 822-87-7
(P) Eb = 82-83/10
(I) P_{fla} = 82
(D) Tr: 36/37/38

8 • Table of organic compounds

Aldehydes - ketones

Name: Empirical formula • Molar mass(g/mol)• Registry number (CAS).
- (P) **Physical data:** ΔH_f^0 (kJ/mol) (physical state) • Eb (°C/mmHg) (if substance volatile) • P_{vap} (mBar) (°C) (if substance volatile).
- (I) **Inflammability data:** LEL/UEL (%) • P_{fla} (°C) • AIT (°C).
- (T) **Toxicity:** Safety Code (Tr) • NFPA • Transport Code (TR) • LVE (MVE) (ppm) F, USA, D • STEL (ppm); IDLH (ppm).
- (D) **Lethal and toxic doses:** LC (mg/l/duration of exposure); LD (mg/kg).

Tetrachlorobenzoquinone: $C_6Cl_4O_2$ • 245.88 • 118-75-2;
- (P) $\Delta H_f^0 = -288.7(s)$ • Eb = sublimation at 290
- (T) Tr: 36/38
- (D) LD50 o-r: 4000; MLD ip-r: 500

2-Chloroacetophenone: C_8H_7ClO • 154.60 • 2142-68-9
- (P) Eb = 227-230
- (I) P_{fla} = 88; 92
- (T) Tr: 22-36/37/38; 22-36/37
- (D) LD50 o-r: 1820; LD50 o-m: 880

3-Chloroacetophenone: C_8H_7ClO • 154.60 • 99-02-5
- (P) Eb = 227-229
- (I) P_{fla} = 105

4-Chloroacetophenone: C_8H_7ClO • 154.60 • 99-91-2
- (P) Eb = 232; 237-247; 124/30; 124-126/24 • P_{vap} = 0.016(0); 10.7(90)
- (I) P_{fla} = 89; 118
- (T) Tr: 22-36/37/38; 22-36/37

α-Chloroacetophenone: C_8H_7ClO • 154.60 • 532-27-4
Eb = 244-247 • P_{vap} = 0.0053; 0.0072(20); 0.019(30); 0.1(50)
- (I) P_{fla} = 118
- (T) Tr: 23/24/25-36/37/38 • NFPA: 3 • TR: 60 • LVE(MVE) (F): – (0.05) • TWA (USA): 0.05
- (D) MLC r: 0.417/15'; 8.75/1'; MLC m: 0.600/15'; MLC ra: 0.465/20'; LD50 o-r: 127; LD50 o-m: 139; LD50 o-ra: 118; LD50 ip-r: 36; LD50 ip-m: 60; LD50 iv-r: 41; LD50 iv-m: 81; LD50 iv-ra: 30

ALDEHYDES - KETONES

• Acetals

Dimethoxymethane: $C_3H_8O_2$ • 76.10 • 109-87-5
- (P) Eb = 41-43; 46; 41.6; 41.5/754 • P_{vap} = 440(20); 610(30); 1200(50); 1549(55)
- (I) LEL/UEL = 1.6/17.6; 3.6/12.6 • P_{fla} = –18cc; –18Toc • AIT = 235; 237
- (T) Tr: 36/37/38; – • NFPA: 2 • TR: 33 • LVE(MVE) (F): –(1000); TWA (USA): 1000; MAK(D): 1000
- (D) LC50 r: 46.687/4h; LC50 m: 57/7h; LD50 o-r: 5700; LD50 o-ra: 5708; LD50 cu-ra: > 5000

Diethoxymethane: $C_5H_{12}O_2$ • 104.15 • 462-95-3
- (P) Eb = 55; 85-88 • P_{vap} =80(25)
- (I) P_{fla} = –5 • AIT = 174
- (T) Tr: 36/37/38

1,1-Dimethoxyethane: $C_4H_{10}O_2$ • 90.12 • 534-15-6
- (P) Eb = 62-65
- (I) P_{fla} = –17; 1oc; 10; 21cc; ~27; 40
- (T) NFPA: 2 • TR: 33
- (D) LC50 r: 11.06/4h; LD50 o-r: 6500; LD50 o-ra: 4507; LD50 cu-ra: 20000

1,1-Diethoxyethane: $C_6H_{14}O_2$ • 118.18 • 105-57-7
- (P) Eb = 100-103; 102.7; 66.3/200; 68/200; 39.8/60; 32/40; 19.6/20; 8/10; –2.5/5; –23/1 • P_{vap} = 13.3(8); 20.3; 26.7(20); 29(21)
- (I) LEL/UEL 1.65/10.4; 1.6/10.4 • P_{fla} = –21cc; 3; 36cc • AIT = 230
- (T) Tr: 36/38 • NFPA: 2
- (D) LC50 r: 19.33/4h; LD50 o-r: 4570; 4600; LD50 o-m: 3500; LD50 o-ra: 3545; LD50 ip-r: 900; LD50 ip-m: 500

1,1-Dimethoxypropane: $C_5H_{12}O_2$ • 104.15 • 4744-10-9
- (P) Eb = 95
- (I) P_{fla} = –7;< 10; 10; 68

2.2-Dimethoxypropane: $C_5H_{12}O_2$ • 104.15 • 77-76-9
- (P) $\Delta H_f^0 = -455.72(l)$ • Eb = 79-83 • P_{vap} = 80(15.8); 33(26); 133(26.1)
- (I) LEL/UEL = 6(27)/31(58) • P_{fla} = –11; –7
- (D) LD50 ip-m: 125

1,1-Diethoxypropane: $C_7H_{16}O_2$ • 132.20 • 4744-08-5
- (P) $\Delta H_f^0 = -539.02(l)$; Eb = 120-123
- (I) P_{fla} = 12

2,2-Diethoxypropane: $C_7H_{16}O_2$ • 132.20 • 126-84-1
- (P) Eb = 45/60
- (I) P_{fla} = 7

1,3-Dioxolane: $C_3H_6O_2$ • 74.08 • 646-06-0
- (P) $\Delta H_f^0 = -376.52(l)$ • Eb = 74-76; 78 • P_{vap} = 93.3; 133(20)
- (I) P_{fla} = –6; –4; –2; 1; 2oc • AIT = 273
- (T) NFPA: 2 • TR: 33
- (D) LC50 r: 20.65/4h; LC50 m: 10.5/2h; MLC ra: 96.96/4h; LD50 o-r: 3000; LD50 o-m: 3200; MLD ip-r: 500; LD50 cu-ra: 8480

Trioxane: $C_3H_6O_3$ • 90.08 • 110-88-3
- (P) Eb = 112-115 • P_{vap} = 16.2; 17.3(25)
- (I) LEL/UEL = 3.6/28.7 • P_{fla} = 44oc • AIT = 410; 414
- (T) Tr: 22 • NFPA: 2 • TR: –
- (D) LD50 o-r: 800; MLDcu-ra: 10000

Paraldehyde: $C_6H_{12}O_3$ • 132.16 • 123-63-7
- (P) $\Delta H_f^0 = -687.01(l)$ • Eb = 109-122; 123-124; 124.4/752; 128 • P_{vap} = 13.3; 64(20)
- (I) LEL/UEL = 1.3/- • P_{fla} = 17; 24cc; 36oc • AIT = 235; 238
- (T) NFPA: 2 • TR: 30 • LVE(MVE) (F): ; TWA (USA):
- (D) MLC r: 10.81/4h; LD50 o-r: 1530; 1650; LD50 o-m: 2750; LD50 o-ra: 3304; MLD scu-r: 1650; LD50 cu-ra: 14000

8 • Table of organic compounds — Carboxylic acids

Name: Empirical formula • Molar mass(g/mol)• Registry number (CAS).
(P) **Physical data:** ΔH_f^0 (kJ/mol) (physical state) • Eb (°C/mmHg) (if substance volatile) • P_{vap} (mBar) (°C) (if substance volatile).
(I) **Inflammability data:** LEL/UEL (%) • P_{fla} (°C) • AIT (°C).
(T) **Toxicity:** Safety Code (Tr) • NFPA • Transport Code (TR) • LVE (MVE) (ppm) F, USA, D • STEL (ppm); IDLH (ppm).
(D) **Lethal and toxic doses:** LC (mg/l/duration of exposure); LD (mg/kg).

CARBOXYLIC ACIDS
• **Aliphatic saturated acids**

Formic acid : CH_2O_2 • 46.03 • 64-18-6
(P) ΔH_f^0 = –378.61(g); –424.72(l) • Eb = 100-101; 44/100; 2.1/10; 103(94%) • P_{vap} = 28; 31(85-95%); 36(94%); 43(90%) ; 44.6; 60(20); 53(24); 500(80);
(I) LEL/UEL = 18/51; 14.3/34 (90%); 18/57(90%); 14.9/47.6(85%);13.9/38.1(94%); 15/47; 10/45.5 • P_{fla} = 46; 50(90%); 54 (94%); 59; 64cc(90%); 69oc • AIT = 434(90%); 490(90%);500; 505(94%); 520; 539; 601
(T) Tr: 34(25-90%); 35(90-100%) • NFPA: 3 • TR: 80
• LVE(MVE) (F): 5 (-); TWA (USA): 5; MAK(D): 5
• STEL: 10; IDLH: 100
(D) LC50 r: 7.4/4h; 15/30'; LC50 m: 6.2/15'; LD50 o-r: 1100; ~1200; 1210-1830; LD50 o-m: 700; 1100; LD50 ip-m: 940; MLD iv-m: 147

Acetic acid : $C_2H_4O_2$ • 60.05 • 64-19-7
(P) ΔH_f^0 = –434.84(g); –484.13(l) • Eb = 116-118; 80/202; 47/50; 18/10; 0/4 • P_{vap} = 14.7; 15.2, 15.3; 15.6; 16(20); 500(100)
(I) LEL/UEL = 4(59)/16(92); 4/16; 4/17; 5.4/16(100); 5.3/16.6; 4.0/17(100) • P_{fla} = 40cc; 43cc; 43oc; 43Tcc; 44Toc
• AIT = 426; 465; 485
(T) Tr: 34(25-90%); 35(90-100%) • NFPA: 2 • TR: 80(25-90%); 83(90-100%) • LVE(MVE) (F): 10 (-); TWA (USA): 10; MAK(D): 10 • STEL: 15; IDLH: 1000
(D) LC50 r: 39.296/4h; LC50 m: 13.80/2h; LD50 o-r: 3310; 3530; MLD o-ra: 1200; LD50 iv-m: 525; LD50 cu-ra: 1060; 1100

Propionic acid : $C_3H_6O_2$ • 74.08 • 79-09-4
(P) ΔH_f^0 = –510.74(l) • Eb = 141-142; 122/400; 85.8/100; 71/50; 41/10; 4.6/1 • P_{vap} = 2.9; 3.2; 3.9; 5; 77(20); 13.3(39.7)
(I) LEL/UEL = 2.1/12; 2.9/12.1; 2.9/14.8 • P_{fla} = 49; 51cc; 54Tcc; 57Toc; 58oc • AIT = 465; 485; 513
(T) Tr: 34(≥ 25%); 36/37/38(10-25%) • NFPA: 2 • TR: 80
• LVE(MVE) (F): –(10); TWA (USA): 10; MAK(D): 10
• STEL: 15; 30
(D) LD50 o-r: 2600; 3500;3500-4200; 4290; LD50 ip-r: 200; LD50 iv-m: 625; LD50 cu-ra: 500

Butyric acid : $C_4H_8O_2$ • 88.11 • 107-92-6
(P) ΔH_f^0 = –533.84(l) • Eb = 162-164; 145/400; 108/100; 75/25; 88/40; 61/10 • P_{vap} = 0.57; 0.67; 1.12(20)
(I) LEL/UEL = 2.0/10.0; 2.2/13.4 • P_{fla} = 67; 71cc; 75oc; 77cc; 77oc • AIT = 425; 440; 452
(T) Tr: 34 • NFPA: 2 • TR: 80
(D) LC50 r: > 5.1/4h; LD50 o-r: 2000; 2940; 8790; 8800; MLD o-m: 500; LD50 ip-m: 3180; LD50 iv-m: 800; LD50 scu-m: 3180; LD50 cu-ra: 530; 2100

Isobutyric acid: $C_4H_8O_2$ • 88.11 • 79-31-2
(P) Eb = 151-155; 98/100; 51/10 • P_{vap} = 1.33(14.7); 2; 12(20)
(I) LEL/UEL = 2/10 • P_{fla} = 55Toc; 62; 74; 77Toc • AIT = 420; 439; 502
(T) Tr: 21/22 • NFPA: 1 • TR: 80
(D) LD50 o-r: 280; LD50 cu-ra: 500

Pentanoic acid: $C_5H_{10}O_2$ • 102.13 • 109-52-4
(P) ΔH_f^0 = –559.44(l) • Eb = 184-187; 112/50; 96/23; 87/15; 110-111/10; 80/10 • P_{vap} = 0.11; 0.2(20)
(I) LEL/UEL = 1.6/7.6 • P_{fla} = 86; 88; 96oc • AIT = 374
(T) Tr: 34 • NFPA: 1 • TR: –
(D) LC50 m: 4.1/2h; LD50 o-m: 600; LD50 ip-m: 3590; LD50 iv-m: 1290 ± 53; LD50 scu-m: 3590

2-Methylbutyric acid: $C_5H_{10}O_2$ • 102.13 • 600-07-7
(P) Eb = 173-176 • P_{vap} = 0.67(20)
(I) P_{fla} = 73; 83
(T) Tr: 34

3-Methylbutyric acid: $C_5H_{10}O_2$ • 102.13 • 503-74-2
(P) Eb = 173-177 • P_{vap} = 0.19; 0.51(20)
(I) LEL/UEL = 1.5/- • P_{fla} = 70; 74 • AIT = 439
(T) Tr: 22-24-34
(D) LD50 o-r: 3560; LD50 o-m: 2580; LD50 iv-m: 1120 ±30; LD50 cu-ra: > 3200

2,2-Dimethylpropionic acid: $C_5H_{10}O_2$ • 102.13 • 75-98-9
(P) Eb = 164
(D) LD50 o-r: 900; LD50 cu-r: 1900

Hexanoic acid: $C_6H_{12}O_2$ • 116.16 • 142-62-1
(P) ΔH_f^0 = –584.55(l) • Eb = 202-205; 146/100; 100/1
• P_{vap} = < 0.01; 0.02; 0.24(20)
(I) P_{fla} = 102cc; 102Coc; 104 • AIT = 380
(T) Tr: 21-34 • NFPA: 2 • TR: 80
(D) LC50 m: 4.1/2h; LD50 o-r: 2050; 2850; 3000; LD50 o-m: 5000; LD50 ip-m: 3180; LD50 iv-m: 1725; LD50 scu-m: 3180; LD50 cu-ra: 630

2-Methylpentanoic acid: $C_6H_{12}O_2$ • 116.16 • 97-61-0
(P) Eb = 196-197; 122/50; 92/10 • P_{vap} = 0.027; 0.27(20)
(I) P_{fla} = 91; 108oc
(D) LD50 o-r: 2040; LD50 cu-ra: 2500

3-Methylpentanoic acid: $C_6H_{12}O_2$ • 116.16 • 105-43-1
(P) Eb = 196-198
(I) P_{fla} = 85
(T) Tr: 34

4-Methylpentanoic acid: $C_6H_{12}O_2$ • 116.16 • 646-07-1
(P) Eb = 199-201; 208; 110-111/25
(I) P_{fla} = 97
(T) Tr: 21-34
(D) LC50 r: LD50 o-r: 2050; LD50 cu-ra: 1050

8 • Table of organic compounds
Carboxylic acids

Name: Empirical formula • Molar mass(g/mol)• Registry number (CAS).
(P) Physical data: ΔH_f^0 (kJ/mol) (physical state) • Eb (°C/mmHg) (if substance volatile) • P_{vap} (mBar) (°C) (if substance volatile).
(I) Inflammability data: LEL/UEL (%) • P_{fla} (°C) • AIT (°C).
(T) Toxicity: Safety Code (Tr) • NFPA • Transport Code (TR) • LVE (MVE) (ppm) F, USA, D • STEL (ppm); IDLH (ppm).
(D) Lethal and toxic doses: LC (mg/l/duration of exposure); LD (mg/kg).

2,2-Dimethylbutyric acid: $C_6H_{12}O_2$ • 116.16 • 595-37-9
(P) Eb = 187; 80/11; 94/5
(I) P_{fla} = 79; 89

3,3-Dimethylbutyric acid: $C_6H_{12}O_2$ • 116.16 • 1070-83-3
(P) Eb = 185-190; 183-184/750; 94-96/26
(I) P_{fla} = 83; 89 • AIT = 495
(T) Tr: 34

2-Ethylbutyric acid: $C_6H_{12}O_2$ • 116.16 • 88-09-5
(P) Eb = 193-196; 120/50; 99-101/18; 89/10 • P_{vap} = 13.3(15.3); 0.053; 0.11(20)
(I) P_{fla} = 26cc(?); 87; 99 • AIT = 463
(T) Tr: 36/37/38 • NFPA: 2
(D) LD50 o-r: 2200; LD50 cu-ra: 520

Heptanoic acid: $C_7H_{14}O_2$ • 130.19 • 11-14-8
(P) ΔH_f^0 = –609.42(l) • Eb = 221.9; 223; 187.5/256; 188/250; 150.8/64; 112-114/10 • P_{vap} = < 0.13(20); 0.1(38)
(I) LEL/UEL = 1.1/10.1 • P_{fla} = 110oc; > 112
(T) Tr: 34 • TR: 80
(D) LC50 r: LD50 o-r: 7000; LD50 o-m: 6400; LD50 iv-m: 1200 ± 56

Octanoic acid: $C_8H_{16}O_2$ • 144.22 • 124-07-2
(P) ΔH_f^0 = –635.68(l) • Eb = 220; 237-240; 147.9/30; 124/10 • P_{vap} = 0.05; 0.067(20); 1.3(78); 1.33(92); 13.3(124)
(I) P_{fla} = 110cc; > 110; 132oc; 135PMcc • AIT = 440
(T) Tr: 36/37/38; 34 • NFPA: – • TR: 80
(D) LC50 r: LD50 o-r: 10080; LD50 iv-m: 600; LD50 cu-ra: 5000; > 5000

2-Ethylhexanoic acid: $C_8H_{16}O_2$ • 144.22 • 149-57-5
(P) Eb = 222; 226-229; 167/100; 149/50; 113-118/11; 114/10 • P_{vap} = < 0.013; 0.04; 0.11(20); 13.3(115)
(I) LEL/UEL = 0.93/6.71; 1/6.2; 1.04(135)/8.64(188) • P_{fla} = 98; 114cc; 124Toc; 127oc • AIT = 310; 370
(T) Tr: 34-63
(D) LD50 o-r: 3000; 3640; LD50 cu-ra: 1260; > 2000

Nonanoic acid: $C_9H_{18}O_2$ • 158.24 • 112-05-0
(P) Eb = 252-253.4; 255.6; 268-269; 143-145/14; 132-133/6.3 • P_{vap} = < 0.13; 0.04(20); 1.33(108)
(I) LEL/UEL = 0.8/9; 1.2/- • P_{fla} = 99; 129Tcc • AIT = 405
(T) Tr: 34-63; 34 • TR: 80
(D) LC50 r: 0.46-3.8/4h; MLD o-r: 3200; > 9000; LD50 o-m: 15000; LD50 iv-m: 224 ± 4.6

Decanoic acid: $C_{10}H_{20}O_2$ • 172.27 • 334-48-5
(P) Eb = 268-270; 172.6/30; 123-124/0.1 • P_{vap} = < 1.33(20); 20(160)
(I) P_{fla} = > 109; 152PMcc
(T) Tr: 36/38
(D) LD50 o-r: 3730; LD50 iv-m: 129 ± 5.4

Dodecanoic acid: $C_{12}H_{24}O_2$ • 200.32.•143-07-7
(P) ΔH_f –774.63(s) • Eb = 299; 225/100; 160-165/20 • P_{vap} = 1.33(121); 66.7(210)
(I) P_{fla} = > 109; 165; 168PMcc
(T) Tr: 36/37/38
(D) LD50 o-r: 12000; LD50 iv-m: 131 ± 5.7

CARBOXYLIC ACIDS
• **Aliphatic unsaturated acids and aromatic acids**

Acrylic acid : $C_3H_4O_2$ • 72.06 • 79-10-7
(P) ΔH_f^0 = –336.23(g); –384.09(l) • Eb = 139-142; 122/400; 103.3/200; 86.1/100; 69/50; 66.2/40; 39/10; 27.3/5 • P_{vap} = 4; 4.13; 5.33(20); 53.3(39.9); 53.3(60)
(I) LEL/UEL = 2/13.7; 2.0/8.0; 2.4/8; 5.3/26 • P_{fla} = 46; 50Tcc; 52oc; 54Toc; 62; 68Coc • AIT = 374; 395; 412; 429; 541
(T) Tr: 34 • NFPA: 3 • TR: 89 • LVE(MVE) (F): 10; TWA (USA): 2 (skin)
(D) LC50 r: 8.34/4h; 11.788/4h; LC50 m: 5.3/2h; LD50 o-r: 34; 235; 340; 353; 355; 2590 (Merck); LD50 o-m: 2400; LD50 ip-r: 22; LD50 ip-m: 144; LD50 scu-m: 1590; LD50 cu-ra: 280; 298

Crotonic acid (E): $C_4H_6O_2$ • 86.09 • 107-93-7 (3724-65-0)
(P) ΔH_f^0 = E: –430.53(s); Z: –347.27(l) • Eb = 180-182; 185; 189; 165.5/400; 146/200; 128/100; 116.7/60; 107./40; 93/20; 80/10 • P_{vap} = 0.25(20)
(I) LEL/UEL = 2.2(135)/15.1(171) • P_{fla} = 88Coc • AIT = 395; 490
(T) Tr: 22-34; 21/22-36/37/38 • NFPA: 3 • TR: –
(D) LD50 o-r: 1000; LD50 o-m: 4800; LD50 ip-r: 100; LD50 ip-m: 25; LD50 scu-m: 3590; LD50 cu-ra: 600

Methacrylic acid: $C_4H_6O_2$ • 86.09 • 79-41-4
(P) Eb = 158-161; 163; 142/400; 90/50; 81/30; 63/12; 60/12 • P_{vap} = 0.81; 0.87; 1; 1.33(20); 1.33(25.5)
(I) LEL/UEL = 1.6(65)/8.1(96) • P_{fla} = 70; 73cc; 77Coc • AIT = 365; 399
(T) Tr: 34 • NFPA: 3 • TR: 89 • LVE(MVE) (F): – (20); TWA (USA): 20
(D) LD50 o-r: 60; 1060; 1600-2200; LD50 o-m: 1250; LD50 o-ra: 1200; LD50 ip-m: 48; LD50 cu-ra: 500

2-Methylcrotonic acid(E): $C_5H_8O_2$ • 100.12 • 80-59-1
(P) Eb = 198-199; 95-96/12; 95-96/11.5
(T) Tr: 36/37/38

3-Methylcrotonic acid: $C_5H_8O_2$ • 100.12 • 541-47-9
(P) Eb = 194-195; 199
(I) P_{fla} = 93
(T) Tr: 36/37/38
(D) LD50 o-r: 3560; LD50 o-m: 2580

435

8 • Table of organic compounds — Carboxylic acids

Name: Empirical formula • Molar mass(g/mol) • Registry number (CAS).
(P) Physical data: ΔH_f^0 (kJ/mol) (physical state) • Eb (°C/mmHg) (if substance volatile) • P_{vap} (mBar) (°C) (if substance volatile).
(I) Inflammability data: LEL/UEL (%) • P_{fla} (°C) • AIT (°C).
(T) Toxicity: Safety Code (Tr) • NFPA • Transport Code (TR) • LVE (MVE) (ppm) F, USA, D • STEL (ppm); IDLH (ppm).
(D) Lethal and toxic doses: LC (mg/l/duration of exposure); LD (mg/kg).

Oleic acid: $C_{18}H_{34}O_2$ • 282.47 • 112-80-1
- (P) $\Delta H_f^0 = -783.24(s)$ • Eb = 360 (dec); 286/100; 228-229/15; 225/10; 194-195/1.2; 194-195/1 • $P_{vap} = 1.33(176.5)$
- (I) $P_{fla} = > 109$; 189cc • AIT = 363
- (T) NFPA: 0
- (D) LD50 o-r: 74000; LD50 iv-r: 2; LD50 ip-m: 282; LD50 iv-m: 23018

Benzoic acid: $C_7H_6O_2$ • 122.12 • 65-85-0
- (P) $\Delta H_f^0 = -385.05(s)$ • Eb = 249; 227/400; 205.8/200; 186.2/100; 172.8/60; 162.6/40; 146.7/20; 132.1/10; 132.5/10 • $P_{vap} = 0.1(60)$; 1.33(96); 8.5(122); 13.3(132)
- (I) $P_{fla} = 121cc$; 121Coc; 121-131 • AIT = 571; 574
- (T) Tr: 22-36/37/38-42/43; - • NFPA: 1; 2
- (D) LD50 o-r: 1700; 2350; 2530; LD50 o-m: 1940; 2370 MLD o-ra: 2000; LD50 ip-r: 1600; LD50 ip-m: 1460; MLD scu-m: 2000; LD50 cu-ra: > 10000

Cinnamic acid(E): $C_9H_8O_2$ • 148.16 • 140-10-3
- (P) $\Delta H_f^0 = E: -336.94(s)$; Z: $-301.25(s)$ • Eb = 300 (dec: $-CO_2$)
- (I) $P_{fla} = > 100$
- (T) Tr: 36/37/38
- (D) LD50 o-r: 2500; LD50 o-m: 5000; LD50 ip-r: 1600; LD50 ip-m: 160

CARBOXYLIC ACIDS
• Diacids

Oxalic acid: $C_2H_2O_4$ • 0.04 • 144-62-7 • 6153-56-6
- (P) $\Delta H_f^0 = -827.60(s)$; $-1427.43(s, 2H_2O)$ • Eb = 149-160; 157 (subl)
- (T) Tr: 21/22 • LVE(MVE) (F): – (1 mg/m³); TWA (USA): 1 mg/m³ • STEL: 2 mg/m³ ; IDLH: 500 mg/m³
- (D) LD50 o-r: 375; 7500 (solution à 5%); LD50 ip-m: 270 ; LD50 cu-ra: 20000

Malonic acid: $C_3H_4O_4$ • 104.06 • 141-82-2
- (P) $\Delta H_f^0 = -891.02(s)$ • Eb = dec at ~153
- (I) $P_{fla} = \sim 172$
- (T) Tr: 22-37/38
- (D) LC50 r: > 8.9/1h; LD50 o-r: 1310; LD50 o-m: 4000 LD50 ip-r: 1500; LD50 ip-m: 300

Succinic acid: $C_4H_6O_4$ • 118.09 • 110-15-6
- (P) $\Delta H_f^0 = -940.52(s)$ • Eb = 235 (dec → anhydride)
- (T) Tr: 36/37/38 • NFPA: 2 ;
- (D) LD50 o-r: 2260; LD50 ip-m: 2702

Glutaric acid: $C_5H_8O_4$ • 132.12 • 110-94-1
- (P) $\Delta H_f^0 = -959.98(s)$ • Eb = 303-304 (dec); 200/20; 195-198/10 • $P_{vap} = 26.7(20)$ (?)
- (T) Tr: 38
- (D) LD50 ip-m: 6000

Adipic acid: $C_6H_{10}O_4$ • 146.14 • 124-04-9
- (P) $\Delta H_f^0 = -985.37(l)$; $-994.12(s)$ • Eb = 337-338; 265/100; 240.5/40; 222/20; 216/15; ~205/10; 205.5/10; 191/5; 159.5/1 • $P_{vap} = 0.00056(85)$; 1.33(159.5)
- (I) $P_{fla} = 191$; 196cc; 198cc • AIT = 405; 419; 422
- (T) Tr: 36 • NFPA: 1 • TR: – • TWA (USA): 5 mg/m³ (proposed)
- (D) LD50 o-r: 3600; ~5700; > 11000; LD50 o-m: 1900; LD50 ip-r: 275; LD50 ip-m: 275; LD50 iv-m: 680; LD50 cu-ra: > 11000

Pimelic acid: $C_7H_{12}O_4$ • 160.17 • 111-16-0
- (P) $\Delta H_f^0 = -1009.39(s)$ • Eb = 272/100; 251.5/50; 223/15; 212/10
- (D) LD50 o-r: 7000; LD50 o-m: 4800

Fumaric acid : $C_4H_4O_4$ • 116.07 • 110-17-8
- (P) $\Delta H_f^0 = -811.03(s)$ • Eb = 290 (subl); 165/1.7 (subl)
- (I) LEL/UEL = 3/40 • AIT = 739
- (T) Tr: 36 • NFPA: 1
- (D) LD50 o-r: 9300; 10700; MLD o-ra: 5000; MLD ip-r: 587; LD50 ip-m: 100; LD50 cu-ra: 20000

Maleic acid: $C_4H_4O_4$ • 116.07 • 110-16-7
- (P) $\Delta H_f^0 = -790.52(s)$ • Eb = 135; 160 (dec)
- (I) $P_{fla} = \sim 100$
- (T) Tr: 22-36/37/38 • NFPA: 2
- (D) LC50 r: > 0.720/1h; LD50 o-r: 708; LD50 o-m: 2400; LD50 cu-ra: 1560

o-Phthalic acid: $C_8H_6O_4$ • 166.13 • 88-99-3
- (P) Eb = 155 (dec); 191 (dec); 289
- (I) $P_{fla} = 167cc$
- (T) Tr: 36/37/38; – • NFPA: 0 • TR: –;
- (D) LD50 o-r: 7900; LD50 o-m: 2530; LD50 ip-m: 550

Terephthalic acid: $C_8H_6O_4$ • 166.13 • 100-21-0
- (P) Eb = > 300 (subl); 402 (subl)
- (T) Tr: 36/37/38; –
- (D) LD50 o-r: 18800; MLD o-m: 10000; LD50 ip-m: 1430

CARBOXYLIC ACIDS
• Acids with complex groups

Glycolic acid: $C_2H_4O_3$ • 76.05 • 79-14-1
- (P) $\Delta H_f^0 = -663.58(s)$ • Eb = (dec); 100; 112 (70%); 116 (65-70%) • $P_{vap} = 15(20)$
- (T) Tr: 34; 22-34 • NFPA: 2 • TR: –;
- (D) LD50 o-r: 1950

Lactic acid : $C_3H_6O_3$ • 90.08 • 50-21-5
- (P) $\Delta H_f^0 = -694.08(s)$; 674.46(l) • Eb = 122/14 to 15; 82-85/0.5 to 1
- (I) $P_{fla} = > 109$
- (T) Tr: 36/38
- (D) LD50 o-r: 3543; 3730; LD50 o-m: 4875; MLD o-ra: 500; LD50 scu-m: 4500; LD50 cu-ra: > 2000

8 • Table of organic compounds

Carboxylic acids

Name: Empirical formula • Molar mass(g/mol)• Registry number (CAS).
(P) Physical data: ΔH_f^0 (kJ/mol) (physical state) • Eb (°C/mmHg) (if substance volatile) • P_{vap} (mBar) (°C) (if substance volatile).
(I) Inflammability data: LEL/UEL (%) • P_{fla} (°C) • AIT (°C).
(T) Toxicity: Safety Code (Tr) • NFPA • Transport Code (TR) • LVE (MVE) (ppm) F, USA, D • STEL (ppm); IDLH (ppm).
(D) Lethal and toxic doses: LC (mg/l/duration of exposure); LD (mg/kg).

Salicylic acid: $C_7H_6O_3$ • 138.12 • 69-72-7
(P) $\Delta H_f^0 = -589.53(s)$ • Eb = 211/20 • $P_{vap} = 1.33(114)$
(I) LEL/UEL = 1.1/- • $P_{fla} = > 109$; 157cc • AIT = 545
(T) Tr: 22-36/37/38 • NFPA: 0; 1 • TWA (USA): 10 mg/m^3;ceiling value 15 mg/m^3
(D) LC50 r: > 0.9/1h; LD50 o-r: 891; LD50 o-m: 480; LD50 o-ra: 1300; LD50 ip-r: 157; LD50 ip-m: 300; 1500; LD50 iv-m: 184; 500; MLD scu-m: 6000; LD50 cu-ra: > 10000

2-Furoic acid: $C_5H_4O_3$ • 112.08 • 88-14-2
(P) $\Delta H_f^0 = -498.40(s)$ • Eb = 230-232; 141-144/20 • $P_{vap} = 4.12(105)$
(I) $P_{fla} = 137$
(T) Tr: 38; 21/22-34
(D) LD50 ip-m: 100

Fluoroacetic acid: $C_2H_3FO_2$ • 78.04 • 144-49-0
(P) $\Delta H_f^0 = -688.27(s)$ • Eb = 165-168 • $P_{vap} = 5.32(20)$
(T) Tr: 28 • NFPA: 2
(D) LD50 o-r: 5; LD50 o-m: 7; LD50 ip-m: 6; LD50 iv-m: 13; LD50 iv-ra: 250; LD50 scu-m: 281

Chloroacetic acid: $C_2H_3ClO_2$ • 94.50 • 79-11-8
(P) $\Delta H_f^0 = -511.70(s)$ • Eb = 186-189; 160/300; 132/100; 101/20; 94/11 • $P_{vap} = 0.23$; 1(20); 1.3(43); 4(55)
(I) LEL/UEL = 8/- • $P_{fla} = 126$; 129cc • AIT = 470
(T) Tr: 25-34 • NFPA: 3 • TR: 80
(D) LC50 r: 0.18/?; LD50 o-r: 76; 580; LD50 o-m: 165; LD50 ip-r: 16.6; LD50 iv-r: 5; 55; LD50 scu-r: 5; LD50 scu-m: 250

Bromoacetic acid: $C_2H_3BrO_2$ • 138.95 • 79-08-3
(P) Eb = 208; 127/30 • $P_{vap} = 0.07(20)$; 0.17(30); 0.95(50)
(I) $P_{fla} = > 109$
(T) Tr: 23/24/25-35 • NFPA: 2 • TR: 80
(D) MLC r: 114/30'; LD50 o-m: 100; LD50 ip-r: 50; LD50 ip-m: 66; MLD iv-ra: 45

2-Chloropropionic acid: $C_3H_5ClO_2$ • 108.52 • 598-78-7
(P) $\Delta H_f^0 = -523.00(l)$ • Eb = 170-190; 178-190; 183-187; 186-190 • $P_{vap} = 5.33(20)$
(I) $P_{fla} = 101$; 107 • AIT = 550
(T) Tr: 22-35 • TR: 80
(D) LD50 cu-guinea pig: 126

3-Chloropropionic acid: $C_3H_5ClO_2$ • 108.52 • 107-94-8
(P) Eb = 203-205; 200/765; 127/35; 108/12
(I) $P_{fla} = 107$; > 109
(T) Tr: 35
(D) MLD su-m: 1040

2-Bromopropionic acid: $C_3H_5BrO_2$ • 152.98 • 598-72-1
(P) Eb = 203
(I) $P_{fla} = 99$
(T) Tr: 22-34
(D) LD50 o-m: 250

3-Bromopropionic acid: $C_3H_5BrO_2$ • 152.98 • 590-92-1
(P) Eb = 140/60; 112/44 • $P_{vap} = 0.5(20)$
(I) $P_{fla} = 65$
(T) Tr: 34
(D) MLD ip-m: 500

2-Bromohexanoic acid: $C_6H_{11}BrO_2$ • 195.06 • 616-05-7
(P) Eb = 240; 148-153/30; 141/23; 136/24; 136-138/18; 128-136/17; 116-125/8; 98-100/1
(I) $P_{fla} = 63$; > 109
(T) Tr: 34

Dichloroacetic acid: $C_2H_2Cl_2O_2$ • 128.94 • 79-43-6
(P) $\Delta H_f^0 = -497.90(l)$ • Eb = 192-195 • $P_{vap} = 0.2$; 0.25(20); 0.43(30); 1.33(44); 2.1(50)
(I) $P_{fla} = > 109$; > 112
(T) Tr: 35 • NFPA: 3 • TR: 80
(D) LD50 o-r: 2820; LD50 cu-ra: 510

2,2-Dichloropropionic acid : $C_3H_4Cl_2O_2$ • 142.97 • 75-99-0
(P) Eb = 98-99/20 • $P_{vap} = 4(50)$
(I) $P_{fla} = > 110$
(T) Tr: 22-38-41
(D) LD50 o-r: 970

Ttrifluoroacetic acid: $C_2HF_3O_2$ • 114.02 • 76-05-1
(P) $\Delta H_f^0 = -1068.59(l)$ • Eb = 71-73; 71.1/734 • $P_{vap} = 11(?)$; 130(20); 142.6(25)
(T) Tr: 20-35 • NFPA: 3 • TR: 88
(D) LC50 r: 10/?; LD50 o-r: 200; LC50 m: 13.5/?; MLD ip-m: 150; LD50 iv-m: 1200

Trichloroacetic acid: $C_2HCl_3O_2$ • 163.39 • 76-03-9
(P) $\Delta H_f^0 = -505.01(s)$ • Eb = 198; 141-142/25 • $P_{vap} = 0.1(20)$; 1.33(51)
(I) $P_{fla} = n$; > 109
(T) Tr: 35 • NFPA: 3 • TR: 80 • LVE(MVE) (F): – (1); TWA (USA): 1
(D) LD50 o-r: 400; 5000; MLD ip-m: 500; LD50 scu-m: 270

2,4-Dichlorophenoxyacetic acid: $C_8H_6Cl_2O_3$ • 221.04 • 94-75-7
(P) Eb = 160/0.4 • $P_{vap} = < 10^{-7}$; $10^{-7}(20)$
(I) AIT = 375
(T) Tr: 22-36/37/38 • TR: 60 • LVE(MVE) (F): – (10 mg/m^3); TWA (USA): 10 mg/m^3; MAK(D): 10 mg/m^3
(D) LD50 o-r: 370; 375; 699; LD50 o-m: 347; 368; MLD o-ra: 800; MLD ip-r: 666; MLD ip-m: 125; LD50 ip-ra: 400 ; LD50 iv-ra: 400; LD50 cu-r: 1500; LD50 cu-ra: > 2000; 1400

Nicotinic acid $C_6H_5NO_2$ • 123.11 • 59-67-6
(P) Eb = sublimation > 236 • $P_{vap} = < 0.1(20)$
(I) $P_{fla} = 193Toc$
(D) LD50 o-r: 7000; LD50 o-m: 3720; LD50 o-ra: 4550; LD50 ip-r: 730; LD50 ip-m: 358; MLD iv-r: 3500; LD50 iv-m: 5000; LD50 scu-r: 5000; LD50 scu-m: 3500

8 • Table of organic compounds — Esters - Lactones

Name: Empirical formula • Molar mass(g/mol)• Registry number (CAS).
(P) Physical data: ΔH_f^0 (kJ/mol) (physical state) • Eb (°C/mmHg) (if substance volatile) • P_{vap} (mBar) (°C) (if substance volatile).
(I) Inflammability data: LEL/UEL (%) • P_{fla} (°C) • AIT (°C).
(T) Toxicity: Safety Code (Tr) • NFPA • Transport Code (TR) • LVE (MVE) (ppm) F, USA, D • STEL (ppm) • IDLH (ppm).
(D) Lethal and toxic doses: LC (mg/l/duration of exposure); LD (mg/kg).

Pyruvic acid: $C_3H_4O_3$ • 88.06 • 127-17-3
- **(P)** $\Delta H_f^0 = -584.50(l)$; Eb = 165(with dec.); 106.5/100; 85.3/40; 70.8/20; 57.9/10; 54/10; 45.8/5; 21.4/1
- **(I)** $P_{fla} = 83; 91$
- **(T)** Tr: 34
- **(D)** LD50 scu-m: 3533

ESTERS - LACTONES
• **Aliphatic saturated monoesters**

• *Formates*

Methyl formate: $C_2H_4O_2$ • 60.05 • 107-31-3
- **(P)** $\Delta H_f^0 = -350.20(g); -379.07(l)$ • Eb = 31-34; -49/10 • $P_{vap} = 533(16); 635; 640; 644(20); 935(30); 1900(50); 2262(55)$
- **(I)** LEL/UEL = 5/23; 5.9/20 • $P_{fla} = -32; -28; -26; -19; -17$ • AIT = 440; 450; 456; 465
- **(T)** NFPA: 2 • TR: 33 • VLE/VME(F): – (100); TWA (USA): 100; MAK(D): 100 • STEL: 150
- **(D)** LD50 o-r: ~1500; LD50 o-ra: 1622

Ethyl formate: $C_3H_6O_2$ • 74.08 • 109-94-4
- **(P)** Eb = 52-55 • $P_{vap} = 133(5.4); 256; 259; 261; 267(20); 390(30); 400(30.2); 875(50); 1045(55)$
- **(I)** LEL/UEL = 2.7/13.5; 2.7/16.5; 2.8/13.5; 2.8/16 • $P_{fla} = -34; -20cc$ • AIT = 440; 455
- **(T)** NFPA: 2 • TR: 33 • TWA (USA): 100; MAK(D):100
- **(D)** MLC r: 24.239/4h; LD50 o-r: 1850; LD50 o-ra: 2075; LD50 scu-m: 1000; LD50 cu-ra: 20000

Propyl formate: $C_4H_8O_2$ • 88.11 • 110-74-7
- **(P)** Eb = 81 • $P_{vap} = 66.7; 79; 84(20); 133(29.5); 135(30); 320(50)$
- **(I)** LEL/UEL = 2.3/- ; 2.4/7. • $P_{fla} = -5; -3cc; -3Coc$ • AIT = 450; 454
- **(T)** NFPA: 2 • TR: 33
- **(D)** LD50 o-r: 3980; LD50 o-m: 3400

Isopropyl formate: $C_4H_8O_2$ • 88.11 • 625-55-8
- **(P)** Eb = 68; 71 • $P_{vap} = 133(17.); 150(20); 240(30); 520(50)$
- **(I)** $P_{fla} = 6cc$ • AIT = 480; 485
- **(T)** NFPA: 2 • TR: 33

Butyl formate: $C_5H_{10}O_2$ • 102.13 • 592-84-7
- **(P)** Eb = 105-107 • $P_{vap} = 29(20); 49(30); 53.3(31.6); 128(50)$
- **(I)** LEL/UEL = 1.7/8; 1.7/8.2 • $P_{fla} = 13; 18cc$ • AIT = 320; 322
- **(T)** NFPA: 2 • TR: 33
- **(D)** LD50 o-ra: 2656

Isobutyl formate: $C_5H_{10}O_2$ • 102.13 • 542-55-2
- **(P)** Eb = 98
- **(I)** LEL/UEL = 2/8; 2/8.9; ~2/~8 • $P_{fla} = 5; 9; < 21$ • AIT = 319
- **(T)** NFPA: – • TR: 33
- **(D)** LD50 o-ra: 3064

Amyl formate: $C_6H_{12}O_2$ • 116.16 • 638-49-3
- **(P)** Eb = 130
- **(I)** LEL/UEL = 1.7/10 • $P_{fla} = 26$
- **(T)** NFPA: 1 • TR: –
- **(D)** LD50 o-r: > 5000; LD50 cu-ra: > 5000

Isoamyl formate: $C_6H_{12}O_2$ • 116.16 • 542-55-2
- **(P)** Eb = 121-124 • $P_{vap} = 13.3(17.1)$
- **(I)** LEL/UEL = 1.2/8; 1.7/10 • $P_{fla} = 23; 26; 29; 53 (?)$ • AIT = 320
- **(T)** NFPA: 1 • TR: 30
- **(D)** LD50 o-r: 9840; LD50 o-ra: 3020; LD50 cu-ra: > 5000

• *Acetates*

Methyl acetate: $C_3H_6O_2$ • 74.08 • 79-20-9
- **(P)** $\Delta H_f^0 = -445.18(l)$ • Eb = 56-58 • $P_{vap} = 133(9.4); 220; 227(20); 266(24); 500(40)$
- **(I)** LEL/UEL = 3.1/16 • $P_{fla} = -16; -13cc; -9cc$ • AIT = 453; 475; 502; 506
- **(T)** NFPA: 1 • TR: 33 • VLE/VME(F): 250 (200); TWA (USA): 200; MAK(D): 200 • STEL: 250 • IDLH: 10000
- **(D)** LC50 r: 96.955/?; MLC m: 34/4h; LD50 o-r: > 5000; 6970; LD50 o-ra: 3705; MLD scu-r: 8000; LD50 cu-ra: > 5000

Ethyl acetate: $C_4H_8O_2$ • 88.11 • 540-88-5
- **(P)** $\Delta H_f^0 = -442.92(g); -479.03(l)$ • Eb = 76-77 • $P_{vap} = 97; 101(20); 120(25); 133(27); 300(50)$
- **(I)** LEL/UEL = 2/11.4; 2.1/11.5; 2.2/9; 2.2/11; 2.2(38)/11.5(38); 2.5/9 • $P_{fla} = -4Tcc; -2; 7oc$ • AIT = 426; 430; 460; 482; 484
- **(T)** NFPA: 1 • TR: 33 • VLE/VME(F): – (400); TWA (USA): 400; MAK(D): 400 • IDLH: 10000
- **(D)** LC50 r: 0.2/?; 5.76/8h; LC50 m: 45/2h; MLC m: 31/2h; LD50 o-r: 5600; 5620; 6100; 11300; LD50 o-m: 4100; LD50 o-ra: 4935; LD50 ip-m: 709; MLD scu-r: 5000; LD50 cu-ra: > 20000

Propyl acetate: $C_5H_{10}O_2$ • 102.13 • 109-60-4
- **(P)** Eb = 101-102 • $P_{vap} = 33.3(20); 43(25); 53(28.8)$
- **(I)** LEL/UEL = 1.7(37)/8; 2.0/8 • $P_{fla} = 13cc$ • AIT = 430; 450
- **(T)** NFPA: 1 • TR: 33 • VLE/VME(F): – (200); TWA (USA): 200; MAK(D): 200 • STEL: 250
- **(D)** MLC r: 33.413/4h; LD50 o-r: 9370; LD50 o-m: 8300; LD50 o-ra: 6640; LD50 ip-m: 1420; LD50 cu-ra: > 20000

8 • Table of organic compounds

Esters - Lactones

Name: Empirical formula • Molar mass(g/mol)• Registry number (CAS).
(P) Physical data: ΔH_f^0 (kJ/mol) (physical state) • Eb (°C/mmHg) (if substance volatile) • P_{vap} (mBar) (°C) (if substance volatile).
(I) Inflammability data: LEL/UEL (%) • P_{fla} (°C) • AIT (°C).
(T) Toxicity: Safety Code (Tr) • NFPA • Transport Code (TR) • LVE (MVE) (ppm) F, USA, D • STEL (ppm); IDLH (ppm).
(D) Lethal and toxic doses: LC (mg/l/duration of exposure); LD (mg/kg).

Isopropyl acetate: $C_5H_{10}O_2$ • 102.13 • 108-21-4
- **(P)** $\Delta H_f^0 = -518.86(l)$ • Eb = 87-90; 93 • P_{vap} = 53.3(17); 57.3; 61; 63(20); 81(25); 250(50)
- **(I)** LEL/UEL = 1/8; 1.8(37)/8; 1.8/7 • P_{fla} = 2cc; 4Tcc; 4oc; 6Tcc; 10oc; 17Toc • AIT = 460; 478
- **(T)** NFPA: 1 • TR: 33 • VLE/VME(F): 300 (250); TWA (USA): 250; MAK(D): 200 • STEL: 310; IDLH: 16000
- **(D)** LC50 r: 133.67/4h; 50.6/8h; LD50 o-r: 3000; 6750; LD50 o-ra: 6946; LD50 cu-ra: 17500; > 20000

Butyl acetate $C_6H_{12}O_2$ • 116.16 • 123-86-4
- **(P)** $\Delta H_f^0 = -529.36(l)$ • Eb = 124-127 • P_{vap} = 10.7; 11; 11.6; 13.3; 17; 17.3(20); 20(25)
- **(I)** LEL/UEL = 1.2/7.5; 1.38/7.6; 1.4/7.6; 1.7/7.6; 3/10.4 • P_{fla} = 22cc; 24cc; 27Tcc; 32Toc; 37Toc • AIT = 360; 370; 399; 420; 425
- **(T)** NFPA: 1 • TR: 30 • VLE/VME(F): 200 (150); TWA (USA): 150; MAK(D): 200
- **(D)** LC50 r: 9.5/4h; LC50 m: 6/2h; LD50 o-r: 10768; 13100; 14000; 14130; ; LD50 o-m: 6000; 7600; LD50 o-ra: 3200; LD50 ip-m: 1230; LD50 cu-ra: > 17600

s-Butyl acetate: $C_6H_{12}O_2$ • 116.16 • 105-46-4
- **(P)** Eb = 112 • P_{vap} = 21.6(20); 32(25)
- **(I)** LEL/UEL = 1.3/7.5; 1.7/- • P_{fla} = 16; 18; 29oc; 31oc
- **(T)** NFPA: 1 • TR: 33 • VLE/VME(F): – (200); TWA (USA): 200; MAK(D): 200

Isobutyl acetate: $C_6H_{12}O_2$ • 116.16 • 110-19-0
- **(P)** Eb = 116-118 • P_{vap} = 13.3(12.8); 17; 18.7; 20; 21(20); 26(25)
- **(I)** LEL/UEL = 1.1/-; 1.3/7.5; 1.3/10.5; 2.4/10.5; 4.0/13.0 • P_{fla} = 18cc; 21cc; 36 • AIT = 405; 423
- **(T)** NFPA: 1 • TR: 33 • VLE/VME(F): 200 (150); TWA (USA): 150; MAK(D): 200
- **(D)** MLC r: 38/4h; LD50 o-r: 13400; 15000; LD50 o-ra: 4763; LD50 cu-ra: 17400

t-Butyl acetate: $C_6H_{12}O_2$ • 116.16 • 540-88-5
- **(P)** Eb = 94-98 • P_{vap} = 53(20); 53(25)
- **(I)** P_{fla} = -2; 1; 15
- **(T)** NFPA: 1 • TR: - • VLE/VME(F): – (200); TWA (USA): 200 MAK(D): 200

Amyl acetate: $C_7H_{14}O_2$ • 130.19 • 628-63-7
- **(P)** Eb = 142; 147-149; 148/737 • P_{vap} = 5.3; 6(20); 8(25)
- **(I)** LEL/UEL = 1/7.5; 1.1/- ; 1.1/7.5 • P_{fla} = 23; 25cc; 37; 39 • AIT = 359; 375; 379
- **(T)** NFPA: 1 • TR: 30 • VLE/VME(F): 150 (100); TWA (USA): 100; MAK(D): 100
- **(D)** LC50 r: 27.69/8h; LD50 o-r: 6500; LD50 o-ra: 7400

s-Amyl acetate: $C_7H_{14}O_2$ • 130.19 • 626-38-0
- **(P)** Eb = 120-121; 130-131
- **(I)** LEL/UEL = 1/7.5; 1.1/7.5; 1.1/10 • P_{fla} = 23cc; 32(impure)
- **(T)** NFPA: 1 • TR: 30 • VLE/VME(F): – (125); TWA (USA): 125; MAK(D): 100

Isoamyl acetate: $C_7H_{14}O_2$ • 130.19 • 123-92-2
- **(P)** Eb = 140-143; 142/756 • P_{vap} = 1.33; 5.33(20); 6; 6.67(25)
- **(I)** LEL/UEL = 1(100)/7.5; 1/10; 1.1/7.0(100) • P_{fla} = 23(impure); 25; 27; 33cc; 36; 38oc; 41 • AIT = 359; 380
- **(T)** NFPA: 1 • TR: 30 • VLE/VME(F): – (100); TWA (USA): 100
- **(D)** LD50 o-r: 7422; 16600; LD50 o-ra: 7422

Hexyl acetate: $C_8H_{16}O_2$ • 144.21 • 142-92-7
- **(P)** Eb = 167-170; 172; 62/12 • P_{vap} = 5(20)
- **(I)** P_{fla} = 37; 41; 43
- **(T)** NFPA: 1 • TR: 30
- **(D)** LD50 o-r: 41500; 42000; LD50 cu-ra: > 5000

s-Hexyl acetate: $C_8H_{16}O_2$ • 144.21 • 108-84-9
- **(P)** Eb = 141; 146.3 • P_{vap} = 4; 5.1(20)
- **(I)** LEL/UEL = -/18.7 • P_{fla} = 43oc; 45Coc
- **(T)** NFPA: 1 • TR: 30 • VLE/VME(F): –(50); TWA (USA): 50; MAK(D): 50
- **(D)** LC50 r: 11.79/2h; LD50 o-r: 6160; LD50 cu-ra: 20000

Cyclohexyl acetate: $C_8H_{14}O_2$ • 142.20 • 622-45-7
- **(P)** Eb = 172-173; 175-177; 58/10 • P_{vap} = 4; 20(20)
- **(I)** LEL/UEL = 4.2/- • P_{fla} = 58; 60cc; 68oc • AIT = 330; 334
- **(T)** NFPA: 1 • TR: 30
- **(D)** LD50 o-r: 6730; LD50 cu-ra: 10000

• *Superior terms*

Methyl propionate : $C_4H_8O_2$ • 88.11 • 554-12-1
- **(P)** Eb = 78-80 • P_{vap} = 53(11); 87; 90(20); 114(25); 140(30); 333(50)
- **(I)** LEL/UEL = 2.4/13; 2.50/13 • P_{fla} = 2cc; 6 • AIT = 465; 468
- **(T)** NFPA: 1 • TR: 33
- **(D)** LC50 m: 27/?; LD50 o-r: 5000; LD50 o-m: 3460; LD50 o-ra: 2550; LD50 cu-ra: > 5000

Ethyl propionate : $C_5H_{10}O_2$ • 102.13 • 105-37-3
- **(P)** $\Delta H_f^0 = -511.12(l)$ • Eb = 96-99 • P_{vap} = 36; 38(20); 53(27.2); 61(30); 190(50)
- **(I)** LEL/UEL = 1.8/11; 1.9/11 • P_{fla} = 5; 12cc • AIT = 440; 475
- **(T)** NFPA: – • TR: 33
- **(D)** LD50 o-r: 8732; LD50 o-ra: 3500; LD50 ip-r: 1200; LD50 ip-m: 1158; 1300

Propyl propionate : $C_6H_{12}O_2$ • 116.16 • 106-36-5
- **(P)** Eb = 120-124 • P_{vap} = 13; 15(20); 24(30); 67(50)
- **(I)** P_{fla} = 19; 22; 79oc (? SAX)
- **(T)** NFPA: 1 • TR: –
- **(D)** LD50 o-r: 10331; LC50 m: 24/2h; LD50 o-ra: 3950; LD50 cu-ra: > 14128

Butyl propionate : $C_7H_{14}O_2$ • 130.19 • 590-01-2
- **(P)** Eb = 145-146.8; 145/756 • P_{vap} = 4(20)
- **(I)** P_{fla} = 32; 38 • AIT = 425; 427
- **(T)** NFPA: 2 • TR: 30
- **(D)** LD50 o-r: 5000; LD50 cu-ra: > 14000

8 • Table of organic compounds — Esters - Lactones

Name: Empirical formula • Molar mass(g/mol) • Registry number (CAS).
(P) **Physical data:** ΔH_f^0 (kJ/mol) (physical state) • Eb (°C/mmHg) (if substance volatile) • P_{vap} (mBar) (°C) (if substance volatile).
(I) **Inflammability data:** LEL/UEL (%) • P_{fla} (°C) • AIT (°C).
(T) **Toxicity:** Safety Code (Tr) • NFPA • Transport Code (TR) • LVE (MVE) (ppm) F, USA, D • STEL (ppm); IDLH (ppm).
(D) **Lethal and toxic doses:** LC (mg/l/duration of exposure); LD (mg/kg).

Isobutyl propionate : $C_7H_{14}O_2$ • 130.19 • 540-42-1
(P) Eb = 137-138; 66.5/60
(I) P_{fla} = 26; 38
(T) NFPA: – • TR: 30
(D) LD50 o-ra: 5599

Methyl butyrate : $C_5H_{10}O_2$ • 102.13 • 623-42-7
(P) Eb = 100-103 • P_{vap} = 32(20); 53(29.6); 140(50)
(I) LEL/UEL = 0.9/3.5 • P_{fla} = 11; 14cc
(T) Tr: –; 36/37/38 • NFPA: 2 • TR: 33
(D) LC50 m: 18/2h; LD50 o-r: > 5000; LD50 o-ra: 3380; LD50 cu-ra: 3560

Ethyl butyrate : $C_6H_{12}O_2$ • 116.16 • 105-54-4
(P) Eb = 119-121.6; 45/50 • P_{vap} = 13; 17; 20.7(20); 28(30); 80(50)
(I) P_{fla} = 19; 26cc; 29oc • AIT = 460; 463
(T) Tr: –; 36/37/38 • NFPA: 2 • TR: 30
(D) LD50 o-r: 13000; 13050; LD50 o-ra: 5228

Isopropyl butyrate : $C_7H_{14}O_2$ • 130.19 • 638-11-9
(P) Eb = 128-131 • P_{vap} = 8(20); 65(50); 115(65)
(I) P_{fla} = 14; 25; 29
(T) TR: 30

Butyl butyrate : $C_8H_{16}O_2$ • 144.21 • 109-21-7
(P) Eb = 164-167; 165.7/736; 55/53 • P_{vap} = 4(20)
(I) P_{fla} = 49; 53oc; AIT = 385
(T) NFPA: 2 • TR: 30
(D) LD50 o-r: 2300; LD50 o-ra: 9520; LD50 ip-m: 8900

Methyl isobutyrate : $C_5H_{10}O_2$ • 102.13 • 547-63-7
(P) Eb = 90-93 • P_{vap} = 49(20); 84(30); 210(50)
(I) P_{fla} = 3; 12; 13oc; 14 • AIT = 482
(D) LD50 o-r: 16000; LC50 m: 25.5/2h; MLD ip-r: 3200

Ethyl isobutyrate : $C_6H_{12}O_2$ • 116.16 • 97-62-1
(P) Eb = 107-110; 110-113 • P_{vap} = 27(20); 44(30); 53.3(33.8); 112(50)
(I) P_{fla} = 13; < 18; 20 • AIT = 463
(T) NFPA: – • TR: 33
(D) LD50 ip-m: 800

Methyl pentanoate : $C_6H_{12}O_2$ • 116.16 • 624-24-8
(P) ΔH_f^0 = –514.17(l) • Eb = 126-128; 87-88/15.75
(I) P_{fla} = 22; 27
(D) LC50 m: 6.6/2h

Ethyl pentanoate : $C_7H_{14}O_2$ • 130.19 • 539-82-2
(P) ΔH_f^0 = –553.12(l) • Eb = 142-146
(I) P_{fla} = 34; 38

Ethyl isopentanoate : $C_7H_{14}O_2$ • 130.19 • 108-64-5
(P) ΔH_f^0 = –571.12(l) • Eb = 131-135 • P_{vap} = 10(20)
(I) P_{fla} = 22; 26; 30
(D) LD50 o-ra: 7031; LD50 ip-r: 1200

Methyl pivalate : $C_6H_{12}O_2$ • 116.16 • 598-98-1
(P) Eb = 99-102
(I) P_{fla} = 6

Ethyl pivalate : $C_7H_{14}O_2$ • 130.19 • 3938-95-2
(P) Eb = 116-118
(I) P_{fla} = 12; 16

Methyl hexanoate : $C_7H_{14}O_2$ • 130.19 • 106-70-7
(P) ΔH_f^0 = –540.20(l) • Eb = 150-151; 63/30; 52/15 • P_{vap} = 3.7(20); 6.7(30); 20(50)
(I) P_{fla} = 41; 44
(T) TR: 30
(D) LC50 m: 14/2h; LD50 o-r: > 5000; MLD iv-m: 48

Ethyl hexanoate : $C_8H_{16}O_2$ • 144.21 • 123-66-0
(P) Eb = 163; 165-168
(I) P_{fla} = 49; 53oc
(T) TR: 30

Methyl heptanoate : $C_8H_{16}O_2$ • 144.21 • 106-73-0
(P) ΔH_f^0 = –567.10(l) • Eb = 171-174
(I) P_{fla} = 52
(T) TR: 30
(D) LD50 o-r: > 5000

Ethyl heptanoate : $C_9H_{18}O_2$ • 158.24 • 106-30-9
(P) Eb = 186-189; 78/14; 68/8
(I) P_{fla} = 65; 74
(D) LD50 o-r: 25960; > 34640

ESTERS - LACTONES
• Aliphatic unsaturated monoesters

Vinyl acetate : $C_4H_6O_2$ • 86.09 • 108-05-4
(P) Eb = 71-73; 9/50; 18/10 • P_{vap} = 113; 117; 118.2(20); 133(21.5)
(I) LEL/UEL = 2.6/13.4 • P_{fla} = –8cc to –5; –1Toc • AIT = 384; 425; 427
(T) Tr: –; 40 • NFPA: 2 • TR: 339 • VLE/VME(F): –(10); TWA (USA): 10; MAK(D): 10 • STEL: 15; 20
(D) LC50 r: 14.317/2h; 11.4/4h; LC50 m: 5.548/4h; LC50 ra: 8.949/4h; LD50 o-r: 2920; LD50 o-m: 1613; LD50 cu-ra: 2335

Allyl acetate : $C_5H_8O_2$ • 100.12 • 591-87-7
(P) Eb = 99-104
(I) P_{fla} = –11; 6; 21oc • AIT = 374
(T) Tr: 23/24/25 • NFPA: 3 • TR: 336
(D) LC50 r: 7.163/1h; LD50 o-r: 130; LD50 o-m: 170; LD50 cu-ra: 1021

Isopropenyl acetate: $C_5H_8O_2$ • 100.12 • 108-22-5
(P) ΔH_f^0 = –386.23(l) • Eb = 93; 94; 95-97.4; 96.6/746; 58-60/200 • P_{vap} = 39.6(20); 210(50); 375(65)
(I) LEL/UEL = 1.6/10; 1.9/- • P_{fla} = 4; 9; 16oc; 18 • AIT = 395; 431
(T) TR: 33
(D) LD50 o-r: 3000

440

8 • Table of organic compounds

Esters - Lactones

Name: Empirical formula • Molar mass(g/mol)• Registry number (CAS).
(P) Physical data: ΔH_f^0 (kJ/mol) (physical state) • Eb (°C/mmHg) (if substance volatile) • P_{vap} (mBar) (°C) (if substance volatile).
(I) Inflammability data: LEL/UEL (%) • P_{fla} (°C) • AIT (°C).
(T) Toxicity: Safety Code (Tr) • NFPA • Transport Code (TR) • LVE (MVE) (ppm) F, USA, D • STEL (ppm); IDLH (ppm).
(D) Lethal and toxic doses: LC (mg/l/duration of exposure); LD (mg/kg).

Methyl acrylate : $C_4H_6O_2$ • 86.09 • 96-33-3
- (P) $\Delta H_f^0 = -293.30(g)$ • Eb = 78-81; 70/608; 60/428; 50/298; 40/200; 28/100; 20/88; 10/54; 5/41.5; 10/41; 0/32; -5/24.5; -10/18.5 • P_{vap} = 86.6; 90; 90.6; 93(20); 133(28); 144(30); 330(50); 500(60)
- (I) LEL/UEL = 2.1/14.5; 2.5/12; 2.6/18.6; 2.8/18.6; 2.8/25 • P_{fla} = -4Toc; -3Tcc; -3oc; -2Toc; 6 • AIT = 390; 392; 415; 463
- (T) Tr: 20/22-36/37/38 • NFPA: 2 • TR: 339 • VLE/VME(F): 15 (10); TWA (USA): 10; MAK(D): 5 • IDLH: 1000
- (D) LC50 r: 4.832/4h; LC50 m: 12.8/?; MLC: 9.3/?; MLC ra: 9.028/1h; LD50 o-r: 277; 300; LD50 o-m: 827; MLD o-ra: 280; LD50 ip-r: 325; LD50 ip-m: 254; LD50 cu-ra: 1243

Ethyl acrylate : $C_5H_8O_2$ • 100.12 • 140-88-3
- (P) Eb = 98-100; 44/100; 20/39.2 • P_{vap} = 39.1; 41.3; 52(20); 150(50)
- (I) LEL/UEL = 1.4/14; 1.4/15.8; 1.8/- 1.8/12.1; 1.8/14 • P_{fla} = 7; 8Tcc; 16oc; 17Toc; 19Toc; AIT = 273; 350; 382; 399
- (T) Tr: 20/22-36/37/38-43; 20/21/22-36/37/38-40-43 • NFPA: 2 • TR: 339 • VLE/VME(F): -(5); TWA (USA): 5; MAK(D): 5 • STEL: 15; IDLH: 2000
- (D) LC50 r: 9.75/4h; LC50 m: 16.2/?; MLC ra: 5.012/7h; LD50 o-r: 760-1020; 800; 1000; LD50 o-m: 1799; LD50 o-ra: 400; LD50 ip-r: 450; LD50 ip-m: 599; LD50 cu-ra: 1234; 1834; ~2000

Butyl acrylate : $C_7H_{12}O_2$ • 128.17 • 141-32-2
- (P) Eb = 145-149; 84-86/101; 69/50; 59/25; 38-39/10; 35/8 • P_{vap} = 4; 4.3; 4.4; 4.8; 5.3(20); 13.3(35.5); 28.3(50)
- (I) LEL/UEL = 1.2/8; 1.3/9.9; 1.5/2.1; 1.5/9.9 • P_{fla} = 36cc; 39Tcc; 41Tcc; 47Toc; 48Toc; 49oc • AIT = 275; 292; 297
- (T) Tr: 36/37/38-43 • NFPA: 2 • TR: 39 • VLE/VME(F): -(10); TWA (USA): 10; MAK(D): 10
- (D) LC50 r: 14.549/4h; LC50 m: 7./2h; LD50 o-r: 900; 3700; 3730; LD50 o-m: 5880; 7561; LC50 ip-r: 550; LD50 ip-m: 835; 853; LD50 cu-r: 2000; MLDcu-r: 1700; LD50 cu-ra: 2000

Methyl methacrylate : $C_5H_8O_2$ • 100.12 • 80-62-6
- (P) Eb = 98-100; 50/120 • P_{vap} = 38.7(20); 53.3(25.5); 66.7(30); 166.7(50)
- (I) LEL/UEL = 1.7/12.5; 1.8/8.2; 2.1/12.5 • P_{fla} = 2; 8-10cc; 10oc • AIT = 421; 430; 434
- (T) Tr: 36/37/38-43 • NFPA: 2 • TR: 339 • VLE/VME(F): 200 (100); TWA (USA): 100; MAK(D): 50 • IDLH: 4000
- (D) LC50 r: 15.61/?; LC50 m: 18.5/2h; MLC m: 13/?; LD50 o-r: 7872; 8400; 8900; LD50 o-m: 3625; 5204; LD50 ip-r: 1328; LD50 ip-m: 945; 1000; LD50 scu-r: 7088; 7500; LD50 scu-m: 5954; 6300

Ethyl methacrylate : $C_6H_{10}O_2$ • 114.14 • 97-63-2
- (P) Eb = 116-120 • P_{vap} = 16; 20; 21.3(20); 24(30); 56(50)
- (I) LEL/UEL = 1.8/- • P_{fla} = 15; 18; 20oc; 26cc • AIT = 329; 395; 410
- (T) Tr: 36/37/38-43 • NFPA: 2 • TR: 339
- (D) LC50 r: 39.39/4h; LD50 o-r: 14800; LD50 o-m: 7836; MLD o-ra: 3630; LD50 ip-r: 1223; LD50 ip-m: 1369; MLD scu-r: 25000

Butyl methacrylate : $C_8H_{14}O_2$ • 142.20 • 97-88-1
- (P) Eb = 162-165; 120/200; 101/100; 62/20; 34/5 • P_{vap} = 2.7; 3; 6.53(20); 5.5(30); 13(50)
- (I) LEL/UEL = 2.0/8.0 • P_{fla} = 41Toc; 49cc; 50; 52Toc; 54oc; 65Coc • AIT = 290; 294
- (T) Tr: 36/37/38-43 • NFPA: 2 • TR: 39
- (D) LC50 r: 29.03/4h; LD50 o-r: 16000; 23000; LD50 o-m: 12900; 14000; LD50 o-ra: 25000; LD50 ip-r: 2304; LD50 ip-m: 1490; LD50 cu-ra: 11300

Isobutyl methacrylate : $C_8H_{14}O_2$ • 142.20 • 97-86-9
- (P) Eb = 155; 93/100; 46/12 • P_{vap} = 4; 4.7(20); 15(45)
- (I) P_{fla} = 41; 43; 49oc • AIT = 390
- (T) Tr: 36/37/38-43; • TR: 39
- (D) LD50 o-r: 6400; LD50 o-m: 11990; LD50 ip-m: 1340

Allyl methacrylate : $C_7H_{10}O_2$ • 126.16 • 96-05-9
- (P) Eb = 139-142; 150; 59-61/43; 55/40; 55/30
- (I) P_{fla} = 33; 38
- (T) Tr: 21/22-36/37/38
- (D) LC50 r: 1.8/?; MLC r: 2.62/?; LC50 m: 5.5/?; LD50 o-r: 70; 430; LD50 o-m: 57; LD50 cu-ra: 500

Methyl crotonate : $C_5H_8O_2$ • 100.12 • 623-43-8
- (P) $\Delta H_f^0 = -382.84(l)$ • Eb = 116-121
- (I) P_{fla} = -1; 4
- (T) Tr: 36/37/38; 22
- (D) LD50 o-r: > 3200; LD50 o-m: 1600; LD50 cu-ra: > 5000

Ethyl crotonate : $C_6H_{10}O_2$ • 114.14 • 623-70-1
- (P) $\Delta H_f^0 = -420.07(l)$ • Eb = 134-139; 142-147 • P_{vap} = 65(50); 110(65)
- (I) P_{fla} = 2; 28
- (T) Tr: 36/37/38; 36 • TR: 33
- (D) LD50 o-r: 3000

ESTERS - LACTONES
• **Aliphatic diesters**

Ethyleneglycol diacetate: $C_6H_{10}O_4$ • 146.14 • 111-55-7
- (P) Eb = 186-191; 79-81/11 • P_{vap} = 0.27; 0.33; 0.4; 0.6(20); 1.33(38.3)
- (I) LEL/UEL = 1.6/8.4 • P_{fla} = 82; 86; 94; 96oc • AIT = 481
- (T) NFPA: 1 • TR: –
- (D) LD50 o-r: 6850; 6860; LD50 o-m: 1070; 1190; LD50 cu-ra: 8480

Dimethyl oxalate: $C_4H_6O_4$ • 118.09 • 553-90-2
- (P) $\Delta H_f^0 = -757.30(l)$ • Eb = 163-164
- (I) P_{fla} = 74cc
- (T) Tr: 36/38
- (D) LD50 o-r: 500

441

8 • Table of organic compounds

Esters - Lactones

Name: Empirical formula • Molar mass(g/mol)• Registry number (CAS).
(P) Physical data: ΔH_f^0 (kJ/mol) (physical state) • Eb (°C/mmHg) (if substance volatile) • P_{vap} (mBar) (°C) (if substance volatile).
(I) Inflammability data: LEL/UEL (%) • P_{fla} (°C) • AIT (°C).
(T) Toxicity: Safety Code (Tr) • NFPA • Transport Code (TR) • LVE (MVE) (ppm) F, USA, D • STEL (ppm); IDLH (ppm).
(D) Lethal and toxic doses: LC (mg/l/duration of exposure); LD (mg/kg).

Diethyl oxalate : $C_6H_{10}O_4$ • 146.14 • 95-92-1
- (P) ΔH_f^0 = –805.46(l) • Eb = 182-186; 130.8/100; 96.8/20; 47/1
  • P_{vap} = 0.27; 1.33(20); 1.33(47)
- (I) LEL/UEL = 0.8/- • P_{fla} = 67; 76cc; 76oc
- (T) Tr: 22-36 • NFPA: 3; 1; – • TR: 60
- (D) LD50 o-r: 400; LD50 o-m: 2000

Dibutyl oxalate : $C_{10}H_{18}O_4$ • 202.25 • 2050-60-4
- (P) Eb = 239-246; 123-124/10; 96/2
- (I) P_{fla} = 93; 104; 108; 129oc
- (T) Tr: 36/37/38 • NFPA: 3; 0

Dimethyl malonate : $C_5H_8O_4$ • 132.12 • 108-59-8
- (P) ΔH_f^0 = –795.80(l) • Eb = 180-181; 204-205 • P_{vap} = 0.15(20)
- (I) P_{fla} = 85; 90
- (T) NFPA: 0
- (D) LD50 o-r: 5331; LD50 cu-ra: > 5000

Diethyl malonate : $C_7H_{12}O_4$ • 160.17 • 105-53-3
- (P) Eb = 199; 95/20; 94-95/11 • P_{vap} = 0.35(20); 1.33(40)
- (I) P_{fla} = 73; 89cc; 93oc; 99
- (T) Tr: 36 • NFPA: 0
- (D) LD50 o-r: 15000; LD50 o-m: 6400

Dimethyl succinate : $C_6H_{10}O_4$ • 146.14 • 106-65-0
- (P) ΔH_f^0 = –835.13(l) • Eb = 190-193; 196; 200; 216; 103.5/25
  • P_{vap} = 0.4(20)
- (I) LEL/UEL = 1/8.5 • P_{fla} = 80; 84; 90 • AIT = 364
- (D) LD50 o-r: > 5000

Diethyl succinate : $C_8H_{14}O_4$ • 174.20 • 123-25-1
- (P) Eb = 216-218; 97/10; 55/1 • P_{vap} = 0.04(20)
- (I) P_{fla} = 90; 110oc
- (T) NFPA: 1
- (D) LD50 o-r: 8530

Diallyl succinate : $C_{10}H_{14}O_4$ • 198.22 • 925-16-6
- (P) Eb = 249; 105/3
- (I) P_{fla} = > 109

Dimethyl glutarate : $C_7H_{12}O_4$ • 160.17 • 1119-40-0
- (P) ΔH_f^0 = –861.49(l) • Eb = 214; 150/50; 109/21; 96-103/15; 93-95/13
- (I) P_{fla} = 97; 103

Diethyl glutarate : $C_9H_{16}O_4$ • 188.22 • 818-38-2
- (P) Eb = 237
- (I) P_{fla} = 96

Dimethyl adipate : $C_8H_{14}O_4$ • 174.20 • 627-93-0
- (P) ΔH_f^0 = –886.59(l) • Eb = 109/19; 109-110/14; 115/12; 112/10; 105/7 • P_{vap} = 0.08(20)
- (I) LEL/UEL = 0.8/8.1; • P_{fla} = 107; 110; AIT = 359
- (D) LD50 ip-r: 1809

Diethyl adipate : $C_{10}H_{18}O_4$ • 202.25 • 141-28-6
- (P) Eb = 240-245; 251; 127/13 ; 124-125/11
- (I) P_{fla} = 107; > 109
- (D) LD50 ip-m: 2190

Dimethyl fumarate : $C_6H_8O_4$ • 144.13 • 624-49-7
- (P) ΔH_f^0 = –729.27(l) • Eb = 192-193
- (T) Tr: 21-36/37/38
- (D) LD50 o-r: 2240; LD50 cu-ra: 1250

Diethyl fumarate : $C_8H_{12}O_4$ • 172.18 • 623-91-6
- (P) Eb = 218-219 • P_{vap} = 1.33(53.2)
- (I) P_{fla} = 91; 94; 96; 104
- (T) Tr: 22 • NFPA: 1
- (D) LD50 o-r: 1780

Dimethyl maleate : $C_6H_8O_4$ • 144.13 • 624-48-6
- (P) ΔH_f^0 = –703.75 • Eb = 200-205; 102/17; 86-89/12
  • P_{vap} = 1.33(45.7)
- (I) P_{fla} = 91; 95; 101; 113oc
- (T) Tr: 22 • NFPA: 1
- (D) MLD o-r: 1410; LD50 cu-ra: 530

Diethyl maleate : $C_8H_{12}O_4$ • 172.18 • 141-05-9
- (P) Eb = 220-225; 105/14 • P_{vap} = 1.33(14); 1.33(20); 1.33(57.3)
- (I) P_{fla} = 93Coc; 121oc • AIT = 349
- (T) Tr: 36/37/38 • NFPA: 1
- (D) LD50 o-r: 300; 3200; LD50 ip-r: 3070; LD50 cu-ra: 4000

ESTERS - LACTONES
•Aliphatic and aromatic carbonates

Dimethyl carbonate : $C_3H_6O_3$ • 90.08 • 616-38-6
- (P) Eb = 86-91 • P_{vap} = 24(21.1); 96(37.)
- (I) P_{fla} = 16; 19oc
- (T) NFPA: – • TR: 33
- (D) LD50 o-r: 13000; LD50 o-m: 6000; LD50 ip-r: 1600; LD50 ip-m: 800; LD50 cu-ra: > 5000

Diethyl carbonate : $C_5H_{10}O_3$ • 118.13 • 105-58-8
- (P) Eb = 125-128 • P_{vap} = 13.3(23.8); 78.7(37.)
- (I) P_{fla} = 25cc; 25oc; 31
- (T) NFPA: 2; –
- (D) MLD o-r: 15000; LD50 scu-r: 8500

Diphenyl carbonate : $C_{13}H_{10}O_3$ • 214.22 • 102-09-0
- (P) ΔH_f^0 = –401.37(s) • Eb = 301-306; 168/15
- (I) P_{fla} = 168
- (D) LD50 o-r: 1800; LD50 cu-ra: 2000

8 • Table of organic compounds

Esters - Lactones

Name: Empirical formula • Molar mass(g/mol)• Registry number (CAS).
(P) Physical data: ΔH_f^0 (kJ/mol) (physical state) • Eb (°C/mmHg) (if substance volatile) • P_{vap} (mBar) (°C) (if substance volatile).
(I) Inflammability data: LEL/UEL (%) • P_{fla} (°C) • AIT (°C).
(T) Toxicity: Safety Code (Tr) • NFPA • Transport Code (TR) • LVE (MVE) (ppm) F, USA, D • STEL (ppm); IDLH (ppm).
(D) Lethal and toxic doses: LC (mg/l/duration of exposure); LD (mg/kg).

Ethylene carbonate : $C_3H_4O_3$ • 88.06 • 96-49-1
(P) ΔH_f^0 = –581.16(s) • Eb = 238; 244/740; 246; 248; 177/100; 156/50; 116/10 • P_{vap} = 0.013(20); 0.027(36.4)
(I) P_{fla} = 143oc; 150cc; 160co • AIT = 465
(T) Tr: 36/38 • NFPA: 2
(D) LD50 o-r: 10000; > 10000; MLD ip-m: 500

Propylene carbonate : $C_4H_6O_3$ • 102.09 • 108-32-7
(P) Eb = 239-242 • P_{vap} = 0.027; 0.04; 0.17(20); 1.31(50)
(I) LEL/UEL = 1.8/14.3; 1.9/-; • P_{fla} = 123; 132oc; 134cc; 135oc • AIT = 430; 454
(T) Tr: 36 • NFPA: 1
(D) LD50 o-r: 29000; LD50 o-m: 20700; LD50 scu-r: 11100; LD50 scu-m: 15800; LD50 cu-ra: > 20000

ESTERS - LACTONES
• Aromatic esters, diesters

Phenyl acetate : $C_8H_8O_2$ • 136.15 • 122-79-2
(P) ΔH_f^0 = –334.80(l) • Eb = 193-196; 75/8 • P_{vap} = 0.25(20)
(I) P_{fla} = 76; 80; 94; 104
(T) NFPA: 1
(D) LD50 o-r: 1630; LD50 cu-ra: 8000

Benzyl acetate : $C_9H_{10}O_2$ • 150.18 • 140-11-4
(P) Eb = 206; 213-214; 216; 134/102; 100/14 • P_{vap} = 0.25; 1.33(20); 1.9(25); 0.6(30); 1.33(45); 2.8(50); 31(110)
(I) P_{fla} = 95; 102cc • AIT = 461
(T) NFPA: 1
(D) MLC r: 1.3/22h; LD50 o-r: 2490; LD50 o-m: 830; LD50 o-ra: 2200; MLD scu-m: 3000; LD50 cu-ra: > 5000

Methyl benzoate : $C_8H_8O_2$ • 136.15 • 93-58-3
(P) ΔH_f^0 = –333.88(l) • Eb = 197-202; 96/24 • P_{vap} = 1.33(39)
(I) LEL/UEL = 8.6/20 • P_{fla} = 82-85; 82cc • AIT = 518
(T) Tr: 22 • NFPA: 0 • TR: 60
(D) LD50 o-r: 1177; 1350; 3430; LD50 o-m: 3330; LD50 o-ra: 2170

Ethyl benzoate : $C_9H_{10}O_2$ • 150.18 • 93-89-0
(P) Eb = 211-214 • P_{vap} = 0.24(20); 1.33(44)
(I) LEL/UEL = 1.0/– • P_{fla} = 84; 87cc; 91; 93; 96; > 96 • AIT = 489; 510
(T) NFPA: 1
(D) LD50 o-r: 2100; 6480; LD50 o-ra: 2630

Propyl benzoate : $C_{10}H_{12}O_2$ • 164.20 • 2315-68-6
(P) Eb = 229-231; 229.5/766 • P_{vap} = 0.16(20)
(I) P_{fla} = 98

Isopropyl benzoate : $C_{10}H_{12}O_2$ • 164.20 • 938-48-0
(P) Eb = 218-219 • P_{vap} = 0.16(20)
(I) P_{fla} = 99oc
(T) NFPA: 1
(D) LD50 o-r: 3730; LD50 cu-ra: 20000

Butyl benzoate : $C_{11}H_{14}O_2$ • 178.23 • 136-60-7
(P) Eb = 249-250; 155/50; 118/10
(I) P_{fla} = 107oc • AIT = 440
(T) Tr: 36/37/38 • NFPA: 1
(D) LD50 o-r: 735; 5140; LD50 cu-ra: 4000

Benzyl benzoate : $C_{14}H_{12}O_2$ • 212.25 • 120-51-4
(P) Eb = 323-324; 189-191/16; 173/13; 156/4.5 • P_{vap} = 1.33(125); 6(156)
(I) P_{fla} = 147cc; ~158 • AIT = 480
(T) Tr: 22 • NFPA: 1 • TR: –
(D) LD50 o-r: 500; 1700; LD50 o-m: 1400; LD50 o-ra: 1680; 1800; LD50 cu-ra: 4000

Ethyl phenylacetate : $C_{10}H_{12}O_2$ • 164.20 • 101-97-3
(P) Eb = 226-229; 135/32; 121/20
(I) P_{fla} = 77; 97; 100
(D) LC50 r: LD50 o-r: 3300

Methyl cinnamate: $C_{10}H_{10}O_2$ • 162.19 • 103-26-4
(P) Eb = 260-263; 261.9; 132.5-134/15; 127/10
(I) P_{fla} = 100; > 109
(D) LD50 o-r: 2610

Ethyl cinnamate : $C_{11}H_{12}O_2$ • 176.22 • 103-36-6
(P) Eb = 272; 144/15; 132-135/10
(I) P_{fla} = 100; 109; 135
(D) LD50 o-r: 4000; LD50 o-m: 4000

Dimethyl phthalate : $C_{10}H_{10}O_4$ • 194.19 • 131-11-3
(P) ΔH_f^0 = –677.1(l) • Eb = 282-285; 257./400; 232.7/200; 210/100; 194/60; 182.8/40; 164/20; 147.6/10; 134/5; 131.8/5; 100.3/1 • P_{vap} = 0.0023; 0.13(20); 1.33(100.3)
(I) LEL/UEL = 0.94(181)/8.03(229) • P_{fla} = 145cc; 149oc • AIT = 489; 556
(T) Tr: 36/37/38;– • NFPA: 0 • TR: – • VLE/VME(F): – (5 mg/m^3); TWA (USA): 5 mg/m^3
(D) LD50 o-r: 6800; 6900; LD50 o-m: 6800; 7200; LD50 o-ra: 4400; LD50 ip-r: 3375; LD50 ip-m: 1380; 1500; MLD scu-m: 6500; LD50 cu-ra: > 20000

Diethyl phthalate : $C_{12}H_{14}O_4$ • 222.24 • 84-66-2
(P) ΔH_f^0 = –778.22(l) • Eb = 296-302; 295/734; 220/100; 156/10; 109/1 • P_{vap} = 0.00044(20); 0.00144(30); 0.013(50); 1.33(100); 18.7(163); 133(220)
(I) LEL/UEL = 0.75(187)/- • P_{fla} = 117; 156; 159oc; 163oc • AIT = 456
(T) Tr: 36/37/38 • NFPA: 0 • TR: – • VLE/VME(F): – (5 mg/m^3); TWA (USA): 5 mg/m^3
(D) LC50 r: 7.51/?; LD50 o-r: 8600; LD50 o-m: 6172; MLD o-ra: 1000; LD50 ip-r: 5058; LD50 ip-m: 2749; MLD iv-ra: 100

443

8 • Table of organic compounds — Esters - Lactones

Name: Empirical formula • Molar mass(g/mol)• Registry number (CAS).
- (P) **Physical data:** ΔH_f^0 (kJ/mol) (physical state) • Eb (°C/mmHg) (if substance volatile) • P_{vap} (mBar) (°C) (if substance volatile).
- (I) **Inflammability data:** LEL/UEL (%) • P_{fla} (°C) • AIT (°C).
- (T) **Toxicity:** Safety Code (Tr) • NFPA • Transport Code (TR) • LVE (MVE) (ppm) F, USA, D • STEL (ppm); IDLH (ppm).
- (D) **Lethal and toxic doses:** LC (mg/l/duration of exposure); LD (mg/kg).

Dibutyl phthalate : $C_{16}H_{22}O_4$ • 278.35 • 84-74-2
- (P) ΔH_f^0 = –840.98(s) • Eb = 340; 287/200; 227-235/37; 182/5; 152-155/1 • P_{vap} = 0.1(100); 1.33(147); 1.48(150); 2.6(162); 17.6(200); 53.3(232)
- (I) LEL/UEL = 0.5(236)/- ; 0.1/1.97 • P_{fla} = 157cc; 171Coc; 175oc; 180Coc • AIT = 400; 403; 410
- (T) Tr: 36/37/38-40; 36/37 • NFPA: 0 • TR: –
  • VLE/VME(F): – (5 mg/m^3); TWA (USA): 5 mg/m^3
- (D) LC50 r: 4.25/?; 7.9/?; LC50 m: 25/2h; LD50 o-r: 8000; LD50 o-m: 5289; LD50 ip-r: 3050; LD50 ip-m: 3500; 3570; LD50 iv-m: 720; LD50 cu-ra: > 20000

Bis(2-ethylhexyl) phthalate: $C_{24}H_{38}O_4$ • 390.56 • 117-81-7
- (P) Eb = 384-386; 391; 231/5 • P_{vap} = 1.6(93); 6.7(180); 1.76(200); 26.7(213)
- (I) LEL/UEL = 0.28/- ; 0.3/- ; 0.3/2.4 • P_{fla} = 199; 207; 215oc; 218Coc
  • AIT = 389; 400; 410
- (T) Tr: 36/37/38; 40 • NFPA: 0 • TR: – • VLE/VME(F): –(5 mg/m^3); TWA (USA): 5 mg/m^3; MAK(D): 10 mg/m^3 • STEL: 10 mg/m^3;
- (D) LD50 o-r: 30600; 31000; LD50 o-m: 30000; LD50 o-ra: 34000; LD50 ip-r: 30700; LD50 iv-r: 250; LD50 ip-m: 14000; LD50 iv-m: 1060; LD50 cu-ra: 15000; 25000

Dimethyl terephthalate : $C_{10}H_{10}O_4$ • 194.19 • 121-61-6
- (P) ΔH_f^0 = –711.28(s); Eb = 288; 300(subl) • P_{vap} = 1.53(93); 17.33(150)
- (I) LEL/UEL = 0.24/- • P_{fla} = 141 • AIT = 569
- (D) LD50 o-r: > 3200; 4390; LD50 ip-r: 3900

Diethyl terephthalate : $C_{12}H_{14}O_4$ • 222.24 • 636-09-9
- (P) Eb = 296; 302
- (I) P_{fla} = 117; 163
- (T) NFPA: 0
- (D) MLD ip-m: 1111

ESTERS - LACTONES

• Lactones

β-Propiolactone: $C_3H_4O_2$ • 72.06 • 57-57-8
- (P) Eb = 155; 162(dec.); 150/750; 61/20; 51/10 • P_{vap} = 5.6(25); 12(50)
- (I) LEL/UEL = 2.9/- • P_{fla} = 69; 74oc
- (T) Tr: 45-26-36/38 • NFPA: 4; – • TR: – • TWA (USA): 0.5
- (D) LC50 r: 0.075/6h; LD50 ip-m: 405

Diketen: $C_4H_4O_2$ • 84.07 • 674-82-8
- (P) ΔH_f^0 = –233.13(l) • Eb = 127.4; 69-70/100 • P_{vap} = 10.53(20); 68(55)
- (I) P_{fla} = 34Toc; 46oc • AIT = 309
- (T) NFPA: 2
- (D) MLC r: 69.912/1h; LD50 o-r: 560; LD50 o-m: 800; LD50 cu-ra: 2830

γ-Butyrolactone: $C_4H_6O_2$ • 86.09 • 96-48-0
- (P) Eb = 204-206; 89/12; 80-81/11; 89/10 • P_{vap} = 1.6; 2(20); 13(25)
- (I) LEL/UEL = 0.25/- ; 1.4/16; 3.6/16 • P_{fla} = 98oc; 102; ~104cc
  • AIT = 454
- (T) Tr: 22-36 • NFPA: 1; – • TR: –
- (D) LD50 o-r: 1540; 1580; 19400; LD50 o-m: 1710; LD50 ip-r: 1000; LD50 ip-m: 1100; LD50 cu-ra: > 5000

δ-Valerolactone: $C_5H_8O_2$ • 100.12 • 542-28-9
- (P) Eb = 206; 226-230; 58-60/0.5 • P_{vap} = 95(50)
- (I) LEL/UEL = 1.7/9 • P_{fla} = 96; 99; 112cc • AIT = 390
- (D) LD50 o-r: > 2000

ε-Caprolactone: $C_6H_{10}O_2$ • 114.14 • 502-44-3
- (P) Eb = 146/50; 96-97/15; 108/10; 98-99/2 • P_{vap} = 0.013(20)
- (I) P_{fla} = 109
- (D) LD50 o-r: 4290; LD50 ip-m: 1300; LD50 cu-ra: 5990

ESTERS - LACTONES
• Esters with complex groups

• Alcohol-esters

Ethyl lactate : $C_5H_{10}O_3$ • 118.13 • 97-64-3
- (P) Eb = 150-154; P_{vap} = 2.7; 3.73(20); 6.7(30)
- (I) LEL/UEL = 1.55(100)/30(100); 1.5/11.4 • P_{fla} = 46cc; 48; 55 (impur) • AIT = 400
- (T) NFPA: 2
- (D) LD50 o-r: 5000; > 5000; LD50 o-m: 2500; MLD ip-r: 1000; LD50 iv-m: 600; LD50 scu-r: 2500; LD50 scu-m: 2500; LD50 cu-ra: 5000; > 5000

• Phenol-esters

Methyl salicylate : $C_8H_8O_3$ • 152.15 • 119-36-8
- (P) ΔH_f^0 = –531.79(l) • Eb = 219-223 • P_{vap} = 0.013(20); 1.33(54)
- (I) P_{fla} = 96; 99cc; 101cc; > 109 • AIT = 454
- (T) Tr: 22 • NFPA: 1
- (D) LD50 o-r: 887; LD50 o-m: 1110; LD50 o-ra: 1300; MLD scu-m: 4250

• Ether-esters

2-methoxyethyl acetate : $C_5H_{10}O_3$ • 118.13 • 110-49-6
- (P) Eb = 140-145 • P_{vap} = 2.7; 3.2; 4.4; 9.3(20); 31(50); 200(100)
- (I) LEL/UEL = 1.5/8.2; 1.7/8.2 • P_{fla} = 44cc; 46; 49Tcc; 52cc; 57Toc; 60oc
  • AIT = 380; 394
- (T) Tr: 60-61-20/21/ 22 • NFPA: 3;1 • TR: 30
  • VLE/VME(F): –(5); TWA (USA): 5(male); 2(female); MAK(D): 5
- (D) MLC r: 34.383/4h; LD50 o-r: 3390; 3400; MLD ip-r: 1200; LD50 cu-ra: 5250; 5285

8 • Table of organic compounds

Esters - Lactones

Name: Empirical formula • Molar mass(g/mol)• Registry number (CAS).
(P) Physical data: ΔH_f^0 (kJ/mol) (physical state) • Eb (°C/mmHg) (if substance volatile) • P_{vap} (mBar) (°C) (if substance volatile).
(I) Inflammability data: LEL/UEL (%) • P_{fla} (°C) • AIT (°C).
(T) Toxicity: Safety Code (Tr) • NFPA • Transport Code (TR) • LVE (MVE) (ppm) F, USA, D • STEL (ppm); IDLH (ppm).
(D) Lethal and toxic doses: LC (mg/l/duration of exposure); LD (mg/kg).

2-Ethoxyethyl acetate: $C_6H_{12}O_3$ • 132.16 • 111-15-9
- (P) Eb = 156-159 • P_{vap} = 1.6; 2.67(20); 5.1(30); 9.5(50); 150(100)
- (I) LEL/UEL = 1.2/12.7; 1.7/6.7; 1.7/10.1; 1.7/12.5; 1.7/13
  • P_{fla} = 47Coc; 49; 52cc; 54; 55oc; 57 • AIT = 380; 382
- (T) Tr: 60-61-20/21/ 22 • NFPA: 2 • TR: 30
  • VLE/VME(F): –(5 skin); TWA (USA): 5 skin;
  MAK(D): 20 skin • IDLH: 2500
- (D) LC50 r: 12.1/8h; 66.49/?; LC50 ra: > 11/4h; LD50 o-r: 2700; 2900; 5100; LD50 o-ra: 1950; LD50 ip-m: 1420; LD50 cu-ra: 10500

2-Butoxyethyl acetate: $C_8H_{16}O_3$ • 160.21 • 112-07-2
- (P) Eb = 188-192.3; 195 • P_{vap} = 0.31; 0.39(20); 150(140)
- (I) LEL/UEL = 0.88(33)/8.54(135); 0.5/3.7; 1.6/8.4; 1.7/8.4
  • P_{fla} = 68; 71cc; 74Tcc; 76; 82; 88 • AIT = 280; 339; 355; 385
- (T) Tr: 20/21 • NFPA: 1 • TR: –; MAK(D): 20
- (D) LD50 o-r: 2400; LD50 o-m: 3200; LD50 cu-ra: 1500

2-Methoxypropyl acetate: $C_6H_{12}O_3$ • 132.16 • 108-65-6
- (P) Eb = 145-146 • P_{vap} = 3.2; 4.67; 4.93(20); 3.8(25)
- (I) LEL/UEL = 1.3/13.1; 1.5/10; 1.7/5.8; 1.7/8.2 • P_{fla} = 43; 46; 47.2PMcc; 51Coc • AIT = 272; 380
- (T) Tr: 36
- (D) LD50 o-r: 8532; LD50 ip-m: 750; LD50 cu-ra: > 5000

3-Methoxybutyl acetate: $C_7H_{14}O_3$ • 146.19 • 4435-53-4
- (P) Eb = 169 • P_{vap} = 1.5(20); 25(50); 45(65)
- (I) LEL/UEL = 0.8/4.7 • P_{fla} = 62; 77 • AIT = 410
- (T) Tr: 36 • TR: 30
- (D) LD50 o-r: 4210

2-Methoxyethyl acrylate: $C_6H_{10}O_3$ • 130.14 • 3121-61-7
- (P) Eb = 156; 61/17; 56/12
- (I) P_{fla} = 59; 82oc
- (T) Tr: 36/37/38
- (D) MLC r: 2.71/4h; LD50 o-r: 810; LD50 cu-ra: 250

• Halogenated esters

Ethyl fluoroacetate: $C_4H_7FO_2$ • 106.10 • 459-72-3
- (P) Eb = 115-118; 122; 119/753 • P_{vap} = 3.7(20)
- (I) P_{fla} = 29; 31
- (T) Tr: 26/27/28 • NFPA: 4 • TR: 663
- (D) LD50 ip-m: 19

Methyl trifluoroacetate: $C_3H_3F_3O_2$ • 128.50 • 431-47-0
- (P) Eb = 42-44 • P_{vap} = 398(20); 1525(55)
- (I) P_{fla} = 20; 7
- (T) Tr: 34

Methyl chloroacetate: $C_3H_5ClO_2$ • 108.52 • 96-34-4
- (P) Eb = 128-132; 130/740 • P_{vap} = 7; 7.2(20); 13.3(25)
- (I) LEL/UEL = 7.5/18.5 • P_{fla} = 47; 51oc; 54 • AIT = 464
- (T) Tr: 23/25-37/38-41 • NFPA: – • TR: 63
- (D) LC50 m: 1/2h; LD50 o-m: 240

Ethyl chloroacetate: $C_4H_7ClO_2$ • 122.55 • 105-39-5
- (P) Eb = 141-145 • P_{vap} = 4.4; 4.8(20); 8.8(30); 13.4(37.5); 30(50)
- (I) P_{fla} = 38; 54; 66
- (T) Tr: 23/24/25 • NFPA: 2 • TR: 63
- (D) LD50 o-r: 235; LD50 o-m: 350; LD50 scu-m: 250; LD50 cu-ra: 230

Methyl 2-chloropropionate: $C_4H_7ClO_2$ • 122.55 • 73246-45-4 (77287-29-7)
- (P) Eb = 127; 132-134; 80-82/110; 49/36 • P_{vap} = 7(20); 13(26)
- (I) LEL/UEL = 2.5/19 • P_{fla} = 37
- (T) Tr: 36/37/38
- (D) MLD ip-m: 250

Ethyl 2-chloropropionate: $C_5H_9ClO_2$ • 136.58 • 535-13-7
- (P) Eb = 144-148; 52/18
- (I) P_{fla} = 38; 42; 50; 65

Ethyl 3-chloropropionate: $C_5H_9ClO_2$ • 136.58 • 623-71-2
- (P) Eb = 163; 50-51/12
- (I) P_{fla} = 54; 61; 75

Methyl dichloroacetate: $C_3H_4Cl_2O_2$ • 142.97 • 116-54-1
- (P) Eb = 141-145 • P_{vap} = 115.7(20)
- (I) P_{fla} = 60; 64; 79 • AIT = 484
- (T) Tr: 20-36/37/38
- (D) MLC cat: 11.89/30'

Ethyl dichloroacetate: $C_4H_6Cl_2O_2$ • 157.00 • 535-15-9
- (P) Eb = 158; 54-55/11
- (I) P_{fla} = 60; 62
- (T) Tr: 20/21-36/37

Methyl trichloroacetate: $C_3H_3Cl_3O_2$ • 177.41 • 598-99-2
- (P) Eb = 152-153
- (I) P_{fla} = 72

Methyl bromoacetate: $C_3H_5BrO_2$ • 152.98 • 96-32-2
- (P) Eb = 144-147; 51-52/15; (265dec)
- (I) P_{fla} = 62
- (T) Tr: 34-37
- (D) MLD iv-m: 16

Ethyl bromoacetate: $C_4H_7BrO_2$ • 167.00 • 105-36-2
- (P) Eb = 158-159; 168; 58/15 • P_{vap} = 4.5(20); 3.47(25); 6.5(30); 15(50)
- (I) P_{fla} = 47cc; 51 • AIT = 565
- (T) Tr: 26/27/28 • NFPA: 3; – • TR: 63

Ethyl 2-bromopropionate: $C_5H_9BrO_2$ • 181.03 • 535-11-5
- (P) Eb = 156-160; 160-165; 71/26; 51-53/10
- (I) P_{fla} = 51; 61; 66; 81
- (T) Tr: 34

8 • Table of organic compounds
Acid anhydrides — Acid chlorides

Name: Empirical formula • Molar mass(g/mol) • Registry number (CAS).
- **(P) Physical data:** ΔH_f^0 (kJ/mol) (physical state) • Eb (°C/mmHg) (if substance volatile) • P_{vap} (mBar) (°C) (if substance volatile).
- **(I) Inflammability data:** LEL/UEL (%) • P_{fla} (°C) • AIT (°C).
- **(T) Toxicity:** Safety Code (Tr) • NFPA • Transport Code (TR) • LVE (MVE) (ppm) F, USA, D • STEL (ppm); IDLH (ppm).
- **(D) Lethal and toxic doses:** LC (mg/l/duration of exposure); LD (mg/kg).

Ethyl 3-bromopropionate : $C_5H_9BrO_2$ • 181.03 • 539-74-2
- (P) Eb = 179; 135-136/50; 122/44; 70/12; 69/10
- (I) P_{fla} = 61; 70; 79
- (T) Tr: 36/37/38

• Ketonic esters

Methyl acetoacetate : $C_5H_8O_3$ • 116.12 • 105-45-3
- (P) Eb = 169-171; 81/28; 73/12 • P_{vap} = 0.93; 2.27(20); 8(25)
- (I) LEL/UEL = 1.8/8.0; 3.1/16 • P_{fla} = 62; 70; 74oc; 77; 82 • AIT = 280
- (T) Tr: 36/37/38; 36 • NFPA: 2 • TR: 30
- (D) LD50 o-r: 3000; 3228

Ethyl acetoacetate : $C_6H_{10}O_3$ • 130.14 • 141-97-9
- (P) Eb = 179-181; 158.2/400; 138/200; 119/100; 106/60; 96/40; 81.1/20; 54/5; 28.5/1 • P_{vap} = 1; 1.07(20); 3(25); 1.33(28.5)
- (I) LEL/UEL = < 1/- ; 1.4/9.5; 1.6/10.4; 2.0/15.2 • P_{fla} = 63; 84cc; 84oc • AIT = 295; 350
- (T) Tr: 36/38; – • NFPA: 2 • TR: 30
- (D) LD50 o-r: 3980; LD50 o-m: 5105

Butyl acetoacetate : $C_8H_{14}O_3$ • 158.20 • 591-60-6
- (P) Eb = 214 • P_{vap} = 0.25(20)
- (I) P_{fla} = 60; 85
- (T) NFPA: 1
- (D) LD50 o-r: 11260

t-Butyl acetoacetate : $C_8H_{14}O_3$ • 158.20 • 1694-31-1
- (P) Eb = 76/35; 56/15; 72/13; 71/11
- (I) P_{fla} = 52; 58; 60
- (T) Tr: 36/37/38

Allyl acetoacetate : $C_7H_{10}O_3$ • 142.15 • 1118-84-9
- (P) Eb = 194-195/737; 84/13
- (I) P_{fla} = 67; 75

Methyl 2-chloroacetoacetate: $C_5H_7ClO_3$ • 150.56 • 4755-81-1
- (P) Eb = 137; 58/15
- (I) P_{fla} = 71; 85 • AIT = 220

Ethyl 2-chloroacetoacetate : $C_6H_9ClO_3$ • 164.59 • 609-15-4
- (P) Eb = 196-199; 107/14; 77/15
- (I) P_{fla} = 50; 54; 87
- (T) Tr: 36/37/38

Ethyl 4-chloroacetoacetate : $C_6H_9ClO_3$ • 164.59 • 638-07-3
- (P) Eb = 209-211; 115/14; 103/16
- (I) P_{fla} = 96; 105
- (T) Tr: 36/37/38
- (D) LD50 ip-r: 108; LD50 ip-m: 88

ACID ANHYDRIDES
ACID CHLORIDES

• Acid anhydrides

Acetic anhydride : $C_4H_6O_3$ • 102.09 • 108-24-7
- (P) ΔH_f^0 = –575.72(g); –624.00(l) • Eb = 138-140; 65/50; 36/10 • P_{vap} = 4.7; 5; 5.33(20); 13.3(36); 300(100)
- (I) LEL/UEL = 2/10; 2/10.2; 2.7/ 10.3; 2.8/12.4; 2.9/10.3; 3/10 • P_{fla} = 49Tcc; 52; 54cc; 64Toc • AIT = 315; 331; 380; 385; 389; 392
- (T) Tr: 34 • NFPA: 2 • TR: 83 • LVE(MVE) (F): 5 (-); TWA (USA): 5; MAK(D): 5 • IDLH: 1000
- (D) LC50 r: 4.17/4h; LD50 o-r: 1780; 1800; LD50 cu-ra: 4000

Propionic anhydride : $C_6H_{10}O_3$ • 130.14 • 123-62-6
- (P) ΔH_f^0 = –675.84(l) • Eb = 165-169; 146/400; 127./200; 107.2/100; 94.5/60; 90/50; 85.6/40; 70.4/20; 57.7/10; 45.3/5; 20.6/1 • P_{vap} = 1.16; 1.33(20); 13.3(57.7)
- (I) LEL/UEL = 1.3/9.5; 1.48/11.9 • P_{fla} = 62cc; 64; 74oc • AIT = 284; 316
- (T) Tr: 34 • NFPA: 2 • TR: 80
- (D) LD50 o-r: 2360; LD50 cu-ra: 10000

Butyric anhydride : $C_8H_{14}O_3$ • 158.20 • 106-31-0
- (P) Eb = 198-201.4; 116/50; 81/10 • P_{vap} = 0.3; 0.4(20); 13.3(79.5)
- (I) LEL/UEL = 1.1(104)/7.6(144) • P_{fla} = 82; 88oc • AIT = 279; 307
- (T) Tr: 34 • NFPA: 3 • TR: 80
- (D) LD50 o-r: 8790; LD 30 o-m: 2000; LD50 cu-ra: 6400

Crotonic anhydride: $C_8H_{10}O_3$ • 154.17 • 78957-07-0
- (P) Eb = 246-250 • P_{vap} = 1(20)
- (I) P_{fla} = 93; 110
- (D) LD50 o-r: 2830

Isobutyric anhydride : $C_8H_{14}O_3$ • 158.20 • 97-72-3
- (P) Eb = 179-183 • P_{vap} = 13.3(67)
- (I) LEL/UEL = 1/6.2; 1.1(87)/ 7.7(127) • P_{fla} = 59; 67 • AIT = 329; 352

Methacrylic anhydride : $C_8H_{10}O_3$ • 154.17 • 760-93-0
- (P) Eb = 87/13
- (I) P_{fla} = 84
- (T) Tr: 26-34
- (D) LC50 m: 0.45/2h

Succinic anhydride: $C_4H_4O_3$ • 100.07 • 108-30-5
- (P) Eb = 261; 237/400; 212/200; 189/100; 174/60; 163/40; 145/20; 130/19; 128/10; 115/5(subl); 92/1(subl) • P_{vap} = 1.33(92)
- (I) P_{fla} = 157
- (T) Tr: 36/37 • NFPA: 2 • TR: –;
- (D) LD50 o-r: 1510

8 • Table of organic compounds

Acid anhydrides
Acid chlorides

Name: Empirical formula • Molar mass(g/mol)• Registry number (CAS).
- (P) **Physical data:** ΔH_f^0 (kJ/mol) (physical state) • Eb (°C/mmHg) (if substance volatile) • P_{vap} (mBar) (°C) (if substance volatile).
- (I) **Inflammability data:** LEL/UEL (%) • P_{fla} (°C) • AIT (°C).
- (T) **Toxicity:** Safety Code (Tr) • NFPA • Transport Code (TR) • LVE (MVE) (ppm) F, USA, D • STEL (ppm); IDLH (ppm).
- (D) **Lethal and toxic doses:** LC (mg/l/duration of exposure); LD (mg/kg).

Maleic anhydride : $C_4H_2O_3$ • 98.06 • 108-31-6
- (P) ΔH_f^0 = –471.96(s) • Eb = 198-202; 179.5/400; 155.9/200; 135.8; 137/100; 122/60; 111.8/40; 95/20; 78.7/10; 63.4/5 • P_{vap} = 0.21; 0.27(20); 1.33(44)
- (I) LEL/UEL = 1.4/7.1 • P_{fla} = 102cc • AIT = 380; 465; 477
- (T) Tr: 22-36/37/38-42 • NFPA: 3 • TR: 80
  • LVE(MVE) (F): 0.25 (?); TWA (USA): 0.25; MAK(D): 0.1; 0.2
- (D) LD50 o-r: 400; 481; LD50 o-m: 465; LD50 o-ra: 875; LD50 ip-r: 97; LD50 cu-ra: 2620

Pentanoic anhydride: $C_{10}H_{18}O_3$ • 186.25 • 2082-59-9
- (P) Eb = 215; 218; 228-230; 111-112/16
- (I) P_{fla} = 99; 101

Isopentanoic anhydride: $C_{10}H_{18}O_3$ • 186.25; –
- (P) ΔH_f^0 = –468.94(l) • Eb = 215; 102/15

Pivalic anhydride: $C_{10}H_{18}O_3$ • 186.25 • 1538-75-6
- (P) ΔH_f^0 = –780.27(l) • Eb = 192-194
- (I) P_{fla} = 57

Glutaric anhydride: $C_5H_6O_3$ • 114.10 • 108-55-4
- (P) Eb = 303; 173/25; 144-146/13; 150/10
- (I) P_{fla} = 173
- (T) Tr: 21-36/37/38-41
- (D) MLD o-r: 4460; MLDcu-ra: 1780

Hexanoic anhydride: $C_{12}H_{22}O_3$ • 214.30 • 2051-49-2
- (P) Eb = 241-245; 246-248; 256-259
- (I) P_{fla} = > 109
- (T) Tr: 34

Benzoic anhydride: $C_{14}H_{10}O_3$ • 226.23 • 93-97-0
- (P) ΔH_f^0 = –430.95(s) • Eb = 360; 299.1/200; 252.7/60; 239.8/40; 218/20; 198/10; 180/5; 143.8/1
- (I) P_{fla} = > 109

Phthalic anhydride: $C_8H_4O_3$ • 148.12 • 85-44-9
- (P) ΔH_f^0 = –460.88(s) • Eb = 284; 295(subl); 202/100; 152/20; 97/5
  • P_{vap} = 0.00027; 0.002; 0.027(20); 0.0013(30); 0.02(50); 1.33(96.5); 8(F: 131); 130(200)
- (I) LEL/UEL = 1.7/10.4; 1.7/10.5 • P_{fla} = 151cc; 152Tcc; 155cc; 165Toc
  • AIT = 538; 570; 580; 584
- (T) Tr: 36/37/38 • NFPA: 2 • TR: 80
  • LVE(MVE)(F): 6.1 mg/m³ (–); TWA (USA): 6.1 mg/m³; MAK(D): 1 mg/m³ • IDLH: 61 mg/m³
- (D) LC50 r: > 0.21/1h; LD50 o-r: 4000; 4020; LD50 o-m: 1500; 2000; LD50 cu-ra: > 10000

Chloroacetic anhydride: $C_4H_4Cl_2O_3$ • 170.98 • 541-88-8
- (P) Eb = 120-123/20
- (I) P_{fla} = 143
- (T) Tr: 25-34-37

ACID ANHYDRIDES
ACID CHLORIDES

• **Acid chlorides**

• *Carboxylic acid chlorides*

Acetyl chloride : C_2H_3ClO • 78.50 • 75-36-5
- (P) ΔH_f^0 = –243.93(g); –273.80(l) • Eb = 50-52 • P_{vap} = 320; 806(20); 2229(55)
- (I) LEL/UEL = 5.0/-; 7.3/19 • P_{fla} = 4cc • AIT = 390; 733
- (T) Tr: 14-34 • NFPA: 3 • TR: X338
- (D) LD50 o-r: 910

Propionyl chloride: C_3H_5ClO • 92.53 • 79-03-8
- (P) Eb = 77-80 • P_{vap} = 390(50)
- (I) LEL/UEL = 3.6/11.9 • P_{fla} = 6; 12 • AIT = 270
- (T) Tr: 34-14 • NFPA: 3 • TR: 338

Acryloyl chloride: C_3H_3ClO • 90.51 • 814-68-6
- (P) Eb = 72-76 • P_{vap} = 133(20); 521(55)
- (I) P_{fla} = –4; –1; 14; 16
- (T) Tr: 34-36/37 • TR: 338
- (D) MLC r: 0.093/4h; LC50 m: 0.092/2h; LD50 iv-m: 180

Butyryl chloride: C_4H_7ClO • 106.55 • 141-75-3
- (P) Eb = 98-102; 107 • P_{vap} = 39(20)
- (I) P_{fla} = 8; 18; < 21; 21
- (T) Tr: 34 • NFPA: 3 • TR: 338

Crotonyl chloride : C_4H_5ClO • 104.54 • 625-35-4
- (P) Eb = 120-123
- (I) P_{fla} = 34
- (T) Tr: 34-36/37; 34 • TR: 83

Isobutyryl chloride : C_4H_7ClO • 106.55 • 79-30-1
- (P) Eb = 91-93
- (I) P_{fla} = 1; 8
- (T) Tr: 35 • TR: 338

Methacryloyl chloride: C_4H_5ClO • 104.54 • 920-46-7
- (P) Eb = 95-99 • P_{vap} =1 04(20); 518.5(55)
- (I) P_{fla} = 2; 22
- (T) Tr: 26-34-37
- (D) LC50 r: 0.060/4h; LC50 m: 0.115/2h

Pentanoyl chloride: C_5H_9ClO • 120.58 • 638-29-9
- (P) Eb = 126-129
- (I) P_{fla} = 32; 36
- (T) Tr: 34

Isopentanoyl chloride: C_5H_9ClO • 120.58 • 108-12-3
- (P) Eb = 110-117
- (I) P_{fla} = 18; 31; 33
- (T) Tr: 34-37

8 • Table of organic compounds

Acid anhydrides
Acid chlorides

Name: Empirical formula • Molar mass(g/mol)• Registry number (CAS).
(P) **Physical data:** ΔH_f^0 (kJ/mol) (physical state) • Eb (°C/mmHg) (if substance volatile) • P_{vap} (mBar) (°C) (if substance volatile).
(I) **Inflammability data:** LEL/UEL (%) • P_{fla} (°C) • AIT (°C).
(T) **Toxicity:** Safety Code (Tr) • NFPA • Transport Code (TR) • LVE (MVE) (ppm) F, USA, D • STEL (ppm); IDLH (ppm).
(D) **Lethal and toxic doses:** LC (mg/l/duration of exposure); LD (mg/kg).

Pivaloyl chloride: C_5H_9ClO • 120.58 • 3282-30-2
(P) Eb = 103-108; 48/100 • P_{vap} = 48(20)
(I) P_{fla} = 8; 19
(T) Tr: 34-36/37

Hexanoyl chloride ; $C_6H_{11}ClO$ • 134.61 • 142-61-0
(P) Eb = 150-153; 47-49/15
(I) P_{fla} = 49; 62; 79; 82
(T) Tr: 34 ; 34-37 • TR: 80

Cyclohexanecarbonyl chloride: $C_7H_{11}ClO$ • 146.62 • 2719-27-9
(P) Eb = 180-188
(I) P_{fla} = 66; 90
(T) Tr: 34-37

Benzoyl chloride : C_7H_5ClO • 140.57 • 98-88-4
(P) ΔH_f^0 = –163.80(l) • Eb = 196-200; 130/108; 100/35; 82.3/15; 74/11; 75/10; 71/9; 49/3 • P_{vap} = 0.5; 1; 1.33(20); 1.33(32)
(I) LEL/UEL = 1.2/4.9 • P_{fla} = 68; 72cc; 88; > 109 • AIT = 568
(T) Tr: 34 • NFPA: 3; 2 • TR: 80
(D) LC50 r: 1.87/4h; 1.87/2h; LD50 o-r: 1900

Oxalyl chloride: $C_2Cl_2O_2$ • 126.93 • 79-37-8
(P) ΔH_f^0 = –358.15(l) • Eb = 62-65 • P_{vap} = 200(20)
(T) Tr: 14-23/24/25-34 • NFPA: 3

Fumaryl chloride: $C_4H_2Cl_2O_2$ • 152.96 • 627-63-4
(P) Eb = 161; 56-58/11 • P_{vap} = 1.86(20)
(I) P_{fla} = 73; 77
(T) Tr: 20/21/22-34-37
(D) MLC r: 3.13/4h; LD50 o-r: 810; LD50 cu-ra: 1410

Chloroacetyl chloride: $C_2H_2Cl_2O$ • 112.94 • 79-04-9
(P) ΔH_f^0 = –284.51(l) • Eb = 101; 105-108 • P_{vap} = 25.3; 27(20); 80(41.5); 267(68.4)
(T) Tr: 34-37 • NFPA: 3 • TR: X80 • LVE(MVE) (F): – (0.05); TWA (USA): 0.05 • STEL: 0.15
(D) MLC r: 4.62/4h; LC50 m: 6.01/2h; LD50 o-r: 120; 208; LD50 o-m: 220; LD50 iv-m: 32; LD50 r-l: 662

Dichloroacetyl chloride: C_2HCl_3O • 147.39 • 79-36-7
(P) Eb = 102-108;107-108/760; 106-107/739 • P_{vap} = 30.6(20)
(I) LEL/UEL = 11.9/- • P_{fla} = 66 • AIT = 585
(T) Tr: 35 • NFPA: 3 • TR: X80
(D) MLC r: 11.98/4h; LD50 o-r: 2460; LD50 cu-ra: 650

Trichloroacetyl chloride: C_2Cl_4O • 181.83 • 76-02-8
(P) ΔH_f^0 = –277.2(l) • Eb = 114-119 • P_{vap} = 1.33; 21(20); 115(55)
(T) Tr: 14-22-26-35 • NFPA: 3
(D) LC50 r: 0.475/4h; LC50 m: 0.445/?; LD50 o-r: 600

2-Chloropropionyl chloride: $C_3H_4Cl_2O$ • 126.97 • 7623-09-8
(P) Eb = 109-113; 45/27 • P_{vap} = 5.2(20)
(I) P_{fla} = 26; 31
(T) Tr: 35-37

4-Chlorobutyryl chloride: $C_4H_6Cl_2O$ • 141.00 • 4635-59-0
(P) Eb = 173-174; 68-70/12; 60/12 • P_{vap} = 4(20)
(I) P_{fla} = 72; 80; 85; > 109 • AIT = 439
(T) Tr: 22-23-35

o-Chlorobenzoyl chloride: $C_7H_4Cl_2O$ • 175.01 • 609-65-4
(P) Eb = 225-238; 11/0.5
(I) P_{fla} = > 109; 110; 124 • AIT = 450
(T) Tr: 34-36/37; 34 • TR: 80

• *Phosgene and chloroformates*

Phosgene: Cl_2CO • 98.92 • 75-44-5
(P) ΔH_f^0 = –220.92(g) • Eb = 8-9 • P_{vap} = 1500; 1573; 1620(20); 2200(30); 4000(50)
(T) Tr: 26; NFPA: 3; – • TR: 266 • LVE(MVE) (F): 0.1; TWA (USA): 0.1; MAK(D): 0.1 • IDLH: 2
(D) LC50 r: 1.4/30'; LC50 m: 1.8/30'; LC50 ra: 1/30'

Methyl chloroformate: $C_2H_3ClO_2$ • 94.50 • 79-22-1
(P) Eb = 70-72 • P_{vap} = 137; 331(20); 839(55)
(I) LEL/UEL = 7./23.3 • P_{fla} = 5cc; 10; 12; 15-17; 24oc • AIT = 484; 504
(T) Tr: 23-36/37/38 • NFPA: 3 • TR: 336
(D) LC50 r: 0.06/4h; 0.345/1h; LC50 m: 0.185/2h; LD50 o-r: 25-200; 60; LD50 o-m: 67; LD50 ip-m: 40; LD50 su-m: 1750; LD50 cu-ra: 890; 7120

Ethyl chloroformate: $C_3H_5ClO_2$ • 108.52 • 541-41-3
(P) Eb = 92-95 • P_{vap} = 236(20); 495(55)
(I) P_{fla} = 2; 10; 13; < 16; 16cc; 18; 26oc • AIT = 449; 500
(T) Tr: 23-36/37/38 • NFPA: 3 • – • TR: 336
(D) LC50 r: 0.654//1h; MLC m: 2.26/10'; LD50 o-r: 270; 411; LD50 cu-ra: > 2000

Allyl chloroformate: $C_4H_5ClO_2$ • 120.54 • 2937-50-0
(P) Eb = 106-114; 57/97; 27/22; ~27/11 • P_{vap} = 253(20); 437(55)
(I) P_{fla} = 31cc
(T) Tr: 22-23-34; 22-26-34 • NFPA: 3 • TR: 88
(D) LC50 r: 0.032/?; LC50 m: 0.023/?; LD50 o-r: 244; LD50 o-m: 210

Benzyl chloroformate : $C_8H_7ClO_2$ • 170.60 • 501-53-1
(P) Eb = 178-180; 103/20; 85-87/7 • P_{vap} = 96(20); 195(55)
(I) P_{fla} = 91
(T) Tr: 34-37 • NFPA: 2 • TR: 88

8 • Table of organic compounds — Nitriles

Name: Empirical formula • Molar mass(g/mol) • Registry number (CAS).
- (P) **Physical data:** ΔH_f^0 (kJ/mol) (physical state) • Eb (°C/mmHg) (if substance volatile) • P_{vap} (mBar) (°C) (if substance volatile).
- (I) **Inflammability data:** LEL/UEL (%) • P_{fla} (°C) • AIT (°C).
- (T) **Toxicity:** Safety Code (Tr) • NFPA • Transport Code (TR) • LVE (MVE) (ppm) F, USA, D • STEL (ppm); IDLH (ppm).
- (D) **Lethal and toxic doses:** LC (mg/l/duration of exposure); LD (mg/kg).

NITRILES

• Aliphatic saturated nitriles and dinitriles

Hydrogen cyanide: CHN • 27.03 • 74-90-8
- (P) $\Delta H_f^0 = 108.87(l); 135.14(g); 130.15(g)$ • Eb = 25.6 • P_{vap} =480(7); 500; 811; 832(20); 877(22); 2500(50)
- (I) LEL/UEL = 5.6/40; 5.6/46.6; 6/41 • P_{fla} = –18Tcc; 15 • AIT = 425; 438; 535; 538
- (T) Tr: 26/27/28 • NFPA: 4 • TR: 663 • LVE(MVE) (F): 10 (2) ; TWA (USA): 10 (skin); MAK(D): 10 • IDLH: 60
- (D) LC50 r: 0.611/5'; 0.534/4h; LC50 m: 0.189/30'; 0.536 /4h; LD50 o-m: 3.7; LD50 ip-m: 3 ; LD50 iv-r: 810; LD50 iv-m: 990; LD50 scu-m: 3

Acetonitrile: C_2H_3N • 41.05 • 75-05-8
- (P) $\Delta H_f^0 = 87.6(g); 53.56(l); 53.1(l)$ • Eb = 81; 13/50; –15/10 • P_{vap} = 97(5); 97(20); 118(25); 133(27); 400(55)
- (I) LEL/UEL = 3/16; 4.4/16; 6/41 • P_{fla} = 2; 5oc; 5cc; 5.5; 6oc; 13 • AIT = 522; 525
- (T) Tr: 23/24/25 • NFPA: 3; 2 • TR: 336 • LVE(MVE) (F): – (40) ; TWA (USA): 40 ; MAK(D): 40 • STEL: 60; IDLH: 4000
- (D) LC50 r: 12.66/4h; LC50 m: 4.52/4h; LC50 ra: 4.74; LD50 o-r: 2460; 2730; 3800; LD50 o-m: 2690 ; LD50 o-ra: 50; LD50 ip-r: 850 ; LD50 ip-m: 175 ; LD50 iv-r: 1680; LD50 iv-m: 612; LD50 scu-r: 3500 ; LD50 scu-m: 4480; LD50 cu-ra: 1250

Propionitrile: C_3H_5N • 55.08 • 107-12-0
- (P) $\Delta H_f^0 = 14.64(l); 50.63(g)$ • Eb = 95-97; 77.7/400; 58.2/200; 41.4/100; 30.1/60; 22/40; 8.8/20 ; –3/10; –13.6/5; –35/1 • P_{vap} = 48; 52(20); 80(30); 190(50)
- (I) LEL/UEL = 3.1/– • P_{fla} = 2; 6; 9
- (T) Tr: 23/24/25 • NFPA: 4 • TR: –;
- (D) LC50 r: 1.12; LC50 m: 0.367; LD50 o-r: 39; LD50 o-m: 36; LD50 ip-r: 25; LD50 ip-m: 28; LD50 iv-ra: 50; LD50 cu-ra: 210

Butyronitrile: C_4H_7N • 69.11 • 109-74-0
- (P) $\Delta H_f^0 = 34.06(g)$ • Eb = 115-117; 117; 96.8/400; 76.7/200; 59/100; 47.3/60; 38.4/40; 25.7/20; 13.4/10; 2.1/5; –20/1 • P_{vap} = 13.3;20(20); 30.7(25); 36(30); 90(50)
- (I) LEL/UEL : 1.65/–; 1.60/– • P_{fla} = 16; 18; 21; 26oc; 29oc • AIT = 487; 501
- (T) Tr: 23/24/25 • TR: 336 • TWA (USA): 8
- (D) LC50 r: 2.82; ; LD50 o-r: 50; 140 ; LD50 o-m: 28; LD50 ip-r: 50 ; LD50 ip-m: 38; LD50 ip-ra: 1250 ; LD50 iv-ra: 980; LD50 scu-r: 703; LD50 scu-m: 10; LD50 cu-ra: 500

Isobutyronitrile: C_4H_7N • 69.11 • 78-82-0
- (P) $\Delta H_f^0 = 25.4(g)$ • Eb = 101; 107-108; 100-103 • P_{vap} = 200(50); 133(55); 350(65)
- (I) LEL/UEL %: 1.6/12.7 • P_{fla} = 8; 3 • AIT: 482
- (T) Tr: 23/24/25
- (D) LC50 r: 2.82; LD50 o-r: 50; 102; LD50 o-m: 25; LD50 o-ra: 14; LD50 ip-m: 25; LD50 scu-m: 9; LD50 cu-ra: 200; 310

Valeronitrile: C_5H_9N • 83.13 • 110-59-8
- (P) Eb = 139-141 • P_{fla} = 34; 40
- (T) Tr: 25
- (D) LD50 o-m: 191

Isovaleronitrile: C_5H_9N • 83.13 • 625-28-5
- (P) Eb = 128
- (I) P_{fla} = 25; 28
- (D) LD50 o-m: 233

Pivalonitrile: C_5H_9N • 83.13 • 630-18-2
- (P) Eb = 101; 105-106; 104-105
- (I) P_{fla} = 4; 21
- (T) Tr: 23/24/25

Capronitrile: $C_6H_{11}N$ • 7.16 • 628-73-9
- (P) Eb = 161; 164
- (I) P_{fla} = 43; 52
- (D) LD50 o-m: 463; LD50 iv-ra: 42

4-Methylvaleronitrile: $C_6H_{11}N$ • 97.16 • 942-54-1
- (P) Eb = 153-156
- (I) P_{fla} = 31; 45
- (D) LD50 o-r: 450; LD50 o-m: 129; 488; LD50 ip-m: 63; LD50 scu-m: 90

Hexanecarbonitrile: $C_7H_{13}N$ • 111.19 • 629-08-3
- (P) Eb = 187
- (I) P_{fla} = 58; 71

Cyclopropanecarbonitrile: C_4H_5N • 67.09 • 5500-21-0
- (P) Eb = 134-135
- (I) P_{fla} = 32; 38; 40

Cyclohexanecarbonitrile: $C_7H_{11}N$ • 109.17 • 762-05-2
- (P) Eb = 188; 75/16
- (I) P_{fla} = 65

Cyanogen: C_2N_2 • 52.04 • 460-19-5
- (P) $\Delta H_f^0 = 307.9(g)$ • Eb = –21 • P_{vap} = 4700(20); 6500(30); 11500(50)
- (I) LEL/UEL = 6/32; 6.6/32
- (T) Tr: 23 • NFPA: 4 • TR: – • LVE(MVE) (F): 10 (2); TWA (USA): 10; MAK(D): 10
- (D) LC50 r: 0.74; LD50 scu-ra: 13

8 • Table of organic compounds — Nitriles

Name: Empirical formula • Molar mass(g/mol)• Registry number (CAS).
(P) Physical data: ΔH_f^0 (kJ/mol) (physical state) • Eb (°C/mmHg) (if substance volatile) • P_{vap} (mBar) (°C) (if substance volatile).
(I) Inflammability data: LEL/UEL (%) • P_{fla} (°C) • AIT (°C).
(T) Toxicity: Safety Code (Tr) • NFPA • Transport Code (TR) • LVE (MVE) (ppm) F, USA, D • STEL (ppm); IDLH (ppm).
(D) Lethal and toxic doses: LC (mg/l/duration of exposure); LD (mg/kg).

Malononitrile: $C_3H_2N_2$ • 66.06 • 109-77-3
- **(P)** $\Delta H_f^0 = 186.61(l)$ • Eb = 198; 220; 218-219; 120/9 • P_{vap} = 1(50); 27(109)
- **(I)** P_{fla} = 62; 86; 112; 130oc
- **(T)** Tr: 23/24/25 • TR: 60 • LVE(MVE)(F): 5 mg/m³(CN); TWA (USA): 5 mg/m³
- **(D)** LD50 o-r: 61; LD50 o-m: 19; LD50 ip-r: 10; 21; LD50 ip-m: 12.9; LD50 iv-m: 32 ; LD50 iv-ra: 28; LD50 scu-r: 7; 32; LD50 scu-m: 8; LD50 scu-m: 6; LD50 cu-r: 350

Succinonitrile: $C_4H_4N_2$ • 80.09 • 110-61-2
- **(P)** Eb = 265; 185/60; 158-160/20; 90/2 • P_{vap} = 2.66(100)
- **(I)** P_{fla} = > 109; 132
- **(T)** Tr: 25 • NFPA: 3; – • TR: – • TWA (USA): 6

Glutaronitrile: $C_5H_6N_2$ • 94.12 • 544-13-8
- **(P)** Eb = 277-280; 285-288; 206/100; 190/60; 176/40; 157/20; 140/10; 124/5; 91.3/1
- **(I)** P_{fla} = > 109; 112
- **(T)** Tr: 20/21/22
- **(D)** LD50 o-m: 266; 2660 (?)

Tetramethylsuccinonitrile: $C_8H_{12}N_2$ • 136, 20 • 3333-52-6
- **(T)** NFPA: 3 • TWA (USA): 0.5 (skin)
- **(D)** LD50 o-r: 450; LD50 o-m: 129; LD50 ip-m: 63; LD50 scu-m: 36

Adiponitrile: $C_6H_8N_2$ • 108.14 • 111-69-3
- **(P)** Eb = 295; 180/20; 161-162/11; 154/10 • P_{vap} = 0.013(40)
- **(I)** LEL/UEL = 1.7/5 • P_{fla} = 93; 93oc; >109; 110; 159; 163 • AIT = 549
- **(T)** Tr: 23/24/25 • NFPA:3 • TR: 60 • TWA (USA): 2 (skin)
- **(D)** LC50 r: 1.71 ; LD50 o-r: 155; LD50 o-m: 172; LD50 o-ra: 22; LD50 ip-m: 40 ; LD50 iv-m:

NITRILES
•Unsaturated aliphatic and aromatic nitriles

Acrylonitrile: C_3H_3N • 53.06 • 107-13-1
- **(P)** $\Delta H_f^0 = 184.93(g)$; +151.04(l) • Eb = 77-78; 64.7/500; 45.5/250; 23.6/100; 8.7/50 • P_{vap} = 110; 111; 115; 120(20); 133(23)
- **(I)** LEL/UEL = 2.8/28; 3.05/17 • P_{fla} = –5; 0; 0oc;1 • AIT = 480
- **(T)** Tr: 45-23/24/25-38 • NFPA: 4 • TR: 336 • LVE(MVE) (F): 15 (2); TWA (USA): 2
- **(D)** LC50 r: 1.08; LC50 m: 0.68; LD50 o-r: 78; 93; LD50 o-m: 27; LD50 ip-r: 65 ; LD50 ip-m: 46; LD50 iv-ra: 69; LD50 scu-r: 75 ; LD50 scu-m: 25; 35; LD50 cu-ra: 250 ; LD50 cu-r: 148

Cyclopropanecarbonitrile: C_4H_5N • 67.09 • 5500-21-0
- **(T)** Tr: 23/24/25-36/ 37/38
- **(P)** Eb = 134-135
- **(I)** P_{fla} = 32; 38; 40

Methacrylonitrile: C_4H_5N • 67.09 • 126-98-7
- **(P)** Eb = 90-92 • P_{vap} = 85(20); 303(55); 86; 120(30); 267(50)
- **(I)** P_{fla} = –1; 12; 13oc
- **(T)** Tr: 23/24/25-43 • LVE(MVE) (F): 1; TWA (USA): 1
- **(D)** LC50 r: 0.899; ; LC50 m: 0.098; LC50 ra: 0.101; LD50 o-r: 120; 250; LD50 o-m: 17; 15; LD50 o-ra: 16; LD50 cu-ra: 320; LD50 cu-r: 2080

Crotonitrile: C_4H_5N • 67.09 • 627-26-9 (4786-20-3)
- **(P)** $\Delta H_f^0 = +149.66(g)$ • Eb = 122; 105-110
- **(I)** P_{fla} = 16; 19
- **(T)** Tr: 23/24/25
- **(D)** LD50 o-r: 501; LD50 o-m: 396

Allyl cyanide: C_4H_5N • 67.09 • 109-75-1
- **(P)** Eb = 116-121; 119; 98/400; 78/200; 60.2/100; 48.8/60; 40/40; 26.6/20; 14.1/10; 2.9/5; 19.6/1
- **(I)** P_{fla} = 18; 19; 23; 24
- **(T)** Tr: 21-23/25-36/38
- **(D)** LC50 r. 1.37; LC50 m: 1 ; LD50 o-r: 115; LD50 o-m: 67; LD50 scu-r: 150; LD50 cu-ra: 1410

Fumaronitrile: $C_4H_2N_2$ • 78.07 • 764-42-1
- **(P)** $\Delta H_f^0 = 268.24(l)$; ΔH_d = 0.454 kJ/g at 340-380°C • Eb = 186
- **(T)** Tr: 23/25
- **(D)** LD50 o-r: 132; LD50 ip-m: 38 ; LD50 iv-m: 56

Benzonitrile: C_7H_5N • 103.12 • 100-47-0
- **(P)** $\Delta H_f^0 = 218.82(g)$ • Eb = 191; 123.5/100; 100/40; 69.2/10; 28.2/1 • P_{vap} = 0.72; 1.33(20); 1.5(30); 4.8(50)
- **(I)** LEL/UEL = 0.9/12; 1.4/7.2 • P_{fla} = 70; 71; Fire point : 75 • AIT = 550
- **(T)** Tr: 21/22 • NFPA: 3 • TR: 60
- **(D)** LC50 r: 4/4h; LC50 m: 6/4h; LD50 o-r: 720 ; LD50 o-m: 971; LD50 o-ra: 800; LD50 ip-r: 740 ; LD50 ip-m: 400; LD50 scu-m: 180; LD50 scu-m: 200; LD50 cu-ra: 1250; ; LD50 cu-r: 1200

Benzyl cyanide: C_8H_7N • 117.15 • 140-29-4
- **(P)** Eb = 231-234; 161.8/100; 119.4/20; 100-103/10; 60/1 • P_{vap} = 0.07(20); 0.7(50); 1.3(60)
- **(I)** P_{fla} = 101; 102
- **(T)** Tr: 20/22 • TR: 60
- **(D)** LC50 r:0.430 ; LC50 m: 0.100; LD50 o-r: 270 ; LD50 o-m: 46; LD50 ip-r: 75 ; LD50 scu-m: 50; LD50 cu-ra: 270 ; LD50 cu-r: 2000

o-Tolunitrile: C_8H_7N • 117.15 • 529-19-1
- **(P)** Eb = 205.2; 135/100; 110/40; 93/20; 77.9/10; 64/5; 36.7/1
- **(I)** P_{fla} = 84; 85; 91; 82
- **(D)** LD50 o-r: 3200 (o-tolunitrile); LD50 ip-r: 700 (o-tolunitrile); LD50 scu-m: 600 (o-tolunitrile); LD50 o-r: 4200 (m-tolunitrile)

8 • Table of organic compounds — Nitriles

Name: Empirical formula • Molar mass(g/mol) • Registry number (CAS).
(P) Physical data: ΔH_f^0 (kJ/mol) (physical state) • Eb (°C/mmHg) (if substance volatile) • P_{vap} (mBar) (°C) (if substance volatile).
(I) Inflammability data: LEL/UEL (%) • P_{fla} (°C) • AIT (°C).
(T) Toxicity: Safety Code (Tr) • NFPA • Transport Code (TR) • LVE (MVE) (ppm) F, USA, D • STEL (ppm); IDLH (ppm).
(D) Lethal and toxic doses: LC (mg/l/duration of exposure); LD (mg/kg).

p-Tolunitrile: C_8H_7N • 117.15 • 104-85-8
- (P) Eb = 217.6; 199-202; 145.2/100; 130/60; 109.5/40; 106/20; 101.7/20; 85.8/10; 71.3/5; 42.5/1
- (I) P_{fla} = 78; 84; 86; 88; 92
- (T) Tr: 20/21
- (D) LD50 o-r: 3800; 4060; LD50 ip-m: 1300; 512; LD50 scu-m: 1080

Phthalonitriles (o-, m-, p-): $C_8H_4N_2$ • 128.13 • 91-15-6; 626-17-5; 623-26-7
- (I) P_{fla} = 162 (o-)

o-Phthalonitrile:
- (T) Tr: 25
- (D) LD50 o-m: 65; LD50 ip-r: 62; LD50 ip-m: 25; LD50 scu-m: 46

m-Phthalonitrile:
- (D) LD50 o-r: 1860; LD50 o-m: 178; LD50 ip-m: 481; LD50 o-ra: 350

p-Phthalonitrile:
- (D) LD50 o-r: 2080; LD50 ip-m: 699

NITRILES

Nitriles with complex groups

Glycolonitrile: C_2H_3NO • 57.05 • 107-16-4
- (P) Eb = 119/24 • P_{vap} = 1(20); 1.5(30); 3(50)
- (I) P_{fla} = 56; 69
- (T) Tr: 26/27/28 • TWA (USA): 2
- (D) LC50 r: 0.063/4h ; LC50 m: 0.06/4h; LD50 o-r: 8; 16; LD50 o-m: 10; LD50 ip-m: 10; 3; LD50 scu-m: 15; LD50 cu-ra: 5

Ethylenecyanohydrin: C_3H_5NO • 71.08 • 109-78-4
- (P) Eb = 228; 178/200; 157.7/100; 144.7/60; 134.1/40; 117.9/20; 106-108/11; 88/11; 102/10; 87./5; 58.7/1 • P_{vap} = 0.1(20); 0.2(30); 0.8(50)
- (I) P_{fla} = 128; 129oc
- (T) Tr: 36/37/38 • NFPA: 2 • TR: –;
- (D) LD50 o-r: 10000; LD50 o-m: 1800; LD50 o-ra: 900; LD50 ip-m: 500; LD50 cu-ra: 5000

Lactonitrile: C_3H_5NO • 71.08 • 42492-95-5
- (P) Eb = 182(dec); 103/50; 90/17 • P_{vap} = < 0.27(25)
- (I) P_{fla} = 77cc
- (T) Tr: 27-23/25 • NFPA: 3 • TR: –
- (D) LC50 r: 0.363/4h; LD50 o-r: 87; LD50 ip-m: 15; LD50 scu-m: 5; LD50 cu-ra: 20

Acetonecyanohydrin: C_4H_7NO • 85.11 • 75-86-5
- (P) Eb = 82(?); 95(?); 120(dec); 170(dec); 82/23; 88-90/20; 81/15; 68-70/11 • P_{vap} = 20.7; 2.67(20); 51.7(55)
- (I) LEL/UEL = 2.25/11; 2.25/12 • P_{fla} = 63; 73; 75 • AIT = 668; 688
- (T) Tr: 26/27/28 • NFPA: 4 • TR: 66 • LVE(MVE) (F): – (50)
- (D) LC50 r: 0.22; LC50 m: 0.52; LD50 o-r: 17.8; 19.3; 170; LD50 o-m: 14; LD50 o-ra: 13.5; LD50 scu-r: 85; LD50 cu-ra: 17

3-Ethoxypropionitrile: C_5H_9NO • 99.13 • 2141-62-0 (14631-45-9)
- (P) Eb = 171; 172-174; 58/13 • P_{vap} = 20(65)
- (I) P_{fla} = 63; 66; 75
- (D) LD50 o-r: 2860; 3200; LD50 o-m: 2200; LD50 ip-m: 2000

Methoxyacetonitrile: C_3H_5NO • 71.08 • 1738-36-9
- (P) Eb = 118-121
- (I) P_{fla} = 30; 32
- (T) Tr: 20/21/22

Chloroacetonitrile: C_2H_2ClN • 75.5 • 107-14-2
- (P) Eb = 123; 124-6; 123-125; 30-2/15 • P_{vap} = 11.5(20); 19(30); 205(55); 138(50)
- (I) P_{fla} = 47; 54; 56
- (T) Tr: 23/24/25 • NFPA: 3 • TR: 60
- (D) LC50 r: 0.78; LD50 o-r: 220; LD50 o-m: 139; LD50 ip-m: 100; LD50 cu-ra: 71

Dichloroacetonitrile: C_2HCl_2N • 109.94 • 3018-12-0
- (P) Eb = 110-112; 112-114
- (I) P_{fla} = 35
- (D) LD50 o-r: 330; LD50 o-m: 270

Trichloroacetonitrile: C_2Cl_3N • 144.39 • 545-06-2
- (P) Eb = 84; 86; 83-84 • P_{vap} = 77.3(20); 355(55)
- (I) P_{fla} = None > 90; ≥ 195
- (T) Tr: 23/24/25 • TR: 60
- (D) LC50 r: 1.46; LD50 o-r: 250; LC50 ra: 1.83; LD50 iv-m: 56; LD50 cu-ra: 900

3-Chloropropionitrile: C_3H_4ClN • 89.52 • 542-76-7
- (P) Eb = 174-176; 176(dec); 95/50; 46/5 • P_{vap} = 8(50)
- (I) P_{fla} = 76; 83; 87
- (T) Tr: 23/25
- (D) LD50 o-r: 100; LD50 o-m: 9; LD50 ip-m: 100; 25 ; LD50 iv-m: 56

4-Chlorobutyronitrile: C_4H_6ClN • 103.55 • 628-20-6
- (P) Eb = 195-197; 73-75/12
- (I) P_{fla} = 84; 91
- (T) Tr: 25
- (D) LD50 o-m: 53

2-Chloroacrylonitrile: C_3H_2ClN • 87.51 • 920-37-6
- (P) Eb = 89
- (I) P_{fla} = 1; 6
- (T) Tr: 23/24/25 • NFPA: – • TR: 336
- (D) LC50 m: 0.105; LD50 o-r: 230; LD50 o-m:128; LD50 iv-m: 100

2-Chlorobenzonitrile: C_7H_4ClN • 137.57 • 873-32-5
- (P) Eb = 232
- (I) P_{fla} = 108; 122
- (T) Tr: 21/22-36 • TR: 60
- (D) LD50 o-r: 435; LD50 ip-m: 150

8 • Table of organic compounds — Amides - isocyanates

Name: Empirical formula • Molar mass(g/mol)• Registry number (CAS).
(P) Physical data: ΔH_f^0 (kJ/mol) (physical state) • Eb (°C/mmHg) (if substance volatile) • P_{vap} (mBar) (°C) (if substance volatile).
(I) Inflammability data: LEL/UEL (%) • P_{fla} (°C) • AIT (°C).
(T) Toxicity: Safety Code (Tr) • NFPA • Transport Code (TR) • LVE (MVE) (ppm) F, USA, D • STEL (ppm); IDLH (ppm).
(D) Lethal and toxic doses: LC (mg/l/duration of exposure); LD (mg/kg).

o–Chlorobenzalmalonitrile: $C_{10}H_5ClN_2$ • 188.62 • 2698-41-1
- (P) Eb = 313; 126/0.1
- (T) TWA (USA): 0.05
- (D) LC50 r: 2.02 ; LC50 m: 3.08; LC50 ra: 2.014; LD50 o-r: 178; LD50 o-m: 282; LD50 o-ra: 143; LD50 ip-r: 48 ; LD50 ip-m: 32; LD50 iv-r: 28 ; LD50 iv-m: 47; LD50 iv-ra: 8

Cyanoacetic acid: $C_3H_3NO_2$ • 85.06 • 372-09-8
- (P) Eb = 108/15; dec 160 °C $\to NO_x$ + HCN • P_{vap} = 0.133(100)
- (I) P_{fla} = 107
- (T) Tr: 20/22-34
- (D) LD50 o-r: 1500; LD50 ip-m: 200; MLD scu-m: 2000

Methyl cyanoacetate: $C_4H_5NO_2$ • 99.09 • 105-34-0
- (P) Eb = 200; 203; 204-7; 115/48; 85-87/11 • P_{vap} = 0.27 (20)
- (I) P_{fla} = >109; 110
- (T) Tr: 20/22 • NFPA: – • TR: 60
- (D) LD50 ip-m: 200

Ethyl cyanoacetate: $C_5H_7NO_2$ • 113.12 • 105-56-6
- (P) Eb = 203; 207; 208-10 • P_{vap} = 1.33(67)
- (I) P_{fla} = 110; 109 • AIT = 478
- (T) Tr: 20/21/22 • NFPA: 3 • TR: 60
- (D) LD50 ip-m: 500

Methyl 2–cyanoacrylate: $C_5H_5NO_2$ • 111.10
- (P) Eb = 47-49/1.8
- (I) P_{fla} = 75
- (T) LVE(MVE) (F): 4 (2); TWA (USA): 2; MAK(D): 2
- (D) LC50 r: 0.454; LD50 o-r: 1600

AMIDES - ISOCYANATES
• Amides, urea

Formamide: CH_3NO • 45.04 • 75-12-7
- (P) ΔH_f^0 = –186.19(g); 253.97(l) • Eb = 210-216(dec); 175.5/200; 147/60; 122.5/20; 109.5/10; 70.5/1; 60-62/0.1 • P_{vap} = 0.03; 0.11(20); 0.08(30); 0.32(50); 39.6(129.4)
- (I) LEL/UEL = 2.7/19 • P_{fla} = 150; 154cc; 154Coc; 175cc • AIT = 499; > 500
- (T) Tr: 36/37/38-63; NFPA: 2 TR: –; LVE(MVE) (F): – (20); TWA (USA): 10 (skin)
- (D) LC50 r: 7.304/6h; LD50 o-r: 5570; 5800; 6000; LD50 o-m: 3150; LD50 ip-r: 5700; LD50 ip-m: 2450; 4600; LD50 cu-ra: 17000

Acetamide: C_2H_5NO • 59.07 • 60-35-5
- (P) ΔH_f^0 = –317.98(s) • Eb = 221-222; 158/100; 136/40; 120/20; 105/10; 92/5 • P_{vap} = 1.33(65)
- (T) Tr: 40
- (D) LD50 o-r: 7000; LD50 o-m: 12900; LD50 ip-r: 10300; LD50 iv-r: 12500; LD50 ip-m: 1000; 10000; LD50 ip-ra; LD50 iv-ra: 7500; LD50 scu-m: 8300

Dimethylformamide: C_3H_7NO • 73.09 • 68-12-2
- (P) ΔH_f^0 = –239.32(l) • Eb = 153-155; 76/39; 25/3.7 • P_{vap} = 0.9(0); 3.5; 3.6; 3.77(20); 4.9(25); 9(30); 32(50); 200(100)
- (I) LEL/UEL = 2.2(100)/15.2(100); 2.2/16.0 • P_{fla} = 58Tcc; 67oc • AIT = 410; 440; 445
- (T) Tr: 61-20/21-36 • NFPA: 1 • TR: 30 • LVE(MVE) (F): –(10); TWA (USA): 10; MAK(D): 10; 20 • IDLH: 3500
- (D) LC50 m: 9.4/2h; LD50 o-r: 2200; 2800; 7600; LD50 o-m: 3700; 3750; 6800; LD50 ip-r: 1400; 4700; LD50 iv-r: 2000; LD50 ip-m: 650; 6200; LD50 iv-m: 2500; LD50 ip-ra: 1000; LD50 iv-ra: 1800; LD50 scu-r: 3800; LD50 scu-m: 4500; LD50 cu-r: 5000; LD50 cu-ra: ~1500; 4720

Dimethylacetamide: C_4H_9NO • 87.12 • 127-19-5
- (P) Eb = 165-167; 96/80; 85-87/33; 74.5/26; 66-67/15; 62-63/12 • P_{vap} = 3.3(20); 1.73; 2; 2.67(25); 7.4(30); 5.33(38); 44(50)
- (I) LEL/UEL = 1.7/11.5; 1.8/13.8; 1.8(100]/ 11.5(160); 2/11.5(160) • P_{fla} = 62Tcc; 66; 69; 77Toc • AIT = 390; 420; 489
- (T) Tr: 20/21-36 • NFPA: 2 • TR: 30 • LVE(MVE) (F): – (10); TWA (USA): 10; MAK(D): 10 • STEL: 20
- (D) LC50 r: 8.81/1h; LC50 m: 7.2/?; LD50 o-r: 4300; 4930; 5000; 5400; LD50 o-m: 4620; LD50 ip-r: 2750; LD50 iv-r: 2640; LD50 ip-m: 2800; LD50 iv-m: 3020; LD50 su-m: 9600; LD50 cu-ra: 2240

Propionamide: C_3H_7NO • 73.09 • 79-05-0
- (P) ΔH_f^0 = –341.83(s) • Eb = 213; 222
- (D) LCM r: 24.312/?; MLD iv-ra: 230

Butyramide: C_4H_9NO • 87.12 • 541-35-5
- (P) ΔH_f^0 = –366.10(s); Eb = 216

Isobutyramide: C_4H_9NO • 87.12 • 563-83-7
- (P) Eb = 216-220

Pyrrolidone: C_4H_7NO • 85.11 • 616-45-5
- (P) ΔH_f^0 = –285.77(s) • Eb = 245; 127-130/10; 123-125/10; 113-114/9.2; 76/0.2 • P_{vap} = ~0.04(20)
- (I) P_{fla} = 130oc; 135; 138cc • AIT = 390
- (T) Tr: 22-36
- (D) LD50 o-r: 328; 6500; LD50 ip-r: 160; LD50 ip-m: 3700; MLD scu-m: 10000

N-Methylpyrrolidone: C_5H_9NO • 99.13 • 872-50-4
- (P) ΔH_f^0 = –262.09(l) • Eb = 202-204; 81/10 • P_{vap} = 0.29; 0.39(20); 0.65(30); 1.33(40); 2.5(50)
- (I) LEL/UEL = 1.3(81)/9.5(129); 1.3/9.5 • P_{fla} = 86; 91cc; 93PMcc; 96oc • AIT = 269; 346
- (T) Tr: 36/38 • NFPA: 2 • TR: 30 • MAK(D): 100
- (D) LC50 r: > 5.1/4h; LD50 o-r: 3500; 3914; 4200; LD50 o-m: 5130; LD50 ip-r: 2472; LD50 iv-r: 81; 2266; LD50 ip-m: 3050; LD50 iv-m: 55; LD50 cu-ra: < 8000; 8000

8 • Table of organic compounds

Miscellaneous compounds

Name: Empirical formula • Molar mass(g/mol) • Registry number (CAS).
- **P** **Physical data:** ΔH_f^0 (kJ/mol) (physical state) • Eb (°C/mmHg) (if substance volatile) • P_{vap} (mBar) (°C) (if substance volatile).
- **I** **Inflammability data:** LEL/UEL (%) • P_{fla} (°C) • AIT (°C).
- **T** **Toxicity:** Safety Code (Tr) • NFPA • Transport Code (TR) • LVE (MVE) (ppm) F, USA, D • STEL (ppm); IDLH (ppm).
- **D** **Lethal and toxic doses:** LC (mg/l/duration of exposure); LD (mg/kg).

ε-Caprolactam: $C_6H_{11}NO$ • 113.16 • 105-60-2
- **P** $\Delta H_f^0 = -328.61(s)$ • Eb = 268.5; 180/50; 139/12; 136-138/10; 100/3 • $P_{vap} = 0.001(20); 4(100); 8(120); 66.65(180)$
- **I** LEL/UEL = 1.4/8 • P_{fla} = 125oc; 139Tcc • AIT = 374
- **T** Tr: 20/22-36/37/38 • LVE(MVE) (F): –(5) vapour; (1 mg/m^3) dusts; TWA (USA): id; MAK(D): 25 mg/m^3
  • STEL: 10 ppm; 3 mg/m^3
- **D** LC50 r: 0.3/2h ; LC50 m: 0.45/?; LD50 o-r: 1210; 2140;
  LD50 o-m: 930; LD50 ip-m: 650; LD50 iv-m: 480;
  LD50 scu-m: 750; LD50 cu-ra: 1438

Acrylamide: C_3H_5NO • 71.08 • 79-06-1
- **P** Eb = 125/25; 103/5; 87/2 • $P_{vap} = 0.0093(20); 0.04(40); 2.13(84.5)$
- **I** P_{fla} = 137oc
- **T** Tr: 45-46-24/25-48/23/24/25 • TR: 60
  • TWA (USA): 0.03 mg/m^3
- **D** LD50 o-r: 124; LD50 o-m: 107; MLD o-ra: 126; LD50 ip-r: 90;
  LD50 ip-m: 170; LD50 cu-ra: 400

Methacrylamide: C_4H_7NO • 85.11 • 79-39-0
- **T** Tr: 20/21/22-36/ 37/38
- **D** LD50 o-r: 459; LD50 o-m: 451; LD50 ip-m: 200;
  LD50 cu-ra: > 6000

Acetanilide: C_8H_9NO • 135.17 • 103-84-4
- **P** $\Delta H_f^0 = -210.46(s)$ • Eb = 303-305 • $P_{vap} = 1.33(114)$
- **I** P_{fla} = 161; 169oc; 174oc • AIT = 528; 540; 545
- **T** Tr: 22; 20/21/22 • NFPA: 3 • TR: –
- **D** LD50 o-r: 800; LD50 o-m: 1210; MLD o-ra: 1500;
  LD50 ip-r: 540; MLD ip-r: 1000; LD50 ip-m: 500

Benzamide: C_7H_7NO • 121.14 • 55-21-0
- **P** $\Delta H_f^0 = -202.59(s)$ • Eb = 288-290
- **I** P_{fla} = ~180
- **T** Tr: 22
- **D** LD50 o-m: 1160; LD50 ip-r: 781; LD50 ip-m: 1282

Benzanilide: $C_{13}H_{11}NO$ • 197.24 • 93-98-1
- **P** $\Delta H_f^0 = -93.30(s)$ • Eb = 117/10

Urea: CH_4N_2O • 60.06 • 57-13-6
- **P** $\Delta H_f^0 = -333.51(s)$
- **T** NFPA: 0
- **D** LD50 o-r: 8471; 14300; LD50 o-m: 11000; LD50 iv-r: 5300;
  LD50 iv-m: 4600; LD50 scu-r: 8200; LD50 scu-m: 9200

Chloroacetamide: C_2H_4ClNO • 93.51 • 79-07-2
- **P** $\Delta H_f^0 = -338.49(s)$ • Eb = 225 (dec) • $P_{vap} = 0.07(20)$
- **I** P_{fla} = 179
- **T** Tr: 25-36/38 • NFPA: – • TR: 60
- **D** LC50 o-r: 70; LD50 o-m: 155; LD50 o-ra: 122; LD50 ip-m: 100;
  LD50 iv-m: 180

AMIDES - ISOCYANATES

• **Isocyanates**

Methyl isocyanate: C_2H_3NO • 57.05 • 624-83-9
- **P** Eb = 37-39; 44 • $P_{vap} = 267(4.2); 400(13.5); 513; 539(20); 533(20.6); 800(31.2); 1828(55)$
- **I** LEL/UEL = 5.3/26 • P_{fla} = –6 • AIT = 533
- **T** Tr: 23/24/25-36/ 37/38 • NFPA: 3 • TR: –
- **D** LC50 r: 0.011/4h; 0.014/6h; LC50 m: 0.028/6h ;
  LD50 o-r: 51.5; 140; LD50 o-m: 120; LD50 scu-r: 261;
  LD50 scu-m: 82; LD50 scu-m: 126; LD50 cu-ra: 220

Toluene-2,4-diisocyanate: $C_9H_6N_2O_2$ • 174.16 • 91-08-7
- **P** Eb = 251; 126/11; 115-120/10 • $P_{vap} = 0.013(20); 0.04(25)$
- **I** LEL/UEL = 0.9/9.5 • P_{fla} = 120; 132oc • AIT = > 619; 620
- **T** Tr: 23-36/37/38-40-42; 23-36/37/38-42 • NFPA: 2 • TR: 60
  • LVE(MVE) (F): 0.02 (0.01); TWA (USA): 0.005;
  MAK(D): 0.01 • STEL: 0.02
- **D** LC50 r: 0.099/4h; LC50 m: 0.071/4h; LC50 ra: 0.0797/4h;
  LD50 o-r: 5800; LD50 iv-m: 56; LD50 cu-ra: > 16000

Toluene-2,6-diisocyanate: $C_9H_6N_2O_2$ • 174.16 • 584-84-9
- **P** Eb = 129-133/18 • $P_{vap} = 0.013(20)$
- **I** LEL/UEL = 0.9/9.5 • P_{fla} = >109; 127 • AIT = > 600; > 619
- **T** Tr: 23-36/37/38-42 • TR: 60 • LVE(MVE) (F): 0.02 (0.01); TWA (USA): 0.005; MAK(D): 0.01 • STEL: 0.02
- **D** LC50 m: 0.091/4h

Hexamethylenediisocyanate: $C_8H_{12}N_2O_2$ • 168.20 • 822-06-0
- **P** Eb = 255; 82-85/0.1; 61/0.08 • $P_{vap} = 0.007; 0.067(25); 0.14(50)$
- **I** LEL/UEL = 0.9/9.5 • P_{fla} = 100; 130; 140 • AIT = 454
- **T** Tr: 23-36/37/38-42/43 • TR: 60 • LVE(MVE) (F): 0.02 (0.01); TWA (USA): 0.005; MAK(D): 0.01 • STEL: 0.02
- **D** LD50 o-r: 738; LC50 m: 0.030/?; LD50 o-m: 350;
  LD50 iv-m: 5.6; LD50 cu-ra: 593

MISCELLANEOUS COMPOUNDS

• **Sulphur derivatives**

• *Simple groups*

Methanethiol: CH_4S • 48.10 • 74-93-1
- **P** $\Delta H_f^0 = -22.97(g); -22.3(g); -46.4(l)$ • Eb = 6-8; –49/100; –67/10
  • $P_{vap} = 1700; 2048(20); 2000(26); 2400(30); 2136(37.7); 4300(50)$
- **I** LEL/UEL = 3.9/21.8; 4/22; 4.1/21 • P_{fla} = < –18; –18oc
- **T** Tr: 20 • NFPA: 2 • TR: 236 • LVE(MVE) (F): – (0.5);
  TWA (USA): 0.5 • MAK(D): 0.5 • IDLH: 400
- **D** LC50 r: 1.33/?; LC50 m: 0.0065/?

453

8 • Table of organic compounds

Miscellaneous compounds

Name: Empirical formula • Molar mass(g/mol)• Registry number (CAS).
(P) Physical data: ΔH_f^0 (kJ/mol) (physical state) • Eb (°C/mmHg) (if substance volatile) • P_{vap} (mBar) (°C) (if substance volatile).
(I) Inflammability data: LEL/UEL (%) • P_{fla} (°C) • AIT (°C).
(T) Toxicity: Safety Code (Tr) • NFPA • Transport Code (TR) • LVE (MVE) (ppm) F, USA, D • STEL (ppm); IDLH (ppm).
(D) Lethal and toxic doses: LC (mg/l/duration of exposure); LD (mg/kg).

Ethanethiol: C_2H_6S • 62.13 • 75-08-1
- (P) ΔH_f^0 = –46.11(g); –45.3(g); –73.6(l) • Eb = 34-37; 34.7-35.04; 17.7/400; 1.5/200; –13/100; –50.2/10; –76.1/1 • P_{vap} = 587(20); 840(30); 1700(50); 1963(55)
- (I) LEL/UEL = 2.8/18; 2.8/18.2 • P_{fla} = –45; < –18; –17 • AIT = 295; 299 (in O_2: 261)
- (T) Tr: 20 • NFPA: 2 • TR: 336 • LVE(MVE) (F): – (0.5); TWA (USA): 0.5; MAK(D): 0.5 • STEL: 2; IDLH 2500
- (D) LC50 r: 11.418/4h; LC50 m: 2.77/?; 8.16/4h; LD50 o-r: 682; 1960; LD50 o-m: 226; 450; LD50 ip-r: 450

1-Propanethiol: C_3H_8S • 76.16 • 107-03-9
- (P) Eb = 66-68 • P_{vap} = 160; 162.6(20); 250(30); 349.2(37.7); 540(50)
- (I) P_{fla} = –20
- (T) Tr: 22
- (D) LC50 r: 23.117/4h; LC50 m: 12.700/4h; LD50 o-r: 1790; LD50 ip-r: 515

2-Propanethiol: C_3H_8S • 76.16; 75-33-2
- (P) ΔH_f^0 = –76.2(g); –105.90(l); Eb = 57-60; P_{vap} = 606.5(37.)
- (I) P_{fla} = < –34; –34
- (T) Tr: 20

1-Butanethiol: $C_4H_{10}S$ • 90.18 • 109-79-5
- (P) ΔH_f^0 = –88.10(g); –124.70(l) • Eb = 96-98 • P_{vap} = 40(20); 65(30); 110.7(37.7); 140(50)
- (I) P_{fla} = 2; 12
- (T) Tr: 20/21/22; 20/22 • NFPA: 2 • TR: 33 • LVE(MVE) (F): – (0.5); TWA (USA): 0.5; MAK(D): 0.5
- (D) LC50 r: 14.8/4h; LC50 m: 9.2/4h; LD50 o-r: 1500; 3800; LD50 o-m: 3000; LD50 ip-r: 399

Thiophenol: C_6H_6S • 110.17 • 108-98-5
- (P) ΔH_f^0 = +111.60(g); +63.70(l) • Eb = 168-169; 103.6/100; 86.2/50; 69.7/20; 18.6/1 • P_{vap} = 1.5(20); 2.9(30); 10.5(50)
- (I) P_{fla} = 50; 55
- (T) Tr: 23/24/25-34; 23/24/25 • TR: 663 • LVE(MVE) (F): – (0.5); TWA (USA): 0.5
- (D) LC50 r: 0.151/4h; LC50 m: 0.128/4h; LD50 o-r: 46.2; LD50 ip-r: 9.8; 10; LD50 ip-m: 25; LD50 cu-r: 300; LD50 cu-ra: 134

p-Thiocresol: C_7H_8S • 124.20 • 106-45-6
- (P) ΔH_f^0 = +44.25(l) ortho isomer • Eb = 195 • P_{vap} = 13.3(71)
- (I) P_{fla} = 68
- (T) Tr: 34
- (D) LD50 ip-m: 200

Diethyl sulphide: $C_4H_{10}S$ • 90.18 • 352-93-2
- (P) ΔH_f^0 = –83.60(g); –119.40(l) • Eb = 92-93
- (I) P_{fla} = –10
- (D) LD50 o-r: 5930

Dimethyl sulphide: C_2H_6S • 62.13 • 75-18-3
- (P) ΔH_f^0 = –37.5(g); –65.4(l) • Eb = 36-38 • P_{vap} = 532; 537(20); 800(30); 1600(50); 1809(55)
- (I) LEL/UEL = 2.2/19.7 • P_{fla} = –36; –34; –25; < –18; 18 • AIT = 206; 215
- (T) Tr: 20/21/22; 22 • NFPA: 4 • TR: 33
- (D) LC50 m: 0.032/?; LD50 o-r: 535; 3300; LD50 o-m: 3700; LD50 ip-m: 8000; LD50 cu-ra: > 5000

Diphenyl sulphide: $C_{12}H_{10}S$ • 186.27 • 139-66-2
- (P) ΔH_f^0 = +163.59(l) • Eb = 295-297; 96-100/0.1
- (I) P_{fla} = > 109
- (T) Tr: 20/22-38
- (D) LD50 o-r: 800; LD50 cu-ra: 11300

Thiophene: C_4H_4S • 84.14 • 110-02-1
- (P) ΔH_f^0 = +114.00(g); +80.2(l) • Eb = 82-84.4; 64.7/400; 46.5/200; 30.5/100; 20.1/60; 12.5/40; 0/20; –10.9/10; –20.8/5 • P_{vap} = 53.3(12.5); 80(20); 130(30); 187(37.); 310(50)
- (I) LEL/UEL = 1.5/12.5 • P_{fla} = –9; –6; –1 • AIT = 394
- (T) Tr: 20/21/22 • NFPA: 2 • TR: 33
- (D) LC50 m: 9.5/2h ; LD50 o-r: 1400; LD50 o-m: 420; LD50 ip-m: 100; LD50 scu-ra: 830

Dimethyl sulphoxide: C_2H_6OS • 78.13 • 67-68-5 • 69-68-5
- (P) ΔH_f^0 = –203.34(l); –203.44(l) • Eb = 189; 83/17; 56.6/5.11; 57/5; 47.4/2.82; 30/0.79; 20/0.37 • P_{vap} = 0.49; 0.56; 0.61; 2.5(20); 3.5(30); 4.08; 7.5(50)
- (I) LEL/UEL = 1.8/dec; 2.6/28.5; 3/43; 3.0/29-63; 3.5/63 • P_{fla} = 82; 88cc; 95oc • AIT = 215; 270; 300
- (T) Tr: 20/21/22 • NFPA: 1 • TR: 30
- (D) LD50 o-r: 14500; 17500; 17900; LD50 o-m: 7920; LD50 ip-r: 8200; 9800; LD50 iv-r: 5360; LD50 ip-m: 2500; LD50 iv-m: 3100; 3800; LD50 scu-r: 12000; LD50 scu-m: 14000; LD50 cu-r: 40000 ; LD50 cu-m: 50000

Dimethylsulphone: $C_2H_6O_2S$ • 94.13 • 67-71-0
- (P) Eb = 238
- (I) P_{fla} = 143

Sulpholane: $C_4H_8O_2S$ • 120.17 • 126-33-0
- (P) Eb = 285; 104/0.2 • P_{vap} = 0.013(20); 0.035(30); 19.4(150); 113.6(200)
- (I) P_{fla} = 159cc; 165; 177oc
- (T) Tr: 22 • NFPA: 2 • TR: –
- (D) LCM r: 4.7/24h; LD50 o-r: 1941; MLD o-m: 1900; LD50 ip-r: 1600; LD50 ip-m: 1250; LD50 iv-m: 1080; LD50 cu-ra: > 3800; 4009

Methanesulphonic acid: CH_4O_3S • 96.10 • 75-75-2
- (P) Eb = 167/10; 122/1 • P_{vap} = 1.33(20)
- (I) P_{fla} = 108cc; > 109
- (D) MLD o-r: 200; LD50 o-m: 6200; MLD ip-r: 50; LD50 cu-ra: 2000

8 • Table of organic compounds

Miscellaneous compounds

Name: Empirical formula • Molar mass(g/mol)• Registry number (CAS).
(P) Physical data: ΔH_f^0 (kJ/mol) (physical state) • Eb (°C/mmHg) (if substance volatile) • P_{vap} (mBar) (°C) (if substance volatile).
(I) Inflammability data: LEL/UEL (%) • P_{fla} (°C) • AIT (°C).
(T) Toxicity: Safety Code (Tr) • NFPA • Transport Code (TR) • LVE (MVE) (ppm) F, USA, D • STEL (ppm); IDLH (ppm).
(D) Lethal and toxic doses: LC (mg/l/duration of exposure); LD (mg/kg).

Dimethyl sulphate : $C_2H_6O_4S$ • 126.13 • 77-78-1

(P) ΔH_f^0 = –733.16(l); –733.51(l) • Eb = 188-189 (dec); 89/27; 76/15; 75/11 • P_{vap} = 0.13(20); 0.4; 0.93(25); 1.47(38); ~1.3(50)

(I) P_{fla} = 83cc; 83oc; 115oc • AIT = 188; 494

(T) Tr: 45-25-26-34 • NFPA: 4 • TR: 66
• LVE(MVE) (F): – (0.1); TWA (USA): 0.1

(D) LC50 r: 0.045/4h; LC50 m: 0.28/?; LD50 o-r: 205; 440; LD50 o-m: 140; MLD o-ra: 45; MLD iv-ra: 50; MLD scu-r: 100; MLD scu-ra: 53

Diethyl sulphate $C_4H_{10}O_4S$ • 154.18 • 64-67-5

(P) ΔH_f^0 = –812.87(l); –813.26(l) • Eb = 208-209.5 (dec); 143/100; 96/15; 88-91/10; 75/5; 47/1 • P_{vap} = < 0.013; 0.133; 0.16; 0.2; 0.25(20); 0.4(30); 1.33(47); 1.6(50); 2.7(55)

(I) LEL/UEL = 4.1/– • P_{fla} = 78; 104cc; 121oc • AIT = 360; 436

(T) Tr: 45-46-20/21/ 22-34 • NFPA: 3 • TR: 60

(D) LCM r: 1.6/4h; LD50 o-r: 880; LD50 o-m: 647; LD50 scu-r: 350; LD50 cu-ra: 600

• *Complex groups, isothiocyanates*

2-Mercaptoethanol: C_2H_6OS • 78.13 • 60-24-2

(P) Eb = 154-158/742 (dec); 154-161; 53-55/12 • P_{vap} = 1; 1.33; 1.6(20)

(I) LEL/UEL = 2.3/18 • P_{fla} = 70cc; 73Coc; 76cc • AIT = 295

(T) Tr: 22-23/24-34 • NFPA: 2

(D) LC50 m: 13.2/15'; LD50 o-r: 244; 300; 330; LD50 o-m: 190; 345; LD50 ip-m: 200; 322; LD50 iv-m: 480; LD50 cu-ra: 150; < 200

Thiodiethanol: $C_4H_{10}O_2S$ • 122.18 • 111-48-8

(P) Eb = 282; 164-166/20; 168/14; 133/1; 96-98/0.1 • P_{vap} = < 0.013

(I) LEL/UEL = 1.2/5.2 • P_{fla} = 160oc • AIT = 245; 260

(T) Tr: 36 • NFPA: 1 • TR: –

(D) LD50 o-r: 6610; LD50 ip-m: 4000; LD50 iv-ra: 3000; LD50 scu-r: 4000; LD50 cu-ra: 20000

Trichloromethanethiol: $CHCl_3S$ • 151.44 • –

(P) Eb = 147-149 (dec); 51/25 • P_{vap} = 2.36(20); 33(51)

(T) Tr: 23/24/25-34 • NFPA: 3 • TR: 66 • LVE(MVE) (F): – (0.1); TWA (USA): 0.1

(D) LC50 r: 0.073/1h; LC50 m: 0.296/2h; LD50 o-r: 8.3; LD50 iv-m: 56; LD50 cu-ra: 1410

Thiomorpholine: C_4H_9NS • 103.18 • 123-90-0

(P) Eb = 175-178; 169/758; 166-167/743

(I) P_{fla} = 59; > 97

(T) Tr: 34-37

Thioglycoic acid: $C_2H_4O_2S$ • 92.11 • 68-11-1

(P) Eb = 154/100; 123/29; 116/20; 108/15; 105/11; 96/5
• P_{vap} = 0.53(25); 0.2(30)

(I) LEL/UEL = 5.9/– • P_{fla} = > 109; > 129 • AIT = 349

(T) Tr: 23/24/25-34 • NFPA: 3 • TR: 80 • LVE(MVE) (F): – (1); TWA (USA): 1

(D) LD50 o-r: 114; 250; LD50 o-m: 242; LD50 o-ra: 119; LD50 ip-r: 70; LD50 ip-m: 138; LD50 iv-m: 145; LD50 iv-ra: 100; MLD scu-m: 1000; MLDcu-ra: 300

Thioacetamide: C_2H_5NS • 75.13 • 62-55-5

(T) Tr: 45-22-36/38 • NFPA: 2 • TR: 60

(D) LD50 o-r: 301; MLD o-r: 200; LD50 ip-m: 300; LD50 iv-m: 300; MLD scu-m: 2000

Thiourea: CH_4N_2S • 76.12 • 62-56-6

(P) ΔH_f^0 = –88.41(s) • Eb = Decomposition at 200
• P_{vap} = 7.8x10^{-8}(20)

(T) Tr: 22-40 • NFPA: 1 • TR: 60

(D) LD50 o-r: 125; 1830(Norwegian wild rat); LD50 o-m: 8500; LD50 ip-r: 436; LD50 ip-m: 100

Methyl isothiocyanate: C_2H_3NS • 73.11 • 556-61-6

(P) ΔH_f^0 = +131.02(g) • Eb = 118-120 • P_{vap} = 27.6; 28(20); 39(30); 88(50)

(I) P_{fla} = 30; 32

(T) Tr: 23/25-34-43 • NFPA: 3 • TR: 63

(D) LC50 r: 1.9/1h; LD50 o-r: 72; 97; LD50 o-m: 90; 97; LD50 ip-r: 54; LD50 ip-m: 82; LD50 scu-r: 59; LD50 scu-m: 50; LD50 scu-m: 33; LD50 cu-r: 2780; LD50 su-m: 1820

Ethyl isothiocyanate : C_3H_5NS • 87.14 • 542-85-8

(P) Eb = 130-132 • P_{vap} = 13(23)

(I) P_{fla} = 24; 32

(T) Tr: 23/24/25; 23/24/25-36/37/38 • TR: 63

Allyl isothiocyanate: C_4H_5NS • 99.15 • 57-06-7

(P) Eb = 151-154 • P_{vap} = 13.3(38.3)

(I) P_{fla} = 29; 46

(T) Tr: 24/25-33-36/37/38; 23/24/25-36/37/38-43 • NFPA: 3
• TR: 69 • LVE(MVE) (F): – (0.02); TWA (USA): 0.02; MAK(D): 0.01

MISCELLANEOUS COMPOUNDS
• Phosphorus derivatives

Trimethylphosphine: C_3H_9P • 76.08 • 594-09-2

(P) ΔH_f^0 = –122.23(l) • Eb = 38-40 • P_{vap} = 499(20)1643.7(55)

(I) P_{fla} = –29; –17

(T) Tr: 36/37/38

8 • Table of organic compounds

Miscellaneous compounds

Name: Empirical formula • Molar mass(g/mol) • Registry number (CAS).
- (P) **Physical data:** ΔH_f^0 (kJ/mol) (physical state) • Eb (°C/mmHg) (if substance volatile) • P_{vap} (mBar) (°C) (if substance volatile).
- (I) **Inflammability data:** LEL/UEL (%) • P_{fla} (°C) • AIT (°C).
- (T) **Toxicity:** Safety Code (Tr) • NFPA • Transport Code (TR) • LVE (MVE) (ppm) F, USA, D • STEL (ppm); IDLH (ppm).
- (D) **Lethal and toxic doses:** LC (mg/l/duration of exposure); LD (mg/kg).

Triphenylphosphine: $C_{18}H_{15}P$ • 262.29 • 603-35-0
- (P) ΔH_f^0 = +232.21(s) • Eb = 377; 195-205/5.25 • P_{vap} = 0.013(88); 1333(388)
- (I) P_{fla} = 179cc; 180oc
- (T) Tr: 20/22-43-48/20/22 • NFPA: 3; 0
- (D) LC50 r: 3.378/4h; LD50 o-r: 700; 800; > 6400; LD50 o-m: 1000; LD50 scu-m: 3000; LD50 cu-ra: > 4000; > 5000

Dibutyl hydrogenphosphite: $C_8H_{19}O_3P$ • 194.21 • 1809-19-4
- (P) Eb = 118/11; 115-125/10
- (I) P_{fla} = 49; 118; 120
- (T) NFPA: 3

Trimethylphosphite: $C_3H_9O_3P$ • 124.08 • 121-45-9
- (P) ΔH_f^0 = –741.34(l) • Eb = 110-112 • P_{vap} = 22.7; 28(20); 43(30); 99(50)
- (I) P_{fla} = 28; 38Coc; 54oc
- (T) Tr: 22-36/37/38
- (D) LC50 r: < 51.59/4h; LD50 o-r: 1600; LD50 ip-m: 4180; LD50 cu-ra: 2600; MLDcu-ra: 2200

Triethylphosphite: $C_6H_{15}O_3P$ • 166.16 • 122-52-1
- (P) ΔH_f^0 = –861.90(l) • Eb = 153-157
- (I) P_{fla} = 26; 54
- (T) Tr: 36/37/38; 20/21 • TR: 30
- (D) LC50 r: 11.063/6h; LC50 m: 6.203/6h; LD50 o-r: 1840; 3200; LD50 o-m: 3720; LD50 cu-ra: 2800

Tributylphosphite: $C_{12}H_{27}O_3P$ • 250.32 • 102-85-2
- (P) Eb = 125-128/10; 118-120/7
- (I) P_{fla} = 82; 120
- (T) NFPA: 2
- (D) LD50 o-r: 3000; LD50 cu-ra: 2000

Trimethylphosphate: $C_3H_9O_4P$ • 140.08 • 512-56-1
- (P) Eb = 192-194; 197.2
- (I) P_{fla} = > 148; ~150; none
- (T) Tr: 20/21/22-40
- (D) LD50 o-r: 840; LD50 o-m: 1470; LD50 o-ra: 1050; MLD ip-r: 800; LD50 iv-r: 2400; LD50 ip-m: 2250; LD50 cu-ra: 3388

Triethylphosphate: $C_6H_{15}O_4P$ • 182.16 • 78-40-0
- (P) ΔH_f^0 = –1243.24(l) • Eb = 215-216; 95-98/10; 90-95/10 • P_{vap} = 0.2(20); 0.6(30); 1.33(39.6); 2.7(50)
- (I) LEL/UEL = 1.7/10 • P_{fla} = 115cc; 116oc • AIT = 451; 457
- (T) Tr: 22 • TR: 60
- (D) LDM-r: 1600; LD50 o-m: 1500; MLD ip-r: 800; MLD iv-r: 1000; LD50 ip-m: 485

Tributylphosphate: $C_{12}H_{27}O_4P$ • 266.32 • 126-73-8
- (P) ΔH_f^0 = –1456.03(l); 1456.73(l) • Eb = 289-292 (dec); 177-178/27; 183/22; 155-157/10 • P_{vap} = 9.73(150); 36(178)
- (I) P_{fla} = 145cc; 146Coc; 193 • AIT = 409
- (T) Tr: 22 • NFPA: 2 • TR: 60 • LVE(MVE) (F): – (0.2); TWA (USA): 0.2
- (D) LC50 r: 28/1h; LC50 m: 1.3/?; LD50 o-r: > 1400; 3000; LD50 o-m: 1189; MLD iv-r: 100; LD50 ip-m: 159; 251; MLD scu-m: 3000; LD50 cu-ra: > 3100

Triphenylphosphate: $C_{18}H_{15}O_4P$ • 326.29 • 115-86-6
- (P) ΔH_f^0 = –757.67(s) • Eb = 245/11; 244/10 • P_{vap} = 1.33(193.5)
- (I) P_{fla} = 220cc; 223
- (T) Tr: 20/21/22 • NFPA: 2 • TR: –
- (D) LD50 o-r: 3800; LD50 o-m: 1320

Tritolylphosphate: $C_{21}H_{21}O_4P$ • 368.37 • 1330-78-5
- (P) Eb = 275-296; 410; 265/10; 180-200/0.09 • P_{vap} = 0.04(25); 0.03; 0.13(150); 2.7(200)
- (I) P_{fla} = > 109; 110; 210; 225cc; 250 • AIT = 385; 405
- (T) Tr: 39/23/24/25; 21/22 • NFPA: 2 • TR: –
- (D) LD50 o-r: 1160; 3000; LD50 o-m: 3900; LD50 ip-r: 2500; MLD ip-ra: 100; MLD iv-ra: 100; MLD scu ra: 100

MISCELLANEOUS COMPOUNDS
• Silicon derivatives

Tetramethylsilane: $C_4H_{12}Si$ • 88.22 • 75-76-3
- (P) ΔH_f^0 = –239.1(g); –264.0(l) • Eb = 23; 26-28 • P_{vap} = 785; 804; 2428(55)
- (I) P_{fla} = –27cc • AIT = 449
- (T) NFPA: 3

Triethylsilane: $C_6H_{16}Si$ • 116.28 • 617-86-7
- (P) Eb = 105-110
- (I) P_{fla} = –20; –3
- (T) Tr: 36/37/38

Hexamethyldisiloxane: $C_6H_{18}OSi$ • 162.38 • 107-46-0
- (P) ΔH_f^0 = –777.70(g); –815.0(l) • Eb = 98-101; P_{vap} = 37.3(20)
- (I) P_{fla} = –6; –2
- (T) Tr: –; 36/37/38 • TR: 33
- (D) MLD o-r: 8000; LD50 ip-m: 4500; LD50 cu-ra: ~15000; 16000

Trimethoxysilane: $C_3H_{10}O_3Si$ • 122.20 • 2487-90-3
- (P) Eb = 81-84 • P_{vap} = < 9.6(20)
- (I) P_{fla} = –4; –9
- (D) LC50 r: 0.213/4h; 0.635/4h; LD50 o-r: 9330; LD50 cu-ra: 6300

Triethoxysilane: $C_6H_{16}O_3Si$ • 164.28 • 998-30-1
- (P) Eb = 131-135
- (I) P_{fla} = 26
- (T) Tr: 20/21/22; 20/21
- (D) LC50 r: 0.5/2h; LC50 m: 0.5/2h; LD50 iv-m: 180

8 • Table of organic compounds — Miscellaneous compounds

Name: Empirical formula • Molar mass(g/mol)• Registry number (CAS).
(P) **Physical data:** ΔH_f^0 (kJ/mol) (Physical state) • Eb (°C/mmHg) (if substance volatile) • P_{vap} (mBar) (°C) (if substance volatile).
(I) **Inflammability data:** LEL/UEL (%) • P_{fla} (°C) • AIT (°C).
(T) **Toxicity:** Safety Code (Tr) • NFPA • Transport Code (TR) • LVE (MVE) (ppm) F, USA, D • STEL (ppm); IDLH (ppm).
(D) **Lethal and toxic doses:** LC (mg/l/duration of exposure); LD (mg/kg).

Tetraethoxysilane: $C_8H_{20}O_4Si$ • 208.33 • 78-10-4
(P) Eb = 163-168 • P_{vap} = < 1.33; 1.7; 2.7(20); 3.5(30); 15(50)
(I) P_{fla} = 37; 39Tcc; 46; 52; 60Toc • AIT = 237
(T) Tr: 20-36/37 • NFPA: 2 • TR: 30 • LVE(MVE) (F): – (10); TWA (USA): 10
(D) LCM r: 8.51/?; LD50 o-r: 6270; MLD iv-ra: 200; LD50 cu-ra: 5878; 6300

Trichlorosilane: Cl_3HSi • 135.45 • 10025-78-2
(P) ΔH_f^0 = –513.0(g); –539.3(l); Eb = 31-34; P_{vap} = 533(14.5); 667; 672(20); 2123(55)
(I) LEL/UEL = 1.2/90.5; 6.9/70 • P_{fla} = –50; –28oc; –24; –13; < –6 • AIT = 104; 195
(T) NFPA: 3 • TR: X338
(D) LC50 r: 5.632/4h; LC50 m: 1.5/2h; LD50 o-r: 1030

Tetrachlorosilane: Cl_4Si • 169.90 • 10026-04-7
(P) ΔH_f^0 = –657.0(g); –687.0(l) • Eb = 57-58 • P_{vap} = 560(37.8)
(T) NFPA: 3
(D) LC50 r: 56.515/4h

Trimethylchlorosilane: C_3H_9ClSi • 108.64 • 75-77-4
(P) ΔH_f^0 = –352.8(g); –382.8(l) • Eb = 56-59 • P_{vap} = 180; 247(25)
(I) P_{fla} = –28; –23cc; –18
(T) Tr: 34/37-40
(D) LCM s: 0.500/10'; 0.100/?; LD50 o-r: 5660; MLD ip-m: 750; MLDcu-m: 750 ; LD50 cu-ra: 1780

Dimethyldichlorosilane: $C_2H_6Cl_2Si$ • 129.06 • 75-78-5
(P) ΔH_f^0 = –461.3(g) • Eb = 69-70 • P_{vap} = 45; 173(20); 230(30); 490(50)
(I) LEL/UEL = 3.4/9.5; 5.5/10.4 • P_{fla} = –16; –12; –9Coc; –5; –3cc • AIT = 425
(T) Tr: 36/37/38 • NFPA: 3 • TR: X338
(D) LC50 r: 4.99/4h; LC50 m: 0.3/2h; LD50 o-r: 5660

Methyltrichlorosilane: CH_3Cl_3Si • 149.48 • 75-79-6
(P) Eb = 65-66 • P_{vap} = 66.7; 180(20); 200(25); 280(30); 580(50); 693(55)
(I) LEL/UEL = 7.2/11.9; 7.6/– • P_{fla} = –13; 4cc; 8; 11 • AIT = 394; 404
(T) Tr: 36/37/38 • TR: X338
(D) LC50 r: 2.797/4h; LC50 m: 0.18/2h; MLD ip-r: 30

Hexamethyldisilazane: $C_6H_{19}N$ • 105.22 • 999-97-3
(P) Eb = 124-127 • P_{vap} = 26.7; 30.7(20)
(I) LEL/UEL = 0.7(40)/31(120); 0.8/16.3 • P_{fla} = 8cc; 9Tcc; 14; 25 • AIT = 379
(T) Tr: 36/37/38; 36/38 • TR: 33
(D) LC50 r: 8.8/4h; LC50 m: 12/2h; LD50 o-r: 850; LD50 o-m: 850; LD50 o-ra: 1100; p-m: 650; LD50 cu-ra: 710

MISCELLANEOUS COMPOUNDS
• **Boron derivatives**

Trimethyl borate : $C_3H_9BO_3$ • 103.91 • 121-43-7
(P) ΔH_f^0 = –932.61(l); –933.06(l) • Eb = 63-69
(I) P_{fla} = –8; < 27
(T) Tr: 21 • NFPA: 2 • TR: 33
(D) LD50 o-r: 6140; LD50 o-m: 1290; MLD ip-r: 1600; MLD ip-m: 1000; MLD ip-ra: 1600; LD50 cu-ra: 1980

Triethyl borate : $C_6H_{15}BO_3$ • 145.99 • 150-46-9
(P) ΔH_f^0 = –1048.17(l) • Eb = 117-120
(I) P_{fla} = 11
(T) TR: 33
(D) LD50 o-m: 1800

Tributyl borate : $C_{12}H_{27}BO_3$ • 230.15 • 688-74-4
(P) ΔH_f^0 = –1200.13(l) • Eb = 230-235
(I) P_{fla} = 93Coc
(D) LD50 o-m: 174^0; MLD ip-m: 500

9 • TABLES OF EMPIRICAL FORMULAE AND CORRESPONDING NAMES

9.1 Inorganic compounds

1 Empirical formulae/Corresponding names

Empirical formula	Name
Ag	
Ag	Silver
AgBr	Silver bromide
AgCl	Silver chloride
AgClO$_4$	Silver perchlorate
AgCN	Silver cyanide
AgF	Silver fluoride
AgI	Silver iodide
AgMnO$_4$	Silver permangate
AgNO$_3$	Silver nitrate
Ag$_2$CO$_3$	Silver carbonate
Ag$_2$O	Silver oxide
Ag$_2$SO$_4$	Silver sulphate
Ag$_2$S	Silver sulphide
Ag$_3$PO$_4$	Silver phosphate
Al	
Al	Aluminium
AlB$_3$H$_{12}$	Aluminium borohydride
AlCl$_3$	Aluminium chloride
AlCl$_3$O$_9$	Aluminium chlorate
NH$_4$AlCl$_4$	Ammonium tetrachloroaluminate
AlF$_3$	Aluminium fluoride
AlO$_3$H$_3$	Aluminium trihydroxide
AlNH$_4$S$_2$O$_8$	Aluminium ammonium sulphate
AlKS$_2$O$_8$	Aluminium potassium sulphate
AlN	Aluminium nitride
AlN$_3$O$_9$	Aluminium nitrate
NaAlO$_2$	Sodium aluminate
AlNaS$_2$O$_8$	Aluminium sodium aluminate
AlPO$_4$	Aluminium phosphate
AlP	Aluminium phosphide
AlB$_2$O$_6$	Aluminium borate
Al$_2$ClH$_5$O$_5$	Aluminium oxychloride
Al$_2$Si$_3$F$_{18}$	Aluminium fluorosilicate
Al$_2$MgSi$_2$O$_8$	Aluminium magnesium silicate
Al$_2$S$_3$O$_{12}$	Aluminium sulphate

Empirical formula	Name
Al$_2$O$_3$	Aluminium oxide
Al$_2$SiO$_5$	Aluminium silicate
Ar	
Ar	Argon
As	
As	Arsenic
AsCl$_3$	Arsenic trichloride
GaAs	Gallium arsenide
K$_2$HAsO$_4$	Potassium arsenate
H$_3$As	Arsine
H$_3$AsO$_4$	Arsenic acid
Na$_2$HAsO$_4$	Sodium arsenate
NaAsO$_2$	Sodium arsenite
Ca$_3$As$_2$O$_8$	Calcium arsenate
As$_2$O$_3$	Arsenic trioxide
As$_2$O$_5$	Arsenic pentoxide
As$_2$S$_3$	Arsenic trisulphide
As$_2$S$_5$	Arsenic pentasulphide
As$_2$Se	Arsenic hemiselenide
B	
B	Boron
BCl$_3$	Boron trichloride
BF$_3$	Boron trifluoride
HBF$_4$	Fluoroboric acid
KBF$_4$	Potassium tetrafluoroborate
H$_3$BO$_3$	Boric acid
BN	Boron nitride
NaBO$_3$	Sodium perborate
B$_2$H$_6$	Diborane
MgB$_2$O$_4$	Magnesium diborate
MgB$_2$O$_6$	Magnesium perborate
B$_2$O$_3$	Boron oxide
TiB$_2$	Titanium diboride
B$_4$C	Boron carbide

9 • Tables of equivalence

9.1 Inorganic compounds
Empirical formulae/Names

Empirical formula	Name
CaB_4O_7	Calcium borate
$Li_2B_4O_7$	Lithium tetraborate
$Na_2B_4O_7$	Sodium tetraborate
Ba	
Ba	Barium
$BaBr_2$	Barium bromide
BaC_2N_2	Barium cyanide
$BaCl_2$	Barium chloride
$BaCl_2O_6$	Barium chlorate
$BaCO_3$	Barium carbonate
$BaCrO_4$	Barium chromate
BaF_2	Barium fluoride
$BaSiF_6$	Barium fluorosilicate
BaH_2O_2	Barium hydroxide
BaN_2O_6	Barium nitrate
BaO	Barium oxide
BaO_2	Barium peroxide
$BaSO_4$	Barium sulphate
BaS	Barium sulphide
Be	
Be	Beryllium
$BeCl_2$	Beryllium chloride
BeF_2	Beryllium fluoride
BeH_2O_2	Beryllium hydroxide
$BeK_2S_2O_8$	Beryllium potassium double sulphate
BeN_2O_6	Beryllium nitrate
BeO	Beryllium oxide
Be_2C	Beryllium carbide
Bi	
Bi	Bismuth
$BiCl_3$	Bismuth trichloride
BiClO	Bismuth oxychloride
BiN_3O_9	Bismuth nitrate
Bi_2O_3	Bismuth (III) oxide
Br	
HBr	Hydrrobromic acid
HBr	Hydrogen bromide
NH_4Br	Ammonium bromide
KBr	Potassium bromide
$KBrO_3$	Potassium bromate
LiBr	Lithium bromide
NaBr	Sodium bromide
$NaBrO_3$	Sodium bromate
Br_2	Bromine
$CaBr_2$	Calcium bromide
$CdBr_2$	Cadmium bromide
$CoBr_2$	Cobalt bromide
$CuBr_2$	Cupric bromide
$MgBr_2$	Magnesium dibromide
$SnBr_4$	Stannic bromide

Empirical formula	Name
C	
C	Carbon
$CaCN_2$	Calcium cyanamide
$CaCO_3$	Calcium carbonate
$CdCO_3$	Cadmium carbonate
$CoCO_3$	Cobalt carbonate
$Cu_2CH_2O_5$	Basic copper carbonate
CuCN	Cuprous cyanide
NH_4SCN	Ammonium thiocyanate
NH_5CO_3	Ammonium hydrogencarbonate
$N_2H_8CO_3$	Diammonium carbonate
$NaHCO_3$	Sodium hydrogencarbonate
$KHCO_3$	Potassium hydrogencarbonate
KCN	Potassium cyanide
$MgCO_3$	Magnesium carbonate
Na_2CO_3	Sodium carbonate
K_2CO_3	Potassium carbonate
Na_2CS_3	Sodium thiocarbonate
$NiCO_3$	Nickel carbonate
NaCN	Sodium cyanide
CO	Carbon monoxide
CO_2	Carbon dioxide
$PbCO_3$	Lead carbonate
Rb_2CO_3	Rubidium carbonate
$SrCO_3$	Strontium carbonate
$ZnCO_3$	Zinc carbonate
CS_2	Carbon disulphide
SiC	Silicon carbide
TiC	Titanium carbide
VC	Vanadium carbide
CaC_2	Calcium carbide
CaC_2N_2	Calcium cyanide
$CaC_2N_2S_2$	Calcium thiocyanate
CdC_2N_2	Cadmium cyanide
HgC_2N_2	Mercuric cyanide
$HgC_2N_2S_2$	Merurous thiocyanate
$Na_3HC_2O_6$	Sodium sesquicarbonate
NiC_2N_2	Nickel cyanide
ZnC_2N_2	Zinc cyanide
C_3O_2	Carbon suboxide
$K_3FeC_6N_6$	Potassium hexacyanoferrate (III)
$Fe_7C_{18}N_{18}$	Ferric hexacyanoferrate (II)
Ca	
Ca	Calcium
$CaCl_2$	Calcium chloride
$CaCl_2O_2$	Calcium hypochlorite
$CaCl_2O_6$	Calcium chlorate
$CaCrO_4$	Calcium chromate
CaF_2	Calcium fluoride
$CaSiF_6$	Calcium hexafluorosilicate
CaO_2H_2	Calcium hydroxide

9.1 Inorganic compounds
Empirical formulae/Names

Empirical formula	Name
$CaH_4P_2O_4$	Calcium hypophosphite
$CaH_4P_2O_8$	Calcium bis(dihydrogenphosphate)
$CaHPO_4$	Calcium hydrogenphosphate
$CaMn_2O_8$	Calcium permanganate
CaN_2O_6	Calcium nitrate
CaO	Calcium oxide
$CaSO_4$	Calcium sulphate
CaS	Calcium sulphide
$Ca_2P_2O_7$	Calcium pyrophosphate
$Ca_3P_2O_8$	Calcium phosphate
Ca_3P_2	Calcium phosphide
Cd	
Cd	Cadmium
$CdCl_2$	Cadmium chloride
CdF_2	Cadmium fluoride
CdO_2H_2	Cadmium hydroxide
CdI_2	Cadmium iodide
CdN_2O_6	Cadmium nitrate
CdO	Cadmium oxide
$CdSO_4$	Cadmium sulphate
CdS	Cadmium sulphide
Cl	
$CuCl$	Cuprous chloride
HCl	Hydrogen chloride
NH_4Cl	Ammonium chloride
NH_4OCl	Hydroxylamine chlorohydrate
NH_4ClO_4	Ammonium perchlorate
$HClSO_3$	Chlorosulphonic acid
$HClO_4$	Perchloric acid
KCl	Potassium chloride
$KClO_3$	Potassium chlorate
$KClO_4$	Potassium perchlorate
$LiCl$	Lithium chloride
$NaCl$	Sodium chloride
$NaClO$	Sodium hypochlorite
$NaClO_2$	Sodium chlorite
$NaClO_3$	Sodium chlorate
$NaClO_4$	Sodium perchlorate
ClO_2	Chlorine dioxide
$RbCl$	Rubidium chloride
Cl_2	Chlorine
$CoCl_2$	Cobalt chloride
CrO_2Cl_2	Chromyl dichloride
$CuCl_2$	Cupric chloride
Cu_2Cl_2	Cuprous chloride
$FeCl_2$	Iron (II) chloride
$N_2H_6Cl_2$	Hydrazine chlorohydrate
$HgCl_2$	Mercuric chloride
Hg_2Cl_2	Mercurous chloride
$MgCl_2$	Magnesium dichloride
$MgCl_2O_8$	Magnesium perchlorate

Empirical formula	Name
$MnCl_2$	Manganese dichloride
$NiCl_2$	Nickel chloride
SO_2Cl_2	Sulphuryl chloride
$SrCl_2O_6$	Strontium chlorate
$SOCl_2$	Thionyl chloride
$ZrOCl_2$	Zirconyl chloride
$PbCl_2$	Lead chloride
$PdCl_2$	Palladium chloride
SCl_2	Sulphur dichloride
S_2Cl_2	Disulphur dichloride
$SnCl_2$	Stannous chloride
$SrCl_2$	Strontium chloride
$ZnCl_2$	Zinc chloride
NCl_3	Nitrogen trichloride
$CrCl_3$	Chromium trichloride
$FeCl_3$	Ferric chloride
$GaCl_3$	Gallium trichloride
$InCl_3$	Indium trichloride
$POCl_3$	Phosphoryl trichloride
$VOCl_3$	Vanadyl chloride
PCl_3	Phosphorous trichloride
$RhCl_3$	Rhodium trichloride
$RuCl_3$	Ruthenium trichloride
$SbCl_3$	Antinomy trichloride
$TiCl_3$	Titanium trichloride
$GeCl_4$	Germanium tetrachloride
$SnCl_4$	Stannic chloride
$TiCl_4$	Titanium tetrachloride
$ZrCl_4$	Zirconium tetrachloride
$NbCl_5$	Niobium pentachloride
PCl_5	Phosphorus pentachloride
Co	
Co	Cobalt
CoO_2H_2	Cobalt dihydroxide
$K_3N_6CoO_{12}$	Potassium hexanitrocobaltate (III)
CoN_2O_6	Cobalt nitrate
CoO	Cobalt oxide
$CoSO_4$	Cobalt sulphate
Co_3O_4	Tricobalt tetroxide
Cr	
Cr	Chromium
$N_2H_8CrO_4$	Ammonium chromate
K_2CrO_4	Potassium chromate
$CrKS_2O_8$	Chromium potassium sulphate
Li_2CrO_4	Lithium chromate
CrN_3O_9	Chromium trinitrate
Na_2CrO_4	Sodium chromate
CrO_3	Chromium (V) oxide
$PbCrO_4$	Lead chromate
$SrCrO_4$	Strontium chromate
$ZnCrO_4$	Zinc chromate

9 • Tables of equivalence

9.1 Inorganic compounds Empirical formulae/Names

Empirical formula	Name
$N_2H_8Cr_2O_7$	Ammonium dichromate
$K_2Cr_2O_7$	Potassium dichromate
$Na_2Cr_2O_7$	Sodium dichromate
$Cr_2S_3O_{12}$	Chromium (III) sulphate
Cr_2O_3	Chromium (III) oxide
Cu	
Cu	Copper
$CuCl_2O_8$	Cupric perchlorate
CuN_2O_6	Cupric nitrate
CuO	Cupric oxide
$CuSO_4$	Copper sulphate
CuS	Cupric sulphide
Cu_2O	Cuprous oxide
F	
F	Fluorine
HF	Hydrogen fluoride
NH_4F	Ammonium fluoride
KF	Potassium fluoride
LiF	Lithium fluoride
NaF	Sodium fluoride
F_2	Fluorine
NH_5F_2	Ammonium difluoride
KHF_2	Potassium difluoride
$NaHF_2$	Sodium difluoride
MgF_2	Magnesium difluoride
SO_2F_2	Sulphuryl fluoride
SnF_2	Stannous fluoride
ZnF_2	Zinc fluoride
SbF_3	Antinomy trifluoride
H_2SiF_6	Fluorosilicic acid
$N_2H_8SiF_6$	Ammonium hexafluorosilicate
K_2SiF_6	Potassium hexafluorosilicate
$MgSiF_6$	Magnesium fluorosilicate
Na_2SiF_6	Sodium hexafluorosilicate
$PbSiF_6$	Lead hexafluorosilicate
SF_6	Sulphur hexafluoride
SeF_6	Selenium hexafluoride
$ZnSiF_6$	Zinc hexafluorosilicate
Fe	
Fe	Iron
$FeN_2H_8S_2O_8$	Iron (II) ammonium sulphate
FeN_3O_9	Ferric nitrate
FeO	Iron (III) oxide
$FeSO_4$	Ferrous sulphate
$Fe_2S_3O_{12}$	Ferric sulphate
Fe_2O_3	Ferric oxide
Ga	
Ga	Gallium
GaN_3O_9	Gallium nitrate

Empirical formula	Name
Ga_2O_3	Digallium trioxide
Ge	
Ge	Germanium
GeO_2	Germanium dioxide
H	
HI	Hydrogen iodide
K_2HPO_4	Dipotassium hydrogenphosphate
KOH	Potassium hydroxide
$KHSO_4$	Potassium hydrogensulphate
LiOH	Lithium hydroxide
HN_3	Hydrogen azide
Na_2HPO_4	Disodium hydrogenphosphate
NaOH	Sodium hydroxide
$NaHSO_3$	Sodium hydrogensulphite
NaHS	Sodium hydrogen sulphide
$(HPO_3)_n$	Metaphosphoric acid
H_2	Hydrogen
KH_2PO_4	Potassium dihydrogenphosphate
MgO_2H_2	Magnesium dihydroxide
NaH_2PO_4	Sodium dihydrogenphosphate
$NaNH_2$	Sodium amide
H_2O	Water
H_2O_2	Hydrogen peroxide
SrO_2H_2	Strontium hydroxide
H_2SO_3	Sulphurous acid
H_2SO_4	Sulphuric acid
H_2S	Hydrogen sulphide
TiH_2	Titanium dihydride
NH_3	Ammonia
NH_3O	Hydroxylamine
H_3NSO_3	Sulphamic acid
H_3PO_4	Orthophosphoric acid
PH_3	Phosphine
H_3Sb	Stibine
NH_4I	Ammonium iodide
$MgH_4P_2O_8$	Magnesium bis (dihydrogenphosphate)
N_2H_4	Hydrazine
NH_4NO_3	Ammonium nitrate
H_4SiO_4	Silicic acid
ZrO_4H_4	Zirconium hydroxide
SiH_4	Silane
$NaNH_5PO_4$	Sodium ammonium hydrogenphosphate
NH_5S	Ammonium hydrogensulphide
$N_2H_6SO_3$	Ammonium sulphamate
$N_2H_6SO_4$	Hydrazine sulphate
NH_6PO_4	Ammonium phosphate
$N_2H_8SO_3$	Diammonium sulphite
$N_2H_8S_2O_3$	Ammonium thiosulphate
$N_2H_8MoO_4$	Ammonium molybdate
$N_2H_8SO_4$	Ammonium sulphate
$N_2H_8S_2O_8$	Diammonium peroxodisulphate

9 • Tables of equivalence

9.1 Inorganic compounds — Empirical formulae/Names

Empirical formula	Name
N_2H_8S	Ammonium sulphide
$N_2H_9PO_3$	Diammonium phosphite
$N_2H_9PO_4$	Diammonium phosphate
He	
He	Helium
Hg	
Hg	Mercury
HgI_2	Mercuric iodide
HgN_2O_6	Mercuric nitrate
HgO	Mercuric oxide
$HgNO_3$	Mercurous nitrate
$HgSO_4$	Mercuric sulphate
HgS	Mercuric sulphide
Hg_2SO_4	Mercurous sulphate
I	
KI	Potassium iodide
KIO_3	Potassium iodate
LI	Lithium iodide
NaI	Sodium iodide
$NaIO_3$	Sodium iodate
I_2	Iodine
PbI_2	Lead iodide
In	
In	Indium
InSb	Indium antimonide
$In_2S_3O_{12}$	Indium sulphate
In_2O_3	Indium oxide
K	
K	Potassium
$KMnO_4$	Potassium permanganate
KNO_2	Potassium nitrite
KNO_3	Potassium nitrate
K_2SO_3	Potassium sulphite
K_2SnO_3	Potassium stannate (IV)
K_2SO_4	Potassium sulphate
$K_2S_2O_5$	Poptassium metabisulphite
$K_2S_2O_8$	Potassium peroxodisulphate
K_2TeO_3	Potassium tellurate (IV)
$K_4P_2O_7$	Potassium pyrophosphate
Kr	
Kr	Krypton
Li	
Li	Lithium
Li_2SO_4	Lithium sulphate
Li_2CO_3	Lithium carbonate
Mg	
Mg	Magnesium
MgN_2O_6	Magnesium dinitrate

Empirical formula	Name
MgO	Magnesium oxide
$MgSO_3$	Magnesium sulphite
$MgSO_4$	Magnesium sulphate
Mn	
Mn	Manganese
MnO	Manganese (II) oxide
MnO_2	Manganese dioxide
$MnSO_4$	Manganese sulphate
Mn_2O_3	Manganese (III) oxide
Mn_3O_4	Trimanganese tetroxide
Mo	
Mo	Molybdenum
MoO_3	Molybdenum trioxide
MoS_2	Molybdenum sulphide
N	
$NaNO_2$	Sodium nitrite
$NaNO_3$	Sodium nitrate
NO	Nitric oxide
NO_2	Nitrogen dioxide
TiN	Titanium nitride
N_2	Nitrogen
NiN_2O_6	Nickel nitrate
N_2O	Nitrous oxide
N_2O_4	Nitrogen tetroxide
PbN_2O_6	Lead nitrate
SrN_2O_6	Strontium nitrate
ZnN_2O_6	Zinc nitrate
NaN_3	Sodium azide
Si_3N_4	Silicon nitride
PbN_6	Lead nitride
Na	
Na	Sodium
$(NaPO_3)_n$	Sodium polymetaphosphate
$NaVO_3$	Sodium vanadate
Na_2O_2	Sodium peroxide
Na_2SO_3	Sodium sulphite
$Na_2S_2O_3$	Sodium thiosulphate
Na_2SeO_3	Sodium selenite
Na_2SnO_3	Sodium stannate (IV)
$NaMo_2O_4$	Sodium molybdate
Na_2SO_4	Sodium sulphate
$Na_2S_2O_4$	Sodium hydrosulphite
$Na_2S_2O_8$	Sodium peroxodisulphate
Na_2S	Sodium sulphide
Na_3PO_4	Sodium phosphate
$Na_3P_3O_9$	Sodium trimetaphosphate
$Na_5P_3O_{10}$	Sodium tripolyphosphate
$Na_6P_6O_{18}$	Sodium hexametaphosphate

9 • Tables of equivalence

9.1 Inorganic compounds — Empirical formulae/Names

Empirical formula	Name	Empirical formula	Name
Nb		TiS_2O_8	Titanium sulphate
Nb	Niobium	ZrS_2O_8	Zirconium sulphate
Nb_2O_5	Niobium pentoxide	$Rh_2S_3O_{12}$	Rhodium sulphate
Ne		**P**	
Ne	Neon	P	Phosphorus
		P_2S_5	Phosphorus pentasulphide
Ni		Zn_3P_2	Zinc phosphide
Ni	Nickel		
NiO	Nickel oxide	**Pb–Ru**	
$NiSO_4$	Nickel sulphate	Pb	Lead
		PbS	Lead sulphide
O		Pd	Palladium
PbO	Lead oxide	Rb	Rubidium
SrO	Strontium oxide	Rh	Rhodium
ZnO	Zinc oxide	Ru	Ruthenium
O_2	Oxygen		
PbO_2	Lead peroxide	**S**	
SO_2	Sulphur dioxide	S	Sulphur
SeO_2	Selenium dioxide	SnS_2	Tin disulphide
SiO_2	Silicon dioxide	Sb_2S_3	Antimony trisulphide
SnO_2	Tin (IV) oxide	Sb_2S_5	Antimony sulphide
TiO_2	Titanium dioxide	Sb	Antimony
ZnO_2	Zinc peroxide	Sc	Scandium
ZrO_2	Zirconium oxide	Se	Selenium
O_3	Ozone	Si	Silicon
SO_3	Sulphur trioxide	Sn	Tin
Sb_2O_3	Antinomy trioxide	Sr	Strontium
V_2O_3	Divanadium trioxide	SrS	Strontium sulphide
Pb_3O_4	Lead tetroxide	ZnS	Zinc sulphide
$PbSO_4$	Lead sulphate		
RuO_4	Ruthenium oxide	**Tc - Zr**	
$ZrSiO_4$	Zirconium silicate	Tc	Technetium
$SrSO_4$	Strontium sulphate	Te	Tellurium
$ZnSO_4$	Zinc sulphate	Ti	Titanium
P_2O_5	Diphosphorus pentoxide	V	Vanadium
Sb_2O_5	Antinomy pentoxide	Xe	Xenon
VSO_5	Vanadyl sulphate	Zn	Zinc
V_2O_5	Divanadium pentoxide	Zr	Zirconium

9 • Tables of equivalence

9.1 Inorganic compounds CAS N°/Name

9.1.2 Equivalent CAS N° /Name

CAS N°	Name	CAS N°	Name
75-15-0	Carbon disulphide	1306-19-0	Cadmium oxide
75-20-7	Calcium carbide	1306-23-6	Cadmium sulphide
124-38-9	Carbon dioxide	1307-96-6	Cobalt oxide
143-33-9	Sodium cyanide	1308-06-1	Tricobalt tetroxide
144-55-8	Sodium hydrogencarbonate	1308-38-9	Chromium (III) oxide
151-50-8	Potassium cyanide	1309-37-1	Ferric oxide
156-62-7	Calcium cyanamide	1309-42-8	Magnesium dihydroxide
298-14-6	Potassium hydrogencarbonate	1309-48-4	Magnesium oxide
302-01-2	Hydrazine	1309-60-0	Lead peroxide
409-21-2	Silicon carbide	1309-64-4	Antinomy trioxide
471-34-1	Calcium carbonate	1310-53-8	Germanium dioxide
497-19-8	Sodium carbonate	1310-58-3	Potassium hydroxide
504-64-3	Carbon suboxide	1310-65-2	Lithium hydroxide
506-64-9	Silver cyanide	1310-73-2	Sodium hydroxide
506-66-1	Beryllium carbide	1312-41-0	Indium antimonide
513-77-9	Barium carbonate	1312-43-2	Indium oxide
513-78-0	Cadmium carbonate	1312-76-1	Potassium silicate
513-79-1	Cobalt carbonate	1313-13-9	Manganese dioxide
533-96-0	Sodium sesquicarbonate	1313-27-5	Molybdenum trioxide
534-16-7	Silver carbonate	1313-60-6	Sodium peroxide
534-18-9	Sodium thiocarbonate	1313-82-2	Sodium sulphide
542-62-1	Barium cyanide	1313-96-8	Niobium pentoxide
542-83-6	Cadmium cyanide	1313-99-1	Nickel oxide
544-92-3	Copper cyanide	1314-11-0	Strontium oxide
546-93-0	Magnesium carbonate	1314-13-2	Zinc oxide
554-13-2	Lithium carbonate	1314-22-3	Zirconium peroxide
557-19-7	Nickel cyanide	1314-23-4	Zirconium oxide
557-21-1	Zinc cyanide	1314-34-7	Divanadium trioxide
584-08-7	Potassium carbonate	1314-41-6	Lead tetroxide
584-09-8	Rubidium carbonate	1314-56-3	Phosphorus pentoxide
592-01-8	Calcium cyanide	1314-60-9	Antinomy pentoxide
592-04-1	Mercuric cyanide	1314-62-1	Divanadium pentoxide
592-85-8	Mercury thiocyanate	1314-80-3	Diphosphorus pentasulphide
598-63-0	Lead carbonate	1314-84-7	Zinc phosphide
630-08-0	Carbon monoxide	1314-87-0	Lead sulphide
1066-33-7	Ammonium hydrogencarbonate	1314-96-1	Strontium sulphide
1303-00-0	Gallium arsenide	1314-98-3	Zinc sulphide
1303-28-2	Arsenic pentoxide	1315-01-1	Tin disulphide
1303-33-9	Arsenic trisulphide	1315-04-4	Antinomy pentasulphide
1303-34-0	Arsenic pentasulphide	1317-33-5	Molybdenum sulphide
1303-35-1	Arsenic hemiselenide	1317-34-6	Manganese (III) oxide
1303-86-2	Boron oxide	1317-35-7	Trimanganese tetroxide
1304-28-5	Barium oxide	1317-36-8	Lead oxide
1304-29-6	Barium peroxide	1317-38-0	Cupric oxide
1304-56-9	Beryllium oxide	1317-39-1	Cuprous oxide
1304-76-3	Bismuth (III) oxide	1317-40-4	Cupric sulphide
1305-62-0	Calcium hydroxide	1317-65-3	Calcium carbonate
1305-78-8	Calcium oxide	1327-41-9	Aluminium oxychloride
1305-99-3	Calcium phosphide	1327-53-3	Arsenic trioxide

9 • Tables of equivalence

9.1 Inorganic compounds CAS N°/Name

CAS N°	Name	CAS N°	Name
1330-43-4	Sodium tetraborate	7440-37-1	Argon
1332-21-4	Asbestos	7440-38-2	Arsenic
1332-29-2	Tin (IV) oxide	7440-39-3	Barium
1333-82-0	Chromium (VI) oxide	7440-41-7	Beryllium
1333-83-1	Sodium bifluoride	7440-42-8	Boron
1341-49-7	Ammonium bifluoride	7440-43-9	Cadmium
1344-28-1	Aluminium oxide	7440-44-0	Carbon
1344-43-0	Manganese (II) oxide	7440-47-3	Chromium
1344-48-5	Mercuric sulphide	7440-48-4	Cobalt
1344-67-8	Cupric chloride	7440-50-8	Copper
1345-04-6	Antimony trisulphide	7440-55-3	Gallium
1345-25-1	Iron (II) oxide	7440-56-4	Germanium
1762-95-4	Ammonium thiocyanate	7440-59-7	Helium
2092-16-2	Calcium thiocyanate	7440-62-2	Vanadium
2551-62-4	Sulphur hexafluoride	7440-63-3	Xenon
2699-79-8	Sulphuryl fluoride	7440-66-6	Zinc
3251-23-8	Cupric nitrate	7440-67-7	Zirconium
3333-67-3	Nickel carbonate	7440-69-9	Bismuth
3486-35-9	Zinc carbonate	7440-70-2	Calcium
3811-04-9	Potassium chlorate	7440-74-6	Indium
5329-14-6	Sulphamic acid	7446-08-4	Selenium dioxide
5341-61-7	Hydrazine chlorohydrate	7446-09-5	Sulphur dioxide
5470-11-1	Hydroxylamine chlorohydrate	7446-11-9	Sulphur trioxide
6484-52-2	Ammonium nitrate	7446-14-2	Lead sulphate
6834-92-0	Sodium silicate	7446-70-0	Aluminium chloride
7320-34-5	Potassium pyrophosphate	7447-40-7	Potassium chloride
7429-90-5	Aluminium	7447-41-8	Lithium chloride
7439-89-6	Iron	7487-88-9	Magnesium sulphate
7439-90-9	Krypton	7487-94-7	Mercuric chloride
7439-92-1	Lead	7550-35-8	Lithium bromide
7439-93-2	Lithium	7550-45-0	Titanium tetrachloride
7439-95-4	Magnesium	7553-56-2	Iodine
7439-96-5	Manganese	7558-79-4	Disodium hydrogenphosphate
7439-97-6	Mercury	7558-80-7	Sodium dihydrogenphosphate
7439-98-7	Molybdenum	7597-37-2	Nitric acid
7440-02-0	Nickel	7601-54-9	Sodium phosphate
7440-03-1	Niobium	7601-89-0	Sodium perchlorate
7440-05-3	Palladium	7601-90-3	Perchloric acid
7440-09-1	Neon	7631-90-5	Sodium hydrogensulphite
7440-09-7	Potassium	7631-95-0	Sodium molybdate
7440-16-6	Rhodium	7631-99-4	Sodium nitrate
7440-17-7	Rubidium	7632-00-0	Sodium nitrite
7440-18-8	Ruthenium	7637-07-2	Boron trifluoride
7440-20-2	Scandium	7646-78-8	Stannic chloride
7440-21-3	Silicon	7646-79-9	Cobalt chloride
7440-22-4	Silver	7646-85-7	Zinc chloride
7440-23-5	Sodium	7646-93-7	Potassium hydrogensulphate
7440-24-6	Strontium	7647-01-0	Hydrogen chloride, hydrochloric acid
7440-26-8	Technetium	7647-10-1	Palladium chloride
7440-31-5	Tin	7647-14-5	Sodium chloride
7440-32-6	Titanium	7647-15-6	Sodium bromide
7440-36-0	Antimony	7664-38-2	Orthophosphoric acid

9 • Tables of equivalence

9.1 Inorganic compounds CAS N°/Name

CAS N°	Name	CAS N°	Name
7664-39-3	Hydrofluoric acid	7773-06-0	Ammonium sulphamate
7664-39-3	Hydrogen fluoride	7774-29-0	Mercuric iodide
7664-41-7	Ammonia	7775-09-9	Sodium chlorate
7664-93-9	Sulphuric acid	7775-11-3	Sodium chromate
7681-11-0	Potassium iodide	7775-14-6	Sodium hydrosulphite
7681-49-7	Sodium fluoride		
7681-52-9	Sodium hypochlorite	7775-27-1	Sodium peroxodisulphate
7681-55-2	Sodium iodate	7775-41-9	Silver fluoride
7681-82-5	Sodium iodide	7778-18-9	Calcium sulphate
7699-43-6	Zirconyl chloride	7778-39-4	Arsenic acid
7704-34-9	Sulphur	7778-43-0	Sodium arsenate
7704-98-5	Titanium dihydride	7778-44-1	Calcium arsenate
7705-07-9	Titanium trichloride	7778-50-9	Potassium dichromate
7705-08-0	Ferric chloride	7778-54-3	Calcium hypochlorite
7718-54-9	Nickel chloride	7778-74-7	Potassium perchlorate
7719-09-7	Thionyl chloride	7778-77-0	Potassium dihydrogenphosphate
7719-12-2	Phosphorus trichloride	7778-80-5	Potassium sulphate
7720-78-7	Ferrous sulphate	7779-88-6	Zinc nitrate
7722-64-7	Potassium permanganate	7782-41-4	Fluorine
7722-76-1	Ammonium phosphate	7782-44-7	Oxygen
7722-84-1	Hydrogen peroxide	7782-49-2	Selenium
7723-14-0	Phosphorus	7782-50-5	Chlorine
7726-95-6	Bromine	7782-61-8	Ferric nitrate
7727-18-6	Vanadyl chloride	7782-79-8	Hydrogen azide
7727-21-1	Potassium peroxodisulphate	7782-92-5	Sodium amide
7727-43-7	Barium sulphate	7782-99-2	Sulphurous acid
7727-54-0	Diammonium peroxodisulphate	7783-06-4	Hydrogen sulphide
7732-18-5	Water	7783-18-8	Diammonium thiosulphate
7733-02-0	Zinc sulphate	7783-20-2	Diammonium sulphate
7757-79-1	Potassium nitrate	7783-28-0	Diammonium phosphate
7757-82-6	Sodium sulphate	7783-35-9	Mercuric sulphate
7757-83-7	Sodium sulphite	7783-36-0	Mercurous sulphate
7757-88-2	Magnesium sulphite	7783-40-6	Magnesium difluoride
7757-93-9	Calcium hydrogenphosphate	7783-47-3	Stannous fluoride
7758-01-2	Potassium bromate	7783-49-5	Zinc fluoride
7758-02-3	Potassium bromide	7783-56-4	Antinomy trifluoride
7758-05-6	Potassium iodate	7783-79-1	Selenium hexafluoride
7758-09-0	Potassium nitrite	7783-85-9	Iron (II) ammonium sulphate
7758-11-4	Dipotassium hydrogenphosphate	7783-90-6	Silver chloride
7758-19-2	Sodium chlorite	7783-93-9	Silver perchlorate
7758-23-8	Calcium bis(dihydrogenphosphate)	7783-96-2	Silver iodide
7758-29-4	Sodium tripolyphosphate	7783-98-4	Silver permanganate
7758-89-6	Cuprous chloride	7784-09-0	Silver phosphate
7758-94-3	Iron (II) chloride	7784-14-7	Ammonium tetrachloroaluminate
7758-95-4	Lead chloride	7784-18-1	Aluminium fluoride
7758-97-6	Lead chromate	7784-25-0	Aluminium ammonium sulphate
7758-98-7	Copper sulphate	7784-30-7	Aluminium phosphate
7759-02-6	Strontium sulphate	7784-34-1	Arsenic trichloride
7761-88-8	Silver nitrate	7784-41-0	Potassium arsenate
7772-98-7	Sodium thiosulphate	7784-42-1	Arsine
7772-99-8	Stannous chloride	7784-46-5	Sodium arsenite
7773-01-5	Manganese dichloride	7785-23-1	Silver bromide

9 • Tables of equivalence

9.1 Inorganic compounds CAS N°/Name

CAS N°	Name	CAS N°	Name
7785-84-4	Sodium trimetaphosphate	10034-85-2	Hydrogen iodide
7785-87-7	Manganese sulphate	10034-93-2	Hydrazine sulphate
7786-30-3	Magnesium dichloride	10035-10-6	Hydrobromic acid
7786-81-4	Nickel sulphate	10035-10-6	Hydrogen bromide
7787-32-8	Barium fluoride	10038-98-9	Germanium tetrachloride
7787-47-5	Beryllium chloride	10042-76-9	Strontium nitrate
7787-49-7	Beryllium fluoride	10043-01-3	Aluminium sulphate
7787-59-9	Bismuth oxychloride	10043-11-5	Boron nitride
7787-60-2	Bismuth trichloride	10043-35-3	Boric acid
7788-98-9	Ammonium chromate	10043-52-4	Calcium chloride
7788-99-0	Potassium chromium sulphate	10043-67-1	Aluminium potassium sulphate
7789-00-6	Potassium chromate	10045-94-0	Mercuric nitrate
7789-06-2	Strontium chromate	10049-04-4	Chlorine dioxide
7789-09-5	Ammonium dichromate	10049-07-7	Rhodium trichloride
7789-23-3	Potassium fluoride	10049-08-8	Ruthenium trichloride
7789-24-4	Lithium fluoride	10099-74-8	Lead nitrate
7789-29-9	Potassium difluoride	10101-52-7	Zirconium silicate
7789-38-0	Sodium bromate	10101-63-0	Lead iodide
7789-41-5	Calcium bromide	10102-18-8	Sodium selenite
7789-42-6	Cadmium bromide	10102-40-6	Sodium molybdate
7789-43-7	Cobalt bromide	10102-43-9	Nitrogen oxide
7789-45-9	Cupric bromide	10102-44-0	Nitrogen dioxide
7789-48-2	Magnesium dibromide	10102-71-3	Aluminium sodium sulphate
7789-65-5	Stannic bromide	10108-64-2	Cadmium chloride
7789-75-5	Calcium fluoride	10112-91-1	Mercurous chloride
7789-79-9	Calcium hypophosphite	10117-38-1	Potassium sulphite
7790-76-3	Calcium pyrophosphate	10118-76-0	Calcium permanganate
7790-79-6	Cadmium fluoride	10124-36-4	Cadmium sulphate
7790-80-9	Cadmium iodide	10124-37-5	Calcium nitrate
7790-94-5	Chlorosulphonic acid	10124-43-3	Cobalt sulphate
7790-98-9	Ammonium perchlorate	10124-56-8	Sodium hexametaphosphate
7791-10-8	Strontium chlorate	10137-74-3	Calcium chlorate
7791-11-9	Rubidium chloride	10141-05-6	Cobalt nitrate
7791-25-5	Sulphuryl chloride	10193-36-9	Silicic acid
7803-49-8	Hydroxylamine	10196-04-0	Diammonium sulphite
7803-51-2	Phosphine	10294-26-5	Silver sulphate
7803-52-3	Stibine	10294-34-5	Boron trichloride
7803-62-5	Silane	10294-40-3	Barium chromate
10022-31-8	Barium nitrate	10294-46-9	Cupric perchlorate
10024-97-2	Nitrous oxide	10325-94-7	Cadmium nitrate
10025-67-9	Disulphur dichloride	10343-62-1	Metaphosphoric acid
10025-73-7	Chromium trichloride	10361-29-2	Diammonium carbonate
10025-82-8	Indium trichloride	10361-37-2	Barium chloride
10025-85-1	Nitrogen trichloride	10361-44-1	Bismuth nitrate
10025-87-3	Phosphoryl trichloride	10377-48-7	Lithium sulphate
10025-91-9	Antinomy trichloride	10377-51-2	Lithium iodide
10026-11-6	Zirconium tetrachloride	10377-60-3	Magnesium dinitrate
10026-12-7	Niobium pentachloride	10415-75-5	Mercurous nitrate
10026-13-8	Phosphorus pentachloride	10476-85-4	Strontium chloride
10028-15-6	Ozone	10486-00-7	Sodium perborate
10028-22-5	Ferric sulphate	10489-46-0	Rhodium sulphate
10034-81-8	Magnesium perchlorate	10544-72-6	Nitrogen tetroxide (see dioxide)

9 • Tables of equivalence

9.1 Inorganic compounds CAS N° /Name

Cas N°	Name	CAS N°	Name
10545-99-0	Sulphur dichloride	14808-60-7	Silicon
10553-31-8	Barium bromide	14940-68-2	Zirconium silicate
10588-01-9	Sodium dichromate	14977-61-8	Chromyl dichloride
11121-16-7	Aluminium borate	15005-90-0	Chromium sulphate
11138-49-1	Sodium aluminate	15477-33-5	Aluminium chlorate
12007-56-6	Calcium borate	155 71-91-2	Potassium tellurate (IV)
12007-60-2	Lithium tetraborate	1633-05-2	Strontium carbonate
12024-21-4	Digallium trioxide	16721-80-5	Sodium hydrogensulphide
12027-06-4	Diammonium iodide	16731-55-8	Potassium metabisulphite
12033-89-5	Silicon nitride	16871-71-9	Zinc hexafluorosilicate
12045-63-5	Titanium diboride	16871-90-2	Potassium hexafluorosilicate
12058-66-1	Sodium stannate (IV)	16872-11-0	Fluoroboric acid
12069-32-8	Boron carbide	16893-85-9	Sodium hexafluorosilicate
12069-69-1	Basic copper carbonate	16919-19-8	Ammonium hexafluorosilicate
12070-08-5	Titanium carbide	16925-39-6	Calcium hexafluorosilicate
12124-97-9	Ammonium bromide	16961-83-4	Fluorosilicic acid
12124-99-1	Ammonium hydrogensulphide	16962-07-5	Aluminium borohydride
12125-01-8	Ammonium fluoride	17099-70-6	Aluminium fluorosilicate
12125-02-9	Ammonium chloride	17125-80-3	Barium fluorosilicate
12125-03-0	Potassium stannate (IV)	17194-00-2	Barium hydroxide
12135-76-1	Diammonium sulphide	18480-07-4	Strontium hydroxide
12141-46-7	Aluminium silicate	18972-56-0	Magnesium fluorosilicate
12167-74-7	Calcium phosphate	19287-45-7	Diborane
12385-13-6	Hydrogen	20427-56-9	Ruthenium oxide
12511-31-8	Magnesium aluminium silicate	20548-54-3	Calcium sulphide
13011-54-6	Sodium ammonium hydrogenphosphate	20667-12-3	Silver oxide
13092-66-5	Magnesium bis (dihydrogenphosphate)	20859-73-8	Aluminium phosphide
13106-76-8	Ammonium molybdate	21041-93-0	Cobalt dihydroxide
13138-45-9	Nickel nitrate	21041-95-2	Cadmium hydroxide
13327-32-7	Beryllium hydroxide	21109-95-5	Barium sulphide
13424-46-9	Lead nitride	21548-73-2	Silver sulphide
13450-90-3	Gallium trichloride	21645-51-2	Aluminium trihydroxide
13463-67-7	Titanium dioxide	21908-53-2	Mercuric oxide
13464-82-9	Indium sulphate	24304-00-5	Aluminium azide
13473-90-0	Aluminium nitrate	25583-20-4	Titanium nitride
13477-00-4	Barium chlorate	25808-74-6	Lead hexafluorosilicate
13494-80-9	Tellurium	26628-22-8	Sodium azide
13494-90-1	Gallium nitrate	27774-13-6	Vanadyl sulphate
13530-65-9	Zinc chromate	34806-73-0	Zirconium sulphate
13548-38-4	Chromium trinitrate	50813-16-6	Sodium polymetaphosphate
13597-99-4	Beryllium nitrate	51503-61-8	Diammonium phosphite
13703-82-7	Magnesium diborate	53684-48-3	Double beryllium potassium sulphate
13718-26-8	Sodium vanadate	82597-01-1	Magnesium carbonate
13746-66-2	Poatssium hexacyanoferrate (III)		
13755-29-8	Poatssium tetrafluoroborate		
13765-19-0	Calcium chromate		
13782-01-9	Potassium hexanitrocobaltate (III)		
13825-74-6	Titanium sulphate		
14038-43-8	Ferric hexacyanoferrate (II)		
14242-05-8	Silver perchlorate		
14307-35-8	Lithium chromate		
14475-63-9	Zirconium hydroxide		
14635-87-1	Magnesium perborate		

9.2 Organic compounds

1 Empirical formulae/Corresponding names

Empirical formula	Name
C	
CBr_2F_2	Dibromodifluoromethane
CBr_4	Tetrabromomethane
$CBrCl_3$	Bromotrichloromethane
$CBrF_3$	Bromotrifluoromethane
CCl_2F_2	Dichlorodifluoromethane
CCl_2O	Phosgene
CCl_3F	Fluorotrichloromethane
CCl_3NO_2	Chloropicrin
CCl_4	Tetrachloromethane
CF_4	Tetrafluoromethane
$CHBr_3$	Tribromomethane
$CHCl_3$	Trichloromethane
$CHCl_3S$	Trichloromethanethiol
CHF_3	Trifluoromethane
HCN	Hydrogen cyanide
CH_2Br_2	Dibromomethane
CH_2BrCl	Bromochloromethane
CH_2Cl_2	Dichloromethane
CH_2O	Formol
CH_2O_2	Formic acid
CH_3Br	Bromomethane
CH_3Cl	Chloromethane
CH_3Cl_3Si	Methyltrichlorosilane
CH_3I	Iodomethane
CH_3NO	Formamide
CH_3NO_2	Nitromethane
CH_3NO_2	Methyl nitrite
CH_3NO_3	Methyl nitrate
CH_4	Methane
CH_4N_2O	Urea
CH_4N_2S	Thiourea
CH_4O	Methanol
CH_4O_3S	Methanesulphonic acid
CH_4S	Methanethiol
CH_5N	Methylamine
CH_6N_2	Methylhydrazine
Cl_3HSi	Trichlorosilane
Cl_4Si	Tetrachlorosilane
CN_4O_8	Tetranitromethane
C_2	
$C_2Br_2F_4$	1,2-Dibromotetrafluoroethane
C_2Cl_2	Dichloroacetylene
$C_2Cl_2F_4$	1,2-Dichlorotetrafluoroethane
$C_2Cl_2O_2$	Oxalyl chloride

Empirical formula	Name
$C_2Cl_3F_3$	1,1,2-Trichlorotrifluoroethane
C_2Cl_3N	Trichloroacetonitrile
C_2Cl_4	Tetrachloroethylene
$C_2Cl_4F_2$	1,1,1,2-Tetrachlorodifluoroethane
$C_2Cl_4F_2$	1,1,2,2-Tetrachlorodifluoroethane
C_2Cl_4O	Trichloroacetyl chloride
C_2ClF_3	Chlorotrifluoroethylene
C_2ClF_5	Chloropentafluoroethane
C_2HCl	Chloroacetylene
C_2HCl_2N	Dichloroacetonitrile
C_2HCl_3	Trichloroethylene
C_2HCl_3O	Chloral
C_2HCl_3O	Dichloroacetyl acid
$C_2HCl_3O_2$	Trichloroacetic acid
$C_2HF_3O_2$	Trifluoroacetic acid
C_2H_2	Acetylene
$C_2H_2Br_2$	1,2-Dibromoethylene
$C_2H_2Br_4$	1,1,2,2-Tetrabromoethane
$C_2H_2Cl_2$	1,1-Dichloroethylene
$C_2H_2Cl_2$	1,2-Dichloroethylene (E)
$C_2H_2Cl_2$	1,2-Dichloroethylene (Z)
$C_2H_2Cl_2F_2$	1,2-Dichloro-1,1-difluoroethane
$C_2H_2Cl_2O$	Chloroacetyl chloride
$C_2H_2Cl_2O_2$	Dichloroacetic acid
$C_2H_2Cl_4$	1,1,2,2-Tetrachloroethane
$C_2H_2Cl_4$	1,1,1,2-Tetrachloroethane
C_2H_2ClN	Chloroacetonitrile
C_2H_2O	Ketene
$C_2H_2O_2$	Glyoxal
$C_2H_2O_4$	Oxalic acid
C_2H_3Br	Bromoethylene
$C_2H_3Br_3O$	2,2,2-Tribromoethanol
$C_2H_3BrO_2$	Bromoacetic acid
C_2H_3Cl	Chloroethylene
$C_2H_3Cl_2NO_2$	1,1-Dichloro-1-nitroethane
$C_2H_3Cl_3$	1,1,1-Trichloroethane
$C_2H_3Cl_3O$	2,2,2-Trichloroethanol
$C_2H_3Cl_3O_2$	Chloral hydrate
$C_2H_3ClF_2$	1-Chloro-1,1-difluoroethane
C_2H_3ClO	Chloroacetaldehyde
C_2H_3ClO	Acetyl chloride
$C_2H_3ClO_2$	Chloroacetic acid
$C_2H_3ClO_2$	Methyl chloroformate
C_2H_3F	Fluoroethylene
$C_2H_3F_3O$	2,2,2-Trifluoroethanol

9.2 Organic compounds — Empirical formulae/Names

Empirical formula	Name
$C_2H_3FO_2$	Fluoroacetic acid
C_2H_3N	Acetonitrile
C_2H_3NO	Glycolonitrile
C_2H_3NO	Methyl isocyanate
C_2H_3NS	Methyl isothiocyanate
C_2H_4	Ethylene
$C_2H_4Br_2$	1,2-Dibromoethane
C_2H_4BrCl	1-Bromo-2-chloroethane
$C_2H_4Cl_2$	1,1-Dichloroethane
$C_2H_4Cl_2$	1,2-Dichloroethane
$C_2H_4Cl_2O$	Methyl dichloromethyl oxide
C_2H_4ClNO	Chloroacetamide
$C_2H_4N_2O_6$	Ethyleneglycol dinitrate
C_2H_4O	Ethylene oxide
C_2H_4O	Acetaldehyde
$C_2H_4O_2$	Acetic acid
$C_2H_4O_2$	Methyl formate
$C_2H_4O_2S$	Thioglycolic acid
$C_2H_4O_3$	Glycolic acid
C_2H_5Br	Bromoethane
C_2H_5BrO	2-Bromoethanol
C_2H_5Cl	Chloroethane
C_2H_5ClO	2-Chloroethanol
C_2H_5ClO	Chloromethyl ether
C_2H_5F	Fluoroethane
C_2H_5FO	2-Fluoroethanol
C_2H_5I	Iodoethane
C_2H_5N	Aziridine
C_2H_5NO	Acetamide
$C_2H_5NO_2$	Nitroethane
$C_2H_5NO_2$	Ethyl nitrite
$C_2H_5NO_3$	Ethy nitrate
$C_2H_5NO_3$	2-Nitroethanol
C_2H_5NS	Thioacetamide
C_2H_6	Ethane
$C_2H_6Cl_2Si$	Dimethyldichlorosilane
C_2H_6O	Ethanol
C_2H_6O	Dimethyl ether
$C_2H_6O_2$	Ethylene glycol
$C_2H_6O_2S$	Dimethylsulphone
$C_2H_6O_4S$	Dimethyl sulphate
C_2H_6OS	Dimethylsulphoxide
C_2H_6OS	2-Mercaptoethanol
C_2H_6S	Ethanethiol
C_2H_6S	Dimethyl sulphide
C_2H_7N	Ethylamine
C_2H_7N	Dimethylamine
C_2H_7NO	Ethanolamine
$C_2H_8N_2$	Ethylenediamine
$C_2H_8N_2$	1,1-Dimethylhydrazine
$C_2H_8N_2$	1,2-Dimethylhydrazine
C_2N_2	Cyanogen

Empirical formula	Name
C_3	
C_3H_2ClN	2–Chloroacrylonitrile
$C_3H_2N_2$	Malononitrile
C_3H_3Br	1-Bromo-2-propyne
C_3H_3Cl	3-Chloro-1-propyne
$C_3H_3Cl_3O_2$	Methy trichloroacetate
C_3H_3ClO	Acryloyl chloride
$C_3H_3F_3O_2$	Methyl trifluoroacetate
C_3H_3N	Acrylonitrile
$C_3H_3NO_2$	Cyanoacetic acid
C_3H_4	Allene
C_3H_4	Propyne
$C_3H_4Cl_2$	1,3-Dichloropropene
$C_3H_4Cl_2$	2,3-Dichloropropene
$C_3H_4Cl_2O$	1,1-Dichloroacetone
$C_3H_4Cl_2O$	2-Chloropropionyl chloride
$C_3H_4Cl_2O_2$	2,2-Dichloropropionic acid
$C_3H_4Cl_2O_2$	Methyl dicloroacetate
C_3H_4ClN	3–Chloropropionitrile
$C_3H_4N_2$	Pyrazole
C_3H_4O	Propargyl alcohol
C_3H_4O	Acrolein
$C_3H_4O_2$	Acrylic acid
$C_3H_4O_2$	ß-Propiolactone
$C_3H_4O_3$	Pyruvic acid
$C_3H_4O_3$	Ethylene carbonate
$C_3H_4O_4$	Malonic acid
C_3H_5Br	1-Bromo-2-propene
$C_3H_5BrO_2$	2-Bromopropionic acid
$C_3H_5BrO_2$	3-Bromopropionic acid
$C_3H_5BrO_2$	Methyl bromoacetate
C_3H_5Cl	3-Chloro-1-propene
$C_3H_5Cl_3$	1,1,1-Trichloropropane
$C_3H_5Cl_3$	1,2,3-Trichloropropane
C_3H_5ClO	Epiclorhydrin
C_3H_5ClO	Chloroacetone
C_3H_5ClO	Propionyl chloride
$C_3H_5ClO_2$	2-Chloropropionic acid
$C_3H_5ClO_2$	3-Chloropropionic acid
$C_3H_5ClO_2$	Methyl chloroacetate
$C_3H_5ClO_2$	Ethyl chloroformate
C_3H_5I	Allyl iodide
C_3H_5N	Propionitrile
$C_3H_5N_3O_9$	Nitroglycerine
C_3H_5NO	Ethylene cyanohydrin
C_3H_5NO	Lactonitrile
C_3H_5NO	Methoxyacetonitrile
C_3H_5NO	Acrylamide
C_3H_5NS	Ethyl isothiocyanate
C_3H_6	Cyclopropane
C_3H_6	Propylene
$C_3H_6Br_2$	1,2-Dibromopropane

9.2 Organic compounds — Empirical formulae/Names

Empirical formula	Name
$C_3H_6Br_2$	1,3-Dibromopropane
C_3H_6BrCl	1-Bromo-3-chloropropane
$C_3H_6Cl_2$	1,1-Dichloropropane
$C_3H_6Cl_2$	1,2-Dichloropropane
$C_3H_6Cl_2$	1,3-Dichloropropane
$C_3H_6Cl_2$	2,2-Dichloropropane
$C_3H_6ClNO_2$	1-Chloro-1-nitropropane
$C_3H_6N_6O_6$	Hexogene
C_3H_6O	Allyl alcohol
C_3H_6O	Propylene oxide
C_3H_6O	Propionaldehyde
C_3H_6O	Acetone
$C_3H_6O_2$	Glycidol
$C_3H_6O_2$	1,3-Dioxolane
$C_3H_6O_2$	Propionic acid
$C_3H_6O_2$	Ethyl formate
$C_3H_6O_2$	Methyl acetate
$C_3H_6O_3$	Trioxane
$C_3H_6O_3$	Lactic acid
$C_3H_6O_3$	Dimethyl carbonate
C_3H_7Br	1-Bromopropane
C_3H_7Br	2-Bromopropane
C_3H_7BrO	3-Bromo-1-propanol
C_3H_7Cl	1-Chloropropane
C_3H_7Cl	2-Chloropropane
C_3H_7ClO	3-Chloro-1-propanol
C_3H_7ClO	2-Methoxychloroethane
C_3H_7N	Allylamine
C_3H_7NO	Dimethylformamide
C_3H_7NO	Propionamide
$C_3H_7NO_2$	1-Nitropropane
$C_3H_7NO_2$	2-Nitropropane
$C_3H_7NO_3$	Propyl nitrate
$C_3H_7NO_3$	Isopropyl nitrate
$C_3H_7NO_3$	2-Nitro-1-propanol
C_3H_8	Propane
C_3H_8O	1-Propanol
C_3H_8O	2-Propanol
C_3H_8O	Methyl ethyl ether
$C_3H_8O_2$	1,2-diol propane
$C_3H_8O_2$	1,3-diol propane
$C_3H_8O_2$	2-Methoxyethanol
$C_3H_8O_2$	Dimethoxymethane
$C_3H_8O_3$	Glycerol
C_3H_8S	1-Propanethiol
C_3H_8S	2-Propanethiol
$C_3H_9BO_3$	Trimethyl borate
C_3H_9ClSi	Trimethylchlorosilane
C_3H_9N	Propylamine
C_3H_9N	Isopropylamine
C_3H_9N	Trimethylamine
C_3H_9NO	1-Amino-2-propanol

Empirical formula	Name
C_3H_9NO	3-Amino-1-propanol
C_3H_9NO	2-Methoxyethylamine
$C_3H_9O_3P$	Trimethylphosphite
$C_3H_9O_4P$	Trimethylphosphate
C_3H_9P	Trimethylphosphine
$C_3H_{10}N_2$	1,2-Diaminopropane
$C_3H_{10}N_2$	1,3-Diaminopropane
$C_3H_{10}O_3Si$	Trimethoxysilane
C_4	
$C_4H_2Cl_2O_2$	Fumaryl chloride
$C_4H_2N_2$	Fumaronitrile
$C_4H_2O_3$	Maleic anhydride
$C_4H_4Cl_2O_3$	Chloroactic anhydride
$C_4H_4N_2$	Succinonitrile
C_4H_4O	Furan
$C_4H_4O_2$	Dicetene
$C_4H_4O_3$	Succinic anhydride
$C_4H_4O_4$	Fumaric acid
$C_4H_4O_4$	Maleic acid
C_4H_4S	Thiophene
C_4H_5Cl	2-Chloro-1,3-butadiene
C_4H_5ClO	Crotonoyl chloride
C_4H_5ClO	Methacryloyl chloride
$C_4H_5ClO_2$	Allyl chloroformate
C_4H_5N	Pyrrole
C_4H_5N	Cyclopropane carbonitrile
C_4H_5N	Methacrylonitrile
C_4H_5N	Crotononitrile
C_4H_5N	Allyl cyanide
$C_4H_5NO_2$	Methyl cyanoacetate
C_4H_5NS	Allyl isothiocyanate
C_4H_6	1,3-Butadiene
C_4H_6	1-Butyne
C_4H_6	2-Butyne
$C_4H_6Cl_2$	1,4-Dichloro-2-butene
$C_4H_6Cl_2$	1,3-Dichloro-2-butene
$C_4H_6Cl_2O$	4-Chlorobutyrylchloride
$C_4H_6Cl_2O_2$	Ethyl dichloroacetate
C_4H_6ClN	4-Chlorobutyronitrile
C_4H_6O	3-Butyn-2-ol
C_4H_6O	Divinyl ether
C_4H_6NO	Ethoxyacetylene
C_4H_6O	2,5-Dihydrofuran
C_4H_6O	Crotonaldehyde
C_4H_6O	Methacrolein
C_4H_6O	Methylvinylketone
$C_4H_6O_2$	2-Butyne-1,4-diol
$C_4H_6O_2$	Butane-2,3-dione
$C_4H_6O_2$	Crotonic (E) acid
$C_4H_6O_2$	Methacrylic acid
$C_4H_6O_2$	Vinyl acetate
$C_4H_6O_2$	Methyl acrylate

9.2 Organic compounds — Empirical formulae/Names

Empirical formula	Name
$C_4H_6O_2$	γ-Butyrolactone
$C_4H_6O_3$	Propylene carbonate
$C_4H_6O_3$	Acetic anhydride
$C_4H_6O_4$	Succinic acid
$C_4H_6O_4$	Dimethyl oxalate
$C_4H_7BrO_2$	Ethyl bromoacetate
C_4H_7Cl	1-Chloro-2-butene
C_4H_7Cl	1-Chloro-3-butene
C_4H_7Cl	3-Chloro-2-methyl-1-propene
C_4H_7Cl	1-Chloro-2-methyl-2-propene
C_4H_7ClO	3-Chloro-2-butanone
C_4H_7ClO	Butyl chloride
C_4H_7ClO	Isobutyl chloride
$C_4H_7ClO_2$	Ethyl chloroacetate
$C_4H_7ClO_2$	Methyl 2-chloropropionate
$C_4H_7FO_2$	Ethyl fluoroacetate
C_4H_7N	Butyronitrile
C_4H_7N	Isobutyronitrile
C_4H_7NO	Acetonecyanohydrin
C_4H_7NO	Pyrrolidone
C_4H_7NO	Methacrylamide
C_4H_8	1-Butene
C_4H_8	2-Butene (E)
C_4H_8	2-Butene (Z)
C_4H_8	2-Butene (Z+E)
C_4H_8	Isobutene
$C_4H_8Br_2$	1,4-Dibromobutane
C_4H_8BrCl	1-Bromo-3-chloro-2-methylpropane
$C_4H_8Cl_2$	1,4-Dichlorobutane
$C_4H_8Cl_2O$	Bis(2-chloroethyl)ether
C_4H_8O	2-Buten-1-ol (Z+E)
C_4H_8O	3-Buten-2-ol
C_4H_8O	2-Methyl-2-prope n-1-ol
C_4H_8O	Ethyl vinyl ether
C_4H_8O	1,2-Epoxybutane
C_4H_8O	Tetrahydrofuran
C_4H_8O	Butyraldehyde
C_4H_8O	Isobutyraldehyde
C_4H_8O	Butanone
$C_4H_8O_2$	2-Butene-1,4-diol (Z)
$C_4H_8O_2$	1,4-dioxan
$C_4H_8O_2$	3-Hydroxybutyraldehyde
$C_4H_8O_2$	1-Hydroxy-2-butanone
$C_4H_8O_2$	3-Hydroxy-2-butanone
$C_4H_8O_2$	Butyric acid
$C_4H_8O_2$	Isobutyric acid
$C_4H_8O_2$	Propyl formate
$C_4H_8O_2$	Isopropyl formate
$C_4H_8O_2$	Ethyl acetate
$C_4H_8O_2$	Methyl propionate
$C_4H_8O_2S$	Sulpholane
C_4H_9Br	1-Bromobutane
C_4H_9Br	2-Bromobutane
C_4H_9Br	1-Bromo-2-methylpropane
C_4H_9Br	2-Bromo-2-methylpropane
C_4H_9C	2-Chlorobutane
C_4H_9Cl	1-Chlorobutane
C_4H_9Cl	1-Chloro-2-methylpropane
C_4H_9Cl	2-Chloro-2-methylpropane
C_4H_9I	1-Iodobutane
C_4H_9I	2-Iodobutane
C_4H_9N	Pyrrolidine
C_4H_9NO	Morpholine
C_4H_9NO	Dimethylacetamide
C_4H_9NO	Butyramide
C_4H_9NO	Isobutyramide
$C_4H_9NO_2$	1-Nitrobutane
$C_4H_9NO_2$	2-Nitrobutane
$C_4H_9NO_2$	Butyl nitrite
$C_4H_9NO_2$	s-Butyl nitrite
$C_4H_9NO_2$	Isobutyl nitrite
$C_4H_9NO_2$	t-Butyl nitrite
C_4H_9NS	Thiomorpholine
C_4H_{10}	Butane
C_4H_{10}	2-Methylpropane
$C_4H_{10}N_2$	Piperazine
$C_4H_{10}N_2O$	Nitrosodiethylamine
$C_4H_{10}O$	1-Butanol
$C_4H_{10}O$	2-Butanol
$C_4H_{10}O$	Isobutanol
$C_4H_{10}O$	t-Butanol
$C_4H_{10}O$	Diethyl ether
$C_4H_{10}O_2$	Butane-1,2-diol
$C_4H_{10}O_2$	Butane-1,3-diol
$C_4H_{10}O_2$	Butane-1,4-diol
$C_4H_{10}O_2$	Butane-2,3-diol
$C_4H_{10}O_2$	1,2-Dimethoxyethane
$C_4H_{10}O_2$	2-Ethoxyethanol
$C_4H_{10}O_2$	1,1-Dimethoxyethane
$C_4H_{10}O_2S$	Thiodiethanol
$C_4H_{10}O_3$	Diethylene glycol
$C_4H_{10}O_4S$	Ethyl sulphate
$C_4H_{10}S$	1-Butanethiol
$C_4H_{10}S$	Diethyl sulphide
$C_4H_{11}N$	Butylamine
$C_4H_{11}N$	s-Butylamine
$C_4H_{11}N$	t-Butylamine
$C_4H_{11}N$	Isobutylamine
$C_4H_{11}N$	Diethylamine
$C_4H_{11}N$	N,N-Dimethylethylamine
$C_4H_{11}NO$	2-Dimethylaminoethanol
$C_4H_{11}NO$	2-Ethylaminoethanol
$C_4H_{11}NO$	2-Amino-1-butanol
$C_4H_{11}NO_2$	Diethanolamine

9.2 Organic compounds Empirical formulae/Names

Empirical formula	Name
$C_4H_{12}N_2$	1,4-Diaminobutane
$C_4H_{12}Si$	Tetramethylsilane
$C_4H_{13}N_3$	Diethylenetriamine
C_5	
$C_5H_4O_2$	Furfural
$C_5H_4O_3$	2-Furoic acid
C_5H_5N	N-Methylpyrrole
C_5H_5N	Pyridine
$C_5H_5NO_2$	Methyl 2-cyanoacrylate
C_5H_6	Cyclopentadiene
$C_5H_6N_2$	2-Aminopyridine
$C_5H_6N_2$	3-Aminopyridine
$C_5H_6N_2$	4-Aminopyridine
$C_5H_6N_2$	Glutaronitrile
C_5H_6O	2-Methylfuran
C_5H_6O	2-Cyclopenten-1-one
$C_5H_6O_2$	Furfurylic alcohol
$C_5H_6O_3$	Glutaric anhydride
C_5H_7ClO	2-Chlorocyclopentanone
$C_5H_7ClO_3$	Methyl 2-chloroacetoacetate
$C_5H_7NO_2$	Ethyl cyanoacetate
C_5H_8	Isoprene
C_5H_8	1,3-Pentadiene (E)
C_5H_8	Cyclopentene
C_5H_8	1-Pentyne
C_5H_8	2-Pentyne
$C_5H_8N_4O_{12}$	Pentaerythritol tetranitrate
C_5H_8O	2-Methyl-3-butyn-2-ol
C_5H_8O	Dihydropyran
C_5H_8O	3-Methylcrotonaldehyde
C_5H_8O	Cyclopentanone
$C_5H_8O_2$	Glutaraldehyde
$C_5H_8O_2$	Acetylacetone
$C_5H_8O_2$	Pentane-2,3-dione
$C_5H_8O_2$	2-Methylcrotonic(E) acid
$C_5H_8O_2$	3-Methylcrotonic acid
$C_5H_8O_2$	Allyl acetate
$C_5H_8O_2$	Isopropenyl acetate
$C_5H_8O_2$	Ethyl acrylate
$C_5H_8O_2$	Methyl methacrylate
$C_5H_8O_2$	Methyl crotonate
$C_5H_8O_2$	δ-Valerolactone
$C_5H_8O_3$	Methyl acetoacetate
$C_5H_8O_4$	Glutaric acid
$C_5H_8O_4$	Dimethyl malonate
$C_5H_9BrO_2$	Ethyl 2-bromopropionate
$C_5H_9BrO_2$	Ethyl 3-bromopropionate
C_5H_9ClO	1-Chloro-3-pentanone
C_5H_9ClO	5-Chloro-2-pentanone
C_5H_9ClO	Pentanoyl chloride
C_5H_9ClO	Isopentanoyl chloride
C_5H_9ClO	Pivaloyl chloride
$C_5H_9ClO_2$	Ethyl 2-Chloropropionate
$C_5H_9ClO_2$	Ethyl 3-Chloropropionate
C_5H_9N	Valeronitrile
C_5H_9N	Isovaleronitrile
C_5H_9N	Pivalonitrile
C_5H_9NO	3-Ethoxypropionitrile
C_5H_9NO	N-Methylpyrrolidone
C_5H_{10}	Cyclopentane
C_5H_{10}	1-Pentene
C_5H_{10}	2-Pentene (E)
C_5H_{10}	2-Pentene (Z)
C_5H_{10}	2-Pentene (E+Z)
C_5H_{10}	2-Methyl-1-butene
C_5H_{10}	3-Methyl-1-butene
C_5H_{10}	2-Methyl-2-butene
$C_5H_{10}O$	Cyclopentanol
$C_5H_{10}O$	Tetrahydropyran
$C_5H_{10}O$	Pentanal
$C_5H_{10}O$	2-Methylbutyraldehyde
$C_5H_{10}O$	3-Methylbutyraldehyde
$C_5H_{10}O$	2-Pentanone
$C_5H_{10}O$	3-Pentanone
$C_5H_{10}O$	Methyl isopropylketone
$C_5H_{10}O_2$	Tetrahydrofurfurylic alcohol
$C_5H_{10}O_2$	Pentanoic acid
$C_5H_{10}O_2$	2-Methylbutyric acid
$C_5H_{10}O_2$	3-Methylbutyric acid
$C_5H_{10}O_2$	2,2-Dimethylpropionic acid
$C_5H_{10}O_2$	Butyl formate
$C_5H_{10}O_2$	Isobutyl formate
$C_5H_{10}O_2$	Propyl acetate
$C_5H_{10}O_2$	Isopropyl acetate
$C_5H_{10}O_2$	Ethyl propionate
$C_5H_{10}O_2$	Methyl butyrate
$C_5H_{10}O_2$	Methyl isobutyrate
$C_5H_{10}O_3$	Diethyl carbonate
$C_5H_{10}O_3$	Ethyl lactate
$C_5H_{10}O_3$	2-Methoxyethyl acetate
$C_5H_{11}Br$	1-Bromopentane
$C_5H_{11}Br$	2-Bromopentane
$C_5H_{11}Br$	1-Bromo-3-methylbutane
$C_5H_{11}Cl$	1-Chloropentane
$C_5H_{11}Cl$	1-Chloro-3-methylbutane
$C_5H_{11}Cl$	2-Chloro-2-methylbutane
$C_5H_{11}N$	Cyclopentylamine
$C_5H_{11}N$	N,N-Dimethylallylamine
$C_5H_{11}N$	Piperidine
$C_5H_{11}NO$	N-Methylmorpholine
$C_5H_{11}NO_2$	1-Nitropentane
$C_5H_{11}NO_2$	Pentyl nitrite
$C_5H_{11}NO_2$	Isopentyl nitrite
$C_5H_{11}NO_3$	Pentyl nitrate

9.2 Organic compounds — Empirical formulae/Names

Empirical formula	Name	Empirical formula	Name
$C_5H_{11}NO_3$	Isopentyl nitrate	C_6H_5Br	Bromobenzene
C_5H_{12}	Pentane	C_6H_5Cl	Chlorobenzene
C_5H_{12}	2-Methylbutane	$C_6H_5Cl_2N$	2,3-Dichloroaniline
C_5H_{12}	2,2-Dimethylpropane	$C_6H_5Cl_2N$	2,4-Dichloroaniline
$C_5H_{12}O$	1-Pentanol	$C_6H_5Cl_2N$	2,5-Dichloroaniline
$C_5H_{12}O$	2-Pentanol	$C_6H_5Cl_2N$	2,6-Dichloroaniline
$C_5H_{12}O$	3-Pentanol	$C_6H_5Cl_2N$	3,4-Dichloroaniline
$C_5H_{12}O$	Isoamylic alcohol	$C_6H_5Cl_2N$	3,5-Dichloroaniline
$C_5H_{12}O$	t-Amylic alcohol	C_6H_5ClO	2-Chlorophenol
$C_5H_{12}O$	2-Methyl-1-butanol	C_6H_5ClO	3-Chlorophenol
$C_5H_{12}O$	3-Methyl-2-butanol	C_6H_5ClO	4-Chlorophenol
$C_5H_{12}O$	2,2-Dimethyl-1-propanol	C_6H_5F	Fluorobenzene
$C_5H_{12}O$	Butyl methyl ether	C_6H_5I	Iodobenzene
$C_5H_{12}O$	Methyl tert-butylether (MTBE)	$C_6H_5N_3O_4$	2,4-Dinitroaniline
$C_5H_{12}O_2$	Pentane-1,5-diol	$C_6H_5NO_2$	Nitrobenzene
$C_5H_{12}O_2$	2,2-Dimethylpropane-1,3-diol	$C_6H_5NO_2$	Nicotinic acid
$C_5H_{12}O_2$	2-Isopropoxyethanol	$C_6H_5NO_3$	2-Nitrophenol
$C_5H_{12}O_2$	Diethoxymethane	$C_6H_5NO_3$	3-Nitrophenol
$C_5H_{12}O_2$	1,1-Dimethoxypropane	$C_6H_5NO_3$	4-Nitrophenol
$C_5H_{12}O_2$	2,2-Dimethoxypropane	C_6H_6	Benzene
$C_5H_{12}O_3$	Diethylene glycol monomethyl ether	C_6H_6ClN	2-Chloroaniline
$C_5H_{12}O_4$	Pentaerythritol	C_6H_6ClN	3-Chloroaniline
$C_5H_{13}N$	n-Pentylamine	C_6H_6ClN	4-Chloroaniline
$C_5H_{13}N$	Isopentylamine	$C_6H_6N_2O_2$	2-Nitroaniline
$C_5H_{13}N$	Neopentylamine	$C_6H_6N_2O_2$	3-Nitroaniline
$C_5H_{13}N$	N-Methylbutylamine	$C_6H_6N_2O_2$	4-Nitroaniline
$C_5H_{13}NO$	1-Dimethylamino-2-propanol	C_6H_6O	Phenol
$C_5H_{13}NO$	3-Dimethylamino-1-propanol	$C_6H_6O_2$	Pyrocatechol
C_6		$C_6H_6O_2$	Resorcinol
$C_6Cl_4O_2$	Tetrachlorobenzoquinone	$C_6H_6O_2$	Hydroquinone
C_6Cl_6	Hexachlorobenzene	$C_6H_6O_3$	Pyrogallol
C_6HCl_5O	Pentachlorophenol	C_6H_6S	Thiophenol
$C_6H_2Cl_4$	1,2,4,5-Tetrachlorobenzene	C_6H_7N	Aniline
$C_6H_3Cl_3$	1,2,3-Trichlorobenzene	C_6H_7N	2-Methylpyridine
$C_6H_3Cl_3$	1,2,4-Trichlorobenzene	C_6H_7N	3-Methylpyridine
$C_6H_3Cl_3$	1,3,5-Trichlorobenzene	C_6H_7N	4-Methylpyridine
$C_6H_3Cl_3O$	2,4,5-Trichlorophenol	C_6H_7NO	2-Aminophenol
$C_6H_3ClN_2O_4$	1-Chloro-2,4-dinitrobenzene	C_6H_7NO	3-Aminophenol
$C_6H_3N_3O_6$	1,3,5-Trinitrobenzene	C_6H_7NO	4-Aminophenol
$C_6H_3N_3O_7$	2,4,6-Trinitrophenol	$C_6H_8N_2$	p-Phenylenediamine
$C_6H_4Cl_2$	1,2-Dichlorobenzene	$C_6H_8N_2$	Phenylhydrazine
$C_6H_4Cl_2$	1,3-Dichlorobenzene	$C_6H_8N_2$	Adiponitrile
$C_6H_4Cl_2$	1,4-Dichlorobenzene	C_6H_8O	2-Cyclohexen-1-one
$C_6H_4Cl_2O$	2,4-Dichlorophenol	$C_6H_8O_4$	Dimethyl fumarate
$C_6H_4ClNO_2$	1-Chloro-2-nitrobenzene	$C_6H_8O_4$	Dimethyl maleate
$C_6H_4ClNO_2$	1-Chloro-3-nitrobenzene	C_6H_9ClO	2-Chlorocyclohexanone
$C_6H_4ClNO_2$	1-Chloro-4-nitrobenzene	$C_6H_9ClO_3$	2-Ethyl chloroacetoacetate
$C_6H_4N_2O_4$	1,2-Dinitrobenzene	$C_6H_9ClO_3$	4-Ethyl chloroacetoacetate
$C_6H_4N_2O_4$	1,3-Dinitrobenzene	C_6H_{10}	Cyclohexene
$C_6H_4N_2O_4$	1,4-Dinitrobenzene	$C_6H_{10}O$	Mesityl oxide
$C_6H_4N_2O_5$	2,4-Dinitrophenol	$C_6H_{10}O$	Cyclohexanone
$C_6H_4O_2$	Benzoquinone	$C_6H_{10}O_2$	Hexane-2,3-dione

9.2 Organic compounds — Empirical formulae/Names

Empirical formula	Name
$C_6H_{10}O_2$	Hexane-2,5-dione
$C_6H_{10}O_2$	Hexane-3,4-dione
$C_6H_{10}O_2$	Ethyl methacrylate
$C_6H_{10}O_2$	Ethyl crotonate
$C_6H_{10}O_2$	ε-Caprolactone
$C_6H_{10}O_3$	Diglycidyl ether
$C_6H_{10}O_3$	2-Methoxyethyl acrylate
$C_6H_{10}O_3$	Ethyl acetoacetate
$C_6H_{10}O_3$	Propionic anhydride
$C_6H_{10}O_4$	Adipic acid
$C_6H_{10}O_4$	Ethylene glycol diacetate
$C_6H_{10}O_4$	Diethyl oxalate
$C_6H_{10}O_4$	Dimethyl succinate
$C_6H_{11}BrO_2$	`2-Bromohexanoic acid
$C_6H_{11}ClO$	Hexanoyl chloride
$C_6H_{11}N$	Diallylamine
$C_6H_{11}N$	Capronitrile
$C_6H_{11}N$	4-Methylvaleronitrile
$C_6H_{11}NO$	ε-Caprolactam
C_6H_{12}	Cyclohexane
C_6H_{12}	Methylcyclopentane
C_6H_{12}	1-Hexene
C_6H_{12}	2-Hexene
C_6H_{12}	2,3-Dimethyl-1-butene
C_6H_{12}	3,3-Dimethyl-1-butene
C_6H_{12}	2,3-Dimethyl-2-butene
$C_6H_{12}N_4$	Hexamethylenetetramine
$C_6H_{12}O$	Cyclohexanol
$C_6H_{12}O$	Butyl vinyl ether
$C_6H_{12}O$	Isobutyl vinyl ether
$C_6H_{12}O$	2,5-Dimethyltetrahydrofuran
$C_6H_{12}O$	Hexanal
$C_6H_{12}O$	2-Methylpentanal
$C_6H_{12}O$	2-Ethylbutyraldehyde
$C_6H_{12}O$	2-Hexanone
$C_6H_{12}O$	3-Hexanone
$C_6H_{12}O$	Methylisobutylketone
$C_6H_{12}O$	Ethylisopropylketone
$C_6H_{12}O$	3,3-Dimethyl-2-butanone
$C_6H_{12}O_2$	Propylglycidyl ether
$C_6H_{12}O_2$	Isopropylglycidyl ether
$C_6H_{12}O_2$	Diaketone achohol
$C_6H_{12}O_2$	Hexanoic acid
$C_6H_{12}O_2$	2-Methylpentanoic acid
$C_6H_{12}O_2$	3-Methylpentanoic acid
$C_6H_{12}O_2$	4-Methylpentanoic acid
$C_6H_{12}O_2$	2,2-Dimethylbutyric acid
$C_6H_{12}O_2$	3,3-Dimethylbutyric acid
$C_6H_{12}O_2$	2-Ethylbutyric acid
$C_6H_{12}O_2$	Amyl formate
$C_6H_{12}O_2$	Isoamyl formate
$C_6H_{12}O_2$	Butyl acetate
$C_6H_{12}O_2$	s-Butyl acetate
$C_6H_{12}O_2$	Isobutyl acetate
$C_6H_{12}O_2$	t-Butyl acetate
$C_6H_{12}O_2$	Propyl propionate
$C_6H_{12}O_2$	Ethyl butyrate
$C_6H_{12}O_2$	Ethyl isobutyrate
$C_6H_{12}O_2$	Methyl pentanoate
$C_6H_{12}O_2$	Methyl pivalate
$C_6H_{12}O_3$	Paraldehyde
$C_6H_{12}O_3$	2-Ethoxyethyl acetate
$C_6H_{12}O_3$	2-Methoxypropyl acetate
$C_6H_{13}Br$	1-Bromohexane
$C_6H_{13}Br$	2-Bromohexane
$C_6H_{13}N$	Cyclohexylamine
$C_6H_{13}N$	N-Methylpiperidine
$C_6H_{13}N$	2-Methylpiperidine
$C_6H_{13}N$	3-Methylpiperidine
$C_6H_{13}N$	4-Ethylpiperidine
$C_6H_{13}NO$	N-Ethylmorpholine
$C_6H_{13}NO_2$	1-Nitrohexane
C_6H_{14}	Hexane
C_6H_{14}	2-Methylpentane
C_6H_{14}	3-Methylpentane
C_6H_{14}	2,2-Dimethylbutane
$C_6H_{14}O$	1-Hexanol
$C_6H_{14}O$	2-Hexanol
$C_6H_{14}O$	3-Hexanol
$C_6H_{14}O$	4-Methyl-2-pentanol
$C_6H_{14}O$	3,3-Dimethyl-2-butanol
$C_6H_{14}O$	Dipropyl ether
$C_6H_{14}O$	Diisopropyl ether
$C_6H_{14}O_2$	Hexane-1,6-diol
$C_6H_{14}O_2$	2,3-Dimethylbutane-2,3-diol
$C_6H_{14}O_2$	1,2-Diethoxyethane
$C_6H_{14}O_2$	2-Butoxyethanol
$C_6H_{14}O_2$	1,1-Diethoxyethane
$C_6H_{14}O_3$	Diethylene glycol monoethyl ether
$C_6H_{14}O_4$	Triethylene glycol
$C_6H_{15}BO_3$	Triethyl borate
$C_6H_{15}N$	Hexylamine
$C_6H_{15}N$	Dipropylamine
$C_6H_{15}N$	Diisopropylamine
$C_6H_{15}N$	Triethylamine
$C_6H_{15}NO$	2-Diethylaminoethanol
$C_6H_{15}NO_3$	Triethanolamine
$C_6H_{15}O_3P$	Triethylphosphite
$C_6H_{15}O_4P$	Triethylphosphate
$C_6H_{16}N_2$	1,6-Diaminohexane
$C_6H_{16}O_3Si$	Triethoxysilane
$C_6H_{16}Si$	Triethylsilane
$C_6H_{18}N_4$	Triethylenetetramine
$C_6H_{18}OSi$	Hexamethyldisiloxane
$C_6H_{19}N$	Hexamethyldisilazane

9.2 Organic compounds — Empirical formulae/Names

Empirical formula	Name	Empirical formula	Name
C_7		C_7H_9N	Benzylamine
$C_7H_4Cl_2O$	o-Chlorobenzoyl chloride	C_7H_9N	o-Toluidine
C_7H_4ClN	2-Chlorobenzonitrile	C_7H_9N	m-Toluidine
$C_7H_5Cl_3$	Trichloromethylbenzene	C_7H_9N	p-Toluidine
C_7H_5ClO	2-Chlorobenzaldehyde	C_7H_9N	N-Methylaniline
C_7H_5ClO	3-Chlorobenzaldehyde	C_7H_9N	2-Ethylpyridine
C_7H_5ClO	4-Chlorobenzaldehyde	C_7H_9N	3-Ethylpyridine
C_7H_5ClO	Benzoyl chloride	C_7H_9N	4-Ethylpyridine
$C_7H_5F_3$	Trifluoromethylbenzene	C_7H_9NO	o-Anisidine
C_7H_5N	Benzonitrile	C_7H_9NO	m-Anisidine
$C_7H_5N_3O_6$	2,4,6-Trinitrotoluene	C_7H_9NO	p-Anisidine
$C_7H_6Cl_2$	2,4-Dichlorotoluene	$C_7H_{10}O_2$	Allyl methacrylate
$C_7H_6Cl_2$	2,6-Dichlorotoluene	$C_7H_{10}O_3$	Allyl acetoacetate
$C_7H_6Cl_2$	Benzylidene chloride	$C_7H_{11}ClO$	Cyclohexanecarbonyl chloride
$C_7H_6N_2O_4$	2,4-Dinitrotoluene	$C_7H_{11}N$	Cyclohexanecarbonitrile
$C_7H_6N_2O_4$	2,6-Dinitrotoluene	C_7H_{12}	4-Methylcyclohexene
$C_7H_6N_2O_5$	4,6-Dinitro-o-cresol	C_7H_{12}	Cycloheptene
C_7H_6O	Benzaldehyde	$C_7H_{12}O$	Cyclohexanecarboxaldehyde
$C_7H_6O_2$	Salicylaldehyde	$C_7H_{12}O$	Cycloheptanone
$C_7H_6O_2$	Benzoic acid	$C_7H_{12}O$	2-Methylcyclohexanone
$C_7H_6O_3$	Salicylic acid	$C_7H_{12}O$	3-Methylcyclohexanone
C_7H_7Br	2-Bromotoluene	$C_7H_{12}O$	4-Methylcyclohexanone
C_7H_7Br	3-Bromotoluene	$C_7H_{12}O_2$	Butyl acrylate
C_7H_7Br	4-Bromotoluene	$C_7H_{12}O_4$	Pimelic acid
C_7H_7Br	Benzyl bromide	$C_7H_{12}O_4$	Diethyl malonate
C_7H_7Cl	2-Chlorotoluene	$C_7H_{12}O_4$	Dimethyl glutarate
C_7H_7Cl	3-Chlorotoluene	$C_7H_{13}N$	Hexanecarbonitrile
C_7H_7Cl	4-Chlorotoluene	C_7H_{14}	Cycloheptane
C_7H_7Cl	Benzyl chloride	C_7H_{14}	Methylcyclohexane
C_7H_7ClO	p-Chloro-m-cresol	C_7H_{14}	1-Heptene
C_7H_7ClO	4-Chloroanisole	C_7H_{14}	2-Heptene (E)
C_7H_7F	2-Fluorotoluene	C_7H_{14}	3-Heptene (E)
C_7H_7F	3-Fluorotoluene	$C_7H_{14}O$	2-Methylcyclohexanol
C_7H_7F	4-Fluorotoluene	$C_7H_{14}O$	3-Methylcyclohexanol
C_7H_7N	2-Vinylpyridine	$C_7H_{14}O$	4-Methylcyclohexanol
C_7H_7N	4-Vinylpyridine	$C_7H_{14}O$	Cycloheptanol
C_7H_7NO	Benzamide	$C_7H_{14}O$	Heptanal
$C_7H_7NO_2$	2-Nitrotoluene	$C_7H_{14}O$	2-Heptanone
$C_7H_7NO_2$	3-Nitrotoluene	$C_7H_{14}O$	3-Heptanone
$C_7H_7NO_2$	4-Nitrotoluene	$C_7H_{14}O$	4-Heptanone
$C_7H_7NO_3$	2-Nitroanisole	$C_7H_{14}O$	Diisopropylketone
$C_7H_7NO_3$	3-Nitroanisole	$C_7H_{14}O$	5-Methyl-2-hexanone
$C_7H_7NO_3$	4-Nitroanisole	$C_7H_{14}O_2$	Butyl glycidyl ether
C_7H_8	Toluene	$C_7H_{14}O_2$	Heptanoic acid
C_7H_8O	Benzyl alcohol	$C_7H_{14}O_2$	Amyl acetate
C_7H_8O	o-Cresol	$C_7H_{14}O_2$	s-amyl acetate
C_7H_8O	m-Cresol	$C_7H_{14}O_2$	Isoamyl acetate
C_7H_8O	p-Cresol	$C_7H_{14}O_2$	Butyl propionate
C_7H_8O	Anisole	$C_7H_{14}O_2$	Isobutyl propionate
$C_7H_8O_2$	Gaiacol	$C_7H_{14}O_2$	Isopropyl butyrate
$C_7H_8O_2$	4-Methoxyphenol	$C_7H_{14}O_2$	Ethyl pentanoate
C_7H_8S	p-Thiocresol	$C_7H_{14}O_2$	Ethyl isopentanoate

9.2 Organic compounds — Empirical formulae/Names

Empirical formula	Name
$C_7H_{14}O_2$	Ethyl pivalate
$C_7H_{14}O_2$	Methyl hexanoate
$C_7H_{14}O_3$	3-Methoxybutyl acetate
$C_7H_{15}N$	N-Methylcyclohexylamine
$C_7H_{15}N$	N-Ethylpiperidine
$C_7H_{15}N$	2-Ethylpiperidine
C_7H_{16}	Heptane
$C_7H_{16}O$	1-Heptanol
$C_7H_{16}O$	2-Heptanol
$C_7H_{16}O$	3-Heptanol
$C_7H_{16}O$	4-Heptanol
$C_7H_{16}O$	2,4-Dimethyl-3-pentanol
$C_7H_{16}O_2$	1,1-Diethoxypropane
$C_7H_{16}O_2$	2,2-Diethoxypropane
$C_7H_{17}N$	Heptylamine
C_8	
$C_8H_4N_2$	o-Phthalonitrile
$C_8H_4N_2$	m-Phthalonitrile
$C_8H_4N_2$	p-Phthalonitrile
$C_8H_4O_3$	Phthalic anhydride
C_8H_6	Phenylacetylene
$C_8H_6Cl_2O_3$	2,4-Dichlorophenoxyacetic acid
C_8H_6O	Benzofuran
$C_8H_6O_4$	o-Phthalic acid
$C_8H_6O_4$	Terephthalic acid
C_8H_7Cl	4-Chlorostyrene
C_8H_7ClO	2-Chloroacetophenone
C_8H_7ClO	3-Chloroacetophenone
C_8H_7ClO	4-Chloroacetophenone
C_8H_7ClO	α-Chloroacetophenone
$C_8H_7ClO_2$	Benzyl chloroformate
C_8H_7N	Benzyl cyanide
C_8H_7N	o-Tolunitrile
C_8H_7N	p-Tolunitrile
C_8H_8	Styrene
C_8H_8O	2-Methylbenzaldehyde
C_8H_8O	3-Methybenzaldehyde
C_8H_8O	4-Methybenzaldehyde
C_8H_8O	Acetophenone
$C_8H_8O_2$	2-Anisaldehyde
$C_8H_8O_2$	3-Anisaldehyde
$C_8H_8O_2$	4-Anisaldehyde
$C_8H_8O_2$	Phenyl acetate
$C_8H_8O_2$	Methyl benzoate
$C_8H_8O_3$	Methyl salicylate
C_8H_9NO	Acetanilide
$C_8H_{12}N_2$	Tetramethylsuccinonitrile
C_8H_{10}	Xylenes (o-, m-, p-)
C_8H_{10}	o-Xylene
C_8H_{10}	m-Xylene
C_8H_{10}	p-Xylene
C_8H_{10}	Ethyl benzene

Empirical formula	Name
$C_8H_{10}O$	1-Phenylethanol
$C_8H_{10}O$	2-Phenylethanol
$C_8H_{10}O$	2-Ethylphenol
$C_8H_{10}O$	3-Ethylphenol
$C_8H_{10}O$	4-Ethylphenol
$C_8H_{10}O$	2,3-Xylenol
$C_8H_{10}O$	2,4-Xylenol
$C_8H_{10}O$	2,5-Xylenol
$C_8H_{10}O$	2,6-Xylenol
$C_8H_{10}O$	3,4-Xylenol
$C_8H_{10}O$	2-Methylanisole
$C_8H_{10}O$	3-Methylanisole
$C_8H_{10}O$	4-Methylanisole
$C_8H_{10}O_2$	1,2-Dimethoxybenzene
$C_8H_{10}O_2$	1,3-Dimethoxybenzene
$C_8H_{10}O_2$	1,4-Dimethoxybenzene
$C_8H_{10}O_3$	Crotonic anhydride
$C_8H_{10}O_3$	Metacrylic anhydride
$C_8H_{11}N$	2-Ethylaniline
$C_8H_{11}N$	4-Ethylaniline
$C_8H_{11}N$	N-Ethylaniline
$C_8H_{11}N$	N,N-Dimethylaniline
$C_8H_{11}NO$	2-Anilinoethanol
C_8H_{12}	4-Vinylcyclohexene
$C_8H_{12}N_2$	Tetramethylsuccinonitrile
$C_8H_{12}N_2O_2$	Hexamethylenediisocyanate
$C_8H_{12}O_4$	Diethyl fumarate
$C_8H_{12}O_4$	Diethyl maleate
$C_8H_{14}O$	2-Ethyl-2-hexene-1-al
$C_8H_{14}O$	Methylcyclohexylketone
$C_8H_{14}O_2$	Cyclohexyl acetate
$C_8H_{14}O_2$	Butyl methacrylate
$C_8H_{14}O_2$	Isobutyl methacrylate
$C_8H_{14}O_3$	Butyl acetoacetate
$C_8H_{14}O_3$	t-Butyl acetoacetate
$C_8H_{14}O_3$	Butyric anhydride
$C_8H_{14}O_3$	Isobutyric anhydride
$C_8H_{14}O_4$	Diethyl succinate
$C_8H_{14}O_4$	Dimethyl adipate
C_8H_{16}	Cyclo-octane
C_8H_{16}	Ethylcyclohexane
C_8H_{16}	1-Octene
C_8H_{16}	2-Octene (E)
C_8H_{16}	3-Octene (E)
C_8H_{16}	4-Octene (E)
C_8H_{16}	2,4,4-Trimethyl-1-pentene
C_8H_{16}	2,4,4-Trimethyl-2-pentene
$C_8H_{16}O$	Octanal
$C_8H_{16}O$	5-Methyl-3-heptanone
$C_8H_{16}O_2$	Octanoic acid
$C_8H_{16}O_2$	2-Ethylhexanoic acid
$C_8H_{16}O_2$	Hexyl acetate

9.2 Organic compounds — Empirical formulae/Names

Empirical formula	Name
$C_8H_{16}O_2$	s-Hexyl acetate
$C_8H_{16}O_2$	Butyl butyrate
$C_8H_{16}O_2$	Ethyl hexanoate
$C_8H_{16}O_2$	Methyl heptanoate
$C_8H_{16}O_3$	2-Butoxyethyl acetate
C_8H_{18}	Octane
C_8H_{18}	2,2,4-Trimethylpentane
$C_8H_{18}O$	1-Octanol
$C_8H_{18}O$	2-Octanol
$C_8H_{18}O$	3-Octanol
$C_8H_{18}O$	2-Ethyl-1-hexanol
$C_8H_{18}O$	Dibutyl ether
$C_8H_{18}O$	Diisobutyl ether
$C_8H_{19}N$	Dibutylamine
$C_8H_{19}N$	Di-s-butylamine
$C_8H_{19}N$	N-Ethyl diisopropylamine
$C_8H_{19}O_3P$	Dibutyl hydrogenphosphite
$C_8H_{20}O_4Si$	Tetraethoxysilane

C_9

Empirical formula	Name
$C_9H_6N_2O_2$	Toluene-2,4-diisocyanate
$C_9H_6N_2O_2$	Toluene-2,6-diisocyanate
C_9H_7N	Quinoline
C_9H_7N	Isoquinoline
C_9H_7NO	8-Quinolinol
C_9H_8	Indene
C_9H_8O	1-Phenyl-2-propyn-1-ol
C_9H_8O	Cinnamaldehyde
$C_9H_8O_2$	Cinnamic acid(E)
C_9H_{10}	Allylbenzene
C_9H_{10}	α-Methylstyrene
C_9H_{10}	2-Methylstyrene
C_9H_{10}	3-Methylstyrene
C_9H_{10}	4-Methylstyrene
$C_9H_{10}O$	3-Phenyl-2-propen-1-ol
$C_9H_{10}O$	2-Allylphenol
$C_9H_{10}O$	2,4-Dimethylbenzaldehyde
$C_9H_{10}O_2$	Phenyl glycidylether
$C_9H_{10}O_2$	Benzyl acetate
$C_9H_{10}O_2$	Ethyl benzoate
C_9H_{12}	Propyl benzene
C_9H_{12}	Cumene
C_9H_{12}	2-Ethyltoluene
C_9H_{12}	3-Ethyltoluene
C_9H_{12}	4-Ethyltoluene
C_9H_{12}	Mesitylene
$C_9H_{12}O$	1-Phenyl-1-propanol
$C_9H_{12}O$	2-Phenyl-1-propanol
$C_9H_{12}O$	3-Phenyl-1-propanol
$C_9H_{13}N$	N,N-Dimethyl-p-toluidine
$C_9H_{14}O$	Phorone
$C_9H_{14}O$	Isophorone
$C_9H_{15}N$	Triallylamine

Empirical formula	Name
$C_9H_{16}O_4$	Diethyl glutarate
$C_9H_{18}O$	Nonanal
$C_9H_{18}O$	Diisobutylketone
$C_9H_{18}O_2$	Nonanoic acid
$C_9H_{18}O_2$	Ethyl heptanoate
$C_9H_{19}N$	N-Isopropylcyclohexylamine
C_9H_{20}	Nonane
$C_9H_{21}N$	Tripropylamine

C_{10}

Empirical formula	Name
$C_{10}H_5ClN_2$	o–Chlorobenzalmalonitrile
$C_{10}H_7Cl$	1-Chloronaphthalene
$C_{10}H_7Cl$	2-Chloronaphthalene
$C_{10}H_7NO_2$	1-Nitronaphthalene
$C_{10}H_7NO_2$	2-Nitronaphthalene
$C_{10}H_8$	Naphthalene
$C_{10}H_8O$	1-Naphthol
$C_{10}H_8O$	2-Naphthol
$C_{10}H_9N$	1-Naphthylamine
$C_{10}H_9N$	2-Naphthylamine
$C_{10}H_9N$	Quinaldine
$C_{10}H_{10}$	Divinyl benzene
$C_{10}H_{10}O_2$	Methyl cinnamate
$C_{10}H_{10}O_4$	Dimethyl phthalate
$C_{10}H_{10}O_4$	Dimethyl terephthalate
$C_{10}H_{12}$	Dicyclopentadiene
$C_{10}H_{12}$	Tetrahydronaphthalene
$C_{10}H_{12}O_2$	Propyl benzoate
$C_{10}H_{12}O_2$	Isopropyl benzoate
$C_{10}H_{12}O_2$	Ethyl phenylacetate
$C_{10}H_{14}$	p-Cymene
$C_{10}H_{14}$	n-Butylbenzene
$C_{10}H_{14}$	s-Butylbenzene
$C_{10}H_{14}$	t-Butylbenzene
$C_{10}H_{14}$	Isobutylbenzene
$C_{10}H_{14}$	1,2-Diethylbenzene
$C_{10}H_{14}$	1,3-Diethylbenzene
$C_{10}H_{14}$	1,4-Diethylbenzene
$C_{10}H_{14}N_2$	Nicotine
$C_{10}H_{14}O_4$	Diallyl succinate
$C_{10}H_{15}N$	N,N-Diethylaniline
$C_{10}H_{16}$	α-Pinene
$C_{10}H_{16}$	β-Pinene
$C_{10}H_{16}$	Essence of turpentine
$C_{10}H_{16}$	Limonene
$C_{10}H_{16}O$	Camphor (natural isomer)
$C_{10}H_{18}$	Decahydronaphthalene(Z+E)
$C_{10}H_{18}$	Decahydronaphthalene(Z)
$C_{10}H_{18}$	Decahydronaphthalene(E)
$C_{10}H_{18}O_3$	Pentanoic anhydride
$C_{10}H_{18}O_3$	Isopentanoic anhydride
$C_{10}H_{18}O_3$	Pivalic anhydride
$C_{10}H_{18}O_4$	Dibutyl oxalate

9.2 Organic compounds — Empirical formulae/Names

Empirical formula	Name
$C_{10}H_{18}O_4$	Diethyl adipate
$C_{10}H_{20}O$	Decanal
$C_{10}H_{20}O$	Isodecanal
$C_{10}H_{20}O_2$	Decanoic acid
$C_{10}H_{22}$	Decane
$C_{10}H_{22}O$	Diamyl ether
$C_{10}H_{22}O$	Diisoamyl ether

C_{11}

Empirical formula	Name
$C_{11}H_{10}$	1-Methylnaphthalene
$C_{11}H_{10}$	2-Methylnaphthalene
$C_{11}H_{12}O_2$	Ethyl cinnamate
$C_{11}H_{14}O_2$	Butyl benzoate
$C_{11}H_{16}$	n-Pentylbenzene
$C_{11}H_{24}$	Undecane

C_{12}

Empirical formula	Name
$C_{12}H_9N$	Carbazol
$C_{12}H_9NO_2$	o-Nitrodiphenyl
$C_{12}H_9NO_2$	p-Nitrodiphenyl
$C_{12}H_{10}$	Diphenyl
$C_{12}H_{10}Cl_2N_2$	3,3'-Dichlorobenzidine
$C_{12}H_{10}O$	o-Phenylphenol
$C_{12}H_{10}O$	Diphenyl ether
$C_{12}H_{10}S$	Diphenyl sulphide
$C_{12}H_{11}N$	2-Diphenylamine
$C_{12}H_{11}N$	Diphenylamine
$C_{12}H_{12}N_2$	Benzidine
$C_{12}H_{14}O_4$	Diethyl phthalate
$C_{12}H_{14}O_4$	Diethyl terephthalate
$C_{12}H_{16}$	Cyclohexylbenzene
$C_{12}H_{22}O_3$	Hexanoic anhydride
$C_{12}H_{23}N$	Dicyclohexylamine
$C_{12}H_{24}$	Cyclododecane
$C_{12}H_{24}O_2$	Dodecanoic acid
$C_{12}H_{26}$	Dodecane
$C_{12}H_{26}O$	Hexyl ether
$C_{12}H_{27}BO_3$	Tributyl borate
$C_{12}H_{27}N$	Dihexylamine
$C_{12}H_{27}N$	Tributylamine
$C_{12}H_{27}N$	Triisobutylamine
$C_{12}H_{27}O_3P$	Tributylphosphite
$C_{12}H_{27}O_4P$	Tributylphosphate

C_{13}-C_{24}

Empirical formula	Name
$C_{13}H_{10}$	Fluorene
$C_{13}H_{10}O_3$	Diphenyl carbonate
$C_{13}H_{11}NO$	Benzanilide
$C_{13}H_{12}$	Diphenylmethane
$C_{14}H_{10}$	Diphenylacetylene
$C_{14}H_{10}$	Anthracene
$C_{14}H_{10}$	Phenanthrene
$C_{14}H_{10}O_3$	Benzoic anhydride
$C_{14}H_{12}$	1,1-Diphenylethylene
$C_{14}H_{12}$	1,2-Diphenylethylene (E)
$C_{14}H_{12}O_2$	Benzyl benzoate
$C_{14}H_{14}O$	Dibenzyl ether
$C_{14}H_{15}N$	Dibenzylamine
$C_{14}H_{16}N_2O_2$	3,3'-Dimethoxybenzidine
$C_{14}H_8O_2$	Anthraquinone
$C_{15}H_{12}$	9-Methylanthracene
$C_{15}H_{24}O$	4-Nonylphenol
$C_{15}H_{33}N$	Triamylamine
$C_{16}H_{22}O_4$	Dibutyl phthalate
$C_{18}H_{14}$	o-Terphenyl
$C_{18}H_{14}$	p-Terphenyl
$C_{18}H_{15}O_4P$	Triphenylphosphate
$C_{18}H_{15}P$	Triphenylphosphine
$C_{18}H_{34}O_2$	Oleic acid
$C_{21}H_{21}O_4P$	Tritolylphosphate
$C_{24}H_{38}O_4$	Bis (2-ethylhexyl) phthalate

9.2.2 CAS N°/Corresponding names

CAS N°	Name
50-00-0	Formol
50-21-5	Lactic acid
51-28-5	2,4-Dinitrophenol
54-11-5	Nicotine
55-18-5	Nitrosodiethylamine
55-21-0	Benzamide
55-63-0	Nitroglycerine
56-23-5	Tetrachloromethane
56-81-5	Glycerol
57-06-7	Allyl isothiocyanate
57-13-6	Urea
57-14-7	1,1-Dimethylhydrazine
57-55-6	Propane-1,2-diol
57-57-8	ß-Propiolactone
59-50-7	p-Chloro-m-cresol
59-67-6	Nicotinic acid
60-12-8	2-Phenylethanol
60-24-2	2-Mercaptoethanol
60-29-7	Diethyl ether
60-34-4	Methylhydrazine
60-35-5	Acetamide
62-53-3	Aniline
62-55-5	Thioacetamide
62-56-6	Thiourea
64-17-5	Ethanol
64-18-6	Formic acid
64-19-7	Acetic acid
64-67-5	Diethyl sulphate
65-85-0	Benzoic acid
66-25-1	Hexanal
67-56-1	Methanol
67-63-0	2-Propanol
67-64-1	Acetone
67-66-3	Trichloromethane
67-68-5	Dimethylsulphether
67-71-0	Dimethylsulphone
68-11-1	Thioglycolic acid
68-12-2	Dimethylformamide
69-72-7	Salicylic acid
71-23-8	1-Propanol
71-36-3	1-Butanol
71-41-0	1-Pentanol
71-43-2	Benzene
71-55-6	1,1,1-Trichloroethane
74-82-8	Methane
74-83-9	Bromomethane
74-84-0	Ethane
74-85-1	Ethylene
74-86-2	Acetylene

CAS N°	Name
74-87-3	Chloromethane
74-88-4	Iodomethane
74-89-5	Methylamine
74-90-8	Hydrogen cyanide
74-93-1	Methanethiol
74-95-3	Dibromomethane
74-96-4	Bromoethane
74-97-5	Bromochloromethane
74-98-6	Propane
74-99-7	Propyne
75-00-3	Chloroethane
75-01-4	Chloroethylene
75-02-5	Fluoroethylene
75-03-6	Iodoethane
75-04-7	Ethylamine
75-05-8	Acetonitrile
75-07-0	Acetaldehyde
75-08-1	Ethanethiol
75-09-2	Dichloromethane
75-12-7	Formamide
75-18-3	Dimethyl sulphide
75-19-4	Cyclopropane
75-21-8	Ethylene ether
75-25-2	Tribromomethane
75-26-3	2-Bromopropane
75-28-5	2-Methylpropane
75-29-6	2-Chloropropane
75-31-0	Isopropylamine
75-33-2	2-Propanethiol
75-34-3	1,1-Dichloroethane
75-35-4	1,1-Dichloroethylene
75-36-5	Acetyl chloride
75-44-5	Phosgene
75-46-7	Trifluoromethane
75-50-3	Trimethylamine
75-52-5	Nitromethane
75-56-9	Propylene oxide
75-61-6	Dibromodifluoromethane
75-62-7	Bromotrichloromethane
75-63-8	Bromotrifluoromethane
75-64-9	t-Butylamine
75-65-0	t-Butanol
75-68-3	1-Chloro-1,1-difluoroethane
75-69-4	Fluorotrichloromethane
75-71-8	Dichlorodifluoromethane
75-73-0	Tetrafluoromethane
75-75-2	Methanesulphonic acid
75-76-3	Tetramethylsilane
75-77-4	Trimethylchlorosilane

9 • Tables of equivalence
9.2 Organic compounds CAS N°/Name

N° CAS	Name	N° CAS	Name
75-78-5	Dimethyldichlorosilane	79-05-0	Propionamide
75-79-6	Methyltrichlorosilane	79-06-1	Acrylamide
75-80-9	2,2,2-Tribromoethanol	79-07-2	Chloroacetamide
75-83-2	2,2-Dimethylbutane	79-08-3	Bromoacetic acid
75-84-3	2,2-Dimethyl-1-propanol	79-09-4	Propionic acid
75-85-4	t-Amylic alcohol	79-10-7	Acrylic acid
75-86-5	Acetonecyanohydrin	79-11-8	Chloroacetic acid
75-87-6	Chloral	79-14-1	Glycolic acid
75-89-8	2,2,2-Trifluoroethanol	79-20-9	Methyl acetate
75-97-8	3,3-Dimethyl-2-butanone	79-22-1	Methyl chloroformate
75-98-9	2,2-Dimethylpropionic acid	79-24-3	Nitroethane
75-99-0	2,2-Dichloropropionic acid	79-27-6	1,1,2,2-Tetrabromoethane
76-02-8	Trichloroacetyl chloride	79-30-1	Isobutyryl chloride
76-03-9	Trichloroacetic acid	79-31-2	Isobutyric acid
76-05-1	Trifluoroacetic acid	79-34-5	1,1,2,2-Tetrachloroethane
76-06-2	Chloropicrin	79-36-7	Dichloroacetyl chloride
76-09-5	2,3-Dimethylbutane-2,3-diol	79-37-8	Oxalyl chloride
76-11-9	1,1,1,2-Tetrachlorodifluoroethane	79-38-9	Chlorotrifluoroethylene
76-12-0	1,1,2,2-Tetrachlorodifluoroethane	79-39-0	Methacrylamide
76-13-1	1,1,2-Trichlorotrifluoroethane	79-41-4	Methacrylic acid
76-14-2	1,2-Dichlorotetrafluoroethane	79-43-6	Dichloroacetic acid
76-15-3	Chloropentafluoroethane	79-46-9	2-Nitropropane
77-73-6	Dicyclopentadiene	80-56-8	ß-Pinene
77-76-9	2,2-Dimethoxypropane	80-59-1	2-methylcrotonic acid (E)
77-78-1	Dimethyl sulphate	80-62-6	Methyl methacrylate
78-10-4	Tetraethoxysilane	84-15-1	o-Terphenyl
78-11-5	Pentaerythritol tetranitrate	84-65-1	Anthraquinone
78-40-0	Triethylphosphate	84-66-2	Diethyl phthalate
78-59-1	Isophorone	84-74-2	Dibutyl phthalate
78-75-1	1,2-Dibromopropane	85-01-8	Phenanthrene
78-76-2	2-Bromobutane	85-44-9	Phthalic anhydride
78-77-3	1-Bromo-2-methylpropane	86-00-0	o-Nitrodiphenyl
78-78-4	2-Methylbutane	86-57-7	1-Nitronaphthalene
78-79-5	Isoprene	86-73-7	Fluorene
78-81-9	Isobutylamine	86-74-8	Carbazole
78-82-0	Isobutyronitrile	87-61-6	1,2,3-Trichlorobenzene
78-83-1	Isobutanol	87-66-1	Pyrogallol
78-84-2	Isobutyraldehyde	87-86-5	Pentachlorophenol
78-85-3	Methacrolein	88-09-5	2-Ethylbutyric acid
78-86-4	2-Chlorobutane	88-14-2	2-Furoic acid
78-87-5	1,2-Dichloropropane	88-72-2	2-Nitrotoluene
78-88-6	2,3-Dichloropropene	88-73-3	1-Chloro-2-nitrobenzene
78-90-0	1,2-Diaminopropane	88-74-4	2-Nitroaniline
78-92-2	2-Butanol	88-75-5	2-Nitrophenol
78-93-3	Butanone	88-89-1	2,4,6-Trinitrophenol
78-94-4	Methylvinylketone	88-99-3	o-Phthalic acid
78-95-5	Chloroacetone	89-98-5	2-Chlorobenzaldehyde
78-96-6	1-Amino-2-propanol	90-00-6	2-Ethylphenol
78-99-9	1,1-Dichloropropane	90-02-8	Salicylaldehyde
79-01-6	Trichloroethylene	90-04-0	o-Anisidine
79-03-8	Propionyl chloride	90-05-1	Gaiacol
79-04-9	Chloroacetyl chloride	90-12-0	1-Methylnaphthalene

9.2 Organic compounds CAS N°/Name

CAS N°	Name	CAS N°	Name
90-13-1	1-Chloronaphthalene	96-32-2	Methyl bromoacetate
90-15-3	1-Naphthol	96-33-3	Methyl acrylate
90-41-5	2-Diphenylamine	96-34-4	Methyl chloroacetate
90-43-7	o-Phenylphenol	96-37-7	Methylcyclopentane
91-08-7	Toluene-2,4-diisocyanate	96-41-3	Cyclopentanol
91-15-6	o-Phthalonitrile	96-48-0	γ-Butyrolactone
91-16-7	1,2-Dimethoxybenzene	96-49-1	Ethylene carbonate
91-17-8	Decahydronaphthalene(Z+E)	96-54-8	N-Methylpyrrole
91-20-3	Naphthalene	97-00-7	1-Chloro-2,4-dinitrobenzene
91-22-5	Quinoline	97-02-9	2,4-Dinitroaniline
91-23-6	2-Nitroanisole	97-61-0	2-Methylpentanoïc acid
91-57-6	2-Methylnaphthalene	97-62-1	Ethyl isobutyrate
91-58-7	2-Chloronaphthalene	97-63-2	Ethyl methacrylate
91-59-8	2-Naphthylamine	97-64-3	Ethyl lactate
91-63-4	Quinaldine	97-72-3	Isobutyric anhydride
91-66-7	N,N-Diethylaniline	97-86-9	Isobutyl methacrylate
91-94-1	3,3'-Dichlorobenzidine	97-88-1	Butyl methacrylate
92-52-4	Diphenyl	97-96-1	2-Ethylbutyraldehyde alcohol
92-87-5	Benzidine	97-99-0	Tetrahydrofurfurylic alcohol
92-93-3	p-Nitrodiphenyl	98-00-0	Furfurylic alcohol
92-94-4	p-Terphenyl	98-01-1	Furfural
93-54-9	1-Phenyl-1-propanol	98-06-6	t-Butylbenzene
93-58-3	Methyl benzoate	98-07-7	Trichloromethylbenzene
93-89-0	Ethyl benzoate	98-08-8	Trifluoromethylbenzene
93-97-0	Benzoic anhydride	98-82-8	Cumene
93-98-1	Benzanilide	98-83-9	α-Methylstyrene
94-14-0	3-Methylpentane	98-85-1	1-Phenylethanol
94-75-7	2,4-Dichlorophenoxyacetic acid	98-86-2	Acetophenone
95-13-6	Indene	98-87-3	Benzylidene chloride
95-46-5	2-Bromotoluene	98-88-4	Benzoyl chloride
95-47-6	o-Xylene	98-95-3	Nitrobenzene
95-48-7	o-Cresol	99-02-5	3-Chloroacetophenone
95-49-8	2-Chlorotoluene	99-08-1	3-Nitrotoluene
95-50-1	1,2-Dichlorobenzene	99-09-2	3-Nitroaniline
95-51-2	2-Chloroaniline	99-35-4	1,3,5-Trinitrobenzene
95-52-3	2-Fluorotoluene	99-65-0	1,3-Dinitrobenzene
95-53-4	o-Toluidine	99-87-6	p-Cymene
95-55-6	2-Aminophenol	99-91-2	4-Chloroacetophenone
95-57-8	2-Chlorophenol	99-97-8	N,N-Dimethyl-p-toluidine
95-65-8	3,4-Xylenol	99-99-0	4-Nitrotoluene
95-73-8	2,4-Dichlorotoluene	100-00-5	1-Chloro-4-nitrobenzene
95-76-1	3,4-Dichloroaniline	100-01-6	4-Nitroaniline
95-82-9	2,5-Dichloroaniline	100-02-7	4-Nitrophenol
95-87-4	2,5-Xylenol	100-17-4	4-Nitroanisole
95-92-1	Diethyl oxalate	100-21-0	Terephthalic acid
95-94-3	1,2,4,5-Tetrachlorobenzene	100-25-4	1,4-Dinitrobenzene
95-95-4	2,4,5-Trichlorophenol	100-37-8	2-Diethylaminoethanol
96-05-9	Allyl methacrylate	100-39-0	Benzyl bromide
96-17-3	2-Methylbutyraldehyde	100-40-3	4-Vinylcyclohexene
96-18-4	1,2,3-Trichloropropane	100-41-4	Ethyl benzene
96-20-8	2-Amino-1-butanol	100-42-5	Styrene
96-22-0	3-Pentanone	100-43-6	4-Vinylpyridine

9 • Tables of equivalence

9.2 Organic compounds CAS N°/Name

CAS N°	Name	CAS N°	Name
100-44-7	Benzyl chloride	105-54-4	Ethyl butyrate
100-46-9	Benzylamine	105-56-6	Ethyl cyanoacetate
100-47-0	Benzonitrile	105-57-7	1,1-Diethoxyethane
100-51-6	Benzyl alcohol	105-58-8	Diethyl carbonate
100-52-7	Benzaldehyde	105-60-2	ε-Caprolactam
100-60-7	N-Methyl cyclohexylamine	105-67-9	2,4-Xylenol
100-61-8	N-Methylaniline	106-30-9	Ethyl heptanoate
100-63-0	Phenylhydrazine	106-31-0	Butyric anhydride
100-66-3	Anisole	106-35-4	3-Heptanone
100-69-6	2-Vinylpyridine	106-36-5	Propyl propionate
100-71-0	2-Ethylpyridine	106-38-7	4-Bromotoluene
100-74-3	N-Ethylmorpholine	106-42-3	p-Xylene
100-80-1	3-Methylstyrene	106-43-4	4-Chlorotoluene
100-84-5	3-Methylanisole	106-44-5	p-Cresol
100-97-0	Hexamethylenetetramine	106-45-6	p-Thiocresol
101-81-5	Diphenylmethane	106-46-7	1,4-Dichlorobenzene
101-83-7	Dicyclohexylamine	106-47-8	4-Chloroaniline
101-84-8	Diphenyl ether	106-48-9	4-Chlorophenol
101-97-3	Ethyl phenylacetate	106-49-0	p-Toluidine
102-09-0	Diphenyl carbonate	106-50-3	p-Phenylenediamine
102-69-2	Tripropylamine	106-51-4	Benzoquinone
102-70-5	Triallylamine	106-65-0	Dimethyl succinate
102-71-6	Triethanolamine	106-70-7	Methyl hexanoate
102-82-9	Tributylamine	106-73-0	Methyl heptanoate
102-85-2	Tributylphosphite	106-88-7	1,2-Epoxybutane
103-26-4	Methyl cinnamate	106-89-8	Epiclorhydrin
103-30-0	1,2-Diphenylethylene (E)	106-93-4	1,2-Dibromoethane
103-36-6	Ethyl cinnamate	106-94-5	1-Bromopropane
103-49-1	Dibenzylamine	106-95-6	1-Bromo-2-propene
103-50-4	Dibenzyl ether	106-96-7	1-Bromo-2-propyne
103-65-1	Propylbenzene	106-97-8	Butane
103-69-5	N-Ethylaniline	106-98-9	1-Butene
103-84-4	Acetanilide	106-99-0	1,3-Butadiene
104-40-5	4-Nonylphenol	107-00-6	1-Butyne
104-51-8	n-Butylbenzene	107-01-7	2-Butene (Z+E)
104-54-1	3-Phenyl-2-propen-1-ol	107-02-8	Acrolein
104-55-2	Cinnamaldehyde	107-03-9	1-Propanethiol
104-76-7	2-Ethyl-1-hexanol	107-04-0	1-Bromo-2-chloroethane
104-85-8	p-Tolunitrile	107-05-1	3-Chloro-1-propene
104-87-0	4-Methybenzaldehyde	107-06-2	1,2-Dichloroethane
104-88-1	4-Chlorobenzaldehyde	107-07-3	2-Chloroethanol
104-93-8	4-Methylanisole	107-10-8	Propylamine
104-94-9	p-Anisidine	107-11-9	Allylamine
105-05-5	1,4-Diethylbenzene	107-12-0	Propionitrile
105-34-0	Methyl cyanoacetate	107-13-1	Acrylonitrile
105-36-2	Ethyl bromoacetate	107-14-2	Chloroacetonitrile
105-37-3	Ethyl propionate	107-15-3	Ethylenediamine
105-39-5	Ethyl chloroacetate	107-16-4	Glycolonitrile
105-43-1	3-Methylpentanoic acid	107-18-6	Allyl alcohol
105-45-3	Methyl acetoacetate	107-19-7	Propargyl alcohol
105-46-4	s-Butyl acetate	107-20-0	Chloroacetaldehyde
105-53-3	Diethyl malonate	107-21-1	Ethylene glycol

9.2 Organic compounds CAS N°/Name

CAS N°	Name	CAS N°	Name
107-22-2	Glyoxal	108-90-7	Chlorobenzene
107-30-2	Chloromethyl ether	108-91-8	Cyclohexylamine
107-31-3	Methyl formate	108-93-0	Cyclohexanol
107-39-1	2,4,4-Trimethyl-1-pentene	108-94-1	Cyclohexanone
107-40-4	2,4,4-Trimethyl-2-pentene	108-95-2	Phenol
107-46-0	Hexamethyldisiloxane	108-98-5	Thiophenol
107-81-3	2-Bromopentane	108-99-6	3-Methylpyridine
107-82-4	1-Bromo-3-methylbutane	109-02-4	N-Methylmorpholine
107-83-5	2-Methylpentane	109-05-7	2-Methylpiperidine
107-84-6	1-Chloro-3-methylbutane	109-06-8	2-Methylpyridine
107-85-7	Isopentylamine	109-21-7	Butyl butyrate
107-86-8	3-Methylcrotonaldehyde	109-52-4	Pentanoic acid
107-87-9	2-Pentanone	109-53-5	Isobutyl vinyl ether
107-88-0	Butane-1,3-diol	109-59-1	2-Isopropoxyethanol
107-89-1	3-Hydroxybutyraldehyde	109-60-4	Propyl acetate
107-92-6	Butyric acid	109-64-8	1,3-Dibromopropane
107-93-7	Crotonic (E) acid	109-65-9	1-Bromobutane
107-94-8	3-Chloropropionic acid	109-66-0	Pentane
108-01-0	2-Dimethylaminoethanol	109-68-2	2-Pentene (E+Z)
108-03-2	1-Nitropropane	109-69-3	1-Chlorobutane
108-05-4	Vinyl acetate	109-70-6	1-Bromo-3-chloropropane
108-10-1	Methyl isobutylketone	109-73-9	Butylamine
108-11-2	4-Methyl-2-pentanol	109-74-0	Butyronitrile
108-12-3	Isopentanoyl chloride	109-75-1	Allyl cyanide
108-16-7	1-Dimethylamino-2-propanol	109-76-1	1-Pentene
108-18-9	Diisopropylamine	109-76-2	1,3-Diaminopropane
108-20-3	Diisopropyl ether	109-77-3	Malo(no)nitrile
108-21-4	Isopropyl acetate	109-78-4	Ethylene cyanohydrin
108-22-5	Isopropenyl acetate	109-79-5	1-Butanethiol
108-24-7	Acetic anhydride	109-85-3	2-Methoxyethylamine
108-30-5	Succinic anhydride	109-86-4	2-Methoxyethanol
108-31-6	Maleic anhydride	109-87-5	Dimethoxymethane
108-32-7	Propylene carbonate	109-89-7	Diethylamine
108-38-3	m-Xylene	109-92-2	Ethyl vinyl ether
108-39-4	m-Cresol	109-93-3	Divinyl ether
108-41-8	3-Chlorotoluene	109-94-4	Ethyl formate
108-42-9	3-Chloroaniline	109-95-5	Ethyl nitrite
108-43-0	3-Chlorophenol	109-97-7	Pyrrole
108-44-1	m-Toluidine	109-99-9	Tetrahydrofuran
108-46-3	Resorcinol	110-00-9	Furan
108-55-4	Glutaric anhydride	110-02-1	Thiophene
108-59-8	Dimethyl malonate	110-12-3	2-Heptanone
108-64-5	Ethyl isopentanoate	110-12-3	5-Methyl-2-hexanone
108-65-6	2-Methoxypropyl acetate	110-13-4	Hexane 2,5-dione
108-67-8	Mesitylene	110-15-6	Succinic acid
108-70-3	1,3,5-Trichlorobenzene	110-16-7	Maleic acid
108-83-8	Diisobutylketone	110-17-8	Fumaric acid
108-84-9	s-Hexyl acetate	110-19-0	Isobutyl acetate
108-86-1	Bromobenzene	110-46-3	Isopentyl nitrite
108-87-2	Methylcyclohexane	110-49-6	2-Methoxyethyl acetate
108-88-3	Toluene	110-52-1	1,4-Dibromobutane
108-89-4	4-Methylpyridine	110-53-2	1-Bromopentane

9 • Tables of equivalence

9.2 Organic compounds CAS N°/Name

CAS N°	Name	CAS N°	Name
110-54-3	Hexane	112-05-0	Nonanoic acid
110-56-5	1,4-Dichlorobutane	112-07-2	2-Butoxyethyl acetate
110-57-6	1,4-Dichloro-2-butene	112-24-3	Triethylenetetramine
110-58-7	n-Pentylamine	112-27-6	Triethylene glycol
110-59-8	Valeronitrile	112-31-2	Decanal
110-60-1	1,4-Diaminobutane	112-40-3	Dodecane
110-61-2	Succinonitrile	112-58-3	Dihexyl ether
110-62-3	Pentanal	112-80-1	Oleic acid
110-63-4	Butane-1,4-diol	115-07-1	Propylene
110-65-6	2-Butyne-1,4-diol	115-10-6	Dimethyl ether
110-68-9	N-Methylbutylamine	115-11-7	Isobutene
110-71-4	1,2-Dimethoxyethane	115-19-5	2-Methyl-3-butyn-2-ol
110-73-6	2-Ethylaminoethanol	115-20-8	2,2,2-Trichloroethanol
110-74-7	Propyl formate	115-77-5	Pentaerythritol
110-80-5	2-Ethoxyethanol	115-86-6	Triphenylphosphate
110-82-7	Cyclohexane	116-54-1	Methyl dichloroacetate
110-83-8	Cyclohexene	117-81-7	Bis(2-ethylhexyl) phthalate
110-85-0	Piperazine	118-74-1	Hexachlorobenzene
110-86-1	Pyridine	118-75-2	Tetrachlorobenzoquinone
110-87-2	Dihydropyran	118-96-7	2,4,6-Trinitrotoluene
110-88-3	Trioxan	119-36-8	Methyl salicylate
110-89-4	Piperidine	119-64-2	Tetrahydronaphthalene
110-91-8	Morpholine	119-65-3	Isoquinoline
110-94-1	Glutaric acid	119-90-4	3,3'-Dimethoxybenzidine
111-14-8	Heptanoic acid	120-12-7	Anthracene
111-15-9	2-Ethoxyethyl acetate	120-51-4	Benzyl benzoate
111-16-0	Pimelic acid	120-80-9	Pyrocatechol
111-25-1	1-Bromohexane	120-82-1	1,2,4-Trichlorobenzene
111-26-2	Hexylamine	120-83-2	2,4-Dichlorophenol
111-27-3	1-Hexanol	120-92-3	Cyclopentanone
111-29-5	Pentane-1,5-diol	121-14-2	2,4-Dinitrotoluene
111-30-8	Glutaraldehyde	121-43-7	Trimethyl borate
111-34-2	Butyl vinyl ether	121-44-8	Triethylamine
111-40-0	Diethylenetriamine	121-45-9	Trimethyl phosphite
111-42-2	Diethanolamine	121-61-6	Dimethyl terephthalate
111-43-3	Dipropyl ether	121-69-7	N,N-Dimethylaniline
111-44-4	Bis(2-Chloroethyl)ether	121-73-3	1-Chloro-3-nitrobenzene
111-46-0	Diethyleneglycol	121-82-4	Hexogene
111-48-8	Thiodiethanol	122-39-4	Diphenylamine
111-55-7	Ethyleneglycol diacetate	122-52-1	Triethylphosphite
111-65-9	Octane	122-60-1	Phenyl glycidylether
111-66-0	1-Octene	122-79-2	Phenyl acetate
111-68-2	Heptylamine	122-97-4	3-Phenyl-1-propanol
111-69-3	Adiponitrile	122-98-5	2-Anilinoethanol
111-70-6	1-Heptanol	123-07-9	4-Ethylphenol
111-71-7	Heptanal	123-11-5	4-Anisaldehyde
111-76-2	2-Butoxyethanol	123-15-9	2-Methylpentanal
111-77-3	Diethylene glycol monomethylether	123-19-3	4-Heptanone
111-84-2	Nonane	123-25-1	Diethyl succinate
111-87-5	1-Octanol	123-30-8	4-Aminophenol
111-90-0	Diethylene glycol monoethylether	123-31-9	Hydroquinone
111-92-2	Dibutylamine	123-38-6	Propionaldehyde

9.2 Organic compounds CAS N°/Name

CAS N°	Name
123-42-2	Diacetone achohol
123-51-3	Isoamyl alcohol
123-54-6	Acetylacetone
123-62-6	Propionic anhydride
123-63-7	Paraldehyde
123-66-0	Ethyl hexanoate
123-72-8	Butyraldehyde
123-73-9	Crotonaldehyde
123-75-1	Pyrrolidine
123-86-4	Butyl acetate
123-90-0	Thiomorpholine
123-91-1	1,4-Dioxan
123-92-2	Isoamyl acetate
123-96-6	2-Octanol
124-02-7	Diallylamine
124-04-9	Adipic acid
124-07-2	Octanoic acid
124-09-4	1,6-Diaminohexane
124-13-0	Octanal
124-18-5	Decane
124-19-6	Nonanal
124-40-3	Dimethylamine
124-73-2	1,2-Dibromotetrafluoroethane
126-30-7	2,2-Dimethylpropane-1,3-diol
126-33-0	Sulpholane
126-73-8	Tributylphosphate
126-84-1	2,2-Diethoxypropane
126-98-7	Methacrylonitrile
126-99-8	2-Chloro-1,3-butadiene
127-17-3	Pyruvic acid
127-18-4	Tetrachloroethylene
127-19-5	Dimethylacetamide
131-11-3	Dimethyl phthalate
134-32-7	1-Naphthylamine
135-01-3	1,2-Diethylbenzene
135-02-4	2-Anisaldehyde
135-19-3	2-Naphthol
135-98-8	s-Butylbenzene
136-60-7	Butyl benzoate
137-32-6	2-Methyl-1-butanol
138-86-3	Limonene
139-66-2	Diphenyl sulphide
140-10-3	Cinnamic(E) acid
140-11-4	Benzyl acetate
140-29-4	Benzyl cyanide
140-88-3	Ethyl acrylate
141-05-9	Diethyl maleate
141-28-6	Diethyl adipate
141-32-2	Butyl acrylate
141-43-5	Ethanolamine
141-75-3	Butyryl chloride
141-79-7	Dimesityl ether
141-82-2	Malonic acid
141-93-5	1,3-Diethylbenzene
141-97-9	Ethyl acetoacetate
142-28-9	1,3-Dichloropropane
142-29-0	Cyclopentene
142-61-0	Hexanoyl chloride
142-62-1	Hexanoic acid
142-68-7	Tetrahydropyran
142-82-5	Heptane
142-84-7	Dipropylamine
142-92-7	Hexyl acetate
142-96-1	Butyl acid
143-07-7	Dodecanoic acid
143-16-8	Dihexylamine
144-49-0	Fluoroacetic acid
144-62-7	Oxalic acid
148-24-3	8-Quinolinol
149-57-5	2-Ethylhexanoic acid
150-46-9	Triethyl borate
150-76-5	4-Methoxyphenol
150-78-7	1,4-Dimethoxybenzene
151-10-3	1,3-Dimethoxybenzene
151-56-4	Aziridine
156-59-2	1,2-Dichloroethylene (Z)
156-60-5	1,2-Dichloroethylene (E)
156-87-6	3-Amino-1-propanol
271-89-6	Benzofuran
287-92-3	Cyclopentane
288-13-1	Pyrazole
291-64-5	Cycloheptane
292-64-8	Cyclo-octane
294-62-2	Cyclododecane
300-57-2	Allylbenzene
302-17-0	Chloral hydrate
334-48-5	Decanoic acid
352-32-9	4-Fluorotoluene
352-70-5	3-Fluorotoluene
352-93-2	Diethyl sulphide
353-36-6	Fluoroethane
371-62-0	2-Fluoroethanol
372-09-8	Cyanoacetic acid
431-03-8	Butane-2,3-dione
431-47-0	Methyl trifluoroacetate
459-72-3	Ethyl fluoroacetate
460-19-5	Cyanogen
462-06-6	Fluorobenzene
462-08-8	3-Aminopyridine
462-95-3	Diethoxymethane
463-04-7	Pentyl nitrite
463-49-0	Allene
463-51-4	Cetene
463-82-1	2,2-Dimethylpropane

9 • Tables of equivalence
9.2 Organic compounds CAS N°/Name

CAS N°	Name	CAS N°	Name
464-07-3	3,3-Dimethyl-2-butanol	540-73-8	1,2-Dimethylhydrazine
464-49-3	Camphor (natural isomer)	540-80-7	t-Butyl nitrite
493-01-6	Decahydronaphthalene (Z)	540-84-1	2,2,4-Trimethylpentane
493-02-7	Decahydronaphthalene (E)	540-88-5	Ethyl acetate
501-53-1	Benzyl chloroformate	540-88-5	t-Butyl acetate
501-65-5	Diphenylacetylene	541-35-5	Butyramide
502-41-0	Cycloheptanol	541-41-3	Ethyl chloroformate
502-42-1	Cycloheptanone	541-47-9	3-Methylcrotonic acid
502-44-3	ε-Caprolactone	541-73-1	1,3-Dichlorobenzene
503-17-3	2-Butyne	541-85-5	5-Methyl-3-heptanone
503-74-2	3-Methylbutyric acid	541-88-8	Chloroacetic anhydride
504-20-1	Phorone	542-28-9	δ-Valerolactone
504-24-5	4-Aminopyridine	542-55-2	Isobutyl formate
504-29-0	2-Aminopyridine	542-55-2	Isoamyl formate
504-63-2	Propane-1,3-diol	542-56-3	Isobutyl nitrite
507-19-7	2-Bromo-2-methylpropane	542-69-8	1-Iodobutane
507-20-0	2-Chloro-2-methylpropane	542-75-6	1,3-Dichloropropene
509-14-8	Tetranitromethane	542-76-7	3-Chloropropionitrile
512-56-1	Trimethylphosphate	542-85-8	Ethyl isothiocyanate
513-35-9	2-Methyl-2-butene	542-92-7	Cyclopentadiene
513-36-0	1-Chloro-2-methylpropane	543-49-7	2-Heptanol
513-37-1	1-Chloro-2-methyl-2-propene	543-59-9	1-Chloropentane
513-42-8	2-Methyl-2-propen-1-ol	543-87-3	Isopentyl nitrate
513-48-4	2-Iodobutane	544-01-4	Diisoamyl ether
513-85-9	Butane-2,3-diol	544-13-8	Glutaronitrile
513-86-0	3-Hydroxy-2-butanone	544-16-1	Butyl nitrite
513-88-2	1,1-Dichloroacetone	545-06-2	Trichloroacetonitrile
526-75-0	2,3-Xylenol	547-63-7	Methyl Isobutyrate
528-29-0	1,2-Dinitrobenzene	553-90-2	Dimethyl oxalate
529-19-1	o-Tolunitrile	554-00-7	2,4-Dichloroaniline
529-20-4	2-Methylbenzaldehyde	554-12-1	Methyl propionate
530-48-3	1,1-Diphenylethylene	554-84-7	3-Nitrophenol
532-27-4	α-Chloroacetophenone	555-03-3	3-Nitroanisole
534-15-6	1,1-Dimethoxyethane	556-52-5	Glycidol
534-22-5	2-Methylfuran	556-56-9	Allyl iodide
534-52-1	4,6-Dinitro-o-cresol	556-61-6	Methyl isothiocyanate
535-11-5	Ethyl 2-Bromopropionate	558-13-4	Tetrabromomethane
535-13-7	Ethyl 2-Chloropropionate	558-37-2	3,3-Dimethyl-1-butene
535-15-9	Ethyl dichloroacetate	563-45-1	3-Methyl-1-butene
536-74-3	Phenylacetylene	563-46-2	2-Methyl-1-butene
536-75-4	4-Ethylpyridine	563-47-3	3-Chloro-2-methyl-1-propene
536-78-7	3-Ethylpyridine	563-52-0	1-Chloro-3-butene
536-90-3	m-Anisidine	563-78-0	2,3-Dimethyl-1-butene
538-68-1	n-Pentylbenzene	563-79-1	2,3-Dimethyl-2-butene
538-93-2	Isobutylbenzene	563-80-4	Methyl isopropylketone
539-74-2	Ethyl 3-bromopropionate	563-83-7	Isobutyramide
539-82-2	Ethyl pentanoate	565-69-5	Ethyl isopropylketone
540-42-1	Isobutyl propionate	565-80-0	Diisopropylketone
540-49-8	1,2-Dibromoethylene	576-26-1	2,6-Xylenol
540-51-2	2-Bromoethanol	578-54-1	2-Ethylaniline
540-54-5	1-Chloropropane	578-58-5	2-Methylanisole
540-67-0	Methyl ethyl ether	581-89-5	2-Nitronaphthalene

9 • Tables of equivalence

9.2 Organic compounds CAS N°/Name

Cas N°	Name	CAS N°	Name
583-59-5	2-Methylcyclohexanol	603-35-0	Triphenylphosphine
583-60-8	2-Methylcyclohexanone	606-20-2	2,6-Dinitrotoluene
584-02-1	3-Pentanol	608-27-5	2,3-Dichloroaniline
584-03-2	Butane-1,2-diol	608-31-1	2,6-Dichloroaniline
584-84-9	Toluene-2,6-diisocyanate	609-15-4	Ethyl 2-chloroacetoacetate
587-04-2	3-Chlorobenzaldehyde	609-65-4	o-Chlorobenzoyl chloride
589-16-2	4-Ethylaniline	611-14-3	2-Ethyltoluene
589-38-8	3-Hexanone	611-15-4	2-Methylstyrene
589-55-9	4-Heptanol	616-05-7	2-Bromohexanoic acid
589-82-2	3-Heptanol	616-38-6	Dimethyl carbonate
589-91-3	4-Methylcyclohexanol	616-45-5	Pyrrolidone
589-92-4	4-Methylcyclohexanone	617-86-7	Triethylsilane
589-98-0	3-Octanol	620-14-4	3-Ethyltoluene
590-01-2	Butyl propionate	620-17-7	3-Ethylphenol
590-18-1	2-Butene (Z)	620-23-5	3-Methylbenzaldehyde
590-86-3	3-Methylbutyraldehyde	621-77-2	Triamylamine
590-92-1	3-Bromopropionic acid	622-45-7	Cyclohexyl acetate
591-17-3	3-Bromotoluene	622-96-8	4-Ethyltoluene
591-23-1	3-Methylcyclohexanol	622-97-9	4-Methylstyrene
591-24-2	3-Methylcyclohexanone	623-12-1	4-Chloroanisole
591-27-5	3-Aminophenol	623-26-7	p-Phthalonitrile
591-31-1	3-Anisaldehyde	623-37-0	3-Hexanol
591-47-9	4-Methylcyclohexene	623-42-7	Methyl butyrate
591-50-4	Iodobenzene	623-43-8	Methyl crotonate
591-60-6	Butyl acetoacetate	623-70-1	Ethyl crotonate
591-78-6	2-Hexanone	623-71-2	Ethyl 3-chloropropionate
591-87-7	Allyl acetate	623-91-6	Diethyl fumarate
591-97-9	1-Chloro-2-butene	624-24-8	Methyl pentanoate
592-41-6	1-Hexene	624-48-6	Dimethyl maleate
592-43-8	2-Hexene	624-49-7	Dimethyl fumarate
592-76-7	1-Heptene	624-64-6	2-Butene (E)
592-84-7	Butyl formate	624-65-7	3-Chloro-1-propyne
593-60-2	Bromoethylene	624-83-9	Methyl isocyanate
593-63-5	Chloroacetylene	624-91-9	Methyl nitrite
594-09-2	Trimethylphosphine	625-28-5	Isovaleronitrile
594-20-7	2,2-Dichloropropane	625-35-4	Crotonoyl chloride
594-36-5	2-Chloro-2-methylbutane	625-48-9	2-Nitroethanol
594-72-9	1,1-Dichloro-1-nitroethane	625-55-8	Isopropyl formate
595-37-9	2,2-Dimethylbutyric acid	625-58-1	Ethyl nitrate
598-32-3	3-Buten-2-ol	626-17-5	m-Phthalonitrile
598-56-1	N,N-Dimethylethylamine	626-38-0	s-Amyl acetate
598-58-3	Methyl nitrate	626-43-7	3,5-Dichloroaniline
598-72-1	2-Bromopropionic acid	626-56-2	3-Methylpiperidine
598-74-3	Isopentylamine	626-58-4	4-Methylpiperidine
598-75-4	3-Methyl-2-butanol	626-67-5	N-Methylpiperidine
598-78-7	2-Chloropropionic acid	626-93-7	2-Hexanol
598-98-1	Methyl pivalate	627-05-4	1-Nitrobutane
598-99-2	Methyl trichloroacetate	627-13-4	Propyl nitrate
600-07-7	2-Methylbutyric acid	627-18-9	3-Bromo-1-propanol
600-14-6	Pentane-2,3-dione	627-19-0	1-Pentyne
600-25-9	1-Chloro-1-nitropropane	627-20-3	2-Pentene (Z)
600-36-2	2,4-Dimethyl-3-pentanol	627-21-4	2-Pentyne

9 • Tables of equivalence

9.2 Organic compounds CAS N°/Name

CAS N°	Name	CAS N°	Name
627-26-9	Crotononitrile	930-30-3	2-Cyclopenten-1-one
627-27-0	2-Buten-1-ol (Z+E)	930-68-7	2-Cyclohexen-1-one
627-30-5	3-Chloro-1-propanol	938-48-0	Isopropyl benzoate
627-42-9	2-Methoxychloroethane	942-54-1	4-Methylvaleronitrile
627-63-4	Fumaryl chloride	998-30-1	Triethoxysilane
627-93-0	Methyl adipate	999-97-3	Hexamethyldisilazane
628-05-7	1-Nitropentane	1002-16-0	Pentyl nitrate
628-20-6	4–Chlorobutyronitrile	1003-03-8	Cyclopentylamine
628-28-4	Butyl methyl ether	1003-38-9	2,5-Dimethyltetrahydrofuran
628-55-7	Diisobutyl ether	1070-83-3	3,3-Dimethylbutyric acid
628-63-7	Amyl acetate	1073-67-2	4-Chlorostyrene
628-73-9	Capronitrile	1116-40-1	Triisobutylamine
628-92-2	Cycloheptene	1118-84-9	Allyl acetoacetate
628-96-6	Ethyleneglycol dinitrate	1119-40-0	Dimethyl glutarate
629-08-3	Hexanecarbonitrile	1120-21-4	Undecane
629-11-8	Hexane-1,6-diol	1195-42-2	N-Isopropylcyclohexylamine
629-14-1	1,2-Diethoxyethane	1321-74-0	Divinylbenzene
630-18-2	Pivalonitrile	1330-20-7	Xylenes (o-, m-, p-)
630-20-6	1,1,1,2-Tetrachloroethane	1330-78-5	Tritolylphosphate
636-09-9	Diethyl terephthalate	1484-80-6	2-Ethylpiperidine
638-07-3	Ethyl 4-chloroacetoacetate	1538-75-6	Pivalic anhydride
638-11-9	Isopropyl butyrate	1634-04-4	Methyltert-butylether (MTBE)
638-29-9	Pentanoyl chloiride	1649-08-7	1,2-Dichloro-1,1-difluoroethane
638-49-3	Amyl formate	1678-91-7	Ethylcyclohexane
645-62-5	2-Ethyl-2-hexen-1-al	1694-31-1	t-Butyl acetoacetate
646-04-8	2-Pentene (E)	1708-29-8	2,5-Dihydrofuran
646-06-0	1,3-Dioxolane	1712-64-7	Isopropyl nitrate
646-07-1	4-Methylpentanoic acid	1738-36-9	Methoxyacetonitrile
646-14-0	1-Nitrohexane	1745-81-9	2-Allylphenol
674-82-8	Diketene	1809-19-4	Dibutyl hydrogenphosphite
688-74-4	Tributyl borate	2004-70-8	1,3-Pentadiene (E)
693-65-2	Diamyl ether	2028-63-9	3-Butyn-2-ol
694-28-0	2-Chlorocyclopentanone	2043-61-0	Cyclohexanecarboxaldehyde
760-93-0	Methacryl anhydride	2050-60-4	Dibutyl oxalate
762-05-2	Cyclohexanecarbonitrile	2051-49-2	Hexanoic anhydride
764-42-1	Fumaronitrile	2082-59-9	Pentanoic anhydride
766-09-6	N-Ethylpiperidine	2141-62-0	3–Ethoxypropionitrile
779-02-2	9-Methylanthracene	2142-68-9	2-Chloroacetophenone
814-68-6	Acryloyl chloride	2155-94-4	N,N-Dimethylallylamine
818-38-2	Diethyl glutarate	2238-07-5	Diglycidyl ether
822-06-0	Hexamethylnediisocyanate	2315-68-6	Propyl benzoate
822-87-7	2-Chlorocyclohexanone	2426-08-6	Butylglycidyl ether
823-76-7	Methylcyclohexylketone	2487-90-3	Trimethoxysilane
827-52-1	Cyclohexylbenzene	2698-41-1	o–Chlorobenzalmalonitrile
872-50-4	N-Methylpyrrolidone	2719-27-9	Cyclohexanecarbonyl chloride
873-32-5	2-Chlorobenzonitrile	2902-96-7	2-Nitro-1-propanol
920-37-6	2-Chloroacrylonitrile	2937-50-0	Allyl chloroformate
920-46-7	Methacryloyl chloride	3018-12-0	Dichloroacetonitrile
924-43-6	s-Butyl nitrite	3121-61-7	2-Methoxyethyl acetate
925-16-6	Diallyl succinate	3179-63-3	3-Dimethylamino-1-propanol
926-57-8	1,3-Dichloro-2-butene	3282-30-2	Pivaloyl chloride
927-80-0	Ethoxyacetylene	3333-52-6	Tetramethylsuccinonitrile

9 • Tables of equivalence

9.2 Organic compounds CAS N°/Name

CAS N°	Name	CAS N°	Name
3724-65-0	Crotonic (E) acid	10026-04-7	Tetrachlorosilane
3848-24-6	Hexane-2,3-dione	13054-87-0	2-Amino-1-butanol
3938-95-2	Ethyl pivalate	13323-81-4	1-Phenylethanol
4016-14-2	Isopropylglycidylether	13389-42-9	2-Octene (E)
4091-39-8	3-Chloro-2-butanone	13952-84-6	s-Butylamine
4128-31-8	2-Octanol	14371-10-9	Cinnamaldehyde
4170-30-3	Crotonaldehyde	14631-45-9	3–Ethoxypropionitrile
4435-53-4	3-Methoxybutyl acetate	14686-13-6	2-Heptene (E)
4437-51-8	Hexane-3,4-dione	14686-14-7	3-Heptene (E)
4444-67-1	Di-s-butylamine	14850-23-8	4-Octene (E)
4635-59-0	4-Chlorobutyryl chloride	14919-01-8	3-Octene (E)
4744-08-5	1,1-Diethoxypropane	15764-16-6	2,4-Dimethylbenzaldehyde
4744-10-9	1,1-Dimethoxypropane	15892-23-6	2-Butanol
4755-81-1	Methyl 2-chloro ether	19902-08-0	β-Pinene
4786-20-3	Crotononitrile	20296-29-1	3-Octanol
4885-02-3	Methyl dichloromethyl ether	22083-74-5	Nicotine
5077-67-8	1-Hydroxy-2-butanone	25167-67-3	1-Butene
5500-21-0	Cyclopropanecarbonitrile	25167-70-8	2,4,4-Trimethyl-1-pentene
5813-64-9	Neopentylamine	25340-17-4	1,2-Diethylbenzene
5891-21-4	5-Chloro-2-pentanone	26171-83-5	Butane-1,2-diol
6032-29-7	2-Pentanol	26760-64-5	2-Methyl-2-butene
6117-80-2	2-Butene-1,4-diol (Z)	32830-97-0	1-Chloro-3-pentanone
6117-91-5	2-Buten-1-ol (Z+E)	33966-50-6	s-Butylamine
6153-56-6	Oxalic acid	34713-94-5	2-Methyl-1-butanol
7087-68-5	N-Ethyldiisopropylamine	42492-95-5	Lactonitrile
7572-29-4	Dichloroacetylene	64344-45-2	2-Ethyl-2-hexen-1-al
7623-09-8	2-Chloropropionyl acid chloride	64599-56-0	1-Phenyl-2-propyn-1-ol
7785-70-8	α-Pinene	73246-45-4	Methyl 2-chloropropionate
7789-89-1	1,1,1-Trichloropropane	77287-29-7	Methyl 2-chloropropionate
8006-64-2	Essence of turpentine	78957-07-0	Crotonic anhydride
9004-70-0	Cellulose nitrate	98103-87-8	2-Phenyl-1-propanol
10025-78-2	Trichlorosilane		

Publications and Internet Sites Bibliography

Books and CD-ROMs

American Society for Testing and Materials, ASTM 215569 Astm Std,17, Nov 1970

Bretherick's Reactive Chemical Hazards, database version 1, 4th edition CD-ROM, Butterworth-Heinemann, 1994

Catalogue de Produits Chimiques, Fuka 1995-1996

CCPs Guideline Series

Comment Utilisier en Toute Sécurité les Produits Chimiques Dangereux, (by subscription), Weka, 1995

Constantes, Techniques de L'ingénieur, updated in 1989

Les Explosifs Occasionnels,Vol 1, L Médard, 1987

Fiches de Sécurité Produits Dangereux, ed Weka, updated 1995

Fire and Materials, PF Thorne,1976

Fire Protection Handbook, NFPA, 13th edition, 1969

Guide de la Chimie International, Chimedit, 1994-1995

Guide for Safety in the Chemical Laboratory, Manufacturing Chemists Association, 1971

Guidelines for Chemical Process Quantitative Risk Analysis, American Institute for Chemical Engineers, 1999

Handbook of Chemistry and Physics, CRC Press, DR Lide, 79th edition,1998

Handbook of Laboratory Safety, RB Haas and RF Newton, CRC Press, 2nd edition, 1971

Health and Safety Manual, Lawrence Livermore National Laboratory, 1991-1996

Lange's Handbook of Chemistry, JA Dean, J Dean, McGraw Hill, 13th edition, 1985

Loss Prevention in the Process Industries, FP Lees, Butterworth-Heinemmann, 1992

Material Safety Data Sheets (MSDS) CD-ROM, Sigma-Aldrich-Fluka, 1995 (subscription)

Major Chemical Hazards, VC Marshall, Ellis Horwood, 1987

Les Melanges Explosifs, Brochure No 335, IRNS, 1980

The Merck Index, An Encyclopedia of Chemicals Drugs and Biologicals (annual) 11th edition, Merck and Co Inc, 1989

MSDS, Canadian Centre for Work Hygiene and Safety, A1 CD-ROM version, 1995 (subscription)

PUCK, Produits Chemiques, Ugine-Kuhlmann

Risk Assessment in the Process Industries, (ed) R Turney, R Pitblado, Gulf Publishing Company, 1996

SAX's Dangerous Properties of Dangerous Materials, CD-ROM version of 8th edition, Van Nostrand Rheinhold; hardcover, Chapman and Hall

Sécuridisque, CD-ROM version of Code du Travail, Cabinet Beugnette Publishing, 19 rue Poincare, 88210 Senones, 1995 (by subscription)

Traité Pratique de Sécurité, Produits Dangereuse pour L'homme et L'environnment, Centre National de Prevention et de Protection, 5th edition.

The American Institute of Chemical Engineers (AIChE) list, in their publications catalogue (Next Century Tools for Today's Process Industries), a number of titles relevant to chemical risk and chemical risk analysis.

Internet Sites

Bmartel@imaginet.fr
http://hazard.com/msds/
http://www.chemsafe.com
http://www.ps.uga.edu/rtk/msds.htm
http://odin.chemistry.uakron.edu/erd/
gopher://ecosys.drdr.Virginia.EDU:70/11/library/gen/toxics
http://www.wco.com/jray/pyro/safety/msds
gopher://gaia.ucs.orst.edu:70/11/osu-i+s/osu-d+o/ehs/msds/Product
http://www.chemexper.be
http://ace.orst.edu/info/extonet/ghindex.html
http://www.orcbs.msu.edu/chemical/nfpa/flammability.
http://www.orcbs.msu.edu/chemical/nfpa/nfpa.html.
http://www.nist.gov/srd/o_sspec16.htm;e-mail: srdp@enh.nist.gov.
http://www.orcbs.msu.edu/chemical/nfpa/reactivity.

US Sites

http://www.epa.gov/ceppo (EPA Chemical and Emergency Preparedness and Prevention Office)
http://www.osha.gov (Occupational Safety and health Aministration)
http://www.osha-slc.gov/SLTC (OSHA, Technical Center)
http://process-safety.tamu.edu
http://www.chemsafety.org<http://www.chemsafety.org

EU European Agency for Health and Saftey at Work
http://osha.eu.int

CHEMICALS INDEX

– A –

Acetaldehyde 39 ; 89 ; 121 ; 123 ; 128 ; 307 ; 308 ; 310 ;311 ;312 ; 314 ; 323 426

Acetamide 452

Acetanilide 453

Acetic acid 39 ; 40 ; 128 ;147 ; 236 ; 245 ; 263 ; 298 ; 310 ; 316 ; 317 ; 319 ; 320 ; 322 : 338 ; 434

Acetic anhydride 39 ; 41 ; 42 ; 66 ; 121 ; 123 ; 128 ; 159 ; 245 ; 295 ; 327 ; 329 ; 330 ; 332 ; 333 ; 446

Acetone 39 ; 40 ; 72 ; 151 ; 295 ; 308 ; 309 ; 310 ; 312 ; 313 ; 314 ; 429

Acetonecyanohydrin 129 ; 337 ; 340 ; 451

Acetonitrile 97 ; 128 ; 135 ; 139 ; 140 ; 335 ; 337 ; 340 ; 350 ; 449

Acetophenone 39 ; 432

Acetyl chloride 329 ; 333 ; 346 ; 447

Acetyl nitrate 330

Acetyl peroxide 98

Acetylacetone 430

Acetylene 50 ; 97 ; 121 ; 96 ; 123 ; 149 ; 150 ; 155 ; 197 ; 236 ; 238 ; 240 ; 241 ; 247 ; 246 ; 286 ; 308 ; 387

Acrolein 128 ; 135 ; 140 ; 310 ; 311 ; 314 ; 427

Acrylamide 453

Acrylic acid 40 ; 65 ; 133 ; 319 ; 320 435

Acrylonitrile 97 ; 129 ; 336 ; 337 ; 340 ; 450

Acryloyl chloride 322 ; 333 ; 447

Adipic acid 436

Adiponitrile 450

Alkyl nitrates 306

Alkyl nitrites 306

Allene 97 ; 149 ; 236 ; 246 ; 387

Allyl acetate 324 ; 440

Allyll acetoacetate 446

Allyl alcohol 98 ; 111 ; 114 ; 115 ; 115 ; 117 ; 118 ; 255 ; 256 ; 257 ; 285 ; 395

Allyl bromide 281

Allyl chloroformate 448

Allyl cyanide 450

Allyl iodide 409

Allyl isothiocyanate 349 ; 351 ; 455

Allyl methacrylate 441

Allyl oxide 255 ; 264

Allylamine 413

Allylbenzene 390

2-Allylphenol 398

Aluminium 148 ; 159 ; 176 ; 177 ; 178 ; 181 ; 182 ; 183 ; 186 ; 187 ; 189 ; 191 ; 193 ; 197 ; 199 ; 201 ; 203 ; 204 ; 206 ; 207 ; 209 ; 210 ; 211 ; 213 ; 220 ; 221 ; 223 ; 224 ; 225 ; 229 ; 230 ; 232 ; 233 ; 242 ; 248 ; 249 ; 267 ; 269 ; 276 ; 278 ; 307 ; 315 ; 348 ;364

Aluminium and magnesium silicate 365

Aluminium and ammonium sulphate 365

Aluminium and potassium sulphate 365

Aluminium and sodium sulphate 365

Aluminium borate 364

Aluminium borohydride 176 ; 364

Index A

Aluminium bromide 285
Aluminium chlorate 364
Aluminium chloride 177; 194; 267; 281; 283; 285; 289; 299; 300; 334
Aluminium fluoride 189; 194; 224; 364
Aluminium fluorosilicate 364
Aluminium nitrate 364
Aluminium nitride 365
Aluminium oxide 176; 177; 242; 264; 307
Aluminium oxychloride 365
Aluminium perchlorate 325
Aluminium phosphate 365
Aluminium phosphide 177; 365
Aluminium silicate 365
Aluminium sulphate 365
Aluminium triacetylide 285
Aluminium trihydroxide 365
o-Aminobenzoic acid 287; 268
2-Amino-1-butanol 420
2-Aminophenol 420
3-Aminophenol 420
4-Aminophenol 420
1-Amino-2-propanol 420
3-Amino-1-propanol 420
3-Aminopyraldehyde 419
2-Aminopyridine 418
4-Aminopyridine 419
Ammonia 128; 165; 166; 168; 169; 170; 171; 172; 173; 175; 182; 186; 187; 191; 194; 196; 202; 204; 205; 207; 211; 213; 219; 221; 224; 227; 230; 267; 297; 298; 302; 348; 361
Ammonium bromide 216; 215; 377
Ammonium chloride 169; 173; 187; 189; 191; 208; 229; 328; 369
Ammonium chromate 372
Ammonium dichromate 112; 114; 115; 115; 117; 118; 201; 254; 372
Ammonium difluoride 362
Ammonium fluoride 362
Ammonium hexacyanoferrate (II) 205
Ammonium hexacyanoferrate 207
Ammonium hexafluorosilicate 362
Ammonium hydrogencarbonate 361
Ammonium iodide 227; 226; 382
Ammonium metavanadate 290
Ammonium molybdate 379
Ammonium nitrate 112; 114; 115; 115; 117; 169; 168; 170; 173; 175; 176; 178; 194; 200; 201; 203; 203; 205; 206; 207; 208; 222; 224; 231; 233; 341; 362
Ammonium perchlorate 191; 190; 203; 207; 369
Ammonium permanganate 203
Ammonium phosphate 366
Ammonium sulphamate 184; 185; 368
Ammonium tetrachloroaluminate 365
Ammonium thiocyanate 164; 368
s-Amyl acetate 439
t-Amyl alcohol 394
Amyl formate 438
Aniline 54; 285; 289; 290; 293; 297; 312; 414
2-Anilinoethanol 420
o-Anisaldehyde 421
3-Anisaldehyde 428
4-Anisaldehyde 428
Anisaldene 288
m-Anisidine 421
p-Anisidine 421
2-Anisidine 428
Anisole 262; 400

Anthracene 39; 41; 239; 392
Anthraquinone 39; 432
Antimony 148; 223; 225; 380
Antimony pentasulphide 224; 228; 381
Antimony pentoxide 381
Antinomy trichloride 224; 327; 381
Antinomy trifluoride 381
Antinomy trioxide 225; 381
Antinomy trisulphide 224; 228; 381
Argon 192; 369
Arsenic 148; 210; 211; 219; 224; 376
Arsenic acid 375
Arsenic hemiselenide 377
Arsenic pentafluoride 247
Arsenic pentasulphide 211; 212; 376
Arsenic pentoxide 376
Arsenic trichloride 212; 211; 377
Arsenic trifluoride 317
Arsenic trioxide 212; 377
Arsenic trisulphide 212; 211; 228; 377
Arsine 97; 212; 211; 225; 376
Asbestos 365
Aziraldehyde 286; 288; 289; 290; 293; 417
Azobis(isobutyronitrile) 336; 338

– B –

Barium 148; 228; 229; 278; 382
Barium bromide 382
Barium carbonate 382
Barium chlorate 229; 382
Barium chloride 382
Barium chlorite 348
Barium chromate 382
Barium cyanide 382
Barium fluoride 382

Barium fluorosilicate 382
Barium hydroxide 343; 383
Barium nitrate 164; 229; 383
Barium oxide 228; 383
Barium perchlorate 251
Barium peroxide 229; 245; 309
Barium sulphate 229; 383
Barium sulphide 229; 232; 383
Basic copper carbonate 375
Bismuth (III) oxide 232; 384
Bismuth chlorate 232
Bicyclo[3,1,0]hexane 103
Bicyclo[4,1,0]heptane 103
Benzaldehyde 39; 309; 310; 314
Benzamide 453
Benzanilide 453
Benzene 40; 90; 97; 128; 135; 140; 239; 242; 243; 219; 244; 245; 247; 246; 259; 275; 389
Benzenediazonium chloride 288
Benzenesulphonyl chloride 324
Benzofuran 402
Benzoic anhydride 447
Benzoic acid 39; 315; 320; 436
Benzonitrile 39; 450
Benzophenone 39
Benzoquinone 314; 431
Benzoyl chloride 329; 333; 324
Benzoyl peroxide 100; 101; 237; 290; 323; 334
Benzyl acetate 282 443
N-Benzyl-4-aminophenol 263
Benzyl benzoate 443
Benzyl bromide 282; 284; 411
Benzyl chloride 149; 410
Benzyl chloroformate 448
Benzyl cyanide 40; 338; 340; 450
Benzyl oxide 264; 400
Benzylamine 291; 293 414
Benzyl alcohol 39; 255; 257
Beryllium carbide 359

Index B

Beryllium 129; 148; 164; 165; 171; 175; 178; 187; 278; 359
Beryllium chloride 359
Beryllium fluoride 175; 360
Beryllium hydroxide 360
Beryllium nitrate 360
Beryllium oxide 171; 175; 360
Bicyclo[1,1,0]butane 103
Bicyclo[2,1,0]pentane 103
Bicyclo[5,1,0]octane 103
Bis(2-chloroethyl)ether 412
Bismuth 233; 252; 384
Bismuth nitrate 384
Bismuth oxychloride 384
Bismuth perchlorate 233
Bismuth tribromide 233
Bismuth trichloride 233; 384
Bismuth triiodide 233
Boric acid 330; 360
Boron 164; 167; 169; 170; 171; 174; 181; 189; 197; 201; 207; 213; 220; 225; 360
Boron carbide 164; 360
Boron nitride 171; 174; 360
Boron oxide 171; 241; 360
Boron trichloride 167; 170; 171; 178; 285; 360
Boron trifluoride 164; 170; 194; 196; 217; 228; 268; 281; 299 300; 338
Bromine 129; 149; 150; 216; 215; 217; 224; 230; 239; 249; 264; 285; 312; 336; 339; 377
Bromoacetic acid 415
Bromobenzene 276; 284; 411
1-Bromobutane 276; 406
Bromobutane 284
Bromochlorodifluoromethane 279
1-Bromo-3-chloroethane 407
1-Bromo-3-chloro-2- ethylpropane 407
1-Bromo-3-chloromethane 407
Bromochloromethane 407
B romoethane 306
1-Bromohexane 406
Bromomethane 277; 279; 284; 347; 405
1-Bromo-3-methylbutane 406
1-Bromo-2-methylpropane 406
2-Bromo-2-methylpropane 406
1-Bromopentane 406
1-Bromopropane 405
3-Bromo-1-propanol 412
1-Bromo-2-propene 129; 409
3-Bromopropionic acid 437
1-Bromo-2-propyne 284; 409
N-Bromosuccinimide 291; 337
3-Bromotoluene 411
4-Bromotoluene 411
Bromotrichlororomethane 407
Bromotrifluoroethylene 280
Butadiene 262
1,3-Butadiene 74; 121; 123; 152; 236; 237; 238; 242; 246; 250; 262; 311; 387
Butane 62; 236; 295;
Butane-1,4-diol 396
Butane-1,3-diol 396
Butane-2,3-diol 397
Butane-2,3-dione 429
Butanethiol 323; 328
1-Butanethiol 454
2-Butanol 46; 153; 253; 257; 262; 393
1-Butanol 46; 59; 67; 69; 128
Butanols 255
Butanone 39; 65; 121; 123; 253; 308; 309; 312; 314; 429
Butenes 324
1-Butene 387
2-Butene-1,4-diol 397
2-Butene-1-ol 395
2-Butenes 387

Index B-C

2-Butoxyethanol 64; 251; 262; 400
2-Butoxyethyl acetate 445
s-Butyl acetate 439
Butyl acetate 50; 53; 439
Butyl acetoacetate 446
Butyl acrylate 441
s-Butylamine 413
Butylamine 413
s-Butylbenzene 391
n-Butylbenzene 391
Butyl benzoate 443
Butyl butyrate 440
Butyl formate 438
Butylglycidylether 402
Butyllithium 238
Butyl methacrylate 324; 441
Butyl methyl ether 262; 399
Butyl nitrite 422
s-Butyl nitrite 423
Butyl peroxide 98
Butyl propionate 439
Butyl vinyl ether 262; 399
2-Butyn-1,4-diol 397
3-Butyn-2-ol 395
1-Butyne 387
2-Butyne 387
Butyne-1,4-diol 255; 257
Butyraldehyde 311; 427
Butyramide 452
Butyric acid 39; 317; 434
Butyric anhydride 329; 446
γ-Butyrolactone 325; 326; 444
Butyronitrile 97; 449
Butyryl chloride 329; 447

– C –

Cadmium 222; 225; 380
Cadmium bromide 222; 380
Cadmium carbonate 197; 380
Cadmium chloride 222; 380
Cadmium cyanide 222; 28
Cadmium fluoride 380
Cadmium hydroxide 380
Cadmium iodide 222; 380
Cadmium nitrate 380
Cadmium oxide 222; 380
Cadmium sulphate 380
Cadmium sulphide 380
Cadmium thiocyanate 99
Cadmium thiocyanate 371
Calcium amide 213
Calcium arsenate 376
Calcium 148; 195; 195; 199; 174; 229; 230; 232; 370
Calcium borate 370
Calcium bromide 197; 370
Calcium carbide 197; 205; 207; 213; 221; 223; 227; 370
Calcium carbonate 175; 197; 370
Calcium chlorate 196; 370
Calcium chloride 197; 209; 370
Calcium chromate 196; 370
Calcium cyanamide 197; 370
Calcium cyanide 370
Calcium dis(hydrogenphosphate) 370
Calcium disilicide 204; 229; 251
Calcium fluoride 370
Calcium hexafluorosilicate 370
Calcium hydrogenphosphate 370
Calcium hydroxide 273; 275; 289; 370
Calcium hypochlorite 159; 189; 196; 204; 240; 250; 259; 280; 303; 339; 370
Calcium hypophosphite 370

Index C

Calcium nitrate 331; 341; 370
Calcium oxide 195; 197; 349
Calcium permanganate 196; 371
Calcium phosphate 371
Calcium phosphide 197; 371
Calcium pyrophosphate 371
Calcium silicide 196
Calcium sulphate 187; 196; 371
Calcium sulphide 197; 200; 232; 371
Camphor 39; 432
ε-Caprolactam 343; 344; 453
ε-Caprolactone 444
Capronitrile 449
Carbazole 419
Carbon 154; 165; 167; 170; 171; 172; 173; 173; 174; 178; 181; 183; 186; 188; 191; 193; 195; 197; 202; 202; 204; 206; 209; 215; 219; 220; 221; 231; 232; 232; 239; 242; 264; 295; 317
Carbon dioxide 152; 165; 166; 173; 175; 175; 176; 186; 193; 196; 200; 202; 223; 229; 239; 291; 360
Carbon disulphide 97; 128; 185; 187; 191; 194; 209; 367
Carbon monoxide 39; 165; 167; 170; 171; 171; 172146; 176; 193; 204; 221; 295 ;317 361
Carbon suboxide 361
Cellulose nitrate 294; 297; 423
Cerium 224; 230; 223
Chloral 429
Chloral hydrate 313; 314; 429
Chloric acid 190; 211; 224; 232
Chlorine 150; 152; 181; 186; 187; 188; 190; 194; 195; 197; 201; 203; 209; 210; 211; 215; 217; 218; 224; 224; 225; 230; 231; 239; 249; 264; 303; 325; 339; 350
Chlorine dioxide 147; 188; 189; 191; 190; 195; 197; 229; 230; 369
Chloroacetaldehyde 429
Chloroacetamide 453
Chloroacetic acid 129; 437
Chloroacetic anhydride 447
Chloroacetone 312; 314; 432
Chloroacetonitrile 451
α-Chloroacetophenone 433
p-Chloroacetophenone 313
4-Chloroacetophenone 314; 433
3-Chloroacetophenone 433
2-Chloroacetophenone 433
Chloroacetyl chloride 329; 448
Chloroacetylene 240; 274; 279; 282; 284
2-Chloroacrylonitrile 451
4-Chloroaniline 288; 421
3-Chloroaniline 421
2-Chloroaniline 288; 421
Chloroanilines 39
4-Chloroanisole 413
4-Chlorobenzaldehyde 429
3-Chlorobenzaldehyde 429
2-Chlorobenzaldehyde 429
o-Chlorobenzalmalonitrile 452
Chlorobenzene 284; 409
2-Chlorobenzonitrile 451
o-Chlorobenzoyl chloride 448
2-Chloro-1,3-butadiene 279; 284; 409
Chlorobutane 284
2-Chlorobutane 405
1-Chlorobutane 276; 405
3-Chloro-2-butanone 432
1-Chloro-3-butene 408
1-Chloro-2-butene 408
4-Chlorobutyronitrile 338; 340; 451
4-Chlorobutyryl chloride 426
Chlorocresols 284
p-Chloro-m-cresol 412

500

Index — C

p-Chloro-o-cresol 283
2-Chlorocyclohexanone 432
2-Chlorocyclopentanone 432
1-Chloro-1,1-difluoroethane 50; 407
1-Chloro-2,4-dinitrobenzene 294; 306; 426
1-Chloro-2,4-dinitrophenol 302
Chloroethane 276; 284; 403
2-Chloroethanol 411
Chloroethylene 149; 241; 284; 408
Chloromethane 128; 276; 284; 403
2-Chloro-2-methylbutane 405
1-Chloro-3-methylbutane 405
Chloromethyl ether 412
1-Chloro-2-methylpropane 405
2-Chloro-2-methylpropane 405
1-Chloro-2-methyl-2-propene 409
3-Chloro-2-methyl-1-propene 408
2-Chloronaphthalene 411
1-Chloronaphthalene 411
1-Chloro-2-nitrobenzene 296; 302; 425
1-Chloro-3-nitrobenzene 425
1-Chloro-4-nitrobenzene 275; 279; 425
Chloronitrobenzenes 294; 306
1-Chloro-1-nitropropane 294; 425
Chloropentafluoroethane 407
1-Chloropentane 405
1-Chloro-3-pentanone 432
5-Chloro-2-pentanone 432
Chorophenols 259
2-Chlorophenol 412
3-Chlorophenol 412
4-Chlorophenol 412
Chloropicrin 129; 294; 296; 298; 306; 425
1-Chloropropane 404
2-Chloropropane 404
3-Chloro-1-propanol 411
3-Chloro-1-propene 326; 408
1-Chloro-1-propene 281
2-Chloropropionic acid 437
3-Chloropropionic acid 437
3-Chloropropionitrile 451
3-Chloro-1-propyne 408
4-Chlorostyrene 411
N-Chlorosuccinimide 291
Chlorosulphonic acid 182; 185; 221; 270; 302
2-Chlorotoluene 410
3-Chlorotoluene 410
4-Chlorotoluene 410
Chlorotoluenes 128; 240
Chlorotrifluoroethylene 279; 280; 281; 284; 409
Chromium (II) sulphate 200
Chromium (III) sulphate 372
Chromium (III) oxide 200; 201; 208; 210; 328; 331; 342; 372
Chromium 148; 174; 372
Chromium and potassium sulphate 372
Chromium dichloride 174
Chromium oxide (VI) 147; 205; 211; 254; 255; 316; 372
Chromium tribromide ???
Chromium trichloride 201; 372
Chromium trifluoride 201
Chromium trinitrate 372
Chromyl dichloride 200; 201; 245; 254; 270; 329; 372
Cinnamaldehyde 428
Cinnamic acid 436
Cobalt 205; 283; 374
Cobalt acetate 308
Cobalt bromide 374
Cobalt carbonate 374
Cobalt dichloride 206; 374
Cobalt dihydroxide 374

Cobalt nitrate 206 ; 374
Cobalt oxide 205 ; 295 ; 374
Cobalt sulphate 374
Copper 206 ; 232 ; 237 ; 247 ; 282 ; 348 ; 374
Copper acetylide 207
Copper chromite 208 ;
Copper nitride 48
Copper sulphate 208 ; 343 ; 375
Cresols 39
m-Cresol 397
o-Cresol 397
p-Cresol 397
Crotonic acid 435
Crotonic anhydride 446
Crotonidine 308 ; 311 ; 427
Crotonitrile 450
Crotonyl chloride 447
Cumene 390
Cupric bromide 375
Cupric chloride 208 ; 375
Cupric cyanide 208
Cupric nitrate 207 ; 223 ; 331 ; 375
Cupric oxide 159 ; 207 ; 218 ; 290 ; 295 ; 350 ; 375
Cupric perchlorate 375
Cupric sulphide 208 ; 228 ; 375
Cuprous chloride 208 ; 264 ; 375
Cuprous cyanide 375
Cuprous oxide 208 ; 375
Cyanoacetic acid 338 ; 340 452
Cyanogen 39 ; 97 ; 129 ; 230 ; 449
Cyanogen chloride 39
Cyanuryl chloride 338 ; 341 ; 324
Cyclobutane 103
Cyclobutene 103
Cyclodedocane 387
Cyclohepta-1,3,5,7-tetraene 103
Cyclohepta-1,3,5-triene 103
Cyclohepta-1,3-diene 103
Cycloheptane 103 ; 386

Cycloheptanol 395
Cycloheptanone 431
Cycloheptene 103 ; 389
Cyclohexa-1,3-diene 103
Cyclohexa-1,4-diene 103
Cyclohexane 40 ; 62 ; 89 ; 103 ; 121 ; 123 ; 128 ; 135 ; 140 ; 219 ; 247 ; 246 ; 386
Cyclohexanecarbonitrile 449
Cyclohexanecarbonyl chloride 426
Cyclohexanecarboxaldehyde 428
Cyclohexanol 55 ; 59 ; 62 ; 67 ; 69 ; 252 ; 309
Cyclohexanone 59 ; 67 ; 309 ; 431
Cyclohexene 103 ; 262 ; 388
2-Cyclohexen-1-one 431
Cyclohexyl acetate 439
Cyclohexylaniline 290 ; 293 ; 414
Cyclohexylbenzene 392
Cyclononane 103
E-Cyclononene 103
Cyclooctane 386
Cyclopentadiene 97 ; 103 ; 238 ; 243 ; 246 ; 388
Cyclopentane 62 ; 103 ; 385
Cyclopentanol 62 ; 394
Cyclopentanone 309 ; 430
Cyclopentene 97 ; 103 ; 388
2-Cyclopenten-1-one 430
Cyclopentylaniline 413
Cyclopropane 62 ; 103 ; 385
Cyclopropanecarbonitrile 449
Cyclopropene 103
p-Cymene 59 ; 67 ; 390

– D –

Decaborane 251
Decahydronaphthalene 262 ; 386
Decanal 428
Decane 386
Decanoic acid 435

Diacetate ethyleneglycol 39; 441
Diacetonealcohol 432
Diallyl succinate 442
Diallylaniline 415
1,4-Diaminobutane 416
1,6-Diaminohexane 416
1,3-Diaminopropane 416
Diammonium carbonate 361
Diammonium hydrogensulphide 367
Diammonium peroxodisulphate 185; 186; 367
Diammonium phosphate 366
Diammonium phosphite 366
Diammonium sulphate 191; 368
Diammonium sulphide 209; 368
Diammonium sulphite 368
Diammonium thiosulphate 185; 186; 368
1,2-Diaminopropane 416
Diamyl ether 87; 400   437
5
Dibenzylaniline 415
Dibenzylketone 39
Diborane 97; 164; 167; 170; 187; 251; 324; 360
1,4-Dibromobutane 406
Dibromodifluoromethane 407
1,2-Dibromoethane 405
1,2-Dibromoethylene 409
Dibromomethane 405
1,2-Dibromopropane 406
1,3-Dibromopropane 406
1,2-Dibromotetrafluoroethane 407
Dibutylamine 415
Di-s-butylaniline 41
Dibutyl ether 262; 263; 399
Dibutyl hydrogenphosphite 326; 328; 456
Dibutyl oxalate 442
Dibutyl phthalate 325; 326

Dichloroacetic acid 437
1,1-Dichloroacetone 432
Dichloroacetonitrile 451
Dichloroacetyl chloride 329; 448
Dichloroacetylene 240; 274; 282; 284
Dichloroaniline 421
2,4-Dichloroaniline 421
2,5-Dichloroaniline 421
2,6-Dichloroaniline 422
3,4-Dichloroaniline 422
3,5-Dichloroaniline 422
3,3'-Dichlorobenzaldehyde 422
1,2-Dichlorobenzene 112; 114; 115; 115; 117; 118; 410
1,3-Dichlorobenzene 410
1,4-Dichlorobenzene 410
1,2-Dichlorobutane 80
1,4-Dichlorobutane 405
1,3-Dichloro-2-butene 409
1,4-Dichloro-2-butene 409
1,2-Dichloro-1,1-difluoroethane 404
Dichlorodifluoromethane 407
1,2-Dichloroethane 341; 404
1,1-Dichloroethylene 280; 281; 284
1,2-Dichloroethylene 274
1,2-Dichloroethylenes 408
Dichloromaleic anhydride 318
Dichloromethane 128; 135; 141; 268; 274; 279; 280; 284; 285; 297; 341; 403
1,1-Dichloro-1-nitroethane 425
1,1-Dichloro-1-nitropropane 294
Dichlorophenols 259
2,4-Dichlorophenol 412
2,4-Dichlorophenoxyacetic acid
Dichloropropane 284
1,1-Dichloropropane 404
1,2-Dichloropropane 404
1,3-Dichloropropane 404
2,2-Dichloropropane 251; 404

Index D

1,3-Dichloropropene 408
2,3-Dichloropropene 408
Dichloropropionic acid 437 478
1',2'-Dichlorostyrene 281
1,2-Dichlorotetrafluoroethane 407
2,4-Dichlorotoluene 68 ; 410
2,6-Dichlorotoluene 410
Dichlorous oxide 194 ; 196 ; 197 ; 210 ; 223 ; 224 ; 229 ; 231
Dicobalt trioxide 205
Dicyclohexylaniline 415
Dicyclopentadiene 262 ; 389
Diethanolaniline 420
Diether styrene 51
Diethoxyethane 313 ; 314 ; 433
1,1-Diethoxyethane 313 ; 314 ; 411
1,2-Diethoxyethane 400
Diethoxymethane 433
1,1-Diethoxypropane 433
2,2-Diethoxypropane 433
Diethyl adipate 442
Diethylamine 287 ; 293
2-Diethylaminoethanol 420
N,N-Diethylaniline 417
Diethylaniline 128 ; 415
1,2-Diethylbenzene 391
1,3-Diethylbenzene 391
1,4-Diethylbenzene 391
Diethyl carbonate 325 ; 442
Diethyleneglycol 262 ; 400
Diethyleneglycolmonomethylether 400
Diethylenetriamine 263 ; 297 ; 416
Diethyl ether 87 ; 89 ; 262 ; 263 ; 264 ; 267 ; 268 ; 269 ; 270 ; 271 ; 272 ; 282 ; 295 ; 399
Diethyl fumarate 442
Diethyl glutarate 442
Diethyl hydrogenphosphite 349 ; 351
Diethyl maleate 442
Diethyl malonate 59 ; 442
Diethyl oxalate 442
Diethyl phthalate 443
Diethyl succinate 322 ; 326 ; 442
Diethyl sulphate 348 ; 351 ; 455
Diethyl sulphide 454
Diethyl zinc 249 ; 275
Digallium trioxide 376
Diglycidyl ether 402
Dihexylaniline 415
Dihexyl ether 400
Dihydrofuran 262
2,5-Dihydrofuran 402
Dihydropyran 262 ; 402
Diiodoacetylene 227
Diiodomethane 275
Diisoamyl ether 41 ; 400
Diisobutyl vinyl ether 399
Diisobutylketone 432
Diisocyanate 2,4-toluene 344 ; 345
Diisopropylaniline 415
Diisopropyl ether 40 ; 121 ; 123 ; 261 ; 262 ; 263 ; 271 ; 399
Diisopropylketone 431
Diketene 322 ; 326
Dimanganese heptoxide 222
Dimesityl ether 430
3,3'-Dimethoxybenzaldehyde 421
1,2-Dimethoxybenzene 400
1,3-Dimethoxybenzene 400
1,4-Dimethoxybenzene 400
1,1-Dimethoxyethane 433
1,2-Dimethoxyethane 41 ; 43 ; 262 ; 264 ; 271 ; 272 ; 400
1,1-Dimethoxypropane 433
N,N-Dimethyl-p-toluadehyde 266 ; 417
2,2-Dimethoxypropane 313 ; 314 433
Dimethylacetamide 139 ; 341 ; 342 ;

Index D

344 ; 452
Dimethyl adipate 442
N,N-Dimethylallylamine 416
Dimethylamine 39 ; 128 ; 313 ; 414
N,N-Dimethylthylamine 416
2-Dimethylaminoethanol 420
1-Dimethylamino-2-propanol 420
3-Dimethylamino-1-propanol 420
N,N-Dimethylaniline 328 ; 417
2,4-Dimethylbenzaldehyde 428
3,5-Dimethylbenzoic acid 245
2,3-Dimethyl-1,3-butadiene 219
2,2-Dimethylbutane 386
2,3-Dimethylbutane-2,3-diol 397
3,3-Dimethyl-2-butanol 394
3,3-Dimethyl-2-butanone 430
2,3-Dimethyl-1-butene 388
2,3-Dimethyl-2-butene 388
3,3-Dimethyl-1-butene 388
2,2-Dimethylbutyric acid 435
3,3-Dimethylbutyric acid 435
Dimethyl carbonate 325 ; 326 ; 442
Dimethylchlorosilane 43
Dimethyldichlorosilane 457
Dimethyl ether 39 ; 295 ; 399
Dimethylformamide(DMF) 339 ; 341
 342 ; 343 ; 344 ; 367 ; 452
Dimethyl fumarate 442
Dimethyl glutarate 442
1,1-Dimethylhydrazine 97 ; 419
Dimethyl maleate 442
Dimethyl malonate 442
Dimethyl oxalate 39 ; 441
1,2-Dimethylhydrazine 97 ; 419
2,3-Dimethylpentane 51
2,4-Dimethyl-3-pentanol 394
Dimethyl phthalate 443
2,2-Dimethyl-1-propanol 394
2,2-Dimethylpropane 385
2,2-Dimethylpropane-1,3-diol 397

2,2-Dimethylpropionic acid 434 ; 478
Dimethyl succinate 442
Dimethyl sulphate 129 ; 348 ; 351 ; 455
Dimethyl sulphide 40 ; 129 ; 346 ; 351 ; 454
Dimethylsulphone 454
Dimethylsulphoxide 346 ; 347 ; 348 ; 351 ; 454
Dimethyl terephthalate ??
Dimethyltetrahydrofuran 402
Dimethylzinc 251
Dinitroanilines 294 ; 303
2,4-Dinitroaniline 426
Dinitrobenzenes 294 ; 304 ; 306
m-Dinitrobenzene 94 ; 95 ; 98 ; 129
1,2-Dinitrobenzene 424
1,3-Dinitrobenzene 294 ; 295 ; 424
1,4-Dinitrobenzene 424
4,6-Dinitro-m-cresol 425
4,6-Dinitro-o-cresol 294
Dinitrogen pentoxide 182 ; 194 ; 244 ; 268 ; 290 ; 312
Dinitrogen tetroxide 148 ; 149 ; 166 ; 167 ; 168 ; 175 ; 176 ; 178 ; 194 ; 196 ; 173 ; 201 ; 203 ; 227 ; 229 ; 219 ; 244 ; 280 ; 290 ; 303 ; 336 ; 346
Dinitrogen trioxide 343
Dinitrophenols 306
2,4-Dinitrophenol 294 ; 296 ; 298 ; 425
Dinitrotoluenes 294 ; 296 ; 306
2,4-Dinitrotoluene 298 424
2,6-Dinitrotoluene 424
Dioxan 111 ; 114 ; 115 ; 117 ; 118 ; 128 ; 133 ; 135 ; 141 ; 262 ; 264 ; 265 ; 268 ; 271 ; 346 ; 401
Dioxolane 313
1,3-Dioxolane 433
Diphenyl 55 ; 392

505

Diphenylacetylene 390
2-Diphenylamidene 414
Diphenylamine 291 ; 415
Diphenyl carbonate 325 ; 442
Diphenyl ether 270 ; 271 ; 400
1,1-Diphenylethylene 242 ; 392
1,2-Diphenylethylene 392
Diphenylmethane 89 ; 392
Diphenyl sulphide 454
Diphenylsulphoxide 350
Diphosphorus pentasulphide 179 ; 180
Diphosphorus pentoxide 179 ; 180 ; 194 ; 196 ; 227 ; 229 ; 317 ; 366
Diphosphorus tetrahydride 179
Diphosphorus trioxide 343 ; 348
Dipotassium hydrogenphosphate 366
Dipropyl ether 262 ; 399
Dipropylamine 415
Disodium hydrogenphosphate 366
Disulphur dichloride 181 ; 182 ; 185 ; 194 ; 200 ; 367
Divanadium pentoxide 290 ; 371
Divanadium trioxide 174 ; 372
Divinylacetylene 261
Divinylbenzene 391
Dodecane 387
Dodecanethiol 345
Dodecanoic acid 435
Double beryllium and potassium sulphate 360

– E –

Epichlorhydrin 265 ; 266 ; 268 ; 270 ; 271 ; 274 ; 260 ; 263 ; 412
1,2-Epoxybutane 402
Ethane 62 ; 150 ; 239 ; 240 ; 246 ; 385
Ethanethiol 454
Ethanol 39 ; 40 ; 50 ; 54 ; 69 ; 72 ; 149 ; 151 ; 153 ; 192 ; 249 ; 250 ; 250 ; 252 ; 254 ; 255 ; 257 ; 307 ; 329 ; 392
Ethanolamine 64 ; 291 ; 293 ; 419
Ethers 260
   halogenation 264
   peroxidation 261
Ethoxyacetylene 270 ; 271
2-Ethoxyethanol 400
2-Ethoxyethyl acetate 325 ; 226
3-Ethoxypropionitrile 451
Ethyl acetate 69 ; 121 ; 123 ; 128 321 ; 322 ; 325 ; 326 ; 438
Ethyl acetoacetate 326 ; 327 ; 446
Ethyl acrylate 128 ; 323 ; 325 ; 326 ; 419
Ethylaluminium dichloride 275
Ethylamine 39 ; 413
2-Ethylaminoethanol 420
2-Ethylaniline 414
3-Ethylaniline 414
4-Ethylaniline 414
N-Ethylaniline 290 ; 415
Ethylbenzene 89 ; 390
Ethyl benzoate 443
Ethyl bromoacetate 445
Ethyl 2-bromopropionate 445
Ethyl 3-bromopropionate 446
2-Ethylbutylaldehyde 427
Ethyl butyrate 440
2-Ethylbutyric acid 435
Ethyl chloroacetate 445
Ethyl 2-chloroacetoacetate 446
Ethyl 4-chloroacetoacetate 446
Ethyl chloroformate 448
2-Ethyl chloropropionate 445
3-Ethyl chloropropionate 445
Ethyl cinnamate 443
Ethyl crotonate 324 ; 441
Ethyl cyanoacetate 328 ; 452
Ethylcyclohexane 386

Index — F

N-Ethyldiisopropylamine 417
Ethyldithiocarbamate 263
Ethylene 97; 111; 113; 115; 115; 117; 118; 121; 123; 150; 152; 235; 237; 238; 240; 241; 247; 246; 248; 281; 299; 387
Ethylene carbonate 325; 443
Ethylenecyanohydrin 337; 340; 451
Ethylenediamine 286; 290; 293; 416
Ethylene glycol 39; 249; 250; 251; 254; 257; 325; 396
Ethylene glycol dinitrate 268; 294; 423
Ethylene oxide 39; 121; 123; 128; 149; 150; 152; 155; 262; 265; 265; 265; 246; 269; 271; 272; 279; 286; 303; 400
Ethyl fluoroacetate 129; 445
Ethyl formate 438
Ethyl heptanoate 440
2-Ethylhexanal 311
2-Ethylhexane-1,3-diol 51
Ethyl hexanoate 440
2-Ethylhexanoic acid 435
2-Ethyl-1-hexanol 55; 395
2-Ethyl-2-hexen-1-al 311; 428
Ethylidene norbornene 319
Ethyl isobutyrate 440
Ethyl isopentanoate 440
Ethylisopropylketone 430
Ethyl isothiocyanate 455
Ethyl lactate 444
Ethylmagnesium iodide 248
Ethylmagnesium bromide 248
Ethyl methacrylate 324; 441
Ethyl methyl ether 262
N-Ethylmorpholine 421
Ethyl nitrate 121; 123; 252; 268; 294; 299; 303; 423
Ethyl nitrite 294; 422
Ethyl pentanoate 440
Ethyl peracetate 322
Ethyl peroxide 267
2-Ethylphenol 398
3-Ethylphenol 398
4-Ethylphenol 398
Ethyl phenylacetate 443
2-Ethylpiperidine 418
N-Ethylpiperidine 418
Ethyl pivalate 440
Ethyl propionate 439
2-Ethylpyridine 418
3-Ethylpyridine 418
4-Ethylpyridine 418
2-Ethyltoluene 390
3-Ethyltoluene 390
4-Ethyltoluene 47; 390
Ethyl trifluoroacetate 322
Ethyl vinyl ether 262; 271; 399

– F –

Ferric chloride 187; 205; 334; 373
Ferric hexacyanoferrate (II) 272; 373
Ferric nitrate 348; 373
Ferric oxide 153; 169; 183; 203; 204; 290; 295; 307; 373
Ferric perchlorate 336; 313
Ferric sulphate 374
Ferrocene 295
Ferrous sulphate 210; 289; 373
Fluorene 392
Fluorine 129; 148; 171; 172; 174; 175; 177; 178; 179; 181; 183; 187; 188; 191; 194; 195; 202; 203; 206; 209; 211; 215; 218; 220; 223; 227; 227; 230; 232; 239; 290; 362
Fluoroacetic acid 437
Fluorobenzene 409
Fluoroboric acid 328; 362

Fluoroethane 403
2-Fluoroethanol 411
Fluoroethylene 281 ; 408
Fluoromethane 39
Fluorosilicic acid 362
2-Fluorotoluene 409
3-Fluorotoluene 409
4-Fluorotoluene 409
Fluorotrichloroethylene 281
Fluorotrichloromethane 407
Formamide 342 ; 344 ; 452
Formic acid 39 ; 128 ; 299 ; 309 ; 315 ; 316 ; 317 ; 318 ; 320 ; 342 ; 434
Formol 309 ; 310 ; 312 ; 314 ; 426
Fumaric acid 318 ; 436
Fumaronitrile 335 ; 450
Fumaryl chloride 448
Furan 262 ; 402
Furfural 128 ; 310 ; 311 ; 314 ; 428
Furfurylic alcohol 128 ; 262 ; 269 ; 270 ; 271 ; 338 ; 403
2-Furoic acid 319 ; 320 ; 437

– G –

Gaiacol 400
Gallium 210 ; 213 ; 376
Gallium arsenide 212 ; 376
Gallium nitrate 376
Gallium trichloride 376
Germanium 210 ; 213 ; 376
Germanium dioxide 376
Germanium tetrachloride 210 ; 376
Glutaraldehyde 427
Glutaric acid 436
Glutaric anhydride 447
Glutaronitrile 335 ; 450
Glycerol 149 ; 153 ; 249 ; 250 ; 251 ; 252 ; 257 ; 267 ; 329 ; 397
Glyceryl triacetate 80

Glycidol 265 ; 402
Glycidyl acrylate 80
Glycolic acid 436
Glycolonitrile 335 ; 340 ; 451
Glyoxal 307 ; 310 ; 314 ; 426
Graphite 218 ; 237

– H –

Hafnium tetrachloride 264
Helium 359
Heptanal 427
Heptane 207 ; 386
Heptanoic acid 39 ; 435
1-Heptanol 46 ; 63 ; 394
2-Heptanol 46 ; 63 ; 394
3-Heptanol 394
4-Heptanol 394
2-Heptanone 431
3-Heptanone 431
4-Heptanone 431
1-Heptene 389
2-Heptene 389
3-Heptene 389
Heptylamine 414
Hexabromoethane 275
Hexachlorobenzene 410
Hexachlorocyclohexane 341
Hexachlorocyclopentadiene 277
Hexachloroethane 341
Hexahexyl dilead 275
Hexamethylbenzene 245
Hexamethyldilazane 457
Hexamethyldisiloxane 456
Hexamethylenediamine 286 ; 293
Hexamethylenediisocyanate 344 ; 345 ; 453
Hexamethylenetetramine 419
Hexanal 311 ; 427
Hexane 62 ; 311 ; 385
Hexanecarbonitrile 449

Index H-I

Hexane-1,6-diol 397
Hexane-2,3-dione 430
Hexane-2,5-dione 430
Hexane-3,4-dione 430
Hexanethiol 345; 351
Hexanitrobenzene 272
Hexanoic acid 434
Hexanoic anhydride 447
1-Hexanol 46; 63; 394
2-Hexanol 46; 394
3-Hexanol 394
2-Hexanone 430
3-Hexanone 430
Hexanoyl chloride 306; 448
1-Hexene 39; 219; 388
2-Hexene 388
Hexogene 299; 306; 426
Hexyl acetate 439
s-Hexyl acetate 417
Hexylamine 414
Hydrazine 97; 166; 167; 168; 170; 171; 173; 186; 187; 19; 200 201; 203; 208; 219; 231; 291 ;297; 302; 331; 361
Hydrazine chlorohydrate 361
Hydrazines 293
Hydrazine sulphate 362
Hydriodic acid 382
Hydrobromic acid 376
Hydrochloric acid 328; 369
Hydrofluoric acid 202; 362
Hydrogen 164; 167; 169; 170; 171; 173; 175; 179; 186; 188; 195; 196; 203; 206; 207; 208; 209; 213; 215; 217; 218; 219; 228; 247; 251; 269; 300; 305 317; 359
Hydrogen bromide 213; 216; 215; 256; 264; 282; 337; 377
Hydrogen chloride 154; 159; 176; 182; 187; 188; 194; 197; 201; 202; 205; 210; 219; 240; 241; 274; 335; 337; 340; 369
Hydrogen cyanide 39; 97; 128; 129; 135; 140; 230; 335 ;337; 340; 449
Hydrogen fluoride 129; 147; 153; 172; 173; 179; 197173; 202; 210; 217; 218; 252; 317; 349; 362
Hydrogen iodide 97; 227; 226; 239; 382
Hydrogen nitride 97; 166; 167; 168; 361
Hydrogen peroxide 147; 151; 153; 164; 165; 172; 173; 174; 176; 178; 179; 184; 186; 193; 196; 200; 203; 204; 205; 205; 207; 210; 211; 220; 223; 231; 245; 253; 269; 270; 289; 308; 309 ;315; 316; 322; 331; 346; 359
Hydrogen sulphide 128; 159; 179; 183; 185; 200; 203; 206; 207; 209; 221; 229; 232; 368
Hydroquinone 149; 150; 259; 260; 310 ;323; 398
1-Hydroxy-2-butanone 432
3-Hydroxy-2-butanone 432
3-Hydroxybutyraldehyde 428
2-Hydroxyethyl perchlorate 267
Hydroxylamine 166; 168; 173; 179; 187; 191; 201; 203; 208; 208; 209; 228; 229; 232; 313 331; 362
Hydroxylamine chlorohydrate 362
2-Hydroxy-2-methylpropanoic acid 146
Hypochlorous acid 330

– I –

Indene 391
Indium 230; 336; 380
Indium antimonide 379
Indium oxide 380
Indium sulphate 380
Indium trichloride 380

Iodine 148; 213; 225; 226; 239; 250; 286; 343; 382
Iodobenzene 411
1-Iodobutane 406
2-Iodobutane 407
Iodoethane 406
Iodomethane 276; 280; 284; 406
Iridium 208; 222
Iron 148; 153; 154; 177; 179; 187; 203; 204; 240; 300; 304; 307; 310; 339; 348; 373
Iron (II) ammonium sulphate 373
Iron (II) chloride 205; 373
Iron oxalate 204
Iron (II) oxide 204; 373
Iron pentacarbonyl 204
Iron sulphide 204
Iron trichloride 154; 240; 267; 274; 290
Isoamyl acetate 55; 439
Isoamyl alcohol 393
Isoamyl formate 438
Isobutanol 393
Isobutene 243; 387
Isobutyl oxide 399
Isobutyl acetate 439
Isobutylamine 413
Isobutylbenzene 391
Isobutyl chloride 447
Isobutyl formate 438
Isobutyl methacrylate 324; 441
Isobutyl nitrite 423
Isobutyl propionate 440
Isobutyraldehyde 311; 427
Isobutyric acid 39; 434
Isobutyric anhydride 446
Isobutyronitrile 97; 112; 114; 115; 117; 118; 449
Isocaproic acid 39
Isodecanal 428
Isopentanoic anhydride 447

Isopentanoyl chloride 447
Isopentylamine 413
Isopentyl nitrate 294; 423
Isophorone 128; 432
Isoprene 97; 238; 242; 308; 388
Isopropenyl acetate 440
2-Isopropoxyethanol 400
Isopropyl acetate 128; 439
Isopropylamine 287; 293; 413
Isopropyl and vinyl ether 80
Isopropyl benzoate 443
Isopropyl butyrate 440
N-Isopropylcyclohexylamine 415
isopropyl formate 438
Isopropylglycidylether 262; 403
Isopropyl nitrate 252; 295; 423
p-Isopropyltoluene 59; 67
Isoquinoline 419
Isovaleronitrile 449

– K –

Ketene 310; 314; 429
Krypton 377

– L –

Lactic acid 317; 320; 436
Lactonitrile 337; 451
Lead 231; 304; 383
Lead azide 231
Lead carbonate 233; 384
Lead chloride 233 384
Lead chromate 232; 384
Lead dioxide 292; 350
Lead hexafluorosilicate 384
Lead iodide 384
Lead nitrate 232; 384
Lead nitride 97; 99; 232; 383
Lead oxide 231; 232; 240; 251; 350; 384
Lead peroxide 232; 384

Index L-M

Lead styphnate 270
Lead sulphate 233 ; 384
Lead sulphide 384
Lead tartrate 232
Lead tetroxide 384
Lead thiocyanate 97
Lewis acids 264
Limonene 389
Lithium 151; 152; 164; 165; 166; 169; 170; 171; 173; 173; 176; 177; 181; 182; 192; 195; 198; 199; 205; 207; 213; 217; 223; 225; 230; 231; 232; 247; 270; 277; 278; 359
Lithium acetylide 247
Lithium azide 166
Lithium bromide 359
Lithium carbonate 171 ; 218 ; 359
Lithium chloride 359
Lithium chromate 359
Lithium fluoride 171 359
Lithium fulminate 305
Lithium hydroxide
Lithium iodide 336
Lithium nitrate 237
Lithium nitride 208
Lithium oxide 171
Lithium perchlorate 191 ; 305
Lithium sulphate 359
Lithium tetraborate 359
Lithium tetrahydrogenaluminate 153 ; 272 ; 305 ; 320 ; 322

– M –

Magnesium 148; 174; 176; 177; 181; 187; 189; 194; 196; 199; 204; 206; 208; 209; 218; 221; 222; 222; 225; 227; 229; 230; 231; 232; 248; 277; 278; 304 ; 364
Magnesium bis(dihydrogenphosphate) 364
Magnesium carbonate 364
Magnesium chlorate 208
Magnesium chloride 203
Magnesium diborate 360
Magnesium dibromide 364
Magnesium dichloride 364
Magnesium difluoride 364
Magnesium dihydroxide 364
Magnesium dinitrate 199 ; 319 ; 364
Magnesium fluorosilicate 364
Magnesium hydroxycarbonate 296
Magnesium hypochlorite 196
Magnesium oxide 178 ; 196 ; 364
Magnesium perborate 360
Magnesium perchlorate 175 ; 178 ; 188 ; 250 ; 348 ; 364
Magnesium silicate 206
Magnesium sulphate 364
Magnesium sulphite 364
Maleic acid 436
Maleic anhydride 332 ; 333 ; 447
Malonic acid 318 ; 340 ; 436
Malononitrile 97 ; 335 ; 340 ; 450
Manganese 148 ; 174 ; 201 202 ; 373
Manganese dichloride 203 ; 209 ; 373
Manganese dioxide 154 ; 191 ; 202 ; 295 ; 316 ; 373
Manganese (II) oxide 203
Manganese (III) oxide 351
Manganese perchlorate 313
Manganese sulphate 373
2-Mercaptoethanol 455
Mercuric bromide 230
Mercuric chloride 230 ; 383
Mercuric cyanide 230 ; 383
Mercuric iodide 383
Mercuric nitrate 231 ; 383
Mercuric oxide 231 ; 239 ; 250 ; 296 ; 345 ; 383

Mercuric sulphate 258 ; 383
Mercuric sulphide 231 ; 383
Mercurous chloride 230 ; 383
Mercurous nitrate 231 ; 383
Mercurous perchlorate 310 ; 325
Mercurous sulphate 383
Mercury 230 ; 239 ; 282 ; 383
Mercury chlorate 310
Mercury oxide 239
Mercury fulminate 232 ; 253
Mercury nitrite 231
Mercury peroxide 231
Mercury thiocyanate 99 ; 230 ; 383
Mesitylene 68 ; 90 ; 149 ; 244 ; 246 ; 390
Metaboric acid 316
Methacrolein 311 ; 427
Methacrylamide 453
Methacrylic acid 319 ; 320 435
Methacrylic anhydride 446
Methacrylonitrile 450
Methacryloyl chloride 447
Methane 39 ; 150 ; 239 ; 242 ; 246 ; 385
Methanesulphonic acid 349 ; 351 ; 454
Methanethiol 453
Methanol 39 ; 128 ; 135 ; 141 ; 159 ; 192 ; 248 ; 250 ; 252 ; 253 ; 255 ; 257 ; 273 ; 277 ; 279 ; 295 ; 297 ; 322 ; 392
Metaphosphoric acid 365
Methoxyacetonitrile 451
3-Methoxybutyl acetate 445
2-Methoxychloroethane 412
2-Methoxyethanol 262 ; 269 ; 271
2-Methoxyethyl acetate 444
2-Methoxyethyl acrylate 423
2-Methoxyethylamine 288 ; 421
4-Methoxyphenol 400
2-Methoxypropyl acetate 445

Methyl acetate 72 ; 128 ; 438
Methyl acetoacetate 446
Methyl acrylate 128 ; 323 ; 324 ; 326 ; 441
Methylamine 128 ; 297 ; 413
N-Methylaniline 112 ; 114 ; 115 ; 115 ; 117 ; 415
2-Methylanisole 400
3-Methylanisole 400
4-Methylanisole 151 ; 158 ; 159 ; 271 ; 400
9-Methylanthracene 392
2-Methylbenzaldehyde 428
3-Methylbenzaldehyde 428
4-Methylbenzaldehyde 428
Methyl benzoate 39 ; 443
Methyl bromoacetate 445
2-Methylbutane 62 ; 385
2-Methyl-1-butanol 394
3-Methyl-1-butanol 39
3-Methyl-2-butanol 394
2-Methyl-1-butene 388
3-Methyl-1-butene 388
2-Methyl-2-butene 388
N-Methylbutylamine 415
2-Methyl-3-butyn-2-ol 395
2-Methylbutyraldehyde 427
3-Methylbutyraldehyde 427
Methyl butyrate 440
2-Methylbutyric acid 434
3-Methylbutyric acid 434
Methyl chloroacetate 445
Methyl 2-chloroacetoacetate 446
Methyl chloroformate 448
2-Methyl chloropropionate 445
Methyl cinnamate 443
3-Methylcrotonaldehyde 427
Methyl crotonate 441
2-Methylcrotonic acid 435
3-Methylcrotonic acid 435
Methyl cyanoacetate 452

Methyl 2-cyanoacrylate 452
Methylcyclohexane 386
2-Methylcyclohexanol 395
3-Methylcyclohexanol 395
4-Methylcyclohexanol 395
2-Methylcyclohexanone 431
3-Methylcyclohexanone 431
4-Methylcyclohexanone 431
4-Methylcyclohexene 388
N-Methylcyclohexylamine 415
Methylcyclohexylketone 431
Methylcyclopentane 386
Methyl dichloroacetate 445
Methyl dichloromethyl ether 412
Methyldichlorosilane 350; 351
Methylenecyclopropane 103
Methyl ethyl ether 39; 399
Methyl formate 39; 321; 326; 438
2-Methylfuran 262; 402
Methyl heptanoate 440
5-Methyl-3-heptanone 431
5-Methyl-2-hexanone 431
Methylhydrazine 97; 269; 419
Methyl hydroperoxide 322
Methyl hypochlorite 249
Methylisobutylketone 308; 309; 314; 430
Methyl isobutyrate 440
Methyl isocyanate 343; 344; 453
Methylisopropylketone 430
Methyl isothiocyanate 455
Methylmagnesium bromide 277
Methyl methacrylate 128; 324; 326; 441
N-Methylmorpholine 421
1-Methylnaphthalene 392
2-Methylnaphthalene 392
Methyl nitrate 294; 295; 299; 305; 423
Methyl nitrite 422
Methyl orthoformate 273

Methylparathion 129
2-Methylpentane 386
3-Methylpentane 386
2-Methylpentanoic acid 80; 434
3-Methylpentanoic acid 39; 434
4-Methylpentanoic acid 434
Methyl pentanoate 440
2-Methylpentanol 311; 427
4-Methyl-2-pentanol 394
Methyl perchlorate 251
Methyl periodate 280
N-Methylpiperaldehyde 418
2-Methylpiperaldehyde 418
4-Methylpiperaldehyde 418
3-Methylpiperidine 418
Methyl pivalate 440
2-Methylpropane 385
2-Methyl-1-propanol 39
2-Methyl-2-propen-1-ol 395
Methyl propionate 439
2-Methylpyraldehyde 112; 114; 115; 115; 117; 118; 418
Methylpyridines 293
3-Methylpyridine 418
4-Methylpyridine 418
N-Methylpyrrole 417
N-Methylpyrrolidone 452
Methyl silicate 39
Methyl salicylate 39
α-Methylstyrene 238; 391
2-Methylstyrene 391
3-Methylstyrene 238; 391
4-Methylstyrene 238; 391
Methyltert-butylether 399
Methyl trichloroacetate 326; 327; 445
Methyltrichlorosilane 457
Methyl trifluoroacetate 445
4-Methylvaleronitrile 449
Methyl vinyl ether 349

Methylvinylketone 310; 311; 314; 429

Molybdenum 218; 232; 283; 378

Molybdenum oxide 218

Molybdenum sulphide 218; 379

Molybdenum triether 218; 218; 379

Morpholine 297; 421

– N –

Naphthalene 97; 244; 311; 391

1-Naphthol 398

2-Naphthol 398

Naphthols 3

9 Naphthoquinone 308

2-Naphthylamidene 414

1-Naphthylamine 414

Neon 172; 362

Neopentylamine 413

Nickel 206; 211; 251; 317; 374

Nickel carbonate 374

Nickel chloride 206; 374

Nickel cyanide 206; 374

Nickel fluoride 206

Nickel nitrate 374

Nickel oxide 295; 374

Nickel perchlorate 313

Nickel sulphate 374

Nickel tetracarbonyl 129

Nicotine 129; 419

Nicotinic acid 437

Niodium 218; 378

Niodium pentachloride 378

Niodium pentoxide 218; 378

Nitric acid 147; 148; 149; 150; 151; 153; 159; 167; 168; 170; 171; 173; 175; 178; 179; 182; 183; 194; 200; 202; 204; 209; 210; 211; 219; 224; 227; 233; 244; 248; 251; 259; 260; 267; 270; 280; 290; 295; 299; 303 ;304; 309; 317; 330; 336; 342; 345; 361

p-Nitroacetanilide 344

Nitroalkanes 306

Nitroanilines 294; 306

2-Nitroaniline 111; 114; 115; 117; 303; 426

3-Nitroaniline 280; 426

4-Nitroaniline 302; 426

2-Nitroanisole 305; 306; 425

3-Nitroanisole 425

4-Nitroanisole 425

Nitrobenzene 39; 40; 48; 98; 294 295; 297; 299; 300; 303 ; 304; 305; 423

3-Nitrobenzenesulphonic acid 278

Nitrobutane 98

1-Nitrobutane 422

2-Nitrobutane 422

o-Nitrodiphenyl 424

p-Nitrodiphenyl 424

Nitroethane 98; 121; 123; 295; 297; 297; 304; 422

2-Nitroethanol 298; 306; 424

Nitrogen 152; 165; 167; 168; 169; 170;175; 175; 192; 195; 197; 218

Nitrogen diiodide 227

Nitrogen dioxide 97; 128; 148; 166;167; 168; 178; 194; 202; 243; 361

Nitrogen oxide 97; 148;167; 168; 172152; 194; 196; 200; 201; 236; 247; 362

Nitrogen oxides 195; 303; 310

Nitrogen trichloride 166; 168; 170; 183; 187; 189; 195; 196; 213; 215; 227; 338 ; 339; 362

Nitrogen triiodide 227

Nitroglycerine 100; 197; 252; 270; 423

1-Nitrohexane 422

Nitromethane 39; 121; 123; 294 295; 297; 298; 299; 303; 305;

312 ; 422
1-Nitronaphthalene 270 ; 271 ; 281 ; 283 ; 424
2-Nitronaphthalene 424
1-Nitropentane 422
Nitrophenols 294 ; 306
2-Nitrophenol 298 ; 302 ; 349 ; 425
3-Nitrophenol 425
4-Nitrophenol 298 ; 425
1-Nitropropane 121 ; 123 ; 297 422
2-Nitropropane 422
2-Nitro-1-propanol 424
Nitropropanes 98 ; 295 ; 296
Nitrosodiethylamine 426
Nitrotolualdehydes 39
3-Nitrotoluene 424
Nitrotoluenes 39 ; 295
4-Nitrotoluene 298 ; 301 ; 306 ; 424
2-Nitrotoluene 298 ; 301 ; 424
Nitrous oxide 97 ; 166 ; 167 ; 168 ; 178 ; 201 ; 362
Nitroxylene 295
Nonanal 428
Nonane 40 ; 386
Nonanoic acid 435
4-Nonylphenol 398

– O –

Octanal 428
Octane 62 ; 386
Octanoic acid 435
3-Octanol 53 ; 395
2-Octanol 395
1-Octanol 251 ; 395
1-Octene 389
2-Octene 389
3-Octene 389
Oleic acid 315 ; 320 ; 436
Oleum 301
Orthophosphoric acid 153 ; 180 ; 254 ; 317 ; 328 ; 335 ; 365
Osmium tetroxide 323
Oxalic acid 317 ; 318 ; 343 ; 436
Oxalyl chloride 329 ; 347 ; 448
Oxidation 279 ; 289 ; 308 ; 315
Oxygen 149 ; 150 ; 151 ; 155 ; 158 ; 159 ; 169 ; 169 ; 170 ; 171 ; 173 ; 175 ; 176 ; 177 ; 178 ; 179 ; 181 ; 183 ; 192 ; 195 ; 197 ; 198 ; 199 ; 201 ; 204 ; 204 ; 206 ; 207 ; 208 ; 210 ; 215 ; 217 ; 219 ; 221 ; 222 ; 227 ; 228 ; 242 ; 244 ; 253 ; 260 ; 269 ; 272
Oxygen difluoride 346 ; 349
Ozone 97 ; 170 ; 177 ; 213 ; 215 ; 242 ; 267 ; 280 ; 325 ; 362
Ozonide 325

– P –

Palladium (II) perchlorate 349
Palladium 220 ; 317 ; 379
Palladium chloride 379
Paraldehyde 313 ; 433
Parathion 129
Pentachlorophenol 412
1,3-Pentadiene 238 ; 388
Pentaerythritol 149 ; 250 ; 251 ; 257 ; 397
Pentaerythritol tetranitrate 294 ; 423
Pentanal 427
Pentane-2,3-dione 430
Pentane 62 ; 87 ; 89 ; 385
Pentane-1,5-diol 397
Pentane-2,3-dione 430
Pentanethiol 345
Pentanoic acid 39 ; 434
Pentanoic anhydride 447
3-Pentanol 393
2-Pentanol 45 ; 46 ; 393
1-Pentanol 39 ; 46 ; 59 ; 62 ; 67 ; 69 ; 393

Index

3-Pentanone 309 ; 291 ; 430

2-Pentanone 429

Pentanoyl chloride 447

2-Pentenes 388

1-Pentene 387

Pentyl acetate 128

n-Pentylamine 413

257 ; 397

n-Pentylbenzene 391

Pentyl nitrate 423

Pentyl nitrite 294 ; 423

Pentyl oxide 262

Pentyl nitrate 423

Pentyl nitrite 294 ; 423

Pentyl oxide 262

2-Pentyne 388

1-Pentyne 388

Peracetic acid 98 ; 121 ; 123 ; 307 316 ; 322 ; 323 ; 330

Perchloric acid 159 ; 175 ; 188 ; 190 ; 209 ; 225 ; 227 ; 232 ; 247 ; 250 ; 251 ; 267 ; 280 ;312 ; 328 ;336 ; 347 ; 348 ; 349 ; 369

Perfluoroiodohexane 152

Performic acid 98 ; 289 ; 309 ; 315

Periodic acid 347

Permanganic acid 202

Peroxodisulphuric acid 245 ; 253 ; 260

Peroxomonosulphuric acid 184 ; 202 ; 203 ; 220 ; 245 ; 253 ; 260 ; 384 ; 309

Perstearic acid 98

Petroleum ether 149

1,3-Phenylenediamene 270

p-Phenylenediamine 298 ; 293 ; 416

2-Phenylethanol 396

1-Phenylethanol 396

o-Phenylphenol 398

3-Phenyl-1-propanol 151 ; 250 ; 257 ; 396

2-Phenyl-1-propanol 396

1-Phenyl-1-propanol 396

3-Phenyl-2-propen-1-ol 396; 364

1-Phenyl-2-propyn-1-ol 396

Pharaon 230

Phenanthrene 39 ; 392

Phenol 39 ; 133 ; 258 ; 260 ; 299 ; 300 ;311 ; 397

Phenothiazine 332

Phenyl acetate 443

Phenylacetylene 159 ; 247 ; 246 ; 390

p-Phenylenediamine 298 ; 293 ; 416

Phenylglycidylether 265 ; 402

Phenylhydrazine 97 ; 269 ; 419

N-Phenyl-1-naphthylamine 264

Phorone 432

Phosgene 39 ; 177 ; 318 ; 448

Phosphine 97 ; 148 ; 149 ; 176 ; 178 ; 178 ; 180 ; 187 ; 188 ; 194 ; 200 ; 209 ; 221 ; 227 ; 231 ; 250 ; 319 ; 366

Phosphorus 148 ; 178 ; 180 ; 181 ; 182 ; 184 ; 191 ; 194 ; 196 ; 200 ; 201 ; 203 ; 211 ; 213 ; 218 ; 221 ; 221 ; 227 ; 227 ; 229 ; 231 ; 249 ; 365

Phosphorous acid 319

Phosphorus pentachloride 178 ; 180 ; 188 ; 194 ; 197 ; 306 ; 339 ; 366

Phosphorus pentoxide 195 ;250 ; 256

Phosphorus tribromide 151 ; 219 ; 250

Phosphorus trichloride 149 ; 179 ; 180 ; 194 ; 200 ; 232 ; 276 ;319 ; 344

Phosphorus triiodide 250

Phthalic acid 317 ; 320

o-Phthalic acid 436

Phthalic anhydride 329 ; 330 ; 332 ; 333 ; 425

Phthalonitriles 451

Phthaloyl chloride 328 ; 333
Pimelic acid 436
α-Pinene 389
Piperaldehyde 417
Piperazine 418
Pivalic anhydride 447
Pivalonitrile 449
Pivaloyl chloride 448
Platinum 188 ; 208 ; 253
Plutonium 278
Polyacrylamide 269
Polyethylene 241
Polyisobutene 239
Polypropylene 153 ; 239
Polypropyleneglycol 269
Potassium 148 ; 192 ; 198 ; 199 ; 200 ; 201 ; 203 ; 205 ; 206 ; 207 ; 208 ; 209 ; 211 ; 213 ; 215 ; 218 ; 220 ; 223 ; 224 ; 225 ; 227 ; 230 ; 231 ; 233 ; 237 ; 247 ; 254 ; 267 ; 277 ; 278 ; 280 ; 305 ; 347
Potassium acetylide 247
Potassium amide 207
Potassium arsenate 376
Potassium azide 177
Potassium bromate 215 ; 377
Potassium bromide 197 ; 377
Potassium carbonate 193 ; 194 ; 218 ; 369
Potassium chlorate 189 ; 191 ; 197 ; 209 ; 211 ; 218 ; 219 ; 222 ; 224 ; 227 ; 229 ; 304 ; 369
Potassium chloride 138 ; 369
Potassium chromate 372
Potassium cyanide 149 ; 166 ; 361
Potassium dichromate 147 ; 201 ; 254 ; 304 ; 372
Potassium dihydrogenphosphate 365
Potassium fluoride 171 ; 362
Potassium hexacyanoferrate (III) 205 ; 373

Potassium hexacyanoferrate 207
Potassium hexafluorosilicate 171 ; 362
Potassium hexanitrocobaltate (III) 205 ; 374
Potassium hydride 282
Potassium hydrogencarbonate 370
Potassium hydrogensulphate 367
Potassium hydroxide 194 ; 195 ; 210 ; 218 ; 238 ; 273 ; 274 ; 297 ; 298 ; 370
Potassium iodate 227 ; 226 ; 382
Potassium iodide 227 ; 226 ; 281 ; 382
Potassium nitrate 169 ; 168 ; 178 ; 191 ; 197 ; 198 ; 210 ; 211 ; 229 ; 362
Potassium nitrite 362
Potassium oxide 193
Potassium perchlorate 191 ; 203 ; 206 ; 218 ; 251 ; 369
Potassium permanganate 147 ; 149 ; 153 ; 202 ; 203 ; 210 ; 224 ; 254 ; 270 ; 309 ; 317 ; 332 ; 342 ; 346 ; 350 ; 373
Potassium peroxide 193
Potassium peroxodisulphate 186 ; 195 ; 368
Potassium pyrophosphate 366
Potassium silicate 365
Potassium stannate (IV) 381
Potassium sulphate 368
Potassium sulphide 274
Potassium sulphite 368
Potassium superoxide 193 ; 207 ; 209 ; 210 ; 223 ; 254
Potassium tellurate (IV) 382
Potassium tert-fluoroborate 362
Potassium tertiobutylate 192 ; 249 ; 274 ; 318 ; 325 ; 347
Potassium thiocyanate 228
Propane 89 ; 236 ; 245 ; 295 ; 385
Propane-1,2-diol 47 ; 252 ; 396

Propane-1,3-diol 396
Propanethiol 323 ; 328
2-Propanethiol 454
2-Propanol 47 ; 128 ; 151 ; 249 ; 252 ; 253 ; 254 ; 255 ; 257 ; 308 ; 393
1-Propanol 39 ; 46 ; 98 ; 249 ; 257 ; 260 ; 393
Propargyl alcohol 256 ; 256 ; 395
Propene 97
Propionaldehyde 138 ; 309 ; 311 314 ; 324 ; 426
Propionamide 452
Propionic acid 39 ; 319 ; 320 ; 434
Propionic anhydride 329 ; 446
α-Propiolactone 129
Propionitrile 97 ; 121 ; 123 ; 129 ; 133 ; 337 ; 340 ; 449
Propionyl chloride 329 ; 333 ; 334 ; 447
Propyl acetate 438
Propyl benzoate 443
Propyl formate 325 ; 326 ; 438
Propyl nitrate 294 ; 295 ; 423
Propyl propionate 439
Propylamine 413
Propylbenzene 89 ; 390
Propylene 98 ; 121 ; 123 ; 149 ; 150 ; 152 ; 236 ; 237 ; 238 ; 246 ; 248 ; 387
Propylene carbonate 325 ; 443
Propylene oxide 139 ; 262 ; 266 ; 271 ; 400
Propylglycidylether 402
Propyne 97 ; 236 ; 238 ; 238 ; 247 ; 246 ; 262 ; 387
Pyrazole 417
Pyridine 128 ; 286 ; 290 ; 293 ; 297 ; 343 ; 418
Pyrites 205
Pyrocatechol 398
Pyrogallol 263 ; 399

Pyromellitic acid 245
Pyrrolaldehyde 417
Pyrrole 337 ; 417
Pyrrolidone 452
Pyruvic acid 318 ; 320 ; 438

– Q –

Quinaldine 419
Quinoline 39 ; 289 ; 291 ; 293 ; 419
8-Quinolinol 421

– R –

Raney cobalt 205
Raney nickel 176 ; 206
Resorcinol 259 ; 398
Rhodium 208 ; 219 ; 379
Rhodium sulphate 379
Rhodium trichloride 379
Rubidium 215 ; 225 ; 230 ; 378
Rubidium carbonate 378
Rubidium chloride 378
Ruthenium 219 ; 379
Ruthenium oxide 379
Ruthenium tetroxide 219 ; 227
Ruthenium trichloride 219 ; 379

– S –

Salicylaldehyde 428
Salicylic acid 159 ; 437
Scandium 213 ; 371
Selenium 211 ; 214 ; 215 ; 221 ; 229 ; 377
Selenium dioxide 214 ; 377
Selenium hexafluoride 377
Silane 97 ; 177 ; 187 ; 213 ; 365
Silica 177 ; 187 ; 365
Silica gel 300

Index — S

Silicic acid 365
Silicon 177; 189; 194; 195; 201; 217; 221; 229; 232; 232; 365
Silicon carbide 365
Silicon nitride 365
Silicon tetrachloride 194; 350; 351; 329
Silicon trichloride 350
Silver 220; 247; 282; 86; 297; 378
Silver bromide 379
Silver carbonate 379
Silver chlorate 255
Silver chloride 220; 379
Silver chlorite 274
Silver cyanide 220; 379
Silver fluoride 220; 379
Silver fulminate 252; 305
Silver iodide 349; 379
Silver nitrate 182; 210; 220; 222; 247; 252; 305; 336; 379
Silver oxide 221; 224; 231; 240; 295; 299; 350; 379
Silver perchlorate 247; 272; 280; 379
Silver permanganate 222; 379
Silver phosphate 379
Silver sulphide 222; 380
Silver tetrafluoroborate 299
Siver sulphate 380
Soda lime 183
Sodium 151; 152; 153; 155; 172; 173; 173; 176; 179; 181; 182; 183; 184; 187; 195; 198; 200; 200; 205; 207; 211; 213; 218; 222; 223; 225; 228; 230; 231; 232; 232; 237; 247; 249; 263; 269; 273; 276; 278; 305; 342; 350; 329; 362
Sodium acetate 324; 328
Sodium acetylide 256
Sodium aluminate 362
Sodium amide 140; 168; 170; 191; 200; 213; 361
Sodium ammonium hydrogenphosphate 366
Sodium arsenate 376
Sodium arsenite 376
Sodium azide 97; 99; 166; 167; 168; 183; 200; 213; 338; 361
Sodium borohydride 219; 220; 249; 320
Sodium bromate 216; 215; 377
Sodium bromide 377
Sodium carbonate 174; 177; 184; 196; 198; 218; 296; 297; 346; 362
Sodium chlorate 189; 190; 198; 207; 210; 224; 369
Sodium chloride 181; 187; 191; 190; 369
Sodium chlorite 189; 190; 255; 317
Sodium chromate 200; 218; 372
Sodium cyanide 159; 168; 171; 175; 191; 200; 228; 231; 338; 361
Sodium dichromate 200; 201; 203; 218; 372
Sodium difluoride 362
Sodium dihydrogenphosphate 366
Sodium fluoride 362
Sodium hexacyanoferrate (III) 204; 232
Sodium hexafluorosilicate 362
Sodium hexametaphosphate 366
Sodium hydride 153; 249; 297; 300; 342; 347
Sodium hydrogencarbonate 174; 312; 362
Sodium hydrogensulphate 329
sodium hydrogensulphide 367
Sodium hydrogensulphite 367
Sodium hydrosulphite 184; 185; 189; 367
Sodium hydroxide 151; 174; 176; 178; 186; 187; 204; 209; 213;

218; 221; 238; 249; 259; 266; 273; 274; 279; 283; 298; 305; 325; 338; 341

Sodium hypochlorite 147; 189; 190; 240; 250; 291; 310; 317 338; 339; 369

Sodium hypophosphite 188

Sodium iodate 227; 226; 382

Sodium iodide 227; 226; 263; 382

Sodium isopropylate 347

Sodium metavanadate 290

Sodium molybdate 219; 379

Sodium nitrate 169; 168; 173; 175; 177; 178; 184; 191; 198; 200; 204; 210; 223; 224; 362

Sodium nitrite 159; 169; 184; 194; 230; 244; 287; 303; 317; 342 362

Sodium oxide 179; 195; 298

Sodium perborate 164; 360

Sodium perchlorate 191; 369

Sodium peroxide 147; 173; 175; 177; 178; 179; 181; 182; 183; 186; 193; 197; 203; 210; 211; 218; 223; 245; 270; 289; 315; 316; 364

Sodium peroxodisulphate 185; 186; 270; 368

Sodium phosphate 366

Sodium phosphite 178

Sodium polyphosphate 366

Sodium selenite 377

Sodium sesquicarbonate 364

Sodium silicate 365

Sodium stannate (IV) 381

Sodium sulphate 159; 185; 186; 218; 368

Sodium sulphide 183; 185; 197; 200; 368

Sodium sulphite 368

Sodium tetraborate 218; 360

Sodium thiocarbonate 368

Sodium thiocyanate 349

Sodium thiosulphate 184; 185; 368

Sodium trimetaphosphate 366

Sodium tripolyphosphate 366

Sodium vanadate 372

Spiropentane 103

Stannic bromide 381

Stannic chloride 381

Stannous chloride 223; 381

Stannous fluoride 223; 381

Starch 263

Stibine 97; 224; 381

Strontium 148; 215; 379

Strontium carbonate 378

Strontium chlorate 378

Strontium chloride 378

Strontium chromate 378

Strontium hydroxide 378

Strontium sulphate 378

Strontium sulphide 232; 378

Styrene 89; 97; 121; 123; 237; 240; 241; 242; 246; 390

Succinic acid 436

Succinic anhydride 322; 323; 446

Succinonitrile 335; 340; 450

Sulphamic acid 184; 185; 187; 367

Sulphides 183

Sulpholane 454

Sulphur 181; 185; 191; 194; 195; 196; 196; 200; 200; 203; 208; 206; 209; 215; 220; 221; 221; 222; 229; 232; 366

Sulphur dichloride 181; 182; 185; 194; 223; 231; 240; 313; 397

Sulphur dioxide 128; 182; 184 185; 191; 194; 200; 202 203; 229; 232; 236; 237; 310; 367

Sulphur hexafluoride 367

Sulphur trioxide 184; 185; 229; 265; 301; 335; 343; 348

Sulphuric acid 149; 151; 182;

184; 185; 187; 191; 194; 202; 207; 210; 222; 229; 232; 238; 244; 247; 252; 253; 254; 256; 258; 267; 268; 270; 281; 287 ;290 ; 295; 300; 301 ;303 ; 309; 310 ; 317; 330; 335 ; 337; 343; 348; 367

Sulphurous acid 367

Sulphuryl chloride 182; 185; 232; 264; 347 367

Sulphuryl fluoride 367

– T –

Technetium 379

Tellurium 381

Terephthalic acid 244; 436

Terpentine spirit 244; 245; 246; 388

o-Terphenyl 392

p-Terphenyl 392

Tert-bromomethane 240; 284; 405
  Tert-butyl peroxide 98; 121; 123; 253;

Tertiobutylpyrocatechol 238

1,2,2-Tetrabromoethane 405

1,2,4,5-Tetrachlorobenzene 259; 410

Tetrachlorobenzene 260

Tetrachlorobenzoquinone 433 1,1,2,

2-Tetrachlorodibenzodioxin 283

1,1,1,2-Tetrachlorodifluoroethane 407

Tetrachlorodifluoroethane 407

1,1,2,2-Tetrachloroethane 273; 274; 284; 404

1,1,1,2-Tetrachloroethane 404

Tetrachloroethylene 277; 278; 284; 408

Tetrachloromethane 128; 152; 240; 274; 251; 253; 256; 260; 261; 262; 318; 403

Tetrachlorosilane 457

Tetraethylenepentamine 286

Tetrafluoroethylene 280

Tetrafluoromethane 403

Tetrahydrocarbazole 337

Tetrahydrofuran 40; 153; 213; 262; 263; 265; 268; 271; 272; 282; 297; 402

Tetrahydrofurfurylic alcohol 238; 403

Tetrahydronaphthhalene 55; 89; 242; 262; 392

Tetrahydropyran 402

Tetramethylsilane 351; 456

Tetramethylsuccinonitrile 338; 450

Tetranitrimethane 422

Tetranitromethane 39; 294; 295; 297; 304; 305; 306

Tetranitronaphthalene 294; 304

Tetryl 97

Thallium trinitrate 317

Thioacetamide 455

p-Thiocresol 454

Thiodiethanol 455

Thioglycolic acid 455

Thiomorpholine 455

Thionyl chloride 182; 186; 269; 339; 347; 367

Thiophene 40; 112; 114; 115; 117; 118; 351; 454

Thiophenol 454

Thiourea 346; 351; 455

Tin (IV) oxide 381

Tin (II) oxide 223

Tin 222; 225; 381

Tin disulphide 223; 381

Tin sulphide 228

Tin tertrachloride 223; 300

Titanium 187; 198; 203; 207; 213; 215; 220; 225; 232; 232; 278;

Titanium carbide 198; 199; 371

Titanium diboride 371

Index   T

Titanium hydride 198 ; 198 ; 371

Titanium dioxide 189 ; 192 ; 199 ;

Titanium nitride 371

Titanium sulphate 371

Titanium tetrachloride 264 ; 339 ; 371

Titanium trichloride 198 ; 199 ; 371

p-Tolualdehyde 4

Toluene 70 ; 72 ; 89 ; 90 ;91 ; 97 ; 181 ; 240 ; 244 ; 247 ; 246 ; 269 ; 276 ;305 ; 389

Toluene-2,4-diisocyanate 453

Toluene-2,6-diisocyanate 453

m-Toluidine 414

o-Toluidine 290 ; 414

o-Tolunitrile 450

p-Tolunitrile 451

1,Triamylamine 417

Tribromomethane 273 ; 273 ; 284

2,2,2-Tribromoethanol 412

2,2,2-Tribromoethanol 412

Tributylamine 417

Tributyl borate 457

Tributylphosphate 456

Tributylphosphine 349 ; 328

Tributylphosphite 456

Trichloroacetic acid 348

Trichloroacetonitrile 338 ;340 ; 451

Trichloroacetyl chloride 329 ; 448

1,2,4-Trichlorobenzene 410

1,2,3-Trichlorobenzene 410

1,3,5-Trichlorobenzene 410

Trichlorethane 280 ; 283

1,1,1-Trichloroethane 280 ; 281 ;

2,2,2-Trichloroethanol 411

Trichloroethylene 39 ; 128 ; 135 ; 140 ; 150 ; 154 ; 274 ; 277 ; 278 280 ; 282 ; 284 ; 408

1,1,1-Trichloromethane 279

Trichloromethane 128 ; 135 ; 140 ; 151 ; 273 ; 274 ; 273 ; 284 ; 305

Trichloromethanethiol 455

Trichloromethylbenzene 411

2,4,5-Trichlorophenol 412

1,2,3-Trichloropropane 404

1,1,1-Trichloropropane 404

Trichlorosilane 350 ; 457

1,2-Trichlorotrifluoroethane 407

Trichlorotrifluoroethane 247

Tricobalt tetroxide 374 ; 283 ;404

Triethanolamine 420

Triethoxysilane 456

Triethyl borate 457

Triethylaluminium 249 ; 276

Triethylamine 128 ; 290 ; 416

Triethyleneglycol 262 ; 400

Triethylenetetramine 45 ; 266 ; 416

Triethylphosphate 456

Triethylphosphine 349 ; 328

Triethylphosphite 456

Triethylsilane 456

Triifluoroacetic acid 316 ; 320

Trifluoroacetic anhydride 347

Trifluoromethane 403

Trifluoromethylbenzene 129 ; 409

Triiodomethane 280

Triisobutylamine 417

Triisopropylaluminium 308

Triisopropylphosphine 274

Trimanganese tetroxide 373

Trimethoxysilane 456

Trimethylamine 285 ; 327 ; 416

1,2,4-Trimethylbenzene 59 ; 67 ; 91 ; 92 ;

Trimethyl borate 457

Trimethylchlorosilane 349 ; 350

2,4-Trimethylpentane 386

2,3,4-Trimethyl-1-pentene 51

2,4,4-Trimethyl-1-pentene 389

Trimethylphosphate 349 ; 351 ; 456

Trimethylphosphine 349 ; 351 ; 455

Trimethyl phosphite 349 ; 456
2,4,6-Trinitroanisole 294
,3,5-Trinitrobenzene 294 ; 300 ; 424
Trinitrocresol 294
1,3,5-Trinitromethylbenzene 149 ; 245
Trinitrophenol 296
2,4,6-Trinitrophenol 98 ; 294 ; 296 ; 425
2,4,6-Trinitrophenol and salts 306
Trinitroresorcinol 294
2,4,6-Trinitrotoluene 98 ; 171 ; 272 ; 424 405
Trioxan 433
Trinitroxylene 294
Triphenylphosphate 456
Triphenylphosphine 456
Tripropylamine 417
Triron tetroxide 203 ; 204
Tritolylphosphate 456
Tungsten 232

– U –

Undecane 387 ;
Urea 339 ; 342 ; 344 ; 453

– V –

δ-Valerolactone 422
Valeronitrile 449 ;
Vanadium carbide 371
Vanadium 198 ; 371 ;
Vanadyl chloride 200 ; 215 ; 371
Vanadyl sulphate 371
Vaniliene 317
Vinyl acetate 323 ; 324 ; 326 ; 440
Vinylacetylene 237 ; 246
N-Vinyl bromide 236
4-Vinylcyclohexene 388
Vinyl oxide 262 ; 270 ; 271 ; 399

2-Vinylpyraldehyde 418
4-Vinylpyridine 418

– X –

Xenon 382
Xenon difluoride 346
Xenon trioxide 228
Xylenes 128 ; 135 ; 141 ; 153 ; 246
o-Xylene 389
m-Xylene 90 ; 247 ; 390
p-Xylene 244 ; 390
3,4-Xylenol 398
2,6-Xylenol 398
2,5-Xylenol 398
2,4-Xylenol 398
2,3-Xylenol 398
Xylidines 290 ; 293

– Z –

Z-Cyclononene 103
Zeolites 237 ; 295
Zinc 203 ; 204 ; 208 ; 210 ; 211 ; 219 ; 221 ; 222 ; 225 ; 230 ; 232 ; 267 ; 277 ; 278 ; 305 ; 327 ; 339 ; 375
Zinc carbonate 375
Zinc chlorate 208
Zinc chloride 208 ; 239 ; 334 ; 375
Zinc chromate 375
Zinc cyanide 209 ; 375
Zinc fluoride 375
Zinc hexafluorosilicate 375
Zinc nitrate 209 ; 375
Zinc nitride 232
Zinc oxide 209 ; 239 ; 278 ; 346 ; 375
Zinc peroxide 209 ; 375
Zinc phosphide 209 ; 375
Zinc sulphate 376

Zinc sulphide 376
Zinc thiocyanate 99
Zirconium oxide 378
Zirconium 217; 219; 232; 278; 378
Zirconium hydroxide 218; 378
Zirconium silicate 378
Zirconium sulphate 378
Zirconium tetrachloride 218; 264; 378
Zirconyl chloride 378

SUBJECT INDEX

– A –

Analysis of risk 18
Antoine's equation 36
Autoignition temperatures (AIT) 35 ; 71
    additivity method 74 ; 79
    apparatus 71
    chemical structure, effect of 73
        chain length 73
        unsaturation 74
    definition 71
    estimation 74
    physical variables, effect of 72

– B –

Ballistic mortar test 95
Benson's tables 101 ; 102-108
Boiling point 35
    estimation 41 ; 45

– C –

Calculated limiting value of exposure 127
card gap test 95
CAS Nos, chemical names 481 et seq
Chain length and substitution, effect of 98
CHETAH 23
CHETAH criteria 101 ; 117
    criterion C1 : enthalpy of decomposition 110
    criterion C2 : propensity to combustion 113
    criterion C3 : `internal redox' measure ; oxygen balance 115
    criterion C4 : effect of mass 116
CHETAH programme 93
    instability 97
Clapeyron's equation 38
Critical diameter, determination of 95
Cyclic tension, effect of 98

– D –

Dangerous materials, transport code 146
Dangerous reactions
    Labour regulations 145
    NFPA Code 145
    parameters 143 et seq
    qualitative approach, et seq 146
        `external' risk factors 148
            apparatus material effect 153
            concentration effects 150
            contact time effects 153
            effect of light 152
            mass effects 152
            pressure effects 150
            product confusion 154
            reagent viscosity effect 153
            thermal effects 149
    structural risk factors 146
    quantitative estimation 155 et seq
        CHETAH programme 157
        physical factors 155
        Stull method 157

Subject Index — E-H

Dangerous reactions, inorganic chemicals 163 et seq
Dangerous reactions, organic chemicals 235 et seq
Degree of fire risk calculation 87 ; 89
Degree of risk 158
Detonatability test 95
Dose, means of penetration 125
Dow Chemical Co 85
Dow Chemical method 17

– E –

Empirical formulae, and chemical names 459 et seq
    inorganic compounds 459 et seq
    organic compounds 470 et seq
Endothermic compounds 97
Enthalpy of formation 97
Enthalpy of formation, estimation of 101
    examples of 108
Enthalpy of reaction
Entropic factor 38
    calculation of 43
    estimation of 41
EU Framework Directive 21
Experimental uncertainty 36
Explosophoric (plosophoric) groups 93 ; 99
Exposure, limiting, mean values 127

– F –

Fire explosion risk (FER) 156
Fire hazard 80
    criteria and classification 86
    index 87
    labour regulations 81 ; 83
    methods of classification 80
    new classification 87
    physical and other features 84
    SAX code 81 ; 84
Fire hazard 87
    degree 87
Fire hazard index 87
Fire risk, calculation of degree of 87
    classes of 87
    degree of 89
Flammability 35 et seq
Flammability code 83
    attribution criteria 84
Flammability, physical factors 86
Flashpoint 35 ; 56
    apparatus effect 58
    applicability 56
    closed cup (cc) 57
    cup effect 58
    definition 56
    environmental effect 60
    measurement 56
    methods of estimation 61
        from boiling points 61
        from Hass equation 63
        mixtures 68 ; 70
        Thorne and Gmehling models 68
    open cup (oc) 57
    operator effect 58
Friction sensitivity tests 94

– G –

Group repetition and close groups, effect of 98

– H –

Hass HB 38
    method 38
Heating-under-containment tests 95
Heteroscedasticity 36
    of vapour pressures 37
Hilado, C.J. 71

Subject Index   I-O

– I –

Immediately dangerous to life and health (IDLH) 127
Impact sensitivity tests 94
Index of reactivity (IR) 158
Inflammability (flammability),
    applicability 50
    definition 50
Inflammability risk 36
Inflammation index 87
Inorganic chemicals, dangerous reactions 163 et seq
Inorganic compounds, tables 359 et seq
Instability index 118 ; 119
Instability risk, classification 120
    indicators 123
    labour code 122
    labelling symbol 122
    NFPA grading ; 'reactivity' 120
        reactivity codes 121
Instability, qualitative approach to 96 et seq
    'external' factors 100
        dilution, desensitisation 100
        confinement 100
        quantity 100
    'internal' factors 96
        endothermic property 97
        low energy linkages 96
Instability, quantitative approach 101 et seq
Institut National Recherche et Securite (INRS) 127
Institution of Chemical Engineers (UK) 22
Internal redox property 98

– L –

Labour Code 18

Le Chatelier's equation 69
Lethal concentration (LC50) 126
Lethal dose (LD50) 126
Limiting value of exposure (LVE) 127
    limits of 50
        estimation of 51
        inflammable mixtures 55
        measurement 57
        symbol 83
Lower explosive limit (LEL) 35 ; 50 ; 67
    estimation of 67
    influence of temperature 55
    published experimental data 67

– M –

Mass effects, ionic 99
Mean value of exposure (MVE) 127
Mechanical sensitivity tests 94
Minimum lethal concentration, dose (MLC, MLD) 126

– N –

NFPA 83
    code 93
    code classification 85
    danger symbol 83

– O –

Occupational Safety & Health Administration (OSHA) 21
    Process Safety Management Programs 21
    Risk Management Programs 21
OECD initiatives 22
Organic chemicals, dangerous reactions 235 et seq
Organic chemicals, tables 385 et seq
Oxygen balance 98

– P –

Plosphoric (explosophoric) groups 93 ; 99
Prevention 29
Product confusion 154
Protection 29

– R –

Rat and mouse, correlation of toxicity data 137
Risk assessment methodologies 23
Risk map 30
Risk parameters 30
Risk profile 30
Risk, classes of 89

– S –

Safety code, flammability danger 81
    Labour code 83
    NFPA code 81
    SAX code 84
Safety sheets 36
SAX code 84
Sevesco Directives 21
Shockwave sensitivity tests 95
Short term exposure limit (STEL) 127
Stability et seq 93 ; 97
    experimental methods of determination 94
Stability, `internal' structural factors 96
Stull, DR 119

– T –

Thermal sensitivity tests 94
Thorne and Gmehling models 68
Toxic risk, establishing level 129 et seq
    classification criteria 132
    Hodge-Sterner Code 130
    Labour Code 130
    NFPA Code 129
    problems in determining 133
    quantitative index 134 ; 136
    SAX Code 130
Toxicity 125 et seq
    classification criteria 132
    evaluation parameters 126
    index of (IT) 126
    quantitative estimation methods 119
    safety and risk factor 135
    safety factor (SF) 135
Toxicity hazard 36
Transport code, dangerous materials 146
Transport of dangerous materials, toxicity code 133

– U –

UNIFAC 69
Upper explosive limit (UEL) 35
Upper explosive limit (UEL) 50

– V –

Vapour pressure 35 ; 36
    estimation of 37
violent heating tests 94

– W –

Weka 36